Kathleen Miranda

Student Solutions Manual
To Accompany
Mathematics:
An Applied Approach
8th Edition

Michael Sullivan
Chicago State University

Abe Mizrahi
Indiana University Northwest

WILEY
JOHN WILEY & SONS, INC.

Cover photo: © Photodisc /Getty Images

To order books or for customer service call 1-800-CALL-WILEY (225-5945).

ISBN 0-471-33379-4

Printed in the United States of America

10 9 8 7 6 5 4 3 2 1

Printed and bound by Malloy, Inc.

TABLE OF CONTENTS

Chapter 1

Linear Equations

1.1 Rectangular Coordinates; Lines $(15/3-31)$ ODD

1. $A = (4, 2)$ $B = (6, 2)$ $C = (5, 3)$ $D = (-2, 1)$
    $E = (-2, -3)$ $F = (3, -2)$ $G = (6, -2)$ $H = (5, 0)$

3. The set of points of the form, $(2, y)$, where y is a real number, is a vertical line passing through 2 on the x-axis.

The equation of the line is $x = 2$.

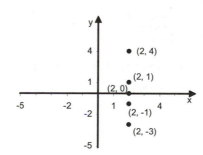

5. $y = 2x + 4$

x	0	-2	2	-2	4	-4
y	**4**	**0**	**8**	**0**	**12**	**-4**

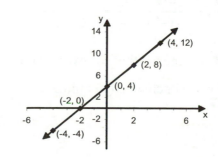

7. $2x - y = 6$

x	0	3	2	–2	4	–4
y	–6	0	–2	–10	2	–14

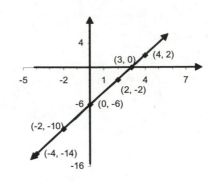

9.   (a)   The vertical line containing the point $(2, -3)$ is $x = 2$.

  (b)   The horizontal line containing the point $(2, -3)$ is $y = -3$.

11.   (a)   The vertical line containing the point $(-4, 1)$ is $x = -4$.

  (b)   The horizontal line containing the point $(-4, 1)$ is $y = 1$.

13. $m = \dfrac{y_2 - y_1}{x_2 - x_1} = \dfrac{1 - 0}{2 - 0} = \dfrac{1}{2}$

We interpret the slope to mean that for every 2 unit change in x, y changes 1 unit. That is, for every 2 units x increases, y increases by 1 unit.

15. $m = \dfrac{y_2 - y_1}{x_2 - x_1} = \dfrac{3 - 1}{-1 - 1} = -1$

We interpret the slope to mean that for every 1 unit change in x, y changes by (-1) unit. That is, for every 1 unit increase in x, y decreases by 1 unit.

17. $m = \dfrac{y_2 - y_1}{x_2 - x_1} = \dfrac{3 - 0}{2 - 1} = \dfrac{3}{1} = 3$

A slope of 3 means that for every 1 unit change in x, y will change 3 units.

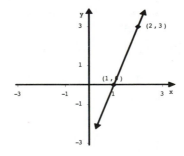

19. $m = \dfrac{y_2 - y_1}{x_2 - x_1} = \dfrac{1 - 3}{2 - (-2)} = \dfrac{-2}{4} = -\dfrac{1}{2}$

A slope of $-\dfrac{1}{2}$ means that for every 2 unit change in x, y will change (-1) unit.

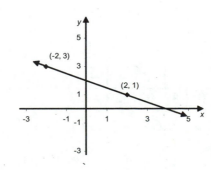

21. $m = \dfrac{y_2 - y_1}{x_2 - x_1} = \dfrac{(-1) - (-1)}{2 - (-3)} = \dfrac{0}{5} = 0$

A slope of zero indicates that regardless of how *x* changes, *y* remains constant.

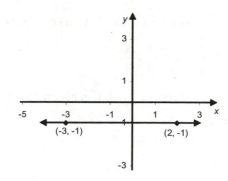

23. $m = \dfrac{y_2 - y_1}{x_2 - x_1} = \dfrac{(-2) - 2}{(-1) - (-1)} = \dfrac{-4}{0}$

The slope is not defined.

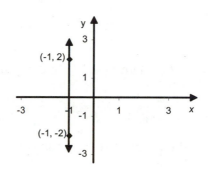

25.

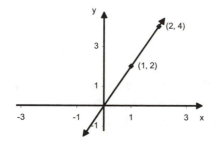

27.

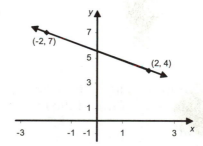

29.

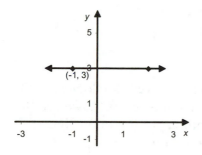

31.

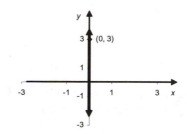

Eq. of Lines 16/33-72 (ODD)

33. Use the points $(0, 0)$ and $(2, 1)$ to compute the slope of the line:

$$m = \frac{y_2 - y_1}{x_2 - x_1} = \frac{1 - 0}{2 - 0} = \frac{1}{2}$$

Since the y-intercept, $(0, 0)$, is given, use the slope-intercept form of the equation of the line:

$$y = mx + b$$
$$y = \frac{1}{2}x + 0$$
$$y = \frac{1}{2}x$$
$$2y = x$$
$$x - 2y = 0$$

35. Use the points $(1, 1)$ and $(-1, 3)$ to compute the slope of the line:

$$m = \frac{y_2 - y_1}{x_2 - x_1} = \frac{3 - 1}{(-1) - 1} = \frac{2}{-2} = -1$$

We now use the point $(1, 1)$ and the slope $m = -1$ to write the point-slope form of the equation of the line:

$$y - y_1 = m(x - x_1)$$
$$y - 1 = (-1)(x - 1)$$
$$y - 1 = -x + 1$$
$$x + y = 2$$

37. Since the slope and a point are given, use the point-slope form of the line:

$$y - y_1 = m(x - x_1)$$
$$y - 1 = 2(x - (-4))$$
$$y - 1 = 2x + 8$$
$$2x - y = -9$$

39. Since the slope and a point are given, use the point-slope form of the line:

$$y - y_1 = m(x - x_1)$$
$$y - (-1) = -\frac{2}{3}(x - 1)$$
$$3y + 3 = -2(x - 1)$$
$$3y + 3 = -2x + 2$$
$$2x + 3y = -1$$

41. Since we are given two points, $(1, 3)$ and $(-1, 2)$, first find the slope.

$$m = \frac{3 - 2}{1 - (-1)} = \frac{1}{2}$$

Then with the slope and one of the points, $(1, 3)$, we use the point-slope form of the line:

$$y - y_1 = m(x - x_1)$$
$$y - 3 = \frac{1}{2}(x - 1)$$
$$2y - 6 = x - 1$$
$$x - 2y = -5$$

43. Since we are given the slope $m = -2$ and the y-intercept $(0, 3)$, we use the slope-intercept form of the line:

$$y = mx + b$$
$$y = -2x + 3$$
$$2x + y = 3$$

45. Since we are given the slope $m = 3$ and the x-intercept $(-4, 0)$, we use the point-slope form of the line:

$$y - y_1 = m(x - x_1)$$
$$y - 0 = 3 \left(x - (-4) \right)$$
$$y = 3x + 12$$
$$3x - y = -12$$

47. We are given the slope $m = \dfrac{4}{5}$ and the point $(0, 0)$, which is the y-intercept. So, we use the slope-intercept form of the line:

$$y = mx + b$$
$$y = \frac{4}{5}x + 0$$
$$5y = 4x$$
$$4x - 5y = 0$$

49. We are given two points, the x-intercept $(2, 0)$ and the y-intercept $(0, -1)$, so we need to find the slope and then to use the slope-intercept form of the line to get the equation.

$$\text{slope} = \frac{0 - (-1)}{2 - 0} = \frac{1}{2}$$

$$y = mx + b$$
$$y = \frac{1}{2}x - 1$$
$$2y = x - 2$$
$$x - 2y = 2$$

51. Since the slope is undefined, the line is vertical. The equation of the vertical line containing the point $(1, 4)$ is:

$$x = 1$$

53. Since the slope $= 0$, the line is horizontal. The equation of the horizontal line containing the point $(1, 4)$ is:

$$y = 4$$

55. $y = 2x + 3$,
slope: $m = 2$; y-intercept: $(0, 3)$

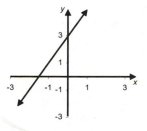

57. To obtain the slope and *y*-intercept, we transform the equation into its slope-intercept form. To do this we solve for *y*.

$$\frac{1}{2}y = x - 1$$
$$y = 2x - 2$$

slope: *m* = 2; *y*-intercept: (0, –2)

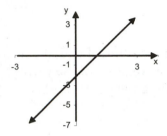

59. To obtain the slope and *y*-intercept, we transform the equation into its slope-intercept form. To do this we solve for *y*.

$$2x - 3y = 6$$
$$y = \frac{2}{3}x - 2$$

slope: $m = \frac{2}{3}$; *y*-intercept: (0, –2)

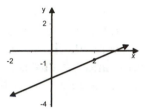

61. To obtain the slope and *y*-intercept, we transform the equation into its slope-intercept form. To do this we solve for *y*.

$$x + y = 1$$
$$y = -x + 1$$

slope: *m* = –1; *y*-intercept: (0, 1)

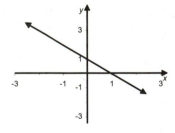

63. $x = -4$

The slope is not defined; there is no *y*-intercept.

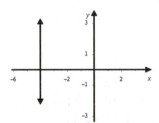

65. $y = 5$

slope: $m = 0$; y-intercept: $(0, 5)$

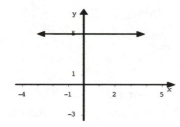

67. To obtain the slope and y-intercept, we transform the equation into its slope-intercept form. To do this we solve for y.

$$y - x = 0$$
$$y = x$$

slope: $m = 1$; y-intercept = $(0, 0)$

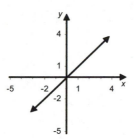

69. To obtain the slope and y-intercept, we transform the equation into its slope-intercept form. To do this we solve for y.

$$2y - 3x = 0$$
$$2y = 3x$$
$$y = \frac{3}{2}x$$

slope: $m = \frac{3}{2}$; y-intercept = $(0, 0)$

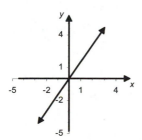

71. The equation of a horizontal line containing the point $(-1, -3)$ is given by

$y = -3$.

Problem Solving 17/73-83, (84)

73. The average cost of operating the car is given as $0.122 per mile. This is the slope of the equation. So the equation will be

$C = 0.122x$, where x is the number of miles the car is driven.

75. The fixed cost of electricity for the month is \$7.58. In addition, the electricity costs \$0.08275 (8.275 cents) for every kilowatt-hour (KWH) used. Then if x represents the number of KWH of electricity used in a month,

    (a) the total monthly is represented by the equation:

$$C = 0.08275x + 7.58, \quad 0 \le x \le 400$$

    (b)

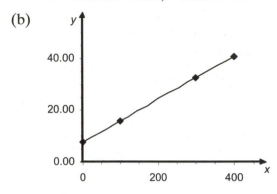

    (c) The charge for using 100 KWH of electricity is found by substituting 100 for x in part (a):

$$C = 0.08275(100) + 7.58$$
$$= 8.275 + 7.58$$
$$= 15.855$$
$$= \$15.86$$

    (d) The charge for using 300 KWH of electricity is found by substituting 300 for x in part (a):

$$C = 0.08275(300) + 7.58$$
$$= 24.825 + 7.58$$
$$= 32.405$$
$$= \$32.41$$

    (e) The slope of the line, $m = 0.08275$, indicates that for every extra KWH used (up to 400 KWH), the electric bill increases by 8.275 cents.

77. Two points are given, $(h_1, w_1) = (67, 139)$ and $(h_2, w_2) = (70, 151)$, and we are told they are linearly related. So we will first compute the slope of the line:

$$m = \frac{w_2 - w_1}{h_2 - h_1} = \frac{151 - 139}{70 - 67} = \frac{12}{3} = 4$$

We use the point $(151, 70)$ and the fact that the slope $m = 4$ to get the point-slope form of the equation of the line:

$$w - 151 = 4(h - 70)$$
$$w - 151 = 4h - 280$$
$$w = 4h - 129$$
$$w = 4h - 129$$

79. The delivery cost of the Sunday *Chicago Tribune* is \$1,070,000 plus \$0.53 for each of the x copies delivered. The total cost of delivering the papers is :

$$C = 0.53x + 1,070,000$$

81. Since we are told the relationship is linear, we will use the two points to get the slope of the line:

$$m = \frac{C_2 - C_1}{F_2 - F_1} = \frac{100 - 0}{212 - 32} = \frac{100}{180} = \frac{5}{9}$$

We use the point (0, 32) and the fact that the slope $m = \frac{5}{9}$ to get the point-slope form of the equation:

$$C - C_1 = m(F - F_1)$$
$$C - 0 = \frac{5}{9}(F - 32)$$
$$C = \frac{5}{9}(F - 32)$$

To find the Celsius measure of 68 °F we substitute 68 for F in the equation and simplify:

$$C = \frac{5}{9}(68 - 32)$$
$$= 20°$$

83. Since the problem states that the *rate* of the loss of water remains constant, we can assume that the relationship is linear.

(a) We are given two points, the amount of water (in billions of gallons) on November 8, 2002 ($t = 8$) and the amount of water on December 8, 2002 ($t = 38$). We use these two points and the fact that the relation is linear to find the slope of the line:

$$m = \frac{y_2 - y_1}{t_2 - t_1} = \frac{52.5 - 52.9}{38 - 8} = \frac{-0.4}{30} = -\frac{1}{75}$$

We use the point (8, 52.9) and the slope $m = -\frac{1}{75}$, to get the point-slope form of the equation of the line:

$$y - y_1 = m(t - t_1)$$
$$y - 52.9 = -\frac{1}{75}(t - 8)$$
$$y = -\frac{1}{75}t + 53.007$$

(b) To find the amount of water in the reservoir on November 20, let $t = 20$ in the equation from part (a).

$$y = -\frac{1}{75}(20) + 53.007$$
$$y = 52.74 \text{ billion gallons of water.}$$

(c) The slope tells us that the reservoir loses one billion gallons of water every 75 days.

(d) To find the amount of water in the reservoir on December 31, 2002, let $t = 61$ in the equation from (a).

$$y = -\frac{1}{75}(61) + 53.007$$
$$y = 52.194 \text{ billion gallons of water.}$$

(e) To determine when the reservoir will be empty, assume $y = 0$ gallons and solve for t.

$$y = -\frac{1}{75}(t) + 53.007$$
$$0 = -\frac{1}{75}(t) + 53.007$$
$$t = (53.007)(75)$$
$$= 3975.525 \text{ days, or about 10}$$
years, 10 months and 22 days.

(f) Answers vary.

85. To graph an equation on a graphing utility, first solve the equation for *y*.

$$1.2x + 0.8y = 2$$
$$0.8y = -1.2x + 2$$
$$y = -1.5x + 2.5$$

Window: Xmin = –10; Xmax = 10
Ymin = –10; Ymax = 10

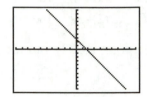

The *x*-intercept is (1.67, 0).
The *y*-intercept is (0, 2.50).

87. To graph an equation on a graphing utility, first solve the equation for *y*.

$$21x - 15y = 53$$
$$15y = 21x - 53$$
$$y = \frac{21}{15}x - \frac{53}{15}$$
$$y = \frac{7}{5}x - \frac{53}{15}$$

Window: Xmin = –10; Xmax = 10
Ymin = –10; Ymax = 10

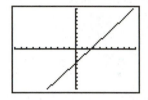

The *x*-intercept is (2.52, 0).
The *y*-intercept is (0, –3.53).

89. To graph an equation on a graphing utility, first solve the equation for *y*.

$$\frac{4}{17}x + \frac{6}{23}y = \frac{2}{3}$$
$$\frac{6}{23}y = -\frac{4}{17}x + \frac{2}{3}$$
$$y = \frac{23}{6}\left(-\frac{4}{17}x + \frac{2}{3}\right)$$
$$y = -\frac{46}{51}x + \frac{23}{9}$$

Window: Xmin = –10; Xmax = 10
Ymin = –10; Ymax = 10

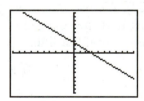

The *x*-intercept is (2.83, 0).
The *y*-intercept is (0, 2.56).

91. To graph an equation on a graphing utility, first solve the equation for *y*.

$$\pi x - \sqrt{3}y = \sqrt{6}$$
$$\sqrt{3}y = \pi x - \sqrt{6}$$
$$y = \frac{\pi}{\sqrt{3}}x - \sqrt{2}$$

Window: Xmin = –10; Xmax = 10
Ymin = –10; Ymax = 10

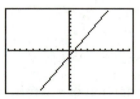

The *x*-intercept is (0.78, 0).
The *y*-intercept is (0, –1.41).

93. The graph passes through the points $(0, 0)$ and $(4, 8)$. We use the points to find the slope of the line:

$$m = \frac{y_2 - y_1}{x_2 - x_1} = \frac{8 - 0}{4 - 0} = \frac{8}{4} = 2$$

The y-intercept $(0, 0)$ is given, so we use the y-intercept and the slope $m = 2$, to get the slope-intercept form of the line:

$$y = mx + b$$
$$y = 2x + 0$$
$$y = 2x \quad \text{which is answer (b).}$$

95. The graph passes through the points $(0, 0)$ and $(2, 8)$. We use the points to find the slope of the line:

$$m = \frac{y_2 - y_1}{x_2 - x_1} = \frac{8 - 0}{2 - 0} = \frac{8}{2} = 4$$

The y-intercept $(0, 0)$ is given, so we use the y-intercept and the slope $m = 4$, to get the slope-intercept form of the line:

$$y = mx + b$$
$$y = 4x + 0$$
$$y = 4x \quad \text{which is answer (d).}$$

97. From the graph we can read the two intercepts, $(-2, 0)$ and $(0, 2)$. Use these points to compute the slope of the line:

$$m = \frac{y_2 - y_1}{x_2 - x_1} = \frac{2 - 0}{0 - (-2)} = \frac{2}{2} = 1$$

We use the y-intercept $(0, 2)$ and the slope $m = 1$, to find the slope-intercept form of the equation:

$$y = mx + b$$
$$y = 1x + 2$$
$$y = x + 2$$

The general form of the equation is:

$$x - y = -2$$

99. From the graph we can read the two intercepts, $(3, 0)$ and $(0, 1)$. Use these points to compute the slope of the line:

$$m = \frac{y_2 - y_1}{x_2 - x_1} = \frac{1 - 0}{0 - 3} = \frac{1}{-3} = -\frac{1}{3}$$

We use the y-intercept $(0, 2)$ and the slope $m = -\frac{1}{3}$, to find the slope-intercept form of the equation:

$$y = mx + b$$
$$y = -\frac{1}{3}x + 1$$

The general form of the equation is:

$$x + 3y = 3$$

101. From the graph we can see that the line has a positive slope and a y-intercept of the form $(0, b)$ where b is a positive number. Put each of the equations into slope-intercept form and choose those with both positive slope and positive y-intercept.

$$\text{(b)} \quad y = \frac{2}{3}x + 2 \qquad \text{(c)} \quad y = \frac{3}{4}x + 3 \qquad \text{(e)} \quad y = x + 1 \qquad \text{(g)} \quad y = 2x + 3$$

103. The x-axis is a horizontal line; its slope is zero. The general equation of the x-axis is $y = 0$.

105. Answers vary.

107. Not every line has two distinct intercepts. A line passing through the origin has the point $(0, 0)$ as both its x- and y-intercept. Usually vertical lines have only an x-intercept and horizontal lines have only a y-intercept.

109. If two lines have the same x-intercept and the same y-intercept, and x-intercept is not $(0, 0)$, then the two lines have equal slopes. Lines that have equal slopes and equal y-intercepts have equivalent equations and identical graphs.

111. Two lines can have the same y-intercept but different slopes only if their y-intercept is the point $(0, 0)$.

Parallel & Inter. Lines 25/1-24 (ODD), 25-43 (ODD)

1.2 Pairs of Lines

1. To determine whether the pair of lines is parallel, coincident, or intersecting, rewrite each equation in slope-intercept form, compare their slopes, and, if necessary, compare their y-intercepts.

$L:$ $x + y = 10$ $M:$ $3x + 3y = 6$
        $y = -x + 10$ $3y = -3x + 6$
                                                          $y = -x + 2$

slope: $m = -1$; y-intercept: $(0, 10)$ slope: $m = -1$; y-intercept: $(0, 2)$

Since the slope of the two lines are the same, but the y-intercepts are different, the lines are parallel.

3. To determine whether the pair of lines is parallel, coincident, or intersecting, rewrite each equation in slope-intercept form, compare their slopes, and, if necessary, compare their y-intercepts.

$L:$ $2x + y = 4$ $M:$ $2x - y = 8$
        $y = -2x + 4$ $-y = -2x + 8$
                                                          $y = 2x - 8$

slope: $m = -2$ slope: $m = 2$

Since the slopes of the two lines are different, the lines intersect.

5. To determine whether the pair of lines is parallel, coincident, or intersecting, rewrite each equation in slope-intercept form, compare their slopes, and, if necessary, compare their y-intercepts.

$L:$ $-x + y = 2$ $M:$ $2x - 2y = -4$
        $y = x + 2$ $-2y = -2x - 4$
                                                          $y = x + 2$

slope: $m = 1$; y-intercept: $(0, 2)$ slope: $m = 1$; y-intercept: $(0, 2)$

Since the slopes of the two lines are the same and the y-intercepts are the same, the lines are coincident.

7. To determine whether the pair of lines is parallel, coincident, or intersecting, rewrite each equation in slope-intercept form, compare their slopes, and, if necessary, compare their *y*-intercepts.

L: $\quad 2x - 3y = -8$
$\qquad\quad -3y = -2x - 8$
$\qquad\qquad\quad y = \dfrac{2}{3}x + \dfrac{8}{3}$

M: $\quad 6x - 9y = -2$
$\qquad\quad -9y = -6x - 2$
$\qquad\qquad\quad y = \dfrac{2}{3}x + \dfrac{2}{9}$

slope: $m = \dfrac{2}{3}$; *y*-intercept: $\left(0, \dfrac{8}{3}\right)$ $\qquad$ slope: $m = \dfrac{2}{3}$; *y*-intercept: $\left(0, \dfrac{2}{9}\right)$

Since the slopes of the two lines are the same, but the *y*-intercepts are different, the lines are parallel.

9. To determine whether the pair of lines is parallel, coincident, or intersecting, rewrite each equation in slope-intercept form, compare their slopes, and, if necessary, compare their *y*-intercepts.

L: $\quad 3x - 4y = 1$
$\qquad\quad -4y = -3x + 1$
$\qquad\qquad\quad y = \dfrac{3}{4}x - \dfrac{1}{4}$

M: $\quad x - 2y = -4$
$\qquad\quad -2y = -x - 4$
$\qquad\qquad\quad y = \dfrac{1}{2}x + 2$

slope: $m = \dfrac{3}{4}$ $\qquad\qquad$ slope: $m = \dfrac{1}{2}$

Since the slopes of the two lines are different, the lines intersect.

11. L: $\quad x = 3$ $\qquad\qquad\qquad\qquad$ M: $\quad y = -2$

slope: not defined; no *y*-intercept $\qquad$ slope: $m = 0$; *y*-intercept: $(0, -2)$

Since the slopes of the two lines are different, the lines intersect.

13. To find the point of intersection of two lines, first put the lines in slope-intercept form.

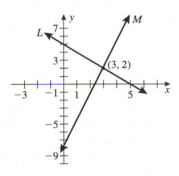

L: $\quad x + y = 5$ $\qquad\qquad$ M: $\quad 3x - y = 7$
$\qquad\quad y = -x + 5$ $\qquad\qquad\qquad\quad y = 3x - 7$

Since the point of intersection, (x_0, y_0), must be on both L and M, we set the two equations equal to each other and solve for x_0. Then we substitute the value of x_0 into the equation of one of the lines to find y_0.

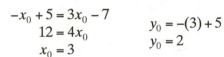

$-x_0 + 5 = 3x_0 - 7$
$\qquad 12 = 4x_0$
$\qquad\; x_0 = 3$

$y_0 = -(3) + 5$
$y_0 = 2$

The point of intersection is $(3, 2)$.

15. To find the point of intersection of two lines, first put the lines in slope-intercept form.

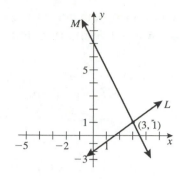

$$L: \quad x - y = 2 \qquad\qquad M: \quad 2x + y = 7$$
$$y = x - 2 \qquad\qquad\qquad y = -2x + 7$$

Since the point of intersection, (x_0, y_0), must be on both L and M, we set the two equations equal to each other and solve for x_0. Then we substitute the value of x_0 into the equation of one of the lines to find y_0.

$$x_0 - 2 = -2x_0 + 7 \qquad\qquad y_0 = 3 - 2$$
$$3x_0 = 9 \qquad\qquad\qquad\quad y_0 = 1$$
$$x_0 = 3$$

The point of intersection is $(3, 1)$.

17. To find the point of intersection of two lines, first put the lines in slope-intercept form.

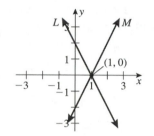

$$L: \quad 4x + 2y = 4 \qquad\qquad M: \quad 4x - 2y = 4$$
$$y = -2x + 2 \qquad\qquad\qquad y = 2x - 2$$

Since the point of intersection, (x_0, y_0), must be on both L and M, we set the two equations equal to each other and solve for x_0. Then we substitute the value of x_0 into the equation of one of the lines to find y_0.

$$-2x_0 + 2 = 2x_0 - 2 \qquad\qquad y_0 = -2(1) + 2$$
$$4 = 4x_0 \qquad\qquad\qquad\quad y_0 = 0$$
$$x_0 = 1$$

The point of intersection is $(1, 0)$.

19. To find the point of intersection of two lines, first put the lines in slope-intercept form.

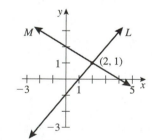

$$L: 3x - 4y = 2 \qquad\qquad M: \ x + 2y = 4$$
$$y = \frac{3}{4}x - \frac{1}{2} \qquad\qquad\qquad y = -\frac{1}{2}x + 2$$

Since the point of intersection, (x_0, y_0), must be on both L and M, we set the two equations equal to each other and solve for x_0. Then we substitute the value of x_0 into the equation of one of the lines to find y_0.

$$\frac{3}{4}x_0 - \frac{1}{2} = -\frac{1}{2}x_0 + 2 \qquad\qquad y_0 = -\frac{1}{2}(2) + 2$$
$$3x_0 - 2 = -2x_0 + 8 \qquad\qquad\quad y_0 = 1$$
$$5x_0 = 10$$
$$x_0 = 2$$

The point of intersection is $(2, 1)$.

21. To find the point of intersection of two lines, first put the lines in slope-intercept form.

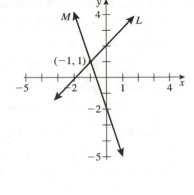

$$L: \ 3x - 2y = -5$$
$$y = \frac{3}{2}x + \frac{5}{2}$$

$$M: \ 3x + y = -2$$
$$y = -3x - 2$$

Since the point of intersection, (x_0, y_0), must be on both L and M, we set the two equations equal to each other and solve for x_0. Then we substitute the value of x_0 into the equation of one of the lines to find y_0.

$$\frac{3}{2}x_0 + \frac{5}{2} = -3x_0 - 2$$
$$3x_0 + 5 = -6x_0 - 4$$
$$9x_0 = -9$$
$$x_0 = -1$$

$$y_0 = -3(-1) - 2$$
$$y_0 = 1$$

The point of intersection is $(-1, 1)$.

23. L is the vertical line on which the x-value is always 4.

M is the horizontal line on which y-value is always -2.

The point of intersection is $(4, -2)$.

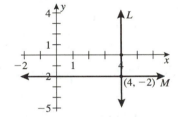

25. To show two lines are perpendicular we show that the product of their slopes equals negative one. To find the slope of each line, write both equations in slope-intercept form.

$$-x + 3y = 2$$
$$y = \frac{1}{3}x + \frac{2}{3}$$

$$6x + 2y = 5$$
$$2y = -6x + 5$$
$$y = -3x + \frac{5}{2}$$

$$m_1 = \frac{1}{3}$$

$$m_2 = -3$$

Since $m_1 m_2 = \frac{1}{3} \cdot (-3) = -1$, the lines are perpendicular.

27. To show two lines are perpendicular we show that the product of their slopes equals negative one. To find the slope of each line, write both equations in slope-intercept form.

$$x + 2y = 7 \qquad\qquad\qquad -2x + y = 15$$
$$2y = -x + 7 \qquad\qquad\qquad y = 2x + 15$$
$$y = -\frac{1}{2}x + \frac{7}{2}$$
$$m_1 = -\frac{1}{2} \qquad\qquad\qquad m_2 = 2$$

Since $m_1 m_2 = -\frac{1}{2} \cdot (2) = -1$, the lines are perpendicular.

29. To show two lines are perpendicular we show that the product of their slopes equals negative one. To find the slope of each line, write both equations in slope-intercept form.

$$3x + 12y = 2 \qquad\qquad\qquad 4x - y = -2$$
$$12y = -3x + 2 \qquad\qquad\qquad -y = -4x - 2$$
$$y = -\frac{1}{4}x + \frac{1}{6} \qquad\qquad\qquad y = 4x + 2$$
$$\qquad\qquad\qquad\qquad\qquad m_2 = 4$$
$$m_1 = -\frac{1}{4}$$

Since $m_1 m_2 = -\frac{1}{4} \cdot (4) = -1$, the lines are perpendicular.

31. L is parallel to $y = 2x$, so the slope of L, $m = 2$.
On the graph we are given the point $(3, 3)$ on line L.
Use the point-slope form of the line.

$$y - y_1 = m(x - x_1)$$
$$y - 3 = 2(x - 3)$$
$$y - 3 = 2x - 6$$

slope-intercept form: $\qquad y = 2x - 3$
general form: $\qquad 2x - y = 3$

33. L is perpendicular to $y = 2x$, so the slope of L, $m = -\frac{1}{2}$. On the graph we are given the point $(3, 3)$ on the line L.
Use the point-slope form of the line.

$$y - y_1 = m(x - x_1)$$
$$y - 3 = -\frac{1}{2}(x - 3)$$
$$y - 3 = -\frac{1}{2}x + \frac{3}{2}$$

slope-intercept form: $\qquad y = -\frac{1}{2}x + \frac{9}{2}$
general form: $\qquad x + 2y = 9$

35. We want a line parallel to $y = 4x$. So our line will have a slope $m = 4$. It must also contain the point $(-1, 2)$.

Use the point slope form of the equation of a line:

$$y - y_1 = m(x - x_1)$$
$$y - 2 = 4(x + 1)$$
$$y - 2 = 4x + 4$$

slope-intercept form: $\qquad y = 4x + 6$
general form: $\quad 4x - y = -6$

37. We want a line parallel to $2x - y = -2$.

Find the slope of the line and use the given point, $(0, 0)$ to obtain the equation.
Since the y-intercept $(0, 0)$ is given, use the slope-intercept form of the equation of a line.

Original line: $\qquad\qquad\qquad\qquad$ Parallel line:
$$2x - y = -2$$
$$\qquad y = 2x + 2 \qquad\qquad\qquad y = mx + b$$
$$\qquad m = 2 \qquad\qquad\qquad\qquad\quad y = 2x$$

general form: $\quad 2x - y = 0$

39. We want a line parallel to the line $x = 3$. This is a vertical line so the slope is not defined. A parallel line will also be vertical, and it must contain the point $(4, 2)$. The parallel line will have the equation:

$$x = 4$$

41. We want a line perpendicular to $y = 2x - 5$ which has a slope $m = 2$.

So the line we are seeking will have a slope $m = -\dfrac{1}{2}$.

Use the slope and the point $(-1, -2)$ to find the point-slope form of the equation.

$$y - y_1 = m(x - x_1)$$
$$y + 2 = -\frac{1}{2}(x + 1)$$
$$y + 2 = -\frac{1}{2}x - \frac{1}{2}$$

slope-intercept form: $\qquad y = -\dfrac{1}{2}x - \dfrac{5}{2}$
general form: $\quad x + 2y = -5$

43.   We want a line perpendicular to the line $y = 2x - 5$. The perpendicular line must contain the point $\left(-\frac{1}{3}, \frac{4}{5}\right)$ and will have a slope $m = -\frac{1}{2}$. Use the point-slope form of the equation.

$$y - y_1 = m(x - x_1)$$
$$y - \frac{4}{5} = -\frac{1}{2}\left(x + \frac{1}{3}\right)$$

slope-intercept form: $\qquad\qquad y = -\frac{1}{2}x + \frac{19}{30}$

general form: $15x + 30y = 19$

45.   If the equations

$tx - 4y = -3$ and $2x + 2y = 5$

are perpendicular, then the product of their slopes will be –1. So first find the slope of each equation by rewriting the equation in slope-intercept form.

Equation one: Equation two:

$tx - 4y = -3$
$\quad -4y = -tx - 3$ $2x + 2y = 5$
$\qquad\qquad\qquad\qquad 2y = -2x + 5$

$\qquad y = \frac{t}{4}x + \frac{3}{4}$

$\qquad\qquad\qquad\qquad y = -x + \frac{5}{2}$

$\quad m_1 = \frac{t}{4}$ $m_2 = -1$

For the lines to be perpendicular, the product of their slopes must be –1.

$$m_1 m_2 = -1$$
$$\frac{t}{4} \cdot (-1) = -1$$
$$t = 4$$

47. To find the equation of the line, we must first find the slope of the line containing the points $(-2, 9)$ and $(3, -10)$.

$$m = \frac{y_2 - y_1}{x_2 - x_1} = \frac{9 - (-10)}{(-2) - 3} = \frac{19}{-5} = -\frac{19}{5}$$

The slope of a line perpendicular to the line containing these points is $\frac{5}{19}$. Use the point $(-2, -5)$ and the slope $m = \frac{5}{19}$ to get the point-slope form of the perpendicular line

$$y - y_1 = m(x - x_1)$$
$$y + 5 = \frac{5}{19}(x + 2)$$

Solving for y we can get the slope-intercept form:

$$y = \frac{5}{19}x - \frac{85}{19}$$

Multiplying both sides by 19 and rearranging terms will give the general form of the equation: $5x - 19y = 85$

49. The two lines in the graph are parallel. Parallel lines have equal slopes, so to determine which set of equations is parallel we must compare their slopes. Do this by writing each equation in slope-intercept form.

   Also notice that the slopes of the graphed equations are positive, and one line has a positive y-intercept and the other has a negative y-intercept.

   Use all this information to answer the question.

(c)
$$\begin{array}{ll} x - y = -2 & \quad x - y = 1 \\ y = x + 2 & \quad y = x - 1 \end{array}$$

both equations have slope $m = 1$. The first has y-intercept $(0, 2)$ and the second has y-intercept $(0, -1)$.

1.3 Applications: Prediction; Break-Even Point; Mixture Problems; Economics

1. **Predicting Sales** The company's sales are given by, $s = \$5000x + \$80,000$, where x stands for years and $x = 0$ corresponds to the year 2002.

   (a) When $x = 0$, that is in 2002,

   $$S = 5000(0) + 80,000$$
   $$S = \$80,000$$

   (b) When $x = 3$, that is in 2005,

   $$S = 5000(3) + 80,000$$
   $$S = \$95,000$$

   (c) If the trend continues, sales in 2007 should be equal to S when $x = 5$.

   $$S = 5000(5) + 80,000$$
   $$S = \$105,000$$

   (d) If the trend continues, sales in 2010 should be equal to S when $x = 8$.

   $$S = 5000(8) + 80,000$$
   $$S = \$120,000$$

3. **Predicting the Cost of a Compact Car** Since the relationship is assumed to be linear we will first find the slope between (2000, 12,500) and (2003, 14,450).

$$m = \frac{y_2 - y_1}{x_2 - x_1} = \frac{14,450 - 12,500}{2003 - 2000} = \frac{1950}{3} = 650$$

We use the slope $m = 650$ and the point (2000, 12,500) to write the point-slope form of the equation.

$$y - y_1 = m(x - x_1)$$
$$y - 12,500 = 650(x - 2000)$$

Simplifying, we get the slope-intercept form of the equation.

$$y - 12,500 = 650x - 1,300,000$$
$$y = 650x - 1,287,500$$

To predict the average cost of a compact car in 2005, evaluate y with $x = 2005$.

$$y = 650(2005) - 1,287,500$$
$$y = 15,750$$

So, we can predict that a compact car will cost $15,750 in 2005.
The slope of the prediction equation, $m = 650$, can be interpreted as the average yearly increase in price of a compact car.

5. **SAT Scores**   Since the problem states that the rate of increase is constant, we use a linear equation to predict future SAT scores.

(a)   We need to find the slope of the line containing the points (1987, 468) and (2001, 488).

$$m = \frac{S_2 - S_1}{t_2 - t_1} = \frac{488 - 468}{2001 - 1987} = \frac{20}{14} = \frac{10}{7}$$

Using the point (2001, 488) and the slope $m = \dfrac{10}{7}$, we get the point-slope form of the equation of a line.

$$S - S_1 = m(t - t_1)$$
$$S - 488 = \frac{10}{7}(t - 2001)$$

Simplifying, we get the slope-intercept form of the equation.

$$S - 488 = \frac{10}{7}t - \frac{20010}{7}$$
$$S = \frac{10}{7}t - \frac{16594}{7}$$

(b)   To find the predicted average SAT score in South Carolina in 2004, evaluate S with $t = 2004$,

$$S = \frac{10}{7}(2004) - \frac{16594}{7}$$
$$\approx 492.29$$

Since SAT scores are reported as whole numbers, we round S to 492. So we would expect the average score to be 492.

7.   **Percent of Population with Bachelor's Degrees**   Since the problem states that the rate of increase is constant, we use a linear equation to predict the percent of the population older than 25 holding Bachelor's Degrees. We will use P for the percent of the population and t for the year.

(a)   We find the slope of the equation containing the points (1990, 20.3) and (2000, 25.6).

$$m = \frac{P_2 - P_1}{t_2 - t_1} = \frac{25.6 - 20.3}{2000 - 1990} = \frac{5.3}{10} = 0.53$$

We use the point (2000, 25.6) and the slope $m = 0.53$ to get the point-slope form of the equation of the line.

$$P - P_1 = m(t - t_1)$$
$$P - 25.6 = 0.53(t - 2000)$$

Simplifying, we get the slope-intercept form of the equation.

$$P - 25.6 = 0.53t - 1060$$
$$P = 0.53t - 1034.4$$

(b)   To determine the percentage of people over 25 who will have a bachelor's degree or higher by 2004, let $t = 2004$ in the equation and simplify.

$$P = 0.53(2004) - 1034.4$$
$$= 27.72$$

This suggests that in 2004, 27.7% of the population over 25 will hold a bachelor's degree or higher.

(c)   The slope of the equation $m = 0.53$ is the annual percentage increase of the population over 25 who hold bachelor's degrees or higher.

9.   The break-even point is the point where the revenue and the cost are equal.

Setting $R = C$, we find

$$30x = 10x + 600$$
$$20x = 600$$
$$x = 30$$

That is, 30 units must be sold to break even.
Break-even point (30, 900).

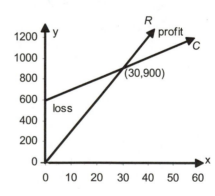

11.   The break-even point is the point where the revenue and the cost are equal.

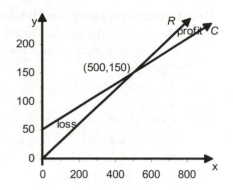

Setting $R = C$, we find

$0.30x = 0.20x + 50$
$0.10x = 50$
$\quad x = 500$

That is, 500 units must be sold to break even.
Break-even point: (500, 150)

13.   **Break-Even Point**   The break-even point is the point where the revenue and the cost are equal.

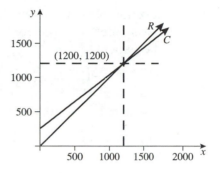

Cost is given by the variable cost of producing x units at \$0.75 per unit, plus the fixed operational overhead of \$300 per day.

$C = \$0.75x + \$300$

Revenue is the product of price of each item (\$1) and the number of items sold.

$R = \$1x$

Setting $R = C$, we find

$\quad 1x = 0.75x + 300$
$0.25x = 300$
$\quad x = 1200$

That is, 1200 items must be sold each day to break even.

15. Profit from Sunday Home Delivery

(a) Revenue is the selling price per copy of the paper times the number of copies sold.

$R = \$1.79x$

(b) The cost of delivery is the sum of the variable cost of delivering the newspapers and the fixed cost.

$C = \$0.53x + \$1,070,000$

(c) Profit is the difference between the revenue obtained from selling the newspapers and the cost of delivering the papers.

$$P = 1.79x - \left(0.53x + 1,070,000\right)$$
$$= 1.79x - 0.53x - 1,070,000$$
$$= 1.26x - 1,070,000$$

(d) The break-even point is the point where the revenue and the cost are equal, or it can be considered the point where the profit equals zero.

$$0 = 1.26x - 1,070,000$$
$$1,070,000 = 1.26x$$
$$849,206.35 = x$$

The Tribune has to deliver 849,207 Sunday papers to break even.

(e) Graph of the break-even point:

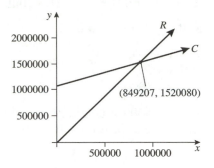

(f) Graph of profit:

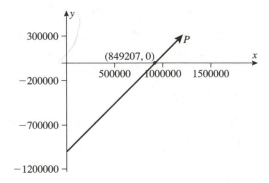

(g) The *x*-value of the break-even point and the *x*-intercept of the profit graph are the same, because the break-even point, is the point where profit equals zero.

17. **Mixture Problem** We will let
x = the number of caramels the box of candy, and
y = the number of creams in the box of candy.

Since there are a total of 50 pieces of candy in a box, we have

$x + y = 50$, or
$\quad y = 50 - x$

Each caramel costs \$0.10 to make, and each cream costs \$0.20 to make. Therefore, the cost of making a box of candy is given by the equation:

$C = 0.1x + 0.2y$
$\quad = 0.1x + 0.2(50 - x)$

The box of candy sells for \$8.00. So to break even, we need

$R = C$
$8 = 0.1x + 0.2(50 - x)$
$8 = 0.1x + 10 - 0.2x$
$8 = 10 - 0.1x$
$0.1x = 2$
$\quad x = 20$

If 20 caramels and 30 creams are put into the box, neither a profit nor a loss will occur.

If the candy shop owner increases the number of caramels to more than 20 (and decreases the number of creams) the owner will obtain a profit since the caramels cost less to produce than the creams.

19. **Investment Problem** Investment problems are simply mixture problems involving money. We will use a table to organize the information.

Investment	Amount Invested	Interest Rate	Interest Earned
AA Bonds	x	10% = 0.10	$0.10x$
S & L Certificates	$y = 150{,}000 - x$	5% = 0.05	$0.05y = 0.05(150{,}000 - x)$
Total	$x + y = 150{,}000$		\$10,000

The last column gives the equation we need to solve since the sum of the interest earned on the two investments must equal the total interest earned.

$0.1x + 0.05(150{,}000 - x) = 10{,}000$
$0.1x + 7500 - 0.05x = 10{,}000$
$0.05x + 7500 = 10{,}000$
$0.05x = 2500$
$x = 50{,}000$

Mr. Nicholson should invest \$50,000 in AA Bonds and \$100,000 in Savings and Loan Certificates in order to earn \$10,000 per year.

21. **Mixture Problem**   We will use the hint and assume that the total weight of the blend is 100 pounds.

Coffee	Amount Mixed	Price per Pound	Total Value
Kona	x	$22.95	$22.95x$
Columbian	$y = 100 - x$	$6.75	$6.75y = 6.75(100 - x)$
Mixture	$x + y = 100$	$10.80	$10.80(100) = \$1080$

The last column gives the information necessary to write the equation, since the sum of the values of each of the two individual coffees must equal the total value of the mixture.

$$22.95x + 6.75(100 - x) = 10.80(100)$$
$$22.95x + 675 - 6.75x = 1080$$
$$16.2x = 405$$
$$x = 25$$

Mix 25 pounds of Kona coffee with 75 pounds of Columbian coffee to obtain 100 pounds of coffee worth $10.80 per pound.

23. **Mixture Problem**   This is a classical mixture problem. Let's use a table to organize the information.

Ingredients	Amount Used	Acidity	Solution
Acid A	x	$15\% = 0.15$	$0.15x$
Acid B	$y = 100 - x$	$5\% = 0.05$	$0.05y = 0.05(100 - x)$
Total	$x + y = 100$	$8\% = 0.08$	$0.08(100) = 8$

The last column gives the information necessary to determine how much of each of the two ingredients should be mixed, since the sum of the two individual solutions must equal the total solution.

$$0.15x + 0.05(100 - x) = 8$$
$$0.15x + 5 - 0.05x = 8$$
$$0.1x = 3$$
$$x = 30$$

This indicates that we should mix 30 cubic centimeters of 15% solution with 70 cubic centimeters of 5% solution to obtain a 100 cubic centimeters of 8% solution.

25. The market price is the price at which the supply and the demand are equal.

$$S = D$$
$$p + 1 = 3 - p$$
$$2p = 2$$
$$p = 1$$

At a price of $1.00 supply and demand are equal, so $1.00 is the market price.

27. The market price is the price at which the supply and the demand are equal.

$$S = D$$
$$20p + 500 = 1000 - 30p$$
$$50p = 500$$
$$p = 10$$

At a price of $10.00 supply and demand are equal, so $10.00 is the market price.

29. **Market Price of Sugar** The market price is the price at which the supply and the demand are equal.

$$S = D$$
$$0.7p + 0.4 = -0.5p + 1.6$$
$$1.2p = 1.2$$
$$p = 1$$

The market price is $1.00.

To find the amount supplied when the market price is $1.00, let $p = 1$ and solve for S:

$$S = 0.7(1) + 0.4$$
$$= 1.1$$

So 1.1 units of supply are demanded at the market price of $1.00.

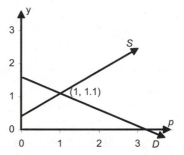

The point of intersection is the market price and the quantity supplied and demanded at the market price.

31. **Supply and Demand Problem** At the market price of $3.00,

$$S = 2(3) + 5 = 11$$

So 11 units of the commodity are supplied. Since at the market price the supply and demand are equal, the point (3, 11) satisfies the demand equation. In addition, we are told that at $1.00, there are 19 units of the commodity demanded. To find the demand equation, use the points (3, 11) and (1, 19) to find the slope.

$$m = \frac{D_2 - D_1}{p_2 - p_1} = \frac{19 - 11}{1 - 3} = \frac{8}{-2} = -4$$

Use the point (1, 19) and the slope $m = -4$ to get the point-slope form of the equation.

$$D - D_1 = m(p - p_1)$$
$$D - 19 = -4(p - 1)$$
$$D = -4p + 23$$

1.4 Scatter Diagrams; Linear Curve Fitting

1. A relation exists, and it appears to be linear.

3. A relation exists, and it appears to be linear.

5.   No relation exists.

7.   (a) and (c)

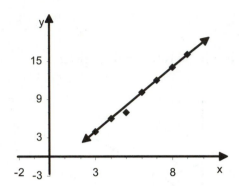

(b)   Answers will vary. We select points
       (3, 4) and (9, 16). The slope of the line
       containing these points is:

$$m = \frac{y_2 - y_1}{x_2 - x_1} = \frac{16 - 4}{9 - 3} = \frac{12}{6} = 2$$

The equation of the line is:

$$y - y_1 = m(x - x_1)$$
$$y - 4 = 2(x - 3)$$
$$y - 4 = 2x - 6$$
$$y = 2x - 2$$

(c)   The line on the scatter diagram on
       the left will vary depending on your
       choice of points in part (b).

(d)   Window: Xmin = –2; Xmax = 10
                    Ymin = –3; Ymax = 20

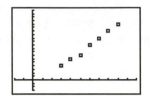

(e)   Using the LinReg program, the line
       of best fit is:

$$y = 2.0357x - 2.357$$

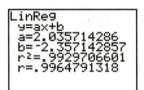

(f)

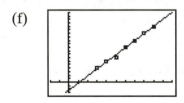

9. (a) and (c)

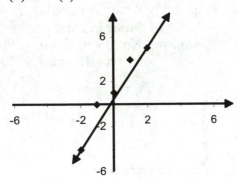

(b) Answers will vary. We select points $(-2, -4)$ and $(2, 5)$. The slope of the line containing these points is:

$$m = \frac{y_2 - y_1}{x_2 - x_1} = \frac{5 - (-4)}{2 - (-2)} = \frac{9}{4}$$

The equation of the line is:

$$y - y_1 = m(x - x_1)$$
$$y - 5 = \frac{9}{4}(x - 2)$$
$$y - 5 = \frac{9}{4}x - \frac{9}{2}$$
$$y = \frac{9}{4}x + \frac{1}{2}$$

(c) The line on the scatter diagram on the left will vary depending on your choice of points in part (b).

(d) Window: Xmin = –6; Xmax = 6
　　　　　Ymin = –6; Ymax = 7

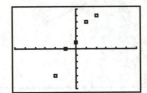

(e) Using the LinReg program the line of best fit is:

$$y = 2.2x + 1.2$$

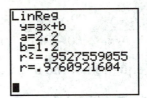

(f)

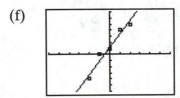

11.   (a) and (c)

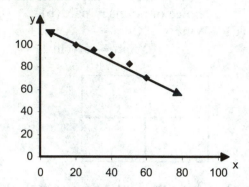

(b)   Answers will vary. We select points (20, 100) and (60, 70). The slope of the line containing these points is:

$$m = \frac{y_2 - y_1}{x_2 - x_1} = \frac{70 - 100}{60 - 20} = \frac{-30}{40} = -\frac{3}{4}$$

The equation of the line is:

$$y - y_1 = m(x - x_1)$$
$$y - 100 = -\frac{3}{4}(x - 20)$$
$$y - 100 = -\frac{3}{4}x + 15$$
$$y = -\frac{3}{4}x + 115$$

(c)   The line on the scatter diagram on the left will vary depending on your choice of points in part (b).

(d)   Window: Xmin = 0; Xmax = 100
                      Ymin = 1; Ymax = 120

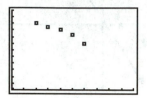

(e)   Using the LinReg function the line of best fit is:

$$y = -0.72x + 116.6$$

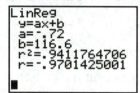

(f)

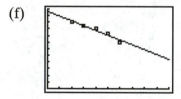

13.　(a) and (c)

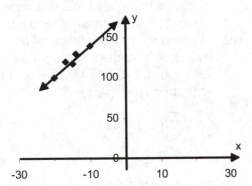

(b)　Answers will vary. We select points (–20, 100) and (–10, 140). The slope of the line containing these points is:

$$m = \frac{y_2 - y_1}{x_2 - x_1} = \frac{140 - 100}{(-10) - (-20)} = \frac{40}{10} = 4$$

The equation of the line is:

$$y - y_1 = m(x - x_1)$$
$$y - 100 = 4(x - (-20))$$
$$y - 100 = 4x + 80$$
$$y = 4x + 180$$

(c)　The line on the scatter diagram on the left will vary depending on your choice of points in part (b).

(d)　Window:　Xmin = –30; Xmax = 10
　　　　　　　Ymin = 0; Ymax = 160

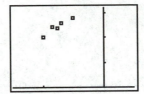

(e)　Using the LinReg function, the line of best fit is:

$$y = 3.86131x + 180.29197$$

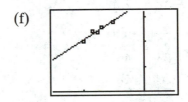

(f)

15.   Consumption and Disposable Income

(a)

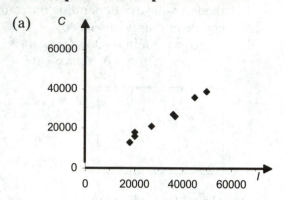

(b)   Answers will vary. We select points (20, 16) and (50, 39) numbers in thousands. The slope of the line containing these points is:

$$m = \frac{C_2 - C_1}{I_2 - I_1} = \frac{39 - 16}{50 - 20} = \frac{23}{30}$$

The equation of the line is:

$$C - C_1 = m(I - I_1)$$
$$C - 16 = \frac{23}{30}(I - 20)$$
$$C - 16 = \frac{23}{30}I - \frac{46}{3}$$
$$C = \frac{23}{30}I + \frac{2}{3}$$

(c)   The slope of this line indicates that a family will spend $23 of every extra $30 of disposable income.

(d)   To find the consumption of a family whose disposable income is $42,000, substitute 42 for x in the equation from part (b).

$$C = \frac{23}{30}(42) + \frac{2}{3}$$
$$= \frac{473}{15}$$
$$= 32.86667$$

So the family will spend $32,867.

(e)   Using the LinReg function on the graphing utility, the line of best fit is:

$$y = 0.75489x + 0.62663$$

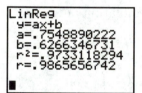

17. Mortgage Qualification   The problem was done with income and loan amounts in thousands of dollars.

(a)   Window: Xmin = 0; Xmax = 75.5
Ymin = 0; Ymax = 236.5

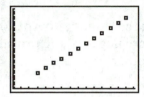

(b)   Using the LinReg function, the line of best fit is:

$y = 2.98140x - 0.07611$

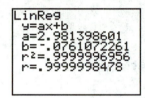

(c)

(d)   The slope indicates that a person can borrow an additional $2.98 for each additional dollar of income.

(e)   To determine the amount for which an individual earning $42,000 would qualify, evaluate the line of best fit at $x = 42$.

$y = 2.9814(42) - 0.07611$
$= 125.14263$

So an individual with an income of $43,000 would qualify for a $125,143 mortgage.

19. Apparent Room Temperature

(a)   window: Xmin = –10; Xmax = 110
Ymin = 50; Ymax = 70

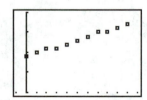

(b)   Using the LinReg function, the line of best fit is:

$y = 0.07818x + 59.0909$

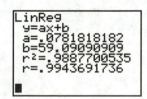

(c)

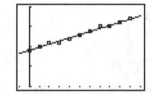

(d)   The slope indicates the apparent change in temperature in a 65°F room for every percent increase in relative humidity.

(e)   To determine the apparent temperature when the relative humidity is 75%, evaluate the equation of the line of best fit when $x = 75$.

$y = 0.07818(75) + 59.0909$
$= 64.95$

So, if the relative humidity is 75%, the temperature of the room will appear to be 65°F.

Chapter 1 Linear Equations; Review

True-False Items

1. False

3. True

5. False; parallel lines never have the same intercept.

7. False; the slopes of perpendicular lines are negative reciprocals of each other.

9. False; if the lines have the same intercepts, they are coincident.

Fill In The Blanks

1. abscissa ... ordinate, or
x-coordinate ... *y*-coordinate

3. negative

5. coincident

7. intersecting

Review Exercises

1.

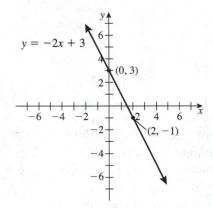

3.

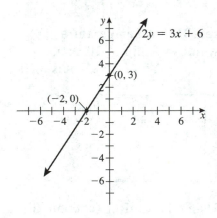

5. (a) $m = \dfrac{y_2 - y_1}{x_2 - x_1} = \dfrac{4 - 2}{(-3) - 1} = \dfrac{2}{-4} = -\dfrac{1}{2}$

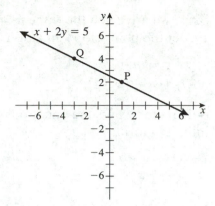

A slope of $-\dfrac{1}{2}$ means that for every 2 unit change in x, y will change (-1) unit. That is, for every 2 units x moves to the right, y will move down 1 unit.

(b) Use the point $(1, 2)$ and the slope to get the point-slope form of the equation of the line:

$$y - y_1 = m(x - x_1)$$
$$y - 2 = -\dfrac{1}{2}(x - 1)$$

Simplifying and solving for y gives the slope-intercept form:

$$y - 2 = -\dfrac{1}{2}x + \dfrac{1}{2}$$
$$y = -\dfrac{1}{2}x + \dfrac{5}{2}$$

Rearranging terms gives the general form of the equation:

$$x + 2y = 5$$

7. (a) $m = \dfrac{y_2 - y_1}{x_2 - x_1} = \dfrac{5 - 3}{(-1) - (-2)} = \dfrac{2}{1} = 2$

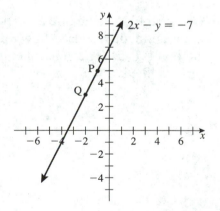

A slope of 2 means that for every 1 unit change in x, y will change by 2 units. That is, for every 1 unit x moves to the right, y will move up 2 units.

(b) Use the point $(-1, 5)$ and the slope to get the point-slope form of the equation:

$$y - y_1 = m(x - x_1)$$
$$y - 5 = 2(x - (-1))$$

Simplifying and solving for y, gives the slope-intercept form:

$$y - 5 = 2x + 2$$
$$y = 2x + 7$$

Rearranging terms gives the general form of the equation:

$$2x - y = -7$$

9.   Since we are given the slope $m = -3$ and a point, we get the point-slope equation of the line:

$$y - y_1 = m(x - x_1)$$
$$y - (-1) = -3(x - 2)$$

Solving for y puts the equation into the slope-intercept form:

$$y + 1 = -3x + 6$$
$$y = -3x + 5$$

Rearranging terms gives the general form of the equation: $3x + y = 5$

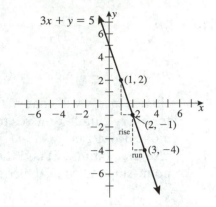

11.   We are given the slope $m = 0$ and a point on the line. We either use the point-slope formula or recognize that this is a horizontal line and the equation of a horizontal line is: $y = b$.

$$y = 4$$

This is the general form of the equation.

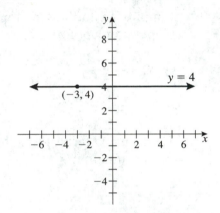

13.   We are told that the line is vertical, so the slope is not defined. We also know the line contains the point $(8, 5)$. The general equation of a vertical line is $x = a$

$$x = 8$$

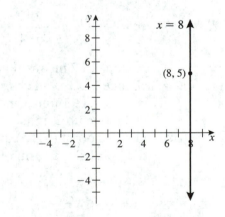

15. We are given the *x*-intercept and a point. First we find the slope of the line containing the two points.

$$m = \frac{y_2 - y_1}{x_2 - x_1} = \frac{(-5) - 0}{4 - 2} = \frac{-5}{2} = -\frac{5}{2}$$

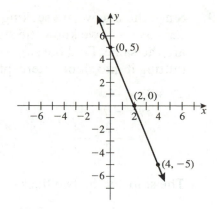

We then use the point $(2, 0)$ and the slope to get the point-slope form of the equation of the line.

$$y - y_1 = m(x - x_1)$$
$$y - 0 = -\frac{5}{2}(x - 2)$$

To get the slope-intercept form, solve for *y*:

$$y = -\frac{5}{2}x + 5$$

Rearrange the terms to get the general form of the equation of the line.

$$5x + 2y = 10$$

17. We are given two points, the *x*-intercept and the *y*-intercept. Use the two points to find the slope of the line.

$$m = \frac{y_2 - y_1}{x_2 - x_1} = \frac{(-4) - 0}{0 - (-3)} = \frac{-4}{3} = -\frac{4}{3}$$

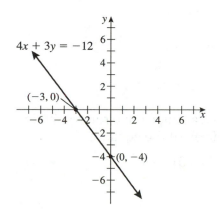

Now since one of the points is the *y*-intercept, use it and the slope to get the slope-intercept form of the equation:

$$y = -\frac{4}{3}x - 4$$

Rearrange the terms to obtain the general form of the equation:

$$4x + 3y = -12$$

19. Since the line we are seeking is parallel to $2x + 3y = -4$; we know the slope of the two lines are the same. Find the slope of the given line by putting it into slope-intercept form:

$$2x + 3y = -4$$
$$3y = -2x - 4$$
$$y = -\frac{2}{3}x - \frac{4}{3}$$

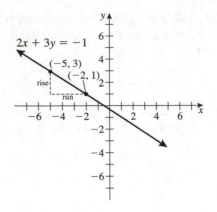

The slope of the two lines is $m = -\frac{2}{3}$. Use the slope and the point $(-5, 3)$ to write the point-slope form of the equation of the parallel line.

$$y - y_1 = m(x - x_1)$$
$$y - 3 = -\frac{2}{3}(x - (-5))$$
$$y - 3 = -\frac{2}{3}(x + 5)$$

To put the equation into slope-intercept form, solve for y.

$$y - 3 = -\frac{2}{3}x - \frac{10}{3}$$
$$y = -\frac{2}{3}x - \frac{1}{3}$$

Rearrange the terms to obtain the general form of the equation:

$$2x + 3y = -1$$

21. We seek a line perpendicular to $2x + 3y = -4$.
We find the slope of this line:

$$3y = -2x - 4$$
$$y = -\frac{2}{3}x - \frac{4}{3}$$
$$m = -\frac{2}{3}$$

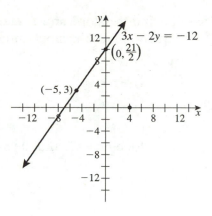

The slope of the perpendicular line obeys

$$m \cdot \left(-\frac{2}{3}\right) = -1$$
$$m = \frac{3}{2}$$

Use the slope and the point $(-5, 3)$ to write the point-slope equation of the perpendicular line:

$$y - y_1 = m(x - x_1)$$
$$y - 3 = \frac{3}{2}(x - (-5))$$

Simplify and solve for y to obtain the slope-intercept form.

$$y - 3 = \frac{3}{2}x + \frac{15}{2}$$
$$y = \frac{3}{2}x + \frac{21}{2}$$

Rearrange the terms for the general form of the equation:

$$3x - 2y = -21$$

23. To find the slope and y-intercept of the line, put the equation into the slope-intercept form.

$$9x + 2y = 18$$
$$2y = -9x + 18$$
$$y = -\frac{9}{2}x + 9$$

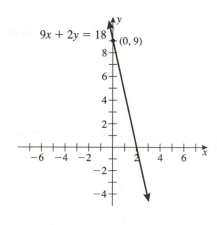

The slope $m = -\frac{9}{2}$, and the y-intercept is $(0, 9)$.

25. To find the slope and y-intercept of the line, put the equation into the slope-intercept form.

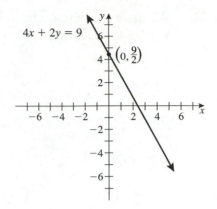

$$4x + 2y = 9$$
$$2y = -4x + 9$$
$$y = -2x + \frac{9}{2}$$

The slope $m = -2$, and the y-intercept is $\left(0, \dfrac{9}{2}\right)$.

27. We put each equation into slope-intercept form:

$$3x - 4y = -12 \qquad\qquad 6x - 8y = -9$$
$$-4y = -3x - 12 \qquad\qquad -8y = -6x - 9$$
$$y = \frac{3}{4}x + 3 \qquad\qquad y = \frac{6}{8}x + \frac{9}{8}$$
$$\qquad\qquad\qquad\qquad y = \frac{3}{4}x + \frac{9}{8}$$

Since both lines have the same slope $m = \dfrac{3}{4}$ but different y-intercepts, the lines are parallel.

29. We put each equation into slope-intercept form:

$$x - y = -2 \qquad\qquad 3x - 4y = -12$$
$$-y = -x - 2 \qquad\qquad -4y = -3x - 12$$
$$y = x + 2 \qquad\qquad\quad y = -\frac{3}{4}x + 3$$

Since the lines have different slopes, they intersect.

31. We put each equation into slope-intercept form:

$$4x + 6y = -12 \qquad\qquad 2x + 3y = -6$$
$$6y = -4x - 12 \qquad\qquad 3y = -2x - 6$$
$$y = -\frac{2}{3}x - 2 \qquad\qquad y = -\frac{2}{3}x - 2$$

Since both lines have the same slope and the same y-intercept, the lines are coincident.

33. To find the point of intersection of two lines, first put the lines in slope-intercept form:

L: $x - y = 4$ $\qquad$ M: $x + 2y = 7$

$\qquad\quad$ $y = x - 4$ $\qquad\qquad\quad$ $y = -\dfrac{1}{2}x + \dfrac{7}{2}$

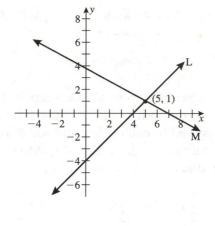

Since the point of intersection, (x_0, y_0), must be on both L and M, we set the two equations equal to each other and solve for x_0. Then we substitute the value of x_0 into the equation of one of the lines to find y_0.

$$x_0 - 4 = -\frac{1}{2}x_0 + \frac{7}{2} \qquad y_0 = x_0 - 4$$
$$2x_0 - 8 = -x_0 + 7 \qquad\quad y_0 = 5 - 4$$
$$3x_0 = 15 \qquad\qquad\quad y_0 = 1$$
$$x_0 = 5$$

The point of intersection is $(5, 1)$.

35. To find the point of intersection of two lines, first put the lines in slope-intercept form.

L: $x - y = -2$ $\qquad$ M: $x + 2y = 7$

$\qquad\quad$ $y = x + 2$ $\qquad\qquad\quad$ $2y = -x + 7$

$\qquad\qquad\qquad\qquad\qquad\qquad$ $y = -\dfrac{1}{2}x + \dfrac{7}{2}$

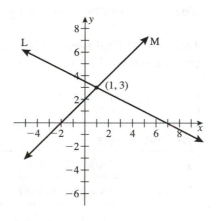

Since the point of intersection, (x_0, y_0), must be on both L and M, we set the two equations equal to each other and solve for x_0. Then we substitute the value of x_0 into the equation of one of the lines to find y_0.

$$x_0 + 2 = -\frac{1}{2}x_0 + \frac{7}{2} \qquad y_0 = x_0 + 2$$
$$2x_0 + 4 = -x_0 + 7 \qquad\quad y_0 = 1 + 2$$
$$3x_0 = 3 \qquad\qquad\quad y_0 = 3$$
$$x_0 = 1$$

The point of intersection is $(1, 3)$.

37. To find the point of intersection of two lines, first put the lines in slope-intercept form.

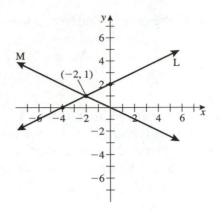

$$L: \quad 2x - 4y = -8 \qquad\qquad M: \quad 3x + 6y = 0$$
$$-4y = -2x - 8 \qquad\qquad\qquad 6y = -3x$$
$$y = \frac{1}{2}x + 2 \qquad\qquad\qquad\quad y = -\frac{1}{2}x$$

Since the point of intersection, (x_0, y_0), must be on both L and M, we set the two equations equal to each other and solve for x_0. Then we substitute the value of x_0 into the equation of one of the lines to find y_0.

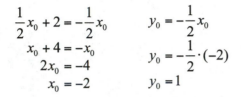

$$\frac{1}{2}x_0 + 2 = -\frac{1}{2}x_0 \qquad\qquad y_0 = -\frac{1}{2}x_0$$
$$x_0 + 4 = -x_0 \qquad\qquad\qquad\quad$$
$$2x_0 = -4 \qquad\qquad\quad y_0 = -\frac{1}{2}\cdot(-2)$$
$$x_0 = -2 \qquad\qquad\qquad y_0 = 1$$

The point of intersection is $(-2, 1)$.

39. Investment Problem   It is often convenient to use a table to organize the information in an investment problem.

	Amount Invested	Interest Rate	Interest Earned
B-Bonds	x	$12\% = 0.12$	$0.12x$
Bank	$y = 90{,}000 - x$	$5\% = 0.05$	$0.05y = 0.05(90{,}000 - x)$
Total	$x + y = 90{,}000$		$\$10{,}000$

The last column gives the information needed for the equation since the sum of the interest earned on the individual investments must equal the total interest earned.

$$0.12x + 0.05\big(90{,}000 - x\big) = 10{,}000$$
$$0.12x + 4500 - 0.05x = 10{,}000$$
$$0.07x = 5500$$
$$x = 78{,}571.429$$

Mr. and Mrs. Byrd should invest $78,571.43 in B-rated bonds and $11,428.57 in the well-known bank in order to achieve their investment goals.

41. Attendance at a Dance

(a) The break-even point is the point where the cost equals the revenue, or when the profit is zero. Before we can find the break-even point we need the equation that describes cost.

We are told the fixed costs, the band and the advertising, and the variable costs. If we let x denote the number of tickets sold, the cost of the dance is described by the equation:

$$C = 500 + 100 + 5x$$
$$= 5x + 600$$

The revenue is given by the equation, $R = 10x$, since each ticket costs $10. Setting $C = R$, and solving for x, will tell how many tickets must be sold to break even.

$$5x + 600 = 10x$$
$$600 = 5x$$
$$x = 120$$

So 120 tickets must be sold to break even.

(b) Profit is the difference between the revenue and the cost. To determine the number of tickets that need to be sold to clear a profit of $900, we will solve the equation:

$$P = R - C$$
$$900 = 10x - (5x + 600)$$
$$900 = 5x - 600$$
$$1500 = 5x$$
$$x = 300$$

The church group must sell 300 tickets to realize a profit of $900.

(c) If tickets cost $12, the break-even point will come from the equation

$$R = C$$
$$12x = 5x + 600$$
$$7x = 600$$
$$x = 85.71$$

To break even, 86 tickets must be sold.
To find the number of ticket sales needed to have a $900 profit, solve the equation:

$$P = R - C$$
$$900 = 12x - (5x + 600)$$
$$900 = 7x - 600$$
$$1500 = 7x$$
$$x = 214.29$$

So the church group needs to sell 215 tickets at $12 each to realize a profit of $900.

43.

This relation is not linear.

45. **Concentration of Carbon Monoxide in the Air**

(a)

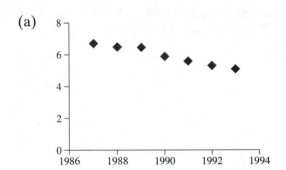

(b) $m = \dfrac{y_2 - y_1}{x_2 - x_1} = \dfrac{5.87 - 6.69}{1990 - 1987} = \dfrac{-0.82}{3} = -0.273$

(c) The slope indicates the average annual decrease in concentration of carbon monoxide between 1987 and 1990.

(d) $m = \dfrac{y_2 - y_1}{x_2 - x_1} = \dfrac{4.88 - 5.87}{1993 - 1990} = \dfrac{-0.99}{3} = -0.33$

(e) The slope indicates the average annual decrease in concentration of carbon monoxide between 1990 and 1993.

(f) Using the LinReg function on a graphing utility, the slope of the line of best fit is

$a = -.30786$

```
LinReg
 y=ax+b
 a=-.3078571429
 b=618.4771429
 r²=.980579603
 r=-.9902421941
■
```

(g) The slope indicates the average annual decrease in the concentration of carbon monoxide between 1987 and 1993.

47. **Value of a Portfolio**

(a)

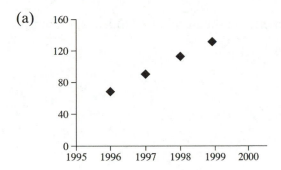

(b) The relation appears to be linear.

(c) $m = \dfrac{135.33 - 69.17}{1999 - 1996} = \dfrac{66.16}{3} = 22.05$

(d) The slope represents the average annual increase in value of a share of the Vanguard 500 Index Fund from 1996 to 1999.

(e) Using the LinReg function on a graphing utility, the line of best fit is:

$y = 22.236x + 44314.28$

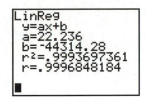

(f) In 2001,

$y = 22.236(2001) - 44314.28$

$y = 179.956$

So if the trend continues, the predicted value of a share of Vanguard 500 Index Fund would be $179.96

49. Answers will vary.

Chapter 1 Project

1. Since Avis charges a flat rate that does not depend on the number of miles, x, driven the equation is that of a horizontal line.

    $A = 64.99$

3.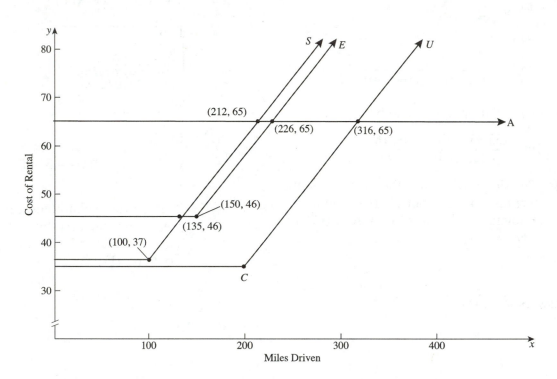

5. By planning your sightseeing carefully and figuring how far you intend to drive, the information in question 4 will help you to decide which car to rent. If you are going to drive more than 226 miles, renting from Avis would be more economical. Otherwise, you would save money by renting from Enterprise.

7. See graph in Problem 3.

9.   Usave Car Rental also has a two tiered pricing policy similar to Enterprise and AutoSave. The equations describing the cost of renting a Usave car are:

$U = 35.99$ when $x \le 200$
$U = 35.99 + 0.25(x - 200)$ when $x > 200$
   $= 35.99 + 0.25x - 50$
   $= 0.25x - 14.01$

Usave Car Rental is always less expensive than Enterprise and AutoSave. Usave's base rate is lower than either of the two, and it offers more free miles than both the other companies. There will be a point, however, where Avis again becomes more economical. To determine at which mileage Avis is the better deal, find the point of intersection of A and U. Consider only $x_0 > 200$.

$$U = A$$
$$0.25x_0 - 14.01 = 64.99$$
$$0.25x_0 = 79$$
$$x_0 = 316$$

Therefore, if you drive more than 316 miles, the Avis car which costs $64.99 is the least expensive to rent.

Mathematical Questions Form Professional Exams

1.   The break-even point is the value of x for which the revenue equals cost. If x units are sold at price of $2.00 each, the revenue is $R = 2x$.
        Cost is the total of the fixed and variable costs. We are told that the fixed costs are $6000, and that the variable cost per item is 40% of the price. So the cost is given by the equation:

$C = (0.40)(2)x + 6000$
   $= 0.80x + 6000$

Setting $R = C$ and solving for x yields,

  $R = C$
 $2x = 0.8x + 6000$
$1.2x = 6000$
   $x = 5000$

Answer: (b) 5000 units

3.   The break-even point is the number of units that must be sold for revenue to equal cost. Using the notation given, we have:

$R = SPx$ and $C = VCx + FC$

To find the sales level necessary to break even, we set $R = C$ and solve for x.

$$R = C$$
$$SPx = VCx + FC$$
$$SPx - VCx = FC$$
$$(SP - VC)x = FC$$
$$x = \frac{FC}{SP - VC}$$

Answer: (d)

5.   Since straight-line depreciation remains the same over the life of the property, its expense over time will be a horizontal line. Sum-of-year's-digits depreciation expense decreases as time increases.
Answer: (c)

7.   Answer: (b); Y is an estimate of total factory overhead.

Chapter 2

Systems of Linear Equations; Matrices

2.1 Systems of Linear Equations: Substitution; Elimination

1. To be a solution to the system of equations, the given values of the variables must solve each equation. So we evaluate each equation at $x = 2$ and $y = -1$.

$$2x - y = 5 \qquad\qquad 5x + 2y = 8$$
$$2(2) - (-1) = 4 + 1 = 5 \qquad 5(2) + 2(-1) = 10 - 2 = 8$$

Since both equations are satisfied, $x = 2$, $y = -1$ is a solution to the system.

3. To be a solution to the system of equations, the given values of the variables must solve each equation. So we evaluate each equation at $x = 2$ and $y = \dfrac{1}{2}$.

$$3x + \quad 4y = 4$$
$$3(2) + 4\left(\frac{1}{2}\right) = 8 \neq 4$$

Since the first equation is not satisfied, $x = 2$, $y = \dfrac{1}{2}$ is not a solution to the system.

5. To be a solution to the system of equations, the given values of the variables must solve each equation. So we evaluate each equation at $x = 4$ and $y = 1$.

$$\begin{aligned} &\qquad\qquad\quad \frac{1}{2}x + y = 3 \\ x - y &= 3 \\ 4 - 1 &= 3 \\ &\qquad\qquad\quad \frac{1}{2}(4) + 1 = 3 \end{aligned}$$

Since both equations are satisfied, $x = 4$, $y = 1$ is a solution to the system.

7.   To be a solution to the system of equations, the given values of the variables must solve each equation. So we evaluate each equation at $x = 1$, $y = -1$, and $z = 2$.

$$\begin{array}{lll} 3x + 3y + 2z = 4 & x - y - z = 0 & 2y - 3z = -8 \\ 3(1) + 3(-1) + 2(2) = 4 & 1 - (-1) - (2) = 0 & 2(-1) - 3(2) = -8 \end{array}$$

Since all three equations are satisfied, $x = 1$, $y = -1$, $z = 2$ is a solution to the system.

9.   To be a solution to the system of equations, the given values of the variables must solve each equation. So we evaluate each equation at $x = 2$, $y = -2$, and $z = 2$.

$$\begin{array}{lll} 3x + 3y + 2z = 4 & x - 3y + z = 10 & 5x - 2y - 3z = 8 \\ 3(2) + 3(-2) + 2(2) = 4 & (2) - 3(-2) + 2 = 10 & 5(2) - 2(-2) - 3(2) = 8 \end{array}$$

Since all three equations are satisfied, $x = 2$, $y = -2$, $z = 2$ is a solution to the system.

11.   Choosing the method of elimination to solve this system,

$$\begin{cases} x + y = 8 & (1) \\ x - y = 4 & (2) \end{cases}$$

$$\begin{array}{ll} 2x = 12 & \text{Add the two equations.} \\ x = 6 & \text{Solve for } x. \\ 6 + y = 8 & \text{Back-substitute 6 for } x \text{ in equation (1).} \\ y = 2 & \text{Solve for } y. \end{array}$$

The solution to the system is $x = 6$, $y = 2$.

13.   Choosing the method of elimination to solve this system,

$$\begin{cases} 5x - y = 13 & (1) \\ 2x + 3y = 12 & (2) \end{cases}$$

$$\begin{array}{ll} 15x - 3y = 39 \quad (1) & \text{Multiply equation (1) by 3.} \\ 17x = 51 & \text{Add (1) to equation (2).} \\ x = 3 & \text{Solve for } x. \\ 5(3) - y = 13 & \text{Back-substitute 3 for } x \text{ in equation (1).} \\ y = 2 & \text{Solve for } y. \end{array}$$

The solution to the system is $x = 3$, $y = 2$.

15.   Choosing the method of substitution to solve this system,

$$\begin{cases} 3x = 24 & (1) \\ x + 2y = 0 & (2) \end{cases}$$

$$\begin{array}{ll} x = 8 \quad (1) & \text{Solve equation (1) for } x. \\ 8 + 2y = 0 \quad (2) & \text{Substitute 8 for } x \text{ in equation (2).} \\ y = -4 & \text{Solve for } y. \end{array}$$

The solution to the system is $x = 8$, $y = -4$.

17. Choosing the method of elimination to solve this system,

$$\begin{cases} 3x - 6y = 2 & (1) \\ 5x + 4y = 1 & (2) \end{cases}$$

$$6x - 12y = 4 \quad (1)$$

Multiply equation (1) by 2.

$$15x + 12y = 13 \quad (2)$$

Multiply equation (2) by 3.

$$21x \quad\quad = 7$$

Add equations (1) and (2).

Solve for x.

$$x = \frac{1}{3}$$

Back-substitute $\dfrac{1}{3}$ for x in equation (1).

$$3\left(\frac{1}{3}\right) - 6y = 2$$

Solve for y.

$$y = -\frac{1}{6}$$

The solution to the system is $x = \dfrac{1}{3}$, $y = -\dfrac{1}{6}$.

19. Choosing the method of elimination to solve this system,

$$\begin{cases} 2x + y = 1 & (1) \\ 4x + 2y = 3 & (2) \end{cases}$$

$$4x + 2y = 2 \quad (1)$$

Multiply equation (1) by 2.

$$0 = 1$$

Subtract (1) from equation (2). This results in a contradiction.

There is no solution to this system of equations. The system is inconsistent.

21. Choosing the method of substitution to solve this system,

$$\begin{cases} 2x - y = 0 & (1) \\ 3x + 2y = 7 & (2) \end{cases}$$

$$y = 2x$$

Solve equation (1) for y.

$$3x + 2(2x) = 7$$

Substitute $2x$ for y in equation (2).

$$7x = 7$$

Simplify.

$$x = 1$$

Solve for x.

$$y = 2$$

Back-substitute 1 for x in equation (1) and solve for y.

The solution to the system of equations is $x = 1$ and $y = 2$.

23. Choosing the method of elimination to solve this system,

$$\begin{cases} x + 2y = 4 & \text{(1)} \\ 2x + 4y = 8 & \text{(2)} \end{cases}$$

$$2x + 4y = 8 \quad \text{(1)}$$
$$0 = 0$$

Multiply equation (1) by 2.
Subtract equation (1) from equation (2).
The equations are dependent.

There are infinitely many solutions to the system of equations. They can be written as $x = 4 - 2y$ and y, where y is any real number, or as $y = -\dfrac{1}{2}x + 2$ and x, where x is any real number.

25. Choosing the method of substitution to solve this system,

$$\begin{cases} 2x - 3y = -1 & \text{(1)} \\ 10x + y = 11 & \text{(2)} \end{cases}$$

$$y = 11 - 10x \quad \text{(2)}$$
$$2x - 3(11 - 10x) = -1$$
$$2x - 33 + 30x = -1$$
$$32x = 32$$
$$x = 1$$
$$y = 11 - 10(1) \quad \text{(2)}$$
$$y = 1$$

Solve equation (2) for y.
Substitute $11 - 10x$ for y in equation (1).
Simplify.

Solve for x.
Back-substitute 1 for x in equation (2).
Simplify.

The solution of the system is $x = 1$, $y = 1$.

27. Choosing the method of elimination to solve this system,

$$\begin{cases} 2x + 3y = 6 & \text{(1)} \\ x - y = \dfrac{1}{2} & \text{(2)} \end{cases}$$

$$2x - 2y = 1 \quad \text{(2)}$$
$$5y = 5$$
$$y = 1$$
$$2x + 3(1) = 6$$
$$2x = 3$$
$$x = \dfrac{3}{2}$$

Multiply equation (2) by 2.
Subtract equation (2) from equation (1).
Solve for y.
Back-substitute 1 for y in equation (1).
Simplify.

Solve for x.

The solution of the system is $x = \dfrac{3}{2}$, $y = 1$.

29. Choosing the method of elimination to solve this system,

$$\begin{cases} \dfrac{1}{2}x + \dfrac{1}{3}y = 3 & \text{(1)} \\[2mm] \dfrac{1}{4}x - \dfrac{2}{3}y = -1 & \text{(2)} \end{cases}$$

$\begin{aligned} 12x + 8y &= 72 \quad \text{(1)} \\ 3x - 8y &= -12 \quad \text{(2)} \\ 15x &= 60 \\ x &= 4 \end{aligned}$	Multiply equation (1) by 24. Multiply equation (2) by 12. Add equations (1) and (2). Solve for x.

$$\dfrac{1}{4}(4) - \dfrac{2}{3}y = -1$$

Back-substitute 4 for x in equation (2).

$$-\dfrac{2}{3}y = -2$$

Simplify.

$$y = 3$$

Solve for y.

The solution of the system is $x = 4$, $y = 3$.

31. Choosing the method of elimination to solve this system,

$$\begin{cases} 3x - 5y = 3 & \text{(1)} \\ 15x + 5y = 21 & \text{(2)} \end{cases}$$

$$18x = 24$$ Add equations (1) and (2).

$$x = \dfrac{4}{3}$$ Solve for x.

$$3\left(\dfrac{4}{3}\right) - 5y = 3$$ Back-substitute $\dfrac{4}{3}$ for x in equation (1).

$$4 - 5y = 3$$ Simplify.

$$y = \dfrac{1}{5}$$ Solve for y.

The solution of the system is $x = \dfrac{4}{3}$, $y = \dfrac{1}{5}$.

33.

$$\begin{cases} x - y = 6 & \text{(1)} \\ 2x - 3z = 16 & \text{(2)} \\ 2y + z = 4 & \text{(3)} \end{cases}$$

$$\begin{aligned} 2x - 2y &= 12 \quad \text{(1)} \\ 2y + 3z &= 4 \quad \text{(2)} \\ 2z &= 0 \\ z &= 0 \end{aligned}$$

Multiply equation (1) by 2.
Subtract equation (1) from (2).
Subtract equation (3) from (2).
Solve for z.
Back-substitute 0 for z in equation (2).

$$2x - 3(0) = 16$$

Solve for x.

$$x = 8$$

Back-substitute 8 for x in equation (1).

$$8 - y = 6$$

Solve for y.

$$y = 2$$

The solution of the system is $x = 8$, $y = 2$, and $z = 0$.

35. $\begin{cases} x - 2y + 3z = 7 & (1) \\ 2x + y + z = 4 & (2) \\ -3x + 2y - 2z = -10 & (3) \end{cases}$

$\begin{array}{ll} x - 2y + 3z = 7 & (1) \\ 2x + y + z = 4 & (2) \end{array}$ Multiply by 2 $\begin{array}{ll} x - 2y + 3z = 7 & (1) \\ 4x + 2y + 2z = 8 & (2) \\ \hline 5x \qquad + 5z = 15 & \text{Add} \end{array}$

$\begin{array}{ll} x - 2y + 3z = 7 & (1) \\ -3x + 2y - 2z = -10 & (2) \\ \hline -2x \qquad + z = -3 & \text{Add} \end{array}$

$\begin{cases} x - 2y + 3z = 7 & (1) \\ 5x \qquad + 5z = 15 & (2) \\ -2x \qquad + z = -3 & (3) \end{cases}$

Working only with equations (2) and (3),

$\begin{array}{ll} 5x + 5z = 15 & (2) \\ -2x + z = -3 & (3) \end{array}$ Multiply by (–5) $\begin{array}{ll} 5x + 5z = 15 & (2) \\ 10x - 5z = 15 & (3) \\ \hline 15x \qquad = 30 & \text{or } x = 2 \end{array}$

Back-substitute 2 for x in equation (3) and solve for z.

$\begin{array}{ll} -2x + z = -3 & (3) \\ -2(2) + z = -3 & (3) \\ z = 1 \end{array}$

Finally back-substitute $x = 2$ and $z = 1$ in equation (1) and solve for y.

$\begin{array}{ll} x - 2y + 3z = 7 & (1) \\ 2 - 2y + 3(1) = 7 \\ y = -1 \end{array}$

The solution of the original system is $x = 2$, $y = -1$, and $z = 1$.

37. $\begin{cases} x - y - z = 1 & (1) \\ 2x + 3y + z = 2 & (2) \\ 3x + 2y = 0 & (3) \end{cases}$

Since (3) has no z term we will eliminate z first.

$\begin{array}{ll} x - y - z = 1 & (1) \\ 2x + 3y + z = 2 & (2) \\ \hline 3x + 2y = 3 & \text{(Add) (2)} \end{array}$

We now use the revised (2) and (3).

$\begin{array}{ll} 3x + 2y = 3 & (2) \\ 3x + 2y = 0 & (3) \\ \hline 0 = 3 & \text{(Subtract) (3)} \end{array}$

Equation (3) has no solution and the system is inconsistent.

39. $\begin{cases} x - y - z = 1 & (1) \\ -x + 2y - 3z = -4 & (2) \\ 3x - 2y - 7z = 0 & (3) \end{cases}$

$\begin{array}{ll} x - y - z = 1 & (1) \\ \underline{-x + 2y - 3z = -4} & (2) \\ y - 4z = -3 & \text{(Add)} \ (2) \end{array}$

$\begin{array}{ll} x - y - z = 1 \quad (1) \\ 3x - 2y - 7z = 0 \quad (3) \end{array}$ Multiply by -3 $\begin{array}{ll} -3x + 3y + 3z = -3 & (1) \\ \underline{3x - 2y - 7z = 0} & (3) \\ y - 4z = -3 & \text{(Add)} \ (3) \end{array}$

We now used revised (2) and (3).

$\begin{array}{ll} y - 4z = -3 & (2) \\ \underline{y - 4z = -3} & (3) \\ 0 = 0 & (3) \end{array}$

The original system is equivalent to a system containing 2 equations, so the equations are dependent and the system has infinitely many solutions. If z represents any real number, then we get $y = 4z - 3$. Substituting into (1) we get

$\begin{array}{l} x - y - z = 1 \\ x - (4z - 3) - z = 1 \ \text{ or } \ x = 5z - 2 \end{array}$

The solution to the system is $\begin{cases} x = 5z - 2 \\ y = 4z - 3 \end{cases}$ where z can be any real number.

41. $\begin{cases} 2x - 2y + 3z = 6 & (1) \\ 4x - 3y + 2z = 0 & (2) \\ -2x + 3y - 7z = 1 & (3) \end{cases}$

$\begin{array}{ll} 2x - 2y + 3z = 6 & (1) \\ 4x - 3y + 2z = 0 & (2) \end{array}$ Multiply by -2 $\begin{array}{ll} -4x + 4y - 6z = -12 & (1) \\ \underline{4x - 3y + 2z = 0} & (2) \\ y - 4z = -12 & \text{(Add)} \ (2) \end{array}$

$\begin{array}{ll} 2x - 2y + 3z = 6 & (1) \\ \underline{-2x + 3y - 7z = 1} & (3) \\ y - 4z = 7 & \text{(Add)} \ (3) \end{array}$

We now use the revised (2) and (3).

$\begin{array}{ll} y - 4z = -12 & (2) \\ \underline{y - 4z = 7} & (3) \\ 0 = -5 & \text{(Subtract)} \ (3) \end{array}$

Equation (3) has no solution and the system is inconsistent.

43.
$$\begin{cases} x + y - z = 6 & (1) \\ 3x - 2y + z = -5 & (2) \\ x + 3y - 2z = 14 & (3) \end{cases}$$

$$\begin{array}{ll} x + y - z = 6 & (1) \\ 3x - 2y + z = -5 & (2) \end{array} \quad \text{Multiply by 3} \quad \begin{array}{ll} 3x + 3y - 3z = 18 & (1) \\ \underline{3x - 2y + z = -5} & (2) \\ 5y - 4z = 23 & \text{(Subtract) (2)} \end{array}$$

$$\begin{array}{ll} x + y - z = 6 & (1) \\ \underline{x + 3y - 2z = 14} & (3) \\ -2y + z = -8 & \text{(Subtract) (3)} \end{array}$$

We now work with revised (2) and (3) and solve for y.

$$\begin{array}{ll} 5y - 4z = 23 & (2) \\ -2y + z = -8 & (3) \end{array} \quad \text{Multiply by 4} \quad \begin{array}{ll} 5y - 4z = 23 & (2) \\ \underline{-8y + 4z = -32} & (3) \\ -3y = -9 & \text{(Add) (3)} \end{array}$$

We solve (3): $-3y = -9$ getting $y = 3$.
Back-substitute $y = 3$ into (2): $5(3) - 4z = 23$ getting $15 - 4z = 23$ or $z = -2$.
Then back substitute $y = 3$ and $z = -2$ into (1): $x + 3 - (-2) = 6$ or $x = 1$.

The solution to the system is $x = 1$, $y = 3$, and $z = -2$.

45.
$$\begin{cases} x + 2y - z = -3 & (1) \\ 2x - 4y + z = -7 & (2) \\ -2x + 2y - 3z = 4 & (3) \end{cases}$$

$$\begin{array}{ll} x + 2y - z = -3 & (1) \\ 2x - 4y + z = -7 & (2) \end{array} \quad \text{Multiply by 2} \quad \begin{array}{ll} 2x + 4y - 2z = -6 & (1) \\ \underline{2x - 4y + z = -7} & (2) \\ 8y - 3z = 1 & \text{(Subtract) (2)} \end{array}$$

$$\begin{array}{ll} x + 2y - z = -3 & (1) \\ -2x + 2y - 3z = 4 & (3) \end{array} \quad \text{Multiply by 2} \quad \begin{array}{ll} 2x + 4y - 2z = -6 & (1) \\ \underline{-2x + 2y - 3z = 4} & (3) \\ 6y - 5z = -2 & \text{(Add) (3)} \end{array}$$

We now work with revised (2) and (3) and solve for z.

$$\begin{array}{ll} 8y - 3z = 1 & (2) \\ 6y - 5z = -2 & (3) \end{array} \quad \begin{array}{l} \text{Multiply by 3} \\ \text{Multiply by 4} \end{array} \quad \begin{array}{ll} 24y - 9z = 3 & (2) \\ \underline{24y - 20z = -8} & (3) \\ 11z = 11 & \text{(Subtract) (3)} \end{array}$$

We solve (3) for z: $11z = 11$ or $z = 1$.

Back-substitute $z = 1$ into (2) to solve for y: $8y - 3(1) = 1$ or $y = \dfrac{1}{2}$.

Last back substitute $y = \dfrac{1}{2}$ and $z = 1$ into (1) to solve for x: $x + 2\left(\dfrac{1}{2}\right) - 1 = -3$ or $x = -3$.

The solution to the system is $x = -3$, $y = \dfrac{1}{2}$, and $z = 1$.

47. We let l represent the length of the floor and w represent its width. We are told the perimeter is 90 feet and that the length is twice the width. We need to solve the system of equations $\begin{cases} 2l + 2w = 90 & (1) \\ l = 2w & (2) \end{cases}$

We solve this system by substitution. Substituting $2w$ for l in (1), we get

$$2(2w) + 2w = 90 \quad (1)$$
$$6w = 90$$
$$w = 15$$

Substituting 15 for w in (2) gives $l = 2(15) = 30$.
The floor has a length of 30 feet and a width of 15 feet.

49. We let x denote the number of acres of corn planted and y denote the number of acres of soybeans planted. We solve the system of equations

$$\begin{cases} x + y = 445 & (1) \\ 246x + 140y = 85,600 & (2) \end{cases}$$

We solve this system by substitution. First solve equation (1) for y, $y = 445 - x$, then substitute $445 - x$ for y in equation (2).

$$246x + 140y = 85,600 \quad (2)$$
$$246x + 140(445 - x) = 85,600$$
$$246x + 62,300 - 140x = 85,600$$
$$106x = 23,300$$
$$x = 219.811 \quad \text{and} \quad y = 445 - 219.811 = 225.189$$

The farmer should plant 219.8 acres of corn and 225.2 acres of soybeans.

51. Let x represent the number of cashews in the mixture and let y represent the total weight of the mix. To find the amount of cashews needed for the mixture, we solve the system of equations

$$\begin{cases} 30 + x = y & (1) \\ 5.00x + 1.50(30) = 3.00y & (2) \end{cases}$$

We solve this system by substitution, replacing y in (2) with $30 + x$.

$$5x + 1.5(30) = 3(30 + x)$$
$$5x + 45 = 90 + 3x$$
$$2x = 45$$
$$x = 22.5 \quad \text{and} \quad y = 30 + x = 55.5$$

The manager should add 22.5 pounds of cashews to the peanuts to make the mixture.

53.   Let x represent the cost of a bowl of noodles and y represent the cost of a carton of fresh milk. To find the cost of each, we solve the system of equations

$$\begin{cases} 3x + 2y = 2153 & \text{(1)} \\ 3y - 89 = x & \text{(2)} \end{cases}$$

We solve this system by substitution, replacing $3y + 89$ for x in equation (1).

$$3(3y - 89) + 2y = 2153$$
$$9y - 267 + 2y = 2153$$
$$11y = 2420$$
$$y = 220 \quad \text{and} \quad x = 3(220) - 89 = 571$$

A bowl of noodles costs 571 yen and a carton of fresh milk costs 220 yen.

55.   Let x represent the cost of a pound of bacon and y represent the cost of a carton of eggs. To find each item's price, we solve the system of equations

$$\begin{cases} 3x + 2y = 7.45 & \text{(1)} \\ 2x + 3y = 6.45 & \text{(2)} \end{cases}$$

We solve this system using the method of elimination.

$3x + 2y = 7.45$	(1)	Multiply by 2	$6x + 4y = 14.90$	(1)
$2x + 3y = 6.45$	(2)	Multiply by 3	$6x + 9y = 19.35$	(2)
			$-5y = -4.45$	(Subtract) (2)
			$y = 0.89$	

Back-substitute 0.89 for y in equation (1) to find x.

$$3x + 2(0.89) = 7.45 \quad \text{(1)}$$
$$3x = 5.67$$
$$x = 1.89$$

Bacon cost \$1.89 per pound and eggs cost \$0.89. So the refund when we return 2 pounds of bacon and 2 cartons of eggs will be $2(1.89) + 2(0.89) = \$5.56$.

57.   Let x represent the milligrams of liquid 1 and y represent the milligrams of liquid 2 necessary to obtain the desired mixture. To learn the amount of each liquid to use, we solve the system of equations

$$\begin{cases} 0.20x + 0.40y = 40 & \text{(1)} \\ 0.30x + 0.20y = 30 & \text{(2)} \end{cases} \quad \text{or} \quad \begin{cases} 2x + 4y = 400 & \text{(1)} \\ 3x + 2y = 300 & \text{(2)} \end{cases}$$

We will solve this system of equations by the method of elimination.

$2x + 4y = 400$	(1)		$2x + 4y = 400$	(1)
$3x + 2y = 300$	(2)	Multiply by 2	$6x + 4y = 600$	(2)
			$-4x = -200$	(Subtract) (2)
			$x = 50$	

Back-substitute 50 for x in equation (1) and solve for y.

$$2(50) + 4y = 400 \quad \text{(1)}$$
$$4y = 300 \quad \text{or} \quad y = 75$$

The pharmacist should mix 50 mg of liquid 1 and 75 mg of liquid 2 to fill the prescription.

59. Let x represent the pounds of rolled oats and y represent the pounds of molasses in the horse's diet. To determine the amount of oats and molasses to feed the horse, we solve the system of equations

$$\begin{cases} 0.41x + 3.35y = 33 & \text{(1)} \\ 1.95x + 0.36y = 21 & \text{(2)} \end{cases}$$

We will solve the system using the method of substitution. First we solve equation (1) for x:

$x = \dfrac{33 - 3.35y}{0.41}$. Then we substitute this value for x in equation (2) and solve for y.

$$1.95\left(\frac{33 - 3.35y}{0.41}\right) + 0.36y = 21$$

$$64.35 - 6.5325y + 0.1476y = 8.61$$
$$64.35 - 6.3849y = 8.61$$
$$6.3849y = 55.74$$
$$y = 8.730$$

Back-substituting into (1) we get $x = \dfrac{33 - 3.35(8.730)}{0.41} = 9.157$

The farmer should feed the horse 9.157 pounds of rolled oats and 8.730 pounds of molasses each day.

61. Let x represent the number of orchestra seats, y represent the number of main seats, and z represent the number of balcony seats in the theater. To find the number of each kind of seat solve the system of equations

$$\begin{cases} x + y + z = 500 & \text{(1)} \\ 50x + 35y + 25z = 17{,}100 & \text{(2)} \\ 50\left(\dfrac{1}{2}x\right) + 35y + 25z = 14{,}600 & \text{(3)} \end{cases}$$

$$\begin{array}{ll} x + y + z = 500 & \text{(1)} \\ 50x + 35y + 25z = 17{,}100 & \text{(2)} \end{array} \quad \text{Multiply by 50} \quad \begin{array}{ll} 50x + 50y + 50z = 25{,}000 & \text{(1)} \\ \underline{50x + 35y + 25z = 17{,}100} & \text{(2)} \\ 15y + 25z = 7900 & \text{(Subtract) (2)} \end{array}$$

$$\begin{array}{ll} x + y + z = 500 & \text{(1)} \\ 25x + 35y + 25z = 14{,}600 & \text{(3)} \end{array} \quad \text{Multiply by 25} \quad \begin{array}{ll} 25x + 25y + 25z = 12{,}500 & \text{(1)} \\ \underline{25x + 35y + 25z = 14{,}600} & \text{(3)} \\ -10y = -2{,}100 & \text{(Subtract) (3)} \end{array}$$

Solve (3) for y: $y = 210$

Back-substitute 210 for y in (2) and solve for z: $15(210) + 25z = 7900$
$$3150 + 25z = 7900$$
$$25z = 4750 \text{ or } z = 190$$

Finally substitute for both y and z in equation (1) and solve for x: $x + 190 + 210 = 500$ or $x = 100$.

There are 100 orchestra seats, 210 main seats, and 190 balcony seats in the theater.

63. Let x represent the amount Kelly invests in treasury bills, y represent the amount she invests in treasury bonds, and z represent the amount she invests in corporate bonds. To determine how much Kelly should invest in each account, we solve the system of equations

$$\begin{cases} x + y + z = 20,000 & (1) \\ 0.05x + 0.07y + 0.10z = 1390 & (2) \\ x - 3000 = z & (3) \end{cases}$$

$$\begin{array}{llll} x + y + z = 20,000 & (1) & \text{Multiply by 7} & 7x + 7y + 7z = 140,000 \quad (1) \\ 0.05x + 0.07y + 0.10z = 1390 & (2) & \text{Multiply by 100} & \underline{5x + 7y + 10z = 139,000} \quad (2) \\ & & & 2x \quad\quad - 3z = 1000 \quad \text{(Subtract) (2)} \end{array}$$

Now we will substitute $z = x + 3000$ in equation (2) to solve for x.

$$2x - 3(x - 3000) = 1000$$
$$2x - 3x + 9000 = 1000$$
$$x = 8000$$

From (3) we find that $z = 8000 - 3000 = 5000$, and from (1) we get $y = 20,000 - 8000 - 5000 = 7000$.

Kelly should invest \$8000 in treasury bills, \$7000 in treasury bonds, and \$5000 in corporate bonds to achieve her investment goals.

65. Answers will vary.

2.2 Systems of Linear Equations: Matrix Method

1. $\begin{bmatrix} 2 & -3 & | & 5 \\ 1 & -1 & | & 3 \end{bmatrix}$

3. $\begin{bmatrix} 2 & 1 & | & -6 \\ 3 & 1 & | & -1 \end{bmatrix}$

5. $\begin{bmatrix} 2 & -1 & -1 & | & 0 \\ 1 & -1 & 1 & | & 1 \\ 3 & -1 & 0 & | & 2 \end{bmatrix}$

7. $\begin{bmatrix} 2 & -3 & 1 & | & 7 \\ 1 & 1 & -1 & | & 1 \\ 2 & 2 & -3 & | & -4 \end{bmatrix}$

9. $\begin{bmatrix} 4 & -1 & 2 & -1 & | & 4 \\ 1 & 1 & 0 & 0 & | & -6 \\ 0 & 2 & -1 & 1 & | & 5 \end{bmatrix}$

11. $\begin{bmatrix} 1 & -1 & 1 & -1 & | & 0 \\ 2 & 3 & -1 & 4 & | & 5 \end{bmatrix}$

13. $\begin{bmatrix} 1 & -3 & | & -2 \\ 2 & -5 & | & 5 \end{bmatrix} \xrightarrow{R_2 = -2r_1 + r_2} \begin{bmatrix} 1 & -3 & | & -2 \\ -2(1)+2 & -2(-3)+(-5) & | & -2(-2)+5 \end{bmatrix} = \begin{bmatrix} 1 & -3 & | & -2 \\ 0 & 1 & | & 9 \end{bmatrix}$

15. (a) $\begin{bmatrix} 1 & -3 & 4 & | & 3 \\ 2 & -5 & 6 & | & 6 \\ -3 & 3 & 4 & | & 6 \end{bmatrix} \xrightarrow{R_2 = -2r_1 + r_2} \begin{bmatrix} 1 & -3 & 4 & | & 3 \\ -2(1)+2 & -2(-3)-5 & -2(4)+6 & | & -2(3)+6 \\ -3 & 3 & 4 & | & 6 \end{bmatrix}$

$= \begin{bmatrix} 1 & -3 & 4 & | & 3 \\ 0 & 1 & -2 & | & 0 \\ -3 & 3 & 4 & | & 6 \end{bmatrix}$

(b) $\begin{bmatrix} 1 & -3 & 4 & | & 3 \\ 2 & -5 & 6 & | & 6 \\ -3 & 3 & 4 & | & 6 \end{bmatrix} \xrightarrow{R_3 = 3r_1 + r_3} \begin{bmatrix} 1 & -3 & 4 & | & 3 \\ 2 & -5 & 6 & | & 6 \\ 3(1)-3 & 3(-3)+3 & 3(4)+4 & | & 3(3)+6 \end{bmatrix} = \begin{bmatrix} 1 & -3 & 4 & | & 3 \\ 2 & -5 & 6 & | & 6 \\ 0 & -6 & 16 & | & 15 \end{bmatrix}$

17. (a) $\begin{bmatrix} 1 & -3 & 2 & | & -6 \\ 2 & -5 & 3 & | & -4 \\ -3 & -6 & 2 & | & 6 \end{bmatrix} \xrightarrow{R_2 = -2r_1 + r_2} \begin{bmatrix} 1 & -3 & 2 & | & -6 \\ -2(1)+2 & -2(-3)-5 & -2(2)+3 & | & -2(-6)-4 \\ -3 & -6 & 2 & | & 6 \end{bmatrix}$

$= \begin{bmatrix} 1 & -3 & 2 & | & -6 \\ 0 & 1 & -1 & | & 8 \\ -3 & -6 & 2 & | & 6 \end{bmatrix}$

(b) $\begin{bmatrix} 1 & -3 & 2 & | & -6 \\ 2 & -5 & 3 & | & -4 \\ -3 & -6 & 2 & | & 6 \end{bmatrix} \xrightarrow{R_3 = 3r_1 + r_3} = \begin{bmatrix} 1 & -3 & 2 & | & -6 \\ 2 & -5 & 3 & | & -4 \\ 3(1)-3 & 3(-3)-6 & 3(2)+2 & | & 3(-6)+6 \end{bmatrix}$

$= \begin{bmatrix} 1 & -3 & 2 & | & -6 \\ 2 & -5 & 3 & | & -4 \\ 0 & -15 & 8 & | & -12 \end{bmatrix}$

19. (a) $\begin{bmatrix} 1 & -3 & 1 & | & -2 \\ 2 & -5 & 6 & | & -2 \\ -3 & 1 & 4 & | & 6 \end{bmatrix} \xrightarrow{R_2 = -2r_1 + r_2} \begin{bmatrix} 1 & -3 & 1 & | & -2 \\ -2(1)+2 & -2(-3)-5 & -2(1)+6 & | & -2(-2)-2 \\ -3 & 1 & 4 & | & 6 \end{bmatrix}$

$= \begin{bmatrix} 1 & -3 & 1 & | & -2 \\ 0 & 1 & 4 & | & 2 \\ -3 & 1 & 4 & | & 6 \end{bmatrix}$

(b) $\begin{bmatrix} 1 & -3 & 1 & | & -2 \\ 2 & -5 & 6 & | & -2 \\ -3 & 1 & 4 & | & 6 \end{bmatrix} \xrightarrow{R_3 = 3r_1 + r_3} \begin{bmatrix} 1 & -3 & 1 & | & -2 \\ 2 & -5 & 6 & | & -2 \\ 3(1)-3 & 3(-3)+1 & 3(1)+4 & | & 3(-2)+6 \end{bmatrix}$

$= \begin{bmatrix} 1 & -3 & 1 & | & -2 \\ 2 & -5 & 6 & | & -2 \\ 0 & -8 & 7 & | & 0 \end{bmatrix}$

21. $\begin{cases} x + 2y = 5 \\ \quad\quad y = -1 \end{cases}$ The system is consistent;
the solution is $x = 7$, $y = -1$.

23. $\begin{cases} x + 2y + 3z = 1 \\ \quad\quad y + 4z = 2 \\ \quad\quad\quad\quad 0 = 3 \end{cases}$ The system is inconsistent.

25. $\begin{cases} x + 2z = -1 \\ y - 4z = -2 \\ \quad\quad 0 = 0 \end{cases}$ The system is consistent; there are an infinite number of solutions. The solutions are: $x = -2z - 1$, $y = 4z - 2$, and z, where z is any real number.

27. $\begin{cases} x_1 + 2x_2 - x_3 + x_4 = 1 \\ \quad\quad x_2 + 4x_3 + x_4 = 2 \\ \quad\quad\quad\quad x_3 + 2x_4 = 3 \end{cases}$ The system is consistent. To find the solutions, express $x_1, x_2,$ and x_3 in terms of x_4.

$$x_4$$
$$x_3 = -2x_4 + 3$$
$$x_2 = -x_4 - 4(-2x_4 + 3) + 2$$
$$= 7x_4 - 10$$
$$x_1 = -x_4 + x_3 - 2x_2 + 1$$
$$= -x_4 + (-2x_4 + 3) - 2(7x_4 - 10) + 1$$
$$= -17x_4 + 24$$

where x_4 is any real number.

29. $\begin{cases} x_1 + 2x_2 + 4x_4 = 2 \\ \quad\quad x_2 + x_3 + 3x_4 = 3 \\ \quad\quad\quad\quad 0 = 0 \end{cases}$ The system is consistent. To find the solutions, express x_1 and x_2 in terms of x_3 and x_4.

$$x_4$$
$$x_3$$
$$x_2 = -3x_4 - x_3 + 3$$
$$x_1 = -4x_4 - 2x_2 + 2$$
$$= -4x_4 - 2(-3x_4 - x_3 + 3) + 2$$
$$= 2x_4 + 2x_3 - 4$$

where x_3, x_4 are any real numbers.

31. $\begin{cases} x_1 - 2x_2 \quad\quad\quad + x_4 = -2 \\ \quad\quad x_2 - 3x_3 + 2x_4 = 2 \\ \quad\quad\quad\quad x_3 - x_4 = 0 \\ \quad\quad\quad\quad\quad\quad 0 = 0 \end{cases}$ The system is consistent. To find the solutions, express x_1 and x_2, and x_3 in terms of x_4.

$$x_4$$
$$x_3 = x_4$$
$$x_2 = -2x_4 + 3x_3 + 2$$
$$= -2x_4 + 3x_4 + 2$$
$$= x_4 + 2$$
$$x_1 = -x_4 + 2x_2 - 2$$
$$= -x_4 + 2(x_4 + 2) - 2$$
$$= x_4 + 2$$

where x_4 is any real number.

33. Write the system as:

$$\begin{bmatrix} 1 & 1 & | & 6 \\ 2 & -1 & | & 0 \end{bmatrix} \xrightarrow{R_2 = -2r_1 + r_2} \begin{bmatrix} 1 & 1 & | & 6 \\ 0 & -3 & | & -12 \end{bmatrix} \xrightarrow{R_2 = -\frac{1}{3}r_2} \begin{bmatrix} 1 & 1 & | & 6 \\ 0 & 1 & | & 4 \end{bmatrix}$$

The row-echelon form of the system is:

$$\begin{cases} x + y = 6 \\ \quad\; y = 4 \end{cases}$$

Back-substitute 4 for y in the first equation.

$x + 4 = 6$ or $x = 2$

The solution of the system of equations is $x = 2$ and $y = 4$.

35. Write the system as

$$\begin{bmatrix} 2 & 1 & | & 5 \\ 1 & -1 & | & 1 \end{bmatrix} \xrightarrow[\text{rows 1 and 2}]{\text{Interchange}} \begin{bmatrix} 1 & -1 & | & 1 \\ 2 & 1 & | & 5 \end{bmatrix} \xrightarrow{R_2 = -2r_1 + r_2} \begin{bmatrix} 1 & -1 & | & 1 \\ 0 & 3 & | & 3 \end{bmatrix} \xrightarrow{R_2 = \frac{1}{3}r_2} \begin{bmatrix} 1 & -1 & | & 1 \\ 0 & 1 & | & 1 \end{bmatrix}$$

The row-echelon form of the system is:

$$\begin{cases} x - y = 1 \\ \quad\; y = 1 \end{cases}$$

Back-substitute 1 for y in the first equation, giving $x - 1 = 1$ or $x = 2$.
The solution of the system of equations is $x = 2$ and $y = 1$.

37. Write the system as

$$\begin{bmatrix} 2 & 3 & | & 7 \\ 3 & -1 & | & 5 \end{bmatrix} \xrightarrow{R_1 = \frac{1}{2}r_1} \begin{bmatrix} 1 & \frac{3}{2} & | & \frac{7}{2} \\ 3 & -1 & | & 5 \end{bmatrix} \xrightarrow{R_2 = -3r_1 + r_2} \begin{bmatrix} 1 & \frac{3}{2} & | & \frac{7}{2} \\ 0 & -\frac{11}{2} & | & -\frac{11}{2} \end{bmatrix} \xrightarrow{R_2 = -\frac{2}{11}r_2} \begin{bmatrix} 1 & \frac{3}{2} & | & \frac{7}{2} \\ 0 & 1 & | & 1 \end{bmatrix}$$

The row-echelon form of the system is:

$$\begin{cases} x + \dfrac{3}{2}y = \dfrac{7}{2} \\ \qquad\; y = 1 \end{cases}$$

Back-substitute 1 for y in the first equation, giving $x = 2$.
The solution of the system of equations is $x = 2$ and $y = 1$.

39. Write the system as

$$\begin{bmatrix} 2 & -3 & | & 6 \\ 6 & -9 & | & 10 \end{bmatrix} \xrightarrow{R_1 = \frac{1}{2}r_1} \begin{bmatrix} 1 & -\frac{3}{2} & | & 3 \\ 6 & -9 & | & 10 \end{bmatrix} \xrightarrow{R_2 = -6r_1 + r_2} \begin{bmatrix} 1 & -\frac{3}{2} & | & 3 \\ 0 & 0 & | & -8 \end{bmatrix}$$

The system is inconsistent.

41.   Write the system as

$$\begin{bmatrix} 2 & -3 & | & 0 \\ 4 & 9 & | & 5 \end{bmatrix} \xrightarrow{R_1 = \frac{1}{2}r_1} \begin{bmatrix} 1 & -\frac{3}{2} & | & 0 \\ 4 & 9 & | & 5 \end{bmatrix} \xrightarrow{R_2 = -4r_1 + r_2} \begin{bmatrix} 1 & -\frac{3}{2} & | & 0 \\ 0 & 15 & | & 5 \end{bmatrix} \xrightarrow{R_2 = \frac{1}{15}r_2} \begin{bmatrix} 1 & -\frac{3}{2} & | & 0 \\ 0 & 1 & | & \frac{1}{3} \end{bmatrix}$$

The row–echelon form of the system is:

$$\begin{cases} x - \dfrac{3}{2}y = 0 \\ \quad\ y = \dfrac{1}{3} \end{cases}$$

Back-substitute $\dfrac{1}{3}$ for y in the first equation, giving $x - \dfrac{3}{2} \cdot \dfrac{1}{3} = 0$ or $x = \dfrac{1}{2}$.

The solution of the system of equations is $x = \dfrac{1}{2}$ and $y = \dfrac{1}{3}$.

43.   Write the system as

$$\begin{bmatrix} 2 & 6 & | & 4 \\ 5 & 15 & | & 10 \end{bmatrix} \xrightarrow{R_1 = \frac{1}{2}r_1} \begin{bmatrix} 1 & 3 & | & 2 \\ 5 & 15 & | & 10 \end{bmatrix} \xrightarrow{R_2 = -5r_1 + r_2} \begin{bmatrix} 1 & 3 & | & 2 \\ 0 & 0 & | & 0 \end{bmatrix}$$

This system has an infinite number of solutions. They are $x = 2 - 3y$ and y, where y is any real number.

45.   Write the system as

$$\begin{bmatrix} \frac{1}{2} & \frac{1}{3} & | & 2 \\ 1 & 1 & | & 5 \end{bmatrix} \xrightarrow[\text{rows 1 and 2}]{\text{Interchange}} \begin{bmatrix} 1 & 1 & | & 5 \\ \frac{1}{2} & \frac{1}{3} & | & 2 \end{bmatrix} \xrightarrow{R_2 = 6r_2} \begin{bmatrix} 1 & 1 & | & 5 \\ 3 & 2 & | & 12 \end{bmatrix}$$

$$\xrightarrow{R_2 = -3r_1 + r_2} \begin{bmatrix} 1 & 1 & | & 5 \\ 0 & -1 & | & -3 \end{bmatrix} \xrightarrow{R_2 = -r_2} \begin{bmatrix} 1 & 1 & | & 5 \\ 0 & 1 & | & 3 \end{bmatrix}$$

The row echelon form of the system of equations is

$$\begin{cases} x = -y + 5 & (1) \\ y = 3 & (2) \end{cases}$$

Back-substitute 3 for y in equation (1), to get $x = -3 + 5 = 2$.
The solution of the system of equations is $x = 2$ and $y = 3$.

47. Write the system as

$$\begin{bmatrix} 1 & 1 & | & 1 \\ 3 & -2 & | & \frac{4}{3} \end{bmatrix} \xrightarrow{R_2 = -3r_1 + r_2} \begin{bmatrix} 1 & 1 & | & 1 \\ 0 & -5 & | & -\frac{5}{3} \end{bmatrix} \xrightarrow{R_2 = -\frac{1}{5}r_2} \begin{bmatrix} 1 & 1 & | & 1 \\ 0 & 1 & | & \frac{1}{3} \end{bmatrix}$$

The row echelon form of the system of equations is

$$\begin{cases} x = -y + 1 & (1) \\ y = \dfrac{1}{3} & (2) \end{cases}$$

Back-substitute $\dfrac{1}{3}$ for y in equation (1), to get $x = -\dfrac{1}{3} + 1 = \dfrac{2}{3}$.

The solution of the system of equations is $x = \dfrac{2}{3}$ and $y = \dfrac{1}{3}$.

49. Write the system as

$$\begin{bmatrix} 2 & 1 & 1 & | & 6 \\ 1 & -1 & -1 & | & -3 \\ 3 & 1 & 2 & | & 7 \end{bmatrix} \xrightarrow[\text{rows 1 and 2}]{\text{Interchange}} \begin{bmatrix} 1 & -1 & -1 & | & -3 \\ 2 & 1 & 1 & | & 6 \\ 3 & 1 & 2 & | & 7 \end{bmatrix} \xrightarrow[R_3 = -3r_1 + r_3]{R_2 = -2r_1 + r_2} \begin{bmatrix} 1 & -1 & -1 & | & -3 \\ 0 & 3 & 3 & | & 12 \\ 0 & 4 & 5 & | & 16 \end{bmatrix}$$

$$\xrightarrow{R_2 = \frac{1}{3}r_2} \begin{bmatrix} 1 & -1 & -1 & | & -3 \\ 0 & 1 & 1 & | & 4 \\ 0 & 4 & 5 & | & 16 \end{bmatrix} \xrightarrow{R_3 = -4r_2 + r_3} \begin{bmatrix} 1 & -1 & -1 & | & -3 \\ 0 & 1 & 1 & | & 4 \\ 0 & 0 & 1 & | & 0 \end{bmatrix}$$

The row echelon form of the system of equations is

$$\begin{cases} x = y + z - 3 & (1) \\ y = -z + 4 & (2) \\ z = 0 & (3) \end{cases}$$

Back-substitute 0 for z in (2) to get $y = 4$.
Then back-substitute $z = 0$ and $y = 4$ in equation (1), to get $x = 4 + 0 - 3 = 1$.

The solution of the system of equations is $x = 1$, $y = 4$, and $z = 0$.

51. Write the system as

$$\begin{bmatrix} 2 & -2 & -1 & | & 2 \\ 2 & 3 & 1 & | & 2 \\ 3 & 2 & 0 & | & 0 \end{bmatrix} \xrightarrow[R_1 = -\frac{1}{2}r_1]{} \begin{bmatrix} 1 & -1 & -\frac{1}{2} & | & 1 \\ 2 & 3 & 1 & | & 2 \\ 3 & 2 & 0 & | & 0 \end{bmatrix} \xrightarrow[\substack{R_2 = -2r_1 + r_2 \\ R_3 = -3r_1 + r_3}]{} \begin{bmatrix} 1 & -1 & -\frac{1}{2} & | & 1 \\ 0 & 5 & 2 & | & 0 \\ 0 & 5 & \frac{3}{2} & | & -3 \end{bmatrix}$$

$$\xrightarrow[R_2 = \frac{1}{5}r_2]{} \begin{bmatrix} 1 & -1 & -\frac{1}{2} & | & 1 \\ 0 & 1 & \frac{2}{5} & | & 0 \\ 0 & 5 & \frac{3}{2} & | & -3 \end{bmatrix} \xrightarrow[R_3 = -5r_2 + r_3]{} \begin{bmatrix} 1 & -1 & -\frac{1}{2} & | & 1 \\ 0 & 1 & \frac{2}{5} & | & 0 \\ 0 & 0 & -\frac{1}{2} & | & -3 \end{bmatrix} \xrightarrow[R_3 = -2r_3]{} \begin{bmatrix} 1 & -1 & -\frac{1}{2} & | & 1 \\ 0 & 1 & \frac{2}{5} & | & 0 \\ 0 & 0 & 1 & | & 6 \end{bmatrix}$$

The row echelon form of the system of equations is

$$\begin{cases} x = y + \dfrac{1}{2}z + 1 & (1) \\[2mm] y = -\dfrac{2}{5}z & (2) \\[2mm] z = 6 & (3) \end{cases}$$

Back-substitute 6 for z in (2) to get $y = -\dfrac{12}{5}$.

Then back-substitute $z = 6$ and $y = -\dfrac{12}{5}$ in equation (1), to get $x = -\dfrac{12}{5} + \dfrac{1}{2}(6) + 1 = \dfrac{8}{5}$.

The solution of the system of equations is $x = \dfrac{8}{5}$, $y = -\dfrac{12}{5}$, and $z = 6$.

53. Write the system as

$$\begin{bmatrix} 2 & 1 & -1 & | & 2 \\ 1 & 3 & 2 & | & 1 \\ 1 & 1 & 1 & | & 2 \end{bmatrix} \xrightarrow[\substack{\text{Interchange} \\ \text{rows 1 and 2}}]{} \begin{bmatrix} 1 & 3 & 2 & | & 1 \\ 2 & 1 & -1 & | & 2 \\ 1 & 1 & 1 & | & 2 \end{bmatrix} \xrightarrow[\substack{R_2 = -2r_1 + r_2 \\ R_3 = -r_1 + r_3}]{} \begin{bmatrix} 1 & 3 & 2 & | & 1 \\ 0 & -5 & -5 & | & 0 \\ 0 & -2 & -1 & | & 1 \end{bmatrix}$$

$$\xrightarrow[R_2 = -\frac{1}{5}r_2]{} \begin{bmatrix} 1 & 3 & 2 & | & 1 \\ 0 & 1 & 1 & | & 0 \\ 0 & -2 & -1 & | & 1 \end{bmatrix} \xrightarrow[R_3 = 2r_2 + r_3]{} \begin{bmatrix} 1 & 3 & 2 & | & 1 \\ 0 & 1 & 1 & | & 0 \\ 0 & 0 & 1 & | & 1 \end{bmatrix}$$

The row echelon form of the system of equations is

$$\begin{cases} x = -3y - 2z + 1 & (1) \\ y = -z & (2) \\ z = 1 & (3) \end{cases}$$

Back-substitute $z = 1$ in equation (2) to get $y = -1$.
Then back-substitute $y = -1$ and $z = 1$ in equation (1), to get $x = -3(-1) - 2(1) + 1 = 2$.

The solution of the system of equations is $x = 2$, $y = -1$, and $z = 1$.

55. Write the system as

$$\begin{bmatrix} 1 & 1 & -1 & | & 0 \\ 4 & 4 & -4 & | & -1 \\ 2 & 1 & 1 & | & 2 \end{bmatrix} \xrightarrow[\substack{R_2 = -4r_1 + r_2 \\ R_3 = -2r_1 + r_3}]{} \begin{bmatrix} 1 & 1 & -1 & | & 0 \\ 0 & 0 & 0 & | & -1 \\ 0 & -1 & 3 & | & 2 \end{bmatrix}$$

The system is inconsistent because the second row of the matrix has $0 = -1$ which is contradictory.

57. Write the system as

$$\begin{bmatrix} 3 & 1 & -1 & | & \dfrac{2}{3} \\ 2 & -1 & 1 & | & 1 \\ 4 & 2 & 0 & | & \dfrac{8}{3} \end{bmatrix} \xrightarrow[R_1 = r_1 - r_2]{} \begin{bmatrix} 1 & 2 & -2 & | & -\dfrac{1}{3} \\ 2 & -1 & 1 & | & 1 \\ 4 & 2 & 0 & | & \dfrac{8}{3} \end{bmatrix} \xrightarrow[\substack{R_2 = -2r_1 + r_2 \\ R_3 = -4r_1 + r_3}]{} \begin{bmatrix} 1 & 2 & -2 & | & -\dfrac{1}{3} \\ 0 & -5 & 5 & | & \dfrac{5}{3} \\ 0 & -6 & 8 & | & 4 \end{bmatrix}$$

$$\xrightarrow[R_2 = -\frac{1}{5}r_2]{} \begin{bmatrix} 1 & 2 & -2 & | & -\dfrac{1}{3} \\ 0 & 1 & -1 & | & -\dfrac{1}{3} \\ 0 & -6 & 8 & | & 4 \end{bmatrix} \xrightarrow[R_3 = 6r_2 + r_3]{} \begin{bmatrix} 1 & 2 & -2 & | & -\dfrac{1}{3} \\ 0 & 1 & -1 & | & -\dfrac{1}{3} \\ 0 & 0 & 2 & | & 2 \end{bmatrix} \xrightarrow[R_3 = -\frac{1}{2}r_3]{} \begin{bmatrix} 1 & 2 & -2 & | & -\dfrac{1}{3} \\ 0 & 1 & -1 & | & -\dfrac{1}{3} \\ 0 & 0 & 1 & | & 1 \end{bmatrix}$$

The row echelon form of the system of equations is

$$\begin{cases} x = -2y + 2z - \dfrac{1}{3} & (1) \\ y = z - \dfrac{1}{3} & (2) \\ z = 1 & (3) \end{cases}$$

Back-substitute $z = 1$ in equation (2) to get $y = 1 - \dfrac{1}{3} = \dfrac{2}{3}$.

Then back-substitute $y = \dfrac{1}{3}$ and $z = 1$ in equation (1), to get $x = -2\left(\dfrac{2}{3}\right) + 2(1) - \dfrac{1}{3} = \dfrac{1}{3}$.

The solution of the system of equations is $x = \dfrac{1}{3}$, $y = \dfrac{2}{3}$, and $z = 1$.

59. Using a TI-83 graphing utility, the row-echelon form is

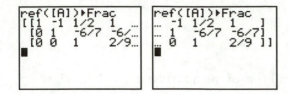

The reduced row-echelon form is

```
rref([A])►Frac
  [[1 0 0 2/9 ]
   [0 1 0 -2/3]
   [0 0 1 2/9 ]]
■
```

The solution to the system is $x = \dfrac{2}{9}$, $y = -\dfrac{2}{3}$, and $z = \dfrac{2}{9}$.

61. Using a TI-83 graphing utility, the row-echelon form is

```
ref([A])
  [[1 1 1 4]
   [0 1 1 2]
   [0 0 1 3]]
■
```

The reduced row-echelon form is

```
rref([A])
  [[1 0 0 2 ]
   [0 1 0 -1]
   [0 0 1 3 ]]
■
```

The solution to the system is $x = 2$, $y = -1$, and $z = 3$.

63. Using a TI-83 graphing utility, the row-echelon form is

```
ref([A])
[[1 1 1 1 20]
 [0 1 1 1 0 ]
 [0 0 1 1 13]
 [0 0 0 1 -4]]
```

The reduced row-echelon form is

$$\begin{bmatrix} 1 & 0 & 0 & 0 & 20 \\ 0 & 1 & 0 & 0 & -13 \\ 0 & 0 & 1 & 0 & 17 \\ 0 & 0 & 0 & 1 & -4 \end{bmatrix}$$

The solution to the system is $x_1 = 20$, $x_2 = -13$, $x_3 = 17$ and $x_4 = -4$.

65. Let x = the price of a mezzanine ticket,
y = the price of a lower balcony ticket, and
z = the price of a middle balcony ticket.

The system of equations representing the problem, can be written:

$$\begin{cases} 4x + 6y = 444 & (1) \\ 2x + 7y + 8z = 614 & (2) \\ 3y + 12z = 474 & (3) \end{cases}$$

We can then represent the system as:

$$\begin{bmatrix} 4 & 6 & 0 & | & 444 \\ 2 & 7 & 8 & | & 614 \\ 0 & 3 & 12 & | & 474 \end{bmatrix} \xrightarrow{R_1 = \frac{1}{4}r_1} \begin{bmatrix} 1 & \frac{3}{2} & 0 & | & 111 \\ 2 & 7 & 8 & | & 614 \\ 0 & 3 & 12 & | & 474 \end{bmatrix} \xrightarrow{R_2 = -2r_1 + r_2} \begin{bmatrix} 1 & \frac{3}{2} & 0 & | & 111 \\ 0 & 4 & 8 & | & 392 \\ 0 & 3 & 12 & | & 474 \end{bmatrix}$$

$$\xrightarrow{R_2 = \frac{1}{4}r_2} \begin{bmatrix} 1 & \frac{3}{2} & 0 & | & 111 \\ 0 & 1 & 2 & | & 98 \\ 0 & 3 & 12 & | & 474 \end{bmatrix} \xrightarrow{R_3 = -3r_2 + r_3} \begin{bmatrix} 1 & \frac{3}{2} & 0 & | & 111 \\ 0 & 1 & 2 & | & 98 \\ 0 & 0 & 6 & | & 180 \end{bmatrix} \xrightarrow{R_3 = \frac{1}{6}r_3} \begin{bmatrix} 1 & \frac{3}{2} & 0 & | & 111 \\ 0 & 1 & 2 & | & 98 \\ 0 & 0 & 1 & | & 30 \end{bmatrix}$$

The row-echelon form of the system is

$$\begin{cases} x = -\dfrac{3}{2}y + 111 & (1) \\ y = -2z + 98 & (2) \\ z = 30 & (3) \end{cases}$$

Back-substitute 30 for z in (2) to get in (2) to get $y = -2(30) + 98 = 38$.

Then back-substitute $y = 38$ in (1) to get $x = -\dfrac{3}{2}(38) + 111 = 54$.

We find that mezzanine tickets cost $54.00 each, lower balcony tickets cost $38.00 each, and middle balcony tickets cost $30.00 each.

67. Let x = number of workstations set up for 2 students each, and
y = number of workstations set up for 3 students each.

The problem can be modeled with the system of equations,

$$\begin{cases} x + y = 16 & \text{(1)} \\ 2x + 3y = 38 & \text{(2)} \end{cases}$$

The system can be represented by the augmented matrices

$$\begin{bmatrix} 1 & 1 & | & 16 \\ 2 & 3 & | & 38 \end{bmatrix} \xrightarrow{R_2 = -2r_1 + r_2} \begin{bmatrix} 1 & 1 & | & 16 \\ 0 & 1 & | & 6 \end{bmatrix}$$

The row-echelon form of the system is

$$\begin{cases} x = -y + 16 & \text{(1)} \\ y = 6 & \text{(2)} \end{cases}$$

There are 10 workstations set up for 2 students each and 6 workstations set up for 3 students each.

69. Let x = amount Carletta invests in treasury bills,
y = amount Carletta invests in treasury bonds, and
z = amount Carletta invests in corporate bonds.

The problem can be modeled with the system of equations,

$$\begin{cases} x + y + x = 10,000 & \text{(1)} \\ 0.06x + 0.07y + 0.08z = 680 & \text{(2)} \\ z = 0.5x & \text{(3)} \end{cases}$$

The system can be represented by the augmented matrices

$$\begin{bmatrix} 1 & 1 & 1 & | & 10000 \\ 6 & 7 & 8 & | & 68000 \\ -5 & 0 & 10 & | & 0 \end{bmatrix} \xrightarrow[R_3 = 5r_2 + r_3]{R_2 = -6r_1 + r_2} \begin{bmatrix} 1 & 1 & 1 & | & 10000 \\ 0 & 1 & 2 & | & 8000 \\ 0 & 5 & 15 & | & 50000 \end{bmatrix} \xrightarrow{R_3 = -5r_2 + r_3} \begin{bmatrix} 1 & 1 & 1 & | & 10000 \\ 0 & 1 & 2 & | & 8000 \\ 0 & 0 & 5 & | & 10000 \end{bmatrix}$$

$$\xrightarrow{R_3 = \frac{1}{5}r_3} \begin{bmatrix} 1 & 1 & 1 & | & 10000 \\ 0 & 1 & 2 & | & 8000 \\ 0 & 0 & 1 & | & 2000 \end{bmatrix}$$

(In this system, we multiplied (2) by 100, and (3) by 10 before writing the matrices. This was done to eliminate decimals.)

The row-echelon form of the system is

$$\begin{cases} x = -y - z + 10,000 & \text{(1)} \\ y = -2z + 8000 & \text{(2)} \\ z = 2000 & \text{(3)} \end{cases}$$

Carletta should invest $2000 in corporate bonds, $4000 in treasury bonds, and $4000 in treasury bills to attain $680 annual income.

71. Let x = number of servings of chicken needed,
y = number of servings of potatoes needed, and
z = number of servings of spinach needed in the diet.

The problem can be modeled with the system of equations,

$$\begin{cases} 14x + y + 6z = 30 & (1) \\ 18y + 8z = 38 & (2) \\ 4.5x + z = 7 & (3) \end{cases}$$

The system can be represented by the augmented matrices

$$\begin{bmatrix} 14 & 1 & 6 & | & 30 \\ 0 & 18 & 8 & | & 38 \\ 4.5 & 0 & 1 & | & 7 \end{bmatrix} \xrightarrow{R_1 = \frac{1}{14}r_1} \begin{bmatrix} 1 & \frac{1}{14} & \frac{3}{7} & | & \frac{15}{7} \\ 0 & 18 & 8 & | & 38 \\ 4.5 & 0 & 1 & | & 7 \end{bmatrix} \xrightarrow{R_3 = -4.5r_1 + r_3} \begin{bmatrix} 1 & \frac{1}{14} & \frac{3}{7} & | & \frac{15}{7} \\ 0 & 18 & 8 & | & 38 \\ 0 & -\frac{9}{28} & -\frac{13}{14} & | & -\frac{37}{14} \end{bmatrix}$$

$$\xrightarrow{R_3 = -28r_3} \begin{bmatrix} 1 & \frac{1}{14} & \frac{3}{7} & | & \frac{15}{7} \\ 0 & 18 & 8 & | & 38 \\ 0 & 9 & 26 & | & 74 \end{bmatrix} \xrightarrow{R_2 = \frac{1}{18}r_2} \begin{bmatrix} 1 & \frac{1}{14} & \frac{3}{7} & | & \frac{15}{7} \\ 0 & 1 & \frac{4}{9} & | & \frac{19}{9} \\ 0 & 9 & 26 & | & 74 \end{bmatrix}$$

$$\xrightarrow{R_3 = -9r_2 + r_3} \begin{bmatrix} 1 & \frac{1}{14} & \frac{3}{7} & | & \frac{15}{7} \\ 0 & 1 & \frac{4}{9} & | & \frac{19}{9} \\ 0 & 0 & 22 & | & 55 \end{bmatrix} \xrightarrow{R_3 = \frac{1}{22}r_3} \begin{bmatrix} 1 & \frac{1}{14} & \frac{3}{7} & | & \frac{15}{7} \\ 0 & 1 & \frac{4}{9} & | & \frac{19}{9} \\ 0 & 0 & 1 & | & \frac{5}{2} \end{bmatrix}$$

The row-echelon form of the system is

$$\begin{cases} x = -\frac{1}{14}y - \frac{3}{7}z + \frac{15}{7} & (1) \\ y = -\frac{4}{9}z + \frac{19}{9} & (2) \\ z = \frac{5}{2} & (3) \end{cases}$$

Back-substituting $z = \frac{5}{2}$ into (2) gives $y = 1$. Then back-substituting $z = \frac{5}{2}$ and $y = 1$ into (1) gives $x = 1$.

The dietician should serve 1 serving of chicken, 1 serving of potatoes, and 2.5 servings of spinach.

73.   Let x = number of cases of orange juice to be prepared,
y = number of cases of tomato juice to be prepared, and
z = number of cases of pineapple juice to be prepared.

The problem can be modeled with the system of equations,

$$\begin{cases} 10x + 12y + 9z = 398 & (1) \\ 4x + 4y + 6z = 164 & (2) \\ 2x + y + z = 58 & (3) \end{cases}$$

The system can be represented by the augmented matrices

$$\begin{bmatrix} 10 & 12 & 9 & | & 398 \\ 4 & 4 & 6 & | & 164 \\ 2 & 1 & 1 & | & 58 \end{bmatrix} \xrightarrow{R_1 = \frac{1}{10}r_1} \begin{bmatrix} 1 & 1.2 & 0.9 & | & 39.8 \\ 4 & 4 & 6 & | & 164 \\ 2 & 1 & 1 & | & 58 \end{bmatrix} \xrightarrow[R_3 = -2r_1 + r_3]{R_2 = -4r_1 + r_2} \begin{bmatrix} 1 & 1.2 & 0.9 & | & 39.8 \\ 0 & -0.8 & 2.4 & | & 4.8 \\ 0 & -1.4 & -0.8 & | & -21.6 \end{bmatrix}$$

$$\xrightarrow{R_2 = -\frac{10}{8}r_2} \begin{bmatrix} 1 & 1.2 & 0.9 & | & 39.8 \\ 0 & 1 & -3 & | & -6 \\ 0 & -1.4 & -0.8 & | & -21.6 \end{bmatrix} \xrightarrow{R_3 = 1.4r_2 + r_3} \begin{bmatrix} 1 & 1.2 & 0.9 & | & 39.8 \\ 0 & 1 & -3 & | & -6 \\ 0 & 0 & -5 & | & -30 \end{bmatrix}$$

$$\xrightarrow{R_3 = -\frac{1}{5}r_3} \begin{bmatrix} 1 & 1.2 & 0.9 & | & 39.8 \\ 0 & 1 & -3 & | & -6 \\ 0 & 0 & 1 & | & 6 \end{bmatrix}$$

The row-echelon form of the system is

$$\begin{cases} x = -1.2y - 0.9z + 39.8 & (1) \\ y = 3z - 6 & (2) \\ z = 6 & (3) \end{cases}$$

The company should produce 6 cases of pineapple juice, 12 cases of tomato juice, and 20 cases of orange juice.

75. Let x = number of packages of package 1 to be ordered,
   y = number of packages of package 2 to be ordered, and
   z = number of packages of package 3 to be ordered.

The problem can be modeled with the system of equations,

$$\begin{cases} 20x + 40z = 200 & (1) \\ 15x + 3y + 30z = 180 & (2) \\ x + y = 12 & (3) \end{cases}$$

The system can be represented by the augmented matrices

$$\begin{bmatrix} 20 & 0 & 40 & | & 200 \\ 15 & 3 & 30 & | & 180 \\ 1 & 1 & 0 & | & 12 \end{bmatrix} \xrightarrow[\text{rows 1 and 3}]{\text{Interchange}} \begin{bmatrix} 1 & 1 & 0 & | & 12 \\ 15 & 3 & 30 & | & 180 \\ 20 & 0 & 40 & | & 200 \end{bmatrix} \xrightarrow[R_3 = \frac{1}{20}r_3]{R_2 = \frac{1}{3}r_2} \begin{bmatrix} 1 & 1 & 0 & | & 12 \\ 5 & 1 & 10 & | & 60 \\ 1 & 0 & 2 & | & 10 \end{bmatrix}$$

$$\xrightarrow[R_3 = -r_1 + r_3]{R_2 = -5r_1 + r_2} \begin{bmatrix} 1 & 1 & 0 & | & 12 \\ 0 & -4 & 10 & | & 0 \\ 0 & -1 & 2 & | & -2 \end{bmatrix} \xrightarrow[\text{rows 2 and 3}]{\text{Interchange}} \begin{bmatrix} 1 & 1 & 0 & | & 12 \\ 0 & -1 & 2 & | & -2 \\ 0 & -4 & 10 & | & 0 \end{bmatrix} \xrightarrow[R_2 = -r_2]{} \begin{bmatrix} 1 & 1 & 0 & | & 12 \\ 0 & 1 & -2 & | & 2 \\ 0 & -4 & 10 & | & 0 \end{bmatrix}$$

$$\xrightarrow[R_3 = 4r_2 + r_3]{} \begin{bmatrix} 1 & 1 & 0 & | & 12 \\ 0 & 1 & -2 & | & 2 \\ 0 & 0 & 2 & | & 8 \end{bmatrix} \xrightarrow[R_3 = \frac{1}{2}r_3]{} \begin{bmatrix} 1 & 1 & 0 & | & 12 \\ 0 & 1 & -2 & | & 2 \\ 0 & 0 & 1 & | & 4 \end{bmatrix}$$

The row-echelon form of the system is

$$\begin{cases} x = -y + 12 & (1) \\ y = 2z + 2 & (2) \\ z = 4 & (3) \end{cases}$$

The teacher should order 4 packages of package 3, 10 packages of package 2, and 2 packages of package 1 paper.

77.   Let x = number of assorted cartons needed,
           y = number of mixed cartons needed, and
           z = number of single cartons needed to fill the recreation center's order.

The problem can be modeled with the system of equations,

$$\begin{cases} 2x + 4y = 40 & (1) \\ 4x + 2y = 32 & (2) \\ x + 2z = 14 & (3) \end{cases}$$

The system can be represented by the augmented matrices

$$\begin{bmatrix} 2 & 4 & 0 & | & 40 \\ 4 & 2 & 0 & | & 32 \\ 1 & 0 & 2 & | & 14 \end{bmatrix} \xrightarrow{R_1 = \frac{1}{2}r_1} \begin{bmatrix} 1 & 2 & 0 & | & 20 \\ 4 & 2 & 0 & | & 32 \\ 1 & 0 & 2 & | & 14 \end{bmatrix} \xrightarrow[R_3 = -r_1 + r_3]{R_2 = -4r_1 + r_2} \begin{bmatrix} 1 & 2 & 0 & | & 20 \\ 0 & -6 & 0 & | & -48 \\ 0 & -2 & 2 & | & -6 \end{bmatrix}$$

$$\xrightarrow{R_2 = -\frac{1}{6}r_2} \begin{bmatrix} 1 & 2 & 0 & | & 20 \\ 0 & 1 & 0 & | & 8 \\ 0 & -2 & 2 & | & -6 \end{bmatrix} \xrightarrow{R_3 = 2r_2 + r_3} \begin{bmatrix} 1 & 2 & 0 & | & 20 \\ 0 & 1 & 0 & | & 8 \\ 0 & 0 & 2 & | & 10 \end{bmatrix} \xrightarrow{R_3 = \frac{1}{2}r_3} \begin{bmatrix} 1 & 2 & 0 & | & 20 \\ 0 & 1 & 0 & | & 8 \\ 0 & 0 & 1 & | & 5 \end{bmatrix}$$

The row-echelon form of the system is

$$\begin{cases} x = -2y + 20 & (1) \\ y = 8 & (2) \\ z = 5 & (3) \end{cases}$$

The supplier should send the recreation center 5 single cartons, 8 mixed cartons, and 4 assorted cartons.

79.   Let x = number of large cans needed,
y = number of mammoth cans needed, and
z = number of giant cans needed to fulfill the order.

The problem can be modeled with the system of equations,

$$\begin{cases} y + z = 5 & (1) \\ 2x + 6y + 4z = 26 & (2) \\ x + 2y + 2z = 12 & (3) \end{cases}$$

The system can be represented by the augmented matrices

$$\begin{bmatrix} 0 & 1 & 1 & | & 5 \\ 2 & 6 & 4 & | & 26 \\ 1 & 2 & 2 & | & 12 \end{bmatrix} \xrightarrow[\text{rows 1 and 3}]{\text{Interchange}} \begin{bmatrix} 1 & 2 & 2 & | & 12 \\ 2 & 6 & 4 & | & 26 \\ 0 & 1 & 1 & | & 5 \end{bmatrix} \xrightarrow{R_2 = -2r_1 + r_2} \begin{bmatrix} 1 & 2 & 2 & | & 12 \\ 0 & 2 & 0 & | & 2 \\ 0 & 1 & 1 & | & 5 \end{bmatrix}$$

$$\xrightarrow{R_2 = \frac{1}{2}r_2} \begin{bmatrix} 1 & 2 & 2 & | & 12 \\ 0 & 1 & 0 & | & 1 \\ 0 & 1 & 1 & | & 5 \end{bmatrix} \xrightarrow{R_3 = -r_2 + r_3} \begin{bmatrix} 1 & 2 & 2 & | & 12 \\ 0 & 1 & 0 & | & 1 \\ 0 & 0 & 1 & | & 4 \end{bmatrix}$$

The row-echelon form of the system is

$$\begin{cases} x = -2y - 2z + 12 & (1) \\ y = 1 & (2) \\ z = 4 & (3) \end{cases}$$

The store should use 4 giant size cans, 1 mammoth size can, and 2 large size cans to fill the order.

81.   Answers will vary.

83.   Answers will vary.

2.3 Systems of *m* Linear Equations Containing *n* Variables

1.   No, rows with all zeros should be at the bottom.

3.   No, there is a 1 above the leftmost 1 in the second row. It should be a zero.

5.   No, the leftmost 1 in the second row is not to the right of the leftmost 1 in the first row.

7.   Yes.

9.   Yes.

11.   Yes.

13.   Infinitely many solutions.
The solutions are $x = 1 - y$ and y, where y is any real number.

15.   One solution.
The solution is $x = 4$ and $y = 5$.

17. Infinitely many solutions. The solutions are $x = 2z + 6$, $y = -3z + 1$, and z, where z is any real number.

19. Infinitely many solutions. The solutions are $x = -2y + 1$, y, and $z = 2$, where y is any real number.

21. Infinitely many solutions. The solutions are $x = 1$, $y = 2$, and z, where z can be any real number.

23. One solution. The solution is $x = -1$, $y = 3$, and $z = 4$.

25. Infinitely many solutions. The solutions are $x = z + 1$, $y = -2z + 1$, and z, where z can be any real number.

27. Infinitely many solutions. The solutions are $x_1 = x_4 + 4$, $x_2 = -2x_3 - 3x_4$, x_3, and x_4 where x_3 and x_4 are any real numbers.

29. Write the system as the augmented matrix, $\begin{bmatrix} 1 & 1 & | & 3 \\ 2 & -1 & | & 3 \end{bmatrix}$.

Then use row operations to find the reduced row-echelon form.

$$\begin{bmatrix} 1 & 1 & | & 3 \\ 2 & -1 & | & 3 \end{bmatrix} \xrightarrow{R_2 = -2r_1 + r_2} \begin{bmatrix} 1 & 1 & | & 3 \\ 0 & -3 & | & -3 \end{bmatrix} \xrightarrow{R_2 = -\frac{1}{3}r_2} \begin{bmatrix} 1 & 1 & | & 3 \\ 0 & 1 & | & 1 \end{bmatrix} \xrightarrow{R_1 = -r_2 + r_1} \begin{bmatrix} 1 & 0 & | & 2 \\ 0 & 1 & | & 1 \end{bmatrix}$$

There is one solution which can be read from the reduced row-echelon matrix. The solution is $x = 2$ and $y = 1$.

31. Write the system as the augmented matrix, $\begin{bmatrix} 3 & -3 & | & 12 \\ 3 & 2 & | & -3 \\ 2 & 1 & | & 4 \end{bmatrix}$.

Then use row operations to find the reduced row-echelon form.

$$\begin{bmatrix} 3 & -3 & | & 12 \\ 3 & 2 & | & -3 \\ 2 & 1 & | & 4 \end{bmatrix} \xrightarrow{R_1 = \frac{1}{3}r_1} \begin{bmatrix} 1 & -1 & | & 4 \\ 3 & 2 & | & -3 \\ 2 & 1 & | & 4 \end{bmatrix} \xrightarrow[R_3 = -2r_1 + r_3]{R_2 = -3r_1 + r_2} \begin{bmatrix} 1 & -1 & | & 4 \\ 0 & 5 & | & -15 \\ 0 & 3 & | & -4 \end{bmatrix}$$

$$\xrightarrow{R_2 = \frac{1}{5}r_2} \begin{bmatrix} 1 & -1 & | & 4 \\ 0 & 1 & | & -3 \\ 0 & 3 & | & -4 \end{bmatrix} \xrightarrow[R_3 = -3r_2 + r_3]{R_1 = r_2 + r_1} \begin{bmatrix} 1 & 0 & | & 1 \\ 0 & 1 & | & -3 \\ 0 & 0 & | & 5 \end{bmatrix}$$

There is no solution. The system is inconsistent.

33. Write the system as the augmented matrix, $\begin{bmatrix} 2 & -4 & | & 8 \\ 1 & -2 & | & 4 \\ -1 & 2 & | & -4 \end{bmatrix}$.

Then use row operations to find the reduced row-echelon form.

$$\begin{bmatrix} 2 & -4 & | & 8 \\ 1 & -2 & | & 4 \\ -1 & 2 & | & -4 \end{bmatrix} \xrightarrow[\text{rows 1 and 2}]{\text{Interchange}} \begin{bmatrix} 1 & -2 & | & 4 \\ 2 & -4 & | & 8 \\ -1 & 2 & | & -4 \end{bmatrix} \xrightarrow[R_3 = r_1 + r_3]{R_2 = -2r_1 + r_2} \begin{bmatrix} 1 & -2 & | & 4 \\ 0 & 0 & | & 0 \\ 0 & 0 & | & 0 \end{bmatrix}$$

There are an infinite number of solutions. They are $x = 2y + 4$ and y, where y can be any real number.

35. Write the system as the augmented matrix, $\begin{bmatrix} 2 & 1 & 3 & | & -1 \\ -1 & 1 & 3 & | & 8 \\ 2 & -2 & -6 & | & -16 \end{bmatrix}$.

Then use row operations to find the reduced row-echelon form.

$$\begin{bmatrix} 2 & 1 & 3 & | & -1 \\ -1 & 1 & 3 & | & 8 \\ 2 & -2 & -6 & | & -16 \end{bmatrix} \xrightarrow[\text{rows 1 and 2}]{\text{Interchange}} \begin{bmatrix} -1 & 1 & 3 & | & 8 \\ 2 & 1 & 3 & | & -1 \\ 2 & -2 & -6 & | & -16 \end{bmatrix} \xrightarrow{R_1 = -r_1} \begin{bmatrix} 1 & -1 & -3 & | & -8 \\ 2 & 1 & 3 & | & -1 \\ 2 & -2 & -6 & | & -16 \end{bmatrix}$$

$$\xrightarrow[R_3 = -2r_1 + r_3]{R_2 = -2r_1 + r_2} \begin{bmatrix} 1 & -1 & -3 & | & -8 \\ 0 & 3 & 9 & | & 15 \\ 0 & 0 & 0 & | & 0 \end{bmatrix} \xrightarrow{R_2 = \frac{1}{3}r_2} \begin{bmatrix} 1 & -1 & -3 & | & -8 \\ 0 & 1 & 3 & | & 5 \\ 0 & 0 & 0 & | & 0 \end{bmatrix} \xrightarrow{R_1 = r_2 + r_1} \begin{bmatrix} 1 & 0 & 0 & | & -3 \\ 0 & 1 & 3 & | & 5 \\ 0 & 0 & 0 & | & 0 \end{bmatrix}$$

There are an infinite number of solutions. They are $x = -3$, $y = -3z + 5$, and z, where z can be any real number.

37. Write the system as the augmented matrix, $\begin{bmatrix} 1 & -1 & 0 & | & 1 \\ 0 & 1 & -1 & | & 6 \\ 1 & 0 & 1 & | & -1 \end{bmatrix}$.

Then use row operations to find the reduced row-echelon form.

$$\begin{bmatrix} 1 & -1 & 0 & | & 1 \\ 0 & 1 & -1 & | & 6 \\ 1 & 0 & 1 & | & -1 \end{bmatrix} \xrightarrow{R_3 = -r_1 + r_3} \begin{bmatrix} 1 & -1 & 0 & | & 1 \\ 0 & 1 & -1 & | & 6 \\ 0 & 1 & 1 & | & -2 \end{bmatrix} \xrightarrow[R_3 = -r_2 + r_3]{R_1 = r_2 + r_1} \begin{bmatrix} 1 & 0 & -1 & | & 7 \\ 0 & 1 & -1 & | & 6 \\ 0 & 0 & 2 & | & -8 \end{bmatrix}$$

$$\xrightarrow{R_3 = \frac{1}{2}r_3} \begin{bmatrix} 1 & 0 & -1 & | & 7 \\ 0 & 1 & -1 & | & 6 \\ 0 & 0 & 1 & | & -4 \end{bmatrix} \xrightarrow[R_2 = r_3 + r_2]{R_1 = r_3 + r_1} \begin{bmatrix} 1 & 0 & 0 & | & 3 \\ 0 & 1 & 0 & | & 2 \\ 0 & 0 & 1 & | & -4 \end{bmatrix}$$

There is one solution. It is $x = 3$, $y = 2$, and $z = -4$.

39.   Write the system as the augmented matrix, $\begin{bmatrix} 1 & 1 & 0 & 0 & | & 7 \\ 0 & 1 & -1 & 1 & | & 5 \\ 1 & -1 & 1 & 1 & | & 6 \\ 0 & 1 & 0 & -1 & | & 10 \end{bmatrix}$.

Then use row operations to find the reduced row-echelon form.

$$\begin{bmatrix} 1 & 1 & 0 & 0 & | & 7 \\ 0 & 1 & -1 & 1 & | & 5 \\ 1 & -1 & 1 & 1 & | & 6 \\ 0 & 1 & 0 & -1 & | & 10 \end{bmatrix} \xrightarrow{R_3 = -r_1 + r_3} \begin{bmatrix} 1 & 1 & 0 & 0 & | & 7 \\ 0 & 1 & -1 & 1 & | & 5 \\ 0 & -2 & 1 & 1 & | & -1 \\ 0 & 1 & 0 & -1 & | & 10 \end{bmatrix}$$

$$\xrightarrow[\substack{R_1 = -r_2 + r_1 \\ R_3 = 2r_2 + r_3 \\ R_4 = -r_2 + r_4}]{} \begin{bmatrix} 1 & 0 & 1 & -1 & | & 2 \\ 0 & 1 & -1 & 1 & | & 5 \\ 0 & 0 & -1 & 3 & | & 9 \\ 0 & 0 & 1 & -2 & | & 5 \end{bmatrix} \xrightarrow{R_3 = -r_3} \begin{bmatrix} 1 & 0 & 1 & -1 & | & 2 \\ 0 & 1 & -1 & 1 & | & 5 \\ 0 & 0 & 1 & -3 & | & -9 \\ 0 & 0 & 1 & -2 & | & 5 \end{bmatrix}$$

$$\xrightarrow[\substack{R_1 = -r_3 + r_1 \\ R_2 = r_3 + r_2 \\ R_4 = -r_3 + r_4}]{} \begin{bmatrix} 1 & 0 & 0 & 2 & | & 11 \\ 0 & 1 & 0 & -2 & | & -4 \\ 0 & 0 & 1 & -3 & | & -9 \\ 0 & 0 & 0 & 1 & | & 14 \end{bmatrix} \xrightarrow[\substack{R_1 = -2r_4 + r_1 \\ R_2 = 2r_4 + r_2 \\ R_3 = 3r_4 + r_3}]{} \begin{bmatrix} 1 & 0 & 0 & 0 & | & -17 \\ 0 & 1 & 0 & 0 & | & 24 \\ 0 & 0 & 1 & 0 & | & 33 \\ 0 & 0 & 0 & 1 & | & 14 \end{bmatrix}$$

There is one solution. It is $x_1 = -17$, $x_2 = 24$, $x_3 = 33$, and $x_4 = 14$.

41. Write the system as the augmented matrix, $\begin{bmatrix} 1 & 2 & 3 & -1 & | & 0 \\ 3 & 0 & 0 & -1 & | & 4 \\ 0 & 1 & -1 & -1 & | & 2 \end{bmatrix}$.

Then use row operations to find the reduced row-echelon form.

$$\begin{bmatrix} 1 & 2 & 3 & -1 & | & 0 \\ 3 & 0 & 0 & -1 & | & 4 \\ 0 & 1 & -1 & -1 & | & 2 \end{bmatrix} \xrightarrow{R_2 = -3r_1 + r_2} \begin{bmatrix} 1 & 2 & 3 & -1 & | & 0 \\ 0 & -6 & -9 & 2 & | & 4 \\ 0 & 1 & -1 & -1 & | & 2 \end{bmatrix} \xrightarrow[\text{rows 2 and 3}]{\text{Interchange}} \begin{bmatrix} 1 & 2 & 3 & -1 & | & 0 \\ 0 & 1 & -1 & -1 & | & 2 \\ 0 & -6 & -9 & 2 & | & 4 \end{bmatrix}$$

$$\xrightarrow[R_3 = 6r_2 + r_3]{R_1 = -2r_2 + r_1} \begin{bmatrix} 1 & 0 & 5 & 1 & | & -4 \\ 0 & 1 & -1 & -1 & | & 2 \\ 0 & 0 & -15 & -4 & | & 16 \end{bmatrix} \xrightarrow{R_3 = -\frac{1}{15}r_3} \begin{bmatrix} 1 & 0 & 5 & 1 & | & -4 \\ 0 & 1 & -1 & -1 & | & 2 \\ 0 & 0 & 1 & \frac{4}{15} & | & -\frac{16}{15} \end{bmatrix}$$

$$\xrightarrow[R_2 = r_3 + r_2]{R_1 = -5r_3 + r_1} \begin{bmatrix} 1 & 0 & 0 & -\frac{1}{3} & | & \frac{4}{3} \\ 0 & 1 & 0 & -\frac{11}{15} & | & \frac{14}{15} \\ 0 & 0 & 1 & \frac{4}{15} & | & -\frac{16}{15} \end{bmatrix}$$

There are an infinite number of solutions. They are $x_1 = \frac{1}{3}x_4 + \frac{4}{3}$, $x_2 = \frac{11}{15}x_4 + \frac{14}{15}$,

$x_3 = -\frac{4}{15}x_4 - \frac{16}{15}$, and x_4 where x_4 is any real number.

43. Write the system as the augmented matrix, $\begin{bmatrix} 1 & -1 & 1 & | & 5 \\ 2 & -2 & 2 & | & 8 \end{bmatrix}$.

Then use row operations to find the reduced row-echelon form.

$$\begin{bmatrix} 1 & -1 & 1 & | & 5 \\ 2 & -2 & 2 & | & 8 \end{bmatrix} \xrightarrow{R_2 = -2r_1 + r_2} \begin{bmatrix} 1 & -1 & 1 & | & 5 \\ 0 & 0 & 0 & | & -2 \end{bmatrix}$$

There is no solution. The system is inconsistent.

45. Write the system as the augmented matrix, $\begin{bmatrix} 3 & -1 & 2 & | & 3 \\ 3 & 3 & 1 & | & 3 \\ 3 & -5 & 3 & | & 12 \end{bmatrix}$.

Then use row operations to find the reduced row-echelon form.

$$\begin{bmatrix} 3 & -1 & 2 & | & 3 \\ 3 & 3 & 1 & | & 3 \\ 3 & -5 & 3 & | & 12 \end{bmatrix} \xrightarrow{R_1 = \frac{1}{3}r_1} \begin{bmatrix} 1 & -\frac{1}{3} & \frac{2}{3} & | & 1 \\ 3 & 3 & 1 & | & 3 \\ 3 & -5 & 3 & | & 12 \end{bmatrix} \xrightarrow[R_3 = -3r_1 + r_3]{R_2 = -3r_1 + r_2} \begin{bmatrix} 1 & -\frac{1}{3} & \frac{2}{3} & | & 1 \\ 0 & 4 & -1 & | & 0 \\ 0 & -4 & 1 & | & 9 \end{bmatrix}$$

$$\xrightarrow{R_2 = \frac{1}{4}r_2} \begin{bmatrix} 1 & -\frac{1}{3} & \frac{2}{3} & | & 1 \\ 0 & 1 & -\frac{1}{4} & | & 0 \\ 0 & -4 & 1 & | & 9 \end{bmatrix} \xrightarrow[R_3 = 4r_2 + r_3]{R_1 = \frac{1}{3}r_2 + r_1} \begin{bmatrix} 1 & 0 & \frac{7}{12} & | & 1 \\ 0 & 1 & -\frac{1}{4} & | & 0 \\ 0 & 0 & 0 & | & 9 \end{bmatrix}$$

There is no solution to the system. It is inconsistent.

47. Write the system as the augmented matrix, $\begin{bmatrix} 1 & 1 & 1 & 1 & | & 4 \\ 2 & -1 & 1 & 0 & | & 0 \\ 3 & 2 & 1 & -1 & | & 6 \\ 1 & -2 & -2 & 2 & | & -1 \end{bmatrix}$.

Then use row operations to find the reduced row-echelon form.

$$\begin{bmatrix} 1 & 1 & 1 & 1 & | & 4 \\ 2 & -1 & 1 & 0 & | & 0 \\ 3 & 2 & 1 & -1 & | & 6 \\ 1 & -2 & -2 & 2 & | & -1 \end{bmatrix} \xrightarrow[\substack{R_2 = -2r_1 + r_2 \\ R_3 = -3r_1 + r_3 \\ R_4 = -r_1 + r_4}]{} \begin{bmatrix} 1 & 1 & 1 & 1 & | & 4 \\ 0 & -3 & -1 & -2 & | & -8 \\ 0 & -1 & -2 & -4 & | & -6 \\ 0 & -3 & -3 & 1 & | & -5 \end{bmatrix}$$

$$\xrightarrow[\substack{R_2 = -r_3 \\ R_3 = -r_2}]{} \begin{bmatrix} 1 & 1 & 1 & 1 & | & 4 \\ 0 & 1 & 2 & 4 & | & 6 \\ 0 & 3 & 1 & 2 & | & 8 \\ 0 & -3 & -3 & 1 & | & -5 \end{bmatrix} \xrightarrow[\substack{R_1 = -r_2 + r_1 \\ R_3 = -3r_2 + r_3 \\ R_4 = 3r_2 + r_4}]{} \begin{bmatrix} 1 & 0 & -1 & -3 & | & -2 \\ 0 & 1 & 2 & 4 & | & 6 \\ 0 & 0 & -5 & -10 & | & -10 \\ 0 & 0 & 3 & 13 & | & 13 \end{bmatrix}$$

$$\xrightarrow{R_3 = -\frac{1}{5}r_3} \begin{bmatrix} 1 & 0 & -1 & -3 & | & -2 \\ 0 & 1 & 2 & 4 & | & 6 \\ 0 & 0 & 1 & 2 & | & 2 \\ 0 & 0 & 3 & 13 & | & 13 \end{bmatrix} \xrightarrow[\substack{R_1 = r_3 + r_1 \\ R_2 = -2r_3 + r_2 \\ R_4 = -3r_3 + r_4}]{} \begin{bmatrix} 1 & 0 & 0 & -1 & | & 0 \\ 0 & 1 & 0 & 0 & | & 2 \\ 0 & 0 & 1 & 2 & | & 2 \\ 0 & 0 & 0 & 7 & | & 7 \end{bmatrix}$$

$$\xrightarrow{R_4 = \frac{1}{7}r_4} \begin{bmatrix} 1 & 0 & 0 & -1 & | & 0 \\ 0 & 1 & 0 & 0 & | & 2 \\ 0 & 0 & 1 & 2 & | & 2 \\ 0 & 0 & 0 & 1 & | & 1 \end{bmatrix} \xrightarrow[\substack{R_4 = r_4 + r_1 \\ R_3 = -2r_4 + r_3}]{} \begin{bmatrix} 1 & 0 & 0 & 0 & | & 1 \\ 0 & 1 & 0 & 0 & | & 2 \\ 0 & 0 & 1 & 0 & | & 0 \\ 0 & 0 & 0 & 1 & | & 1 \end{bmatrix}$$

There is one solution to the system. It is $x_1 = 1$, $x_2 = 2$, $x_3 = 0$ and $x_4 = 1$.

49. Write the system as the augmented matrix, $\begin{bmatrix} 2 & -1 & -1 & | & 0 \\ 1 & -1 & -1 & | & 1 \\ 3 & -1 & -1 & | & 2 \end{bmatrix}$.

Then use row operations to find the reduced row-echelon form.

$$\begin{bmatrix} 2 & -1 & -1 & | & 0 \\ 1 & -1 & -1 & | & 1 \\ 3 & -1 & -1 & | & 2 \end{bmatrix} \xrightarrow[\text{rows 1 and 2.}]{\text{Interchange}} \begin{bmatrix} 1 & -1 & -1 & | & 1 \\ 2 & -1 & -1 & | & 0 \\ 3 & -1 & -1 & | & 2 \end{bmatrix} \xrightarrow[R_3 = -3r_1 + r_3]{R_2 = -2r_1 + r_2} \begin{bmatrix} 1 & -1 & -1 & | & 1 \\ 0 & 1 & 1 & | & -2 \\ 0 & 2 & 2 & | & -1 \end{bmatrix}$$

$$\xrightarrow[R_3 = -2r_2 + r_3]{R_1 = r_2 + r_1} \begin{bmatrix} 1 & 0 & 0 & | & -1 \\ 0 & 1 & 1 & | & -2 \\ 0 & 0 & 0 & | & 3 \end{bmatrix}$$

There is no solution to the system. It is inconsistent.

51. Write the system as the augmented matrix, $\begin{bmatrix} 2 & -1 & 1 & | & 6 \\ 3 & -1 & 1 & | & 6 \\ 4 & -2 & 2 & | & 12 \end{bmatrix}$.

Then use row operations to find the reduced row-echelon form.

$$\begin{bmatrix} 2 & -1 & 1 & | & 6 \\ 3 & -1 & 1 & | & 6 \\ 4 & -2 & 2 & | & 12 \end{bmatrix} \xrightarrow{R_1 = r_2 - r_1} \begin{bmatrix} 1 & 0 & 0 & | & 0 \\ 3 & -1 & 1 & | & 6 \\ 4 & -2 & 2 & | & 12 \end{bmatrix} \xrightarrow[R_3 = -4r_1 + r_3]{R_2 = -3r_1 + r_2} \begin{bmatrix} 1 & 0 & 0 & | & 0 \\ 0 & -1 & 1 & | & 6 \\ 0 & -2 & 2 & | & 12 \end{bmatrix}$$

$$\xrightarrow{R_2 = -r_2} \begin{bmatrix} 1 & 0 & 0 & | & 0 \\ 0 & 1 & -1 & | & -6 \\ 0 & -2 & 2 & | & 12 \end{bmatrix} \xrightarrow{R_3 = 2r_2 + r_3} \begin{bmatrix} 1 & 0 & 0 & | & 0 \\ 0 & 1 & -1 & | & -6 \\ 0 & 0 & 0 & | & 0 \end{bmatrix}$$

There are an infinite number of solutions of the system. They are $x = 0$, $y = z - 6$, and z, where z is any real number.

53. If, as in Example 11, we let x represent the amount invested in EE/E bond, y the amount invested in I bonds, and z the amount invested in HH/H bonds, then the augmented matrix representing the system becomes,

$$\begin{bmatrix} 1 & 1 & 1 & | & 25000 \\ .032 & .04 & .015 & | & 800 \end{bmatrix} \xrightarrow{R_2 = -0.032 r_1 + r_2} \begin{bmatrix} 1 & 1 & 1 & | & 25000 \\ 0 & .008 & -.017 & | & 0 \end{bmatrix}$$

$$\xrightarrow{R_2 = \frac{r_2}{0.008}} \begin{bmatrix} 1 & 1 & 1 & | & 25000 \\ 0 & 1 & -2.125 & | & 0 \end{bmatrix}$$

$$\xrightarrow{R_1 = -r_2 + r_1} \begin{bmatrix} 1 & 0 & 3.125 & | & 25000 \\ 0 & 1 & -2.125 & | & 0 \end{bmatrix}$$

The system of equations can be written as $\begin{cases} x = -3.125z + 25,000 \\ y = 2.125z \end{cases}$ where z is the parameter.

Possible allocations are:

Amount in EE/E	Amount in I	Amount in HH/H
$ 3125	$14,875	$7000
$12,500	$ 8500	$4000
$18,750	$ 4250	$2000
$21,875	$ 2125	$1000

55. If, as in Example 11, we let x represent the amount invested in EE/E bond, y the amount invested in I bonds, and z the amount invested in HH/H bonds, then the augmented matrix representing the system becomes,

$$\begin{bmatrix} 1 & 1 & 1 & | & 25,000 \\ .032 & .035 & .015 & | & 500 \end{bmatrix} \xrightarrow{R_2 = -0.032r_1 + r_2} \begin{bmatrix} 1 & 1 & 1 & | & 25,000 \\ 0 & .003 & -.017 & | & -300 \end{bmatrix}$$

$$\xrightarrow{R_2 = \frac{r_2}{0.003}} \begin{bmatrix} 1 & 1 & 1 & | & 25,000 \\ 0 & 1 & -\dfrac{17}{3} & | & -100,000 \end{bmatrix} \xrightarrow{R_1 = -r_2 + r_1} \begin{bmatrix} 1 & 0 & \dfrac{20}{3} & | & 125,000 \\ 0 & 1 & -\dfrac{17}{3} & | & -100,000 \end{bmatrix}$$

The system of equations can be written as $\begin{cases} x = -\dfrac{20}{3}z + 125,000 \\ y = \dfrac{17}{3}z - 100,000 \end{cases}$ where z is the parameter.

Since we want to invest in each of the three bond types, the conditions $x > 0$, $y > 0$, and $z > 0$ must hold. We determine from the system that

$$-\frac{20}{3}z + 125,000 > 0 \text{ so } z < 18,750$$

$$\frac{17}{3}z - 100,000 > 0 \text{ so } z > 17,647$$

Possible allocations are:

Amount in EE/E	Amount in I	Amount in HH/H
$1666.67	$4833.33	$18,500
$3333.33	$3416.67	$18,250
$5000	$2000	$18,000
$6666.67	$ 583.33	$17,750

57. Let x represent the price of a hamburger, y represent the price of fries, and z represent the price of cola. To find the price of each item, we need to solve the system of equations.

$$\begin{cases} 8x + 6y + 6z = 26.10 \\ 10x + 6y + 8z = 31.60 \end{cases}$$

The system can be represented by the augmented matrix, and then use row operations to write the matrix in reduced row-echelon form.

$$\begin{bmatrix} 8 & 6 & 6 & | & 26.10 \\ 10 & 6 & 8 & | & 31.60 \end{bmatrix} \xrightarrow{R_1 = \frac{r_1}{8}} \begin{bmatrix} 1 & \frac{3}{4} & \frac{3}{4} & | & \frac{261}{80} \\ 10 & 6 & 8 & | & 31.60 \end{bmatrix} \xrightarrow{R_2 = -10r_1 + r_2} \begin{bmatrix} 1 & \frac{3}{4} & \frac{3}{4} & | & \frac{261}{80} \\ 0 & -\frac{3}{2} & \frac{1}{2} & | & -\frac{41}{40} \end{bmatrix}$$

$$\xrightarrow{R_2 = -\frac{2}{3}r_2} \begin{bmatrix} 1 & \frac{3}{4} & \frac{3}{4} & | & \frac{261}{80} \\ 0 & 1 & -\frac{1}{3} & | & \frac{41}{60} \end{bmatrix} \xrightarrow{R_1 = -\frac{3}{4}r_2 + \frac{3}{4}r_1} \begin{bmatrix} 1 & 0 & 1 & | & \frac{11}{4} \\ 0 & 1 & -\frac{1}{3} & | & \frac{41}{60} \end{bmatrix}$$

There is not enough information to determine the price of each food item. We have

$x = -z + \dfrac{11}{4} = -z + 2.75, \ y = \dfrac{1}{3}z + \dfrac{41}{60}$ where z is the parameter. Using these equations and

that we are told $\$1.75 \le x \le \$2.25$, $\$0.75 \le y \le \$1.00$, and $\$0.60 \le z \le \$0.90$, possible prices include

Price of a Hamburger	Price of Fries	Price of Cola
$2.15	$0.88	$0.60
$2.00	$0.93	$0.75
$1.95	$0.95	$0.80
$1.85	$0.98	$0.90

59. Let x represent the amount invested in treasury bills which yield 7%, y represent the amount invested in corporate bonds which yield 9%, and z represent the amount invested in junk bonds which yield 11% in interest.

(a) The first couple have \$20,000 to invest, and require \$2000 in income. To determine how they should allot their funds we need to solve the system of equations:

$$\begin{cases} x+ \quad y+ \quad z=20,000 \\ 0.07x+0.09y+0.11z= \;2,000 \end{cases}$$

We write the system as an augmented matrix and use row operations to put it in reduced row-echelon form.

$$\begin{bmatrix} 1 & 1 & 1 & 20000 \\ 0.07 & 0.09 & 0.11 & 2000 \end{bmatrix} \xrightarrow{R_2 = -0.07r_1 + r_2} \begin{bmatrix} 1 & 1 & 1 & 20000 \\ 0 & 0.02 & 0.04 & 600 \end{bmatrix}$$

$$\xrightarrow{R_2 = 50r_2} \begin{bmatrix} 1 & 1 & 1 & 20000 \\ 0 & 1 & 2 & 30000 \end{bmatrix} \xrightarrow{R_1 = -r_2 + r_1} \begin{bmatrix} 1 & 0 & -1 & -10000 \\ 0 & 1 & 2 & 30000 \end{bmatrix}$$

The solutions to the system are $x = z - 10,000$, $y = -2z + 30,000$, and z, where z is the parameter. Using $x \geq 0$ we get $z - 10,000 \geq 0$ or $z \geq 10,000$. Using $y \geq 0$, we get $-2z + 30,000 \geq 0$ or $z \leq 15,000$.

So the couple could achieve their goal of earning \$2000 in investment income by investing in the following ways.

Treasury bills	Corporate bonds	Junk bonds
\$0	\$10,000	\$10,000
1,000	8,000	11,000
2,000	6,000	12,000
3,000	4,000	13,000
4,000	2,000	14,000
5,000	0	15,000

(b) The second couple have \$25,000 to invest, and require \$2000 in income. To determine how they should allot their funds we need to solve the system of equations:

$$\begin{cases} x+ \quad y+ \quad z=25,000 \\ 0.07x+0.09y+0.11z= \;2,000 \end{cases}$$

We write the system as an augmented matrix and use row operations to put it in reduced row-echelon form.

$$\begin{bmatrix} 1 & 1 & 1 & 25000 \\ 0.07 & 0.09 & 0.11 & 2000 \end{bmatrix} \xrightarrow{R_2 = -0.07r_1 + r_2} \begin{bmatrix} 1 & 1 & 1 & 25000 \\ 0 & 0.02 & 0.04 & 250 \end{bmatrix}$$

$$\xrightarrow[R_2 = 50r_2]{} \begin{bmatrix} 1 & 1 & 1 & | & 25000 \\ 0 & 1 & 2 & | & 12500 \end{bmatrix} \xrightarrow[R_1 = -r_2 + r_1]{} \begin{bmatrix} 1 & 0 & -1 & | & 12500 \\ 0 & 1 & 2 & | & 12500 \end{bmatrix}$$

The solutions to the system are $x = z + 12{,}500$, $y = -2z + 12{,}500$, and z, where z is the parameter. Using $x \geq 0$ we get $z + 12{,}500 \geq 0$ or $z \geq 0$. Using $y \geq 0$, we get $-2z + 12{,}500 \geq 0$ or $z \leq 6250$.

So the couple could achieve their goal of earning $2000 in investment income by investing in the following ways.

Treasury bills	Corporate bonds	Junk bonds
$12,500	$12,500	$0
13,500	10,500	1,000
14,500	8,500	2,000
15,500	6,500	3,000
16,500	4,500	4,000
18,750	0	6,250

(c)   The third couple have $30,000 to invest, and require $2000 in income
To determine how they should allot their funds we need to solve the system of equations:

$$\begin{cases} x + y + z = 30{,}000 \\ 0.07x + 0.09y + 0.11z = 2{,}000 \end{cases}$$

This couple will exceed their needs regardless of how they allot their investment, since $(0.07)(30{,}000) = 2100$. Therefore, even if they put all $30,000 in treasuries (which yield the least) their interest income will be $2100, more than their needs.

61. Let x represent the mg. of liquid 1 to be used, y represent the mg. of liquid 2 to be used and z represent the mg. of liquid 3 to be used.

To create the table we must solve the system of equations:

$$\begin{cases} 0.20x + 0.40y + 0.30z = 40 \\ 0.30x + 0.20y + 0.50z = 30 \end{cases} \text{ or multiplying both equations by 10 } \begin{cases} 2x + 4y + 3z = 400 \\ 3x + 2y + 5z = 300 \end{cases}.$$

We will write the system as an augmented matrix and use row operations to put it in reduced row-echelon form.

$$\begin{bmatrix} 2 & 4 & 3 & | & 400 \\ 3 & 2 & 5 & | & 300 \end{bmatrix} \xrightarrow{R_1 = \frac{1}{2}r_1} \begin{bmatrix} 1 & 2 & \frac{3}{2} & | & 200 \\ 3 & 2 & 5 & | & 300 \end{bmatrix} \xrightarrow{R_2 = -3r_1 + r_2} \begin{bmatrix} 1 & 2 & \frac{3}{2} & | & 200 \\ 0 & -4 & \frac{1}{2} & | & -300 \end{bmatrix}$$

$$\xrightarrow{R_2 = -\frac{1}{4}r_2} \begin{bmatrix} 1 & 2 & \frac{3}{2} & | & 200 \\ 0 & 1 & -\frac{1}{8} & | & 75 \end{bmatrix} \xrightarrow{R_1 = -2r_2 + r_1} \begin{bmatrix} 1 & 0 & \frac{7}{4} & | & 50 \\ 0 & 1 & -\frac{1}{8} & | & 75 \end{bmatrix}$$

So the solutions to the system are $x = -\frac{7}{4}z + 50$, $y = \frac{1}{8}z + 75$, and z where z is the parameter. Using $x \geq 0$, we get $-\frac{7}{4}z + 50 \geq 0$ or $z \leq \frac{200}{7} = 28.571$; using $y \geq 0$, we find $z \geq 0$.

Possible ways the pharmacist can fill the prescription include:

Mg. of liquid of 1	Mg. of liquid of 2	Mg. of liquid of 3
50	75	0
41.25	75.625	5
32.5	76.25	10
23.75	76.875	15
15	77.5	20
6.25	78.125	25
0	78.571	28.571

63. Answers will vary.

2.4 Matrix Algebra: Equality, Addition, Subtraction

1. $\begin{bmatrix} 3 & 2 \\ -1 & 3 \end{bmatrix}$ is a 2×2 matrix. It is square.

3. $\begin{bmatrix} 2 & 1 & -3 \\ 1 & 0 & -1 \end{bmatrix}$ is a 2×3 matrix.

5. $\begin{bmatrix} 4 & 0 \\ -1 & 2 \\ 5 & 8 \end{bmatrix}$ is a 3×2 matrix.

7. $\begin{bmatrix} 1 & 4 \\ -2 & 8 \\ 0 & 0 \end{bmatrix}$ is 3×2 matrix.

9. $\begin{bmatrix} 4 \\ 1 \end{bmatrix}$ is a 2×1 column matrix.

11. $[2]$ is a 1×1 matrix. It is a column matrix, a row matrix, and a square matrix.

13. False, to be equal 2 matrices must have the same dimensions.

15. True

17. True

19. True

21. True

23. False, to be equal 2 matrices must have the same dimensions.

25. $\begin{bmatrix} 3 & -1 \\ 4 & 2 \end{bmatrix} + \begin{bmatrix} -2 & 2 \\ 2 & 5 \end{bmatrix} = \begin{bmatrix} 3+-2 & -1+2 \\ 4+2 & 2+5 \end{bmatrix} = \begin{bmatrix} 1 & 1 \\ 6 & 7 \end{bmatrix}$

27. $3\begin{bmatrix} 2 & 6 & 0 \\ 4 & -2 & 1 \end{bmatrix} = \begin{bmatrix} 3\cdot 2 & 3\cdot 6 & 3\cdot 0 \\ 3\cdot 4 & 3\cdot (-2) & 3\cdot 1 \end{bmatrix} = \begin{bmatrix} 6 & 18 & 0 \\ 12 & -6 & 3 \end{bmatrix}$

29. $2\begin{bmatrix} 1 & -1 & 8 \\ 2 & 4 & 1 \end{bmatrix} - 3\begin{bmatrix} 0 & -2 & 8 \\ 1 & 4 & 1 \end{bmatrix} = \begin{bmatrix} 2 & -2 & 16 \\ 4 & 8 & 2 \end{bmatrix} - \begin{bmatrix} 0 & -6 & 24 \\ 3 & 12 & 3 \end{bmatrix}$

$$= \begin{bmatrix} 2-0 & -2+6 & 16-24 \\ 4-3 & 8-12 & 2-3 \end{bmatrix}$$

$$= \begin{bmatrix} 2 & 4 & -8 \\ 1 & -4 & -1 \end{bmatrix}$$

31. $3\begin{bmatrix} a & 8 \\ b & 1 \\ c & -2 \end{bmatrix} + 5\begin{bmatrix} 2a & 6 \\ -b & -2 \\ -c & 0 \end{bmatrix} = \begin{bmatrix} 3a & 24 \\ 3b & 3 \\ 3c & -6 \end{bmatrix} + \begin{bmatrix} 10a & 30 \\ -5b & -10 \\ -5c & 0 \end{bmatrix} = \begin{bmatrix} 3a+10a & 24+30 \\ 3b-5b & 3-10 \\ 3c-5c & -6+0 \end{bmatrix}$

$$= \begin{bmatrix} 13a & 54 \\ -2b & -7 \\ -2c & -6 \end{bmatrix}$$

33. $A - B = \begin{bmatrix} 2 & -3 & 4 \\ 0 & 2 & 1 \end{bmatrix} - \begin{bmatrix} 1 & -2 & 0 \\ 5 & 1 & 2 \end{bmatrix} = \begin{bmatrix} 2-1 & -3-(-2) & 4-0 \\ 0-5 & 2-1 & 1-2 \end{bmatrix} = \begin{bmatrix} 1 & -1 & 4 \\ -5 & 1 & -1 \end{bmatrix}$

35. $2A - 3C = 2\begin{bmatrix} 2 & -3 & 4 \\ 0 & 2 & 1 \end{bmatrix} - 3\begin{bmatrix} -3 & 0 & 5 \\ 2 & 1 & 3 \end{bmatrix} = \begin{bmatrix} 4 & -6 & 8 \\ 0 & 4 & 2 \end{bmatrix} + \begin{bmatrix} 9 & 0 & -15 \\ -6 & -3 & -9 \end{bmatrix}$

$$= \begin{bmatrix} 4+9 & -6+0 & 8-15 \\ 0-6 & 4-3 & 2-9 \end{bmatrix} = \begin{bmatrix} 13 & -6 & -7 \\ -6 & 1 & -7 \end{bmatrix}$$

37. $(A + B) - 2C = \left(\begin{bmatrix} 2 & -3 & 4 \\ 0 & 2 & 1 \end{bmatrix} + \begin{bmatrix} 1 & -2 & 0 \\ 5 & 1 & 2 \end{bmatrix}\right) - 2\begin{bmatrix} -3 & 0 & 5 \\ 2 & 1 & 3 \end{bmatrix}$

$$= \begin{bmatrix} 3 & -5 & 4 \\ 5 & 3 & 3 \end{bmatrix} - \begin{bmatrix} -6 & 0 & 10 \\ 4 & 2 & 6 \end{bmatrix} = \begin{bmatrix} 3+6 & -5-0 & 4-10 \\ 5-4 & 3-2 & 3-6 \end{bmatrix} = \begin{bmatrix} 9 & -5 & -6 \\ 1 & 1 & -3 \end{bmatrix}$$

39. $3A + 4(B + C) = 3\begin{bmatrix} 2 & -3 & 4 \\ 0 & 2 & 1 \end{bmatrix} + 4\left(\begin{bmatrix} 1 & -2 & 0 \\ 5 & 1 & 2 \end{bmatrix} + \begin{bmatrix} -3 & 0 & 5 \\ 2 & 1 & 3 \end{bmatrix}\right)$

$$= \begin{bmatrix} 6 & -9 & 12 \\ 0 & 6 & 3 \end{bmatrix} + 4\begin{bmatrix} -2 & -2 & 5 \\ 7 & 2 & 5 \end{bmatrix} = \begin{bmatrix} 6 & -9 & 12 \\ 0 & 6 & 3 \end{bmatrix} + \begin{bmatrix} -8 & -8 & 20 \\ 28 & 8 & 20 \end{bmatrix}$$

$$= \begin{bmatrix} 6-8 & -9-8 & 12+20 \\ 0+28 & 6+8 & 3+20 \end{bmatrix} = \begin{bmatrix} -2 & -17 & 32 \\ 28 & 14 & 23 \end{bmatrix}$$

41. $2(A - B) - C = 2\left(\begin{bmatrix} 2 & -3 & 4 \\ 0 & 2 & 1 \end{bmatrix} - \begin{bmatrix} 1 & -2 & 0 \\ 5 & 1 & 2 \end{bmatrix}\right) - \begin{bmatrix} -3 & 0 & 5 \\ 2 & 1 & 3 \end{bmatrix}$

$$= 2\begin{bmatrix} 2-1 & -3-(-2) & 4-0 \\ 0-5 & 2-1 & 1-2 \end{bmatrix} - \begin{bmatrix} -3 & 0 & 5 \\ 2 & 1 & 3 \end{bmatrix}$$

$$= 2\begin{bmatrix} 1 & -1 & 4 \\ -5 & 1 & -1 \end{bmatrix} + \begin{bmatrix} 3 & 0 & -5 \\ -2 & -1 & -3 \end{bmatrix} = \begin{bmatrix} 2+3 & -2-0 & 8-5 \\ -10-2 & 2-1 & -2-3 \end{bmatrix}$$

$$= \begin{bmatrix} 5 & -2 & 3 \\ -12 & 1 & -5 \end{bmatrix}$$

43. $3A - B - 6C = 3\begin{bmatrix} 2 & -3 & 4 \\ 0 & 2 & 1 \end{bmatrix} - \begin{bmatrix} 1 & -2 & 0 \\ 5 & 1 & 2 \end{bmatrix} - 6\begin{bmatrix} -3 & 0 & 5 \\ 2 & 1 & 3 \end{bmatrix}$

$$= \begin{bmatrix} 6 & -9 & 12 \\ 0 & 6 & 3 \end{bmatrix} + \begin{bmatrix} -1 & +2 & 0 \\ -5 & -1 & -2 \end{bmatrix} + \begin{bmatrix} 18 & 0 & -30 \\ -12 & -6 & -18 \end{bmatrix}$$

$$= \begin{bmatrix} 6-1+18 & -9+2-0 & 12-0-30 \\ 0-5-12 & 6-1-6 & 3-2-18 \end{bmatrix} = \begin{bmatrix} 23 & -7 & -18 \\ -17 & -1 & -17 \end{bmatrix}$$

45. The commutative property for addition: $A + B = B + A$.

$$A + B = \begin{bmatrix} 2 & -3 & 4 \\ 0 & 2 & 1 \end{bmatrix} + \begin{bmatrix} 1 & -2 & 0 \\ 5 & 1 & 2 \end{bmatrix} = \begin{bmatrix} 2+1 & -3-2 & 4+0 \\ 0+5 & 2+1 & 1+2 \end{bmatrix} = \begin{bmatrix} 3 & -5 & 4 \\ 5 & 3 & 3 \end{bmatrix}$$

$$B + A = \begin{bmatrix} 1 & -2 & 0 \\ 5 & 1 & 2 \end{bmatrix} + \begin{bmatrix} 2 & -3 & 4 \\ 0 & 2 & 1 \end{bmatrix} = \begin{bmatrix} 1+2 & -2-3 & 0+4 \\ 5+0 & 1+2 & 2+1 \end{bmatrix} = \begin{bmatrix} 3 & -5 & 4 \\ 5 & 3 & 3 \end{bmatrix} = A + B$$

47. The additive inverse property: $A + (-A) = 0$

$$A + (-A) = \begin{bmatrix} 2 & -3 & 4 \\ 0 & 2 & 1 \end{bmatrix} + \begin{bmatrix} -2 & 3 & -4 \\ 0 & -2 & -1 \end{bmatrix} = \begin{bmatrix} 2-2 & -3+3 & 4-4 \\ 0+0 & 2-2 & 1-1 \end{bmatrix} = \begin{bmatrix} 0 & 0 & 0 \\ 0 & 0 & 0 \end{bmatrix}$$

49. Property of scalar multiplication: $(k + h)A = kA + hA$.

$$(2 + 3)B = 5B = 5\begin{bmatrix} 1 & -2 & 0 \\ 5 & 1 & 2 \end{bmatrix} = \begin{bmatrix} 5 & -10 & 0 \\ 25 & 5 & 10 \end{bmatrix}$$

$$2B + 3B = 2\begin{bmatrix} 1 & -2 & 0 \\ 5 & 1 & 2 \end{bmatrix} + 3\begin{bmatrix} 1 & -2 & 0 \\ 5 & 1 & 2 \end{bmatrix} = \begin{bmatrix} 2 & -4 & 0 \\ 10 & 2 & 4 \end{bmatrix} + \begin{bmatrix} 3 & -6 & 0 \\ 15 & 3 & 6 \end{bmatrix}$$

$$= \begin{bmatrix} 2+3 & -4-6 & 0+0 \\ 10+15 & 2+3 & 4+6 \end{bmatrix} = \begin{bmatrix} 5 & -10 & 0 \\ 25 & 5 & 10 \end{bmatrix} = 5B$$

51. Two matrices are equal if they have the same dimension and if corresponding entries are equal. $\begin{bmatrix} x \\ 4 \end{bmatrix}$ and $\begin{bmatrix} -4 \\ z \end{bmatrix}$ have the same dimension so the matrices are equal if $x = -4$ and $z = 4$.

53. Two matrices are equal if they have the same dimension and if corresponding entries are equal. $\begin{bmatrix} x-2y & 0 \\ -2 & 6 \end{bmatrix}$ and $\begin{bmatrix} 3 & 0 \\ -2 & x+y \end{bmatrix}$ have the same dimension so the matrices are equal if

$x - 2y = 3$ and $x + y = 6$. To find the values of x and y, solve the system $\begin{cases} x-2y=3 \\ x+\ y=6 \end{cases}$.

Writing the system as an augmented matrix and then using row operations to put it in reduced row-echelon form, we get

$$\begin{bmatrix} 1 & -2 & | & 3 \\ 1 & 1 & | & 6 \end{bmatrix} \xrightarrow{R_2 = -r_1 + r_2} \begin{bmatrix} 1 & -2 & | & 3 \\ 0 & 3 & | & 3 \end{bmatrix} \xrightarrow{R_2 = \frac{1}{3}r_2} \begin{bmatrix} 1 & -2 & | & 3 \\ 0 & 1 & | & 1 \end{bmatrix} \xrightarrow{R_1 = 2r_2 + r_1} \begin{bmatrix} 1 & 0 & | & 5 \\ 0 & 1 & | & 1 \end{bmatrix}$$

So the matrices are equal when $x = 5$ and $y = 1$.

55. $[2 \quad 3 \quad -4] + [x \quad 2y \quad z] = [x+2 \quad 2y+3 \quad z-4] = [6 \quad -9 \quad 2]$
Two matrices are equal if they have the same dimension and if corresponding entries are equal. $[x+2 \quad 2y+3 \quad z-4]$ and $[6 \quad -9 \quad 2]$ have the same dimension so the matrices are equal if the corresponding entries are equal.

To find the values of x, y, and z, we solve the system $\begin{cases} x+2=6 \\ 2y+3=-9 \\ z-4=2 \end{cases}$

So we find that the sum of the two matrices equals $[6 \quad -9 \quad 2]$ when $x = 4$, $y = -6$, and $z = 6$.

57.

```
[A]+[B]
[[-2    1    7    5 ...
 [4     6    7    5 ...
 [-3.5  8   -4   13 ...
 [12   -1    7    6 ...
```

$$A + B = \begin{bmatrix} -2 & 1 & 7 & 5 \\ 4 & 6 & 7 & 5 \\ -3.5 & 8 & -4 & 13 \\ 12 & -1 & 7 & 6 \end{bmatrix}$$

59.

```
[C]-3*([A]+[B])          [C]-3*([A]+[B])
[[19    -11  -14 ...      ...  -11  -14  -15]
 [-12   -13  -21 ...      ...  -13  -21  -17]
 [15.5  -24   19 ...      ..5  -24   19  -39]
 [-29    10  -14 ...      ...   10  -14  -11]]
```

$$C - 3(A + B) = \begin{bmatrix} 19 & -11 & -14 & -15 \\ -12 & -13 & -21 & -17 \\ 15.5 & -24 & 19 & -39 \\ -29 & 10 & -14 & -11 \end{bmatrix}$$

61.

```
3([B]+[C])-[A]           3([B]+[C])-[A]
[[37   -17   30  1...    ..7   -17   30  15]
 [4     9    13  1...    ...    9    13   1 ]
 [20.5  16   -3  3...    ..0.5  16   -3  31]
 [29    18   22  4...    ..9    18   22  43]]
```

$$3(B + C) - A = \begin{bmatrix} 37 & -17 & 30 & 15 \\ 4 & 9 & 13 & 1 \\ 20.5 & 16 & -3 & 31 \\ 29 & 18 & 22 & 43 \end{bmatrix}$$

63. Before constructing the matrix, we need to find the numbers of males and females in each type of prison.

Local: female: $(0.115)(613,534) = 70,556$; male: $613,534 - 70,556 = 542,978$
State: male: $(0.934)(1,236,476) = 1,154,869$; female: $1,236,476 - 1,154,869 = 81,607$
Federal: female: $(0.070)(145,416) = 10,179$; male: $145,416 - 10,179 = 135,237$

	LOCAL	STATE	FED
MALE	542,978	1,154,869	135,237
FEMALE	70,556	81,607	10,179

65. Before constructing the matrix, we need to find the number of degrees earned by each gender.

Associate: women: $582,000 - 218,000 = 364,000$ degrees
Bachelor: men: $1,251,000 - 714,000 = 537,000$ degrees
Master: men: $442,000 - 261,000 = 181,000$ degrees
Doctoral: women: $47,100 - 26,700 = 20,400$ degrees

	Associate	Bachelor's	Master's	Doctoral
Male	218,000	537,000	181,000	26,700
Female	364,000	714,000	261,000	20,400

The data could be represented with a 4×2 matrix by interchanging the rows and the columns.

67. The matrix will have dimension 2×3.

	DEMOCRATS	REPUBLICANS	INDEPENDENTS
UNDER $25,000	351	271	73
OVER $25,000	203	215	55

69. Before constructing the matrix, we need to determine how many students of each gender majored in each of the three areas.

LAS: females: $(0.50)(500) = 250$; males: $(0.50)(500) = 250$
ENG: males: $(0.75)(300) = 225$; females: $300 - 225 = 75$
EDUC Majors: $1000 - (500 + 300) = 200$
females: $(0.60)(200) = 120$; males: $200 - 120 = 80$

	LAS	ENG	EDUC
MALE	250	225	80
FEMALE	250	75	120

2.5 Multiplication of Matrices

1. $[1 \quad 3] \begin{bmatrix} 2 \\ 4 \end{bmatrix} = [(1)(2) + (3)(4)] = [14]$

3. $[1 \quad -2 \quad 3] \begin{bmatrix} 0 \\ 1 \\ 2 \end{bmatrix} = [-1 \cdot 0 + (-2) \cdot 1 + 3 \cdot 2] = [4]$

5. $[1 \quad 4] \begin{bmatrix} 2 & 0 \\ 4 & -2 \end{bmatrix} = [1 \cdot 2 + 4 \cdot 4 \quad 1 \cdot 0 + 4 \cdot (-2)] = [18 \quad -8]$

7. $\begin{bmatrix} 2 & 0 \\ 4 & -2 \end{bmatrix} \begin{bmatrix} 2 & 1 \\ 3 & -2 \end{bmatrix} = \begin{bmatrix} 2 \cdot 2 + 0 \cdot 3 & 2 \cdot 1 + 0 \cdot (-2) \\ 4 \cdot 2 + (-2) \cdot 3 & 4 \cdot 1 + (-2)(-2) \end{bmatrix} = \begin{bmatrix} 4 & 2 \\ 2 & 8 \end{bmatrix}$

9. $[1 \quad -2 \quad 3] \begin{bmatrix} 0 & 1 \\ 1 & 2 \\ 2 & 3 \end{bmatrix} = [1 \cdot 0 + (-2) \cdot 1 + 3 \cdot 2 \quad 1 \cdot 1 + (-2) \cdot 2 + 3 \cdot 3] = [4 \quad 6]$

11. $\begin{bmatrix} 1 & -2 & 3 \\ 4 & 0 & 6 \end{bmatrix} \begin{bmatrix} 0 & -2 \\ 1 & 0 \\ 2 & -4 \end{bmatrix} = \begin{bmatrix} 1 \cdot 0 + (-2) \cdot 1 + 3 \cdot 2 & 1 \cdot (-2) + (-2) \cdot 0 + 3 \cdot (-4) \\ 4 \cdot 0 + 0 \cdot 1 + 6 \cdot 2 & 4 \cdot (-2) + 0 \cdot 0 + 6 \cdot (-4) \end{bmatrix} = \begin{bmatrix} 4 & -14 \\ 12 & -32 \end{bmatrix}$

13. $\begin{bmatrix} 2 & 0 \\ 4 & -2 \\ 6 & -1 \end{bmatrix} \begin{bmatrix} 2 & 1 \\ 3 & -2 \end{bmatrix} = \begin{bmatrix} 2 \cdot 2 + 0 \cdot 3 & 2 \cdot 1 + 0 \cdot (-2) \\ 4 \cdot 2 + (-2) \cdot 3 & 4 \cdot 1 + (-2)(-2) \\ 6 \cdot 2 + (-1) \cdot 3 & 6 \cdot 1 + (-1)(-2) \end{bmatrix} = \begin{bmatrix} 4 & 2 \\ 2 & 8 \\ 9 & 8 \end{bmatrix}$

15. $\begin{bmatrix} 1 & -1 & 6 \\ 2 & 0 & -1 \\ 3 & 1 & 2 \end{bmatrix} \begin{bmatrix} 3 & 2 \\ 0 & 1 \\ 1 & 0 \end{bmatrix} = \begin{bmatrix} 1 \cdot 3 + (-1) \cdot 0 + 6 \cdot 1 & 1 \cdot 2 + (-1) \cdot 1 + 6 \cdot 0 \\ 2 \cdot 3 + 0 \cdot 0 + (-1) \cdot 1 & 2 \cdot 2 + 0 \cdot 1 + (-1) \cdot 0 \\ 3 \cdot 3 + 1 \cdot 0 + 2 \cdot 1 & 3 \cdot 2 + 1 \cdot 1 + 2 \cdot 0 \end{bmatrix} = \begin{bmatrix} 9 & 1 \\ 5 & 4 \\ 11 & 7 \end{bmatrix}$

17. BA is defined; the dimensions of the product are 3×4.

19. AB is not defined.

21. $(BA)C$ is not defined. (BA is 3×4 and C is 2×3.)

23. $BA + A$ is defined with dimensions 3×4.

25. $DC + B$ is defined. It has dimensions 3×3.

27. $AB = \begin{bmatrix} 1-2 & 2+8 & 3-4 \\ 0-4 & 0+16 & 0-8 \end{bmatrix} = \begin{bmatrix} -1 & 10 & -1 \\ -4 & 16 & -8 \end{bmatrix}$

29. $BC = \begin{bmatrix} 3+8+0 & 1-2+6 \\ -3+16+0 & -1-4-4 \end{bmatrix} = \begin{bmatrix} 11 & 5 \\ 13 & -9 \end{bmatrix}$

31. $(D+I_3)C = \begin{bmatrix} 2 & 0 & 4 \\ 0 & 2 & 2 \\ 0 & -1 & 2 \end{bmatrix}\begin{bmatrix} 3 & 1 \\ 4 & -1 \\ 0 & 2 \end{bmatrix} = \begin{bmatrix} 6+0+0 & 2+0+8 \\ 0+8+0 & 0-2+4 \\ 0-4+0 & 0+1+4 \end{bmatrix} = \begin{bmatrix} 6 & 10 \\ 8 & 2 \\ -4 & 5 \end{bmatrix}$

33. $EI_2 = \begin{bmatrix} 3\cdot1+-1\cdot0 & 3\cdot0+-1\cdot1 \\ 4\cdot1+2\cdot0 & 4\cdot0+2\cdot1 \end{bmatrix} = \begin{bmatrix} 3 & -1 \\ 4 & 2 \end{bmatrix}$

35. $(2E)B = \begin{bmatrix} 6 & -2 \\ 8 & 4 \end{bmatrix}\cdot B = \begin{bmatrix} 6\cdot1+-2\cdot-1 & 6\cdot2+-2\cdot4 & 6\cdot3+-2\cdot-2 \\ 8\cdot1+4\cdot-1 & 8\cdot2+4\cdot4 & 8\cdot3+4\cdot-2 \end{bmatrix} = \begin{bmatrix} 8 & 4 & 22 \\ 4 & 32 & 16 \end{bmatrix}$

37. $-5E + A = \begin{bmatrix} -15 & 5 \\ -20 & -10 \end{bmatrix} + \begin{bmatrix} 1 & 2 \\ 0 & 4 \end{bmatrix} = \begin{bmatrix} -14 & 7 \\ -20 & -6 \end{bmatrix}$

39. $3CB + 4D = 3\begin{bmatrix} 3\cdot1+1\cdot-1 & 3\cdot2+1\cdot4 & 3\cdot3+1\cdot-2 \\ 4\cdot1+-1\cdot-1 & 4\cdot2+-1\cdot4 & 4\cdot3+-1\cdot-2 \\ 0\cdot1+2\cdot-1 & 0\cdot2+2\cdot4 & 0\cdot3+2\cdot-2 \end{bmatrix} + \begin{bmatrix} 4 & 0 & 16 \\ 0 & 4 & 8 \\ 0 & -4 & 4 \end{bmatrix}$

$= \begin{bmatrix} 6+4 & 30+0 & 21+16 \\ 15+0 & 12+4 & 42+8 \\ -6+0 & 24-4 & -12+4 \end{bmatrix} = \begin{bmatrix} 10 & 30 & 37 \\ 15 & 16 & 50 \\ -6 & 20 & -8 \end{bmatrix}$

41. First we will find $D(CB)$, doing parentheses first, and then $(DC)B$ and compare the results.

$$CB = \begin{bmatrix} 3\cdot1+1\cdot-1 & 3\cdot2+1\cdot4 & 3\cdot3+1\cdot-2 \\ 4\cdot1+-1\cdot-1 & 4\cdot2-1\cdot4 & 4\cdot3+-1\cdot-2 \\ 0\cdot1+2\cdot-1 & 0\cdot2+2\cdot4 & 0\cdot3+2\cdot-2 \end{bmatrix} = \begin{bmatrix} 2 & 10 & 7 \\ 5 & 4 & 14 \\ -2 & 8 & -4 \end{bmatrix}$$

$$D(CB) = \begin{bmatrix} 1 & 0 & 4 \\ 0 & 1 & 2 \\ 0 & -1 & 1 \end{bmatrix} \begin{bmatrix} 2 & 10 & 7 \\ 5 & 4 & 14 \\ -2 & 8 & -4 \end{bmatrix}$$

$$= \begin{bmatrix} 1\cdot2+0\cdot5+4\cdot-2 & 1\cdot10+0\cdot4+4\cdot8 & 1\cdot7+0\cdot14+4\cdot-4 \\ 0\cdot2+1\cdot5+2\cdot-2 & 0\cdot10+1\cdot4+2\cdot8 & 0\cdot7+1\cdot14+2\cdot-4 \\ 0\cdot2+-1\cdot5+1\cdot-2 & 0\cdot10+-1\cdot4+1\cdot8 & 0\cdot7+-1\cdot14+1\cdot-4 \end{bmatrix} = \begin{bmatrix} -6 & 42 & -9 \\ 1 & 20 & 6 \\ -7 & 4 & -18 \end{bmatrix}$$

$$DC = \begin{bmatrix} 1\cdot3+0\cdot4+4\cdot0 & 1\cdot1+0\cdot-1+4\cdot2 \\ 0\cdot3+1\cdot4+2\cdot0 & 0\cdot1+1\cdot-1+2\cdot2 \\ 0\cdot3+-1\cdot4+1\cdot0 & 0\cdot1+-1\cdot-1+1\cdot2 \end{bmatrix} = \begin{bmatrix} 3 & 9 \\ 4 & 3 \\ -4 & 3 \end{bmatrix}$$

$$(DC)B = \begin{bmatrix} 3 & 9 \\ 4 & 3 \\ -4 & 3 \end{bmatrix} \begin{bmatrix} 1 & 2 & 3 \\ -1 & 4 & -2 \end{bmatrix}$$

$$= \begin{bmatrix} 3\cdot1+9\cdot-1 & 3\cdot2+9\cdot4 & 3\cdot3+9\cdot-2 \\ 4\cdot1+3\cdot-1 & 4\cdot2+3\cdot4 & 4\cdot3+3\cdot-2 \\ -4\cdot1+3\cdot-1 & -4\cdot2+3\cdot4 & -4\cdot3+3\cdot-2 \end{bmatrix} = \begin{bmatrix} -6 & 42 & -9 \\ 1 & 20 & 6 \\ -7 & 4 & -18 \end{bmatrix} = D(CB)$$

43.

```
[A]*[B]             [A]*[B]
[[.5    16 -30 2…   …5    16 -30 25]
 [21    14 28  6…   …1    14 28 64]
 [19.5 8  -23  3…   …9.5 8  -23 33]
 [-9.5 45 -9   8…   …9.5 45 -9  83]]
```

$$AB = \begin{bmatrix} 0.5 & 16 & -30 & 25 \\ 21 & 14 & 28 & 64 \\ 19.5 & 8 & -23 & 33 \\ -9.5 & 45 & -9 & 83 \end{bmatrix}$$

45.

```
([A]*[B])[C]          ([A]*[B])[C]
[[31.5   251  -31…    …251  -31.5 143]
 [861    350 791…     …350 791  420]
 [369.5 115 206…      …115 206.5 215]
 [412.5 882 451…      …882 451.5 491]]
```

$$(AB)C = \begin{bmatrix} 31.5 & 251 & -31.5 & 143 \\ 861 & 350 & 791 & 420 \\ 369.5 & 115 & 206.5 & 215 \\ 412.5 & 882 & 451.5 & 491 \end{bmatrix}$$

47.

```
[B]*([A]+[C])         [B]*([A]+[C])
[[66   74    94 …     … 74    94  38]
 [71   13   106…      … 13   106 28]
 [165 124.5 79 …      … 124.5 79  52]
 [158 -3    152…      … -3   152 46]]
```

$$B(A + C) = \begin{bmatrix} 66 & 74 & 94 & 38 \\ 71 & 13 & 106 & 28 \\ 165 & 124.5 & 79 & 52 \\ 158 & -3 & 152 & 46 \end{bmatrix}$$

49.

```
[A]*(2[B]-3[C])        [A]*(2[B]-3[C])
 .23    -102 44 ]       .23    -102 44 ]
..-56   -70  122]      ..-56   -70  122]
..-152  -67  36 ]      ..-152  -67  36 ]
..279   -249 187]]     ..279   -249 187]]
■                      ■
```

$$A(2B-3C) = \begin{bmatrix} -5 & 23 & -102 & 44 \\ -108 & -56 & -70 & 122 \\ 108 & -152 & -67 & 36 \\ -346 & 279 & -249 & 187 \end{bmatrix}$$

51. $AB = \begin{bmatrix} 3+2 & 2-4 \\ 6+0 & 4+0 \end{bmatrix} = \begin{bmatrix} 5 & -2 \\ 6 & 4 \end{bmatrix}$ $\qquad BA = \begin{bmatrix} 3+4 & -3+0 \\ -2+8 & 2+0 \end{bmatrix} = \begin{bmatrix} 7 & -3 \\ 6 & 2 \end{bmatrix}$

$AB \neq BA$

53. To find the value of the variable, x, first multiply the matrices.

$$\begin{bmatrix} x & 4 & 1 \end{bmatrix} \begin{bmatrix} 2 & 1 & 0 \\ 1 & 0 & 2 \\ 0 & 2 & 4 \end{bmatrix} \begin{bmatrix} x \\ -7 \\ 5/4 \end{bmatrix} = \begin{bmatrix} 2x+4 & x+2 & 12 \end{bmatrix} \begin{bmatrix} x \\ -7 \\ 5/4 \end{bmatrix}$$

$$= (2x+4)x + (x+2)(-7) + 15$$
$$= 2x^2 - 3x + 1$$

Since the product of the three matrices is **0**, set $2x^2 - 3x + 1$ equal to 0 and solve.

$2x^2 - 3x + 1 = 0$

$(2x-1)(x-1) = 0$

$2x - 1 = 0$ or $x - 1 = 0$

$x = \dfrac{1}{2}$ or $x = 1$

55. When $A = \begin{bmatrix} a & b \\ c & d \end{bmatrix}$ and $B = \begin{bmatrix} 1 & 1 \\ -1 & 1 \end{bmatrix}$, then

$AB = \begin{bmatrix} a-b & a+b \\ c-d & c+d \end{bmatrix}$ and $BA = \begin{bmatrix} a+c & b+d \\ -a+c & -b+d \end{bmatrix}$.

If $AB = BA$ then corresponding entries must be equal, or

$$\begin{cases} a-b = a+c & \text{or} & -b = c \\ a+b = b+d & \text{or} & a = d \\ c-d = -a+c & \text{or} & d = a \\ c+d = -b+d & \text{or} & c = -b \end{cases}$$

So, if A and B are both 2×2 matrices, and A is not the identity matrix, then $AB = BA$ whenever $c = -b$ and $d = a$.

57.
$$A^2 = A \cdot A = \begin{bmatrix} a & 1-a \\ 1+a & -a \end{bmatrix} \begin{bmatrix} a & 1-a \\ 1+a & -a \end{bmatrix}$$

$$= \begin{bmatrix} a^2 + (1-a)(1+a) & a(1-a) + (1-a)(-a) \\ (1+a)a - a(1+a) & (1+a)(1-a) - a(-a) \end{bmatrix}$$

$$= \begin{bmatrix} a^2 + 1 - a^2 & a - a^2 - a + a^2 \\ a + a^2 - a - a^2 & 1 - a^2 + a^2 \end{bmatrix}$$

$$= \begin{bmatrix} 1 & 0 \\ 0 & 1 \end{bmatrix} = I_2$$

59.

	Number of pants bought	Number of shirts bought	Number of jackets bought
Lee	6	8	2
Chan	2	5	3

	Cost per item
Pants	$25.00
Shirts	$18.00
Jackets	$39.00

$$\begin{bmatrix} 6 & 8 & 2 \\ 2 & 5 & 3 \end{bmatrix} \begin{bmatrix} 25 \\ 18 \\ 39 \end{bmatrix} = \begin{bmatrix} (6)(25) + (8)(18) + (2)(39) \\ (2)(25) + (5)(18) + (3)(39) \end{bmatrix} = \begin{bmatrix} 372 \\ 257 \end{bmatrix}$$

Lee spent $372 and Chan spent $257.

61.
$$A^2 = A \cdot A = \begin{bmatrix} 1 & 0 \\ 3 & 2 \end{bmatrix} \begin{bmatrix} 1 & 0 \\ 3 & 2 \end{bmatrix} = \begin{bmatrix} 1 & 0 \\ 9 & 4 \end{bmatrix}; \qquad A^3 = A \cdot A^2 = \begin{bmatrix} 1 & 0 \\ 3 & 2 \end{bmatrix} \begin{bmatrix} 1 & 0 \\ 9 & 4 \end{bmatrix} = \begin{bmatrix} 1 & 0 \\ 21 & 8 \end{bmatrix};$$

$$A^4 = A \cdot A^3 = \begin{bmatrix} 1 & 0 \\ 3 & 2 \end{bmatrix} \begin{bmatrix} 1 & 0 \\ 21 & 8 \end{bmatrix} = \begin{bmatrix} 1 & 0 \\ 45 & 16 \end{bmatrix}$$

63.
$$A^2 = \begin{bmatrix} 1 & 0 \\ 0 & 1 \end{bmatrix} \begin{bmatrix} 1 & 0 \\ 0 & 1 \end{bmatrix} = \begin{bmatrix} 1 & 0 \\ 0 & 1 \end{bmatrix} \qquad A^3 = A \cdot A^2 = \begin{bmatrix} 1 & 0 \\ 0 & 1 \end{bmatrix} \begin{bmatrix} 1 & 0 \\ 0 & 1 \end{bmatrix} = \begin{bmatrix} 1 & 0 \\ 0 & 1 \end{bmatrix}$$

$$A^4 = A \cdot A^3 = \begin{bmatrix} 1 & 0 \\ 0 & 1 \end{bmatrix} \begin{bmatrix} 1 & 0 \\ 0 & 1 \end{bmatrix} = \begin{bmatrix} 1 & 0 \\ 0 & 1 \end{bmatrix}$$

65. Answers will vary. However, $A^n = \begin{bmatrix} 1 & 0 \\ 0 & 1 \end{bmatrix}$.

67.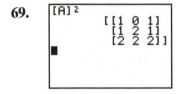

$$A^2 = \begin{bmatrix} -2.95 & -0.5 & -1.21 & 1.6 & -0.49 \\ -2.8 & 2.56 & 2.54 & 1.4 & 1.44 \\ 5.2 & 8.6 & 1.11 & -2.6 & -0.26 \\ -1.5 & -1 & 0.3 & 1 & 0.3 \\ 3.6 & -1.8 & -0.63 & -1.8 & -0.18 \end{bmatrix}$$

$$A^{10} = \begin{bmatrix} 433.0971 & -1583.5617 & -369.8169 & -216.5481 & -141.6383 \\ 1207.6949 & 5998.4376 & 2563.2274 & -603.8474 & 1045.7047 \\ -1423.0228 & 8065.0704 & 4271.7984 & 711.5114 & 2023.3641 \\ 479.4764 & -246.5381 & -231.5431 & -239.7372 & -140.5573 \\ -899.3405 & -3064.3130 & -1627.5015 & 449.6703 & -698.6086 \end{bmatrix}$$

$$A^{15} = \begin{bmatrix} -2247.8449 & -118449.3176 & -69122.3644 & 1123.9224 & -32134.3036 \\ -11094.8095 & 542433.5133 & 249928.4018 & 5547.4047 & 110820.2098 \\ 100366.7830 & 831934.5635 & 386083.7906 & -50183.3915 & 165597.1358 \\ -18782.2970 & -38432.0936 & -18654.0856 & 9391.1485 & -7395.9599 \\ -1633.6655 & -306291.0222 & -130154.5685 & 816.8327 & -56015.4536 \end{bmatrix}$$

69.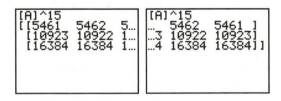

$$A^2 = \begin{bmatrix} 1 & 0 & 1 \\ 1 & 2 & 1 \\ 2 & 2 & 2 \end{bmatrix}$$

$$A^{10} = \begin{bmatrix} 171 & 170 & 171 \\ 341 & 342 & 341 \\ 512 & 512 & 512 \end{bmatrix}$$

$$A^{15} = \begin{bmatrix} 5461 & 5462 & 5461 \\ 10923 & 10922 & 10923 \\ 16384 & 16384 & 16384 \end{bmatrix}$$

71.   Answers will vary.

2.6 The Inverse of a Matrix

1. Since the product $\begin{bmatrix} 1 & 2 \\ 2 & 3 \end{bmatrix} \begin{bmatrix} -3 & 2 \\ 2 & -1 \end{bmatrix} = \begin{bmatrix} -3+4 & 2-2 \\ -6+6 & 4-3 \end{bmatrix} = \begin{bmatrix} 1 & 0 \\ 0 & 1 \end{bmatrix} = I_2$ and the product

$\begin{bmatrix} -3 & 2 \\ 2 & -1 \end{bmatrix} \begin{bmatrix} 1 & 2 \\ 2 & 3 \end{bmatrix} = \begin{bmatrix} -3+4 & -6+6 \\ 2-2 & 4-3 \end{bmatrix} = \begin{bmatrix} 1 & 0 \\ 0 & 1 \end{bmatrix} = I_2$, the matrices are inverses of each other.

3. Since the product $\begin{bmatrix} -1 & -2 \\ 3 & 4 \end{bmatrix} \begin{bmatrix} 2 & 1 \\ -\dfrac{3}{2} & -\dfrac{1}{2} \end{bmatrix} = \begin{bmatrix} -2+3 & -1+1 \\ 6-6 & 3-2 \end{bmatrix} = \begin{bmatrix} 1 & 0 \\ 0 & 1 \end{bmatrix} = I_2$ and the product

$\begin{bmatrix} 2 & 1 \\ -\dfrac{3}{2} & -\dfrac{1}{2} \end{bmatrix} \begin{bmatrix} -1 & -2 \\ 3 & 4 \end{bmatrix} = \begin{bmatrix} 1 & 0 \\ 0 & 1 \end{bmatrix} = I_2$, the matrices are inverses of each other.

5. Since the product $\begin{bmatrix} 1 & 2 & 3 \\ 2 & 3 & 4 \\ 1 & 2 & 1 \end{bmatrix} \begin{bmatrix} -\dfrac{5}{2} & 2 & -\dfrac{1}{2} \\ 1 & -1 & 1 \\ \dfrac{1}{2} & 0 & -\dfrac{1}{2} \end{bmatrix} = \begin{bmatrix} -\dfrac{5}{2}+\dfrac{4}{2}+\dfrac{3}{2} & 2-2+0 & -\dfrac{1}{2}+\dfrac{4}{2}-\dfrac{3}{2} \\ -5+3+2 & 4-3+0 & -1+3-2 \\ -\dfrac{5}{2}+\dfrac{4}{2}+\dfrac{1}{2} & 2-2+0 & -\dfrac{1}{2}+\dfrac{4}{2}-\dfrac{1}{2} \end{bmatrix}$

$= \begin{bmatrix} 1 & 0 & 0 \\ 0 & 1 & 0 \\ 0 & 0 & 1 \end{bmatrix} = I_3$ and the product $\begin{bmatrix} -\dfrac{5}{2} & 2 & -\dfrac{1}{2} \\ 1 & -1 & 1 \\ \dfrac{1}{2} & 0 & -\dfrac{1}{2} \end{bmatrix} \begin{bmatrix} 1 & 2 & 3 \\ 2 & 3 & 4 \\ 1 & 2 & 1 \end{bmatrix} = \begin{bmatrix} 1 & 0 & 0 \\ 0 & 1 & 0 \\ 0 & 0 & 1 \end{bmatrix} = I_3$, the matrices

are inverses of each other.

7. First we augment the matrix with I_2, then we use row operations to obtain the reduced row-echelon form of the matrix.

$\begin{bmatrix} 3 & 7 & | & 1 & 0 \\ 2 & 5 & | & 0 & 1 \end{bmatrix} \xrightarrow[R_1 = r_1 - r_2]{} \begin{bmatrix} 1 & 2 & | & 1 & -1 \\ 2 & 5 & | & 0 & 1 \end{bmatrix} \xrightarrow[R_2 = -2r_1 + r_2]{} \begin{bmatrix} 1 & 2 & | & 1 & -1 \\ 0 & 1 & | & -2 & 3 \end{bmatrix}$

$\xrightarrow[R_1 = -2r_2 + r_1]{} \begin{bmatrix} 1 & 0 & | & 5 & -7 \\ 0 & 1 & | & -2 & 3 \end{bmatrix}$

Since the identity matrix I_2 is on the left side, the matrix on the right is the inverse.

$\begin{bmatrix} 3 & 7 \\ 2 & 5 \end{bmatrix}^{-1} = \begin{bmatrix} 5 & -7 \\ -2 & 3 \end{bmatrix}$

9. First we augment the matrix with I_2, then we use row operations to obtain the reduced row-echelon form of the matrix.

$$\begin{bmatrix} 1 & -1 & | & 1 & 0 \\ 3 & -4 & | & 0 & 1 \end{bmatrix} \xrightarrow{R_2=-3r_1+r_2} \begin{bmatrix} 1 & -1 & | & 1 & 0 \\ 0 & -1 & | & -3 & 1 \end{bmatrix} \xrightarrow{R_2=-r_2} \begin{bmatrix} 1 & -1 & | & 1 & 0 \\ 0 & 1 & | & 3 & -1 \end{bmatrix}$$

$$\xrightarrow{R_1=r_2+r_1} \begin{bmatrix} 1 & 0 & | & 4 & -1 \\ 0 & 1 & | & 3 & -1 \end{bmatrix}$$

Since the identity matrix I_2 is on the left side, the matrix on the right is the inverse.

$$\begin{bmatrix} 1 & -1 \\ 3 & -4 \end{bmatrix}^{-1} = \begin{bmatrix} 4 & -1 \\ 3 & -1 \end{bmatrix}$$

11. First we augment the matrix with I_2, then we use row operations to obtain the reduced row-echelon form of the matrix.

$$\begin{bmatrix} 2 & 1 & | & 1 & 0 \\ 4 & 3 & | & 0 & 1 \end{bmatrix} \xrightarrow{R_1=\frac{1}{2}r_1} \begin{bmatrix} 1 & \frac{1}{2} & | & \frac{1}{2} & 0 \\ 4 & 3 & | & 0 & 1 \end{bmatrix} \xrightarrow{R_2=-4r_1+r_2} \begin{bmatrix} 1 & \frac{1}{2} & | & \frac{1}{2} & 0 \\ 0 & 1 & | & -2 & 1 \end{bmatrix}$$

$$\xrightarrow{R_1=-\frac{1}{2}r_2+r_1} \begin{bmatrix} 1 & 0 & | & \frac{3}{2} & -\frac{1}{2} \\ 0 & 1 & | & -2 & 1 \end{bmatrix}$$

Since the identity matrix I_2 is on the left side, the matrix on the right is the inverse.

$$\begin{bmatrix} 2 & 1 \\ 4 & 3 \end{bmatrix}^{-1} = \begin{bmatrix} \frac{3}{2} & -\frac{1}{2} \\ -2 & 1 \end{bmatrix}$$

13. First we augment the matrix with I_3, then we use row operations to obtain the reduced row-echelon form of the matrix.

$$\begin{bmatrix} 0 & 0 & 1 & | & 1 & 0 & 0 \\ 0 & 1 & 0 & | & 0 & 1 & 0 \\ 1 & 0 & 0 & | & 0 & 0 & 1 \end{bmatrix} \xrightarrow[\text{rows 1 and 3}]{\text{Interchange}} \begin{bmatrix} 1 & 0 & 0 & | & 0 & 0 & 1 \\ 0 & 1 & 0 & | & 0 & 1 & 0 \\ 0 & 0 & 1 & | & 1 & 0 & 0 \end{bmatrix}$$

Since the identity matrix I_3 is on the left side, the matrix on the right is the inverse.

$$\begin{bmatrix} 0 & 0 & 1 \\ 0 & 1 & 0 \\ 1 & 0 & 0 \end{bmatrix}^{-1} = \begin{bmatrix} 0 & 0 & 1 \\ 0 & 1 & 0 \\ 1 & 0 & 0 \end{bmatrix}$$

15. First we augment the matrix with I_3, then we use row operations to obtain the reduced row-echelon form of the matrix.

$$\left[\begin{array}{ccc|ccc} 1 & 1 & -1 & 1 & 0 & 0 \\ 3 & -1 & 0 & 0 & 1 & 0 \\ 2 & -3 & 4 & 0 & 0 & 1 \end{array}\right] \xrightarrow[\substack{R_2=-3r_1+r_2 \\ R_3=-2r_1+r_3}]{} \left[\begin{array}{ccc|ccc} 1 & 1 & -1 & 1 & 0 & 0 \\ 0 & -4 & 3 & -3 & 1 & 0 \\ 0 & -5 & 6 & -2 & 0 & 1 \end{array}\right]$$

$$\xrightarrow[R_2=-\frac{1}{4}r_2]{} \left[\begin{array}{ccc|ccc} 1 & 1 & -1 & 1 & 0 & 0 \\ 0 & 1 & -\dfrac{3}{4} & \dfrac{3}{4} & -\dfrac{1}{4} & 0 \\ 0 & -5 & 6 & -2 & 0 & 1 \end{array}\right] \xrightarrow[\substack{R_1=-r_2+r_1 \\ R_3=5r_2+r_3}]{} \left[\begin{array}{ccc|ccc} 1 & 0 & -\dfrac{1}{4} & \dfrac{1}{4} & \dfrac{1}{4} & 0 \\ 0 & 1 & -\dfrac{3}{4} & \dfrac{3}{4} & -\dfrac{1}{4} & 0 \\ 0 & 0 & \dfrac{9}{4} & \dfrac{7}{4} & -\dfrac{5}{4} & 1 \end{array}\right]$$

$$\xrightarrow[R_3=\frac{4}{9}r_3]{} \left[\begin{array}{ccc|ccc} 1 & 0 & -\dfrac{1}{4} & \dfrac{1}{4} & \dfrac{1}{4} & 0 \\ 0 & 1 & -\dfrac{3}{4} & \dfrac{3}{4} & -\dfrac{1}{4} & 0 \\ 0 & 0 & 1 & \dfrac{7}{9} & -\dfrac{5}{9} & \dfrac{4}{9} \end{array}\right] \xrightarrow[\substack{R_1=\frac{1}{4}r_3+r_1 \\ R_2=\frac{3}{4}r_3+r_2}]{} \left[\begin{array}{ccc|ccc} 1 & 0 & 0 & \dfrac{4}{9} & \dfrac{1}{9} & \dfrac{1}{9} \\ 0 & 1 & 0 & \dfrac{4}{3} & -\dfrac{2}{3} & \dfrac{1}{3} \\ 0 & 0 & 1 & \dfrac{7}{9} & -\dfrac{5}{9} & \dfrac{4}{9} \end{array}\right]$$

Since the identity matrix I_3 is on the left side, the matrix on the right is the inverse.

$$\left[\begin{array}{ccc} 1 & 1 & -1 \\ 3 & -1 & 0 \\ 2 & -3 & 4 \end{array}\right]^{-1} = \left[\begin{array}{ccc} \dfrac{4}{9} & \dfrac{1}{9} & \dfrac{1}{9} \\ \dfrac{4}{3} & -\dfrac{2}{3} & \dfrac{1}{3} \\ \dfrac{7}{9} & -\dfrac{5}{9} & \dfrac{4}{9} \end{array}\right]$$

17. First we augment the matrix with I_3, then we use row operations to obtain the reduced row-echelon form of the matrix.

$$\left[\begin{array}{rrr|rrr} 1 & 1 & -1 & 1 & 0 & 0 \\ 2 & 1 & 1 & 0 & 1 & 0 \\ 1 & 0 & 1 & 0 & 0 & 1 \end{array}\right] \xrightarrow[\substack{R_2=-2r_1+r_2 \\ R_3=-r_1+r_3}]{} \left[\begin{array}{rrr|rrr} 1 & 1 & -1 & 1 & 0 & 0 \\ 0 & -1 & 3 & -2 & 1 & 0 \\ 0 & -1 & 2 & -1 & 0 & 1 \end{array}\right]$$

$$\xrightarrow[R_2=-r_2]{} \left[\begin{array}{rrr|rrr} 1 & 1 & -1 & 1 & 0 & 0 \\ 0 & 1 & -3 & 2 & -1 & 0 \\ 0 & -1 & 2 & -1 & 0 & 1 \end{array}\right] \xrightarrow[\substack{R_1=-r_2+r_1 \\ R_3=r_2+r_3}]{} \left[\begin{array}{rrr|rrr} 1 & 0 & 2 & -1 & 1 & 0 \\ 0 & 1 & -3 & 2 & -1 & 0 \\ 0 & 0 & -1 & 1 & -1 & 1 \end{array}\right]$$

$$\xrightarrow[R_3=-r_3]{} \left[\begin{array}{rrr|rrr} 1 & 0 & 2 & -1 & 1 & 0 \\ 0 & 1 & -3 & 2 & -1 & 0 \\ 0 & 0 & 1 & -1 & 1 & -1 \end{array}\right] \xrightarrow[\substack{R_1=-2r_3+r_1 \\ R_2=3r_3+r_2}]{} \left[\begin{array}{rrr|rrr} 1 & 0 & 0 & 1 & -1 & 2 \\ 0 & 1 & 0 & -1 & 2 & -3 \\ 0 & 0 & 1 & -1 & 1 & -1 \end{array}\right]$$

Since the identity matrix I_3 is on the left side, the matrix on the right is the inverse.

$$\left[\begin{array}{rrr} 1 & 1 & -1 \\ 2 & 1 & 1 \\ 1 & 0 & 1 \end{array}\right]^{-1} = \left[\begin{array}{rrr} 1 & -1 & 2 \\ -1 & 2 & -3 \\ -1 & 1 & -1 \end{array}\right]$$

19. First we augment the matrix with I_4, then we use row operations to obtain the reduced row-echelon form of the matrix.

$$\left[\begin{array}{cccc|cccc} 1 & 1 & 0 & 0 & 1 & 0 & 0 & 0 \\ 0 & 1 & -1 & 1 & 0 & 1 & 0 & 0 \\ 1 & -1 & 1 & 1 & 0 & 0 & 1 & 0 \\ 0 & 1 & 0 & -1 & 0 & 0 & 0 & 1 \end{array}\right] \xrightarrow{R_3=-r_1+r_3} \left[\begin{array}{cccc|cccc} 1 & 1 & 0 & 0 & 1 & 0 & 0 & 0 \\ 0 & 1 & -1 & 1 & 0 & 1 & 0 & 0 \\ 0 & -2 & 1 & 1 & -1 & 0 & 1 & 0 \\ 0 & 1 & 0 & -1 & 0 & 0 & 0 & 1 \end{array}\right]$$

$$\xrightarrow[\substack{R_1=-r_2+r_1 \\ R_3=2r_2+r_3 \\ R_4=-r_2+r_4}]{} \left[\begin{array}{cccc|cccc} 1 & 0 & 1 & -1 & 1 & -1 & 0 & 0 \\ 0 & 1 & -1 & 1 & 0 & 1 & 0 & 0 \\ 0 & 0 & -1 & 3 & -1 & 2 & 1 & 0 \\ 0 & 0 & 1 & -2 & 0 & -1 & 0 & 1 \end{array}\right]$$

$$\xrightarrow{R_3=-r_3} \left[\begin{array}{cccc|cccc} 1 & 0 & 1 & -1 & 1 & -1 & 0 & 0 \\ 0 & 1 & -1 & 1 & 0 & 1 & 0 & 0 \\ 0 & 0 & 1 & -3 & 1 & -2 & -1 & 0 \\ 0 & 0 & 1 & -2 & 0 & -1 & 0 & 1 \end{array}\right]$$

$$\xrightarrow[\substack{R_1=-r_3+r_1 \\ R_2=r_3+r_2 \\ R_4=-r_3+r_4}]{} \left[\begin{array}{cccc|cccc} 1 & 0 & 0 & 2 & 0 & 1 & 1 & 0 \\ 0 & 1 & 0 & -2 & 1 & -1 & -1 & 0 \\ 0 & 0 & 1 & -3 & 1 & -2 & -1 & 0 \\ 0 & 0 & 0 & 1 & -1 & 1 & 1 & 1 \end{array}\right]$$

$$\xrightarrow[\substack{R_1=-2r_4+r_1 \\ R_2=2r_4+r_2 \\ R_3=3r_4+r_3}]{} \left[\begin{array}{cccc|cccc} 1 & 0 & 0 & 0 & 2 & -1 & -1 & -2 \\ 0 & 1 & 0 & 0 & -1 & 1 & 1 & 2 \\ 0 & 0 & 1 & 0 & -2 & 1 & 2 & 3 \\ 0 & 0 & 0 & 1 & -1 & 1 & 1 & 1 \end{array}\right]$$

Since the identity matrix I_4 is on the left side, the matrix on the right is the inverse.

$$\left[\begin{array}{cccc} 1 & 1 & 0 & 0 \\ 0 & 1 & -1 & 1 \\ 1 & -1 & 1 & 1 \\ 0 & 1 & 0 & -1 \end{array}\right]^{-1} = \left[\begin{array}{cccc} 2 & -1 & -1 & -2 \\ -1 & 1 & 1 & 2 \\ -2 & 1 & 2 & 3 \\ -1 & 1 & 1 & 1 \end{array}\right]$$

21. First we augment the matrix with I_2, then we use row operations to obtain the reduced row-echelon form of the matrix.

$$\left[\begin{array}{cc|cc} 4 & 6 & 1 & 0 \\ 2 & 3 & 0 & 1 \end{array}\right] \xrightarrow{R_1=\frac{1}{4}r_1} \left[\begin{array}{cc|cc} 1 & \frac{3}{2} & \frac{1}{4} & 0 \\ 2 & 3 & 0 & 1 \end{array}\right] \xrightarrow{R_2=-2r_1+r_2} \left[\begin{array}{cc|cc} 1 & \frac{3}{2} & \frac{1}{4} & 0 \\ 0 & 0 & -\frac{1}{2} & 1 \end{array}\right]$$

The 0s in row 2 indicate that we cannot get the identity matrix. This tells us that the original matrix has no inverse.

23. First we augment the matrix with I_2, then we use row operations to obtain the reduced row-echelon form of the matrix.

$$\begin{bmatrix} -8 & 4 & | & 1 & 0 \\ -4 & 2 & | & 0 & 1 \end{bmatrix} \xrightarrow{R_1 = -\frac{1}{8}r_1} \begin{bmatrix} 1 & -\dfrac{1}{2} & | & -\dfrac{1}{8} & 0 \\ -4 & 2 & | & 0 & 1 \end{bmatrix} \xrightarrow{R_2 = 4r_1 + r_2} \begin{bmatrix} 1 & -\dfrac{1}{2} & | & -\dfrac{1}{8} & 0 \\ 0 & 0 & | & -\dfrac{1}{2} & 1 \end{bmatrix}$$

The 0s in row 2 indicate that we cannot get the identity matrix. This tells us that the original matrix has no inverse.

25. First we augment the matrix with I_3. Since the original matrix has a row of zeros we cannot get an identity matrix. This tells us that the matrix has no inverse.

$$\begin{bmatrix} 1 & 1 & 1 & | & 1 & 0 & 0 \\ 3 & -4 & 2 & | & 0 & 1 & 0 \\ 0 & 0 & 0 & | & 0 & 0 & 1 \end{bmatrix}$$

27. To find the inverse we augment the matrix with I_2, then use row operations to obtain the reduced row-echelon form.

$$\begin{bmatrix} 1 & 1 & | & 1 & 0 \\ 1 & 2 & | & 0 & 1 \end{bmatrix} \xrightarrow{R_2 = -r_1 + r_2} \begin{bmatrix} 1 & 1 & | & 1 & 0 \\ 0 & 1 & | & -1 & 1 \end{bmatrix} \xrightarrow{R_1 = -r_2 + r_1} \begin{bmatrix} 1 & 0 & | & 2 & -1 \\ 0 & 1 & | & -1 & 1 \end{bmatrix}$$

Since the identity matrix I_2 is on the left side, the matrix on the right is the inverse.

$$\begin{bmatrix} 1 & 1 \\ 1 & 2 \end{bmatrix}^{-1} = \begin{bmatrix} 2 & -1 \\ -1 & 1 \end{bmatrix}$$

29. To find the inverse we augment the matrix with I_2, then use row operations to obtain the reduced row-echelon form.

$$\begin{bmatrix} 3 & -2 & | & 1 & 0 \\ 0 & 4 & | & 0 & 1 \end{bmatrix} \xrightarrow{R_1 = \frac{1}{3}r_1} \begin{bmatrix} 1 & -\dfrac{2}{3} & | & \dfrac{1}{3} & 0 \\ 0 & 4 & | & 0 & 1 \end{bmatrix} \xrightarrow{R_2 = \frac{1}{4}r_2} \begin{bmatrix} 1 & -\dfrac{2}{3} & | & \dfrac{1}{3} & 0 \\ 0 & 1 & | & 0 & \dfrac{1}{4} \end{bmatrix}$$

$$\xrightarrow{R_1 = \frac{2}{3}r_2 + r_1} \begin{bmatrix} 1 & 0 & | & \dfrac{1}{3} & \dfrac{1}{6} \\ 0 & 1 & | & 0 & \dfrac{1}{4} \end{bmatrix}$$

Since the identity matrix I_2 is on the left side, the matrix on the right is the inverse.

$$\begin{bmatrix} 3 & -2 \\ 0 & 4 \end{bmatrix}^{-1} = \begin{bmatrix} \dfrac{1}{3} & \dfrac{1}{6} \\ 0 & \dfrac{1}{4} \end{bmatrix}$$

31. To find the inverse we augment the matrix with I_2, then use row operations to obtain the reduced row-echelon form.

$$\begin{bmatrix} 3 & 2 & | & 1 & 0 \\ 6 & 4 & | & 0 & 1 \end{bmatrix} \xrightarrow{R_1=\frac{1}{3}r_1} \begin{bmatrix} 1 & \dfrac{2}{3} & | & \dfrac{1}{3} & 0 \\ 6 & 4 & | & 0 & 1 \end{bmatrix} \xrightarrow{R_2=-6r_1+r_2} \begin{bmatrix} 1 & \dfrac{2}{3} & | & \dfrac{1}{3} & 0 \\ 0 & 0 & | & -2 & 1 \end{bmatrix}$$

The 0s in the row 2 tell us we cannot get the identity matrix. This indicates that the original matrix has no inverse.

33. To find the inverse we augment the matrix with I_3, then use row operations to obtain the reduced row-echelon form.

$$\begin{bmatrix} 1 & -2 & -1 & | & 1 & 0 & 0 \\ -2 & 5 & 4 & | & 0 & 1 & 0 \\ 3 & -8 & -5 & | & 0 & 0 & 1 \end{bmatrix} \xrightarrow[R_3=-3r_1+r_3]{R_2=2r_1+r_2} \begin{bmatrix} 1 & -2 & -1 & | & 1 & 0 & 0 \\ 0 & 1 & 2 & | & 2 & 1 & 0 \\ 0 & -2 & -2 & | & -3 & 0 & 1 \end{bmatrix}$$

$$\xrightarrow[R_3=2r_2+r_3]{R_1=2r_2+r_1} \begin{bmatrix} 1 & 0 & 3 & | & 5 & 2 & 0 \\ 0 & 1 & 2 & | & 2 & 1 & 0 \\ 0 & 0 & 2 & | & 1 & 2 & 1 \end{bmatrix} \xrightarrow{R=\frac{1}{2}r_3} \begin{bmatrix} 1 & 0 & 3 & | & 5 & 2 & 0 \\ 0 & 1 & 2 & | & 2 & 1 & 0 \\ 0 & 0 & 1 & | & \dfrac{1}{2} & 1 & \dfrac{1}{2} \end{bmatrix}$$

$$\xrightarrow[R_2=-2r_3+r_2]{R_1=-3r_3+r_1} \begin{bmatrix} 1 & 0 & 0 & | & \dfrac{7}{2} & -1 & -\dfrac{3}{2} \\ 0 & 1 & 0 & | & 1 & -1 & -1 \\ 0 & 0 & 1 & | & \dfrac{1}{2} & 1 & \dfrac{1}{2} \end{bmatrix}$$

Since the identity matrix I_3 is on the left side, the matrix on the right is the inverse.

$$\begin{bmatrix} 1 & -2 & -1 \\ -2 & 5 & 4 \\ 3 & -8 & -5 \end{bmatrix}^{-1} = \begin{bmatrix} 7/2 & -1 & -3/2 \\ 1 & -1 & -1 \\ 1/2 & 1 & 1/2 \end{bmatrix}$$

35. To find A^{-1} we use row operations to obtain the reduced row-echelon form of the matrix $[A \mid I_2]$.

$$\begin{bmatrix} 1 & 2 & | & 1 & 0 \\ 2 & -1 & | & 0 & 1 \end{bmatrix} \xrightarrow{R_2 = -2r_1 + r_2} \begin{bmatrix} 1 & 2 & | & 1 & 0 \\ 0 & -5 & | & -2 & 1 \end{bmatrix} \xrightarrow{R_2 = -\frac{1}{5}r_2} \begin{bmatrix} 1 & 2 & | & 1 & 0 \\ 0 & 1 & | & \frac{2}{5} & -\frac{1}{5} \end{bmatrix}$$

$$\xrightarrow{R_1 = -2r_2 + r_1} \begin{bmatrix} 1 & 0 & | & \frac{1}{5} & \frac{2}{5} \\ 0 & 1 & | & \frac{2}{5} & -\frac{1}{5} \end{bmatrix} \quad \text{So, } A^{-1} = \begin{bmatrix} \frac{1}{5} & \frac{2}{5} \\ \frac{2}{5} & -\frac{1}{5} \end{bmatrix}.$$

To find B^{-1} we use row operations to obtain the reduced row-echelon form of the matrix $[B \mid I_2]$.

$$\begin{bmatrix} 1 & 3 & | & 1 & 0 \\ 2 & 1 & | & 0 & 1 \end{bmatrix} \xrightarrow{R_2 = -2r_1 + r_2} \begin{bmatrix} 1 & 3 & | & 1 & 0 \\ 0 & -5 & | & -2 & 1 \end{bmatrix} \xrightarrow{R_2 = -\frac{1}{5}r_2} \begin{bmatrix} 1 & 3 & | & 1 & 0 \\ 0 & 1 & | & \frac{2}{5} & -\frac{1}{5} \end{bmatrix}$$

$$\xrightarrow{R_1 = -3r_2 + r_1} \begin{bmatrix} 1 & 0 & | & -\frac{1}{5} & \frac{3}{5} \\ 0 & 1 & | & \frac{2}{5} & -\frac{1}{5} \end{bmatrix} \quad \text{So, } B^{-1} = \begin{bmatrix} -\frac{1}{5} & \frac{3}{5} \\ \frac{2}{5} & -\frac{1}{5} \end{bmatrix} \text{ and}$$

$$A^{-1} - B^{-1} = \begin{bmatrix} \frac{1}{5} & \frac{2}{5} \\ \frac{2}{5} & -\frac{1}{5} \end{bmatrix} - \begin{bmatrix} -\frac{1}{5} & \frac{3}{5} \\ \frac{2}{5} & -\frac{1}{5} \end{bmatrix} = \begin{bmatrix} \frac{1}{5}+\frac{1}{5} & \frac{2}{5}-\frac{3}{5} \\ \frac{2}{5}-\frac{2}{5} & -\frac{1}{5}+\frac{1}{5} \end{bmatrix} = \begin{bmatrix} \frac{2}{5} & -\frac{1}{5} \\ 0 & 0 \end{bmatrix}$$

37. To solve the system $\begin{cases} x + 3y + 2z = 2 \\ 2x + 7y + 3z = 1 \\ x \quad\quad + 6z = 3 \end{cases}$ we define $A = \begin{bmatrix} 1 & 3 & 2 \\ 2 & 7 & 3 \\ 1 & 0 & 6 \end{bmatrix}$, $X = \begin{bmatrix} x \\ y \\ z \end{bmatrix}$, and $B = \begin{bmatrix} 2 \\ 1 \\ 3 \end{bmatrix}$.

The solution to the system is $X = A^{-1}B$

$$\begin{bmatrix} x \\ y \\ z \end{bmatrix} = \begin{bmatrix} 1 & 3 & 2 \\ 2 & 7 & 3 \\ 1 & 0 & 6 \end{bmatrix}^{-1} \begin{bmatrix} 2 \\ 1 \\ 3 \end{bmatrix} = \begin{bmatrix} 42 & -18 & -5 \\ -9 & 4 & 1 \\ -7 & 3 & 1 \end{bmatrix} \begin{bmatrix} 2 \\ 1 \\ 3 \end{bmatrix} = \begin{bmatrix} 51 \\ -11 \\ -8 \end{bmatrix}$$

or $x = 51$, $y = -11$, and $z = -8$.

39. To solve the system $\begin{cases} 3x + 7y = 10 \\ 2x + 5y = 2 \end{cases}$ we define $A = \begin{bmatrix} 3 & 7 \\ 2 & 5 \end{bmatrix}$, $X = \begin{bmatrix} x \\ y \end{bmatrix}$, and $B = \begin{bmatrix} 10 \\ 2 \end{bmatrix}$.

The solution to the system is $X = A^{-1}B$

$$\begin{bmatrix} x \\ y \end{bmatrix} = \begin{bmatrix} 3 & 7 \\ 2 & 5 \end{bmatrix}^{-1} \begin{bmatrix} 10 \\ 2 \end{bmatrix} = \begin{bmatrix} 5 & -7 \\ -2 & 3 \end{bmatrix} \begin{bmatrix} 10 \\ 2 \end{bmatrix} = \begin{bmatrix} 36 \\ -14 \end{bmatrix}$$

or $x = 36$ and $y = -14$.

41. To solve the system $\begin{cases} 3x+7y=13 \\ 2x+5y=9 \end{cases}$ we define $A=\begin{bmatrix} 3 & 7 \\ 2 & 5 \end{bmatrix}$, $X=\begin{bmatrix} x \\ y \end{bmatrix}$, and $B=\begin{bmatrix} 13 \\ 9 \end{bmatrix}$.

The solution to the system is $X=A^{-1}B$

$$\begin{bmatrix} x \\ y \end{bmatrix} = \begin{bmatrix} 3 & 7 \\ 2 & 5 \end{bmatrix}^{-1} \begin{bmatrix} 13 \\ 9 \end{bmatrix} = \begin{bmatrix} 5 & -7 \\ -2 & 3 \end{bmatrix} \begin{bmatrix} 13 \\ 9 \end{bmatrix} = \begin{bmatrix} 2 \\ 1 \end{bmatrix}$$

or $x=2$ and $y=1$.

43. To solve the system $\begin{cases} 3x+7y=12 \\ 2x+5y=-4 \end{cases}$ we define $A=\begin{bmatrix} 3 & 7 \\ 2 & 5 \end{bmatrix}$, $X=\begin{bmatrix} x \\ y \end{bmatrix}$, and $B=\begin{bmatrix} 12 \\ -4 \end{bmatrix}$.

The solution to the system is $X=A^{-1}B$

$$\begin{bmatrix} x \\ y \end{bmatrix} = \begin{bmatrix} 3 & 7 \\ 2 & 5 \end{bmatrix}^{-1} \begin{bmatrix} 12 \\ -4 \end{bmatrix} = \begin{bmatrix} 5 & -7 \\ -2 & 3 \end{bmatrix} \begin{bmatrix} 12 \\ -4 \end{bmatrix} = \begin{bmatrix} 88 \\ -36 \end{bmatrix}$$

or $x=88$ and $y=-36$.

45. To solve $\begin{cases} x+y-z=3 \\ 3x-y=-4 \\ 2x-3y+4z=6 \end{cases}$ we define $A=\begin{bmatrix} 1 & 1 & -1 \\ 3 & -1 & 0 \\ 2 & -3 & 4 \end{bmatrix}$, $X=\begin{bmatrix} x \\ y \\ z \end{bmatrix}$, and $B=\begin{bmatrix} 3 \\ -4 \\ 6 \end{bmatrix}$.

The solution to the system is $X=A^{-1}B$

$$\begin{bmatrix} x \\ y \\ z \end{bmatrix} = \begin{bmatrix} 1 & 1 & -1 \\ 3 & -1 & 0 \\ 2 & -3 & 4 \end{bmatrix}^{-1} \begin{bmatrix} 3 \\ -4 \\ 6 \end{bmatrix} = \begin{bmatrix} \frac{4}{9} & \frac{1}{9} & \frac{1}{9} \\ \frac{4}{3} & -\frac{2}{3} & \frac{1}{3} \\ \frac{7}{9} & -\frac{5}{9} & \frac{4}{9} \end{bmatrix} \begin{bmatrix} 3 \\ -4 \\ 6 \end{bmatrix} = \begin{bmatrix} \frac{14}{9} \\ \frac{26}{3} \\ \frac{65}{9} \end{bmatrix}$$

or $x=\dfrac{14}{9}$, $y=\dfrac{26}{3}$, and $z=\dfrac{65}{9}$.

47. To solve $\begin{cases} x+y-z=12 \\ 3x-y=-4 \\ 2x-3y+4z=16 \end{cases}$ we define $A=\begin{bmatrix} 1 & 1 & -1 \\ 3 & -1 & 0 \\ 2 & -3 & 4 \end{bmatrix}$, $X=\begin{bmatrix} x \\ y \\ z \end{bmatrix}$, and $B=\begin{bmatrix} 12 \\ -4 \\ 16 \end{bmatrix}$.

The solution to the system is $X=A^{-1}B$

$$\begin{bmatrix} x \\ y \\ z \end{bmatrix} = \begin{bmatrix} 1 & 1 & -1 \\ 3 & -1 & 0 \\ 2 & -3 & 4 \end{bmatrix}^{-1} \begin{bmatrix} 12 \\ -4 \\ 16 \end{bmatrix} = \begin{bmatrix} \frac{4}{9} & \frac{1}{9} & \frac{1}{9} \\ \frac{4}{3} & -\frac{2}{3} & \frac{1}{3} \\ \frac{7}{9} & -\frac{5}{9} & \frac{4}{9} \end{bmatrix} \begin{bmatrix} 12 \\ -4 \\ 16 \end{bmatrix} = \begin{bmatrix} \frac{60}{9} \\ \frac{72}{3} \\ \frac{168}{9} \end{bmatrix} = \begin{bmatrix} \frac{20}{3} \\ 24 \\ \frac{56}{3} \end{bmatrix}$$

or $x=\dfrac{20}{3}$, $y=24$, and $z=\dfrac{56}{3}$.

49. To solve $\begin{cases} x+\ y-\ z=\ 0 \\ 3x-\ y\ \ \ \ \ \ =-8 \\ 2x-3y+4z=-6 \end{cases}$ we define $A=\begin{bmatrix} 1 & 1 & -1 \\ 3 & -1 & 0 \\ 2 & -3 & 4 \end{bmatrix}$, $X=\begin{bmatrix} x \\ y \\ z \end{bmatrix}$, and $B=\begin{bmatrix} 0 \\ -8 \\ -6 \end{bmatrix}$.

The solution to the system is $X=A^{-1}B$

$$\begin{bmatrix} x \\ y \\ z \end{bmatrix}=\begin{bmatrix} 1 & 1 & -1 \\ 3 & -1 & 0 \\ 2 & -3 & 4 \end{bmatrix}^{-1}\begin{bmatrix} 0 \\ -8 \\ -6 \end{bmatrix}=\begin{bmatrix} \frac{4}{9} & \frac{1}{9} & \frac{1}{9} \\ \frac{4}{3} & -\frac{2}{3} & \frac{1}{3} \\ \frac{7}{9} & -\frac{5}{9} & \frac{4}{9} \end{bmatrix}\begin{bmatrix} 0 \\ -8 \\ -6 \end{bmatrix}=\begin{bmatrix} -\frac{14}{9} \\ \frac{10}{3} \\ \frac{16}{9} \end{bmatrix}$$

or $x=-\dfrac{14}{9}$, $y=\dfrac{10}{3}$, and $z=\dfrac{16}{9}$.

51. To find the inverse of an $n \times n$ matrix using a graphing utility, enter the matrix augmented with I_n, and compute the reduced row-echelon (RREF) form.

```
rref([A])              rref([A])
[[1 0 0 .005447...     ...0509   -.0066]
 [0 1 0 .010356...     ...-.0186 .0095 ]
 [0 0 1 -.01933...     ...0116   .0344 ]]
```

The inverse is $\begin{bmatrix} 0.0054 & 0.0509 & -0.0066 \\ 0.0104 & -0.0186 & 0.0095 \\ -0.0193 & 0.0116 & 0.0344 \end{bmatrix}$

53. To find the inverse of an $n \times n$ matrix using a graphing utility, enter the matrix augmented with I_n, and compute the reduced row-echelon (RREF) form.

```
rref([C])              rref([C])              rref([C])              rref([C])
[[1.0000 0.0000...     ... 0.0000 0.0000...   ...0249   -.0360 ...   ...0057 .0059     ...
 [0.0000 1.0000...     ... 0.0000 0.0000...   ...-.0171 .0521 ...    ...292   -.0305   ...
 [0.0000 0.0000...     ... 1.0000 0.0000...   ...0206   .0081 ...    ...0421 4.8764E-4...
 [0.0000 0.0000...     ... 0.0000 1.0000...   ...-.0175 .0570 ...    ...657   .0619    ...
```

The inverse is $\begin{bmatrix} 0.0249 & -0.0360 & -0.0057 & 0.0059 \\ -0.0171 & 0.0521 & 0.0292 & -0.0305 \\ 0.0206 & 0.0081 & -0.0421 & 0.0005 \\ -0.0175 & 0.0570 & 0.0657 & 0.0619 \end{bmatrix}$

55. To find the inverse of an $n \times n$ matrix using a graphing utility, enter the matrix augmented with I_n, and compute the reduced row-echelon (RREF) form.

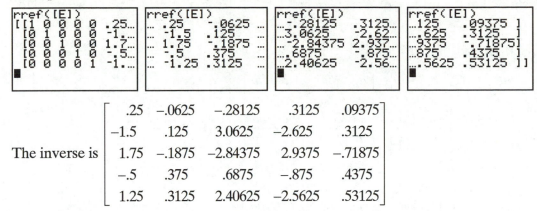

The inverse is $\begin{bmatrix} .25 & -.0625 & -.28125 & .3125 & .09375 \\ -1.5 & .125 & 3.0625 & -2.625 & .3125 \\ 1.75 & -.1875 & -2.84375 & 2.9375 & -.71875 \\ -.5 & .375 & .6875 & -.875 & .4375 \\ 1.25 & .3125 & 2.40625 & -2.5625 & .53125 \end{bmatrix}$

57. To solve the system of equations using a graphing utility (or Excel) first we write the system in the form $AX = B$, where

$$A = \begin{bmatrix} 25 & 61 & -12 \\ 18 & -12 & 7 \\ 3 & 4 & -1 \end{bmatrix}, X = \begin{bmatrix} x \\ y \\ z \end{bmatrix} \text{ and } B = \begin{bmatrix} 10 \\ -9 \\ 12 \end{bmatrix}$$

Enter matrices A and B into the graphing utility's matrix editor. Then calculate $A^{-1}B$.

```
[A]⁻¹*[B]
[[4.566617862 ]
 [-6.44363104 ]
 [-24.07467057]]
■
```

So $x = 4.567$, $y = -6.4436$, and $z = -24.0747$.

59. To solve the system of equations using a graphing utility (or Excel) first we write the system in the form $AX = D$, where

$$A = \begin{bmatrix} 25 & 61 & -12 \\ 18 & -12 & 7 \\ 3 & 4 & -1 \end{bmatrix}, X = \begin{bmatrix} x \\ y \\ z \end{bmatrix} \text{ and } D = \begin{bmatrix} 21 \\ 7 \\ -2 \end{bmatrix}$$

Enter matrices A and D into the graphing utility's matrix editor. Then calculate $A^{-1}D$.

```
[A]⁻¹*[D]
[[-1.187408492]
 [2.456808199 ]
 [8.265007321 ]]
■
```

So $x = -1.1874$, $y = 2.4568$, and $z = 8.2650$.

61. To show that the inverse of A is given by $\begin{bmatrix} \dfrac{d}{\Delta} & \dfrac{-b}{\Delta} \\ \dfrac{-c}{\Delta} & \dfrac{a}{\Delta} \end{bmatrix}$ where $\Delta = ad - bc \neq 0$ we need to

show that $AA^{-1} = A^{-1}A = I_2$.

$$AA^{-1} = \begin{bmatrix} a & b \\ c & d \end{bmatrix} \begin{bmatrix} \dfrac{d}{\Delta} & \dfrac{-b}{\Delta} \\ \dfrac{-c}{\Delta} & \dfrac{a}{\Delta} \end{bmatrix} = \begin{bmatrix} \dfrac{ad}{\Delta} - \dfrac{bc}{\Delta} & -\dfrac{ab}{\Delta} + \dfrac{ab}{\Delta} \\ \dfrac{cd}{\Delta} - \dfrac{cd}{\Delta} & -\dfrac{bc}{\Delta} + \dfrac{ad}{\Delta} \end{bmatrix} = \begin{bmatrix} \dfrac{ad - bc}{\Delta} & 0 \\ 0 & \dfrac{ad - bc}{\Delta} \end{bmatrix} = \begin{bmatrix} 1 & 0 \\ 0 & 1 \end{bmatrix} = I_2$$

$$A^{-1}A = \begin{bmatrix} \dfrac{d}{\Delta} & \dfrac{-b}{\Delta} \\ \dfrac{-c}{\Delta} & \dfrac{a}{\Delta} \end{bmatrix} \begin{bmatrix} a & b \\ c & d \end{bmatrix} = \begin{bmatrix} \dfrac{ad}{\Delta} - \dfrac{bc}{\Delta} & \dfrac{bd}{\Delta} - \dfrac{bd}{\Delta} \\ -\dfrac{ac}{\Delta} + \dfrac{ac}{\Delta} & -\dfrac{bc}{\Delta} + \dfrac{ad}{\Delta} \end{bmatrix} = \begin{bmatrix} \dfrac{ad - bc}{\Delta} & 0 \\ 0 & \dfrac{ad - bc}{\Delta} \end{bmatrix} = \begin{bmatrix} 1 & 0 \\ 0 & 1 \end{bmatrix} = I_2$$

So, $\begin{bmatrix} \dfrac{d}{\Delta} & \dfrac{-b}{\Delta} \\ \dfrac{-c}{\Delta} & \dfrac{a}{\Delta} \end{bmatrix}$ is the inverse of $A = \begin{bmatrix} a & b \\ c & d \end{bmatrix}$.

63. To find the inverse of $\begin{bmatrix} 1 & 5 \\ 2 & 0 \end{bmatrix}$ we first find $\Delta = ad - bc = 1 \cdot 0 - 5 \cdot 2 = -10$, and then write

the inverse $\begin{bmatrix} 1 & 5 \\ 2 & 0 \end{bmatrix}^{-1} = \begin{bmatrix} \dfrac{0}{-10} & \dfrac{-5}{-10} \\ \dfrac{-2}{-10} & \dfrac{1}{-10} \end{bmatrix} = \begin{bmatrix} 0 & \dfrac{1}{2} \\ \dfrac{1}{5} & -\dfrac{1}{10} \end{bmatrix}$

65. To find the inverse of $\begin{bmatrix} 1 & 2 \\ 8 & 15 \end{bmatrix}$ we first find $\Delta = ad - bc = 1 \cdot 15 - 2 \cdot 8 = -1$, and then write

the inverse $\begin{bmatrix} 1 & 2 \\ 8 & 15 \end{bmatrix}^{-1} = \begin{bmatrix} \dfrac{15}{-1} & \dfrac{-2}{-1} \\ \dfrac{-8}{-1} & \dfrac{1}{-1} \end{bmatrix} = \begin{bmatrix} -15 & 2 \\ 8 & -1 \end{bmatrix}$

2.7 Applications

Application 1: Leontief Models

1. We let x represent A's wages, y represent B's wages, and z represent C's wages, which we are told equal \$30,000.

Since the amount paid out by each A, B, and C equals the amount each receives, we get

$$\begin{bmatrix} x \\ y \\ z \end{bmatrix} = \begin{bmatrix} \frac{1}{2} & \frac{1}{3} & \frac{1}{4} \\ \frac{1}{4} & \frac{1}{3} & \frac{1}{4} \\ \frac{1}{4} & \frac{1}{3} & \frac{1}{2} \end{bmatrix} \begin{bmatrix} x \\ y \\ z \end{bmatrix}$$ giving the system $$\begin{cases} x = \frac{1}{2}x + \frac{1}{3}y + \frac{1}{4}z \\ y = \frac{1}{4}x + \frac{1}{3}y + \frac{1}{4}z \\ z = \frac{1}{4}x + \frac{1}{3}y + \frac{1}{2}z \end{cases}$$ or $$\begin{cases} \frac{1}{2}x - \frac{1}{3}y - \frac{1}{4}z = 0 \\ -\frac{1}{4}x + \frac{2}{3}y - \frac{1}{4}z = 0 \\ -\frac{1}{4}x - \frac{1}{3}y + \frac{1}{2}z = 0 \end{cases}$$

Solving the equations for x, y, and z, we find

$$\left[\begin{array}{ccc|c} \frac{1}{2} & -\frac{1}{3} & -\frac{1}{4} & 0 \\ -\frac{1}{4} & \frac{2}{3} & -\frac{1}{4} & 0 \\ -\frac{1}{4} & -\frac{1}{3} & \frac{1}{2} & 0 \end{array} \right] \xrightarrow[\substack{R_1 = 2r_1 \\ R_2 = \frac{1}{4}r_1 + r_2 \\ R_3 = \frac{1}{4}r_1 + r_3}]{} \left[\begin{array}{ccc|c} 1 & -\frac{2}{3} & -\frac{1}{2} & 0 \\ 0 & \frac{1}{2} & -\frac{3}{8} & 0 \\ 0 & -\frac{1}{2} & \frac{3}{8} & 0 \end{array} \right] \xrightarrow[\substack{R_2 = 2r_2 \\ R_1 = \frac{2}{3}r_2 + r_1 \\ R_3 = \frac{1}{2}r_2 + r_3}]{} \left[\begin{array}{ccc|c} 1 & 0 & -1 & 0 \\ 0 & 1 & -\frac{3}{4} & 0 \\ 0 & 0 & 0 & 0 \end{array} \right]$$

or $x = z$ and $y = \frac{3}{4}z$ where z is the parameter. Since we are told C's wages are \$30,000,

we know A's wages are also \$30,000, and B's wages are $\frac{3}{4}(30{,}000) = \$22{,}500$.

3.   We let x represent A's wages, y represent B's wages, and z represent C's wages, which we are told equal \$30,000. Since the amount paid out by each A, B, and C equals the amount each receives, we get $\begin{bmatrix} x \\ y \\ z \end{bmatrix} = \begin{bmatrix} 0.2 & 0.3 & 0.1 \\ 0.6 & 0.4 & 0.2 \\ 0.2 & 0.3 & 0.7 \end{bmatrix} \begin{bmatrix} x \\ y \\ z \end{bmatrix}$ which gives the system

$$\begin{cases} x = 0.2x + 0.3y + 0.1z \\ y = 0.6x + 0.4y + 0.2z \\ z = 0.2x + 0.3y + 0.7z \end{cases} \text{ or } \begin{cases} 0.8x - 0.3y - 0.1z = 0 \\ -0.6x + 0.6y - 0.2z = 0 \\ -0.2x - 0.3y + 0.3z = 0 \end{cases} \text{ or } \begin{cases} 8x - 3y - 1z = 0 \\ -6x + 6y - 2z = 0 \\ -2x - 3y + 3z = 0 \end{cases}$$

Solving the equations for x, y, and z, we find

$$\begin{bmatrix} 8 & -3 & -1 & | & 0 \\ -6 & 6 & -2 & | & 0 \\ -2 & -3 & 3 & | & 0 \end{bmatrix} \xrightarrow[\substack{R_1 = \frac{1}{8}r_1 \\ R_2 = 6r_1 + r_2 \\ R_3 = 2r_1 + r_3}]{} \begin{bmatrix} 1 & -\frac{3}{8} & -\frac{1}{8} & | & 0 \\ 0 & \frac{15}{4} & -\frac{11}{4} & | & 0 \\ 0 & -\frac{15}{4} & \frac{11}{4} & | & 0 \end{bmatrix} \xrightarrow[\substack{R_2 = \frac{4}{15}r_2 \\ R_1 = \frac{3}{8}r_2 + r_1 \\ R_3 = \frac{15}{4}r_2 + r_3}]{} \begin{bmatrix} 1 & 0 & -\frac{2}{5} & | & 0 \\ 0 & 1 & -\frac{11}{15} & | & 0 \\ 0 & 0 & 0 & | & 0 \end{bmatrix}$$

or $x = \dfrac{2}{5}z$ and $y = \dfrac{11}{15}z$ where z is the parameter. Since we are told C's wages are \$30,000, we find A's wages are $\dfrac{2}{5}(30{,}000) = \$12{,}000$, and B's wages are $\dfrac{11}{15}(30{,}000) = \$22{,}000$.

5.   The total output vector X for an open Leontief model is from the system $X = [I_3 - A]^{-1} \cdot D$ where D represents future demand for the goods produced in the system. If $D_2 = \begin{bmatrix} 80 \\ 90 \\ 60 \end{bmatrix}$, then using the information from Table 4, we get

$$X = \begin{bmatrix} 1.6048 & 0.3568 & 0.7131 \\ 0.2946 & 1.3363 & 0.3857 \\ 0.3660 & 0.2721 & 1.4013 \end{bmatrix} \begin{bmatrix} 80 \\ 90 \\ 60 \end{bmatrix} = \begin{bmatrix} 203.28 \\ 166.98 \\ 137.85 \end{bmatrix}$$

The total output of R, S, and T required for the forecast demand D_2 is to produce 203.28 units of product R, 166.98 units of product S, and 137.85 units of product T.

7. We place the information in the 4×4 input/output matrix, A,

$$A: \quad \begin{array}{c} \\ \text{farmer} \\ \text{builder} \\ \text{tailor} \\ \text{rancher} \end{array} \begin{array}{cccc} \text{farmer} & \text{builder} & \text{tailor} & \text{rancher} \\ \begin{bmatrix} 0.3 & 0.3 & 0.3 & 0.2 \\ 0.2 & 0.3 & 0.3 & 0.2 \\ 0.2 & 0.1 & 0.1 & 0.2 \\ 0.3 & 0.3 & 0.3 & 0.4 \end{bmatrix} \end{array}$$

and define the variables x_1: the farmer's income; x_2; the builder's income; x_3: the tailor's income; and x_4: the rancher's income (which we are told is \$25,000). Since in a closed Leontief model the amount paid equals the amount received for each member, we form the system of equations $X = AX$ or $(I_4 - A)X = 0$.

$$\begin{cases} 0.3x_1 + 0.3x_2 + 0.3x_3 + 0.2x_4 = x_1 \\ 0.2x_1 + 0.3x_2 + 0.3x_3 + 0.2x_4 = x_2 \\ 0.2x_1 + 0.1x_2 + 0.1x_3 + 0.2x_4 = x_3 \\ 0.3x_1 + 0.3x_2 + 0.3x_3 + 0.4x_4 = x_4 \end{cases} \quad \text{or} \quad \begin{cases} 0.7x_1 - 0.3x_2 - 0.3x_3 - 0.2x_4 = 0 \\ -0.2x_1 + 0.7x_2 - 0.3x_3 - 0.2x_4 = 0 \\ -0.2x_1 - 0.1x_2 + 0.9x_3 - 0.2x_4 = 0 \\ -0.3x_1 - 0.3x_2 - 0.3x_3 + 0.6x_4 = 0 \end{cases}$$

We solve the system of equations for X.

$$\begin{bmatrix} 7 & -3 & -3 & -2 & | & 0 \\ -2 & 7 & -3 & -2 & | & 0 \\ -2 & -1 & 9 & -2 & | & 0 \\ -3 & -3 & -3 & 6 & | & 0 \end{bmatrix} \xrightarrow[\substack{R_1 = \frac{1}{7}r_1 \\ R_2 = 2r_1 + r_2 \\ R_3 = 2r_1 + r_3 \\ R_4 = 3r_1 + r_4}]{} \begin{bmatrix} 1 & -\frac{3}{7} & -\frac{3}{7} & -\frac{2}{7} & | & 0 \\ 0 & \frac{43}{7} & -\frac{27}{7} & -\frac{18}{7} & | & 0 \\ 0 & -\frac{13}{7} & \frac{57}{7} & -\frac{18}{7} & | & 0 \\ 0 & -\frac{30}{7} & -\frac{30}{7} & \frac{36}{7} & | & 0 \end{bmatrix}$$

$$\xrightarrow[\substack{R_2 = \frac{7}{43}r_2 \\ R_1 = \frac{3}{7}r_2 + r_1 \\ R_3 = \frac{13}{7}r_2 + r_3 \\ R_4 = \frac{30}{7}r_2 + r_4}]{} \begin{bmatrix} 1 & 0 & -\frac{30}{43} & -\frac{20}{43} & | & 0 \\ 0 & 1 & -\frac{27}{43} & -\frac{18}{43} & | & 0 \\ 0 & 0 & \frac{300}{43} & -\frac{144}{43} & | & 0 \\ 0 & 0 & -\frac{300}{43} & \frac{144}{0} & | & 0 \end{bmatrix} \xrightarrow[\substack{R_3 = \frac{43}{300}r_3 \\ R_1 = \frac{30}{43}r_3 + r_1 \\ R_2 = \frac{27}{43}r_3 + r_2 \\ R_4 = \frac{300}{43}r_3 + r_4}]{} \begin{bmatrix} 1 & 0 & 0 & -\frac{4}{5} & | & 0 \\ 0 & 1 & 0 & -\frac{18}{25} & | & 0 \\ 0 & 0 & 1 & -\frac{12}{25} & | & 0 \\ 0 & 0 & 0 & 0 & | & 0 \end{bmatrix}$$

The solutions to the system are $x_1 = \frac{4}{5}x_4$, $x_2 = \frac{18}{25}x_4$, $x_3 = \frac{12}{25}x_4$ where x_4 is the parameter. Since the rancher's income was \$25,000, we get the farmer's income was $\frac{4}{5}(25,000) = \$20,000$; the builder's income was $\frac{18}{25}(25,000) = \$18,000$; and the tailor's income was $\frac{12}{25}(25,000) = \$12,000$.

9.   In an open Leontief model total output must satisfy producer needs and consumer demand.
First we construct matrix A which specifies the material needed per unit of production.

$$A: \begin{array}{c} \\ R \\ S \end{array} \begin{array}{cc} R & S \end{array} \begin{bmatrix} \dfrac{3}{13} & \dfrac{4}{7} \\ \dfrac{2}{13} & \dfrac{1}{7} \end{bmatrix}$$

We define X as the total output of R and S needed to meet future demand. $AX + D = X$ or

$$X = [I-A]^{-1} \cdot D = \begin{bmatrix} \dfrac{10}{13} & -\dfrac{4}{7} \\ -\dfrac{2}{13} & \dfrac{6}{7} \end{bmatrix}^{-1} \begin{bmatrix} 80 \\ 40 \end{bmatrix} = \begin{bmatrix} \dfrac{3}{2} & 1 \\ \dfrac{7}{26} & \dfrac{35}{26} \end{bmatrix} \begin{bmatrix} 80 \\ 40 \end{bmatrix} = \begin{bmatrix} 160 \\ \dfrac{980}{13} \end{bmatrix}$$

To meet future demand, produce 160 units of product R and 75.38 units of product S.

(We got $[I-A]^{-1}$ from the formula from section 2.6 Problem 61.)

$$\Delta = \frac{10}{13} \cdot \frac{6}{7} - \left(-\frac{4}{7}\right)\left(-\frac{2}{13}\right) = \frac{60-8}{7 \cdot 13} = \frac{4}{7} \quad \text{and}$$

$$(I_2 - A)^{-1} = \begin{bmatrix} \dfrac{6}{7} \cdot \dfrac{7}{4} & \dfrac{4}{7} \cdot \dfrac{7}{4} \\ \dfrac{2}{13} \cdot \dfrac{7}{4} & \dfrac{10}{13} \cdot \dfrac{7}{4} \end{bmatrix} = \begin{bmatrix} \dfrac{3}{2} & 1 \\ \dfrac{7}{26} & \dfrac{35}{26} \end{bmatrix}$$

Application 2: Cryptography

1. If $A = \begin{bmatrix} 2 & 3 \\ 1 & 2 \end{bmatrix}$, then $A^{-1} = \begin{bmatrix} 2 & -3 \\ -1 & 2 \end{bmatrix}$. To decode the messages, we form 2×1 column matrices and multiply on the left by A^{-1}.

    (a) $\begin{bmatrix} 2 & -3 \\ -1 & 2 \end{bmatrix}\begin{bmatrix} 64 \\ 36 \end{bmatrix} = \begin{bmatrix} 20 \\ 8 \end{bmatrix} = \begin{matrix} T \\ H \end{matrix}$; $\begin{bmatrix} 2 & -3 \\ -1 & 2 \end{bmatrix}\begin{bmatrix} 75 \\ 47 \end{bmatrix} = \begin{bmatrix} 9 \\ 19 \end{bmatrix} = \begin{matrix} I \\ S \end{matrix}$; $\begin{bmatrix} 2 & -3 \\ -1 & 2 \end{bmatrix}\begin{bmatrix} 75 \\ 47 \end{bmatrix} = \begin{bmatrix} 9 \\ 19 \end{bmatrix} = \begin{matrix} I \\ S \end{matrix}$;

    $\begin{bmatrix} 2 & -3 \\ -1 & 2 \end{bmatrix}\begin{bmatrix} 49 \\ 29 \end{bmatrix} = \begin{bmatrix} 11 \\ 9 \end{bmatrix} = \begin{matrix} K \\ I \end{matrix}$; $\begin{bmatrix} 2 & -3 \\ -1 & 2 \end{bmatrix}\begin{bmatrix} 60 \\ 36 \end{bmatrix} = \begin{bmatrix} 12 \\ 12 \end{bmatrix} = \begin{matrix} L \\ L \end{matrix}$; $\begin{bmatrix} 2 & -3 \\ -1 & 2 \end{bmatrix}\begin{bmatrix} 60 \\ 37 \end{bmatrix} = \begin{bmatrix} 9 \\ 14 \end{bmatrix} = \begin{matrix} I \\ N \end{matrix}$;

    $\begin{bmatrix} 2 & -3 \\ -1 & 2 \end{bmatrix}\begin{bmatrix} 53 \\ 33 \end{bmatrix} = \begin{bmatrix} 7 \\ 13 \end{bmatrix} = \begin{matrix} G \\ M \end{matrix}$; $\begin{bmatrix} 2 & -3 \\ -1 & 2 \end{bmatrix}\begin{bmatrix} 71 \\ 39 \end{bmatrix} = \begin{bmatrix} 25 \\ 7 \end{bmatrix} = \begin{matrix} Y \\ G \end{matrix}$; $\begin{bmatrix} 2 & -3 \\ -1 & 2 \end{bmatrix}\begin{bmatrix} 35 \\ 18 \end{bmatrix} = \begin{bmatrix} 16 \\ 1 \end{bmatrix} = \begin{matrix} P \\ A \end{matrix}$

    The message is: This is killing my GPA.

    (b) $\begin{bmatrix} 2 & -3 \\ -1 & 2 \end{bmatrix}\begin{bmatrix} 76 \\ 49 \end{bmatrix} = \begin{bmatrix} 5 \\ 22 \end{bmatrix} = \begin{matrix} E \\ V \end{matrix}$; $\begin{bmatrix} 2 & -3 \\ -1 & 2 \end{bmatrix}\begin{bmatrix} 64 \\ 41 \end{bmatrix} = \begin{bmatrix} 5 \\ 18 \end{bmatrix} = \begin{matrix} E \\ R \end{matrix}$; $\begin{bmatrix} 2 & -3 \\ -1 & 2 \end{bmatrix}\begin{bmatrix} 95 \\ 55 \end{bmatrix} = \begin{bmatrix} 25 \\ 15 \end{bmatrix} = \begin{matrix} Y \\ 0 \end{matrix}$;

    $\begin{bmatrix} 2 & -3 \\ -1 & 2 \end{bmatrix}\begin{bmatrix} 43 \\ 24 \end{bmatrix} = \begin{bmatrix} 14 \\ 5 \end{bmatrix} = \begin{matrix} N \\ E \end{matrix}$; $\begin{bmatrix} 2 & -3 \\ -1 & 2 \end{bmatrix}\begin{bmatrix} 69 \\ 42 \end{bmatrix} = \begin{bmatrix} 12 \\ 15 \end{bmatrix} = \begin{matrix} L \\ 0 \end{matrix}$; $\begin{bmatrix} 2 & -3 \\ -1 & 2 \end{bmatrix}\begin{bmatrix} 59 \\ 32 \end{bmatrix} = \begin{bmatrix} 22 \\ 5 \end{bmatrix} = \begin{matrix} V \\ E \end{matrix}$;

    $\begin{bmatrix} 2 & -3 \\ -1 & 2 \end{bmatrix}\begin{bmatrix} 77 \\ 45 \end{bmatrix} = \begin{bmatrix} 19 \\ 13 \end{bmatrix} = \begin{matrix} S \\ M \end{matrix}$; $\begin{bmatrix} 2 & -3 \\ -1 & 2 \end{bmatrix}\begin{bmatrix} 27 \\ 15 \end{bmatrix} = \begin{bmatrix} 9 \\ 3 \end{bmatrix} = \begin{matrix} I \\ C \end{matrix}$; $\begin{bmatrix} 2 & -3 \\ -1 & 2 \end{bmatrix}\begin{bmatrix} 37 \\ 21 \end{bmatrix} = \begin{bmatrix} 11 \\ 5 \end{bmatrix} = \begin{matrix} K \\ E \end{matrix}$;

    $\begin{bmatrix} 2 & -3 \\ -1 & 2 \end{bmatrix}\begin{bmatrix} 128 \\ 77 \end{bmatrix} = \begin{bmatrix} 25 \\ 26 \end{bmatrix} = \begin{matrix} Y \\ Z \end{matrix}$

    The message is: Everyone loves Mickey.

3. To decode the message we group the code into form 2×1 column matrices, then we multiply each column matrix on the left by A^{-1}.

    $\begin{bmatrix} 2 & -3 \\ -1 & 2 \end{bmatrix}\begin{bmatrix} 70 \\ 39 \end{bmatrix} = \begin{bmatrix} 23 \\ 8 \end{bmatrix} = \begin{matrix} W \\ H \end{matrix}$; $\begin{bmatrix} 2 & -3 \\ -1 & 2 \end{bmatrix}\begin{bmatrix} 62 \\ 41 \end{bmatrix} = \begin{bmatrix} 1 \\ 20 \end{bmatrix} = \begin{matrix} A \\ T \end{matrix}$; $\begin{bmatrix} 2 & -3 \\ -1 & 2 \end{bmatrix}\begin{bmatrix} 113 \\ 69 \end{bmatrix} = \begin{bmatrix} 19 \\ 25 \end{bmatrix} = \begin{matrix} S \\ Y \end{matrix}$

    $\begin{bmatrix} 2 & -3 \\ -1 & 2 \end{bmatrix}\begin{bmatrix} 93 \\ 57 \end{bmatrix} = \begin{bmatrix} 15 \\ 21 \end{bmatrix} = \begin{matrix} O \\ U \end{matrix}$; $\begin{bmatrix} 2 & -3 \\ -1 & 2 \end{bmatrix}\begin{bmatrix} 51 \\ 28 \end{bmatrix} = \begin{bmatrix} 18 \\ 5 \end{bmatrix} = \begin{matrix} R \\ E \end{matrix}$; $\begin{bmatrix} 2 & -3 \\ -1 & 2 \end{bmatrix}\begin{bmatrix} 29 \\ 15 \end{bmatrix} = \begin{bmatrix} 13 \\ 1 \end{bmatrix} = \begin{matrix} M \\ A \end{matrix}$;

    $\begin{bmatrix} 2 & -3 \\ -1 & 2 \end{bmatrix}\begin{bmatrix} 54 \\ 33 \end{bmatrix} = \begin{bmatrix} 9 \\ 12 \end{bmatrix} = \begin{matrix} I \\ L \end{matrix}$; $\begin{bmatrix} 2 & -3 \\ -1 & 2 \end{bmatrix}\begin{bmatrix} 14 \\ 9 \end{bmatrix} = \begin{bmatrix} 1 \\ 4 \end{bmatrix} = \begin{matrix} A \\ D \end{matrix}$; $\begin{bmatrix} 2 & -3 \\ -1 & 2 \end{bmatrix}\begin{bmatrix} 62 \\ 40 \end{bmatrix} = \begin{bmatrix} 4 \\ 18 \end{bmatrix} = \begin{matrix} D \\ R \end{matrix}$;

    $\begin{bmatrix} 2 & -3 \\ -1 & 2 \end{bmatrix}\begin{bmatrix} 67 \\ 43 \end{bmatrix} = \begin{bmatrix} 5 \\ 19 \end{bmatrix} = \begin{matrix} E \\ S \end{matrix}$; $\begin{bmatrix} 2 & -3 \\ -1 & 2 \end{bmatrix}\begin{bmatrix} 116 \\ 71 \end{bmatrix} = \begin{bmatrix} 19 \\ 26 \end{bmatrix} = \begin{matrix} S \\ Z \end{matrix}$

    The message is: Whats your email address

5.   To encode a message using A, assign each letter in the message a number according to the scheme, then group the numbers into 2×1 column matrices, and multiply each on the left by matrix A.

(a)
I	A	M	G	O	I	N	G	T	O	D	I	S	N	E	Y
9	1	13	7	15	9	14	7	20	15	4	9	19	14	5	25

$$\begin{bmatrix} 2 & 3 \\ 1 & 2 \end{bmatrix}\begin{bmatrix} 9 \\ 1 \end{bmatrix} = \begin{bmatrix} 21 \\ 11 \end{bmatrix}; \quad \begin{bmatrix} 2 & 3 \\ 1 & 2 \end{bmatrix}\begin{bmatrix} 13 \\ 7 \end{bmatrix} = \begin{bmatrix} 47 \\ 27 \end{bmatrix}; \quad \begin{bmatrix} 2 & 3 \\ 1 & 2 \end{bmatrix}\begin{bmatrix} 15 \\ 9 \end{bmatrix} = \begin{bmatrix} 57 \\ 33 \end{bmatrix}; \quad \begin{bmatrix} 2 & 3 \\ 1 & 2 \end{bmatrix}\begin{bmatrix} 14 \\ 7 \end{bmatrix} = \begin{bmatrix} 49 \\ 28 \end{bmatrix};$$

$$\begin{bmatrix} 2 & 3 \\ 1 & 2 \end{bmatrix}\begin{bmatrix} 20 \\ 15 \end{bmatrix} = \begin{bmatrix} 85 \\ 50 \end{bmatrix}; \quad \begin{bmatrix} 2 & 3 \\ 1 & 2 \end{bmatrix}\begin{bmatrix} 4 \\ 9 \end{bmatrix} = \begin{bmatrix} 35 \\ 22 \end{bmatrix}; \quad \begin{bmatrix} 2 & 3 \\ 1 & 2 \end{bmatrix}\begin{bmatrix} 19 \\ 14 \end{bmatrix} = \begin{bmatrix} 80 \\ 47 \end{bmatrix}; \quad \begin{bmatrix} 2 & 3 \\ 1 & 2 \end{bmatrix}\begin{bmatrix} 5 \\ 25 \end{bmatrix} = \begin{bmatrix} 85 \\ 55 \end{bmatrix}$$

The encoded message is: 21 11 47 27 57 33 49 28 85 50 35 22 80 47 85 55

(b)
W	E	S	U	R	F	T	H	E	N	E	T
23	5	19	21	18	6	20	8	5	14	5	20

$$\begin{bmatrix} 2 & 3 \\ 1 & 2 \end{bmatrix}\begin{bmatrix} 23 \\ 5 \end{bmatrix} = \begin{bmatrix} 61 \\ 33 \end{bmatrix}; \quad \begin{bmatrix} 2 & 3 \\ 1 & 2 \end{bmatrix}\begin{bmatrix} 19 \\ 21 \end{bmatrix} = \begin{bmatrix} 101 \\ 61 \end{bmatrix}; \quad \begin{bmatrix} 2 & 3 \\ 1 & 2 \end{bmatrix}\begin{bmatrix} 18 \\ 6 \end{bmatrix} = \begin{bmatrix} 54 \\ 30 \end{bmatrix}; \quad \begin{bmatrix} 2 & 3 \\ 1 & 2 \end{bmatrix}\begin{bmatrix} 20 \\ 8 \end{bmatrix} = \begin{bmatrix} 64 \\ 36 \end{bmatrix};$$

$$\begin{bmatrix} 2 & 3 \\ 1 & 2 \end{bmatrix}\begin{bmatrix} 5 \\ 14 \end{bmatrix} = \begin{bmatrix} 52 \\ 33 \end{bmatrix}; \quad \begin{bmatrix} 2 & 3 \\ 1 & 2 \end{bmatrix}\begin{bmatrix} 4 \\ 20 \end{bmatrix} = \begin{bmatrix} 70 \\ 45 \end{bmatrix}$$

The encoded message is: 61 33 101 61 54 30 64 36 52 33 70 45

(c)
L	E	T	S	G	O	C	L	U	B	B	I	N	G
12	5	20	19	7	15	3	12	21	2	2	9	14	7

$$\begin{bmatrix} 2 & 3 \\ 1 & 2 \end{bmatrix}\begin{bmatrix} 12 \\ 5 \end{bmatrix} = \begin{bmatrix} 39 \\ 22 \end{bmatrix}; \quad \begin{bmatrix} 2 & 3 \\ 1 & 2 \end{bmatrix}\begin{bmatrix} 20 \\ 19 \end{bmatrix} = \begin{bmatrix} 97 \\ 58 \end{bmatrix}; \quad \begin{bmatrix} 2 & 3 \\ 1 & 2 \end{bmatrix}\begin{bmatrix} 7 \\ 15 \end{bmatrix} = \begin{bmatrix} 59 \\ 37 \end{bmatrix}; \quad \begin{bmatrix} 2 & 3 \\ 1 & 2 \end{bmatrix}\begin{bmatrix} 3 \\ 12 \end{bmatrix} = \begin{bmatrix} 42 \\ 27 \end{bmatrix};$$

$$\begin{bmatrix} 2 & 3 \\ 1 & 2 \end{bmatrix}\begin{bmatrix} 21 \\ 2 \end{bmatrix} = \begin{bmatrix} 48 \\ 25 \end{bmatrix}; \quad \begin{bmatrix} 2 & 3 \\ 1 & 2 \end{bmatrix}\begin{bmatrix} 2 \\ 9 \end{bmatrix} = \begin{bmatrix} 31 \\ 20 \end{bmatrix}; \quad \begin{bmatrix} 2 & 3 \\ 1 & 2 \end{bmatrix}\begin{bmatrix} 14 \\ 7 \end{bmatrix} = \begin{bmatrix} 49 \\ 28 \end{bmatrix}$$

The encoded message is: 39 22 97 58 59 37 42 27 48 25 31 20 49 28

Application 3: Accounting

1. We first find the total costs for the 2 service departments by solving the system

$$\begin{cases} x_1 = 2000 + \dfrac{1}{9}x_1 + \dfrac{3}{9}x_2 \\[2mm] x_2 = 1000 + \dfrac{3}{9}x_1 + \dfrac{1}{9}x_2 \end{cases}$$ We will denote $X = \begin{bmatrix} x_1 \\ x_2 \end{bmatrix}$, $C = \begin{bmatrix} \dfrac{1}{9} & \dfrac{3}{9} \\[2mm] \dfrac{3}{9} & \dfrac{1}{9} \end{bmatrix}$, and $D = \begin{bmatrix} 2000 \\ 1000 \end{bmatrix}$.

The system can now be represented by $X = D + CX$, which means the total costs of the 2 service departments can be obtained by solving $X = [I_2 - C]^{-1}D$.

$$[I_2 - C]^{-1} = \begin{bmatrix} \dfrac{8}{9} & -\dfrac{3}{9} \\[2mm] -\dfrac{3}{9} & \dfrac{8}{9} \end{bmatrix}^{-1} = \begin{bmatrix} \dfrac{72}{55} & \dfrac{27}{55} \\[2mm] \dfrac{27}{55} & \dfrac{72}{55} \end{bmatrix}, \text{ so } X = \begin{bmatrix} \dfrac{72}{55} & \dfrac{27}{55} \\[2mm] \dfrac{27}{55} & \dfrac{72}{55} \end{bmatrix}\begin{bmatrix} 2000 \\ 1000 \end{bmatrix} = \begin{bmatrix} 3109.09 \\ 2290.91 \end{bmatrix}$$

The problem has a solution because $[I_2 - C]^{-1}$ exists and because both $[I_2 - C]^{-1}$ and D have only nonnegative entries.

By substituting $x_1 = \$3109.09$ and $x_2 = \$2290.91$ into the table we find all direct and indirect costs:

Dept.	Total Costs	Direct Costs	Indirect Costs S_1	S_2
S_1	$3109.09	$2000	$345.45	$763.64
S_2	$2290.91	$1000	$1036.36	$254.55
P_1	$3354.54	$2500	$345.45	$509.09
P_2	$2790.91	$1500	$1036.36	$254.55
P_3	$3854.54	$3000	$345.45	$509.09
Totals:	$15399.99	$10000	$3109.07	$2290.92

Finally, we show that the total of service charges allocated to production departments P_1, P_2, and P_3 equals the sum of the direct costs of the service departments S_1 and S_2.
Service charges allocated to P_1, P_2, P_3:
$$\$345.45 + \$509.09 + \$1036.36 + \$254.55 + \$345.45 + \$509.09 = \$2999.99$$
Direct costs of S_1 & S_2: $\$2000 + \$1000 = \$3000$

Application 4: The Method of Least Squares

1. To find the transpose A^T of matrix A we interchange the rows and the columns of A.

$$A^T = \begin{bmatrix} 4 & 3 \\ 1 & 1 \\ 2 & 0 \end{bmatrix}$$

3. To find the transpose A^T of matrix A we interchange the rows and the columns of A.

$$A^T = \begin{bmatrix} 1 & 0 & 1 \\ 11 & 12 & 4 \end{bmatrix}$$

5. To find the transpose A^T of matrix A we interchange the rows and the columns of A.

$$A^T = \begin{bmatrix} 8 & 6 & 3 \end{bmatrix}$$

7. (a) To find the least squares line, we define $A = \begin{bmatrix} 3 & 1 \\ 5 & 1 \\ 6 & 1 \\ 7 & 1 \end{bmatrix}$, $Y = \begin{bmatrix} 10 \\ 13 \\ 15 \\ 16 \end{bmatrix}$ and $X = \begin{bmatrix} m \\ b \end{bmatrix}$ and use

these matrices to form the equation $A^T A X = A^T Y$.

$$\begin{bmatrix} 3 & 5 & 6 & 7 \\ 1 & 1 & 1 & 1 \end{bmatrix} \begin{bmatrix} 3 & 1 \\ 5 & 1 \\ 6 & 1 \\ 7 & 1 \end{bmatrix} \begin{bmatrix} m \\ b \end{bmatrix} = \begin{bmatrix} 3 & 5 & 6 & 7 \\ 1 & 1 & 1 & 1 \end{bmatrix} \begin{bmatrix} 10 \\ 13 \\ 15 \\ 16 \end{bmatrix}$$ which simplifies to

$$\begin{bmatrix} 119 & 21 \\ 21 & 4 \end{bmatrix} \begin{bmatrix} m \\ b \end{bmatrix} = \begin{bmatrix} 297 \\ 54 \end{bmatrix}$$ and represents the system $\begin{cases} 119m + 21b = 297 \\ 21m + 4b = 54 \end{cases}$.

We solve the system of equations for m and b.

$$\begin{bmatrix} 119 & 21 & | & 297 \\ 21 & 4 & | & 54 \end{bmatrix} \xrightarrow[\substack{R_1 = \frac{1}{119}r \\ R_2 = -21r_1 + r_2}]{} \begin{bmatrix} 1 & \frac{3}{17} & | & \frac{297}{119} \\ 0 & \frac{5}{17} & | & \frac{27}{17} \end{bmatrix} \xrightarrow[\substack{R_2 = \frac{17}{5}r_2 \\ R_1 = -\frac{3}{17}r_2 + r_1}]{} \begin{bmatrix} 1 & 0 & | & \frac{54}{35} \\ 0 & 1 & | & \frac{27}{5} \end{bmatrix}$$

and we get $m = \dfrac{54}{35}$ and $b = \dfrac{27}{5}$. So the least squares line of best fit is $y = \dfrac{54}{35}x + \dfrac{27}{5}$.

   (b) The predicted supply when the price $8.00 per item is $y = \dfrac{54}{35}(8) + \dfrac{27}{5} = 17.7429$. That is, when the price is $8, the supply is predicted to be about 17,743 units.

9. To find the least squares line , we define $A = \begin{bmatrix} 10 & 1 \\ 17 & 1 \\ 11 & 1 \\ 18 & 1 \\ 21 & 1 \end{bmatrix}$, $Y = \begin{bmatrix} 50 \\ 61 \\ 55 \\ 60 \\ 70 \end{bmatrix}$, and $X = \begin{bmatrix} m \\ b \end{bmatrix}$ and use these matrices to form the equation $A^{\mathrm{T}}AX = A^{\mathrm{T}}Y$.

$$\begin{bmatrix} 10 & 17 & 11 & 18 & 21 \\ 1 & 1 & 1 & 1 & 1 \end{bmatrix} \begin{bmatrix} 10 & 1 \\ 17 & 1 \\ 11 & 1 \\ 18 & 1 \\ 21 & 1 \end{bmatrix} \begin{bmatrix} m \\ b \end{bmatrix} = \begin{bmatrix} 10 & 17 & 11 & 18 & 21 \\ 1 & 1 & 1 & 1 & 1 \end{bmatrix} \begin{bmatrix} 50 \\ 61 \\ 55 \\ 60 \\ 70 \end{bmatrix}$$ which simplifies to

$$\begin{bmatrix} 1275 & 77 \\ 77 & 5 \end{bmatrix} \begin{bmatrix} m \\ b \end{bmatrix} = \begin{bmatrix} 4692 \\ 296 \end{bmatrix}$$ and represents the system $\begin{cases} 1275m + 77b = 4692 \\ 77m + 5b = 296 \end{cases}$.

We solve the system of equations for m and b.

$$\begin{bmatrix} 1275 & 77 & | & 4692 \\ 77 & 5 & | & 296 \end{bmatrix} \xrightarrow[R_1 = \frac{1}{1275}r_1]{} \begin{bmatrix} 1 & \frac{77}{1275} & | & \frac{92}{25} \\ 77 & 5 & | & 296 \end{bmatrix} \xrightarrow[R_2 = -77r_1 + r_2]{} \begin{bmatrix} 1 & \frac{77}{1275} & | & \frac{92}{25} \\ 0 & \frac{446}{1275} & | & \frac{316}{25} \end{bmatrix}$$

$$\xrightarrow[R_2 = \frac{1275}{446}r_2]{} \begin{bmatrix} 1 & \frac{77}{1275} & | & \frac{92}{25} \\ 0 & 1 & | & \frac{8058}{223} \end{bmatrix} \xrightarrow[R_1 = -\frac{77}{1275}r_2 + r_1]{} \begin{bmatrix} 1 & 0 & | & \frac{334}{223} \\ 0 & 1 & | & \frac{8058}{223} \end{bmatrix}$$

We get $m = \dfrac{334}{223}$ and $b = \dfrac{8058}{223}$. So the least squares line of best fit for these data is

$y = \dfrac{334}{223}x + \dfrac{8058}{223}$.

11. (a) $\begin{bmatrix} 1 & 1 & 2 \\ 1 & 0 & 1 \\ 3 & 2 & 3 \end{bmatrix}^{\mathrm{T}} = \begin{bmatrix} 1 & 1 & 3 \\ 1 & 0 & 2 \\ 2 & 1 & 3 \end{bmatrix}$ Since the matrix does not equal its transpose, the matrix is not symmetric.

(b) $\begin{bmatrix} 0 & 1 & 3 \\ 1 & 4 & 7 \\ 3 & 7 & 5 \end{bmatrix}^{\mathrm{T}} = \begin{bmatrix} 0 & 1 & 3 \\ 1 & 4 & 7 \\ 3 & 7 & 5 \end{bmatrix}$ Since the matrix equals its transpose, the matrix is symmetric.

(c) $\begin{bmatrix} 1 & 2 & 3 & 0 \\ 2 & 4 & 5 & 0 \\ 3 & 5 & 1 & 0 \end{bmatrix}^{\mathrm{T}} = \begin{bmatrix} 1 & 2 & 3 \\ 2 & 4 & 5 \\ 3 & 5 & 1 \\ 0 & 0 & 0 \end{bmatrix}$ Since the matrix does not equal its transpose, the matrix is not symmetric.

A symmetric matrix must be square. If A has dimension $n \times m$, then A^{T} has dimension $m \times n$. If A is symmetric, then $A = A^{\mathrm{T}}$. For two matrices to be equal they must have the same dimensions, so $n = m$.

Chapter 2 Review

True-False Items

1. True   **3.** False   **5.** False   **7.** False

Fill in the Blanks

1. 3×2 **3.** rows; columns   **5.** 3×3

Review Exercises

1. We will solve the system by elimination.

$$
\begin{array}{ll}
2x - y = 5 & (1) \\
5x + 2y = 8 & (2)
\end{array}
\qquad \text{Multiply by 2:} \qquad
\begin{array}{ll}
4x - 2y = 10 & (1) \\
5x + 2y = \ 8 & (2) \\
\hline
9x \qquad = 18 & (\text{Add}) \ (2)
\end{array}
$$

Solving (2) for x we get $x = 2$, and back-substituting 2 for x in equation (1) we get $2(2) - y = 5$ or $y = -1$.

The solution to the system is $x = 2$, $y = -1$.

3. We will solve the system by elimination.

$$\begin{array}{ll} x - 2y - 4 = 0 & (1) \\ 3x + 2y - 4 = 0 & (2) \\ \hline 4x \qquad - 8 = 0 & \text{(Add)} \quad \text{or } x = 2 \end{array}$$

Back-substituting $x = 2$ in equation (1) we get $2 - 2y - 4 = 0$ or $-2y = 2$ or $y = -1$. The solution to the system is $x = 2$ and $y = -1$.

5. We will solve the system by elimination.

$$\begin{array}{ll} 3x - 2y = 8 & (1) \\ x - \dfrac{2}{3}y = 12 & (2) \end{array} \qquad \text{Multiply by 3:} \qquad \begin{array}{ll} 3x - 2y = 8 & (1) \\ 3x - 2y = 36 & (2) \\ \hline 0 = -28 & \text{(Subtract) (2)} \end{array}$$

Since equation (2) has no solution, the system is inconsistent.

7.

$$\begin{array}{ll} x + 2y - z = 6 & (1) \\ 2x - y + 3z = -13 & (2) \end{array} \quad \text{Multiply by 2:} \quad \begin{array}{ll} 2x + 4y - 2z = 12 & (1) \\ 2x - y + 3z = -13 & (2) \\ \hline 5y - 5z = 25 & \text{(Subtract) (2)} \end{array}$$

$$\begin{array}{ll} x + 2y - z = 6 & (1) \\ 3x - 2y + 3z = -16 & (3) \end{array} \quad \text{Multiply by 3:} \quad \begin{array}{ll} 3x + 6y - 3z = 18 & (1) \\ 3x - 2y - 3z = -16 & (3) \\ \hline 8y - 6z = 34 & \text{(Subtract) (3)} \end{array}$$

Now we work with the revised (2) and (3) and solve for z.

$$\begin{array}{ll} 5y - 5z = 25 & (2) \\ 8y - 6z = 34 & (3) \end{array} \quad \begin{array}{l} \text{Divide by 5:} \\ \text{Divide by 2:} \end{array} \quad \begin{array}{ll} y - z = 5 & (2) \\ 4y - 3z = 17 & (3) \end{array} \quad \text{Multiply by 4:} \quad \begin{array}{ll} 4y - 4z = 20 & (2) \\ 4y - 3z = 17 & (3) \\ \hline -z = 3 & \text{Subtract (3)} \end{array}$$

Solving (3) for z, we get $z = -3$. Back-substituting into (2) we $y - (-3) = 5$ or $y = 2$. Finally substituting both y and z into equation (1), we solve for x

$$x + 2(2) - (-3) = 6 \text{ or } x + 7 = 6 \text{ or } x = -1.$$

The solution to the system is $x = -1$, $y = 2$, and $z = -3$.

9.

$$2x-4y+z=-15 \quad (1)$$
$$x+2y-4z=27 \quad (2)$$

Multiply by 2:

$$2x-4y+z=-15 \quad (1)$$
$$2x+4y-8z=54 \quad (2)$$
$$\overline{\quad -8y+9z=-69 \quad} \text{(Subtract) (2)}$$

$$2x-4y+z=-15 \quad (1)$$
$$5x-6y-2z=-3 \quad (2)$$

Multiply by 5:
Multiply by 2:

$$10x-20y+5z=-75 \quad (1)$$
$$10x-12y-4z=-6 \quad (3)$$
$$\overline{\quad -8y+9z=-69 \quad} \text{(Subtract) (3)}$$

Now we work with revised (2) and (3).

$$-8y+9z=-69 \quad (2)$$
$$\underline{-8y+9z=-69 \quad (3)}$$
$$0=0 \quad \text{(Subtract)}$$

The system is equivalent to a system with only two equations, so the equations are dependent and the system has infinitely many solutions. We let z be any real number and solve (2) for y: $\quad -8y+9z=-69 \quad (2)$

$$y=\frac{9}{8}z+\frac{69}{8}$$

Substitute this expression into (1) to determine x in terms of z.

$$2x-4y+z=-15 \quad (1)$$
$$2x-4\left(\frac{9}{8}z+\frac{69}{8}\right)+z=-15$$
$$2x-\frac{9}{2}z-\frac{69}{2}+z=-15$$
$$2x=\frac{7}{2}z+\frac{39}{2} \quad \text{or } x=\frac{9}{4}z+\frac{39}{4}$$

The solutions to the system are $x=\frac{9}{4}z+\frac{39}{4}$ and $y=\frac{9}{8}z+\frac{69}{8}$ where z is any real number.

11. The system of equations is

$$\begin{cases} 3x+2y=8 \\ x+4y=-1 \end{cases}$$

13. The system of equations is

$$\begin{cases} x=4 \\ y=6 \\ z=-1 \end{cases}$$

This system has one solution. It is $x=4$, $y=6$, and $z=-1$.

15. We write the system as an augmented matrix and then use row operations to write it in reduced row-echelon form.

$$\begin{bmatrix} -5 & 2 & | & -2 \\ -3 & 3 & | & 4 \end{bmatrix} \xrightarrow{R_1 = -\frac{1}{5}r_1} \begin{bmatrix} 1 & -\frac{2}{5} & | & \frac{2}{5} \\ -3 & 3 & | & 4 \end{bmatrix} \xrightarrow{R_2 = 3r_1 + r_2} \begin{bmatrix} 1 & -\frac{2}{5} & | & \frac{2}{5} \\ 0 & \frac{9}{5} & | & \frac{26}{5} \end{bmatrix}$$

$$\xrightarrow{R_2 = \frac{5}{9}r_2} \begin{bmatrix} 1 & -\frac{2}{5} & | & \frac{2}{5} \\ 0 & 1 & | & \frac{26}{9} \end{bmatrix} \xrightarrow{R_1 = \frac{2}{5}r_2 + r_1} \begin{bmatrix} 1 & 0 & | & \frac{14}{9} \\ 0 & 1 & | & \frac{26}{9} \end{bmatrix}$$

The solution to the system is $x = \dfrac{14}{9}$ and $y = \dfrac{26}{9}$.

17. We write the system as an augmented matrix and then use row operations to write it in reduced row-echelon form.

$$\begin{bmatrix} 1 & 2 & 5 & | & 6 \\ 3 & 7 & 12 & | & 23 \\ 1 & 4 & 0 & | & 25 \end{bmatrix} \xrightarrow[R_3 = r_3 - r_1]{R_2 = r_2 - 3r_1} \begin{bmatrix} 1 & 2 & 5 & | & 6 \\ 0 & 1 & -3 & | & 5 \\ 0 & 2 & -5 & | & 19 \end{bmatrix} \xrightarrow[R_3 = r_3 - 2r_2]{R_1 = r_1 - 2r_2} \begin{bmatrix} 1 & 0 & 11 & | & -4 \\ 0 & 1 & -3 & | & 5 \\ 0 & 0 & 1 & | & 9 \end{bmatrix}$$

$$\xrightarrow[R_2 = r_2 + 3r_3]{R_1 = r_1 - 11r_3} \begin{bmatrix} 1 & 0 & 0 & | & -103 \\ 0 & 1 & 0 & | & 32 \\ 0 & 0 & 1 & | & 9 \end{bmatrix}$$

The solution to the system is $x = -103$, $y = 32$, and $z = 9$.

19. We write the system as an augmented matrix and then use row operations to write it in reduced row-echelon form.

$$\begin{bmatrix} 1 & 2 & 7 & | & 2 \\ 3 & 7 & 18 & | & -1 \\ 1 & 4 & 2 & | & -13 \end{bmatrix} \xrightarrow[R_3 = -r_1 + r_3]{R_2 = -3r_1 + r_2} \begin{bmatrix} 1 & 2 & 7 & | & 2 \\ 0 & 1 & -3 & | & -7 \\ 0 & 2 & -5 & | & -15 \end{bmatrix} \xrightarrow[R_3 = -2r_2 + r_3]{R_1 = -2r_2 + r_1} \begin{bmatrix} 1 & 0 & 13 & | & 16 \\ 0 & 1 & -3 & | & -7 \\ 0 & 0 & 1 & | & -1 \end{bmatrix}$$

$$\xrightarrow[R_2 = 3r_3 + r_2]{R_1 = -13r_3 + r_1} \begin{bmatrix} 1 & 0 & 0 & | & 29 \\ 0 & 1 & 0 & | & -10 \\ 0 & 0 & 1 & | & -1 \end{bmatrix}$$

The solution to the system is $x = 29$, $y = -10$, and $z = -1$.

21. We write the system as an augmented matrix and then use row operations to write it in reduced row-echelon form.

$$
\begin{bmatrix} 2 & -1 & 1 & | & 1 \\ 1 & 1 & -1 & | & 2 \\ 3 & -1 & 1 & | & 0 \end{bmatrix}
\xrightarrow[\text{rows 1 and 2}]{\text{Interchange}}
\begin{bmatrix} 1 & 1 & -1 & | & 2 \\ 2 & -1 & 1 & | & 1 \\ 3 & -1 & 1 & | & 0 \end{bmatrix}
\xrightarrow[R_3=-3r_1+r_3]{R_2=-2r_1+r_2}
\begin{bmatrix} 1 & 1 & -1 & | & 2 \\ 0 & -3 & 3 & | & -3 \\ 0 & -4 & 4 & | & -6 \end{bmatrix}
$$

$$
\xrightarrow[R_2=-\frac{1}{3}r_2]{}
\begin{bmatrix} 1 & 1 & -1 & | & 2 \\ 0 & 1 & -1 & | & 1 \\ 0 & -4 & 4 & | & -6 \end{bmatrix}
\xrightarrow[R_3=4r_2+r_3]{}
\begin{bmatrix} 1 & 1 & -1 & | & 2 \\ 0 & 1 & -1 & | & 1 \\ 0 & 0 & 0 & | & \mathbf{-2} \end{bmatrix}
$$

Row 3 of the final matrix indicates that the third equation has no solution. The system is inconsistent.

23. We write the system as an augmented matrix and then use row operations to write it in reduced row-echelon form.

$$
\begin{bmatrix} 0 & 1 & -2 & | & 6 \\ 3 & 2 & -1 & | & 2 \\ 4 & 0 & 3 & | & -1 \end{bmatrix}
\xrightarrow[\text{rows 1 and 3}]{\text{Interchange}}
\begin{bmatrix} 4 & 0 & 3 & | & -1 \\ 3 & 2 & -1 & | & 2 \\ 0 & 1 & -2 & | & 6 \end{bmatrix}
\xrightarrow[R_1=r_1-r_2]{}
\begin{bmatrix} 1 & -2 & 4 & | & -3 \\ 3 & 2 & -1 & | & 2 \\ 0 & 1 & -2 & | & 6 \end{bmatrix}
$$

$$
\xrightarrow[R_2=-3r_1+r_2]{}
\begin{bmatrix} 1 & -2 & 4 & | & -3 \\ 0 & 8 & -13 & | & 11 \\ 0 & 1 & -2 & | & 6 \end{bmatrix}
\xrightarrow[\text{rows 2 and 3}]{\text{Interchange}}
\begin{bmatrix} 1 & -2 & 4 & | & -3 \\ 0 & 1 & -2 & | & 6 \\ 0 & 8 & -13 & | & 11 \end{bmatrix}
$$

$$
\xrightarrow[R_3=-8r_2+r_3]{R_1=2r_2+r_1}
\begin{bmatrix} 1 & 0 & 0 & | & 9 \\ 0 & 1 & -2 & | & 6 \\ 0 & 0 & 3 & | & -37 \end{bmatrix}
\xrightarrow[R_3=\frac{1}{3}r_3]{}
\begin{bmatrix} 1 & 0 & 0 & | & 9 \\ 0 & 1 & -2 & | & 6 \\ 0 & 0 & 1 & | & -\frac{37}{3} \end{bmatrix}
$$

$$
\xrightarrow[R_2=2r_3+r_2]{}
\begin{bmatrix} 1 & 0 & 0 & | & 9 \\ 0 & 1 & 0 & | & -\frac{56}{3} \\ 0 & 0 & 1 & | & -\frac{37}{3} \end{bmatrix}
$$

The solution to the system is $x = 9$, $y = -\dfrac{56}{3}$, and $z = -\dfrac{37}{3}$.

25. We write the system as an augmented matrix and then use row operations to write it in reduced row-echelon form.

$$\begin{bmatrix} 1 & -3 & 0 & | & 5 \\ 0 & 3 & 1 & | & 0 \\ 2 & -1 & 2 & | & 2 \end{bmatrix} \xrightarrow{R_3=-2r_1+r_3} \begin{bmatrix} 1 & -3 & 0 & | & 5 \\ 0 & 3 & 1 & | & 0 \\ 0 & 5 & 2 & | & -8 \end{bmatrix} \xrightarrow{R_2=\frac{1}{3}r_2} \begin{bmatrix} 1 & -3 & 0 & | & 5 \\ 0 & 1 & \frac{1}{3} & | & 0 \\ 0 & 5 & 2 & | & -8 \end{bmatrix}$$

$$\xrightarrow[R_3=-5r_2+r_3]{R_1=3r_2+r_1} \begin{bmatrix} 1 & 0 & 1 & | & 5 \\ 0 & 1 & \frac{1}{3} & | & 0 \\ 0 & 0 & \frac{1}{3} & | & -8 \end{bmatrix} \xrightarrow{R_3=3r_3} \begin{bmatrix} 1 & 0 & 1 & | & 5 \\ 0 & 1 & \frac{1}{3} & | & 0 \\ 0 & 0 & 1 & | & -24 \end{bmatrix} \xrightarrow[R_2=-\frac{1}{3}r_3+r_2]{R_1=-r_3+r_1} \begin{bmatrix} 1 & 0 & 0 & | & 29 \\ 0 & 1 & 0 & | & 8 \\ 0 & 0 & 1 & | & -24 \end{bmatrix}$$

The solution to the system is $x = 29$, $y = 8$ and $z = -24$.

27. We write the system as an augmented matrix and then use row operations to write it in reduced row-echelon form.

$$\begin{bmatrix} 3 & 1 & -2 & | & 3 \\ 1 & -2 & 1 & | & 4 \end{bmatrix} \xrightarrow[\text{rows 1 and 2}]{\text{Interchange}} \begin{bmatrix} 1 & -2 & 1 & | & 4 \\ 3 & 1 & -2 & | & 3 \end{bmatrix} \xrightarrow{R_2=-3r_1+r_2} \begin{bmatrix} 1 & -2 & 1 & | & 4 \\ 0 & 7 & -5 & | & -9 \end{bmatrix}$$

$$\xrightarrow{R_2=\frac{1}{7}r_2} \begin{bmatrix} 1 & -2 & 1 & | & 4 \\ 0 & 1 & -\frac{5}{7} & | & -\frac{9}{7} \end{bmatrix} \xrightarrow{R_1=2r_2+r_1} \begin{bmatrix} 1 & 0 & -\frac{3}{7} & | & \frac{10}{7} \\ 0 & 1 & -\frac{5}{7} & | & -\frac{9}{7} \end{bmatrix}$$

The system has an infinite number of solutions. If we let z be the parameter, the solutions are $x = \frac{3}{7}z + \frac{10}{7}$ and $y = \frac{5}{7}z - \frac{9}{7}$ where z is any real number.

Sample solutions will vary. We find a sample solution by choosing a z-value and substituting it into the expressions for x and y.

29. We write the system as an augmented matrix and then use row operations to write it in reduced row-echelon form.

$$\begin{bmatrix} 1 & 2 & -1 & | & 5 \\ 2 & -1 & 2 & | & 0 \end{bmatrix} \xrightarrow{R_2=-2r_1+r_2} \begin{bmatrix} 1 & 2 & -1 & | & 5 \\ 0 & -5 & 4 & | & -10 \end{bmatrix} \xrightarrow{R_2=-\frac{1}{5}r_2} \begin{bmatrix} 1 & 2 & -1 & | & 5 \\ 0 & 1 & -\frac{4}{5} & | & 2 \end{bmatrix}$$

$$\xrightarrow{R_1=-2r_2+r_1} \begin{bmatrix} 1 & 0 & \frac{3}{5} & | & 1 \\ 0 & 1 & -\frac{4}{5} & | & 2 \end{bmatrix}$$

The system has an infinite number of solutions. If we let z be the parameter, the solutions are $x = -\frac{3}{5}z + 1$ and $y = \frac{4}{5}z + 2$ where z is any real number.

Sample solutions will vary. We find a sample solution by choosing a z-value and substituting it into the expressions for x and y.

31. We write the system as an augmented matrix and then use row operations to write it in reduced row-echelon form.

$$\begin{bmatrix} 2 & -1 & | & 6 \\ 1 & -2 & | & 0 \\ 3 & -1 & | & 6 \end{bmatrix} \xrightarrow[\text{rows 1 and 2}]{\text{Interchange}} \begin{bmatrix} 1 & -2 & | & 0 \\ 2 & -1 & | & 6 \\ 3 & -1 & | & 6 \end{bmatrix} \xrightarrow[R_3=-3r_1+r_3]{R_2=-2r_1+r_2} \begin{bmatrix} 1 & -2 & | & 0 \\ 0 & 3 & | & 6 \\ 0 & 5 & | & 6 \end{bmatrix} \xrightarrow{R_2=\frac{1}{3}r_2} \begin{bmatrix} 1 & -2 & | & 0 \\ 0 & 1 & | & 2 \\ 0 & 5 & | & 6 \end{bmatrix}$$

$$\xrightarrow{R_3=-5r_2+r_3} \begin{bmatrix} 1 & -2 & | & 0 \\ 0 & 1 & | & 2 \\ 0 & 0 & | & -4 \end{bmatrix}$$

Since equation (3) has no solution, the system is inconsistent.

33. The augmented matrix represents the system of equations

$$\begin{cases} x+4y+3z= \ 4 & \text{(1)} \\ \quad\quad y \quad\quad =-1 & \text{(2)} \\ \quad\quad\quad\quad z=-1 & \text{(3)} \end{cases}$$

We can read two of the solutions, $y=-1$ and $z=-1$ directly from the system. We obtain the third solution, x, by back-substituting y and z in equation (1) and solving for x.

$$x+4(-1)+3(-1)= \ 4$$
$$x-7= \ 4$$
$$x=11$$

The system has one solution; it is $x=11$, $y=-1$, and $z=-1$.

35. The augmented matrix represents the system of equations

$$\begin{cases} x_1 \quad\quad\quad +2x_4=1 & \text{(1)} \\ \quad x_2+x_3+2x_4=2 & \text{(2)} \\ \quad\quad\quad x_3 \quad\quad =3 & \text{(3)} \end{cases}$$

The system has an infinite number of solutions. If we let x_4 be the parameter, we find

$$x_3=3 \quad\quad\quad\quad\quad\quad\quad\quad \text{(3)}$$
$$x_2+3+2x_4=2 \quad \text{or } x_2=-2x_4-1 \quad \text{(2)}$$
$$\text{and} \quad\quad\quad\quad x_1=-2x_4+1 \quad\quad\quad\quad \text{(1)}$$

So the solutions are $x_1=-2x_4+1$, $x_2=-2x_4-1$, and $x_3=3$ where x_4 is any real number.

37. $A+C=\begin{bmatrix} 1 & 0 \\ 2 & 4 \\ -1 & 2 \end{bmatrix}+\begin{bmatrix} 3 & -4 \\ 1 & 5 \\ 5 & -2 \end{bmatrix}=\begin{bmatrix} 4 & -4 \\ 3 & 9 \\ 4 & 0 \end{bmatrix}$ The sum has dimension 3×2.

39. $6A=6\cdot\begin{bmatrix} 1 & 0 \\ 2 & 4 \\ -1 & 2 \end{bmatrix}=\begin{bmatrix} 6 & 0 \\ 12 & 24 \\ -6 & 12 \end{bmatrix}$ The scalar product has dimension 3×2.

41. $AB = \begin{bmatrix} 1 & 0 \\ 2 & 4 \\ -1 & 2 \end{bmatrix} \begin{bmatrix} 4 & -3 & 0 \\ 1 & 1 & -2 \end{bmatrix} = \begin{bmatrix} 4+0 & -3+0 & 0+0 \\ 8+4 & -6+4 & 0-8 \\ -4+2 & 3+2 & 0-4 \end{bmatrix} = \begin{bmatrix} 4 & -3 & 0 \\ 12 & -2 & -8 \\ -2 & 5 & -4 \end{bmatrix}$

The matrix product has dimension 3×3.

43. $CB = \begin{bmatrix} 3 & -4 \\ 1 & 5 \\ 5 & -2 \end{bmatrix} \begin{bmatrix} 4 & -3 & 0 \\ 1 & 1 & -2 \end{bmatrix} = \begin{bmatrix} 12-4 & -9-4 & 0+8 \\ 4+5 & -3+5 & 0-10 \\ 20-2 & -15-2 & 0+4 \end{bmatrix} = \begin{bmatrix} 8 & -13 & 8 \\ 9 & 2 & -10 \\ 18 & -17 & 4 \end{bmatrix}$

The matrix product has dimension 3×3.

45. $(A+C)B = \left(\begin{bmatrix} 1 & 0 \\ 2 & 4 \\ -1 & 2 \end{bmatrix} + \begin{bmatrix} 3 & -4 \\ 1 & 5 \\ 5 & -2 \end{bmatrix} \right) \cdot \begin{bmatrix} 4 & -3 & 0 \\ 1 & 1 & -2 \end{bmatrix} = \begin{bmatrix} 4 & -4 \\ 3 & 9 \\ 4 & 0 \end{bmatrix} \begin{bmatrix} 4 & -3 & 0 \\ 1 & 1 & -2 \end{bmatrix}$

$= \begin{bmatrix} 16-4 & -12-4 & 0+8 \\ 12+9 & -9+9 & 0-18 \\ 16+0 & -12+0 & 0+0 \end{bmatrix} = \begin{bmatrix} 12 & -16 & 8 \\ 21 & 0 & -18 \\ 16 & -12 & 0 \end{bmatrix}$

The resulting matrix has dimension 3×3.

47. $A^{\mathrm{T}} = \begin{bmatrix} 1 & 0 \\ 2 & 4 \\ -1 & 2 \end{bmatrix}^{\mathrm{T}} = \begin{bmatrix} 1 & 2 & -1 \\ 0 & 4 & 2 \end{bmatrix}$ The transpose of A has dimension 2×3.

49. $C + \mathbf{0} = \begin{bmatrix} 3 & -4 \\ 1 & 5 \\ 5 & -2 \end{bmatrix} + \begin{bmatrix} 0 & 0 \\ 0 & 0 \\ 0 & 0 \end{bmatrix} = \begin{bmatrix} 3 & -4 \\ 1 & 5 \\ 5 & -2 \end{bmatrix} = C$ The sum has dimension 3×2.

51. The inverse of a 2×2 matrix, $\begin{bmatrix} a & b \\ c & d \end{bmatrix}$ is given by $\begin{bmatrix} \dfrac{d}{ad-bc} & -\dfrac{b}{ad-bc} \\ -\dfrac{c}{ad-bc} & \dfrac{a}{ad-bc} \end{bmatrix}$ provided

$ad - bc \neq 0$. For the matrix $\begin{bmatrix} 3 & 0 \\ -2 & 1 \end{bmatrix}$ $ad - bc = (3)(1) - (0)(-2) = 3$. So the inverse is

given by $\begin{bmatrix} \dfrac{1}{3} & 0 \\ \dfrac{2}{3} & 1 \end{bmatrix}$.

53. The inverse of a 2 × 2 matrix, $\begin{bmatrix} a & b \\ c & d \end{bmatrix}$ is given by $\begin{bmatrix} \dfrac{d}{ad-bc} & -\dfrac{b}{ad-bc} \\ -\dfrac{c}{ad-bc} & \dfrac{a}{ad-bc} \end{bmatrix}$ provided

$ad - bc \neq 0$. For the matrix $\begin{bmatrix} 4 & 2 \\ 6 & 3 \end{bmatrix}$ $ad - bc = (4)(3) - (2)(6) = 0$. So the inverse does not exist.

55. To find the inverse of a 3 × 3 matrix we augment it with I_3 and use row operations to write the matrix in reduced row-echelon form.

$$\left[\begin{array}{ccc|ccc} 4 & 3 & -1 & 1 & 0 & 0 \\ 0 & 2 & 2 & 0 & 1 & 0 \\ 3 & -1 & 0 & 0 & 0 & 1 \end{array}\right] \xrightarrow{R_1 = r_1 - r_3} \left[\begin{array}{ccc|ccc} 1 & 4 & -1 & 1 & 0 & -1 \\ 0 & 2 & 2 & 0 & 1 & 0 \\ 3 & -1 & 0 & 0 & 0 & 1 \end{array}\right]$$

$$\xrightarrow{R_3 = -3r_1 + r_3} \left[\begin{array}{ccc|ccc} 1 & 4 & -1 & 1 & 0 & -1 \\ 0 & 2 & 2 & 0 & 1 & 0 \\ 0 & -13 & 3 & -3 & 0 & 4 \end{array}\right] \xrightarrow[\substack{R_1 = -4r_2 + r_1 \\ R_3 = 13r_2 + r_3}]{R_2 = \frac{1}{2}r_2} \left[\begin{array}{ccc|ccc} 1 & 0 & -5 & 1 & -2 & -1 \\ 0 & 1 & 1 & 0 & \frac{1}{2} & 0 \\ 0 & 0 & 16 & -3 & \frac{13}{2} & 4 \end{array}\right]$$

$$\xrightarrow{R_3 = \frac{1}{16}r_3} \left[\begin{array}{ccc|ccc} 1 & 0 & -5 & 1 & -2 & -1 \\ 0 & 1 & 1 & 0 & \frac{1}{2} & 0 \\ 0 & 0 & 1 & -\frac{3}{16} & \frac{13}{32} & \frac{1}{4} \end{array}\right] \xrightarrow[\substack{R_2 = -r_3 + r_2}]{R_1 = 5r_3 + r_1} \left[\begin{array}{ccc|ccc} 1 & 0 & 0 & \frac{1}{16} & \frac{1}{32} & \frac{1}{4} \\ 0 & 1 & 0 & \frac{3}{16} & \frac{3}{32} & -\frac{1}{4} \\ 0 & 0 & 1 & -\frac{3}{16} & \frac{13}{32} & \frac{1}{4} \end{array}\right]$$

Since the left side is now I_3, the right side of the matrix is the inverse.

$$\begin{bmatrix} 4 & 3 & -1 \\ 0 & 2 & 2 \\ 3 & -1 & 0 \end{bmatrix}^{-1} = \begin{bmatrix} \frac{1}{16} & \frac{1}{32} & \frac{1}{4} \\ \frac{3}{16} & \frac{3}{32} & -\frac{1}{4} \\ -\frac{3}{16} & \frac{13}{32} & \frac{1}{4} \end{bmatrix}$$

57. We want $AB = BA$ when $A = \begin{bmatrix} x & y \\ z & w \end{bmatrix}$ and $B = \begin{bmatrix} 1 & 1 \\ -1 & 1 \end{bmatrix}$.

$$AB = \begin{bmatrix} x & y \\ z & w \end{bmatrix} \begin{bmatrix} 1 & 1 \\ -1 & 1 \end{bmatrix} = \begin{bmatrix} x-y & x+y \\ z-w & z+w \end{bmatrix} \text{ and } BA = \begin{bmatrix} 1 & 1 \\ -1 & 1 \end{bmatrix} \begin{bmatrix} x & y \\ z & w \end{bmatrix} = \begin{bmatrix} x+z & y+w \\ -x+z & -y+w \end{bmatrix}$$

If the two products are equal then $\begin{bmatrix} x-y & x+y \\ z-w & z+w \end{bmatrix}$ and $\begin{bmatrix} x+z & y+w \\ -x+z & -y+w \end{bmatrix}$ must be equal.

These 2 matrices are equal when

$$\begin{array}{llll} x-y=x+z & x+y=y+w & z-w=-x+z & \text{and} \quad z+w=-y+w \\ \quad -y=z & \quad x=w & \quad w=x & \quad\quad z=-y \end{array}$$

So for $AB = BA$, x must equal z and $y = -z$.

59. Let x represent the number of caramels in the box and y represent the number of creams in the box. To find out how many of each kind of candy to package, we solve the system of equations

$$\begin{cases} x + y = 50 & \text{(1)} \\ 0.05x + 0.10y = 4.00 & \text{(2)} \end{cases}$$

We will solve the system using substitution. First we solve (1) for y, and substitute its value into equation (2).

$$y = 50 - x \quad \text{(1);} \qquad \begin{aligned} 0.05x + 0.10(50 - x) &= 4.00 \quad \text{(2)} \\ 0.05x + 5 - 0.10x &= 4.00 \\ -0.05x &= -1 \\ x &= 20 \end{aligned}$$

Sweet Delight Candies, Inc should pack 20 caramels and 30 creams in each box if they want no profit and no loss.

If they want to increase profit while keeping 50 pieces of candy in the box, Sweet Delight Candies Inc. should increase the number of caramels (and decrease the number of creams) in each box.

61. Let x represent the amount of almonds needed, y represent the amount of cashews needed, and z represent the amount of peanuts needed to make the 100 bags of nuts.
To determine how many pounds of each type of nut the store needs, we solve the system

$$\begin{cases} x + y + z = 100 & \text{(1)} \\ 6x + 5y + 2z = 4.00(100) & \text{(2)} \end{cases}$$

We form an augmented matrix and use row operations to write it in reduced row-echelon form.

$$\begin{bmatrix} 1 & 1 & 1 & | & 100 \\ 6 & 5 & 2 & | & 400 \end{bmatrix} \xrightarrow{R_2 = -2r_1 + r_2} \begin{bmatrix} 1 & 1 & 1 & | & 100 \\ 0 & -1 & -4 & | & -200 \end{bmatrix} \xrightarrow{R_2 = -r_2} \begin{bmatrix} 1 & 1 & 1 & | & 100 \\ 0 & 1 & 4 & | & 200 \end{bmatrix}$$

$$\xrightarrow{R_1 = -r_2 + r_1} \begin{bmatrix} 1 & 0 & -3 & | & -100 \\ 0 & 1 & 4 & | & 200 \end{bmatrix}$$

There are an infinite number of solutions to this system. If z is the parameter, then $x = -3z - 100$ and $y = -4z + 200$ where z is a real number. The physical constraints of this problem limit z. x, y, and z all need to be nonnegative. $y = -4z + 200 \geq 0$ or $z \leq 50$
Possible combinations of the nuts that can be packaged include:

Almonds (pounds)	Cashews (pounds)	Peanuts (pounds)
5	60	35
20	40	40
35	20	45
50	0	50

63. Let x represent the money invested in treasury bills, y represent the money invested in corporate bonds, and z represent the money invested in junk bonds.

The couple invests \$40,000. We find how they should allocate their funds to meet their investment goal by solving a system of equations. We solve the system by writing an augmented matrix which we put into reduced row-echelon form. Before writing the matrix we multiplied equation (2) by 100 to remove decimal points.

(a) The couple require \$2500 in investment income.

$$\begin{cases} x + y + z = 40,000 & (1) \\ 0.06x + 0.08y + 0.10z = 2,500 & (2) \end{cases}$$

$$\begin{bmatrix} 1 & 1 & 1 & | & 40000 \\ 6 & 8 & 10 & | & 250000 \end{bmatrix} \xrightarrow{R_2=-6r_1+r_2} \begin{bmatrix} 1 & 1 & 1 & | & 40000 \\ 0 & 2 & 4 & | & 10000 \end{bmatrix} \xrightarrow{R_2=\frac{1}{2}r_2} \begin{bmatrix} 1 & 1 & 1 & | & 40000 \\ 0 & 1 & 2 & | & 5000 \end{bmatrix}$$

$$\xrightarrow{R_1=-r_2+r_1} \begin{bmatrix} 1 & 0 & -1 & | & 35000 \\ 0 & 1 & 2 & | & 5000 \end{bmatrix}$$

We find that $x = z + 35,000$ and $y = -2z + 5000$. Since x and y must be nonnegative, $z + 35,000 \geq 0$ or $z \geq 0$ and $-2z + 5000 \geq 0$ or $z \leq 2500$.

Possible investments allocations available to the couple include

Treasury Bills	Corporate Bonds	Junk Bonds
\$35,000	\$5000	0
\$36,000	\$3000	\$1000
\$36,500	\$2000	\$1500
\$37,000	\$1000	\$2000
\$37,500	0	\$2500

(b) The couple require \$3000 in investment income.

$$\begin{cases} x + y + z = 40,000 & (1) \\ 0.06x + 0.08y + 0.10z = 3,000 & (2) \end{cases}$$

$$\begin{bmatrix} 1 & 1 & 1 & | & 40000 \\ 6 & 8 & 10 & | & 300000 \end{bmatrix} \xrightarrow{R_2=-6r_1+r_2} \begin{bmatrix} 1 & 1 & 1 & | & 40000 \\ 0 & 2 & 4 & | & 60000 \end{bmatrix} \xrightarrow{R_2=\frac{1}{2}r_2} \begin{bmatrix} 1 & 1 & 1 & | & 40000 \\ 0 & 1 & 2 & | & 30000 \end{bmatrix}$$

$$\xrightarrow{R_1=-r_2+r_1} \begin{bmatrix} 1 & 0 & -1 & | & 10000 \\ 0 & 1 & 2 & | & 30000 \end{bmatrix}$$

We find that $x = z + 10,000$ and $y = -2z + 30,000$. Since x and y must be nonnegative, $z + 10,000 \geq 0$ or $z \geq 0$ and $-2z + 30,000 \geq 0$ or $z \leq 15,000$.

Possible investments allocations available to the couple include

Treasury Bills	Corporate Bonds	Junk Bonds
$10,000	$30,000	0
$15,000	$20,000	$5000
$20,000	$10,000	$10,000
$22,000	$6,000	$12,000
$25,000	0	$15,000

(c)   The couple require \$3500 in investment income.

$$\begin{cases} x + y + z = 40,000 \quad \text{(1)} \\ 0.06x + 0.08y + 0.10z = 3,500 \quad \text{(2)} \end{cases}$$

$$\begin{bmatrix} 1 & 1 & 1 & | & 40000 \\ 6 & 8 & 10 & | & 350000 \end{bmatrix} \xrightarrow{R_2 = -6r_1 + r_2} \begin{bmatrix} 1 & 1 & 1 & | & 40000 \\ 0 & 2 & 4 & | & 110000 \end{bmatrix} \xrightarrow{R_2 = \frac{1}{2}r_2} \begin{bmatrix} 1 & 1 & 1 & | & 40000 \\ 0 & 1 & 2 & | & 55000 \end{bmatrix}$$

$$\xrightarrow{R_1 = -r_2 + r_1} \begin{bmatrix} 1 & 0 & -1 & | & -15000 \\ 0 & 1 & 2 & | & 55000 \end{bmatrix}$$

We find that $x = z - 15,000$ and $y = -2z + 55,000$. Since x and y must be nonnegative, $z - 15,000 \geq 0$ or $z \geq 15,000$ and $-2z + 55,000 \geq 0$ or $z \leq 27,500$.

Possible investments allocations available to the couple include

Treasury Bills	Corporate Bonds	Junk Bonds
0	$25,000	$15,000
$5,000	$15,000	$20,000
$7,500	$10,000	$22,500
$10,000	$5,000	$25,000
$12,500	0	$27,500

65. (a)   The scatter diagram representing U.S. emigration between the years 1991 and 2001 is

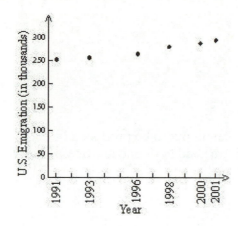

(b)   We define $A = \begin{bmatrix} 1 & 1 \\ 3 & 1 \\ 6 & 1 \\ 8 & 1 \\ 10 & 1 \\ 11 & 1 \end{bmatrix}$, $Y = \begin{bmatrix} 252 \\ 258 \\ 267 \\ 278 \\ 287 \\ 293 \end{bmatrix}$ (expressed in thousands), and $X = \begin{bmatrix} m \\ b \end{bmatrix}$ and use these matrices to form the equation $A^{T}AX = A^{T}Y$.

$$\begin{bmatrix} 1 & 3 & 6 & 8 & 10 & 11 \\ 1 & 1 & 1 & 1 & 1 & 1 \end{bmatrix} \begin{bmatrix} 1 & 1 \\ 3 & 1 \\ 6 & 1 \\ 8 & 1 \\ 10 & 1 \\ 11 & 1 \end{bmatrix} \begin{bmatrix} m \\ b \end{bmatrix} = \begin{bmatrix} 1 & 3 & 6 & 8 & 10 & 11 \\ 1 & 1 & 1 & 1 & 1 & 1 \end{bmatrix} \begin{bmatrix} 252 \\ 258 \\ 267 \\ 278 \\ 287 \\ 293 \end{bmatrix}$$ which simplifies

to $\begin{bmatrix} 331 & 39 \\ 39 & 6 \end{bmatrix} \begin{bmatrix} m \\ b \end{bmatrix} = \begin{bmatrix} 10945 \\ 1635 \end{bmatrix}$. Solving the system of equations for X,

$\begin{bmatrix} 331 & 39 & | & 10945 \\ 39 & 6 & | & 1635 \end{bmatrix} \rightarrow \begin{bmatrix} 1 & 0 & | & 4.09677 \\ 0 & 1 & | & 245.87097 \end{bmatrix}$. We get $m = 4.09677$ and $b = 245.87097$ (both in thousands). So the least squares line of best fit for these data is $y = 4096.77x + 245{,}870.97$.

(c)   The slope is the average annual change in number of persons emigrating from the United States from 1991 through 2001.

(d)   2005 is year 15, so we evaluate $y = 4096.77x + 245{,}870.97$ when $x = 15$. An estimated 307,323 persons will emigrate from the United States in 2005.

(e)   1995 is year 5, so we evaluate $y = 4096.77x + 245{,}870.97$ when $x = 5$. An estimated 266,355 persons emigrated from the United States in 1995.

(f)   On a TI – 83 graphing utility, the line of best fit is given by

```
LinReg
 y=ax+b
 a=4.096774194
 b=245.8709677
 r²=.9842798384
 r=.9921087836
```

Both equations are the same.

67.   This is an open Leontief system. To find the total output to be produced by each corporation to meet predicted future demand, we need to define matrices A and D_5 and to solve the equation $X = [I_3 - A]^{-1} D_5$.

$$A = \begin{bmatrix} \dfrac{100}{250} & \dfrac{50}{100} & \dfrac{40}{200} \\ \dfrac{20}{250} & \dfrac{10}{100} & \dfrac{30}{200} \\ \dfrac{30}{250} & \dfrac{40}{100} & \dfrac{30}{200} \end{bmatrix} = \begin{bmatrix} 0.4 & 0.5 & 0.2 \\ 0.08 & 0.1 & 0.15 \\ 0.12 & 0.4 & 0.15 \end{bmatrix} ; D_5 = \begin{bmatrix} 80 \\ 40 \\ 80 \end{bmatrix}$$

$$X = [I_3 - A]^{-1} D_5 = \begin{bmatrix} 0.6 & -0.5 & -0.2 \\ -0.08 & 0.9 & -0.15 \\ -0.12 & -0.4 & 0.85 \end{bmatrix}^{-1} \begin{bmatrix} 80 \\ 40 \\ 80 \end{bmatrix} = \begin{bmatrix} 275.57 \\ 98.86 \\ 179.55 \end{bmatrix}$$

To meet the predicted future demand 275.57 units of A, 98.86 units of B, and 179.55 units of C should be produced.

69. The total costs for the 2 service departments come from the two first lines of the table. They are given by the system of equations $\begin{cases} x_1 = 800 + 0.2x_1 + 0.1x_2 \\ x_2 = 4000 + 0.1x_1 + 0.3x_2 \end{cases}$

We define the matrices $X = \begin{bmatrix} x_1 \\ x_2 \end{bmatrix}$, $C = \begin{bmatrix} 0.2 & 0.1 \\ 0.1 & 0.3 \end{bmatrix}$, and $D = \begin{bmatrix} 800 \\ 4000 \end{bmatrix}$. We can use the matrices to express the system of equations as $X = D + CX$ and solve for X.

$$X = [I_2 - C]^{-1}D = \begin{bmatrix} 0.8 & -0.1 \\ -0.1 & 0.7 \end{bmatrix}^{-1} \begin{bmatrix} 800 \\ 4000 \end{bmatrix} = \begin{bmatrix} \dfrac{70}{55} & \dfrac{10}{55} \\ \dfrac{10}{55} & \dfrac{80}{55} \end{bmatrix} \begin{bmatrix} 800 \\ 4000 \end{bmatrix} = \begin{bmatrix} 1745.4545 \\ 5963.6363 \end{bmatrix}$$

So we get $x_1 = \$1745.45$ and $x_2 = \$5963.64$.
The table gives all direct and indirect costs.

Dept.	Total Costs	Direct Costs	Indirect Costs S_1	Indirect Costs S_2
S_1	$1745.45	$800	$349.09	$596.36
S_2	$5963.64	$4000	$174.55	$1789.09
P_1	$2445.45	$1500	$349.09	$596.36
P_2	$2216.37	$500	$523.64	$1192.73
P_3	$3338.18	$1200	$349.09	$1789.09
Totals:	$15709.09	$8000	$1745.46	$5963.63

Chapter 2 Project

1. Matrix A represents the amounts of materials needed to produce 1 unit of a product. The entries in the matrix are quotients. Column 1 is found by dividing each entry in the agriculture column of the table by total gross output of agriculture. Column 2 entries are the quotients found by dividing each entry in the manufacturing column by total gross output of manufacturing, and column 3 entries are the quotients found by dividing entries by total gross outcome of services. D is the consumer demand. It is labeled Open Sector in Table 1.

$$A = \begin{bmatrix} 0.410 & 0.030 & 0.026 \\ 0.062 & 0.378 & 0.105 \\ 0.124 & 0.159 & 0.192 \end{bmatrix} \qquad D = \begin{bmatrix} 39.24 \\ 60.02 \\ 130.65 \end{bmatrix}$$

3. To calculate the new X_1, we need to solve the equation $X_1 = [I_3 - A]^{-1}D_1$

$$X_1 = \begin{bmatrix} 1.720 & 0.100 & 0.068 \\ 0.223 & 1.676 & 0.225 \\ 0.308 & 0.345 & 1.292 \end{bmatrix} \begin{bmatrix} 40.24 \\ 60.02 \\ 130.65 \end{bmatrix} = \begin{bmatrix} 84.099 \\ 138.963 \\ 201.901 \end{bmatrix}$$

5. Using the interpretation from Problem 4, we find that service production would need to increase by 0.308 unit for a one unit increase in agriculture.

7. $A =$
$$\begin{bmatrix} 0.2424 & 0.0005 & 0.0058 & 0.0366 & 0.0001 & 0.0012 & 0.0045 & 0.0354 & 0.0005 \\ 0.0013 & 0.2131 & 0.0073 & 0.0207 & 0.0419 & 0.0000 & 0.0000 & 0.0000 & 0.0025 \\ 0.049 & 0.0318 & .0009 & 0.0073 & 0.0379 & 0.0082 & 0.0261 & 0.0083 & 0.0214 \\ 0.1744 & 0.0982 & 0.2976 & 0.3492 & 0.0564 & 0.0438 & 0.0076 & 0.0980 & 0.0145 \\ 0.0446 & 0.0856 & 0.0247 & 0.0455 & 0.1607 & 0.0439 & 0.0207 & 0.0347 & 0.0189 \\ 0.0492 & 0.0237 & 0.0812 & 0.0583 & 0.0121 & 0.0211 & 0.0019 & 0.0196 & 0.0022 \\ 0.0729 & 0.2251 & 0.0164 & 0.0180 & 0.0322 & 0.0698 & 0.1749 & 0.0701 & 0.0066 \\ 0.0318 & 0.0396 & 0.1031 & 0.0607 & 0.1156 & 0.1412 & 0.0751 & 0.1526 & 0.0112 \\ 0.0006 & 0.0002 & 0.0011 & 0.0035 & 0.0026 & 0.0072 & 0.0111 & 0.0071 & 0.0025 \end{bmatrix}$$

$$D_0 = \begin{bmatrix} 34940 \\ -39241 \\ 787208 \\ 1611520 \\ 586248 \\ 1103110 \\ 1520718 \\ 2214382 \\ 1032052 \end{bmatrix}$$

9. $D_1 = \begin{bmatrix} 34940 \\ -39241 \\ 787209 \\ 1611520 \\ 586248 \\ 1103110 \\ 1520718 \\ 2214382 \\ 1032052 \end{bmatrix}$ $X_1 = [I_9 - A]^{-1}D_1 = \begin{bmatrix} 434735.5291 \\ 135766.6225 \\ 1009114.196 \\ 3906797.921 \\ 1300135.743 \\ 1566308.417 \\ 2531704.148 \\ 3715940.758 \\ 119064.343 \end{bmatrix}$

If demand for construction increases by 1 million dollars, demand for transportation, communication, and utilities increases by $80,000.00

1300135.74–1300135.66 = 0.08

11.   An increase of $1 million in demand for construction, produces an composite increase of $2.1 million in the economy.

Mathematical Questions From Professional Exams

1.   (b)   **3.**   (d)

Chapter 3

Linear Programming: Geometric Approach

3.1 Linear Inequalities

1. The corresponding linear equation is $x = 0$. We graph a solid line, since the inequality is nonstrict. The test point $(1, 1)$ satisfies the inequality, so we shade the region containing it.

$x \geq 0$
$1 > 0$

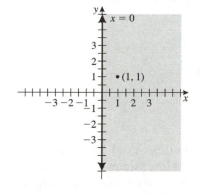

3. The corresponding linear equation is $x = 4$. We graph a dashed line, since the inequality is strict. The test point $(0, 0)$ satisfies the inequality, so we shade the region containing it.

$x < 4$
$0 < 4$

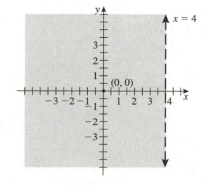

5. The corresponding linear equation is $y = 1$. We graph a solid line, since the inequality is non-strict. The test point $(0, 0)$ does not satisfy the inequality, so we shade the region opposite it.

$y \geq 1$
$0 < 1$

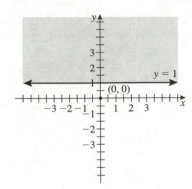

7. The corresponding linear equation is $2x + 3y = 6$. We graph a solid line, since the inequality is nonstrict. The test point $(0, 0)$ satisfies the inequality, so we shade the region containing it.

$$2x + 3y \leq 6$$
$$2 \cdot 0 + 3 \cdot 0 = 0$$
$$0 < 6$$

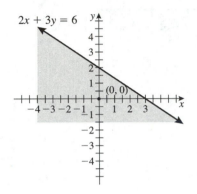

9. The corresponding linear equation is $5x + y = 10$. We graph a solid line, since the inequality is nonstrict. The test point $(0, 0)$ does not satisfy the inequality, so we shade the region opposite it.

$$5x + y \geq 10$$
$$5 \cdot 0 + 0 = 0$$
$$0 < 10$$

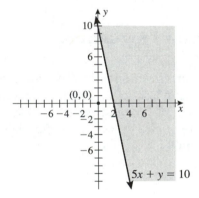

11. The corresponding linear equation is $x + 5y = 5$. We graph a dashed line, since the inequality is strict. The test point $(0, 0)$ satisfies the inequality, so we shade the region containing it.

$$x + 5y < 5$$
$$0 + 5 \cdot 0 = 0$$
$$0 < 5$$

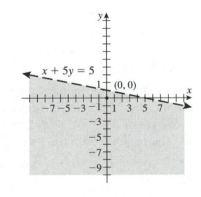

13. To determine which points are part of the graph of the system, check each point to see if it satisfies all of the inequalities:

Point	Inequality 1	Inequality 2	Conclusion
$P_1 = (3, 8)$	Is $3 + 3 \cdot 8 \geq 0$? $27 > 0$	Is $-3 \cdot 3 + 2 \cdot 8 \geq 0$? $7 > 0$	Both inequalities are satisfied; $(3, 8)$ is part of the graph.
$P_2 = (12, 9)$	Is $12 + 3 \cdot 9 \geq 0$? $39 > 0$	Is $-3 \cdot 12 + 2 \cdot 9 \geq 0$? $-18 < 0$	Since the second inequality was not satisfied, $(12, 9)$ is not part of the graph.
$P_3 = (5, 1)$	Is $5 + 3 \cdot 1 \geq 0$? $8 > 0$	Is $-3 \cdot 5 + 2 \cdot 1 \geq 0$? $-13 < 0$	Since the second inequality was not satisfied, $(5, 1)$ is not part of the graph.

15. To determine which points are part of the graph of the system, check each point to see if it satisfies all of the inequalities:

$P_1 = (2, 3)$	Is $3 \cdot 2 + 2 \cdot 3 \geq 0$? $12 > 0$	Is $2 + 3 \leq 15$? $5 < 15$	Both inequalities are satisfied; $(2, 3)$ is part of the graph.
$P_2 = (10, 10)$	Is $3 \cdot 10 + 2 \cdot 10 \geq 0$? $50 > 0$	Is $10 + 10 \leq 15$? $20 > 15$	Since the second inequality was not satisfied, $(10, 10)$ is not part of the graph.
$P_3 = (5, 1)$	Is $3 \cdot 5 + 2 \cdot 1 \geq 0$? $17 > 0$	Is $5 + 1 \leq 15$? $6 < 15$	Both inequalities are satisfied; $(2, 3)$ is part of the graph.

17. The region that represents the graph of the system is the set of points common to the solutions of each individual inequality. Use the test point $(0, 0)$.

$$5x - 4y \leq 8 \qquad 2x + 5y \leq 23$$
$$5 \cdot 0 - 4 \cdot 0 = 0 \qquad 2 \cdot 0 + 5 \cdot 0 = 0$$
$$0 < 8 \qquad\qquad 0 < 23$$

The test point $(0, 0)$ satisfies both inequalities, so the region (**b**) represents the graph of the system.

19. The region that represents the graph of the system is the set of points common to the solutions of each individual inequality. Use the test point $(0, 0)$.

$$2x - 3y \geq -3 \qquad 2x + 3y \leq 16$$
$$2 \cdot 0 - 3 \cdot 0 = 0 \qquad 2 \cdot 0 + 6 \cdot 0 = 0$$
$$0 > -3 \qquad\qquad 0 < 16$$

The test point $(0, 0)$ satisfies both inequalities, so the region (**c**) represents the graph of the system.

21. The region that represents the graph of the system is the set of points common to the solutions of each individual inequality. Use the test point $(0, 0)$.

$$5x - 3y \geq 3 \qquad 2x + 6y \geq 30$$
$$5 \cdot 0 - 3 \cdot 0 = 0 \qquad 2 \cdot 0 + 6 \cdot 0 = 0$$
$$0 < 3 \qquad\qquad 0 < 30$$

The test point $(0, 0)$ satisfies neither inequality, so the region representing each inequality is on the opposite side of the line from $(0, 0)$. This means (**d**) represents the graph of the system.

23. The region that represents the graph of the system is the set of points common to the solutions of each individual inequality. Use the test point $(0, 2)$.

$$5x - 4y \leq 0 \qquad 2x + 4y \leq 28$$
$$5 \cdot 0 - 5 \cdot 2 = -10 \qquad 2 \cdot 0 + 4 \cdot 2 = 8$$
$$-10 < 0 \qquad\qquad 8 < 28$$

The test point $(0, 2)$ satisfies both inequalities, so the region (**c**) represents the graph of the system.

25. Graph is unbounded.
Corner points: $(2, 0), (0, 2)$

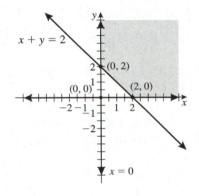

27. Graph is bounded.
Corner points: $(2, 0), (3, 0), (0, 2)$

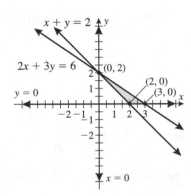

29. Graph is bounded.

Corner points: $(2, 0)$, $(5, 0)$, $(0, 2)$, $(0, 8)$, $(x_0, y_0) = (2, 6)$

To find the last corner point, solve the system of equations:

$$\begin{cases} 2x + y = 10 & (1) \\ x + y = 8 & (2) \end{cases}$$

We will use elimination. Subtracting (2) from (1) gives $x = 2$. Back substituting in (2), you get $y = 6$.
These give the fifth corner (x_0, y_0).

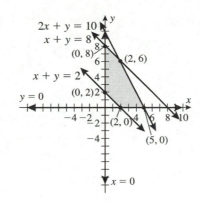

31. Graph is bounded.

Corner points: $(2, 0)$, $(4, 0)$, $(0, 2)$, $(0, 4)$, $(x_0, y_0) = \left(\dfrac{24}{7}, \dfrac{12}{7} \right)$

To find (x_0, y_0), find the intersection of lines

$$\begin{cases} 3x + y = 12 & (1) \\ 2x + 3y = 12 & (2) \end{cases}$$

Using the substitution method we solve (1) for y; $y = -3x + 12$, and substitute $-3x + 12$ for y in equation (2).

$$\begin{aligned} 2x + 3(-3x + 12) &= 12 \quad (2) \\ 2x - 9x + 36 &= 12 \\ -7x &= -24 \\ x &= \frac{24}{7} \end{aligned}$$

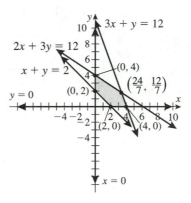

Back substituting into (2) we get

$$y = -3 \left(\frac{24}{7} \right) + 12 \quad (1)$$

$$y = -\frac{72}{7} + \frac{84}{7}$$

$$y = \frac{12}{7}$$

These give the last corner point (x_0, y_0).

33. Graph is unbounded.

Corner points: $(1, 0), \left(0, \dfrac{1}{2}\right)$, and $(0, 4)$

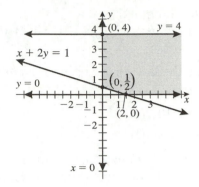

35. Graph is bounded.

Corner points: $(0, 1), (0, 3)$, and

$$(x_0, y_0) = \left(\dfrac{4}{5}, \dfrac{3}{5}\right)$$

To find (x_0, y_0), find the intersection of

$$\begin{cases} x + 2y = 2 & (1) \\ 3x + y = 3 & (2) \end{cases}$$

Using the substitution method, we solve (1) for x and substitute the result in (2)

$$\begin{aligned} x + 2y &= 2 \qquad (1) \\ x &= 2 - 2y \\ 3(2 - 2y) + y &= 3 \qquad (2) \\ 6 - 6y + y &= 3 \\ -5y &= -3 \\ y &= \dfrac{3}{5} \end{aligned}$$

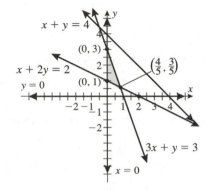

Back substituting $\dfrac{3}{5}$ for y in (1) gives:

$$\begin{aligned} x + 2\left(\dfrac{3}{5}\right) &= 2 \qquad (1) \\ x + \dfrac{6}{5} &= 2 \\ x &= \dfrac{4}{5} \end{aligned}$$

37. As in Example 11, we let x = the number of 1 pound packages of low-grade mix, and
y = the number of packages of high-grade mix to be prepared.
There are 60 pounds (960 ounces) of cashews and 90 pounds (1440 ounces) of peanuts
available.

The system of inequalities representing the system is:

$$
\begin{cases}
4x + 8y \le 960 & (1) \\
12x + 8y \le 1440 & (2) \\
x \ge 0 & (3) \\
y \ge 0 & (4)
\end{cases}
\quad \text{or} \quad
\begin{cases}
x + 2y \le 240 & (1) \\
3x + 2y \le 360 & (2) \\
x \ge 0 & (3) \\
y \ge 0 & (4)
\end{cases}
$$

From the graph we see that three of the
corner points are $(0, 0)$, $(120, 0)$, and $(0, 120)$.
The fourth corner $(60, 90)$, is found by
solving the system of equations

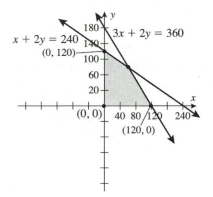

$$
\begin{cases}
x + 2y = 240 & (1) \\
3x + 2y = 360 & (2)
\end{cases}
$$

Using the elimination method, we subtract
(1) from (2) and get $2x = 120$ or $x = 60$. Back
substituting into (1) we find

$$
\begin{aligned}
60 + 2y &= 240 \quad (1) \\
2y &= 180 \\
y &= 90
\end{aligned}
$$

39. First we note that meaningful values for x
and y are nonnegative values.

Next we note that there are only limited
numbers of grinding and finishing hours.
There are

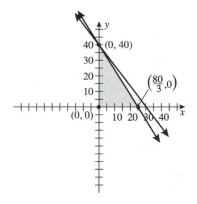

$80 = 2 \cdot 40$ grinding hours available, and
$120 = 3 \cdot 40$ finishing hours available.

Using these constraints, the system is:

$$
\begin{cases}
3x + 2y \le 80 & (1) \\
4x + 3y \le 120 & (2) \\
x \ge 0 & (3) \\
y \ge 0 & (4)
\end{cases}
$$

Corner points: $(0, 0)$, $\left(\dfrac{80}{3}, 0\right)$, and $(0, 40)$

41. (a) First we note that meaningful values of x and y are nonnegative. Then we see that there is at most $25,000 to be invested. Using these constraints, the system is

$$\begin{cases} x + y \le 25,000 \\ x \ge 15,000 \\ y \le 10,000 \\ x \ge 0 \\ y \ge 0 \end{cases}$$

(b)

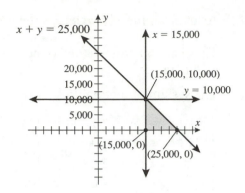

Corner points: (15000, 10000), (15000, 0), and (25000, 0)

(c) The x-value of each corner point represents the amount in dollars to be invested in Treasury Bills, the y-value is the amount in dollars to be invested in corporate bonds.

43. (a) Let x = the units of grain 1 used, and y = the units of grain 2 used.

$$\begin{cases} x + 2y \ge 5 \\ 5x + y \ge 16 \\ x \ge 0 \\ y \ge 0 \end{cases}$$

(b) Corner Points: (5, 0), (3, 1), and (0, 16).

(b)

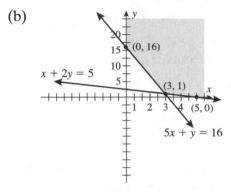

45. (a) Let x = amount of food A to be eaten, and let y = amount of food B to be eaten.

$$\begin{cases} 5x + 4y \ge 85 \\ 3x + 3y \ge 70 \\ 2x + 3y \ge 50 \\ x \ge 0 \\ y \ge 0 \end{cases}$$

(b) Corner Points: $\left(0, \dfrac{70}{3}\right)$, $\left(20, \dfrac{10}{3}\right)$, and (25, 0).

(b)

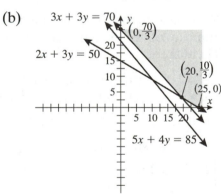

47. Answers will vary.

3.2 A Geometric Approach to Linear Programming Problems

1. Evaluate the objective function at each corner point; choose the maximum and minimum values.

   $z = 2x + 3y$

Point: $(2, 2)$	Point: $(8, 1)$	Point: $(2, 7)$	Point: $(7, 8)$
$z = 2(2) + 3(2)$	$z = 2(8) + 3(1)$	$z = 2(2) + 3(7)$	$z = 2(7) + 3(8)$
$z = 10$	$z = 19$	$z = 25$	$z = 38$

   Maximum: $z = 38$; Minimum: $z = 10$

3. Evaluate the objective function at each corner point; choose the maximum and minimum values.

   $z = x + y$

Point: $(2, 2)$	Point: $(8, 1)$	Point: $(2, 7)$	Point: $(7, 8)$
$z = 2 + 2$	$z = 8 + 1$	$z = 2 + 7$	$z = 7 + 8$
$z = 4$	$z = 9$	$z = 9$	$z = 15$

   Maximum: $z = 15$; Minimum: $z = 4$

5. Evaluate the objective function at each corner point; choose the maximum and minimum values.

   $z = x + 6y$

Point: $(2, 2)$	Point: $(8, 1)$	Point: $(2, 7)$	Point: $(7, 8)$
$z = 2 + 6(2)$	$z = 8 + 6(1)$	$z = 2 + 6(7)$	$z = 7 + 6(8)$
$z = 14$	$z = 14$	$z = 44$	$z = 55$

   Maximum: $z = 55$; Minimum: $z = 14$, all feasible points on line containing the points $(2, 2)$ and $(8, 1)$ will give the minimum value of z.

7. Evaluate the objective function at each corner point; choose the maximum and minimum values.

   $z = 3x + 4y$

Point: $(2, 2)$	Point: $(8, 1)$	Point: $(2, 7)$	Point: $(7, 8)$
$z = 3(2) + 4(2)$	$z = 3(8) + 4(1)$	$z = 3(2) + 4(7)$	$z = 3(7) + 4(8)$
$z = 14$	$z = 28$	$z = 34$	$z = 53$

   Maximum: $z = 53$; Minimum: $z = 14$

9. Evaluate the objective function at each corner point; choose the maximum and minimum values.

   $z = 10x + y$

Point: $(2, 2)$	Point: $(8, 1)$	Point: $(2, 7)$	Point: $(7, 8)$
$z = 10(2) + 2$	$z = 10(8) + 1$	$z = 10(2) + 7$	$z = 10(7) + 8$
$z = 22$	$z = 81$	$z = 27$	$z = 78$

   Maximum: $z = 81$; Minimum: $z = 22$

11. The constraints form the unbounded region that is shaded in the graph.
The corner points are: $(0, 4)$, $(3, 0)$, and $(13, 0)$.

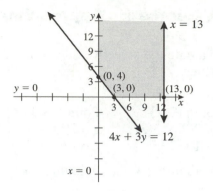

13. The constraints form the region that is shaded in the graph.
Three of the corner points, $(0, 0)$, $(15, 0)$, and $(0, 10)$ are easy to identify. The fourth corner, $(5, 10)$ was found by solving

$$\begin{cases} y = 10 \\ x + y = 15 \end{cases}$$

Substituting $y = 10$ into the second equation, we find $x = 5$.

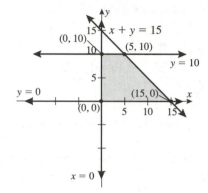

15. The constraints form the region that is shaded in the graph.
Four of the corner points $(3, 0)$, $(10, 0)$, $(0, 4)$, and $(0, 8)$ are easy to identify. The fifth corner, $(10, 8)$ is found at the intersection of the lines $x = 10$ and $y = 8$.

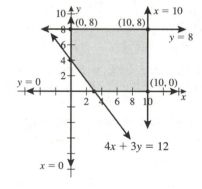

17. To maximize the quantity $z = 5x + 7y$, graph the system of linear inequalities, shade the set of feasible points, and locate the corner points. Then evaluate the objective function at each corner point.

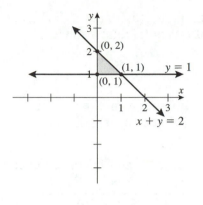

The corner points are $(0, 1)$, $(0, 2)$ and $(1, 1)$.
The third point was obtained by solving

$$\begin{cases} x + y = 2 \\ y = 1 \end{cases}$$

Substituting $y = 1$ into the first equation we find
$x + 1 = 2$ or $x = 1$.

Corner Point (x, y)	Value of the Objective Function $z = 5x + 7y$
$(0, 1)$	$z = 5(0) + 7(1) = 7$
$(0, 2)$	$z = 5(0) + 7(2) = 14$
$(1, 1)$	$z = 5(1) + 7(1) = 12$

The maximum value of z is 14, and it occurs at the point $(0, 2)$.

19. To maximize the quantity $z = 5x + 7y$, graph the system of linear inequalities, shade the set of feasible points, and locate the corner points. Then evaluate the objective function at each corner point.

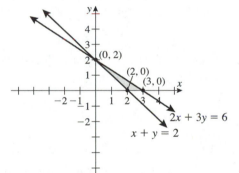

The corner points are $(2, 0)$, $(3, 0)$, and $(0, 2)$.

Corner Point (x, y)	Value of the Objective Function $z = 5x + 7y$
$(2, 0)$	$z = 5(2) + 7(0) = 10$
$(3, 0)$	$z = 5(3) + 7(0) = 15$
$(0, 2)$	$z = 5(0) + 7(2) = 14$

The maximum value of z is 15, and it occurs at the point $(3, 0)$.

21. To maximize the quantity $z = 5x + 7y$, graph the system of linear inequalities, shade the set of feasible points, and locate the corner points. Then evaluate the objective function at each corner point.

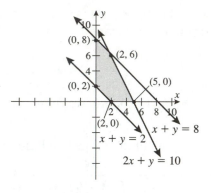

The corner points are $(2, 0)$, $(5, 0)$, $(0, 2)$, $(0, 8)$, and $(2,6)$. The fifth point was obtained by solving

$$\begin{cases} x + y = 8 \\ 2x + y = 10 \end{cases}$$

Subtracting the first equation from the second equation, we find $x = 2$. Back-substituting into equation 1 we find $y = 6$.

Corner Point (x, y)	Value of the Objective Function $z = 5x + 7y$
$(2, 0)$	$z = 5(2) + 7(0) = 10$
$(5, 0)$	$z = 5(5) + 7(0) = 25$
$(0, 2)$	$z = 5(0) + 7(2) = 14$
$(0, 8)$	$z = 5(0) + 7(8) = 56$
$(2, 6)$	$z = 5(2) + 7(6) = 52$

The maximum value of z is 56, and it occurs at the point $(0,8)$.

23. To maximize the quantity $z = 5x + 7y$, graph the system of linear inequalities, shade the set of feasible points, and locate the corner points. Then evaluate the objective function at each corner point.

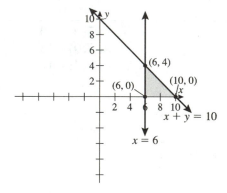

The corner points are $(6, 0)$, $(10, 0)$, and $(6, 4)$. The third point was obtained by solving

$$\begin{cases} x = 6 \\ x + y = 10 \end{cases}$$

Substituting 6 for x in equation 2, we find $y = 4$.

Corner Point (x, y)	Value of the Objective Function $z = 5x + 7y$
$(6, 0)$	$z = 5(6) + 7(0) = 30$
$(10, 0)$	$z = 5(10) + 7(0) = 50$
$(6, 4)$	$z = 5(6) + 7(4) = 58$

The maximum value of z is 58, and it occurs at the point $(6, 4)$.

25. To minimize the quantity $z = 2x + 3y$, graph the system of linear inequalities, shade the set of feasible points, and locate the corner points. Then evaluate the objective function at each corner point.

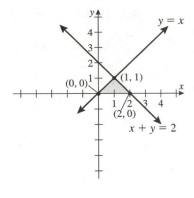

The corner points are $(0, 0)$, $(2, 0)$, and $(1, 1)$. The third point was obtained by solving

$$\begin{cases} x + y = 2 \\ \quad y = x \end{cases}$$

Substituting x for y in equation 1, we find $2x = 2$, or $x = 1$. Back-substituting in equation 2, gives $y = 1$.

Corner Point (x, y)	Value of the Objective Function $z = 2x + 3y$
$(0, 0)$	$z = 2(0) + 3(0) = 0$
$(2, 0)$	$z = 2(2) + 3(0) = 4$
$(1, 1)$	$z = 2(1) + 3(1) = 5$

The minimum value of z is 0, and it occurs at the point $(0, 0)$.

27. To minimize the quantity $z = 2x + 3y$, graph the system of linear inequalities, shade the set of feasible points, and locate the corner points. Then evaluate the objective function at each corner point.

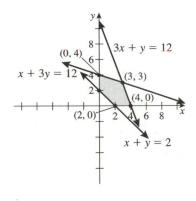

The corner points are $(2, 0)$, $(4, 0)$, $(0, 2)$, $(0, 4)$, and $(3, 3)$. The fifth point was obtained by solving

$$\begin{cases} 3x + y = 12 \\ x + 3y = 12 \end{cases}$$

Subtracting the first equation from three times the second equation gives, $8y = 24$ or $y = 3$. Back-substituting 3 for y in the first equation gives $3x + 3 = 12$ or $3x = 9$ or $x = 3$.

Corner Point (x, y)	Value of the Objective Function $z = 2x + 3y$
$(2, 0)$	$z = 2(2) + 3(0) = 4$
$(4, 0)$	$z = 2(4) + 3(0) = 8$
$(0, 2)$	$z = 2(0) + 3(2) = 6$
$(0, 4)$	$z = 2(0) + 3(4) = 12$
$(3, 3)$	$z = 2(3) + 3(3) = 15$

The minimum value of z is 4, and it occurs at the point $(2, 0)$.

29. To minimize the quantity $z = 2x + 3y$, graph the system of linear inequalities, shade the set of feasible points, and locate the corner points. Then evaluate the objective function at each corner point.

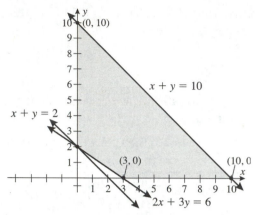

The corner points are $(10, 0)$, $(3, 0)$, $(0, 10)$ and $(0, 2)$.

Corner Point (x, y)	Value of the Objective Function $z = 2x + 3y$
$(0, 10)$	$z = 2(0) + 3(10) = 30$
$(3, 0)$	$z = 2(3) + 3(0) = 6$
$(0, 2)$	$z = 2(0) + 3(2) = 6$
$(10, 0)$	$z = 2(10) + 3(0) = 20$

The minimum value of z is 6, and it occurs at any point on the line segment $2x + 3y = 6$ connecting adjacent corners $(3, 0)$ and $(0, 2)$.

31. To minimize the quantity $z = 2x + 3y$, graph the system of linear inequalities, shade the set of feasible points, and locate the corner points. Then evaluate the objective function at each corner point.

The corner points are $(0, 5)$, $(2, 4)$, $\left(0, \dfrac{1}{2}\right)$,

and $\left(\dfrac{1}{5}, \dfrac{2}{5}\right)$.

The point $(2,4)$ is found by solving

$$\begin{cases} x + 2y = 10 \\ y = 2x \end{cases}$$

Substituting $2x$ for y in the first equation gives $5x = 10$ or $x = 2$. Back-substituting 2 for x in the second equation gives $y = 4$.

The point $\left(\dfrac{1}{5}, \dfrac{2}{5}\right)$ is found by solving

$$\begin{cases} x + 2y = 1 \\ y = 2x \end{cases}$$

Substituting $2x$ for y in the first equation gives

$5x = 1$ or $x = \dfrac{1}{5}$. Back-substituting $\dfrac{1}{5}$ for x in the

second equation gives $y = \dfrac{2}{5}$.

The minimum value of z is $\dfrac{3}{2}$, and it occurs at the point $\left(0, \dfrac{1}{2}\right)$.

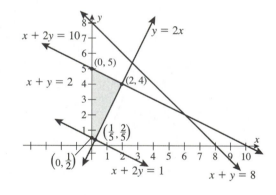

Corner Point (x, y)	Value of the Objective Function $z = 2x + 3y$
$(0, 5)$	$z = 2(0) + 3(5) = 15$
$(2, 4)$	$z = 2(2) + 3(4) = 16$
$\left(0, \dfrac{1}{2}\right)$	$z = 2(0) + 3\left(\dfrac{1}{2}\right) = \dfrac{3}{2}$
$\left(\dfrac{1}{5}, \dfrac{2}{5}\right)$	$z = 2\left(\dfrac{1}{5}\right) + 3\left(\dfrac{2}{5}\right) = \dfrac{8}{5}$

33.

Corner Point (x, y)	Value of Objective Function $z = x + y$
$(10, 0)$	$z = 10 + 0 = 10$
$(0, 10)$	$z = 0 + 10 = 10$
$\left(\dfrac{10}{3}, \dfrac{10}{3}\right)$	$z = \dfrac{10}{3} + \dfrac{10}{3} = \dfrac{20}{3}$

The maximum value of z is 10, and it occurs at the points $(10, 0)$ and $(0, 10)$ as well as at all the points on the line segment connecting them. The minimum value of z is $\dfrac{20}{3}$, and it occurs at the point $\left(\dfrac{10}{3}, \dfrac{10}{3}\right)$.

35.

Corner Point (x, y)	Value of Objective Function $z = 5x + 2y$
$(10, 0)$	$z = 5(10) + 2(0) = 50$
$(0, 10)$	$z = 5(0) + 2(10) = 20$
$\left(\dfrac{10}{3}, \dfrac{10}{3}\right)$	$z = 5\left(\dfrac{10}{3}\right) + 2\left(\dfrac{10}{3}\right) = \dfrac{70}{3}$

The maximum value of z is 50, and it occurs at the point $(10, 0)$. The minimum value of z is 20, and it occurs at the point $(0, 10)$.

37.

Corner Point (x, y)	Value of Objective Function $z = 3x + 4y$
$(10, 0)$	$z = 3(10) + 4(0) = 30$
$(0, 10)$	$z = 3(0) + 4(10) = 40$
$\left(\dfrac{10}{3}, \dfrac{10}{3}\right)$	$z = 3\left(\dfrac{10}{3}\right) + 4\left(\dfrac{10}{3}\right) = \dfrac{70}{3}$

The maximum value of z is 40, and it occurs at the point $(0, 10)$. The minimum value of z is $\dfrac{70}{3}$, and it occurs at the point $\left(\dfrac{10}{3}, \dfrac{10}{3}\right)$.

39.

Corner Point (x, y)	Value of Objective Function $z = 10x + y$
$(10, 0)$	$z = 10(10) + 0 = 100$
$(0, 10)$	$z = 10(0) + 10 = 10$
$\left(\dfrac{10}{3}, \dfrac{10}{3}\right)$	$z = 10\left(\dfrac{10}{3}\right) + \dfrac{10}{3} = \dfrac{110}{3}$

The maximum value of z is 100, and it occurs at the point $(10, 0)$. The minimum value of z is 10, and it occurs at the point $(0, 10)$.

41. To find the maximum and minimum values of z, we will graph the constraints, shade the set of feasible points, find the corner points, and evaluate the objective function, z, at each corner point.

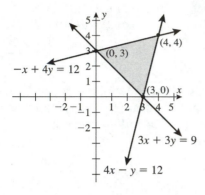

The corner points are $(3, 0)$, $(0, 3)$, and $(4, 4)$.

We find $(4, 4)$ by solving $\begin{cases} -x + 4y = 12 & (1) \\ 4x - y = 12 & (2) \end{cases}$.

Multiplying (1) by 4 and adding it to (2) gives $15y = 60$ or $y = 4$. Back-substituting into (2) gives $4x = 16$ or $x = 4$.

Corner Point	Value of Objective Function $z = 18x + 30y$
$(3, 0)$	$z = 18(3) + 30(0) = 54$
$(0, 3)$	$z = 18(0) + 30(3) = 90$
$(4, 4)$	$z = 18(4) + 30(4) = 192$

The maximum value of $z = 192$; it occurs at the point $(4, 4)$.
The minimum value of $z = 54$; it occurs at the point $(3, 0)$.

43. To find the maximum and minimum values of z, we will graph the constraints, shade the set of feasible points, find the corner points, and evaluate the objective function, z, at each corner point.

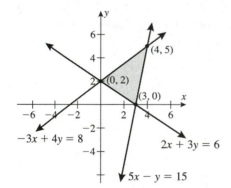

The corner points are $(3, 0)$, $(0, 2)$, and $(4, 5)$.

We find $(4, 5)$ by solving $\begin{cases} -3x + 4y = 8 & (2) \\ 5x - y = 15 & (3) \end{cases}$.

Multiplying (3) by 4 and adding it to (2) gives $17x = 68$ or $x = 4$. Back-substituting into (3) gives $y = 5$.

Corner Point	Value of Objective Function $z = 7x + 6y$
$(3, 0)$	$z = 7(3) + 6(0) = 21$
$(0, 2)$	$z = 7(0) + 6(2) = 12$
$(4, 5)$	$= 7(4) + 6(5) = 58$

The maximum value of $z = 58$; it occurs at the point $(4, 5)$.
The minimum value of $z = 12$; it occurs at the point $(0, 2)$.

45. To find the maximum value of z, we will graph the constraints, shade the set of feasible points, find the corner points, and evaluate the objective function, z, at each corner point.

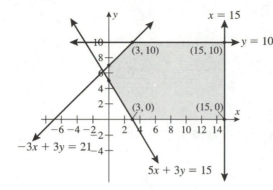

The corner points are $(3, 0)$, $(15, 0)$ $(0, 5)$, $(0, 7)$, $(15, 10)$, and $(3, 10)$.

We find $(3, 10)$ by evaluating $-3x + 3y = 21$ when $y = 10$.

Corner Point	Value of Objective Function $z = -20x + 30y$
$(3, 0)$	$z = -20(3) + 30(0) = -60$
$(15, 0)$	$z = -20(15) + 30(0) = -300$
$(0, 5)$	$z = -20(0) + 30(5) = 150$
$(0, 7)$	$z = -20(0) + 30(7) = 210$
$(15, 10)$	$z = -20(15) + 30(10) = 0$
$(3, 10)$	$z = -20(3) + 30(10) = 240$

The maximum value of $z = 240$; it occurs at the point $(3, 10)$.

47. To find the maximum value of z, we will graph the constraints, shade the set of feasible points, find the corner points, and evaluate the objective function, z, at each corner point.

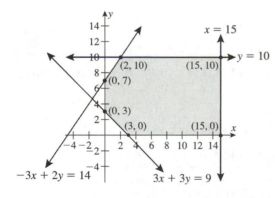

The corner points are $(3, 0)$, $(15, 0)$ $(0, 3)$, $(0, 7)$, $(15, 10)$, and $(2, 10)$.

We find $(2, 10)$ by evaluating $-3x + 2y = 14$ when $y = 10$.

Corner Point	Value of Objective Function $z = -12x + 24y$
$(3, 0)$	$z = -12(3) + 24(0) = -36$
$(15, 0)$	$z = -12(15) + 24(0) = -180$
$(0, 3)$	$z = -12(0) + 24(3) = 72$
$(0, 7)$	$z = -12(0) + 24(7) = 168$
$(15, 10)$	$z = -12(15) + 24(10) = 60$
$(2, 10)$	$z = -12(2) + 24(10) = 216$

The maximum value of $z = 216$; it occurs at the point $(2, 10)$.

49. We will use all the information from Example 5 and form a table to evaluate the objective function $P = (\$0.30)x + (\$0.40)y$ at each corner point.

Corner Point (x, y)	Value of Objective Function $P = (\$0.30)x + (\$0.40)y$
$(0, 0)$	$P = 0.30(0) + 0.40(0) = 0$
$(0, 150)$	$P = 0.30(0) + 0.40(150) = 60$
$(160, 0)$	$P = 0.30(160) + 0.40(0) = 48$
$(90, 105)$	$z = 0.30(90) + 0.40(105) = 69$

The maximum profit $P = \$69$ is obtained when 90 packages of low-grade mix and 105 packages of high-grade mix are made and sold.

3.3 Applications

1. Let x = number of acres of soybeans planted, and
y = number of acres of corn planted.

The objective is to maximize the profit,

$P = 300x + 150y$

The limited amount of land, money for cultivation, and working days available form the constraints:

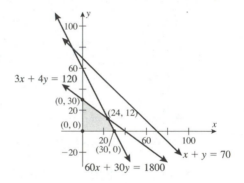

$$\begin{cases} x + y \le 70 & \text{acres available for planting} \\ 60x + 30y \le 1800 & \text{funds available for cultivation} \\ 3x + 4y \le 120 & \text{work days available} \\ x \ge 0 \\ y \ge 0 \end{cases}$$

The constraints are graphed and the set of feasible points is shaded. The corner points are $(0, 0)$, $(30, 0)$, $(0, 30)$, and $(24, 12)$. The fourth point $(24, 12)$ is found by solving the system of equations $\begin{cases} 60x + 30y = 1800 \\ 3x + 4y = 120 \end{cases}$

The objective function P is evaluated at the corner points as shown in the table.

Corner Point (x, y)	Value of Objective Function $P = 300x + 150y$
$(0, 0)$	$P = 0$
$(30, 0)$	$P = 300(30) + 150(0) = 9000$
$(0, 30)$	$P = 300(0) + 150(30) = 4500$
$(24, 12)$	$P = 300(24) + 150(12) = 9000$

The maximum profit of $9000 is made when either 30 acres of soybeans and no acres of corn are planted, when 24 acres of soybeans and 12 acres of corn are planted or when the combination of soybeans and corn planted satisfy $2x + y = 60$ and $24 \le x \le 30$.

3. Let x = amount invested in type A bonds, and
y = amount invested in type B bonds.

The return on the investment is to maximized, which we call P, is

$$P = 0.10x + 0.15y$$

The conditions of the investment form the constraints on the problem.

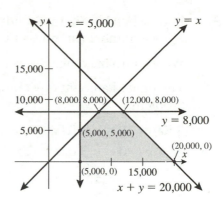

$$\begin{cases} x + y \leq 20{,}000 & \text{maximum to be invested} & (1) \\ x \geq \quad y & \text{at least as much in A as in B} & (2) \\ x \geq \quad 5000 & \text{minimum investment in A} & (3) \\ y \leq \quad 8000 & \text{maximum investment in B} & (4) \\ x \geq \quad 0 & & (5) \\ y \geq \quad 0 & & (6) \end{cases}$$

The constraints are graphed and the set of feasible points is shaded. The corner points are (5000, 0), (20000, 0), (5000, 5000), (8000, 8000), and (12000, 8000). The third point, (5000, 5000), is the intersection of the lines $x = 5000$ and $y = x$. The fourth point, (8000, 8000) is the intersection of the lines $y = 8000$ and $y = x$. The fifth point, (12000, 8000) is found by solving equations (1) and (4).

The objective function P is evaluated at the corner points as shown in the table.

Corner Point (x, y)	Value of Objective Function $P = 0.10x + 0.15y$
(5000, 0)	$P = 0.10(5000) + 0.15(0) = 500$
(20000, 0)	$P = 0.10(20000) + 0.15(0) = 2000$
(5000, 5000)	$P = 0.10(5000) + 0.15(5000) = 1250$
(8000, 8000)	$P = 0.10(8000) + 0.15(8000) = 2000$
(12000, 8000)	$P = 0.10(12000) + 0.15(8000) = 2400$

The maximum return is $2400 and is earned when $12,000 is invested in the type A bond and $8000 is invested in the type B bond.

5.   Let x = number (in thousands) of high potency vitamins produced, and
           y = number (in thousands) of calcium-enriched vitamins produced.

We want to maximize the profit, P, which is given by: $P = [0.10x + 0.05y] \cdot 1000$

The limited amount of vitamin C and calcium available put the following constraints on the problem:

$$\begin{cases} 500x + 100y \le 300,000 & \text{vitamin C constraint} & (1) \\ 40x + 400y \le 220,000 & \text{calcium constraint} & (2) \\ \qquad x \ge 0 & & (3) \\ \qquad y \ge 0 & & (4) \end{cases}$$

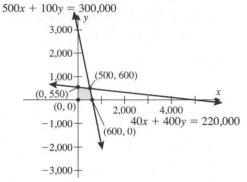

The constraints are graphed and the set of feasible points is shaded. The corner points are $(0, 0)$, $(600, 0)$, $(0, 550)$, and $(500, 500)$. The last point, $(500, 500)$ is found by solving equations (1) and (2).

The objective function P is evaluated at the corner points as shown in the table. Since x and y are in thousands, the profit corresponding to each corner point is:

Corner Point (x, y)	Value of Objective Function $P = (0.10x + 0.05y)1000$
$(0, 0)$	$P = (0.10(0) + 0.05(0))1000 = 0$
$(600, 0)$	$P = (0.10(600) + 0.05(0))1000 = 60,000$
$(0, 550)$	$P = (0.10(0) + 0.05(550))1000 = 27,500$
$(500, 500)$	$P = (0.10(500) + 0.05(500))1000 = 75,000$

The maximum profit of \$75,000 is made when 500,000 high-potency vitamins and 500,000 calcium enriched vitamins are produced.

7.   Let x = number of microwaves in the store, and
           y = number of stoves in the store.

Blink Appliances wants to maximize revenue, R,
which is given by:

$R = 300x + 200y$

The limited amount space and set-up time available
put the following constraints on the problem:

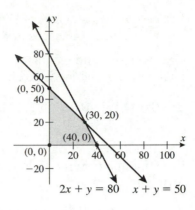

$$\begin{cases} x + y \leq 50 & \text{storeroom limit} & \text{(1)} \\ 2x + y \leq 80 & \text{time constraint} & \text{(2)} \\ \quad x \geq 0 & & \text{(3)} \\ \quad y \geq 0 & & \text{(4)} \end{cases}$$

The constraints are graphed and the set of feasible points is shaded. The corner points are
$(0, 0)$, $(40, 0)$, $(0, 50)$, and $(30, 20)$. The last point is found by solving the system of equations
formed by (1) and (2).

The objective function R is evaluated at the corner points as shown in the table:

Corner Point (x, y)	Value of Objective Function $R = 300x + 200y$
$(0, 0)$	$R = 300(0) + 200(0) = 0$
$(40, 0)$	$R = 300(40) + 200(0) = 12000$
$(0, 50)$	$R = 300(0) + 200(50) = 10000$
$(30, 20)$	$R = 300(30) + 200(20) = 13000$

The maximum revenue is \$13,000, obtained by selling 30 microwaves and 20 stoves.

9.   Let x = number of shares of Duke Energy purchased,
         y = number of shares of Kodak purchased.

The pension fund wants to maximize the yield on
its investment which is given by:

$$P = 0.08(14x) + 0.06(30y)$$
$$= 1.12x + 1.8y$$

The fund's restriction on the allocation of the
investments form the following constraints on the
problem:

$$\begin{cases} 14x + 30y \le 45,000 & \text{investment limit} \\ 14x \ge 11,250 & \text{Duke minimum} \\ 30y \ge 11,250 & \text{Kodak minimum} \\ 14x \le 28,350 & \text{Duke maximum} \\ 30y \le 28,350 & \text{Kodak maximum} \end{cases}$$

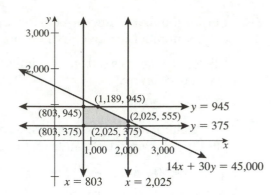

where \$11,250 = 25% of \$45,000, and
      \$28,350 = 63% of \$45,000.

The constraints are graphed and the set of feasible points is shaded. The corner points are
(803, 375), (2025, 375), (2025, 555), (1189, 945), and (803, 945).

The objective function P is evaluated at the corner points as shown in the table:

Corner Point (x, y)	Value of Objective Function $P = 1.12x + 1.8y$
(803, 375)	$P = 1.12(803) + 1.8(375) = 1574.36$
(2025, 375)	$P = 1.12(2025) + 1.8(375) = 2943$
(2025, 555)	$P = 1.12(2025) + 1.8(555) = 3267$
(1189, 945)	$P = 1.12(1189) + 1.8(945) = 3032.68$
(803, 945)	$P = 1.12(803) + 1.8(945) = 2600.36$

The maximum yield is \$3267 and it is obtained if the fund purchases 2025 shares of Duke
Energy Corporation and 375 shares of Eastman Kodak.

11. Let x = number of servings of Gerber granola, and y = number of servings of Gerber juice.

The consumer wants to minimize cost, C, which is given by:

$$C = 0.89x + 0.79y$$

Nutritional guidelines provide the following constraints on the amounts of each food to be consumed.

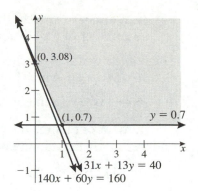

$$\begin{cases} 140x + 60y \geq 160 & \text{minimum calories needed} \\ 31x + 13y \geq 40 & \text{minimum carbohydrates needed} \\ y \geq 0.7 & \text{minimum percent vitamin C} \\ x \geq 0 \\ y \geq 0 \end{cases}$$

The constraints are graphed and the set of feasible points, which is unbounded, is shaded. The corner points are $(0, 3.08)$ and $(1, 0.7)$.

The objective function C is evaluated at the corner points as shown in the table:

Corner Point (x, y)	Value of Objective Function $C = 0.89x + 0.79y$
$(0, 3.08)$	$C = 0.89(0) + 0.79(3.08) = 2.43$
$(1, 0.7)$	$C = 0.89(1) + 0.79(0.7) = 1.44$

The minimum cost is \$1.44 which is obtained by using 1 serving of Gerber granola and 0.7 serving of Gerber mixed fruit carrot juice.

13. Let x = number of months the ad runs on AOL, and
y = number of months the ad runs on Yahoo!.

The company wants to maximize the number of people, P, who see the ad, which is given by:

$P = 76.4x + 66.2y$ (in millions)

The number of months the ads run is limited by the budget and by the minimum time the ad is on Yahoo!.

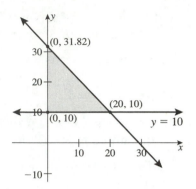

$$\begin{cases} 1200x + 1100y \le 35{,}000 & \text{advertising budget} \\ y \ge \quad 10 & \text{minimum time on Yahoo} \\ x \ge \quad 0 \\ y \ge \quad 0 \end{cases}$$

The constraints are graphed and the set of feasible points is shaded. The corner points are $(0, 10)$, $(0, 31.82)$, and $(20, 10)$.

The objective function P is evaluated at the corner points as shown in the table:

Corner Point (x, y)	Value of Objective Function $P = 76.4x + 66.2y$
$(0, 10)$	$P = 76.4(0) + 66.2(10) = 662$
$(0, 31.82)$	$P = 76.4(0) + 66.2(31.82) = 2106.484$
$(20, 10)$	$P = 76.4(20) + 66.2(10) = 2190$

The maximum number of people exposed to the advertising is 2190 million (2,190,000,000). It is obtained by running the ads on AOL for 20 months and on Yahoo! for 10 months.

15. Let x = number of rolls of high-grade carpet and
y = number of rolls of low-grade carpet made.

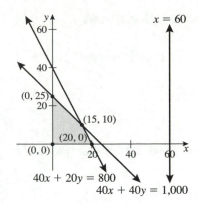

The company wants to maximize income, which is defined to be revenue minus costs.

Income = Revenue – Cost of wool, nylon and labor

The income from manufacturing and selling x rolls of high-grade carpet is:

$$I_{\text{high-grade}} = \$500x - [\$5(20x) + \$2(40x) + \$6(40x)]$$
$$= \$80x$$

The income from manufacturing and selling y rolls of low-grade carpet is:

$$I_{\text{low-grade}} = \$300y - [\$5(0) + \$2(40y) + \$6(20y)]$$
$$= \$100y$$

So the income to be maximized is given by:

$$I = 80x + 100y$$

The limited amounts of materials and labor-hours available form the constraints on the problem.

$$\begin{cases} 20x \le 1200 & \text{wool constraint} & (1) \\ 40x + 40y \le 1000 & \text{nylon constraint} & (2) \\ 40x + 20y \le 800 & \text{labor constraint} & (3) \\ x \ge 0 & & (4) \\ y \ge 0 & & (5) \end{cases}$$

The constraints are graphed and the set of feasible points is shaded. The corner points are $(0, 0)$, $(20, 0)$, $(0, 25)$, and $(15, 10)$. The corner point $(15, 10)$ is found by solving equations (2) and (3).

The objective function I is evaluated at the corner points as shown in the table:

Corner Point (x, y)	Value of Objective Function $I = 80x + 100y$
$(0, 0)$	$I = 80(0) + 100(0) = 0$
$(20, 0)$	$I = 80(20) + 100(0) = 1600$
$(0, 25)$	$I = 80(0) + 100(25) = 2500$
$(15, 10)$	$I = 80(15) + 100(10) = 2200$

The maximum income of $2500 is obtained when 25 rolls of low-grade carpet are manufactured, and no high-grade carpet is made.

Chapter 3 Review

True-False Items (p. 187)

1. True   **3.** True   **5.** True

Fill in the Blanks (p. 187)

1. half-plane   **3.** feasible   **5.** corner point

Review Exercises (p. 187)

1. 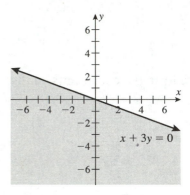   **3.**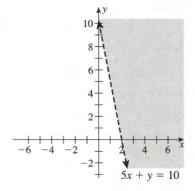

5. To determine which points are part of the graph of the system check each point to see if it satisfies both of the inequalities:

Point	Inequality 1	Inequality 2	Conclusion
$P_1 = (4, -3)$	$x + 2y \leq 8$ Is $4 + 2 \cdot (-3) \leq 8$? $-2 < 8$	$2x - y \geq 4$ Is $2 \cdot 4 - (-3) \geq 4$? $11 > 4$	Both inequalities are satisfied; $(4, -3)$ is part of the graph.
$P_2 = (2, -6)$	Is $2 + 2 \cdot (-6) \leq 8$? $-10 < 8$	Is $2 \cdot 2 - (-6) \geq 4$? $10 > 4$	Both inequalities are satisfied; $(2, -6)$ is part of the graph.
$P_3 = (8, -3)$	Is $8 + 2 \cdot (-3) \leq 8$? $2 < 8$	Is $2 \cdot 8 - (-3) \geq 4$? $19 > 4$	Both inequalities are satisfied; $(8, -3)$ is part of the graph.

7. The region that represents the graph of the system is the set of points common to the solutions of each individual inequality. Use the test point $(0, 0)$.

$$6x - 4y \le 12 \qquad 3x + 2y \le 18$$
$$6 \cdot 0 - 4 \cdot 0 = 0 \qquad 3 \cdot 0 + 2 \cdot 0 = 0$$
$$0 < 12 \qquad\qquad 0 < 18$$

The test point $(0, 0)$ satisfies both inequalities, so the region (*a*) represents the graph of the system.

9. Graph is bounded.
Corner points are $(4, 0)$, $(0, 4)$ and $(0, 6)$.

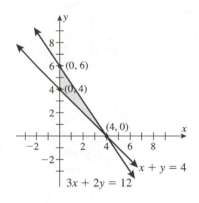

11. Graph is bounded.

Corner points are $(0, 2)$, $(0, 6)$, and $\left(\dfrac{8}{5}, \dfrac{6}{5}\right)$.

To find the third corner point solve

$$\begin{cases} x + 2y = 4 & (1) \\ 3x + y = 6 & (2) \end{cases}$$

Subtracting (1) from twice (2) gives

$5x = 8$ or $x = \dfrac{8}{5}$. Back-substituting in (1) gives

$y = \dfrac{6}{5}$.

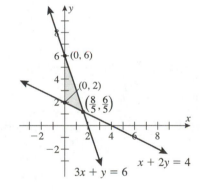

13. Graph is bounded.
Corner points are $(2, 0)$, $(4, 0)$, $(0, 3)$, $(0, 4)$, and $(2, 3)$.
To find $(2, 3)$ solve

$$\begin{cases} x + 2y = 8 & (1) \\ 3x + 2y = 12 & (2) \end{cases}$$

Subtracting (1) from (2) gives $2x = 4$ or $x = 2$.
Back-substituting in (1) gives $2y = 6$ or $y = 3$.

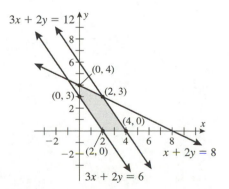

15.

Corner Point (x, y)	Value of Objective Function $z = x + y$
$(10, 0)$	$z = 10 + 0 = 10$
$(20, 0)$	$z = 20 + 0 = 20$
$(0, 10)$	$z = 0 + 10 = 10$
$(0, 20)$	$z = 0 + 20 = 20$
$\left(\dfrac{40}{3}, \dfrac{40}{3}\right)$	$z = \dfrac{40}{3} + \dfrac{40}{3} = \dfrac{80}{3}$

The maximum value of z is $\dfrac{80}{3} \approx 26.667$,

and it occurs at the point $\left(\dfrac{40}{3}, \dfrac{40}{3}\right)$.

17.

Corner Point (x, y)	Value of Objective Function $z = 5x + 2y$
$(10, 0)$	$z = 5(10) + 2(0) = 50$
$(20, 0)$	$z = 5(20) + 2(0) = 100$
$(0, 10)$	$z = 5(0) + 2(10) = 20$
$(0, 20)$	$z = 5(0) + 2(20) = 40$
$\left(\dfrac{40}{3}, \dfrac{40}{3}\right)$	$z = 5\left(\dfrac{40}{3}\right) + 2\left(\dfrac{40}{3}\right) = \dfrac{280}{3}$

The minimum value of z is 20, and it occurs at the point $(0, 10)$.

19

Corner Point (x, y)	Value of Objective Function $z = 2x + y$
$(10, 0)$	$z = 2(10) + 0 = 20$
$(20, 0)$	$z = 2(20) + 0 = 40$
$(0, 10)$	$z = 2(0) + 10 = 10$
$(0, 20)$	$z = 2(0) + 20 = 20$
$\left(\dfrac{40}{3}, \dfrac{40}{3}\right)$	$z = 2\left(\dfrac{40}{3}\right) + \dfrac{40}{3} = \dfrac{120}{3}$ $= 40$

The maximum value of z is 40, and it occurs at the points $(20, 0)$, $\left(\dfrac{40}{3}, \dfrac{40}{3}\right)$ and at all points on the line segment connecting them.

21

Corner Point (x, y)	Value of Objective Function $z = 2x + 5y$
$(10, 0)$	$z = 2(10) + 5(0) = 20$
$(20, 0)$	$z = 2(20) + 5(0) = 40$
$(0, 10)$	$z = 2(0) + 5(10) = 50$
$(0, 20)$	$z = 2(0) + 5(20) = 100$
$\left(\dfrac{40}{3}, \dfrac{40}{3}\right)$	$z = 2\left(\dfrac{40}{3}\right) + 5\left(\dfrac{40}{3}\right) = \dfrac{280}{3}$

The minimum value of z is 20, and it occurs at the point $(10, 0)$.

23. To find the maximum and minimum values of z, we will graph the constraints, shade the set of feasible points, find the corner points, and evaluate the objective function, z, at each corner point.

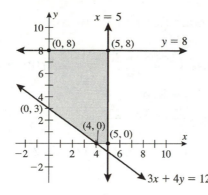

The corner points are $(4, 0)$, $(5, 0)$, $(0, 3)$, $(0, 8)$, and $(5, 8)$.

Point $(5, 8)$ is the intersection of lines $x = 5$ and $y = 8$.

Corner Point	Value of Objective Function $z = 15x + 20y$
$(4, 0)$	$z = 15(4) + 20(0) = 60$
$(5, 0)$	$z = 15(5) + 20(0) = 75$
$(0, 3)$	$z = 15(0) + 20(3) = 60$
$(0, 8)$	$z = 15(0) + 20(8) = 160$
$(5, 8)$	$z = 15(5) + 20(8) = 235$

The maximum value of $z = 235$; it occurs at the point $(5, 8)$.
The minimum value of $z = 60$; it occurs at the points $(4, 0)$, $(0, 3)$, and at all points on the line segment connecting them.

25. To find the maximum and minimum values of z, we will graph the constraints, shade the set of feasible points, find the corner points, and evaluate the objective function, z, at each corner point.

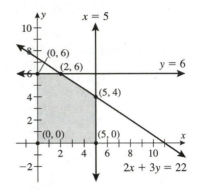

The corner points are $(0, 0)$, $(5, 0)$, $(0, 6)$, $(5, 4)$, and $(2, 6)$.

To find $(5, 4)$ solve $\begin{cases} 2x + 3y = 22 & (1) \\ x = 5 & (2) \end{cases}$. Substitute 5

for x in (1) to get $3y = 12$ or $y = 4$.

To find $(2, 6)$ solve $\begin{cases} 2x + 3y = 22 & (1) \\ y = 6 & (2) \end{cases}$.

Substitute 6 for y in (1) to get $2x = 4$ or $x = 2$.

Corner Point	Value of Objective Function $z = 15x + 20y$
$(0, 0)$	$z = 15(0) + 20(0) = 0$
$(5, 0)$	$z = 15(5) + 20(0) = 75$
$(0, 6)$	$z = 15(0) + 20(6) = 120$
$(5, 4)$	$z = 15(5) + 20(4) = 155$
$(2, 6)$	$z = 15(2) + 20(6) = 150$

The maximum value of $z = 155$; it occurs at the point $(5, 4)$.
The minimum value of $z = 0$; it occurs at the point $(0, 0)$.

27. To solve a linear programming problem graph the constraints, identify the corner points, evaluate the objective function at each corner point, and choose the largest -value of z.

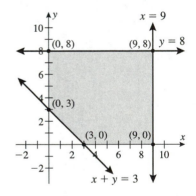

The corner points are $(3, 0)$, $(9, 0)$, $(0, 3)$, $(0, 8)$, and $(9, 8)$.

Corner Point	Value of Objective Function $z = 2x + 3y$
$(3, 0)$	$z = 2(3) + 3(0) = 6$
$(9, 0)$	$z = 2(9) + 3(0) = 18$
$(0, 3)$	$z = 2(0) + 3(3) = 9$
$(0, 8)$	$z = 2(0) + 3(8) = 24$
$(9, 8)$	$z = 2(9) + 3(8) = 42$

The maximum value of $z = 42$; it occurs at the point $(9, 8)$.

29. To solve a linear programming problem graph the constraints, identify the corner points, evaluate the objective function at each corner point, and choose the largest value of z.

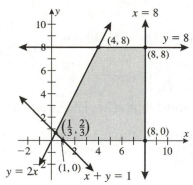

The corner points are $(1, 0)$, $(8, 0)$, $(4, 8)$, $\left(\dfrac{1}{3}, \dfrac{2}{3}\right)$, and $(8, 8)$.

$\left(\dfrac{1}{3}, \dfrac{2}{3}\right)$ is the solution of $\begin{cases} x + y = 1 & (1) \\ y = 2x & (2) \end{cases}$

By substituting $2x$ for y in (1), we get $x = \dfrac{1}{3}$, then back–substituting into (2) gives $y = \dfrac{2}{3}$.

$(4, 8)$ is the intersection of the lines $y = 8$ and $y = 2x$.

Corner Point	Value of Objective Function $z = x + 2y$
$(1, 0)$	$z = 1 + 2(0) = 1$
$(8, 0)$	$z = 8 + 2(0) = 8$
$\left(\dfrac{1}{3}, \dfrac{2}{3}\right)$	$z = \dfrac{1}{3} + 2\left(\dfrac{2}{3}\right) = \dfrac{5}{3}$
$(4, 8)$	$z = 4 + 2(8) = 20$
$(8, 8)$	$z = 8 + 2(8) = 24$

The maximum value of $z = 24$; it occurs at the point $(8, 8)$.

31. To solve a linear programming problem graph the constraints, identify the corner points, evaluate the objective function at each corner point, and choose the smallest value of z.

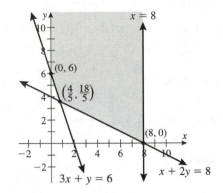

The corner points are $(8, 0)$, $(0, 6)$, and $\left(\dfrac{4}{5}, \dfrac{18}{5}\right)$.

Solve $\begin{cases} x + 2y = 8 & (1) \\ 3x + y = 6 & (2) \end{cases}$ to find $\left(\dfrac{4}{5}, \dfrac{18}{5}\right)$.

Subtracting (1) from 2 times (2) gives $5x = 4$, or $x = \dfrac{4}{5}$. Back-substituting into (1), we find $y = \dfrac{18}{5}$.

Corner Point	Value of Objective Function, $z = 3x + 2y$
$(8, 0)$	$z = 3(8) + 2(0) = 24$
$\left(\dfrac{4}{5}, \dfrac{18}{5}\right)$	$z = 3\left(\dfrac{4}{5}\right) + 2\left(\dfrac{18}{5}\right) = \dfrac{48}{5} = 9.6$
$(0, 6)$	$z = 3(0) + 2(6) = 12$

The minimum value of $z = \dfrac{48}{5}$; it occurs at the point $\left(\dfrac{4}{5}, \dfrac{18}{5}\right)$.

33. Let x = number of downhill skis made, and y = number of cross-country skis made.

The objective is to maximize the profit,

$$P = 70x + 50y$$

The limited amount of time available for manufacturing and finishing put the following constraints on the problem:

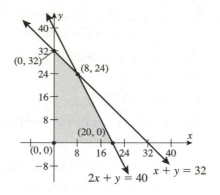

$$\begin{cases} 2x + y \le 40 & \text{manufacturing time} & (1) \\ x + y \le 32 & \text{finishing time} & (2) \\ x \ge 0 & & (3) \\ y \ge 0 & & (4) \end{cases}$$

The constraints are graphed and the set of feasible points is shaded. The corner points are $(0, 0)$, $(20, 0)$, $(0, 32)$, and $(8, 24)$. The fourth point $(8, 24)$ is found by solving the system of equations $\begin{cases} 2x + y = 40 & (1) \\ x + y = 32 & (2) \end{cases}$. When (2) is subtracted from (1), we get $x = 8$. Back-substituting into (2) gives $y = 24$.

The objective function P is evaluated at the corner points as shown in the table.

Corner Point (x, y)	Value of Objective Function $P = 70x + 50y$
$(0, 0)$	$P = 70(0) + 50(0) = 0$
$(20, 0)$	$P = 70(20) + 50(0) = 1400$
$(0, 32)$	$P = 70(0) + 50(32) = 1600$
$(8, 24)$	$P = 70(8) + 50(24) = 1760$

The maximum profit of \$1760 is made when 8 pairs of downhill skis and 24 pairs of cross-country skis are made.

35.   Let x = pounds of Food A purchased, and
y = pounds of Food B purchased.

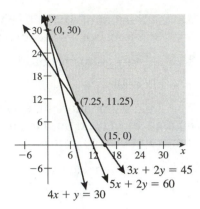

The objective is to minimize the cost of the food, which is given by:

$$C = 1.30x + 0.80y$$

Katy's dietary needs put the following constraints on the problem:

$$\begin{cases} 5x + 2y \geq 60 & \text{carbohydrate needs} & (1) \\ 3x + 2y \geq 45 & \text{protein needs} & (2) \\ 4x + y \geq 30 & \text{fat needs} & (3) \\ \quad x \geq 0 & & (4) \\ \quad y \geq 0 & & (5) \end{cases}$$

The constraints are graphed and the set of feasible points is shaded. The corner points are $(15, 0)$, $(0, 30)$, and $(7.5, 11.25)$. The point $(7.5, 11.25)$ is found by solving the system of equations $\begin{cases} 5x + 2y = 60 & (1) \\ 3x + 2y = 45 & (2) \end{cases}$. When (2) is subtracted from (1), we get $2x = 15$ or

$x = \dfrac{15}{2} = 7.5$. Back-substituting into (2) gives $y = \dfrac{45}{4} = 11.25$.

The objective function C is evaluated at the corner points as shown in the table.

Corner Point (x, y)	Value of Objective Function $C = 1.30x + 0.80y$
$(15, 0)$	$C = 1.30(15) + 0.80(0) = 19.5$
$(7.5, 11.25)$	$C = 1.30(7.5) + 0.80(11.25) = 18.75$
$(0, 30)$	$C = 1.30(0) + 0.80(30) = 24$

The cost of $18.75 is minimized if Katy buys 7.5 pounds of Food A and 11.25 pounds of Food B.

37. Let x = servings of Banana Oatmeal & Peach, and
y = servings of Mixed Fruit Juice consumed.

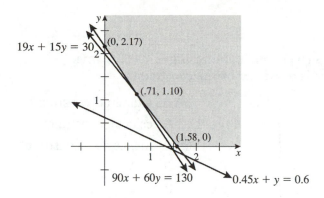

The objective is to minimize the cost, C, of the baby food while maintaining nutritional requirements. The cost is given by:

$$C = 0.79x + 0.65y$$

The nutritional requirement form the constraints:

$$\begin{cases} 90x + 60y \ge 130 & \text{calorie requirement} & (1) \\ 19x + 15y \ge 30 & \text{carbohydrate requirement} & (2) \\ 0.45x + y \ge 0.60 & \text{vitamin C requirement} & (3) \\ x \ge 0 & & (4) \\ y \ge 0 & & (5) \end{cases}$$

The constraints are graphed and the set of feasible points, which is unbounded, is shaded. The corner points are $(1.58, 0)$, $(0.71, 1.10)$, and $(0, 2.17)$. The point $(0.71, 1.10)$ is found by solving the system $\begin{cases} 90x + 60y = 130 & (1) \\ 19x + 15y = 30 & (2) \end{cases}$. When 4 times (2) is subtracted from (1), we get

$14x = 10$ or $x = \dfrac{5}{7} \approx 0.71$. Back-substituting into (2) gives $y = \dfrac{23}{21} \approx 1.10$.

The objective function C is evaluated at the corner points as shown in the table.

Corner Point (x, y)	Value of Objective Function $C = 0.79x + 0.65y$
$(1.58, 0)$	$C = 0.79(1.58) + 0.65(0) = 1.248$
$(0.71, 1.10)$	$C = 0.79(0.71) + 0.65(1.10) = 1.276$
$(0, 2.17)$	$C = 0.79(0) + 0.65(2.17) = 1.411$

The minimum cost is $1.25, and it is attained when 1.58 servings of banana oatmeal and peach cereal and no mixed fruit juice are consumed.

39.   Let x = number (in thousands) of high-potency vitamins to be produced, and
          y = number (in thousands) of calcium-enriched vitamins to be produced.

We want to maximize the profit

$P = 0.10(1000)x + 0.05(1000)y$

The amount of available ingredients is given in kilograms. 1kg = 1,000,000 mg.
The number of vitamins manufactured is restricted by the constraints:

$$\begin{cases} 500,000x + 100,000y \le 300,000,000 \\ 40,000x + 400,000y \le 122,000,000 \\ 100,000x + 40,000y \le 65,000,000 \\ \qquad\qquad x \ge 0 \\ \qquad\qquad y \ge 0 \end{cases} = \begin{cases} 5x + \quad y \le \quad 3000 \quad \text{vitamin C constraint} \\ 4x + 40y \le 12,200 \quad \text{calcium constraint} \\ 10x + 4y \le \quad 6500 \quad \text{magnesium constraint} \\ \qquad x \ge \quad 0 \\ \qquad y \ge \quad 0 \end{cases}$$

The constraints are graphed, and the corner points identified.

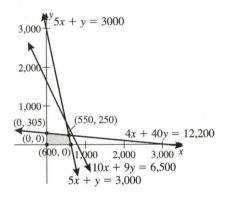

The corner points are $(0, 0)$, $(600, 0)$, $(0, 305)$, and $(550, 250)$. The point $(550, 250)$ is found by solving the system of equations any two of the first three constraints.

The objective function P is evaluated at the corner points as shown in the table.

Corner Point (x, y)	Value of Objective Function $P = 100x + 50y$
$(0, 0)$	$P = 100(0) + 50(0) = 0$
$(600, 0)$	$P = 100(600) + 50(0) = 60,000$
$(0, 305)$	$P = 100(0) + 50(305) = 15,250$
$(550, 250)$	$P = 100(550) + 50(250) = 67,500$

The maximum profit is $67,500. It is attained when 550,000 high-potency vitamins and 250,000 calcium-enriched vitamins are produced.

Chapter 3 Project

1. Objective function: $c = 31.4x + 114.74y$

3.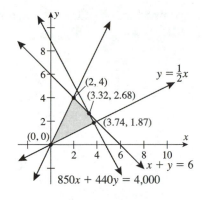

    The corner points are: $(0, 0)$, $(3.74, 1.87)$, $(3.32, 2.68)$, and $(2, 4)$.

    The point $(3.32, 2.68)$ is the solution to the system
    $$\begin{cases} x + y = 6 & (1) \\ 850x + 440y = 4000 & (2) \end{cases}.$$

    Substituting $6 - x$ for y in (2), gives $410x = 1360$ or $x = 3.32$. Back-substitution into (1) gives $y = 2.68$.

    The point $(3.74, 1.87)$ is the solution to the system
    $$\begin{cases} 850x + 440y = 4000 \\ y = 1/2x \end{cases}$$

5. The total mix contains 6 cups, so there are $521.76 \div 6 = 86.96$ grams of carbohydrates per cup.

    Since each cup of raisins contain 440 calories and each cup of peanuts have 850 calories, the calorie count per cup is $(4 \cdot 440 + 2 \cdot 850) \div 6 = 576.67$ calories per cup of mix.

7. To determine the recipe needed to minimize the fat in the mix, construct a new objective function and evaluate it at the corner points.

    Objective function: $f = 72.5x + 0.67y$

Corner Point (x, y)	Value of Objective Function $f = 72.5x + 0.67y$
$(0, 0)$	$f = 72.5(0) + 0.67(0) = 0$
$(3.74, 1.87)$	$f = 72.5(3.74) + 0.67(1.87) = 272.403$
$(3.32, 2.68)$	$f = 72.5(3.32) + 0.67(2.68) = 242.496$
$(2, 4)$	$f = 72.5(2) + 0.67(4) = 147.68$

    Since you intend to eat at least 3 cups of mix, we cannot use the corner point $(0, 0)$. So you would minimize the fat content by making a mix that consists of 2 cups of peanuts and 4 cups of raisins.

9. This low–fat mix has 87.82 grams of protein. See problem 6.

Mathematical Questions from Professional Exams

1.   (b)   **3.**   (c)

5.   Profit is the difference between revenue and cost, so profit for product $Q = 20 - 12 = 8$, and profit for product $P = 17 - 13 = 4$. The objective function is **(d).**

7.   **(c)**

9.   **(b)**

11.   **(b)**

13.   **(d)**

Chapter 4

Linear Programming: Simplex Method

4.1 The Simplex Tableau; Pivoting

Problems 1-10: A maximum problem is in standard form if both of the following conditions are met.

        Condition 1 All the variables are nonnegative.
        Condition 2 Every other constraint is written as a linear expression that is less than or
            equal to a positive constant.

1.   The problem is in standard form since both conditions are met.

3.   The problem is not in standard form since variables x_2 and x_3 are not given as nonnegative.

5.   The problem is not in standard form. The constraint $2x_1 + x_2 + 4x_3 \geq 6$ is not a linear expression that is less than or equal to a positive constant.

7.   The problem is not in standard form. The constraint $2x_1 + x_2 \geq -6$ is not a linear expression that is less than or equal to a positive constant.

9.   The problem is in standard form since both conditions are met.

11.   The maximum problem cannot be modified so to be in standard form.

13. The maximum problem cannot be modified so to be in standard form.

15. The problem can be modified so it is in standard form. It becomes

$$P = 2x_1 + x_2 + 3x_3$$

subject to the constraints

$$
\begin{aligned}
x_1 - x_2 - x_3 &\le 6 \\
-2x_1 + 3x_2 &\le 12 \\
x_3 &\le 2 \\
x_1 \ge 0 \quad x_2 \ge 0 \quad x_3 &\ge 0
\end{aligned}
$$

17. We write the objective function as

$$P - 2x_1 - x_2 - 3x_3 = 0$$

subject to the constraints

$$
\begin{aligned}
5x_1 + 2x_2 + x_3 + s_1 &= 20 \\
6x_1 - x_2 + 4x_3 + s_2 &= 24 \\
x_1 + x_2 + 4x_3 + s_3 &= 16 \\
x_1 \ge 0 \quad x_2 \ge 0 \quad x_3 \ge 0 \quad s_1 \ge 0 \quad s_2 \ge 0 \quad s_3 &\ge 0
\end{aligned}
$$

The initial tableau is

BV	P	x_1	x_2	x_3	s_1	s_2	s_3	RHS
s_1	0	5	2	1	1	0	0	20
s_2	0	6	–1	4	0	1	0	24
s_3	0	1	1	4	0	0	1	16
P	1	–2	–1	–3	0	0	0	0

19. We write the objective function as

$$P - 3x_1 - 5x_2 = 0$$

subject to the constraints

$$
\begin{aligned}
2.2x_1 - 1.8x_2 + s_1 &= 5 \\
0.8x_1 + 1.2x_2 + s_2 &= 2.5 \\
x_1 + x_2 + s_3 &= 0.1 \\
x_1 \ge 0 \quad x_2 \ge 0 \quad s_1 \ge 0 \quad s_2 \ge 0 \quad s_3 &\ge 0
\end{aligned}
$$

The initial tableau is

BV	P	x_1	x_2	s_1	s_2	s_3	RHS
s_1	0	2.2	–1.8	1	0	0	5
s_2	0	0.8	1.2	0	1	0	2.5
s_3	0	1	1	0	0	1	0.1
P	1	–3	–5	0	0	0	0

21. We write the objective function as

$$P - 2x_1 - 3x_2 - x_3 = 0$$

subject to the constraints

$$\begin{aligned} x_1 + x_2 + x_3 + s_1 \quad\quad &= 50 \\ 3x_1 + 2x_2 + x_3 \quad + s_2 &= 10 \end{aligned}$$

$$x_1 \ge 0 \quad x_2 \ge 0 \quad x_3 \ge 0 \quad s_1 \ge 0 \quad s_2 \ge 0$$

The initial tableau is

BV	P	x_1	x_2	x_3	s_1	s_2	RHS
s_1	0	1	1	1	1	0	50
s_2	0	3	2	1	0	1	10
P	1	−2	−3	−1	0	0	0

23. We write the objective function as

$$P - 3x_1 - 4x_2 - 2x_3 = 0$$

subject to the constraints

$$\begin{aligned} 3x_1 + x_2 + 4x_3 + s_1 \quad\quad\quad &= 5 \\ x_1 - x_2 \quad\quad + s_2 \quad &= 5 \\ 2x_1 - x_2 + x_3 \quad\quad + s_3 &= 6 \end{aligned}$$

$$x_1 \ge 0 \quad x_2 \ge 0 \quad x_3 \ge 0 \quad s_1 \ge 0 \quad s_2 \ge 0 \quad s_3 \ge 0$$

The initial tableau is

BV	P	x_1	x_2	x_3	s_1	s_2	s_3	RHS
s_1	0	3	1	4	1	0	0	5
s_2	0	1	−1	0	0	1	0	5
s_3	0	2	−1	1	0	0	1	6
P	1	−3	−4	−2	0	0	0	0

25. We modify the problem by writing each linear constraint as an inequality less than or equal to a positive constant. The modified maximum problem in standard form is

Maximize $P = x_1 + 2x_2 + 5x_3$

subject to the constraints:

$$x_1 - 2x_2 - 3x_3 \leq 10$$
$$-3x_1 - x_2 + x_3 \leq 12$$
$$x_1 \geq 0, \; x_2 \geq 0, \; x_3 \geq 0$$

We then introduce slack variables and set up the initial simplex tableau.

$$P - x_1 - 2x_2 - 5x_3 = 0$$

subject to the constraints

$$x_1 - 2x_2 - 3x_3 + s_1 \quad\;\; = 10$$
$$-3x_1 - x_2 + x_3 \quad\; + s_2 = 12$$
$$x_1 \geq 0, \; x_2 \geq 0, \; x_3 \geq 0, \; s_1 \geq 0, \; s_2 \geq 0$$

The initial tableau is

BV	P	x_1	x_2	x_3	s_1	s_2	RHS
s_1	0	1	–2	–3	1	0	10
s_2	0	–3	–1	1	0	1	12
P	1	–1	–2	–5	0	0	0

27.   We modify the problem by writing each linear constraint as an inequality less than or equal to a positive constant. The modified maximum problem in standard form is

Maximize $P = 2x_1 + 3x_2 + x_3 + 6x_4$

subject to the constraints:

$$-x_1 + x_2 + 2x_3 + x_4 \leq 10$$
$$-x_1 + x_2 - x_3 + x_4 \leq 8$$
$$x_1 + x_2 + x_3 + x_4 \leq 9$$
$$x_1 \geq 0 \quad x_2 \geq 0 \quad x_3 \geq 0 \quad x_4 \geq 0$$

We then introduce slack variables and set up the initial simplex tableau.

$$P - 2x_1 - 3x_2 - x_3 - 6x_4 = 0$$

subject to the constraints:

$$-x_1 + x_2 + 2x_3 + x_4 + s_1 = 10$$
$$-x_1 + x_2 - x_3 + x_4 + s_2 = 8$$
$$x_1 + x_2 + x_3 + x_4 + s_3 = 9$$
$$x_1 \geq 0 \quad x_2 \geq 0 \quad x_3 \geq 0 \quad x_4 \geq 0 \quad s_1 \geq 0 \quad s_2 \geq 0 \quad s_3 \geq 0$$

The initial tableau is

BV	P	x_1	x_2	x_3	x_4	s_1	s_2	s_3	RHS
s_1	0	−1	1	2	1	1	0	0	10
s_2	0	−1	1	−1	1	0	1	0	8
s_3	0	1	1	1	1	0	0	1	9
P	1	−2	−3	−1	−6	0	0	0	0

29.   Pivoting: Step 1: Divide each entry in the *pivot row* by the *pivot element*.

BV	P	x_1	x_2	s_1	s_2	RHS
s_1	0	1	[2]	1	0	300
s_2	0	3	2	0	1	480
P	1	−1	−2	0	0	0

BV	P	x_1	x_2	s_1	s_2	RHS
→ x_2	0	$\frac{1}{2}$	1	$\frac{1}{2}$	0	150
s_2	0	3	2	0	1	480
P	1	−1	−2	0	0	0

Step 2: Obtain 0s elsewhere in the *pivot column* by performing row operations using the revised *pivot row*.

BV	P	x_1	x_2	s_1	s_2	RHS
→ x_2	0	$\frac{1}{2}$	1	$\frac{1}{2}$	0	150
s_2	0	2	0	−1	1	180
P	1	0	0	1	0	300

The system of equations corresponding to the new tableau is

$$\begin{cases} x_2 = -\dfrac{1}{2}x_1 - \dfrac{1}{2}s_1 + 150 \\ s_2 = -2x_1 + s_1 + 180 \\ P = -s_1 + 300 \end{cases}$$

The current values of the objective function and the basic variables are

$$P = 300, \qquad x_2 = 150 \qquad \text{and} \qquad s_2 = 180$$

31. Pivoting: Step 1: Divide each entry in the *pivot row* by the *pivot element*.

BV	P	x_1	x_2	x_3	s_1	s_2	s_3	RHS
s_1	0	1	2	4	1	0	0	24
s_2	0	2	−1	1	0	1	0	32
s_3	0	3	2	$\boxed{4}$	0	0	1	18
P	1	−1	−2	−3	0	0	0	0

BV	P	x_1	x_2	x_3	s_1	s_2	s_3	RHS
s_1	0	1	2	4	1	0	0	24
s_2	0	2	−1	1	0	1	0	32
$\rightarrow x_3$	0	$\frac{3}{4}$	$\frac{1}{2}$	1	0	0	$\frac{1}{4}$	$\frac{9}{2}$
P	1	−1	−2	−3	0	0	0	0

Step 2: Obtain 0s elsewhere in the *pivot column* by performing row operations using the revised *pivot row*.

BV	P	x_1	x_2	x_3	s_1	s_2	s_3	RHS
s_1	0	−2	0	0	1	0	−1	6
s_2	0	$\frac{5}{4}$	$-\frac{3}{2}$	0	0	1	$-\frac{1}{4}$	$\frac{55}{2}$
$\rightarrow x_3$	0	$\frac{3}{4}$	$\frac{1}{2}$	1	0	0	$\frac{1}{4}$	$\frac{9}{2}$
P	1	$\frac{5}{4}$	$-\frac{1}{2}$	0	0	0	$\frac{3}{4}$	$\frac{27}{2}$

The system of equations corresponding to the new tableau is

$$\begin{cases} s_1 = 2x_1 + s_3 + 6 \\[2mm] s_2 = -\dfrac{5}{4}x_1 + \dfrac{3}{2}x_2 + \dfrac{1}{4}s_3 + \dfrac{55}{2} \\[2mm] x_3 = -\dfrac{3}{4}x_1 - \dfrac{1}{2}x_2 - \dfrac{1}{4}s_3 + \dfrac{9}{2} \\[2mm] P = -\dfrac{5}{4}x_1 + \dfrac{1}{2}x_2 - \dfrac{3}{4}s_3 + \dfrac{27}{2} \end{cases}$$

The current values of the objective function and the basic variables are

$$P = \frac{27}{2}, \qquad x_3 = \frac{9}{2} \qquad s_1 = 6 \qquad s_2 = \frac{55}{2}$$

33. Pivoting: Step 1: Divide each entry in the *pivot row* by the *pivot element*.
The pivot element already equals 1, so we move directly to step 2.

$$
\begin{array}{c|ccccccccc|c}
\text{BV} & P & x_1 & x_2 & x_3 & x_4 & s_1 & s_2 & s_3 & s_4 & \text{RHS} \\
\hline
s_1 & 0 & -3 & 0 & 1 & 0 & 1 & 0 & 0 & 0 & 20 \\
s_2 & 0 & 2 & 0 & 0 & \boxed{1} & 0 & 1 & 0 & 0 & 24 \\
s_3 & 0 & 0 & -3 & 1 & 0 & 0 & 0 & 1 & 0 & 28 \\
s_4 & 0 & 0 & -3 & 0 & 1 & 0 & 0 & 0 & 1 & 24 \\
\hline
P & 1 & -1 & -2 & -3 & -4 & 0 & 0 & 0 & 0 & 0
\end{array}
$$

Step 2: Obtain 0s elsewhere in the *pivot column* by performing row operations using the revised *pivot row*.

$$
\begin{array}{c|ccccccccc|c}
\text{BV} & P & x_1 & x_2 & x_3 & x_4 & s_1 & s_2 & s_3 & s_4 & \text{RHS} \\
\hline
s_1 & 0 & -3 & 0 & 1 & 0 & 1 & 0 & 0 & 0 & 20 \\
\rightarrow x_4 & 0 & 2 & 0 & 0 & 1 & 0 & 1 & 0 & 0 & 24 \\
s_3 & 0 & 0 & -3 & 1 & 0 & 0 & 0 & 1 & 0 & 28 \\
s_4 & 0 & -2 & -3 & 0 & 0 & 0 & -1 & 0 & 1 & 0 \\
\hline
P & 1 & 7 & -2 & -3 & 0 & 0 & 4 & 0 & 0 & 96
\end{array}
$$

The system of equations corresponding to the new tableau is

$$
\begin{cases}
s_1 = 3x_1 - x_3 + 20 \\
x_4 = -2x_1 - s_2 + 24 \\
s_3 = 3x_2 - x_3 + 28 \\
s_4 = 2x_1 + 3x_2 + s_2 \\
P = -7x_1 + 2x_2 + 3x_3 - 4s_2 + 96
\end{cases}
$$

The current values of the objective function and the basic variables are

$$P = 96 \quad x_4 = 24 \quad s_1 = 20 \quad s_3 = 28 \quad s_4 = 0$$

4.2 The Simplex Method: Solving Maximum Problems in Standard Form

1. (b) It requires further pivoting.
Since there is still a negative numeral in the objective row, the tableau needs further pivoting.
We find the pivot element by selecting the column containing the smallest, negative entry in the objective row, here it is column x_1.
We select the pivot row by computing the quotients formed by dividing the entry in the right hand side by the corresponding positive entry of the pivot column.
The pivot row has the smallest nonnegative quotient, here we get $20 \div 1 = 20$
and $30 \div \dfrac{1}{2} = 60$
The new pivot element is 1 in row s_1, column x_1.

3. (a) This is the final tableau since there are no negative entries in the objective row.
The maximum value of P is $P = \dfrac{256}{7}$. It occurs when $x_1 = \dfrac{32}{7}$ and $x_2 = 0$.

5. (c) There is no solution to this problem. Although there is a negative entry in the objective row, all entries in the pivot column are negative, so the problem is unbounded and has no solution.

7. (b) This tableau requires further pivoting since there is still a negative numeral in the objective row.
We find the pivot element by selecting the column containing the smallest, negative entry in the objective row; here it is -10 in column s_1.
We select the pivot row by computing the quotients formed by dividing the entry in the right hand side by the corresponding positive entry of the pivot column. The pivot row has the smallest nonnegative quotient. Here we get $6 \div 1 = 6$ and $1 \div 1 = 1$.
The new pivot element is 1 in row x_1, column s_1.

9.   To solve the problem using the simplex method, we first must introduce slack variables and construct the initial tableau.

Maximize $P - 5x_1 - 7x_2 = 0$

subject to

$$2x_1 + 3x_2 + s_1 \qquad = 12$$
$$3x_1 + \ x_2 \qquad + s_2 = 12$$
$$x_1 \geq 0 \ \ x_2 \geq 0 \ \ s_1 \geq 0 \ \ s_2 \geq 0$$

The initial simplex tableau with the pivot element marked is

BV	P	x_1	x_2	s_1	s_2	RHS
$\rightarrow s_1$	0	2	$\boxed{3}$	1	0	12
s_2	0	3	1	0	1	12
P	1	-5	-7	0	0	0

We pivot using row operations

BV	P	x_1	x_2	s_1	s_2	RHS
$\rightarrow s_1$	0	2	$\boxed{3}$	1	0	12
s_2	0	3	1	0	1	12
P	1	-5	-7	0	0	0

$\xrightarrow[\substack{R_2 = \frac{3}{7}r_2 \\ R_1 = -\frac{2}{3}r_2 + r_1 \\ R_3 = \frac{1}{3}r_2 + r_3}]{}$

BV	P	x_1	x_2	s_1	s_2	RHS
x_2	0	$\frac{2}{3}$	1	$\frac{1}{3}$	0	4
$\rightarrow s_2$	0	$\boxed{\frac{7}{3}}$	0	$-\frac{1}{3}$	1	8
P	1	$-\frac{1}{3}$	0	$\frac{7}{3}$	0	28

Since there is still a negative entry in the objective row, the problem needs further pivoting. We choose the pivot element in row s_2, column x_2, and pivot using row operations.

$\xrightarrow[\substack{R_2 = \frac{3}{7}r_2 \\ R_1 = -\frac{2}{3}r_2 + r_1 \\ R_3 = \frac{1}{3}r_2 + r_3}]{}$

BV	P	x_1	x_2	s_1	s_2	RHS
x_2	0	0	1	$\frac{3}{7}$	$-\frac{2}{7}$	$\frac{12}{7}$
x_1	0	1	0	$-\frac{1}{7}$	$\frac{3}{7}$	$\frac{24}{7}$
P	1	0	0	$\frac{16}{7}$	$\frac{1}{7}$	$\frac{204}{7}$

This is the final tableau since all entries in the objective row are positive. The solution is $P = \dfrac{204}{7}$, obtained when $x_1 = \dfrac{24}{7}$ and $x_2 = \dfrac{12}{7}$.

11. To solve the problem using the simplex method, we first must introduce slack variables and construct the initial tableau.

Maximize $P - 5x_1 - 7x_2 = 0$

subject to

$$
\begin{aligned}
x_1 + 2x_2 + s_1 \quad\;\; &= 2 \\
2x_1 + \;\; x_2 \quad + s_2 &= 2 \\
x_1 \geq 0 \; x_2 \geq 0 \; s_1 \geq 0 \; s_2 \geq 0
\end{aligned}
$$

BV	P	x_1	x_2	s_1	s_2	RHS
$\to s_1$	0	1	[2]	1	0	2
s_2	0	2	1	0	1	2
P	1	-5	-7	0	0	0

$R_1 = \frac{1}{2}r_1$
$R_2 = -r_1 + r_2$
$R_3 = 7r_1 + r_3$

BV	P	x_1	x_2	s_1	s_2	RHS
x_1	0	$\frac{1}{2}$	[1]	$\frac{1}{2}$	0	1
$\to s_2$	0	$\frac{3}{2}$	0	$-\frac{1}{2}$	1	1
P	1	$-\frac{3}{2}$	0	$\frac{7}{2}$	0	7

BV	P	x_1	x_2	s_1	s_2	RHS
x_2	0	$\frac{1}{2}$	1	$\frac{1}{2}$	0	1
$\to x_1$	0	[$\frac{3}{2}$]	0	$-\frac{1}{2}$	1	1
P	1	$-\frac{3}{2}$	0	$\frac{7}{2}$	0	7

$R_2 = \frac{2}{3}r_2$
$R_1 = -\frac{1}{2}r_2 + r_1$
$R_3 = \frac{3}{2}r_2 + r_3$

BV	P	x_1	x_2	s_1	s_2	RHS
x_2	0	0	1	$\frac{1}{3}$	$-\frac{1}{3}$	$\frac{2}{3}$
$\to x_1$	0	[1]	0	$-\frac{1}{3}$	$\frac{2}{3}$	$\frac{2}{3}$
P	1	0	0	3	1	8

This is the final tableau since all entries in the objective row are nonnegative. The solution is $P = 8$, obtained when $x_1 = \dfrac{2}{3}$ and $x_2 = \dfrac{2}{3}$.

13. To solve the problem using the simplex method, we first must introduce slack variables and construct the initial tableau.

Maximize $P - 3x_1 - x_2 = 0$

subject to

$$
\begin{aligned}
x_1 + \;\; x_2 + s_1 \quad\quad &= 2 \\
2x_1 + 3x_2 \quad + s_2 \quad &= 12 \\
3x_1 + \;\; x_2 \quad\quad + s_3 &= 12 \\
x_1 \geq 0 \; x_2 \geq 0 \; s_1 \geq 0 \; s_2 \geq 0 \; s_3 \geq 0
\end{aligned}
$$

The initial tableau is on the left. The pivot element is 1; x_1 is the entering basic variable.

BV	P	x_1	x_2	s_1	s_2	s_3	RHS
$\to s_1$	0	[1]	1	1	0	0	2
s_2	0	2	3	0	1	0	12
s_3	0	3	1	0	0	1	12
P	1	-3	-1	0	0	0	0

$R_2 = -2r_1 + r_2$
$R_3 = -3r_1 + r_3$
$R_4 = 3r_1 + r_4$

BV	P	x_1	x_2	s_1	s_2	s_3	RHS
$\to x_1$	0	[1]	1	1	0	0	2
s_2	0	0	1	-2	1	0	8
s_3	0	0	-2	-3	0	1	6
P	1	0	2	3	0	0	6

This is the final tableau since all entries in the objective row are nonnegative. The solution is $P = 6$, obtained when $x_1 = 2$ and $x_2 = 0$.

15. To solve the problem using the simplex method, we first must introduce slack variables and construct the initial tableau.

Maximize $P - 2x_1 - x_2 - x_3 = 0$

subject to

$$\begin{aligned} -2x_1 + x_2 - 2x_3 + s_1 \qquad &= 4 \\ x_1 - 2x_2 + x_3 \qquad + s_2 &= 2 \end{aligned}$$

$x_1 \geq 0 \quad x_2 \geq 0 \quad x_3 \geq 0 \quad s_1 \geq 0 \quad s_2 \geq 0$

The initial tableau is on the left. The pivot element is 1; the entering basic variable is x_1.

BV	P	x_1	x_2	x_3	s_1	s_2	RHS
s_1	0	–2	1	–2	1	0	4
→ s_2	0	1	–2	1	0	1	2
P	1	–2	–1	–1	0	0	0

$$\xrightarrow{\substack{R_1 = 2r_2 + r_1 \\ R_3 = 2r_1 + r_3}}$$

BV	P	x_1	x_2	x_3	s_1	s_2	RHS
s_1	0	0	–3	0	1	2	8
→ x_1	0	1	–2	1	0	1	2
P	1	0	–5	1	0	2	4

The new pivot column is x_2, since –5 is the smallest negative entry in the objective row. But all the entries in the pivot column are negative. This problem is unbounded and has no solution.

17. To solve the problem using the simplex method, we first must introduce slack variables and construct the initial tableau.

Maximize $P - 2x_1 - x_2 - 3x_3 = 0$

subject to

$$\begin{aligned} x_1 + 2x_2 + x_3 + s_1 \qquad &= 25 \\ 3x_1 + 2x_2 + 3x_3 \qquad + s_2 &= 30 \end{aligned}$$

$x_1 \geq 0 \quad x_2 \geq 0 \quad x_3 \geq 0 \quad s_1 \geq 0 \quad s_2 \geq 0$

The initial tableau is on the left. The pivot element is 3; the entering basic variable is x_3.

BV	P	x_1	x_2	x_3	s_1	s_2	RHS
s_1	0	1	2	1	1	0	25
→ s_2	0	3	2	3	0	1	30
P	1	–2	–1	–3	0	0	0

$$\xrightarrow{\substack{R_2 = \frac{1}{3}r_2 \\ R_1 = -r_2 + r_1 \\ R_3 = 3r_2 + r_3}}$$

BV	P	x_1	x_2	x_3	s_1	s_2	RHS
s_1	0	0	$\frac{4}{3}$	0	1	$-\frac{1}{3}$	15
→ x_3	0	1	$\frac{2}{3}$	1	0	$\frac{1}{3}$	10
P	1	1	1	0	0	1	30

Since all the entries in the objective row are nonnegative, this is the final tableau. The solution is $P = 30$, obtained when $x_1 = 0$, $x_2 = 0$, and $x_3 = 10$.

19. To solve the problem using the simplex method, we first must introduce slack variables and construct the initial tableau.

Maximize $P - 2x_1 - 4x_2 - x_3 - x_4 = 0$

subject to

$$2x_1 + x_2 + 2x_3 + 3x_4 + s_1 \qquad\quad = 12$$
$$2x_2 + x_3 + 2x_4 \quad + s_2 \qquad = 20$$
$$2x_1 + x_2 + 4x_3 \qquad\qquad + s_3 = 16$$
$$x_1 \ge 0 \;\; x_2 \ge 0 \;\; x_3 \ge 0 \;\; x_4 \ge 0 \;\; s_1 \ge 0 \;\; s_2 \ge 0 \;\; s_3 \ge 0$$

The initial tableau is below. The pivot element is 2 and x_2 is the entering basic variable.

BV	P	x_1	x_2	x_3	x_4	s_1	s_2	s_3	RHS
s_1	0	2	1	2	3	1	0	0	12
$\to s_2$	0	0	2	1	2	0	1	0	20
s_3	0	2	1	4	0	0	0	1	16
P	1	−2	−4	−1	−1	0	0	0	0

$$\begin{array}{l} R_2 = \frac{1}{2} r_2 \\ R_1 = -r_2 + r_1 \\ R_3 = -r_2 + r_3 \\ R_4 = 4r_2 + r_4 \end{array} \longrightarrow$$

BV	P	x_1	x_2	x_3	x_4	s_1	s_2	s_3	RHS
s_1	0	2	0	$\frac{3}{2}$	2	1	$-\frac{1}{2}$	0	2
x_2	0	0	1	$\frac{1}{2}$	1	0	$\frac{1}{2}$	0	10
s_3	0	2	0	$\frac{7}{2}$	−1	0	$-\frac{1}{2}$	1	6
P	1	−2	0	1	3	0	2	0	40

The new tableau needs further pivoting. The new pivot is 2 and x_1 is the entering variable.

BV	P	x_1	x_2	x_3	x_4	s_1	s_2	s_3	RHS
$\to s_1$	0	2	0	$\frac{3}{2}$	2	1	$-\frac{1}{2}$	0	2
x_2	0	0	1	$\frac{1}{2}$	1	0	$\frac{1}{2}$	0	10
s_3	0	2	0	$\frac{7}{2}$	−1	0	$-\frac{1}{2}$	1	6
P	1	−2	0	1	3	0	2	0	40

$$\begin{array}{l} R_1 = \frac{1}{2} r_1 \\ R_3 = -2r_1 + r_3 \\ R_4 = 2r_1 + r_4 \end{array} \longrightarrow$$

BV	P	x_1	x_2	x_3	x_4	s_1	s_2	s_3	RHS
$\to x_1$	0	1	0	$\frac{3}{4}$	1	$\frac{1}{2}$	$-\frac{1}{4}$	0	1
x_2	0	0	1	$\frac{1}{2}$	1	0	$\frac{1}{2}$	0	10
s_3	0	0	0	2	−3	−1	0	1	4
P	1	0	0	$\frac{5}{2}$	5	1	$\frac{3}{2}$	0	42

Since there are no negative entries in the objective row, this is the final tableau. The solution is $P = 42$, obtained when $x_1 = 1$, $x_2 = 10$, $x_3 = 0$, and $x_4 = 0$.

21. To solve the problem using the simplex method, we first must introduce slack variables and construct the initial tableau.

Maximize $P - 2x_1 - x_2 - x_3 = 0$

subject to

$$x_1 + 2x_2 + 4x_3 + s_1 \qquad\quad = 20$$
$$2x_1 + 4x_2 + 4x_3 \quad + s_2 \qquad = 60$$
$$3x_1 + 4x_2 + x_3 \qquad\quad + s_3 = 90$$
$$x_1 \ge 0 \;\; x_2 \ge 0 \;\; x_3 \ge 0 \;\; s_1 \ge 0 \;\; s_2 \ge 0 \;\; s_3 \ge 0$$

The initial tableau is below. The pivot element is 1; the entering variable is x_1.

BV	P	x_1	x_2	x_3	s_1	s_2	s_3	RHS
$\to s_1$	0	1	2	4	1	0	0	20
s_2	0	2	4	4	0	1	0	60
s_3	0	3	4	1	0	0	1	90
P	1	−2	−1	−1	0	0	0	0

$$\begin{array}{l} R_2 = -2r_1 + r_2 \\ R_3 = -3r_1 + r_3 \\ R_4 = 2r_1 + r_4 \end{array} \longrightarrow$$

BV	P	x_1	x_2	x_3	s_1	s_2	s_3	RHS
$\to x_1$	0	1	2	4	1	0	0	20
s_2	0	0	0	−4	−2	1	0	20
s_3	0	0	−2	−11	−3	0	1	30
P	1	0	3	7	2	0	0	40

Since all entries in the new tableau are nonnegative, it is the final tableau. The solution is $P = 40$, obtained when $x_1 = 20$, $x_2 = 0$, and $x_3 = 0$.

23. To solve the problem using the simplex method, we first must introduce slack variables and construct the initial tableau.

Maximize $P - x_1 - 2x_2 - 4x_3 - x_4 = 0$

subject to

$$5x_1 \quad\ + 4x_3 + 6x_4 + s_1 \quad\quad = 20$$
$$4x_1 + 2x_2 + 2x_3 + 8x_4 \quad\ + s_2 = 40$$
$$x_1 \ge 0 \ \ x_2 \ge 0 \ \ x_3 \ge 0 \ \ x_4 \ge 0 \ \ s_1 \ge 0 \ \ s_2 \ge 0$$

The initial tableau is below. The pivot element is 4. The new basic variable will be x_3.

BV	P	x_1	x_2	x_3	x_4	s_1	s_2	RHS
$\rightarrow s_1$	0	5	0	$\boxed{4}$	6	1	0	20
s_2	0	4	2	2	8	0	1	40
P	1	–1	–2	–4	1	0	0	0

$R_1 = \frac{1}{4}r_1$
$R_2 = -2r_1 + r_2$
$R_3 = 4r_1 + r_3$

$\longrightarrow$

BV	P	x_1	x_2	x_3	x_4	s_1	s_2	RHS
$\rightarrow x_3$	0	$\frac{5}{4}$	0	$\boxed{1}$	$\frac{3}{2}$	$\frac{1}{4}$	0	5
s_2	0	$\frac{3}{2}$	2	0	5	$-\frac{1}{2}$	1	30
P	1	4	–2	0	7	1	0	20

Since there is still a negative entry in the objective row, this tableau needs further pivoting. The new pivot will be 2, x_2 will be the entering basic variable.

BV	P	x_1	x_2	x_3	x_4	s_1	s_2	RHS
x_3	0	$\frac{5}{4}$	0	1	$\frac{3}{2}$	$\frac{1}{4}$	0	5
$\rightarrow s_2$	0	$\frac{3}{2}$	$\boxed{2}$	0	5	$-\frac{1}{2}$	1	30
P	1	4	–2	0	7	1	0	20

$R_2 = \frac{1}{2}r_2$
$R_3 = 2r_2 + r_3$

$\longrightarrow$

BV	P	x_1	x_2	x_3	x_4	s_1	s_2	RHS
x_3	0	$\frac{5}{4}$	0	1	$\frac{3}{2}$	$\frac{1}{4}$	0	5
$\rightarrow x_2$	0	$\frac{3}{4}$	$\boxed{1}$	0	$\frac{5}{2}$	$-\frac{1}{4}$	$\frac{1}{2}$	15
P	1	$\frac{11}{2}$	0	0	12	$\frac{1}{2}$	1	50

This is the final tableau. The solution $P = 50$, is obtained when $x_1 = 0$, $x_2 = 15$, $x_3 = 5$, and $x_4 = 0$.

25.   We let x_1, x_2, and x_3 represent the number of type I jeans, type II jeans and type III jeans produced. We want to

Maximize $P = 3x_1 + 4.50x_2 + 6x_3$

subject to the following manufacturing constraints

$$8x_1 + 12x_2 + 18x_3 + \leq 5200 \qquad \text{Cutting constraint}$$
$$12x_1 + 18x_2 + 24x_3 + \leq 6000 \qquad \text{Sewing constraint}$$
$$4x_1 + 8x_2 + 12x_3 + \leq 2200 \qquad \text{Finishing constraint}$$
$$x_1 \geq 0 \quad x_2 \geq 0 \quad x_3 \geq 0$$

Slack variables are added and the initial tableau is constructed and shown below. The pivot element is 12.

BV	P	x_1	x_2	x_3	s_1	s_2	s_3	RHS
s_1	0	8	12	18	1	0	0	5200
s_2	0	12	18	24	0	1	0	6000
$\rightarrow s_3$	0	4	8	12	0	0	1	2200
P	1	−3	−4.5	−6	0	0	0	0

$R_3 = \frac{1}{12}r_3$
$R_1 = -18r_3 + r_1$
$R_2 = -24r_3 + r_2$
$R_4 = 6r_3 + r_4$

BV	P	x_1	x_2	x_3	s_1	s_2	s_3	RHS
s_1	0	2	0	0	1	0	$-\frac{3}{2}$	1900
s_2	0	4	2	0	0	1	−2	1600
$\rightarrow x_3$	0	$\frac{1}{3}$	$\frac{2}{3}$	1	0	0	$\frac{1}{12}$	$\frac{550}{3}$
P	1	−1	$-\frac{1}{2}$	0	0	0	$\frac{1}{2}$	1100

This tableau needs further pivoting, the smallest entry in the objective row is −1; the smallest positive quotient formed by the RHS and the elements in column x_1 is 400. The new pivot element is 4; the entering basic variable is x_1.

$R_2 = \frac{1}{4}r_2$
$R_1 = -2r_2 + r_1$
$R_3 = -\frac{1}{3}r_2 + r_3$
$R_4 = r_2 + r_4$

BV	P	x_1	x_2	x_3	s_1	s_2	s_3	RHS
s_1	0	0	−1	0	1	$-\frac{1}{2}$	$-\frac{1}{2}$	1100
$\rightarrow x_1$	0	1	$\frac{1}{2}$	0	0	$\frac{1}{4}$	$-\frac{1}{2}$	400
x_3	0	0	$\frac{1}{2}$	1	0	$-\frac{1}{12}$	$\frac{1}{4}$	50
P	1	0	0	0	0	$\frac{1}{4}$	0	1500

This is the final tableau. The solution is $P = 1500$, obtained when $x_1 = 400$, $x_2 = 0$, and $x_3 = 50$.

So the manufacturer should produce 400 pairs of type I jeans and 50 pairs of type III jeans for a maximum profit of $1500.

27.   We let x_1, x_2, and x_3 represent the number of products A, B, and C sold. We want to

Maximize $P = x_1 + x_2 + 2x_3$

subject to the following constraints

$$
\begin{aligned}
3x_1 + \ 5x_2 + \ 4x_3 &\le 500 \qquad &\text{cost limit of products}\\
10x_1 + 15x_2 + 12x_3 &\le 1800 \qquad &\text{time available for selling}\\
.50x_1 + \qquad\quad x_3 &\le 75 \qquad &\text{amount available for delivery}\\
x_1 \ge 0 \quad x_2 \ge 0 \quad x_3 &\ge 0
\end{aligned}
$$

Slack variables are added and the initial tableau is constructed and shown below. The pivot element is 1.

BV	P	x_1	x_2	x_3	s_1	s_2	s_3	RHS
s_1	0	3	5	4	1	0	0	500
s_2	0	10	15	12	0	1	0	1800
→s_3	0	.5	0	[1]	0	0	1	75
P	1	−1	−1	−2	0	0	0	0

BV	P	x_1	x_2	x_3	s_1	s_2	s_3	RHS
s_1	0	1	5	0	1	0	−4	200
s_2	0	4	15	0	0	1	−12	900
→x_3	0	.5	0	[1]	0	0	1	75
P	1	0	−1	0	0	0	2	150

Since there is still a negative entry in the objective row, it denotes the new pivot column. The pivot row is determined by choosing the smallest quotient when the RHS is divided by the elements in the pivot column. The new pivot element is 5.

BV	P	x_1	x_2	x_3	s_1	s_2	s_3	RHS
→s_1	0	1	[5]	0	1	0	−4	200
s_2	0	4	15	0	0	1	−12	900
x_3	0	.5	0	1	0	0	1	75
P	1	0	−1	0	0	0	2	150

BV	P	x_1	x_2	x_3	s_1	s_2	s_3	RHS
→x_2	0	$\frac{1}{5}$	[1]	0	$\frac{1}{5}$	0	$-\frac{4}{5}$	40
s_2	0	1	0	0	−3	1	0	300
x_3	0	.5	0	1	0	0	1	75
P	1	$\frac{1}{5}$	0	0	$\frac{1}{5}$	0	$\frac{6}{5}$	190

Since all the entries in the objective row are nonnegative, this is the final tableau. The solution is $P = 190$, obtained when $x_1 = 0$, $x_2 = 40$, and $x_3 = 75$.

So the salesperson maximizes profit by selling no product A, 40 units of product B an 75 units of product C for a maximum profit of $190.

29. We let x_1, x_2, and x_3 represent the number of gallons of regular, premium, and super premium gasoline, respectively to be refined. The company wants to refine amounts that will maximize its revenue subject to its available resources.

Maximize $P = 1.20x_1 + 1.30x_2 + 1.40x_3$

subject to the following constraints

$$
\begin{aligned}
.6x_1 + .7x_2 + .8x_3 &\leq 140,000 \qquad \text{high octane gas available} \\
.4x_1 + .3x_2 + .2x_3 &\leq 120,000 \qquad \text{low octane gas available} \\
x_1 + x_2 + x_3 &\leq 225,000 \qquad \text{plant capacity} \\
x_1 \geq 0 \quad x_2 \geq 0 \quad x_3 &\geq 0
\end{aligned}
$$

Slack variables are added and the initial tableau is constructed and shown below. The pivot element is 0.8.

BV	P	x_1	x_2	x_3	s_1	s_2	s_3	RHS
→ s_1	0	.6	.7	.8	1	0	0	140,000
s_2	0	.4	.3	.2	0	1	0	120,000
s_3	0	1	1	1	0	0	1	225,000
P	1	−1.2	−1.3	−1.4	0	0	0	0

$R_1 = \frac{r_1}{0.8}$
$R_2 = -.2r_1 + r_2$
$R_3 = -r_1 + r_3$
$R_4 = 1.4r_1 + r_4$

BV	P	x_1	x_2	x_3	s_1	s_2	s_3	RHS
x_3	0	.75	.875	1	1.25	0	0	175,000
→ s_2	0	.25	.125	0	−.25	1	0	85,000
s_3	0	.25	.125	0	−1.25	0	1	50,000
P	1	−.15	−.075	0	1.75	0	0	245,000

$R_3 = 4r_3$
$R_1 = -.75r_3 + r_1$
$R_2 = -.25r_3 + r_2$
$R_4 = .15r_3 + r_4$

BV	P	x_1	x_2	x_3	s_1	s_2	s_3	RHS
x_3	0	0	.5	1	5	0	−3	25,000
→ s_2	0	0	0	0	1	1	−1	35,000
x_1	0	1	.5	0	−5	0	4	200,000
P	1	0	0	0	1	0	.6	275,000

Since all the entries in the objective row are nonnegative, this is the final tableau. The solution is $P = 275,000$, obtained when $x_1 = 200,000$, $x_2 = 0$, and $x_3 = 25,000$.

To achieve a maximize revenue of $275,000, the company should refine 200,000 gallons of regular gasoline, no premium gasoline, and 25,000 of super premium gasoline.

31. We let x_1, x_2, and x_3 represent the amount of money to be invested in stocks, corporate bonds, and municipal bonds respectively. The financial planner wants to maximize return subject to an investment strategy.

Maximize $P = 0.1x_1 + 0.08x_2 + 0.06x_3$

subject to the following constraints

$$\begin{aligned}
x_1 + x_2 + x_3 &\leq 90,000 && \text{money available for investment} \\
x_1 &\leq 45,000 && \text{limit to be invested in stocks} \\
x_2 - x_3 &\leq 18,000 && \text{bond allocation} \\
x_1 \geq 0 \quad x_2 &\geq 0 \quad x_3 \geq 0
\end{aligned}$$

Slack variables are added and the initial tableau is constructed and shown below. The pivot element is 1.

BV	P	x_1	x_2	x_3	s_1	s_2	s_3	RHS
s_1	0	1	1	1	1	0	0	90000
→s_2	0	[1]	0	0	0	1	0	45000
s_3	0	0	1	−1	0	0	1	18000
P	1	−.1	−.08	−.06	0	0	0	0

There are several negative entries in the objective row. We choose the smallest number, make its column the new pivot column, compute the quotients formed by dividing the entries in the RHS by the corresponding entries in the pivot column, and make the entry with the smallest positive quotient the new pivot.

$$\begin{aligned} R_1 &= -r_2 + r_1 \\ R_4 &= .1r_2 + r_4 \end{aligned} \longrightarrow$$

BV	P	x_1	x_2	x_3	s_1	s_2	s_3	RHS
s_1	0	0	1	1	1	−1	0	45000
→x_1	0	[1]	0	0	0	1	0	45000
s_3	0	0	1	−1	0	0	1	18000
P	1	0	−.08	−.06	0	.1	0	4500

This still is not the final tableau, we choose 1 in row s_3, column x_2 and pivot again.

$$\begin{aligned} R_1 &= -r_3 + r_1 \\ R_4 &= .08r_3 + r_4 \end{aligned} \longrightarrow$$

BV	P	x_1	x_2	x_3	s_1	s_2	s_3	RHS
s_1	0	0	0	2	1	−1	−1	27000
x_1	0	1	0	0	0	1	0	45000
→x_2	0	0	[1]	−1	0	0	1	18000
P	1	0	0	−.14	0	.1	.08	5940

Pivoting again, we introduce variable x_3 as a basic variable.

$$\begin{aligned} R_1 &= \tfrac{1}{2}r_1 \\ R_3 &= r_1 + r_3 \\ R_4 &= .14r_1 + r_4 \end{aligned} \longrightarrow$$

BV	P	x_1	x_2	x_3	s_1	s_2	s_3	RHS
→x_3	0	0	0	[1]	$\tfrac{1}{2}$	$-\tfrac{1}{2}$	$-\tfrac{1}{2}$	13500
x_1	0	1	0	0	0	1	0	45000
x_2	0	0	1	0	$\tfrac{1}{2}$	$-\tfrac{1}{2}$	$\tfrac{1}{2}$	31500
P	1	0	0	0	.07	.03	.01	7830

Since all the entries in the objective row are nonnegative, this is the final tableau. The solution is $P = 7830$, obtained when $x_1 = 45,000$, $x_2 = 31,500$, and $x_3 = 13,500$.

The financial planner can maximize her client's investment income, while maintaining her investment strategy by investing \$45,000 in stocks, \$31,500 in corporate bonds, and \$13,500 in municipal bonds. The maximum return will be \$7830.

33. We let x_1, x_2, and x_3 represent the acres of crop A, B, and C respectively that are to be planted. The farmer wants to maximize the profit made from growing the crops.

Maximize $\quad P = 70x_1 + 90x_2 + 50x_3$

subject to the following constraints

$$\begin{array}{ll} x_1 + x_2 + x_3 \le 200 & \text{acres available for planting} \\ 40x_1 + 50x_2 + 30x_3 \le 18,000 & \text{funds available for cultivation} \\ 20x_1 + 30x_2 + 15x_3 \le 4,200 & \text{hours available for labor} \\ x_1 \ge 0 \quad x_2 \ge 0 \quad x_3 \ge 0 \end{array}$$

Slack variables are added and the initial tableau is constructed and shown below.

BV	P	x_1	x_2	x_3	s_1	s_2	s_3	RHS
s_1	0	1	1	1	1	0	0	200
s_2	0	40	50	30	0	1	0	18,000
$\rightarrow s_3$	0	20	$\boxed{30}$	15	0	0	1	4,200
P	1	-70	-90	-50	0	0	0	0

BV	P	x_1	x_2	x_3	s_1	s_2	s_3	RHS
$\rightarrow s_1$	0	$\boxed{\frac{1}{3}}$	0	$\frac{1}{2}$	1	0	$-\frac{1}{30}$	60
s_2	0	$\frac{20}{3}$	0	5	0	1	$-\frac{5}{3}$	11,000
x_2	0	$\frac{2}{3}$	1	$\frac{1}{2}$	0	0	$\frac{1}{30}$	140
P	1	-10	0	-5	0	0	3	12,600

BV	P	x_1	x_2	x_3	s_1	s_2	s_3	RHS
x_1	0	1	0	$\frac{3}{2}$	3	0	$-\frac{1}{10}$	180
s_2	0	0	0	-5	-20	1	-1	9,800
x_2	0	0	1	$-\frac{1}{2}$	-2	0	$\frac{1}{10}$	20
P	1	0	0	10	30	0	2	14,400

Since all the entries in the objective row are nonnegative, this is the final tableau. The solution is $P = 14,400$, obtained when $x_1 = 180$, $x_2 = 20$, and $x_3 = 0$.

The farmer will realize a maximum profit of \$14,400 when he plants 180 acres of crop A, 20 acres of crop B, no crop C.

35. We let x_1, x_2, and x_3 represent the number of cans of type I, type II, and type III cans of nuts to be packed. The company wants to maximize revenue with the nuts it has available.

Maximize $P = 28x_1 + 24x_2 + 20x_3$

subject to the following constraints

$$3x_1 + 4x_2 + 5x_3 \leq 500 \qquad \text{pounds of peanuts on hand}$$

$$x_1 + \frac{1}{2}x_2 \qquad \leq 100 \qquad \text{pounds of pecans on hand}$$

$$x_1 + \frac{1}{2}x_2 \qquad \leq 50 \qquad \text{pounds of cashews on hand}$$

$$x_1 \geq 0 \quad x_2 \geq 0 \quad x_3 \geq 0$$

Slack variables are added and the initial tableau is constructed and shown below.

BV	P	x_1	x_2	x_3	s_1	s_2	s_3	RHS
s_1	0	3	4	5	1	0	0	500
s_2	0	1	$\frac{1}{2}$	0	0	1	0	100
→ s_3	0	$\boxed{1}$	$\frac{1}{2}$	0	0	0	1	50
P	1	−28	−24	−20	0	0	0	0

BV	P	x_1	x_2	x_3	s_1	s_2	s_3	RHS
→ s_1	0	0	$\frac{5}{2}$	$\boxed{5}$	1	0	−3	350
s_2	0	0	0	0	0	1	−1	50
x_1	0	1	$\frac{1}{2}$	0	0	0	1	50
P	1	0	−10	−20	0	0	28	1400

BV	P	x_1	x_2	x_3	s_1	s_2	s_3	RHS
x_3	0	0	$\frac{1}{2}$	1	$\frac{1}{5}$	0	$-\frac{3}{5}$	70
s_2	0	0	0	0	0	1	−1	50
x_1	0	1	$\frac{1}{2}$	0	0	0	1	50
P	1	0	0	0	4	0	16	2800

Since all the entries in the objective row are nonnegative, this is the final tableau. The solution is $P = 2800$, obtained when $x_1 = 50$, $x_2 = 0$, and $x_3 = 70$.

Nutt's Nuts Company will maximize its revenue at \$2800 if it makes 50 cans of type I, 70 cans of type III and no cans of type II.

37. We let x_1, x_2, and x_3 represent the television consoles, stereo systems, and radios, respectively to be manufactured each week in order to maximize profit while keeping with the production constraints.

Maximize $P = 10x_1 + 25x_2 + 3x_3$

subject to the following constraints

$3x_1 + 10x_2 + \ x_3 \le 30,000$	hours available for assembly
$5x_1 + \ 8x_2 + \ x_3 \le 40,000$	hours available for decorating
$.1x_1 + .6x_2 + .1x_3 \le \quad 120$	hours available for crating

$x_1 \ge 0 \quad x_2 \ge 0 \quad x_3 \ge 0$

Slack variables are added and the initial tableau is constructed and shown below. Before setting up the initial tableau, we multiplied the crating constraint by 10 to remove decimals.

BV	P	x_1	x_2	x_3	s_1	s_2	s_3	RHS
s_1	0	3	10	1	1	0	0	30,000
s_2	0	5	8	1	0	1	0	40,000
→ s_3	0	1	$\boxed{6}$	1	0	0	1	1,200
P	1	−10	−25	−3	0	0	0	0

BV	P	x_1	x_2	x_3	s_1	s_2	s_3	RHS
s_1	0	$\frac{4}{3}$	0	$-\frac{2}{3}$	1	0	$-\frac{5}{3}$	28,000
s_2	0	$\frac{11}{3}$	0	$-\frac{1}{3}$	0	1	$-\frac{4}{3}$	38,400
→ x_2	0	$\boxed{\frac{1}{6}}$	1	$\frac{1}{6}$	0	0	$\frac{1}{6}$	200
P	1	$-\frac{35}{6}$	0	$\frac{7}{6}$	0	0	$\frac{25}{6}$	5,000

BV	P	x_1	x_2	x_3	s_1	s_2	s_3	RHS
s_1	0	0	−8	−2	1	0	−3	26,400
s_2	0	0	−22	−4	0	1	−5	34,000
x_1	0	1	6	1	0	0	1	1,200
P	1	0	35	7	0	0	10	12,000

Since all the entries in the objective row are nonnegative, this is the final tableau. The solution is $P = 12,000$, obtained when $x_1 = 1200$, $x_2 = 0$, and $x_3 = 0$.

The manufacturer will maximize profit at \$12,000 if it manufactures 1200 television consoles, and neither stereo cabinets nor radios.

39. We let x_1, x_2, and x_3 represent the televisions made in Chicago, New York, and Denver, respectively. The manufacturer wants to maximize profit.

Maximize $P = 50x_1 + 80x_2 + 40x_3$

subject to the following constraints

$$
\begin{array}{ll}
x_1 + x_2 + x_3 \le 400 & \text{maximum number ordered} \\
20x_1 + 20x_2 + 40x_3 \le 10{,}000 & \text{amount alloted for shipping} \\
6x_1 + 8x_2 + 4x_3 \le 3{,}000 & \text{amount alloted for labor} \\
x_1 \ge 0 \quad x_2 \ge 0 \quad x_3 \ge 0 &
\end{array}
$$

Slack variables are added and the initial tableau is constructed and shown below.

BV	P	x_1	x_2	x_3	s_1	s_2	s_3	RHS
s_1	0	1	1	1	1	0	0	400
s_2	0	20	20	40	0	1	0	10,000
$\to s_3$	0	6	$\boxed{8}$	4	0	0	1	3,000
P	1	−50	−80	−40	0	0	0	0

$$
\begin{array}{l}
R_3 = \tfrac{1}{8}r_3 \\
R_1 = -r_3 + r_1 \\
R_2 = -20r_3 + r_2 \\
R_4 = 80r_3 + r_4
\end{array}
$$

BV	P	x_1	x_2	x_3	s_1	s_2	s_3	RHS
s_1	0	$\tfrac{1}{4}$	0	$\tfrac{1}{2}$	1	0	$-\tfrac{1}{8}$	25
s_2	0	5	0	30	0	1	$-\tfrac{5}{2}$	2,500
x_2	0	$\tfrac{3}{4}$	1	$\tfrac{1}{2}$	0	0	$\tfrac{1}{8}$	375
P	1	10	0	0	0	0	10	30,000

Since all the entries in the objective row are nonnegative, this is the final tableau. The solution is $P = 30{,}000$, obtained when $x_1 = 0$, $x_2 = 375$, and $x_3 = 0$.

The manufacturer will obtain a maximum profit of $30,000, if it ships 375 televisions from New York and none from either Chicago or Denver.

4.3 Solving Minimum Problems in Standard Form Using the Duality Principle

There are three conditions to be met for a minimum problem to be in standard form:
CONDITION 1:   All the variables must be nonnegative.
CONDITION 2:   All other constraints are written as linear expressions that are greater than or equal to a constant.
CONDITION 3:   The objective function is a linear expression with nonnegative coefficients.

1.   The minimum problem is in standard form.

3.   The minimum problem is not in standard form. Condition 3 is not met; the objective function has a negative coefficient.

5.   The minimum problem is not in standard form. Condition 2 is not met; one of the constraints is an expression less than or equal to a constant.

7. To write the dual of the minimum problem we first write the matrix that represents the constraints and the objective function and its transpose.

$$x_1 \quad x_2$$

the matrix $\begin{bmatrix} 1 & 1 & | & 2 \\ 2 & 3 & | & 6 \\ 2 & 3 & | & 0 \end{bmatrix}$ the transpose of the matrix $\begin{bmatrix} 1 & 2 & | & 2 \\ 1 & 3 & | & 3 \\ 2 & 6 & | & 0 \end{bmatrix}$

From the transpose we create the maximum problem:

Maximize $P = 2y_1 + 6y_2$

subject to the conditions

$$y_1 + 2y_2 \le 2$$
$$y_1 + 3y_2 \le 3$$
$$y_1 \ge 0 \quad y_2 \ge 0$$

This maximum problem is the dual of the minimum problem.

9. To write the dual of the minimum problem we first write the matrix that represents the constraints and the objective function and its transpose.

$$x_1 \quad x_2 \quad x_3$$

the matrix $\begin{bmatrix} 1 & 1 & 1 & | & 5 \\ 2 & 1 & 0 & | & 4 \\ 3 & 1 & 1 & | & 0 \end{bmatrix}$ the transpose of the matrix $\begin{bmatrix} 1 & 2 & | & 3 \\ 1 & 1 & | & 1 \\ 1 & 0 & | & 1 \\ 5 & 4 & | & 0 \end{bmatrix}$

From the transpose we create the maximum problem:

Maximize $P = 5y_1 + 4y_2$

subject to the conditions

$$y_1 + 2y_2 \le 2$$
$$y_1 + \ y_2 \le 1$$
$$y_1 \qquad \le 1$$
$$y_1 \ge 0 \quad y_2 \ge 0$$

This maximum problem is the dual of the minimum problem.

11. To write the dual of the minimum problem we first write the matrix that represents the constraints and the objective function and its transpose.

$$
\begin{array}{cccc}
x_1 & x_2 & x_3 & x_4
\end{array}
$$

the matrix $\begin{bmatrix} 1 & 1 & 1 & 2 & | & 60 \\ 3 & 2 & 1 & 2 & | & 90 \\ 3 & 4 & 1 & 2 & | & 0 \end{bmatrix}$ the transpose of the matrix $\begin{bmatrix} 1 & 3 & | & 3 \\ 1 & 2 & | & 4 \\ 1 & 1 & | & 1 \\ 2 & 2 & | & 2 \\ 60 & 90 & | & 0 \end{bmatrix}$

From the transpose we create the maximum problem:

Maximize $P = 60y_1 + 90y_2$

subject to the conditions

$$
\begin{aligned}
y_1 + 3y_2 &\leq 3 \\
y_1 + 2y_2 &\leq 4 \\
y_1 + y_2 &\leq 1 \\
2y_1 + 2y_2 &\leq 2 \\
y_1 \geq 0 \quad y_2 &\geq 0
\end{aligned}
$$

This maximum problem is the dual of the minimum problem.

13. To solve a minimum problem using the Duality Principle, we first must write the matrix that represents the constraints and the objective function, its transpose, and the maximum problem.

$$
\begin{array}{cc}
x_1 & x_2
\end{array}
$$

the matrix: $\begin{bmatrix} 1 & 1 & | & 2 \\ 2 & 6 & | & 6 \\ 6 & 3 & | & 0 \end{bmatrix}$; its transpose: $\begin{bmatrix} 1 & 2 & | & 6 \\ 1 & 6 & | & 3 \\ 2 & 6 & | & 0 \end{bmatrix}$;

the dual: Maximize $P = 2y_1 + 6y_2$

subject to the conditions

$$
\begin{aligned}
y_1 + 2y_2 &\leq 6 \\
y_1 + 6y_2 &\leq 3 \\
y_1 \geq 0 \quad y_2 &\geq 0
\end{aligned}
$$

From here we set up the initial tableau and solve the problem using the simplex method.

BV	P	y_1	y_2	s_1	s_2	RHS
s_1	0	1	2	1	0	6
$\rightarrow s_2$	0	1	$\boxed{6}$	0	1	3
P	1	-2	-6	0	0	0

$\longrightarrow$

BV	P	y_1	y_2	s_1	s_2	RHS
s_1	0	$\frac{2}{3}$	0	1	$-\frac{1}{3}$	5
$\rightarrow y_2$	0	$\boxed{\frac{1}{6}}$	1	0	$\frac{1}{6}$	$\frac{1}{2}$
P	1	-1	0	0	1	3

$\longrightarrow$

BV	P	y_1	y_2	s_1	s_2	RHS
s_1	0	0	-4	1	-1	3
y_1	0	1	6	0	1	3
P	1	0	6	0	2	6

This is the final tableau. From it we read that the minimum cost $C = 6$ is obtained when $x_1 = 0$ and $x_2 = 2$.

15. To solve a minimum problem using the Duality Principle, we first must write the matrix that represents the constraints and the objective function, its transpose, and the maximum problem.

$$\begin{array}{cc} x_1 & x_2 \end{array}$$

the matrix: $\begin{bmatrix} 1 & 1 & | & 4 \\ 3 & 4 & | & 12 \\ 6 & 3 & | & 0 \end{bmatrix}$; its transpose: $\begin{bmatrix} 1 & 3 & | & 6 \\ 1 & 4 & | & 3 \\ 4 & 12 & | & 0 \end{bmatrix}$;

the dual: Maximize $P = 4y_1 + 12y_2$

subject to the conditions

$$y_1 + 3y_2 \le 6$$
$$y_1 + 4y_2 \le 3$$
$$y_1 \ge 0 \quad y_2 \ge 0$$

From here we set up the initial tableau and solve the problem using the simplex method.

BV	P	y_1	y_2	s_1	s_2	RHS
s_1	0	1	3	1	0	6
→ s_2	0	1	[4]	0	1	3
P	1	−4	−12	0	0	0

$\longrightarrow$

BV	P	y_1	y_2	s_1	s_2	RHS
s_1	0	$\frac{1}{4}$	0	1	$-\frac{3}{4}$	$\frac{15}{4}$
→ y_2	0	$\boxed{\frac{1}{4}}$	1	0	$\frac{1}{4}$	$\frac{3}{4}$
P	1	−1	0	0	3	9

$\longrightarrow$

BV	P	y_1	y_2	s_1	s_2	RHS
s_1	0	0	−1	1	−1	3
y_1	0	1	4	0	1	3
P	1	0	4	0	4	12

This is the final tableau. From it we read that the minimum cost $C = 12$ is obtained when $x_1 = 0$ and $x_2 = 4$.

17. To solve a minimum problem using the Duality Principle, we first must write the matrix that represents the constraints and the objective function, its transpose, and the maximum problem.

$$
\begin{array}{ccc}
x_1 & x_2 & x_3
\end{array}
$$

the matrix: $\left[\begin{array}{rrr|r} 1 & -3 & 4 & 12 \\ 3 & 1 & 2 & 10 \\ 1 & -1 & -1 & -8 \\ 1 & 2 & 1 & 0 \end{array}\right]$ its transpose: $\left[\begin{array}{rrr|r} 1 & 3 & 1 & 1 \\ -3 & 1 & -1 & 2 \\ 4 & 2 & -1 & 1 \\ 12 & 10 & -8 & 0 \end{array}\right]$

the dual: Maximize $P = 12y_1 + 10y_2 - 8y_3$

subject to the conditions

$$
\begin{aligned}
y_1 + 3y_2 + y_3 &\le 1 \\
-3y_1 + y_2 - y_3 &\le 2 \\
4y_1 + 2y_2 - y_3 &\le 1 \\
y_1 \ge 0 \quad y_2 \ge 0 \quad y_3 &\ge 0
\end{aligned}
$$

From here we set up the initial tableau and solve the problem using the simplex method.

BV	P	y_1	y_2	y_3	s_1	s_2	s_3	RHS
s_1	0	1	3	1	1	0	0	1
s_2	0	-3	1	-1	0	1	0	2
$\to s_3$	0	☐4	2	-1	0	0	1	1
P	1	-12	-10	8	0	0	0	0

$\longrightarrow$

BV	P	y_1	y_2	y_3	s_1	s_2	s_3	RHS
$\to s_1$	0	0	☐$\frac{5}{2}$	$\frac{5}{4}$	1	0	$-\frac{1}{4}$	$\frac{3}{4}$
s_2	0	0	$\frac{5}{2}$	$-\frac{7}{4}$	0	1	$\frac{3}{4}$	$\frac{11}{4}$
y_1	0	1	$\frac{1}{2}$	$-\frac{1}{4}$	0	0	$\frac{1}{4}$	$\frac{1}{4}$
P	1	0	-4	5	0	0	3	3

BV	P	y_1	y_2	y_3	s_1	s_2	s_3	RHS
y_2	0	0	1	$\frac{1}{2}$	$\frac{2}{5}$	0	$-\frac{1}{10}$	$\frac{3}{10}$
$\longrightarrow s_2$	0	0	0	-3	-1	1	1	2
y_1	0	1	0	$-\frac{1}{2}$	$-\frac{1}{5}$	0	$\frac{3}{10}$	$\frac{1}{10}$
P	1	0	0	7	$\frac{8}{5}$	0	$\frac{13}{5}$	$\frac{21}{5}$

This is the final tableau. From it we read that the minimum cost $C = 4.20$ is obtained when $x_1 = \dfrac{8}{5}$, $x_2 = 0$, and $x_3 = \dfrac{13}{5}$.

19. To solve a minimum problem using the Duality Principle, we first must write the matrix that represents the constraints and the objective function, its transpose, and the maximum problem.

$$
\begin{array}{cccc}
x_1 & x_2 & x_3 & x_4
\end{array}
$$

the matrix:
$$
\begin{bmatrix}
1 & 0 & 1 & 0 & | & 1 \\
0 & 1 & 0 & 1 & | & 1 \\
-1 & -1 & -1 & -1 & | & -3 \\
1 & 4 & 2 & 4 & | & 0
\end{bmatrix}
$$

its transpose:
$$
\begin{bmatrix}
1 & 0 & -1 & | & 1 \\
0 & 1 & -1 & | & 4 \\
1 & 0 & -1 & | & 2 \\
0 & 1 & -1 & | & 4 \\
1 & 1 & -3 & | & 0
\end{bmatrix}
$$

the dual: Maximize $P = y_1 + y_2 - 3y_3$

subject to the conditions

$$
\begin{aligned}
y_1 \quad & -y_3 \le 1 \\
y_2 - y_3 & \le 4 \\
y_1 \quad & -y_3 \le 2 \\
y_2 - y_3 & \le 4 \\
y_1 \ge 0 \quad y_2 \ge 0 \quad & y_3 \ge 0
\end{aligned}
$$

From here we set up the initial tableau and solve the problem using the simplex method.

BV	P	y_1	y_2	y_3	s_1	s_2	s_3	s_4	RHS
→s_1	0	**1**	0	-1	1	0	0	0	1
s_2	0	0	1	-1	0	1	0	0	4
s_3	0	1	0	-1	0	0	1	0	2
s_4	0	0	1	-1	0	0	0	1	4
P	1	-1	-1	3	0	0	0	0	0

BV	P	y_1	y_2	y_3	s_1	s_2	s_3	s_4	RHS
y_1	0	1	0	-1	1	0	0	0	1
→s_2	0	0	**1**	-1	0	1	0	0	4
s_3	0	0	0	0	-1	0	1	0	1
s_4	0	0	1	-1	0	0	0	1	4
P	1	0	-1	2	1	0	0	0	1

BV	P	y_1	y_2	y_3	s_1	s_2	s_3	s_4	RHS
y_1	0	1	0	-1	1	0	0	0	1
y_2	0	0	1	-1	0	1	0	0	4
s_3	0	0	0	0	-1	0	1	0	1
s_4	0	0	0	0	0	-1	0	1	0
P	1	0	0	1	1	1	0	0	5

This is the final tableau. From it we read that the minimum cost $C = 5$ is obtained when $x_1 = 1, x_2 = 1, x_3 = 0$ and $x_4 = 0$.

21. Let x_1 and x_2 represent the number of pill Ps and pill Qs, respectively, that Mr. Jones should add to his diet. He wishes to minimize his cost while meeting the nutritional requirements. He wants to

Minimize $C = 3x_1 + 4x_2$ The matrix representing the system and its transpose are

subject to the conditions

$$5x_1 + 10x_2 \geq 50$$
$$2x_1 + x_2 \geq 8$$
$$x_1 \geq 0 \quad x_2 \geq 0$$

$$\text{matrix:} \begin{bmatrix} 5 & 10 & | & 50 \\ 2 & 1 & | & 8 \\ 3 & 4 & | & 0 \end{bmatrix} \qquad \text{transpose:} \begin{bmatrix} 5 & 2 & | & 3 \\ 10 & 1 & | & 4 \\ 50 & 8 & | & 0 \end{bmatrix}$$

The dual of the problem is

Maximize $P = 50y_1 + 8y_2$

subject to the conditions

$$5y_1 + 2y_2 \leq 3$$
$$10y_1 + y_2 \leq 4$$
$$y_1 \geq 0 \quad y_2 \geq 0$$

We set up the initial simplex tableau and solve the maximum problem.

BV	P	y_1	y_2	s_1	s_2	RHS
s_1	0	5	2	1	0	3
→ s_2	0	[10]	1	0	1	4
P	1	−50	−8	0	0	0

$\longrightarrow$

BV	P	y_1	y_2	s_1	s_2	RHS
→ s_1	0	0	$\frac{3}{2}$	1	$-\frac{1}{2}$	1
y_1	0	1	$\frac{1}{10}$	0	$\frac{1}{10}$	$\frac{2}{5}$
P	1	0	−3	0	5	20

$\longrightarrow$

BV	P	y_1	y_2	s_1	s_2	RHS
y_2	0	0	[1]	$\frac{2}{3}$	$-\frac{1}{3}$	$\frac{2}{3}$
y_1	0	1	0	$-\frac{1}{15}$	$\frac{2}{15}$	$\frac{1}{3}$
P	1	0	0	2	4	22

The solution to the minimum problem is minimal $C = 22$ when $x_1 = 2$ and $x_2 = 4$.
Mr. Jones can take his supplements with a minimum cost of \$0.22 per day if he takes 2 of pill P and 4 of pill Q.

23. Let x_1, x_2, and x_3 represent the units of product A, B, and C, respectively, that Argus Company manufactures. The company wants to minimize costs while meeting production requirements.

Minimize $C = 4x_1 + 2x_2 + x_3$

subject to the conditions

$$
\begin{array}{rl}
x_1 & \geq 20 \\
x_2 & \geq 30 \\
x_3 & \geq 40 \\
x_1 + x_2 + x_3 & \geq 200 \\
x_1 \geq 0 \quad x_2 \geq 0 \quad x_3 \geq 0 &
\end{array}
$$

The matrix (on the left) representing the system and its transpose (on the right) are:

$$
\begin{array}{c}
\begin{array}{ccc} x_1 & x_2 & x_3 \end{array} \\
\begin{bmatrix}
1 & 0 & 0 & 20 \\
0 & 1 & 0 & 30 \\
0 & 0 & 1 & 40 \\
1 & 1 & 1 & 200 \\
4 & 2 & 1 & 0
\end{bmatrix}
\end{array}
\qquad
\begin{array}{c}
\begin{array}{cccc} y_1 & y_2 & y_3 & y_4 \end{array} \\
\begin{bmatrix}
1 & 0 & 0 & 1 & 4 \\
0 & 1 & 0 & 1 & 2 \\
0 & 0 & 1 & 1 & 1 \\
20 & 30 & 40 & 200 & 0
\end{bmatrix}
\end{array}
$$

The dual of the problem is

Maximize $P = 20y_1 + 30y_2 + 40y_3 + 200y_4$

subject to the conditions

$$
\begin{array}{rl}
y_1 + & y_4 \leq 4 \\
y_2 + & y_4 \leq 2 \\
y_3 + & y_4 \leq 1 \\
y_1 \geq 0 \quad y_2 \geq 0 \quad y_3 \geq 0 \quad y_4 \geq 0 &
\end{array}
$$

We set up the initial simplex tableau and solve the maximum problem.

BV	P	y_1	y_2	y_3	y_4	s_1	s_2	s_3	RHS
s_1	0	1	0	0	1	1	0	0	4
s_2	0	0	1	0	1	0	1	0	2
→ s_3	0	0	0	1	$\boxed{1}$	0	0	1	1
P	1	−20	−30	−40	−200	0	0	0	0

BV	P	y_1	y_2	y_3	y_4	s_1	s_2	s_3	RHS
s_1	0	1	0	−1	0	1	0	−1	3
→ s_2	0	0	$\boxed{1}$	−1	0	0	1	−1	1
y_4	0	0	0	1	1	0	0	1	1
P	1	−20	−30	160	0	0	0	200	200

BV	P	y_1	y_2	y_3	y_4	s_1	s_2	s_3	RHS
→ s_1	0	[1]	0	−1	0	1	0	−1	3
⟶ y_2	0	0	1	−1	0	0	1	−1	1
y_4	0	0	0	1	1	0	0	1	1
P	1	−20	0	130	0	0	30	170	230

BV	P	y_1	y_2	y_3	y_4	s_1	s_2	s_3	RHS
y_1	0	1	0	−1	0	1	0	−1	3
⟶ y_2	0	0	1	−1	0	0	1	−1	1
y_4	0	0	0	1	1	0	0	1	1
P	1	0	0	110	0	20	30	150	290

Minimum cost is $C = 290$, obtained when $x_1 = 20$, $x_2 = 30$, and $x_3 = 150$.

Argus will spend a minimum of $290 when they produce 20 of product A, 30 of product B, and 150 of product C.

25. Let $x_1, x_2,$ and x_3 represent the number of lunch #1, lunch #2, and lunch #3 orders Mrs. Mintz purchases. She and her friends want to order the foods they need at minimum cost. They want to

Minimize $C = 6.20x_1 + 7.40x_2 + 9.10x_3$

subject to the conditions

$$\begin{aligned}
x_1 & \geq 4 \\
x_1 + x_2 + x_3 &\geq 9 \\
x_1 + x_3 &\geq 6 \\
x_2 + x_3 &\geq 5 \\
x_1 \geq 0 \quad x_2 \geq 0 \quad x_3 &\geq 0
\end{aligned}$$

The matrix (on the left) representing the system and its transpose (on the right) are:

$$
\begin{array}{ccc}
x_1 & x_2 & x_3 \\
\end{array}
\left[\begin{array}{ccc|c}
1 & 0 & 0 & 4 \\
1 & 1 & 1 & 9 \\
1 & 0 & 1 & 6 \\
0 & 1 & 1 & 5 \\
6.2 & 7.4 & 9.1 & 0
\end{array}\right]
\qquad
\begin{array}{cccc}
y_1 & y_2 & y_3 & y_4 \\
\end{array}
\left[\begin{array}{cccc|c}
1 & 1 & 1 & 0 & 6.2 \\
0 & 1 & 0 & 1 & 7.4 \\
0 & 1 & 1 & 1 & 9.1 \\
4 & 9 & 6 & 5 & 0
\end{array}\right]
$$

The dual of the problem is

Maximize $P = 4y_1 + 9y_2 + 6y_3 + 5y_4$

subject to the conditions

$$\begin{aligned}
y_1 + y_2 + y_3 & \leq 6.2 \\
y_2 + y_4 &\leq 7.4 \\
y_2 + y_3 + y_4 &\leq 9.1 \\
y_1 \geq 0 \quad y_2 \geq 0 \quad y_3 \geq 0 \quad y_4 &\geq 0
\end{aligned}$$

We set up the initial simplex tableau and solve the maximum problem.

BV	P	y_1	y_2	y_3	y_4	s_1	s_2	s_3	RHS
→ s_1	0	1	[1]	1	0	1	0	0	6.2
s_2	0	0	1	0	1	0	1	0	7.4
s_3	0	0	1	1	1	0	0	1	9.1
P	1	−4	−9	−6	−5	0	0	0	0

BV	P	y_1	y_2	y_3	y_4	s_1	s_2	s_3	RHS
y_2	0	1	1	1	0	1	0	0	6.2
→ s_2	0	−1	0	−1	[1]	−1	1	0	1.2
s_3	0	−1	0	0	1	−1	0	1	2.9
P	1	5	0	3	−5	9	0	0	55.8

BV	P	y_1	y_2	y_3	y_4	s_1	s_2	s_3	RHS
y_2	0	1	1	1	0	1	0	0	6.2
y_4	0	−1	0	−1	1	−1	1	0	1.2
→ s_3	0	0	0	[1]	0	0	−1	1	1.7
P	1	0	0	−2	0	4	5	0	61.8

BV	P	y_1	y_2	y_3	y_4	s_1	s_2	s_3	RHS
y_3	0	1	1	0	0	0	1	−1	4.5
y_4	0	−1	0	0	1	−1	0	1	2.9
→ s_3	0	0	0	1	0	0	−1	1	1.7
P	1	0	0	0	0	4	3	2	65.2

Minimum cost of lunch is $C = \$65.20$, obtained when Mrs. Mintz buys 4 orders of lunch #1, 3 orders of lunch #2, and 2 orders of lunch #3.

4.4 The Simplex Method with Mixed Constraints

1.   Rewrite the constraints: Introduce slack variables:

$$\begin{aligned} x_1 + x_2 &\le 12 \\ -5x_1 - 2x_2 &\le -36 \\ -7x_1 - 4x_2 &\le -14 \\ x_1 \ge 0,\ x_2 &\ge 0 \end{aligned}$$
$$\begin{aligned} x_1 + x_2 + s_1 &= 12 \\ -5x_1 - 2x_2 \qquad + s_2 \quad\ &= -36 \\ -7x_1 - 4x_2 \qquad\qquad + s_3 &= -14 \\ x_1 \ge 0,\ x_2 \ge 0,\ s_1 \ge 0,\ s_2 \ge 0,\ s_3 \ge 0 \end{aligned}$$

We now set up the initial tableau and use the alternate pivoting method as long as there are negative entries in the RHS. When all the entries in the RHS are positive, we use the standard way of choosing a pivot.

BV	P	x_1	x_2	s_1	s_2	s_3	RHS
s_1	0	1	1	1	0	0	12
s_2	0	$\boxed{-5}$	-2	0	1	0	-36
s_3	0	-7	-4	0	0	1	-14
P	1	-3	-4	0	0	0	0

$\xrightarrow{\text{Alternative Pivoting Strategy}}$

BV	P	x_1	x_2	s_1	s_2	s_3	RHS
s_1	0	0	$\boxed{\frac{3}{5}}$	1	$\frac{1}{5}$	0	$\frac{24}{5}$
x_1	0	1	$\frac{2}{5}$	0	$-\frac{1}{5}$	0	$\frac{36}{5}$
s_3	0	0	$-\frac{6}{5}$	0	$-\frac{7}{5}$	1	$\frac{182}{5}$
P	1	0	$-\frac{14}{5}$	0	$-\frac{3}{5}$	0	$\frac{108}{5}$

$\xrightarrow{\text{Standard Pivoting Strategy}}$

BV	P	x_1	x_2	s_1	s_2	s_3	RHS
x_2	0	0	1	$\frac{5}{3}$	$\frac{1}{3}$	0	8
x_1	0	1	0	$-\frac{2}{3}$	$-\frac{1}{3}$	0	4
s_3	0	0	0	2	-1	1	46
P	1	0	0	$\frac{14}{3}$	$\frac{1}{3}$	0	44

The maximum $P = 44$, obtained when $x_1 = 4$ and $x_2 = 8$.

3. Rewrite the constraints: Introduce slack variables:

$$x_1 + 3x_2 + x_3 \le 9$$
$$-2x_1 - 3x_2 + x_3 \le -2$$
$$-3x_1 + 2x_2 - x_3 \le -5$$
$$x_1 \ge 0, \ x_2 \ge 0, \ x_3 \ge 0$$

$$x_1 + 3x_2 + x_3 + s_1 \qquad\qquad = 9$$
$$-2x_1 - 3x_2 + x_3 \qquad + s_2 \qquad = -2$$
$$-3x_1 + 2x_2 - x_3 \qquad\qquad + s_3 = -5$$
$$x_1 \ge 0, \ x_2 \ge 0, \ x_3 \ge 0, \ s_1 \ge 0, \ s_2 \ge 0, \ s_3 \ge 0$$

We now set up the initial tableau and use the alternate pivoting method as long as there are negative entries in the RHS. When all the entries in the RHS are positive, we use the standard way of choosing a pivot.

BV	P	x_1	x_2	x_3	s_1	s_2	s_3	RHS
s_1	0	1	3	1	1	0	0	9
s_2	0	[−2]	−3	1	0	1	0	−2
s_3	0	−3	2	−1	0	0	1	−5
P	1	−3	−2	1	0	0	0	0

Alternative Pivoting Strategy →

BV	P	x_1	x_2	x_3	s_1	s_2	s_3	RHS
s_1	0	0	$\frac{3}{2}$	$\frac{3}{2}$	1	$\frac{1}{2}$	0	8
x_1	0	1	$\frac{3}{2}$	$-\frac{1}{2}$	0	$-\frac{1}{2}$	0	1
s_3	0	0	$\frac{13}{2}$	$\boxed{-\frac{5}{2}}$	0	$-\frac{3}{2}$	1	−2
P	1	0	$\frac{5}{2}$	$-\frac{1}{2}$	0	$-\frac{3}{2}$	0	3

Alternative Pivoting Strategy →

BV	P	x_1	x_2	x_3	s_1	s_2	s_3	RHS
s_1	0	0	$\frac{27}{5}$	0	1	$-\frac{2}{5}$	$\frac{3}{5}$	$\frac{34}{5}$
x_1	0	1	$\frac{1}{5}$	0	0	$-\frac{1}{5}$	$-\frac{1}{5}$	$\frac{7}{5}$
x_3	0	0	$-\frac{13}{5}$	1	0	$\boxed{\frac{3}{5}}$	$-\frac{2}{5}$	$\frac{4}{5}$
P	1	0	$\frac{6}{5}$	0	0	$-\frac{6}{5}$	$-\frac{1}{5}$	$\frac{17}{5}$

Standard Pivoting Strategy →

BV	P	x_1	x_2	x_3	s_1	s_2	s_3	RHS
s_1	0	0	$\boxed{\frac{11}{3}}$	$\frac{2}{3}$	1	0	$\frac{1}{3}$	$\frac{22}{3}$
x_1	0	1	$-\frac{2}{3}$	$\frac{1}{3}$	0	0	$-\frac{1}{3}$	$\frac{5}{3}$
s_2	0	0	$-\frac{13}{3}$	$\frac{5}{3}$	0	1	$-\frac{2}{3}$	$\frac{4}{3}$
P	1	0	−4	2	0	0	−1	5

→

BV	P	x_1	x_2	x_3	s_1	s_2	s_3	RHS
x_2	0	0	1	$\frac{2}{11}$	$\frac{3}{11}$	0	$\boxed{\frac{1}{11}}$	2
x_1	0	1	0	$\frac{5}{11}$	$\frac{2}{11}$	0	$-\frac{3}{11}$	3
s_2	0	0	0	$\frac{27}{11}$	$\frac{13}{11}$	1	$-\frac{3}{11}$	10
P	1	0	0	$\frac{30}{11}$	$\frac{12}{11}$	0	$-\frac{7}{11}$	13

→

BV	P	x_1	x_2	x_3	s_1	s_2	s_3	RHS
s_3	0	0	11	2	3	0	1	22
x_1	0	1	3	1	1	0	0	9
s_2	0	0	3	3	2	1	0	16
P	1	0	7	4	3	0	0	27

Maximum $P = 27$, obtained when $x_1 = 9$, $x_2 = 0$ and $x_3 = 0$.

5.   We rewrite the constraints: Then we introduce slack variables:

$$2x_1 + x_2 \leq 4$$
$$x_1 + x_2 \leq 3$$
$$-x_1 - x_2 \leq -3$$
$$x_1 \geq 0, \ x_2 \geq 0$$

$$2x_1 + x_2 + s_1 \qquad\qquad = 4$$
$$x_1 + x_2 \qquad + s_2 \qquad = 3$$
$$-x_1 - x_2 \qquad\qquad + s_3 = -3$$
$$x_1 \geq 0, \ x_2 \geq 0, \ s_1 \geq 0, \ s_2 \geq 0, \ s_3 \geq 0$$

We now set up the initial tableau and use the alternate pivoting method as long as there are negative entries in the RHS. When all the entries in the RHS are positive, we use the standard way of choosing a pivot.

BV	P	x_1	x_2	s_1	s_2	s_3	RHS
s_1	0	2	1	1	0	0	4
s_2	0	1	1	0	1	0	3
s_3	0	[−1]	−1	0	0	1	−3
P	1	−3	−2	0	0	0	0

$\xrightarrow{\text{Alternative Pivoting Strategy}}$

BV	P	x_1	x_2	s_1	s_2	s_3	RHS
s_1	0	0	[−1]	1	0	2	−2
s_2	0	0	0	0	1	1	0
x_1	0	1	1	0	0	−1	3
P	1	0	1	0	0	−3	9

$\xrightarrow{\text{Alternative Pivoting Strategy}}$

BV	P	x_1	x_2	s_1	s_2	s_3	RHS
x_2	0	0	1	−1	0	−2	2
s_2	0	0	0	0	1	[1]	0
x_1	0	1	0	1	0	1	1
P	1	0	0	1	0	−1	7

$\xrightarrow{\text{Standard Pivoting Strategy}}$

BV	P	x_1	x_2	s_1	s_2	s_3	RHS
x_2	0	0	1	−1	2	0	2
s_3	0	0	0	0	1	1	0
x_1	0	1	0	1	−1	0	1
P	1	0	0	1	1	0	7

The final tableau is on the right. Maximum $P = 7$, obtained when $x_1 = 1$ and $x_2 = 2$.

7. Maximize $P = -z = -6x_1 - 8x_2 - x_3$.

Rewrite the constraints:

$$-3x_1 - 5x_2 - 3x_3 \le -20$$
$$-x_1 - 3x_2 - 2x_3 \le -9$$
$$-6x_1 - 2x_2 - 5x_3 \le -30$$
$$x_1 + x_2 + x_3 \le 10$$
$$x_1 \ge 0, \ x_2 \ge 0, \ x_3 \ge 0$$

Introduce slack variables:

$$-3x_1 - 5x_2 - 3x_3 + s_1 = -20$$
$$-x_1 - 3x_2 - 2x_3 + s_2 = -9$$
$$-6x_1 - 2x_2 - 5x_3 + s_3 = -30$$
$$x_1 + x_2 + x_3 + s_4 = 10$$
$$x_1 \ge 0, \ x_2 \ge 0, \ x_3 \ge 0, \ s_1 \ge 0, \ s_2 \ge 0, \ s_3 \ge 0, \ s_4 \ge 0$$

We now set up the initial tableau and use the alternate pivoting method as long as there are negative entries in the RHS. When all the entries in the RHS are positive, we use the standard way of choosing a pivot.

BV	P	x_1	x_2	x_3	s_1	s_2	s_3	s_4	RHS
s_1	0	$\boxed{-3}$	-5	-3	1	0	0	0	-20
s_2	0	-1	-3	-2	0	1	0	0	-9
s_3	0	-6	-2	-5	0	0	1	0	-30
s_4	0	1	1	1	0	0	0	1	10
P	1	6	8	1	0	0	0	0	0

Alternative Pivoting Strategy →

BV	P	x_1	x_2	x_3	s_1	s_2	s_3	s_4	RHS
x_1	0	1	$\frac{5}{3}$	1	$-\frac{1}{3}$	0	0	0	$\frac{20}{3}$
s_2	0	0	$\boxed{-\frac{4}{3}}$	-1	$-\frac{1}{3}$	1	0	0	$-\frac{7}{3}$
s_3	0	0	8	1	-2	0	1	0	10
s_4	0	0	$-\frac{2}{3}$	0	$\frac{1}{3}$	0	0	1	$\frac{10}{3}$
P	1	0	-2	-5	2	0	0	0	-40

Alternative Pivoting Strategy →

BV	P	x_1	x_2	x_3	s_1	s_2	s_3	s_4	RHS
x_1	0	1	0	$-\frac{1}{4}$	$-\frac{3}{4}$	$\frac{5}{4}$	0	0	$\frac{15}{4}$
x_2	0	0	1	$\frac{3}{4}$ ·	$\frac{1}{4}$	$-\frac{3}{4}$	0	0	$\frac{7}{4}$
s_3	0	0	0	$\boxed{-5}$	-4	6	1	0	-4
s_4	0	0	0	$\frac{1}{2}$	$\frac{1}{2}$	$-\frac{1}{2}$	0	1	$\frac{9}{2}$
P	1	0	0	$-\frac{7}{2}$	$\frac{5}{2}$	$-\frac{3}{2}$	0	0	$-\frac{73}{2}$

Alternative Pivoting Strategy →

BV	P	x_1	x_2	x_3	s_1	s_2	s_3	s_4	RHS
x_1	0	1	0	0	$-\frac{11}{20}$	$\boxed{\frac{19}{20}}$	$-\frac{1}{20}$	0	$\frac{79}{20}$
x_2	0	0	1	0	$-\frac{7}{20}$	$\frac{3}{20}$	$\frac{3}{20}$	0	$\frac{23}{20}$
x_3	0	0	0	1	$\frac{4}{5}$	$-\frac{6}{5}$	$-\frac{1}{5}$	0	$\frac{4}{5}$
s_4	0	0	0	0	$\frac{1}{10}$	$\frac{1}{10}$	$\frac{1}{10}$	1	$\frac{41}{10}$
P	1	0	0	0	$\frac{53}{10}$	$-\frac{57}{10}$	$-\frac{7}{10}$	0	$-\frac{337}{10}$

BV	P	x_1	x_2	x_3	s_1	s_2	s_3	s_4	RHS
s_2	0	$\frac{20}{19}$	0	0	$-\frac{11}{19}$	1	$-\frac{1}{19}$	0	$\frac{79}{19}$
x_2	0	$-\frac{3}{19}$	1	0	$-\frac{5}{19}$	0	$\boxed{\frac{3}{19}}$	0	$\frac{10}{19}$
x_3	0	$\frac{24}{19}$	0	1	$\frac{2}{19}$	0	$-\frac{5}{19}$	0	$\frac{110}{19}$
s_4	0	$-\frac{2}{19}$	0	0	$\frac{3}{19}$	0	$\frac{2}{19}$	1	$\frac{70}{19}$
P	1	6	0	0	2	0	-1	0	-10

Standard Pivoting Strategy $\longrightarrow x_3$

BV	P	x_1	x_2	x_3	s_1	s_2	s_3	s_4	RHS
s_2	0	1	$\frac{1}{3}$	0	$-\frac{2}{3}$	1	0	0	$\frac{13}{3}$
s_3	0	-1	$\frac{19}{3}$	0	$-\frac{5}{3}$	0	1	0	$\frac{10}{3}$
x_3	0	1	$\frac{5}{3}$	1	$-\frac{1}{3}$	0	0	0	$\frac{20}{3}$
s_4	0	0	$-\frac{2}{3}$	0	$\frac{1}{3}$	0	0	1	$\frac{10}{3}$
P	1	5	$\frac{19}{3}$	0	$\frac{1}{3}$	0	0	0	$-\frac{20}{3}$

Standard Pivoting Strategy $\longrightarrow x_3$

The maximum value of $P = -\dfrac{20}{3}$, so the minimum value of $z = \dfrac{20}{3}$, and is obtained when

$x_1 = 0$, $x_2 = 0$, and $x_3 = \dfrac{20}{3}$.

9. We define the variables $x_1, x_2, x_3,$ and x_4 so that

x_1 = the number of units shipped from M1 to A1,
x_2 = the number of units shipped from M1 to A2,
x_3 = the number of units shipped from M2 to A1, and
x_4 = the number of units shipped from M2 to A2.

The objective is to minimize shipping costs

$C = 400x_1 + 100x_2 + 200x_3 + 300x_4$

Subject to the constraints

$$\begin{aligned}
x_1 + x_2 &\leq 600 \\
x_3 + x_4 &\leq 400 \\
x_1 + x_3 &\geq 500 \\
x_2 + x_4 &\geq 300 \\
x_1 \geq 0, \; x_2 \geq 0, \; x_3 \geq 0, \; x_4 &\geq 0
\end{aligned}$$

We change the problem to a maximization problem and write the constraints as less than or equal to inequalities.

Maximize $P = -C = -400x_1 - 100x_2 - 200x_3 - 300x_4$

subject to

$$\begin{aligned}
x_1 + x_2 &\leq 600 \\
x_3 + x_4 &\leq 400 \\
-x_1 - x_3 &\leq -500 \\
-x_2 - x_4 &\leq -300 \\
x_1 \geq 0, \; x_2 \geq 0, \; x_3 \geq 0, \; x_4 &\geq 0
\end{aligned}$$

We introduce nonnegative slack variables and set up the initial tableau.

BV	P	x_1	x_2	x_3	x_4	s_1	s_2	s_3	s_4	RHS
s_1	0	1	1	0	0	1	0	0	0	600
s_2	0	0	0	1	1	0	1	0	0	400
s_3	0	[-1]	0	-1	0	0	0	1	0	-500
s_4	0	0	-1	0	-1	0	0	0	1	-300
P	1	400	100	200	300	0	0	0	0	0

Since there are negative entries in the RHS, we use the alternate pivoting strategy.

Alternative Pivoting Strategy →

BV	P	x_1	x_2	x_3	x_4	s_1	s_2	s_3	s_4	RHS
s_1	0	0	1	-1	0	1	0	1	0	100
s_2	0	0	0	1	1	0	1	0	0	400
x_1	0	1	0	1	0	0	0	-1	0	500
s_4	0	0	[-1]	0	-1	0	0	0	1	-300
P	1	0	100	-200	300	0	0	400	0	-200,000

Alternative Pivoting Strategy →

BV	P	x_1	x_2	x_3	x_4	s_1	s_2	s_3	s_4	RHS
s_1	0	0	0	[-1]	-1	1	0	1	1	-200
s_2	0	0	0	1	1	0	1	0	0	400
x_1	0	1	0	1	0	0	0	-1	0	500
x_2	0	0	1	0	1	0	0	0	-1	300
P	1	0	0	-200	200	0	0	400	100	-230,000

Alternative Pivoting Strategy →

BV	P	x_1	x_2	x_3	x_4	s_1	s_2	s_3	s_4	RHS
x_3	0	0	0	1	1	-1	0	-1	-1	200
s_2	0	0	0	0	0	[1]	1	1	1	200
x_1	0	1	0	0	-1	1	0	0	1	300
x_2	0	0	1	0	1	0	0	0	-1	300
P	1	0	0	0	400	-200	0	200	-100	-190,000

Standard Pivoting Strategy →

BV	P	x_1	x_2	x_3	x_4	s_1	s_2	s_3	s_4	RHS
x_3	0	0	0	1	1	0	1	0	0	400
s_1	0	0	0	0	0	1	1	1	1	200
x_1	0	1	0	0	-1	0	-1	-1	0	100
x_2	0	0	1	0	1	0	0	0	-1	300
P	1	0	0	0	400	0	200	400	100	-150,000

The maximum $P = -150,000$, so the minimum cost $= 150,000$ obtained when $x_1 = 100$, $x_2 = 300$, $x_3 = 400$ and $x_4 = 0$.

Private motors should ship 100 engines from M1 to A1, 300 engines from M1 to A2, 400 engines from M2 to A1, and no engines from M2 to A2 for a minimum shipping charge of $150,000.

11. We will let x_1, x_2, x_3 represent the number of units of foods I, II, III, respectively, to be put into the mixture. The objective is to minimize the cost of the mixture while meeting or exceeding the nutrient levels. We want to

Minimize $C = 2x_1 + x_2 + 3x_3$

subject to the conditions

$2x_1 + 3x_2 + 4x_3 \geq 20$
$4x_1 + 2x_2 + 2x_3 \geq 15$
$x_1 \geq 0, \ x_2 \geq 0, \ x_3 \geq 0$

We change the problem to a maximization problem and write the constraints as less than or equal to inequalities.

Maximize $P = -C = -2x_1 - x_2 - 3x_3$

subject to

$-2x_1 - 3x_2 - 4x_3 \leq -20$
$-4x_1 - 2x_2 - 2x_3 \leq -15$
$x_1 \geq 0, \ x_2 \geq 0, \ x_3 \geq 0$

We introduce nonnegative slack variables and set up the initial tableau. Since there are negative entries in the RHS, we use the alternate pivoting strategy.

BV	P	x_1	x_2	x_3	s_1	s_2	RHS
s_1	0	$\boxed{-2}$	-3	-4	1	0	-20
s_2	0	-4	-2	-2	0	1	-15
P	1	2	1	3	0	0	0

Alternative Pivoting Strategy $\longrightarrow$

BV	P	x_1	x_2	x_3	s_1	s_2	RHS
x_1	0	1	$\frac{3}{2}$	2	$-\frac{1}{2}$	0	10
s_2	0	0	$\boxed{4}$	6	-2	1	25
P	1	0	-2	-1	1	0	-20

Standard Pivoting Strategy $\longrightarrow$

BV	P	x_1	x_2	x_3	s_1	s_2	RHS
x_1	0	1	0	$-\frac{1}{4}$	$\frac{1}{4}$	$-\frac{3}{8}$	$\frac{5}{8}$
x_2	0	0	1	$\frac{3}{2}$	$-\frac{1}{2}$	$\frac{1}{4}$	$\frac{25}{4}$
P	1	0	0	2	0	$\frac{1}{2}$	$-\frac{15}{2}$

Maximum $P = -\dfrac{15}{2}$, so minimum $C = \dfrac{15}{2}$, which is obtained when $x_1 = \dfrac{5}{8}$ $x_2 = \dfrac{25}{4}$ and $x_3 = 0$.

The most economical mixture that meets the requirements for protein and carbohydrates costs \$7.50 and consists of $\dfrac{5}{8}$ unit of food I, $\dfrac{25}{4}$ units of food II and no units of food III.

13. We define x_1 as the number of television sets shipped from W_1 to R_1;

x_2 as the number shipped from W_1 to R_2;
x_3 as the number shipped from W_2 to R_1; and
x_4 as the number of sets shipped from W_2 to R_2.

The manufacturer wants to fill the orders at the lowest possible cost.

Minimize $\quad C = 8x_1 + 12x_2 + 13x_3 + 7x_4$

subject to the constraints

$$x_1 + x_3 = 55$$
$$x_2 + x_4 = 75$$
$$x_1 + x_2 \le 100$$
$$x_3 + x_4 \le 120$$
$$x_1 \ge 0, \ x_2 \ge 0, \ x_3 \ge 0, \ x_4 \ge 0$$

We rewrite the first two equations as inequalities, add nonnegative slack variables and set up the initial tableau.

$$x_1 + x_3 \le 55$$
$$-x_1 - x_3 \le -55$$
$$x_2 + x_4 \le 75$$
$$-x_2 - x_4 \le -75$$
$$x_1 + x_2 \le 100$$
$$x_3 + x_4 \le 120$$
$$x_1 \ge 0, \ x_2 \ge 0, \ x_3 \ge 0, \ x_4 \ge 0$$

BV	P	x_1	x_2	x_3	x_4	s_1	s_2	s_3	s_4	s_5	s_6	RHS
s_1	0	1	0	1	0	1	0	0	0	0	0	55
s_2	0	−1	0	−1	0	0	1	0	0	0	0	−55
s_3	0	0	1	0	1	0	0	1	0	0	0	75
s_4	0	0	−1	0	−1	0	0	0	1	0	0	−75
s_5	0	1	1	0	0	0	0	0	0	1	0	100
s_6	0	0	0	1	1	0	0	0	0	0	1	120
P	1	8	12	13	7	0	0	0	0	0	0	0

Since the initial tableau has negative entries in the RHS, we use the alternate pivoting strategy.

Alternative Pivoting Strategy →

BV	P	x_1	x_2	x_3	x_4	s_1	s_2	s_3	s_4	s_5	s_6	RHS
s_1	0	0	0	0	0	1	1	0	0	0	0	0
x_1	0	1	0	1	0	0	−1	0	0	0	0	55
s_3	0	0	1	0	1	0	0	1	0	0	0	75
s_4	0	0	−1	0	−1	0	0	0	1	0	0	−75
s_5	0	0	1	−1	0	0	1	0	0	1	0	45
s_6	0	0	0	1	1	0	0	0	0	0	1	120
P	1	0	12	5	7	0	8	0	0	0	0	−440

BV	P	x_1	x_2	x_3	x_4	s_1	s_2	s_3	s_4	s_5	s_6	RHS
s_1	0	0	0	0	0	1	1	0	0	0	0	0
x_1	0	1	0	1	0	0	-1	0	0	0	0	55
s_3	0	0	0	0	0	0	0	1	1	0	0	0
x_2	0	0	1	0	1	0	0	0	-1	0	0	75
s_5	0	0	0	[-1]	-1	0	1	0	1	1	0	-30
s_6	0	0	0	1	1	0	0	0	0	0	1	120
P	1	0	0	5	-5	0	8	0	12	0	0	-1,340

Alternative Pivoting Strategy → x_2

BV	P	x_1	x_2	x_3	x_4	s_1	s_2	s_3	s_4	s_5	s_6	RHS
s_1	0	0	0	0	0	1	1	0	0	0	0	0
x_1	0	1	0	0	-1	0	0	0	1	1	0	25
s_3	0	0	0	0	0	0	0	1	1	0	0	0
x_2	0	0	1	0	1	0	0	0	-1	0	0	75
x_3	0	0	0	1	[1]	0	-1	0	-1	-1	0	30
s_6	0	0	0	0	0	0	1	0	1	1	1	90
P	1	0	0	0	-10	0	13	0	17	5	0	-1,490

Alternative Pivoting Strategy → x_2

BV	P	x_1	x_2	x_3	x_4	s_1	s_2	s_3	s_4	s_5	s_6	RHS
s_1	0	0	0	0	0	1	1	0	0	0	0	0
x_1	0	1	0	1	0	0	-1	0	0	0	0	55
s_3	0	0	0	0	0	0	0	1	1	0	0	0
x_2	0	0	1	-1	0	0	1	0	0	[1]	0	45
x_4	0	0	0	1	1	0	-1	0	-1	-1	0	30
s_6	0	0	0	0	0	0	1	0	1	1	1	90
P	1	0	0	10	0	0	3	0	7	-5	0	-1,190

Standard Pivoting Strategy → x_2

BV	P	x_1	x_2	x_3	x_4	s_1	s_2	s_3	s_4	s_5	s_6	RHS
s_1	0	0	0	0	0	1	1	0	0	0	0	0
x_1	0	1	0	1	0	0	-1	0	0	0	0	55
s_3	0	0	0	0	0	0	0	1	1	0	0	0
s_5	0	0	1	-1	0	0	1	0	0	1	0	45
x_4	0	0	1	0	1	0	0	0	-1	0	0	75
s_6	0	0	-1	1	0	0	0	0	1	0	1	45
P	1	0	5	5	0	0	8	0	7	0	0	-965

Standard Pivoting Strategy → s_5

The maximum $P = -965$, so the minimum $C = 965$ which is obtained when $x_1 = 55$, $x_2 = 0$, $x_3 = 0$, and $x_4 = 75$. The television manufacturer minimizes shipping costs when 55 televisions are shipped from warehouse W_1 to retailer R_1, and 75 televisions from warehouse W_2 to retailer R_2 for a minimum cost of \$965.

15. Let x_1 denote the number of representatives sent from New York to Dallas, x_2 denote the number of representatives sent from New York to Chicago, x_3 denote the number of representatives sent from San Francisco to Dallas, x_4 denote the number of representatives sent from San Francisco to Chicago. We want to

Minimize $C = 280x_1 + 180x_2 + 340x_3 + 180x_4$
subject to the constraints

$$
\begin{aligned}
x_1 \quad\;\; + x_3 \qquad\; &\geq 15 \\
x_2 \qquad\; + x_4 &\geq 10 \\
x_3 \qquad\; &\geq 5 \\
x_1 + x_2 \qquad\qquad &\leq 12 \\
x_3 + x_4 &\leq 18 \\
x_1 \geq 0 \quad x_2 \geq 0 \quad x_3 \geq 0 \quad x_4 &\geq 0
\end{aligned}
$$

This is a minimum problem with mixed constraints.
Step 1 Rewrite the constraints.

$$
\begin{aligned}
-x_1 \quad\;\; - x_3 \qquad\; &\leq -15 \\
- x_2 \qquad\; - x_4 &\leq -10 \\
- x_3 \qquad\; &\leq -5 \\
x_1 + x_2 \qquad\qquad &\leq 12 \\
x_3 + x_4 &\leq 18 \\
x_1 \geq 0 \quad x_2 \geq 0 \quad x_3 \geq 0 \quad x_4 &\geq 0
\end{aligned}
$$

The objective function becomes

Maximize $- z = -280x_1 - 180x_2 - 340x_3 - 180x_4$
Steps 2 and 3 Add slack variables and set up the initial tableau.

BV	P	x_1	x_2	x_3	x_4	s_1	s_2	s_3	s_4	s_4	RHS
s_1	0	−1	0	−1	0	1	0	0	0	0	−15
s_2	0	0	−1	0	−1	0	1	0	0	0	−10
s_3	0	0	0	−1	0	0	0	1	0	0	−5
s_4	0	1	1	0	0	0	0	0	1	0	12
s_5	0	0	0	1	1	0	0	0	0	1	18
P	1	280	180	340	180	0	0	0	0	0	0

Step 4 and 5 Choose the pivot; pivot. There are negative entries in the RHS, we use the alternate strategy. The pivot element is in row s_1, column x_1.

BV	P	x_1	x_2	x_3	x_4	s_1	s_2	s_3	s_4	s_4	RHS
s_1	0	[−1]	0	−1	0	1	0	0	0	0	−15
s_2	0	0	−1	0	−1	0	1	0	0	0	−10
s_3	0	0	0	−1	0	0	0	1	0	0	−5
s_4	0	1	1	0	0	0	0	0	1	0	12
s_5	0	0	0	1	1	0	0	0	0	1	18
P	1	280	180	340	180	0	0	0	0	0	0

Since there are still negative entries in the RHS, we again use the alternate strategy. The pivot element is in row s_2, column x_2.

BV	P	x_1	x_2	x_3	x_4	s_1	s_2	s_3	s_4	s_4	RHS
x_1	0	1	0	1	0	−1	0	0	0	0	15
s_2	0	0	−1	0	−1	0	1	0	0	0	−10
s_3	0	0	0	−1	0	0	0	1	0	0	−5
s_4	0	0	1	−1	0	1	0	0	1	0	−3
s_5	0	0	0	1	1	0	0	0	0	1	18
P	1	0	180	60	180	280	0	0	0	0	−4200

$\rightarrow$

BV	P	x_1	x_2	x_3	x_4	s_1	s_2	s_3	s_4	s_4	RHS
x_1	0	1	0	1	0	−1	0	0	0	0	15
x_2	0	0	1	0	1	0	−1	0	0	0	10
s_3	0	0	0	−1	0	0	0	1	0	0	−5
s_4	0	0	0	−1	−1	1	1	0	1	0	−13
s_5	0	0	0	1	1	0	0	0	0	1	18
P	1	0	0	60	0	280	180	0	0	0	−6000

$\rightarrow$

BV	P	x_1	x_2	x_3	x_4	s_1	s_2	s_3	s_4	s_4	RHS
x_1	0	1	0	0	0	−1	0	1	0	0	10
x_2	0	0	1	0	1	0	−1	0	0	0	10
x_3	0	0	0	1	0	0	0	−1	0	0	5
s_4	0	0	0	0	−1	1	1	−1	1	0	−8
s_5	0	0	0	0	1	0	0	1	0	1	13
P	1	0	0	0	0	280	180	60	0	0	−6300

$\rightarrow$

BV	P	x_1	x_2	x_3	x_4	s_1	s_2	s_3	s_4	s_4	RHS
x_1	0	1	0	0	0	−1	0	1	0	0	10
x_2	0	0	1	0	0	1	0	−1	1	0	2
x_3	0	0	0	1	0	0	0	−1	0	0	5
x_4	0	0	0	0	1	−1	−1	1	−1	0	8
s_5	0	0	0	0	0	1	1	0	1	1	5
P	1	0	0	0	0	280	180	60	0	0	−6300

This is the final tableau. The maximum value of P is − 6300, so the minimum value of $C = 6300$. The company will minimize travel costs at \$6300 while meeting constraints if they send 10 representatives from New York to Dallas, 2 representatives from New York to Chicago, 5 representatives from San Francisco to Dallas and 8 representatives from San Francisco to Chicago.

Actually the company can achieve the minimum cost if they send 10 representatives from New York and 5 from San Francisco to Dallas and any combination of representatives to Chicago. (10 from San Francisco and 0 from New York; or 9 from San Francisco and 1 from New York; or 8 from San Francisco and 2 from New York.)

Chapter 4 Review

True-False Items

1. T 3. T 5. T

Fill in the Blanks

1. slack variables 3. greater than or equal, $\geq$

Review Exercises

Problems 1-8: A maximum problem is in standard form if the following conditions are met.

    Condition 1   All the variables are nonnegative.

    Condition 2   Every other constraint is written as a linear expression that is less than or equal to a positive constant.

1. The problem is in standard form. Both conditions are met.

3. The problem is in standard form. Both conditions are met.

5. This problem is not in standard form. One of the constraints is written as a greater than or equal to inequality. Condition 2 is not met.

7. This problem is not in standard form. One of the variables can be negative. Condition 1 is not met.

9. We add slack variables and rewrite the problem as

Maximize $P - 2x_1 - x_2 - 3x_3 = 0$

subject to the constraints

$$\begin{aligned} 2x_1 + 5x_2 + x_3 + s_1 &= 100 \\ x_1 + 3x_2 + x_3 \quad + s_2 &= 80 \\ 2x_1 + 3x_2 + 3x_3 \quad\quad + s_3 &= 120 \end{aligned}$$

$x_1 \geq 0 \quad x_2 \geq 0 \quad x_3 \geq 0 \quad s_1 \geq 0 \quad s_2 \geq 0 \quad s_3 \geq 0$

We can then write the initial tableau as

BV	P	x_1	x_2	x_3	s_1	s_2	s_3	RHS
s_1	0	2	5	1	1	0	0	100
s_2	0	1	3	1	0	1	0	80
s_3	0	2	3	3	0	0	1	120
P	1	−2	−1	−3	0	0	0	0

11.   We add slack variables and rewrite the problem as

Maximize $P - 6x_1 - 3x_2 = 0$

subject to the constraints

$$
\begin{aligned}
x_1 + 5x_2 + s_1 \qquad\qquad &= 200 \\
5x_1 + 3x_2 \qquad + s_2 \qquad &= 450 \\
x_1 + x_2 \qquad\qquad + s_3 &= 120
\end{aligned}
$$

$$x_1 \geq 0 \qquad x_2 \geq 0 \qquad s_1 \geq 0 \qquad s_2 \geq 0 \qquad s_3 \geq 0$$

We can then write the initial tableau as

BV	P	x_1	x_2	s_1	s_2	s_3	RHS
s_1	0	1	5	1	0	0	200
s_2	0	5	3	0	1	0	450
s_3	0	1	1	0	0	1	120
P	1	−6	−3	0	0	0	0

13.   We add slack variables and rewrite the problem as

Maximize $P - x_1 - 2x_2 - x_3 - 4x_4 = 0$

subject to the constraints

$$
\begin{aligned}
x_1 + 3x_2 + x_3 + 2x_4 + s_1 \qquad &= 20 \\
4x_1 + x_2 + x_3 + 6x_4 \qquad + s_2 &= 80
\end{aligned}
$$

$$x_1 \geq 0 \qquad x_2 \geq 0 \qquad x_3 \geq 0 \qquad x_4 \geq 0 \qquad s_1 \geq 0 \qquad s_2 \geq 0$$

We can then write the initial tableau as

BV	P	x_1	x_2	x_3	x_4	s_1	s_2	RHS
s_1	0	1	3	1	2	1	0	20
s_2	0	4	1	1	6	0	1	80
P	1	−1	−2	−1	−4	0	0	0

15. **(a)** We choose the pivot element by following the steps below.
Step 1 We find the smallest negative entry in the objective row. It identifies the pivot column. The pivot column is column x_2.
Step 2 For each positive entry above the objective row in the pivot column, we form the quotient of the corresponding RHS entry divided by the positive entry.

$$40 \div 5 = 8 \qquad 10 \div 2 = 5$$

Step 3 The smallest nonnegative quotient is 5 so the pivot row is s_2
The original tableau with the pivot element marked is shown on the left. The tableau after pivoting is on the right.

$$\downarrow$$

$$
\begin{array}{c|ccccc|c}
\text{BV} & P & x_1 & x_2 & s_1 & s_2 & \text{RHS} \\
\hline
x_1 & 0 & 1 & \mathbf{5} & 1 & 0 & 40 \\
\rightarrow s_2 & 0 & 0 & \boxed{2} & 2 & 1 & 10 \\
P & 1 & 0 & -1 & 0 & 3 & 120
\end{array}
\qquad
\begin{array}{c|ccccc|c}
\text{BV} & P & x_1 & x_2 & s_1 & s_2 & \text{RHS} \\
\hline
x_1 & 0 & 1 & \mathbf{0} & -4 & -\frac{5}{2} & 15 \\
x_2 & 0 & 0 & 1 & 1 & \frac{1}{2} & 5 \\
P & 1 & 0 & 0 & 1 & \frac{7}{2} & 125
\end{array}
$$

(b) The resulting system of equations is

$$x_1 = 15 + 4s_1 + \frac{5}{2}s_2$$

$$x_2 = 5 - s_1 - \frac{1}{2}s_2$$

$$P = 125 - s_1 - \frac{7}{2}s_2$$

(c) The new tableau is the final tableau. The solution is maximum $P = 125$ obtained when $x_1 = 15$ and $x_2 = 5$.

17.   (a)   We choose the pivot element by following the steps below.
Step 1 We find the smallest negative entry in the objective row. It identifies the pivot column.   The pivot column is column x_3.
Step 2 For each positive entry above the objective row in the pivot column, we form the quotient of the corresponding RHS entry divided by the positive entry.

In this problem there is only one positive entry, so it becomes the pivot.
The original tableau with the pivot element marked is shown below.

$$
\begin{array}{c|cccccc|c}
\text{BV} & P & x_1 & x_2 & x_3 & s_1 & s_2 & \text{RHS} \\
\hline
s_1 & 0 & 1 & 1 & -1 & 1 & 0 & 10 \\
\rightarrow s_2 & 0 & 0 & 1 & \boxed{1} & 0 & 1 & 4 \\
\hline
P & 1 & -2 & -1 & -3 & 0 & 0 & 0
\end{array}
$$

The tableau after pivoting:

$$
\begin{array}{c|cccccc|c}
\text{BV} & P & x_1 & x_2 & x_3 & s_1 & s_2 & \text{RHS} \\
\hline
s_1 & 0 & 1 & 2 & 0 & 1 & 1 & 14 \\
x_3 & 0 & 0 & 1 & 1 & 0 & 1 & 4 \\
\hline
P & 1 & -2 & 2 & 0 & 0 & 3 & 12
\end{array}
$$

(b)   The resulting system of equations is

$$
\begin{aligned}
s_1 &= 14 - x_1 - 2x_1 - s_2 \\
x_3 &= 4 - x_2 - s_2 \\
P &= 12 + 2x_1 - 2x_2 - 3s_2
\end{aligned}
$$

(c)   This tableau needs further pivoting, since there is still a negative entry in the objective row.   We find the new pivot by using the steps from part (a).
Step 1: x_1 forms the pivot column.
Step 2: We examine the quotients found by dividing the RHS entries by the corresponding positive entries in the pivot column. In this tableau there is only one positive entry so row s_1 becomes the new pivot row. The new pivot element is 1.

19. (a) We choose the pivot element by following the steps below.
Step 1 We find the smallest negative entry in the objective row. It identifies the pivot column. The pivot column is column x_1.
Step 2 For each positive entry above the objective row in the pivot column, we form the quotient of the corresponding RHS entry divided by the positive entry.

$$1 \div 0.5 = 2 \qquad 3 \div 1 = 3$$

Step 3 The smallest nonnegative quotient is 2 so the pivot row is s_1
The original tableau with the pivot element marked is

BV	P	x_1	x_2	s_1	s_2	RHS
s_1	0	0.5	0.5	1	0	1
s_2	0	1	1.5	0	1	3
P	1	−2.5	−2	0	0	0

The tableau after pivoting is

BV	P	x_1	x_2	s_1	s_2	RHS
x_1	0	1	1	2	0	2
s_2	0	0	0.5	−2	1	1
P	1	0	0.5	5	0	5

(b) The resulting system of equations is

$$x_1 = 2 - \quad x_2 - 2s_1$$
$$s_2 = 1 - 0.5x_2 + 2s_1$$
$$P = 5 - 0.5x_2 - 5s_2$$

(c) This is the final tableau. The solution to the maximum problem is maximum $P = 5$ when $x_1 = 2$ and $x_2 = 0$.

21. (a)   We choose the pivot element by following the steps below.
         Step 1 We find the smallest negative entry in the objective row. It identifies the pivot column. The pivot column is column x_1.
         Step 2 For each positive entry above the objective row in the pivot column, we form the quotient of the corresponding RHS entry divided by the positive entry. Here there is only one positive entry in the pivot column, so it will be the pivot element.
         Step 3 The pivot row is s_3.
         The original tableau with the pivot element marked is

BV	P	x_1	x_2	x_3	s_1	s_2	s_3	RHS
s_1	0	−1	0	1	1	−1	0	7
x_2	0	−1	1	5	0	1	0	5
s_3	0	$\boxed{1}$	0	3	0	−5	1	3
P	1	−3	0	4	0	0	0	5

The tableau after pivoting is

BV	P	x_1	x_2	x_3	s_1	s_2	s_3	RHS
s_1	0	0	0	4	1	−6	1	10
x_2	0	0	1	8	0	−4	1	8
x_1	0	1	0	3	0	−5	1	3
P	1	0	0	13	0	−15	3	14

(b)   The resulting system of equations is

$$s_1 = 10 - 4x_3 + 6s_2 - s_3$$
$$x_2 = 8 - 8x_3 + 4s_2 - s_3$$
$$x_1 = 3 - 3x_3 + 5s_2 - s_3$$
$$P = 14 - 13x_3 + 15s_2 - 3s_3$$

(c)   This tableau indicates no solution exists for the problem. The pivot column would be column s_3, but every entry in the pivot column is negative.

23.

BV	P	x_1	x_2	x_3	s_1	s_2	s_3	RHS
s_1	0	5	5	10	1	0	0	1000
s_2	0	10	8	5	0	1	0	2000
s_3	0	10	$\boxed{5}$	0	0	0	1	500
P	1	−100	−200	−50	0	0	0	0

$\rightarrow$

BV	P	x_1	x_2	x_3	s_1	s_2	s_3	RHS
s_1	0	−5	0	$\boxed{10}$	1	0	−1	500
s_2	0	−6	0	5	0	1	$-\frac{8}{5}$	1200
x_2	0	2	1	0	0	0	$\frac{1}{5}$	100
P	1	300	0	−50	0	0	40	20000

$\rightarrow$

BV	P	x_1	x_2	x_3	s_1	s_2	s_3	RHS
x_3	0	$-\frac{1}{2}$	0	1	$\frac{1}{10}$	0	$-\frac{1}{10}$	50
s_2	0	$-\frac{7}{2}$	0	0	$-\frac{1}{2}$	1	$-\frac{11}{10}$	950
x_2	0	2	1	0	0	0	$\frac{1}{5}$	100
P	1	275	0	0	5	0	35	22500

The maximum value for P is 22,500, obtained when $x_1 = 0$, $x_2 = 100$ and $x_3 = 50$.

25.

BV	P	x_1	x_2	x_3	s_1	s_2	RHS
s_1	0	2	$\boxed{2}$	1	1	0	8
s_2	0	1	−4	3	0	1	12
P	1	−40	−60	−50	0	0	0

$\rightarrow$

BV	P	x_1	x_2	x_3	s_1	s_2	RHS
x_2	0	1	1	$\frac{1}{2}$	$\frac{1}{2}$	0	4
s_2	0	5	0	$\boxed{5}$	2	1	28
P	1	20	0	−20	30	0	240

BV	P	x_1	x_2	x_3	s_1	s_2	RHS
x_2	0	$\frac{1}{2}$	1	0	$\frac{3}{10}$	$-\frac{1}{10}$	$\frac{6}{5}$
x_3	0	1	0	1	$\frac{2}{5}$	$\frac{1}{5}$	$\frac{28}{5}$
P	1	40	0	0	38	4	352

The maximum value for P is 352, obtained when $x_1 = 0$, $x_2 = \dfrac{6}{5}$ and $x_3 = \dfrac{28}{5}$.

27. The minimum problem is in standard form.

29. The minimum problem is not in standard form. Condition 2 is not met; the constraints are expressions greater than or equal to a constant.

31. The minimum problem is not in standard form. Condition 2 is not met; one of the constraints is an expression less than or equal to a constant.

33. To write the dual of the minimum problem we first write the matrix that represents the constraints and the objective function and its transpose.

$$\text{the matrix } \begin{array}{c} \begin{matrix} x_1 & x_2 \end{matrix} \\ \left[\begin{array}{cc|c} 2 & 2 & 8 \\ 1 & -1 & 2 \\ 2 & 1 & 0 \end{array}\right] \end{array} \quad \text{the transpose of the matrix } \left[\begin{array}{cc|c} 2 & 1 & 2 \\ 2 & -1 & 1 \\ 8 & 2 & 0 \end{array}\right]$$

From the transpose we create the maximum problem

Maximize $P = 8y_1 + 2y_2$

subject to the conditions

$2y_1 + y_2 \le 2$
$2y_1 - y_2 \le 1$
$y_1 \ge 0 \quad y_2 \ge 0$

This maximum problem is the dual of the minimum problem.

35. To write the dual of the minimum problem we first write the matrix that represents the constraints and the objective function and its transpose.

$$\text{the matrix } \begin{array}{c} \begin{matrix} x_1 & x_2 & x_3 \end{matrix} \\ \left[\begin{array}{ccc|c} 1 & 1 & 1 & 100 \\ 2 & 1 & 0 & 50 \\ 5 & 4 & 3 & 0 \end{array}\right] \end{array} \quad \text{the transpose of the matrix } \left[\begin{array}{cc|c} 1 & 2 & 5 \\ 1 & 1 & 4 \\ 1 & 0 & 3 \\ 100 & 50 & 0 \end{array}\right]$$

From the transpose we create the maximum problem

Maximize $P = 100y_1 + 50y_2$

subject to the conditions

$y_1 + 2y_2 \le 5$
$y_1 + \ y_2 \le 4$
$y_1 \qquad \le 3$
$y_1 \ge 0 \quad y_2 \ge 0$

This maximum problem is the dual of the minimum problem.

37. Using the dual found in Problem 33, we get

BV	P	y_1	y_2	s_1	s_2	RHS
s_1	0	2	1	1	0	2
s_2	0	2	-1	0	1	1
P	1	-8	-2	0	0	0

$\rightarrow$

BV	P	y_1	y_2	s_1	s_2	RHS
s_1	0	0	2	1	-1	1
y_1	0	1	$-\frac{1}{2}$	0	$\frac{1}{2}$	$\frac{1}{2}$
P	1	0	-6	0	4	4

$\rightarrow$

BV	P	y_1	y_2	s_1	s_2	RHS
y_2	0	0	1	$\frac{1}{2}$	$-\frac{1}{2}$	$\frac{1}{2}$
y_1	0	1	0	$\frac{1}{4}$	$\frac{1}{4}$	$\frac{3}{4}$
P	1	0	0	3	1	7

The minimum value for C is 7, when $x_1 = 3$ and $x_2 = 1$.

39. Using the dual found in Problem 35, we get

BV	P	y_1	y_2	s_1	s_2	s_3	RHS
s_1	0	1	2	1	0	0	5
s_2	0	1	1	0	1	0	4
s_3	0	$\boxed{1}$	0	0	0	1	3
P	1	-100	-50	0	0	0	0

$\rightarrow$

BV	P	y_1	y_2	s_1	s_2	s_3	RHS
s_1	0	0	2	1	0	-1	2
s_2	0	0	$\boxed{1}$	0	1	-1	1
y_1	0	1	0	0	0	1	3
P	1	0	-50	0	0	100	300

$\rightarrow$

BV	P	y_1	y_2	s_1	s_2	s_3	RHS
s_1	0	0	0	1	-2	-1	0
y_2	0	0	1	0	1	-1	1
y_1	0	1	0	0	0	1	3
P	1	0	0	0	50	50	350

The minimum value for C is 350 when $x_1 = 0$, $x_2 = 50$ and $x_3 = 50$.
If in tableau 2, the pivot row s_1 was used, the minimum solution would be $C = 350$ when $x_1 = 25$, $x_2 = 0$ and $x_3 = 75$.

41.   **Step 1** Write the constraints as less than or equal to inequalities.

$$-x_1 - x_2 \le -2$$
$$2x_1 + 3x_2 \le 12$$
$$3x_1 + 2x_2 \le 12$$

Step 2 Introduce slack variables

$$-x_1 - x_2 + s_1 \qquad\qquad = -2$$
$$2x_1 + 3x_2 \qquad + s_2 \qquad = 12$$
$$3x_1 + 2x_2 \qquad\qquad + s_3 = 12$$

Step 3 Set up the initial tableau

BV	P	x_1	x_2	s_1	s_2	s_3	RHS
s_1	0	−1	−1	1	0	0	−2
s_2	0	2	3	0	1	0	12
s_3	0	3	2	0	0	1	12
P	1	−3	−5	0	0	0	0

Step 4 The RHS has a negative entry, so we use the alternative strategy. The first negative entry is −1 in column x_1. It is the pivot element.
Step 5 Pivot

BV	P	x_1	x_2	s_1	s_2	s_3	RHS
s_1	0	−1	−1	1	0	0	−2
s_2	0	2	3	0	1	0	12
s_3	0	3	2	0	0	1	12
P	1	−3	−5	0	0	0	0

$\rightarrow$

BV	P	x_1	x_2	s_1	s_2	s_3	RHS
x_1	0	1	1	−1	0	0	2
s_2	0	0	1	2	1	0	8
s_3	0	0	−1	3	0	1	6
P	1	0	−2	−3	0	0	6

The new tableau has only nonnegative entries in the RHS, it represents a maximum problem in standard form. Since the objective row has negative entries, we use the standard pivoting strategy. The pivot column is s_1. We form the quotients

$$8 \div 2 = 4 \qquad\qquad 6 \div 3 = 2$$

The smaller of these is 2, so the pivot row is row s_3.

BV	P	x_1	x_2	s_1	s_2	s_3	RHS
x_1	0	1	$\frac{2}{3}$	0	0	$\frac{1}{3}$	4
$\rightarrow$ s_2	0	0	$\frac{5}{3}$	0	1	$-\frac{2}{3}$	4
s_1	0	0	$-\frac{1}{3}$	1	0	$\frac{1}{3}$	2
P	1	0	−3	0	0	1	12

$\rightarrow$

BV	P	x_1	x_2	s_1	s_2	s_3	RHS
x_1	0	1	0	0	$-\frac{2}{5}$	$\frac{3}{5}$	$\frac{12}{5}$
$\rightarrow$ x_2	0	0	1	0	$\frac{3}{5}$	$-\frac{2}{5}$	$\frac{12}{5}$
s_1	0	0	0	1	$\frac{1}{5}$	$\frac{1}{5}$	$\frac{14}{5}$
P	1	0	0	0	$\frac{9}{5}$	$-\frac{1}{5}$	$\frac{96}{5}$

BV	P	x_1	x_2	s_1	s_2	s_3	RHS
s_3	0	$\frac{5}{3}$	0	0	$-\frac{2}{3}$	1	4
$\rightarrow$ x_2	0	$\frac{2}{3}$	1	0	$\frac{1}{3}$	0	4
s_1	0	$-\frac{1}{3}$	0	1	$\frac{1}{3}$	0	2
P	1	$\frac{1}{3}$	0	0	$\frac{5}{3}$	0	20

The maximum value of P is 20, and it is achieved when $x_1 = 0$ and $x_2 = 4$.

43. **Step 1** Write the constraints as less than or equal to inequalities.

$$-x_1 - x_2 \leq -3$$
$$x_1 + x_2 \leq 9$$

Step 2 Introduce slack variables

$$-x_1 - x_2 + s_1 = -3$$
$$x_1 + x_2 + s_2 = 9$$

The objective function is $P = -2x_1 - 3x_2$

Step 3 Set up the initial tableau

BV	P	x_1	x_2	s_1	s_2	RHS
s_1	0	−1	−1	1	0	−3
s_2	0	1	1	0	1	9
P	1	2	3	0	0	0

Step 4 The RHS has a negative entry, so we use the alternative strategy. The first negative entry is −1 in column x_1. It is the pivot element.

Step 5 Pivot

BV	P	x_1	x_2	s_1	s_2	RHS
s_1	0	[−1]	−1	1	0	−3
s_2	0	1	1	0	1	9
P	1	2	3	0	0	0

$\rightarrow$

BV	P	x_1	x_2	s_1	s_2	RHS
x_1	0	1	1	−1	0	3
s_2	0	0	0	1	1	6
P	1	0	1	2	0	−6

This is the final tableau. The maximum value of P is −6, so the minimum value of z is 6. this occurs when $x_1 = 3$ and $x_2 = 0$.

45. **Step 1** Write the constraints as less than or equal to inequalities.

$$4x_1 + 3x_2 + 5x_3 \leq 140$$
$$x_1 + x_2 + x_3 \leq 30$$
$$-x_1 - x_2 - x_3 \leq -30$$

Step 2 Introduce slack variables

$$4x_1 + 3x_2 + 5x_3 + s_1 = 140$$
$$x_1 + x_2 + x_3 + s_2 = 30$$
$$-x_1 - x_2 - x_3 + s_3 = -30$$

The objective function is $P = -300x_1 - 200x_2 - 450x_3$

Step 3 Set up the initial tableau

BV	P	x_1	x_2	x_3	s_1	s_2	s_3	RHS
s_1	0	4	3	5	1	0	0	140
s_2	0	1	1	1	0	1	0	30
s_3	0	−1	−1	−1	0	0	1	−30
P	1	−300	−200	−450	0	0	0	0

Step 4 The RHS has a negative entry, so we use the alternative strategy. The first negative entry is −1 in column x_1. It is the pivot element.

Step 5 Pivot

BV	P	x_1	x_2	x_3	s_1	s_2	s_3	RHS
s_1	0	4	3	5	1	0	0	140
s_2	0	1	1	1	0	1	0	30
s_3	0	$\boxed{-1}$	-1	-1	0	0	1	-30
P	1	-300	-200	-450	0	0	0	0

$\rightarrow$

BV	P	x_1	x_2	x_3	s_1	s_2	s_3	RHS
s_1	0	0	-1	1	1	0	4	20
s_2	0	0	0	0	0	1	1	0
x_1	0	1	1	1	0	0	-1	30
P	1	0	100	-150	0	0	-300	9000

The new tableau has only nonnegative entries in the RHS, it represents a maximum problem in standard form. Since the objective row has negative entries, we use the standard pivoting strategy. The pivot column is s_3. We form the quotients

$$20 \div 4 = 5 \qquad\qquad 0 \div 1 = 0$$

The smaller of these is 0, so the pivot row is row s_2.

Step 5 Pivot

BV	P	x_1	x_2	x_3	s_1	s_2	s_3	RHS
s_1	0	0	-1	1	1	0	4	20
s_2	0	0	0	0	0	1	$\boxed{1}$	0
x_1	0	1	1	1	0	0	-1	30
P	1	0	100	-150	0	0	-300	9000

$\rightarrow$

BV	P	x_1	x_2	x_3	s_1	s_2	s_3	RHS
s_1	0	0	-1	1	1	-4	0	20
s_3	0	0	0	0	0	1	1	0
x_1	0	1	1	1	0	1	0	30
P	1	0	100	-150	0	300	0	9000

The objective row still has a negative entry, we pivot again. The pivot column is x_3. We form the quotients

$$20 \div 0 = 20 \qquad\qquad 30 \div 1 = 30$$

The smaller of these is 20, so the pivot row is row s_1.

BV	P	x_1	x_2	x_3	s_1	s_2	s_3	RHS
s_1	0	0	-1	$\boxed{1}$	1	-4	0	20
s_3	0	0	0	0	0	1	1	0
x_1	0	1	1	1	0	1	0	30
P	1	0	100	-150	0	300	0	9000

$\rightarrow$

BV	P	x_1	x_2	x_3	s_1	s_2	s_3	RHS
x_3	0	0	−1	1	1	−4	0	20
s_3	0	0	0	0	0	$\boxed{1}$	1	0
x_1	0	1	2	0	−1	5	0	10
P	1	0	−50	0	150	−300	0	12,000

$\rightarrow$

BV	P	x_1	x_2	x_3	s_1	s_2	s_3	RHS
x_3	0	0	−1	1	1	0	4	20
s_2	0	0	0	0	0	1	1	0
x_1	0	1	$\boxed{2}$	0	−1	0	−5	10
P	1	0	−50	0	150	0	300	12,000

$\rightarrow$

BV	P	x_1	x_2	x_3	s_1	s_2	s_3	RHS
x_3	0	$\frac{1}{2}$	0	1	$\frac{1}{2}$	0	$\frac{3}{2}$	25
s_2	0	0	0	0	0	1	1	0
x_2	0	$\frac{1}{2}$	1	0	$-\frac{1}{2}$	0	$-\frac{5}{2}$	5
P	1	25	0	0	125	0	175	12,250

This is the final tableau. The maximum value of P is 12,250, obtained when $x_1 = 0$, $x_2 = 5$ and $x_3 = 25$.

47. Let x_1, x_2, and x_3 represent the number of vats of lite, regular, and dark beer, respectively. The objective is to maximize profit

$$P = 10x_1 + 20x_2 + 30x_3$$

subject to the constraints

$$\begin{cases} 6x_1 + 4x_2 + 2x_3 \le 800 \quad \text{or} \quad 3x_1 + 2x_2 + x_3 \le 400 \\ x_1 + 3x_2 + 2x_3 \le 600 \\ x_1 + x_2 + 4x_3 \le 300 \\ x_1 \ge 0,\ x_2 \ge 0,\ x_3 \ge 0 \end{cases}$$

The constraints are imposed by the availability of barley, sugar, and hops.

This is a maximum problem in standard form. The initial tableau is

BV	P	x_1	x_2	x_3	s_1	s_2	s_3	RHS
s_1	0	3	2	1	1	0	0	400
s_2	0	1	3	2	0	1	0	600
s_3	0	1	1	☐4	0	0	1	300
P	1	−10	−20	−30	0	0	0	0

$\rightarrow$

BV	P	x_1	x_2	x_3	s_1	s_2	s_3	RHS
s_1	0	$\frac{11}{4}$	$\frac{7}{4}$	0	1	0	$-\frac{1}{4}$	325
s_2	0	$\frac{1}{2}$	$\frac{5}{2}$	0	0	1	$-\frac{1}{2}$	450
x_3	0	$\frac{1}{4}$	$\frac{1}{4}$	1	0	0	$\frac{1}{4}$	75
P	1	$-\frac{5}{2}$	$-\frac{25}{2}$	0	0	0	$\frac{15}{2}$	2250

$\rightarrow$

BV	P	x_1	x_2	x_3	s_1	s_2	s_3	RHS
s_1	0	$\frac{12}{5}$	0	0	1	$-\frac{7}{10}$	$\frac{1}{10}$	10
x_2	0	$\frac{1}{5}$	1	0	0	$\frac{2}{5}$	$-\frac{1}{5}$	180
x_3	0	$\frac{1}{5}$	0	1	0	$-\frac{1}{10}$	$\frac{3}{10}$	30
P	1	0	0	0	0	5	5	4500

This is the final tableau.
The brewer should brew no lite beer, 180 vats of regular beer, and 30 vats of dark beer to attain a maximum profit of $4500.

49. Let x_1 represent the number of cars shipped from W_1 to D_1; x_2 the number shipped from W_1 to D_2; x_3 the number shipped from W_2 to D_1; and x_4 the number shipped from W_2 to D_2. The objective is to

Minimize cost $C = 180x_1 + 150x_2 + 160x_3 + 170x_4$

subject to the constraints

$$x_1 + x_3 = 40$$
$$x_2 + x_4 = 25$$
$$x_1 + x_2 \le 30$$
$$x_3 + x_4 \le 50$$
$$x_1 \ge 0, x_2 \ge 0, x_3 \ge 0, x_4 \ge 0$$

This is a minimum problem with mixed constraints.

Step 1 Write the constraints as less than or equal to inequalities

$$x_1 + x_3 \le 40$$
$$-x_1 - x_3 \le -40$$
$$x_2 + x_4 \le 25$$
$$-x_2 - x_4 \le -25$$
$$x_1 + x_2 \le 30$$
$$x_3 + x_4 \le 50$$
$$x_1 \ge 0, x_2 \ge 0, x_3 \ge 0, x_4 \ge 0$$

Step 2 Introduce slack variables

$$
\begin{aligned}
x_1 \quad + x_3 \quad + s_1 \qquad\qquad\qquad &= 40 \\
-x_1 \quad - x_3 \quad\quad + s_2 \qquad\qquad &= -40 \\
x_2 \quad + x_4 \quad + s_3 \qquad\qquad &= 25 \\
-x_2 \quad - x_4 \quad\quad + s_4 \qquad &= -25 \\
x_1 + x_2 \qquad\qquad\quad + s_5 \quad &= 30 \\
x_3 + x_4 \qquad\qquad\qquad + s_6 &= 50
\end{aligned}
$$

$$x_1 \ge 0,\ x_2 \ge 0,\ x_3 \ge 0,\ x_4 \ge 0,\ s_1 \ge 0,\ s_2 \ge 0,\ s_3 \ge 0,\ s_4 \ge 0,\ s_5 \ge 0,\ s_6 \ge 0$$

The objective function is

Maximize $P = -C = -180x_1 - 150x_2 - 160x_3 - 170x_4$

Step 3 Set up the initial tableau

BV	P	x_1	x_2	x_3	x_4	s_1	s_2	s_3	s_4	s_5	s_6	RHS
s_1	0	1	0	1	0	1	0	0	0	0	0	40
s_2	0	**-1**	0	-1	0	0	1	0	0	0	0	-40
s_3	0	0	1	0	1	0	0	1	0	0	0	25
s_4	0	0	-1	0	-1	0	0	0	1	0	0	-25
s_5	0	1	1	0	0	0	0	0	0	1	0	30
s_6	0	0	0	1	1	0	0	0	0	0	1	50
P	1	180	150	160	170	0	0	0	0	0	0	0

Step 4 The RHS has negative entries, so we use the alternative strategy. The first negative entry in row s_2 is in column x_1. It becomes the pivot.

Step 5 Pivot

BV	P	x_1	x_2	x_3	x_4	s_1	s_2	s_3	s_4	s_5	s_6	RHS
s_1	0	0	0	0	0	1	1	0	0	0	0	0
x_1	0	1	0	1	0	0	-1	0	0	0	0	40
s_3	0	0	1	0	1	0	0	1	0	0	0	25
s_4	0	0	[-1]	0	-1	0	0	0	1	0	0	-25
s_5	0	0	1	-1	0	0	1	0	0	1	0	30
s_6	0	0	0	1	1	0	0	0	0	0	1	50
P	1	0	150	-20	170	0	180	0	0	0	0	-7200

After the first pivot, there is still a negative entry in the RHS. So we continue to use the alternate strategy. The new pivot is -1 in row s_4; column x_2. It is marked on the tableau above.

BV	P	x_1	x_2	x_3	x_4	s_1	s_2	s_3	s_4	s_5	s_6	RHS
s_1	0	0	0	0	0	1	1	0	0	0	0	0
x_1	0	1	0	[1]	0	0	-1	0	0	0	0	40
s_3	0	0	0	0	0	0	0	1	1	0	0	0
x_2	0	0	1	0	1	0	0	0	-1	0	0	25
s_5	0	0	0	-1	-1	0	1	0	1	1	0	5
s_6	0	0	0	1	1	0	0	0	0	0	1	50
P	1	0	0	-20	20	0	180	0	150	0	0	-10,950

The new tableau has only nonnegative entries in the RHS, it represents a maximum problem in standard form. Since the objective row has a negative entry, we use the standard pivoting strategy. The pivot column is x_3. We form the quotients

$$40 \div 1 = 40 \qquad\qquad 50 \div 1 = 50$$

The smaller of these is 40, so the pivot row is row x_1.

Step 5 Pivot

	BV	P	x_1	x_2	x_3	x_4	s_1	s_2	s_3	s_4	s_5	s_6	RHS
	s_1	0	0	0	0	0	1	1	0	0	0	0	0
	x_3	0	1	0	1	0	0	-1	0	0	0	0	40
	s_3	0	0	0	0	0	0	0	1	1	0	0	0
→	x_2	0	0	1	0	1	0	0	0	-1	0	0	25
	s_5	0	1	0	0	-1	0	0	0	1	1	0	45
	s_6	0	-1	0	0	1	0	1	0	0	0	1	10
	P	1	20	0	0	20	0	160	0	150	0	0	-10,150

This is the final tableau. The maximum value of $P = -10{,}150$, so the minimum value $C = 10{,}150$. This is attained when $x_1 = 0$, $x_2 = 25$, $x_3 = 40$, and $x_4 = 0$.

The manufacturer should ship no cars to dealer 1 and 25 cars to dealer 2 from warehouse 1, and ship 40 cars to dealer 1 and none to dealer 2 from warehouse 2. The minimum cost is $10,150.

51. Let x_1, x_2, and x_3 represent the acres of corn, wheat, soybeans, respectively, planted. The objective is to

Maximize $P = 30x_1 + 40x_2 + 40x_3$

subject to the constraints

$$\begin{aligned} x_1 + \quad x_2 + \quad x_3 &\le 1000 \\ 100x_1 + 120x_2 + 70x_3 &\le 10000 \\ 7x_1 + \quad 10x_2 + \quad 8x_3 &\le 8000 \\ x_1 \ge 0,\ x_2 \ge 0,\ x_3 &\ge 0 \end{aligned}$$

This is a maximization problem in standard form. Add slack variables and set up the initial tableau. For the first pivot column we can use either x_2 or x_3, we chose x_3. We then choose the pivot row by forming the quotients formed by the RHS and the corresponding entries in column x_3 and choosing the row with the smallest quotient.

$$1000 \div 1 = 1000 \qquad 10{,}000 \div 70 = 142.86 \qquad 8000 \div 8 = 1000$$

The pivot row is s_2.

BV	P	x_1	x_2	x_3	s_1	s_2	s_3	RHS
s_1	0	1	1	1	1	0	0	1000
s_2	0	100	120	70	0	1	0	10000
s_3	0	7	10	8	0	0	1	8000
P	1	−30	−40	−40	0	0	0	0

Pivot:

BV	P	x_1	x_2	x_3	s_1	s_2	s_3	RHS
s_1	0	$-\frac{3}{7}$	$-\frac{5}{7}$	0	1	$-\frac{1}{70}$	0	$\frac{6000}{7}$
x_3	0	$\frac{10}{7}$	$\frac{12}{7}$	1	0	$\frac{1}{70}$	0	$\frac{1000}{7}$
s_3	0	$-\frac{31}{7}$	$-\frac{26}{7}$	0	0	$-\frac{4}{35}$	1	$\frac{48000}{7}$
P	1	$\frac{190}{7}$	$\frac{200}{7}$	0	0	$\frac{4}{7}$	0	$\frac{40000}{7}$

This is the final tableau. The maximum profit $P = \dfrac{40000}{7} = \$5714.29$ which is obtained by

planting no corn, no wheat, and $\dfrac{1000}{7} = 142.86$ acres of soybeans.

53. Let x_1 denote the number of shares of Duke Energy purchased,
x_2 denote the number of shares Eastman Kodak purchased,
x_3 denote the number of shares of General Motors purchased,
x_4 denote the number of shares of H.J Heinz purchased.
We want to

Maximize Yield $P = 0.08(14x_1) + 0.06(30x_2) + 0.06(34\ x_3) + 0.05(31.50\ x_4)$
$P = 1.12x_1 + 1.8x_2 + 2.04\ x_3 + 1.575\ x_4$

subject to the constraints

$$14x_1 + 30x_2 + 34x_3 + 31.50x_4 \le 50,000$$
$$31.50x_4 \ge 5,000$$
$$14x_1 \ge 10,000$$
$$30x_2 \ge 10,000$$
$$34x_3 \ge 10,000$$
$$14x_1 + 34x_3 \le 25,000$$
$$x_1 \ge 0 \quad x_2 \ge 0 \quad x_3 \ge 0 \quad x_4 \ge 0$$

This is a maximum problem with mixed constraints. Rewrite the constraints.

$$14x_1 + 30x_2 + 34x_3 + 31.50x_4 \le 50,000$$
$$-31.50x_4 \le -5,000$$
$$-14x_1 \le -10,000$$
$$-30x_2 \le -10,000$$
$$-34x_3 \le -10,000$$
$$14x_1 + 34x_3 \le 25,000$$
$$x_1 \ge 0 \quad x_2 \ge 0 \quad x_3 \ge 0 \quad x_4 \ge 0$$

Add slack variables and set up the initial tableau.

BV	P	x_1	x_2	x_3	x_4	s_1	s_2	s_3	s_4	s_5	s_6	RHS
s_1	0	14	30	34	31.5	1	0	0	0	0	0	50000
s_2	0	0	0	0	−31.5	0	1	0	0	0	0	−5000
s_3	0	−14	0	0	0	0	0	1	0	0	0	−10000
s_4	0	0	−30	0	0	0	0	0	1	0	0	−10000
s_5	0	0	0	−34	0	0	0	0	0	1	0	−10000
s_6	0	14	0	34	0	0	0	0	0	0	1	25000
P	1	−1.12	−1.8	−2.04	−1.575	0	0	0	0	0	0	0

Since the problem is so large, we will use Excel to find the solution.
As illustrated in the text we
1. Enter the variables and their initial values of 0 (since they are nonbasic in the initial tableau).
2. Enter the objective function.
3. Enter the constraints, and the initial RHS
4. Go to the solver, set the target cell equal to the max(imum).
   then enter the constraints including the nonnegativity constraints.
5. Check the options to be sure a "assume linear model" is checked.
6. Solve and highlight answer.

We find that the maximum yield of $3250.00 is obtained when 1071.43 shares of Duke Energy, 666.67 shares of Eastman Kodak, 294.12 shares of General Motors and 158.73 shares of H.J. Heinz are purchased.

Chapter 4 Project

1. The objective is to maximize the carbohydrates in the trail mix. The objective function is maximize $C = 31.4x_1 + 114.74x_2 + 148.12x_3 + 33.68x_4$

3. This is a maximum problem with mixed constraints, so we will rewrite all the constraints but the nonnegativity inequalities to be less than or equal to inequalities.

$$
\begin{aligned}
x_1 + x_2 + x_3 + x_4 &\le 10 \\
-0.9x_1 + 0.1x_2 + 0.1x_3 + 0.1x_4 &\le 0 \\
+0.1x_1 - 0.9x_2 + 0.1x_3 + 0.1x_4 &\le 0 \\
+0.1x_1 + 0.1x_2 - 0.9x_3 + 0.1x_4 &\le 0 \\
+0.1x_1 + 0.1x_2 + 0.1x_3 - 0.9x_4 &\le 0 \\
854x_1 + 435x_2 + 1023.96x_3 + 162.02x_4 &\le 7000 \\
x_1 \ge 0 \quad x_2 \ge 0 \quad x_3 \ge 0 \quad x_4 &\ge 0
\end{aligned}
$$

Since the problem is so large we will use Excel to find the solution.
As illustrated in the text
1. Enter the variables and their initial values of 0 (since they are nonbasic in the initial tableau).
2. Enter the objective function.
3. Enter the constraints, and the initial RHS
4. Go to the solver, set the target cell equal to the max(imum).
   then enter the constraints including the nonnegativity constraints.
5. Check the options to be sure a "assume linear model" is checked.
6. Solve and highlight answer.

We find that the trail mix will maximize carbohydrates while meeting the constraints on the mix if 1 cup of peanuts, 3.7 cups of raisins, 4.3 cups of M&M's, and 1 cup of pretzels are used in the mix. Then the maximum carbohydrates will be 1124.9 grams.

5. To find the mix that will maximize the protein, we change the objective function. We

maximize $P = 34.57x_1 + 4.67x_2 + 9.01x_3 + 3.87x_4$

Again we use Excel, following the steps in the text, and we find that the trail mix which maximizes protein while meeting the constraints will contain 6.5 cups of peanuts, 0.9 cups of raisins, 0.9 cups of M&M's, and 0.9 cups of pretzels. The maximum protein in the mix will be 239.06 grams.

This trail mix contains 9.2 cups of mixture. So there are $\dfrac{231.06}{9.2} = 25.9$ grams of protein per cup.

There are $\dfrac{(854.10)(6.5) + (435)(0.9) + (1023.96)(0.9) + (162.02)(0.9)}{9.2} = 762.01$ calories per cup of mix.

7. The mix that minimizes the fat contains

$$\frac{(72.5)(1)+(0.67)(7)+(43.95)(1)+(1.49)(1)}{10} = 12.26 \text{ grams of fat per cup of mix.}$$

$$\frac{(34.57)(1)+(4.67)(7)+(9.01)(1)+(3.87)(1)}{10} = 8.0 \text{ grams of protein per cup of mix.}$$

$$\frac{(31.4)(1)+(114.74)(7)+(148.12)(1)+(33.68)(1)}{10} = 101.64 \text{ grams of carbohydrates per cup}$$

of mix.

Mathematical Questions from Professional Exams

1. (c)   **3.** (c)

5. (b)   **7.** (c)

9. (d)   **11.** (d)

Chapter 5

Finance

5.1 Interest

1. $0.60 = \dfrac{60}{100} = 60\%$

3. $1.1 = \dfrac{110}{100} = 110\%$

5. $0.06 = \dfrac{6}{100} = 6\%$

7. $0.0025 = \dfrac{25}{10000} = \dfrac{0.25}{100} = 0.25\%$

9. $25\% = \dfrac{25}{100} = 0.25$

11. $100\% = \dfrac{100}{100} = 1$

13. $6.5\% = \dfrac{6.5}{100} = 0.065$

15. $73.4\% = \dfrac{73.4}{100} = 0.734$

17. 15% of $1000 = (0.15) \cdot (1000) = 150$

19. 18% of $100 = (0.18) \cdot (100) = 18$

21. 210% of $50 = (2.10) \cdot (50) = 105$

23. $x\%$ of $80 = 4$

$\quad \dfrac{x}{100} \cdot 80 = 4$

$\qquad 80x = 400$

$\qquad\quad x = 5$

4 is 5% of 80

25. $x\%$ of $5 = 8$

$\quad \dfrac{x}{100} \cdot 5 = 8$

$\qquad 5x = 800$

$\qquad\ x = 160$

8 is 160% of 5

27. $20 = 8\%$ of x
$20 = 0.08x$

$$x = \frac{20}{0.08} = 250$$

20 is 8% of 250

29. $50 = 15\%$ of x
$50 = 0.15x$

$$x = \frac{50}{0.15} = 333.333$$

50 is 15% of 333.33

31. $I = Prt$

$$= (\$1000)(0.04)\left(\frac{3}{12}\right)$$

$$= \$10$$

33. $I = Prt$

$$= (\$500)(0.12)\left(\frac{9}{12}\right)$$

$$= \$45$$

35. $I = Prt$

$$= (\$1000)(0.10)\left(\frac{18}{12}\right)$$

$$= \$150$$

37. $A = P + Prt$

$$1050 = 1000 + 1000r\left(\frac{6}{12}\right)$$

$$50 = 500r$$

$$r = \frac{50}{500} = 0.1$$

The per annum rate of interest is 10%.

39. $A = P + Prt$

$$400 = 300 + 300r\left(\frac{12}{12}\right)$$

$$100 = 300r$$

$$r = \frac{100}{300} = 0.3333$$

The per annum rate of interest is 33.333%.

41. $A = P + Prt$

$$1000 = 900 + 900r\left(\frac{10}{12}\right)$$

$$100 = 750r$$

$$r = \frac{100}{750} = 0.1333$$

The per annum rate of interest is 13.333%.

43. $R = L - Lrt$

$$= 1200 - 1200(0.10)\left(\frac{6}{12}\right)$$

$$= 1140$$

The proceeds of the loan is $1140.

45. $R = L - Lrt$
$$= 2000 - 2000(0.08)(2)$$
$$= 1680$$

The proceeds of the loan is $1680.

47. $R = L - Lrt = L(1 - rt)$

$$1200 = L\left[1 - (0.10)\left(\frac{1}{2}\right)\right]$$

$$L = \frac{1200}{.95} = 1263.16$$

You must repay $1263.16 for the discounted loan.

The equivalent simple interest on the loan

$$A = P + Prt$$

$$1263.16 = 1200 + 1200r\left(\frac{1}{2}\right)$$

$$63.16 = 600r$$

$$r = 0.1052667$$

The simple rate of interest for the loan is 10.53%.

49.

$$R = L - Lrt = L(1 - rt)$$
$$2000 = L\left[1 - (0.08)(2)\right]$$
$$L = \frac{2000}{.84} = 2380.95$$

You must repay $2380.95 for the discounted loan.

The equivalent simple interest on the loan

$$A = P + Prt$$
$$2380.95 = 2000 + 2000r(2)$$
$$380.95 = 4000r$$
$$r = 0.0952375$$

The simple rate of interest for the loan is 9.52%.

51. We know the amount $A = \$500$, $t = 9$ months $= \frac{9}{12}$ of a year, and $r = 3\%$ simple interest.

We want to know P, the principal. $A = P + Prt = P(1 + rt)$

$$500 = P\left[1 + 0.03\left(\frac{9}{12}\right)\right]$$
$$500 = 1.0225P$$
$$P = 489.00$$

Madalyn should invest $489.00 if she wants to buy the stereo.

53. We know $P = \$600$, $I = \$156$, and $r = 8\% = 0.08$ simple interest. We need to find t, the length of the loan. $I = Prt$

$$156 = (600)(0.08)t$$
$$156 = 48t \text{ or } t = 3.25$$

Tami borrowed the money for 3.25 years or 3 years and 3 months.

55. We need to compare both loans and choose the one with the least interest due.

Discounted Loan: $r = 9\% = 0.09$

$$R = L - Lrt = L(1 - rt)$$
$$1000 = L\left[(1 - (0.09)\left(\frac{1}{2}\right)\right]$$
$$1000 = 0.955L$$
$$L = 1047.1204$$

The interest on the loan is
$1047.12 - \$1000 = \$47.12$.

Simple Interest Loan: $r = 10\% = 0.10$

$$I = Prt$$
$$I = 1000(0.10)\left(\frac{1}{2}\right)$$
$$= 50$$

The interest on the loan is $50.00.

You should choose the discounted loan. You will save $2.88.

57. We need to compare both loans and choose the one with the least interest due.

Discounted Loan: $r = 6\% = 0.06$

$$R = L - Lrt = L(1 - rt)$$
$$4000 = L\left[(1 - (0.06)(1)\right]$$
$$4000 = 0.94L$$
$$L = 4255.319$$

The interest on the loan is
$4255.32 - \$4000 = \$255.32$.

Simple Interest Loan: $r = 6.3\% = 0.063$

$$I = Prt$$
$$I = 4000(0.063)(1)$$
$$= 252.0$$

The interest on the loan is $252.00.

You should choose the simple interest loan. You will save $ 3.32.

59. We need to compare both loans and choose the one with the least interest due.

Discounted Loan: $r = 12.1\% = 0.121$ Simple Interest Loan: $r = 12.3\% = 0.123$

$$R = L - Lrt = L(1 - rt)$$
$$2000 = L\left[(1 - (0.121)(1)\right]$$
$$2000 = 0.879L$$
$$L = 2275.313$$

$$I = Prt$$
$$I = 2000(0.123)(1)$$
$$= 246$$

The interest on the loan is $246.00.

The interest on the loan is
$2275.31 - $2000 = $ 275.31.

Ruth should choose the simple interest loan. She will save $ 29.31.

61. This is an example of a discounted loan. The bank's $1million is L; the amount the bank pays (its bid) is R, the proceeds. In this problem $r = 2\% = 0.02$ and $t = 3$ months $= \dfrac{3}{12}$ year. $R = L(1 - rt)$

$$R = 1,000,000\left[1 - 0.02\left(\frac{3}{12}\right)\right] = 1,000,000\,(0.995) = 995,000$$

The bank should bid $995,000 for the treasury bill.

63. Since the price of the T-bills were $993.78 per $1000, and the investor purchased $10,000 worth of bills, the investor paid ($993.78)(10) = $9937.80 for the treasuries.

The interest earned on the investment was $10,000 – $9937.80 = $62.20.

The interest rate of the investment was r, where $R = L(1 - rt)$.

$$9937.80 = 10,000\left[1 - r\left(\frac{1}{2}\right)\right]$$
$$0.993780 = \left[1 - r\left(\frac{1}{2}\right)\right]$$
$$r = 2[1 - 0.993780] = 0.01244$$

The simple interest rate for this investment was 1.244%.

65. Since the price of the T-bills were $999.12 per $1000, and the investor purchased $5,000 worth of bills, the investor paid ($999.12)(5) = $4995.60 for the treasuries.

The interest earned on the investment was $5000 – $4995.60 = $ 4.40.

The interest rate of the investment was r, where $R = L(1 - rt)$.

$$4995.60 = 5000\left[1 - r\left(\frac{1}{12}\right)\right]$$
$$0.99912 = \left[1 - r\left(\frac{1}{12}\right)\right]$$
$$r = 12[1 - 0.99912] = 0.01056$$

The simple interest rate for this investment was 1.056%.

5.2 Compound Interest

1. $P = \$1000;\ i = \dfrac{0.04}{12};\ n = 36$ $A = P(1+i)^n = 1000\left(1 + \dfrac{0.04}{12}\right)^{36} = 1127.2719$

The amount accumulated in the investment is $A = \$1129.27$.

3. $P = \$500;\ i = 0.05;\ n = 3$ $A = P(1+i)^n = 500(1 + 0.05)^3 = 578.8125$

The amount accumulated in the investment is $A = \$578.81$.

5. $P = \$800;\ i = \dfrac{0.06}{365};\ n = 200$ $A = P(1+i)^n = 800\left(1 + \dfrac{0.06}{365}\right)^{200} = 826.7363$

The amount accumulated in the investment is $A = \$826.74$.

7. We need to find the present value P of the investment.

$A = \$100,\ n = 6$ months, $i = \dfrac{0.04}{12}$ $P = A(1+i)^{-n} = 100\left(1 + \dfrac{0.04}{12}\right)^{-6} = 98.0231$

The principal needed for the investment is $\$98.02$.

9. We need to find the present value P of the investment.

$A = \$500,\ n = 1$ year $= 365$ days, $i = \dfrac{0.07}{365}$; $P = A(1+i)^{-n} = 500\left(1 + \dfrac{0.07}{365}\right)^{-365} = 466.20$

The principal needed for the investment is $\$466.20$.

11. We use $A_n = P(1 + i)^n$. The principal $P = \$1000$

    (a)   For annual compounding $i = 0.04$ and $n = 3$. The amount A is

$$A_3 = P(1 + i)^3 = (\$1000)(1 + 0.04)^3 = (\$1000)(1.124864) = \$1124.86$$

         The interest earned is $A - P = \$1124.86 - \$1000.00 = \$124.86$.

    (b)   For semiannual compounding, there are $3 \cdot 2 = 6$ payment periods over 3 years. The interest rate per payment period is $i = \dfrac{0.04}{2}$. The amount A is

$$A_6 = P(1 + i)^6 = (\$1000)\left(1 + \frac{0.04}{2}\right)^6 = (\$1000)(1.12616) = \$1126.16$$

         The interest earned is $A - P = \$1126.16 - \$1000.00 = \$126.16$.

    (c)   For quarterly compounding there are $3 \cdot 4 = 12$ payment periods over 3 years. The interest rate per payment period is $i = \dfrac{0.04}{4}$. The amount A is

$$A_{12} = P(1 + i)^{12} = (\$1000)\left(1 + \frac{0.04}{4}\right)^{12} = (\$1000)(1.12683) = \$1126.83$$

         The interest earned is $A - P = \$1126.83 - \$1000.00 = \$126.83$.

    (d)   For monthly compounding there are $3 \cdot 12 = 36$ payment periods over 3 years. The interest rate per payment period is $i = \dfrac{0.04}{12}$. The amount A is

$$A_{36} = P(1 + i)^{36} = (\$1000)\left(1 + \frac{0.04}{12}\right)^{36} = (\$1000)(1.12727) = \$1127.27$$

         The interest earned is $A - P = \$1127.27 - \$1000.00 = \$127.27$.

13. We use $A_n = P(1 + i)^n$. The principal $P = \$1000$ and $i = \frac{0.06}{4}$.

    (a)   For an investment of 2 years $n = 2 \cdot 4 = 8$. The amount A is

$$A_8 = P(1 + i)^8 = (\$1000)\left(1 + \frac{0.06}{4}\right)^8 = (\$1000)(1.12649) = \$1126.49.$$

    (b)   For an investment of 3 years $n = 3 \cdot 4 = 12$. The amount A is

$$A_{12} = P(1 + i)^{12} = (\$1000)\left(1 + \frac{0.06}{4}\right)^{12} = (\$1000)(1.195618) = \$1195.62.$$

    (c)   For an investment of 4 years $n = 4 \cdot 4 = 16$. The amount A is

$$A_{16} = P(1 + i)^{16} = (\$1000)\left(1 + \frac{0.06}{4}\right)^{16} = (\$1000)(1.2689855) = \$1268.99.$$

15. In this problem we want to find the principal P needed now to have the amount $A = \$5000$ in the future. We need the present value of $\$5000$.

(a) Since the compounding is twice per year for 4 years, $n = 8$ and $i = \dfrac{0.03}{2}$. The present value $P = A(1 + i)^{-n}$ of $\$5000$ is

$$P = (\$5000)\left(1 + \frac{0.03}{2}\right)^{-8} = (\$5000)(0.887711) = \$4438.56.$$

(b) Since the compounding is twice per year for 8 years, $n = 16$ and $i = \dfrac{0.03}{2}$. The present value $P = A(1 + i)^{-n}$ of $\$5000$ is

$$P = (\$5000)\left(1 + \frac{0.03}{2}\right)^{-16} = (\$5000)(0.788031) = \$3940.16.$$

17. The effective rate of interest is the equivalent annual simple rate of interest that yields the same amount as compounding after 1 year. If $i = \dfrac{r}{k}$, then the effective rate of interest is $(1 + i)^k - 1 = (1 + \frac{r}{k})^k - 1$.

(a) The effective rate of interest for 8% compounded semiannually is

$$\left(1 + \frac{0.08}{2}\right)^2 - 1 = 0.0816 = 8.16\%$$

(b) The effective rate of interest for 4% compounded monthly is

$$\left(1 + \frac{0.04}{12}\right)^{12} - 1 = 0.04074 = 4.07\%$$

19. If P is the principal and we want P to double, then amount $A = 2P$. We use the compound interest formula with $n = 3$ to find i.

$$A_3 = P(1 + i)^3$$
$$2P = P(1 + i)^3$$
$$2 = (1 + i)^3$$
$$i = \sqrt[3]{2} - 1 = 0.25992$$

An annual interest rate of 25.992% will double an investment in 3 years.

21. If P is the principal and we want P to triple, then amount $A = 3P$. We use the compound interest formula with $i = 0.10$ to find n.

$$A = P(1 + i)^n$$
$$3P = P(1 + 0.10)^n$$
$$3 = (1.10)^n$$
$$n = \log_{1.1} 3 = \frac{\log 3}{\log 1.10} = 11.527 \text{ years}$$

At 10% interest, it will take an investment approximately 11.5 years to triple in value.

23. We find the interest due on each loan. We have $P = \$1000$, $t = 2$ years.
With a simple interest rate $r = 12\% = 0.12$ the interest due after two years is

$$I = Prt = (\$1000)(0.12)(2) = \$240$$

With an interest rate of $10\% = 0.10$ compounded monthly the interest due after 2 years is

$$I = A - P = (\$1000)\left(1 + \frac{0.10}{12}\right)^{24} - \$1000 = \$220.39$$

Mr. Nielsen will pay less interest with a loan charging 10% interest compounded monthly.

25. We need the present value of $1000. Since the compounding is annually, $n = 1$ and $i = 0.09$.

The present value $P = A(1 + i)^{-n}$ of $1000 after 1 year is

$$P = (\$1000)(1.09)^{-1} = \$917.43$$

The present value of $1000 after 2 years is

$$P = (\$1000)(1.09)^{-2} = \$841.68$$

27. The effective rate of interest is the equivalent annual simple rate of interest that yields the same amount as compounding after 1 year.
The effective rate of interest for $5.25\% = 0.0525$ compounded quarterly is

$$\left(1 + \frac{0.0525}{4}\right)^4 - 1 = 0.05354 = 5.354\%$$

29. To find the interest rate compounded quarterly that is equivalent to a simple interest rate of 7%, we solve $(1 + \frac{i}{4})^4 - 1 = 0.07$ for i.

$$\left(1 + \frac{i}{4}\right)^4 - 1 = 0.07$$

$$\left(1 + \frac{i}{4}\right)^4 = 1.07$$

$$\left(1 + \frac{i}{4}\right) = \sqrt[4]{1.07}$$

$$i = 4[\sqrt[4]{1.07} - 1] = 0.068234 = 6.823\%$$

An interest rate of 6.823% compounded quarterly has an effective annual simple rate of interest of 7%.

31. To decide which rate yields the larger amount in 1 year, we use $A = P(1 + i)^n$ for each rate and compare the results. We use $P = \$10,000$ as suggested in the text.

6% compounded quarterly:

$$A = \$10,000\left(1 + \frac{0.06}{4}\right)^4$$
$$= \$10,613.64$$

$6\frac{1}{4}\%$ compounded annually:

$$A = \$10,000(1 + 0.0625)^1$$
$$= \$10,625.00$$

$6\frac{1}{4}\%$ compounded annually has a greater yield.

33.   To decide which rate yields the larger amount in 1 year, we use $A = P(1 + i)^n$ for each rate and compare the results. We use $P = \$10,000$ as suggested in the text.

9% compounded monthly: 8.8% compounded daily:

$$A = \$10,000\left(1 + \frac{0.09}{12}\right)^{12}$$ $$A = \$10,000\left(1 + \frac{0.088}{365}\right)^{365}$$
$$= \$10,938.07$$ $$= \$10,919.77$$

9% compounded monthly has a greater yield.

35.   We use the compound interest formula, $A = P(1 + i)^n$, with $P = \$90,000$, $i = 5\% = 0.05$, and $n = 4$ years. $A = \$90,000(1 + 0.05)^4$
$$= \$109,395.5625$$

The house will be worth $\$109,400$ in four years.

37.   We use the compound interest formula, $A = P(1 + i)^n$, with $P = \$600$, $i = 1.5\% = 0.015$ per month, and $n = 6$ months. $A = \$600(1 + 0.015)^6$
$$= \$656.07$$

Caryl will owe $\$656.07$ after 6 months.

39.   We do this problem in two parts since the inflation rates for tuition and room and board are different. At the end we sum the results. For each we need the compound interest formula with $n = 5$.

Tuition: $P = \$3506$, $i = 5.09\% = 0.0509$ Room and Board: $P = \$5149$,
$A = P(1 + i)^n$ $i = 2.85\% = 0.0285$.
$\quad = (\$3506)(1 + 0.0509)^5$ $A = P(1 + i)^n$
$\quad = \$4493.85$ $\quad = (\$5149)(1 + 0.0285)^5$
                                                            $\quad = \$5925.76$

The cost of college in the year 2005-06 is projected to be $\$4493.85 + \$5925.76 = \$10,419.61$.

41.   We need the present value of $\$40,000$ after 4 years. Since the compounding is 4 times per year for 4 years, $n = 16$ and $i = \dfrac{0.08}{4}$. The present value $P = A(1 + i)^{-n}$ is

$$P = (\$40,000)\left(1 + \frac{0.08}{4}\right)^{-16} = \$29,137.83$$

Tami and Todd should deposit $\$29,137.83$ now if they want the $\$40,000$ down payment for the house.

43.   We want the accumulated value of the $\$6000$ investment after 25 years. Since the compounding is 2 times per year for 25 years, $n = 2 \cdot 25 = 50$ and $i = \dfrac{0.08}{2}$. The amount $A = P(1 + i)^n$ is,

$$A = (\$6000)\left(1 + \frac{0.08}{2}\right)^{50} = \$42,640.10$$

The child will have $\$42,640.10$ when she is 25 years old.

45. To decide if Jack's investment is in line with his goal of earning at least 7% interest compounded quarterly, we use the compound interest formula, $A = P(1 + i)^n$, with $A =$ \$20, $P =$ \$15, $n = 16$ (4 compounding periods per year for 4 years), and solve for i,

where $i = \dfrac{r}{4}$. $\$20 = \$15(1 + i)^{16}$

$$\frac{4}{3} = (1 + i)^{16}$$

$$i = \sqrt[16]{\frac{4}{3}} - 1 = 0.0181427; \quad r = 4i = (4)(0.0181427) = 0.07257$$

The growth of the stock is more than 7% per year, so Jack should invest.

47. We use the compound interest formula, $A = P(1 + i)^n$ with $P = \$2000$. Since $i = 9\% = 0.09$ compounded quarterly, there are $n = (4)(25) = 100$ interest periods over the life of the investment.

$$A = \$2000\left(1 + \frac{0.09}{4}\right)^{100} = \$18{,}508.09$$

There will be \$18,508.09 in the IRA account after 25 years.

49. To solve the first part of the question we use the compound interest formula, $A = P(1 + i)^n$ with $A = \$6.406$ trillion, $P = \$4.8$ trillion, and $n = 8$. We solve for i.

$\$6.406 = \$4.8(1 + i)^8$

$1.33458333 = (1 + i)^8$

$i = \sqrt[8]{1.33458333} - 1 = 0.036736$

The national debt grew at an annual rate of 3.67% over the years from 1995 through 2003. If the current rate continues on January 1, 2010 the national debt will have grown to

$A = \$6.406(1 + 0.367)^7 = 8.246$ trillion dollars.

51. We want the present value of Tami's grandparents' investment.

$A = \$40,000$ (the face value of the bond), $n = 17$, $i = 8\% = 0.08$.
$P = A(1 + i)^{-n} = \$40,000(1 + 0.08)^{-17} = \$10,810.758$

Tami's grandparents' should pay \$10,810.76 for the bond.

53. We will use the compound interest formula $A = P(1 + i)^n$ to find i, the rate of return. We have A (the face value) = \$25,000, P (the purchase price) = \$12,485.52, and n (the time until maturity) = 8 years. $\$25,000 = \$12,485.52(1 + i)^8$

$$2.002319 = (1 + i)^8$$
$$\sqrt[8]{2.002319} = 1 + i$$
$$i = 1.090666 - 1 = 0.090666$$

The annual compound rate of return is 9.067%.

55. We use the compound interest formula, $A = P(1 + i)^n$ with $A = \$25{,}000$, $P = \$10{,}000$, and $i = 6\%$ compounded daily. We solve for t, the number of years.

$$\$25{,}000 = \$10{,}000\left(1 + \frac{0.06}{365}\right)^{365t}$$

$$2.5 = \left(1 + \frac{0.06}{365}\right)^{365t}$$

$$2.5 = (1.000164384)^{365t}$$

$$365t = \log_{1.000164384}(2.5)$$

$$t = \left(\frac{1}{365}\right) \cdot \frac{\log 2.5}{\log(1.000164384)}$$

$$t = 15.2727$$

It will take about 15.27 years for \$10,000 to grow to \$25,000 at 6% compounded daily.

57. We use the formula $A = P(1 - r)^n$ to measure the purchasing power of $P = \$1000$ after $n = 2$ years. The inflation rate $r = 3\% = 0.03$.

$$A = \$1000(1 - 0.03)^2 = 940.9$$

With a 3% inflation rate, in 2 years \$1000 will buy \$940.90 worth of goods and services.

59. We use the formula $A = P(1 - r)^n$ to measure the purchasing power of $P = \$1000$ after $n = 5$ years. The inflation rate $r = 3\% = 0.03$.

$$A = \$1000(1 - 0.03)^5 = 858.7340$$

With a 3% inflation rate, in 5 years \$1000 will buy \$858.73 worth of goods and services.

61. We use the formula $A = P(1 - r)^n$ to measure when the purchasing power of $P = \$1000$ is reduced to \$500. The inflation rate $r = 3\% = 0.03$.

$$500 = \$1000(1 - 0.03)^n$$

$$0.5 = (0.97)^n$$

$$n = \log_{0.97}(0.5)$$

$$= \frac{\log 0.5}{\log 0.97} = 22.7566$$

With a 3% inflation rate, purchasing power will be halved in approximately 22.75 years.

63. We use the formula $A = P(1 - r)^n$ to measure when the purchasing power of $P = \$1000$ is reduced to \$500. The inflation rate $r = 6\% = 0.06$.

$$500 = \$1000(1 - 0.06)^n$$

$$0.5 = (0.94)^n$$

$$n = \log_{0.94}(0.5)$$

$$= \frac{\log 0.5}{\log 0.94} = 11.2023$$

With a 6% inflation rate, purchasing power will be halved in approximately 11.20 years.

5.3 Annuities; Sinking Funds

1. The deposit is $P = \$100$. The number of deposits is $n = 10$ and the interest per payment period is $i = 0.10$. We use the formula

$$A = P\frac{(1+i)^n - 1}{i} = \$100 \cdot \frac{(1+0.10)^{10} - 1}{0.10} = (\$100)(15.93742) = \$1593.74$$

There is $1593.74 in the account after 10 years.

3. The deposit is $P = \$400$. The number of deposits is $n = 12$ and the interest per payment period is $i = \frac{0.12}{12} = 0.01$. We use the formula

$$A = P\frac{(1+i)^n - 1}{i} = \$400 \cdot \frac{(1+0.01)^{12} - 1}{0.01} = (\$400)(12.682503) = \$5073.00$$

There is $5073.00 in the account after 12 months (1 year).

5. The deposit is $P = \$200$. The number of deposits is $n = 36$ and the interest per payment period is $i = \frac{0.06}{12} = 0.005$. We use the formula

$$A = P\frac{(1+i)^n - 1}{i} = \$200 \cdot \frac{(1+0.005)^{36} - 1}{0.005} = (\$200)(39.3361) = \$7867.22$$

There is $7867.22 in the account after 36 months (3 years).

7. The deposit is $P = \$100$. The number of deposits is $n = 60$ and the interest per payment period is $i = \frac{0.06}{12} = 0.005$. We use the formula

$$A = P\frac{(1+i)^n - 1}{i} = \$100 \cdot \frac{(1+0.005)^{60} - 1}{0.005} = (\$100)(69.77003) = \$6977.00$$

There is $6977.00 in the account after 60 months (5 years).

9. The deposit is $P = \$9000$. The number of deposits is $n = 10$ and the interest per payment period is $i = 0.05$. We use the formula

$$A = P\frac{(1+i)^n - 1}{i} = \$9000 \cdot \frac{(1+0.05)^{10} - 1}{0.05} = (\$9000)(12.57789) = \$113,201.03$$

There is $113,201.03 in the account after 10 years.

11. The amount required is $A = \$10,000$ after $n = 12 \cdot 5 = 60$ monthly payments. The interest rate per payment period is $i = \frac{0.05}{12}$. To find the monthly payment we use $A = P\dfrac{(1+i)^n - 1}{i}$ and solve for P.

$$\$10,000 = P\,\frac{\left(1 + \dfrac{0.05}{12}\right)^{60} - 1}{\dfrac{0.05}{12}}$$

$$\$10,000 = P(68.0061)$$
$$P = \$147.05$$

Sixty monthly payments of \$147.05 are needed to accumulate \$10,000.

13. The amount required is $A = \$20,000$ after $n = 4 \cdot 2.5 = 10$ quarterly payments. The interest rate per payment period is $i = \dfrac{0.06}{4} = 0.015$. To find the quarterly payment we use

$$A = P\frac{(1+i)^n - 1}{i} \quad \text{and solve for } P.$$

$$\$20,000 = P\,\frac{(1 + 0.015)^{10} - 1}{0.015}$$

$$\$20,000 = P(10.7027)$$
$$P = \$1868.68$$

Ten quarterly payments of \$1868.68 are needed to accumulate \$20,000.

15. The amount required is $A = \$25,000$ after $n = 6$ monthly payments. The interest rate per payment period is $i = \dfrac{0.055}{12}$. To find the monthly payment we use $A = P\dfrac{(1+i)^n - 1}{i}$ and solve for P.

$$\$25,000 = P\,\frac{\left(1 + \dfrac{0.055}{12}\right)^{6} - 1}{\dfrac{0.055}{12}}$$

$$\$25,000 = P(6.06917)$$
$$P = \$4119.178$$

Six monthly payments of \$4119.18 are needed to accumulate \$25,000.

17. The amount required is $A = \$5000$ after $n = 12 \cdot 2 = 24$ monthly payments. The interest rate per payment period is $i = \dfrac{0.04}{12}$. To find the monthly payment we

use $A = P\dfrac{(1+i)^n - 1}{i}$ and solve for P.

$$\$5000 = P\,\frac{\left(1 + \dfrac{0.04}{12}\right)^{24} - 1}{\dfrac{0.04}{12}}$$

$$\$5000 = P(24.94289)$$
$$P = \$200.458$$

Twenty-four monthly payments of $200.46 are needed to accumulate $5000.

19. The amount required is $A = \$9000$ after $n = 4$ annual payments. The interest rate per payment period is $i = 0.05$. To find the monthly payment we use $A = P\dfrac{(1+i)^n - 1}{i}$ and solve for P.

$$\$9000 = P\,\frac{(1 + 0.05)^4 - 1}{0.05}$$
$$\$9000 = P(4.31013)$$
$$P = \$2088.106$$

Four annual payments of $2088.11 are needed to accumulate $9000.

21. Al's investment is an example of an annuity. The deposit is $P = \$2500$, the number of deposits is $n = 15$ and the interest per payment period is $i = 0.07$. To find the value of the fund after 15 deposits we use the formula

$$A = P\frac{(1+i)^n - 1}{i} = \$2500\,\frac{(1 + 0.07)^{15} - 1}{0.07} = \$62822.555$$

The mutual fund is worth $62,822.56.

23. Todd and Tami have set up an annuity. The deposit is $P = \$300$, the number of deposits is $n = 4 \cdot 6 = 24$, and the interest per payment period is $i = \dfrac{0.08}{4} = 0.02$. To find the value of the fund after 24 quarterly deposits we use the formula

$$A = P\frac{(1+i)^n - 1}{i} = \$300\,\frac{(1 + 0.02)^{24} - 1}{0.02} = \$9126.559$$

Todd and Tami's annuity will be worth $9126.56.

25. Dan's pension fund can be thought of as a sinking fund. The amount required is $A = \$350,000$ after $n = 12 \cdot 20 = 240$ monthly payments. The interest rate per payment period is $i = \dfrac{0.09}{12} = 0.0075$. To find the monthly payment we use $A = P\dfrac{(1+i)^n - 1}{i}$ and solve for P.

$$\$350,000 = P\frac{(1+0.0075)^{240} - 1}{0.0075}$$
$$\$350,000 = P(667.88687)$$
$$P = \$524.04$$

Dan needs to save $524.04 per month to have $350,000 in 20 years.

27. In this sinking fund the amount required is $A = \$100,000$ after $n = 4$ annual payments. The interest rate per payment period is $i = 0.08$. To find the annual payment we use $A = P\dfrac{(1+i)^n - 1}{i}$ and solve for P.

$$\$100,000 = P\frac{(1+0.08)^4 - 1}{0.08}$$
$$\$100,000 = P(4.506112)$$
$$P = \$22,192.08$$

An annual payment of $22,192.08 is necessary to accumulate the $100,000 needed. The table shows the growth of the sinking fund over time.

Payment Number	Deposit $	Cumulative Deposits	Accumulated Interest $	Total $
1	22,192.08	22,192.08	0	22,192.08
2	22,192.08	44,384.16	1775.37	46,159.53
3	22,192.08	66,576.24	5468.13	72,044.37
4	22,192.08	88,768.32	11,231.68	100,000.00

29. Let x denote the price to be paid for the oil well. Then $0.14x$ represents a 14% annual Return on Investment (ROI). The annual Sinking Fund Contribution (SFC) needed to recover the purchase price x in 30 years can be calculated letting $A = x$, $n = 30$, $i = 0.10$, and $P = \text{SFC}$.

$$A = P\frac{(1+i)^n - 1}{i}$$
$$x = \text{SFC}\frac{(1+0.10)^{30} - 1}{0.10}$$
$$x = \text{SFC}(164.4940227)$$
$$\text{SFC} = 0.0060792483x$$

The required annual ROI plus the annual SFC equals the annual net income of $30,000.

$$0.14x + 0.006079x = \$30,000$$
$$0.146079x = \$30,000$$
$$x = \$205,368.33$$

The investor should pay $205,368.33 for the oil well.

31. In this sinking fund the amount required is $A = \$1,000,000$ after $n = 4 \cdot 10 = 40$ quarterly payments. The interest rate per payment period is $i = \dfrac{0.08}{4} = 0.02$. To find the quarterly sinking fund payment we use $A = P\dfrac{(1+i)^n - 1}{i}$ and solve for P.

$$\$1,000,000 = P\,\frac{(1 + 0.02)^{40} - 1}{0.02}$$

$$\$1,000,000 = P(60.40198)$$

$$P = \$16,555.75$$

Forty quarterly payments of \$16,555.75 are necessary to accumulate \$1,000,000.

33. (a) The projected cost of the roof in 20 years is given by $A = P(1 + i)^n$. We have $P = \$100,000$, $i = 0.03$, and $n = 20$. The roof will cost $A = \$100,000(1.03)^{20} = \$180,611.12$.

(b) If the Condo Association makes $n = 2 \cdot 20 = 40$ payments at an interest rate of $i = \dfrac{0.06}{2} = 0.03$ per payment period, then each payment should be

$$A = P\frac{(1+i)^n - 1}{i}$$

$$\$180,611.12 = P\,\frac{(1 + 0.03)^{40} - 1}{0.03}$$

$$\$180,611.12 = P(75.40126)$$

$$P = \$2395.33$$

35. $A = \$1,000,000$, $P = \$600$, $i = \dfrac{0.07}{12}$. We will use the formula (1) to find n.

$$A = P\frac{(1+i)^n - 1}{i}$$

$$\$1,000,000 = \$600\,\frac{\left(1 + \dfrac{0.07}{12}\right)^n - 1}{\dfrac{0.07}{12}}$$

$$\frac{1,000,000}{600}\left(\frac{0.07}{12}\right) + 1 = \left(1 + \frac{0.07}{12}\right)^n$$

$$10.72222 = (1.0058333)^n$$

$$\log_{1.0058333}(10.72222) = n = \frac{\log 10.72222}{\log 1.0058333} = 407.868$$

It will take approximately 407.868 months or 33.989 or 34 years to accumulate \$1,000,000.

37. The projected cost of a Honda in 2007 ($n = 4$) assuming an annual inflation rate $i = 0.0158$, is given by $A = P(1 + i)^n$. $A = \$21,600(1 + 0.0158)^4 = \$22,997.82$.

Since the sales tax is 9.25%, the projected cost of the car including tax is

$\$22,997.82(1.0925) = \$25,125.11$

The sinking fund has $A = \$25,125.11$, $n = 12 \cdot 4 = 48$ payments, and $i = \dfrac{0.0275}{12}$. We use

$A = P\dfrac{(1+i)^n - 1}{i}$ to find P, the monthly payment.

$\$25,125.11 = P\dfrac{\left(1 + \dfrac{0.0275}{12}\right)^{48} - 1}{\dfrac{0.0275}{12}}$

$\$25,125.11 = P(50.67822)$
$\qquad P = \$495.78$

39. The projected annual cost of college tuition assuming a constant inflation rate of 5.09% is given by $A = \$3506(1.0509)^n$. The projected total cost of future tuition is

Year	n	$A = \$3506(1.0509)^n$
2012 – 13	12	$6361.34
2013 – 14	13	$6685.13
2014 – 15	14	$7025.41
2015 – 16	15	$7383.00

The projected cost of 4 years tuition at a public institution is $27,454.88.

Since the accumulated amount of the sinking fund continues to earn 4.2% annual interest compounded quarterly after payments have ceased, we need to find the present value of the 2nd through 4th years tuition at 4.2% compounded quarterly on December 2012. The sum of these values and the first year's tuition will be A, the value of the sinking fund.

Present value of $6685.13 on 12/12: $P = \$6685.13\left(1 + \dfrac{0.042}{4}\right)^{-4} = \$6411.45$

Present value of $7025.41 on 12/12: $P = \$7025.41\left(1 + \dfrac{0.042}{4}\right)^{-8} = \$6462.21$

Present value of $7383.00 on 12/12: $P = \$7383.00\left(1 + \dfrac{0.042}{4}\right)^{-12} = \$6513.24$

The quarterly payments to the sinking fund must accumulate to
$\$6361.34 + \$6411.45 + \$6462.21 + \$6513.24 = \$25,748.24$

The quarterly payments needed to the sinking fund are

$A = P\dfrac{(1+i)^n - 1}{i}$

$\$25,748.24 = P\dfrac{\left(1 + \dfrac{0.042}{4}\right)^{40} - 1}{\dfrac{0.042}{4}}$

$\$25748.24 = P(49.39358)$
$\qquad P = \$521.29$

5.4 Present Value of an Annuity; Amortization

1. $P = \$500; n = 36; i = \dfrac{0.10}{12}$

$$V = P\frac{1-(1+i)^{-n}}{i} = \$500\frac{1-\left(1+\dfrac{0.10}{12}\right)^{-36}}{\dfrac{0.10}{12}} = \$500\,(30.991236) = \$15,\!495.62$$

3. $P = \$100; n = 9; i = \dfrac{0.12}{12} = 0.01$

$$V = P\frac{1-(1+i)^{-n}}{i} = \$100\frac{1-(1+.01)^{-9}}{0.01} = \$100\,(8.56602) = \$856.60$$

5. $P = \$10,\!000; n = 20; i = 0.10$

$$V = P\frac{1-(1+i)^{-n}}{i} = \$10,\!000\frac{1-(1+0.10)^{-20}}{0.10} = \$10,\!000\,(8.513564) = \$85,\!135.64$$

7. First we find the amount of the annuity using $A = P\dfrac{(1+i)^n - 1}{i}$. The couple save $P =$

$\$4000$ per year for $n = 20$ years at $i = 0.10$ interest. $A = \$4000\dfrac{(1+0.10)^{20} - 1}{0.10} = \$229,\!100$

The value of their IRA is $229,100.

The $229,100 becomes the present value of the future annuity. To find how much the couple can withdraw each year for $n = 25$ years at 0.10 compounded annually, we solve

$$P = V\frac{i}{1-(1+i)^{-n}} = \$229,\!100\,\frac{0.10}{1-(1+0.10)^{-25}} = \$25,\!239.51$$

The couple can withdraw $25,239.51 a year for 25 years.

9. For the 2 year loan $V = \$10,000$, $n = 12 \cdot 2 = 24$, and $i = \dfrac{0.12}{12} = 0.01$. The monthly payment P is

$$P = V\frac{i}{1-(1+i)^{-n}} = \$10,000\ \frac{0.01}{1-(1+0.01)^{-24}} = \$470.73$$

11. For the 20 year mortgage $V = \$240,000$, $n = 12 \cdot 20 = 240$, and $i = \dfrac{0.06}{12} = 0.005$. The monthly payment P is

$$P = V\frac{i}{1-(1+i)^{-n}} = \$240,000\ \frac{0.005}{1-(1+0.005)^{-240}} = \$1719.43$$

13. For the 20 year inheritance $V = \$15,000$, $n = 20$, and $i = 0.12$. The monthly payment P is

$$P = V\frac{i}{1-(1+i)^{-n}} = \$15,000\ \frac{0.12}{1-(1+0.12)^{-20}} = \$2008.18$$

15. Mr. Doody is interested in finding the present value of an annuity that has an interest rate of $i = \dfrac{0.10}{12}$ and from which he withdraws $P = \$250$ each month for $n = 12 \cdot 20 = 240$ months.

$$V = P\frac{1-(1+i)^{-n}}{i} = \$250\ \frac{1-\left(1+\dfrac{0.10}{12}\right)^{-240}}{\dfrac{0.10}{12}} = \$250(103.62462) = \$25,906.15$$

Mr. Doody needs \$25,906.15 now to guarantee his future income.

17. If the couple chooses to finance their home with a 20 year mortgage at 8% interest, they will have $n = 12 \cdot 20 = 240$ monthly payments, the interest rate per payment period is $i = \dfrac{0.08}{12}$, and their monthly payments will be

$$P = V\frac{i}{1-(1+i)^{-n}} = \$160{,}000\ \frac{\dfrac{0.08}{12}}{1-\left(1+\dfrac{0.08}{12}\right)^{-240}} = \$1338.30$$

The total interest paid on this loan is $(\$1338.30)(240) - \$160{,}000 = \$161{,}192$.

The couple's equity in the house is the sum of their down payment and the amount paid on the loan. After 10 years the couple still owes 120 payments. The amount still owed on the loan is the present value of these payments

$$V = P\frac{1-(1+i)^{-n}}{i} = \$1338.30\ \frac{1-\left(1+\dfrac{0.08}{12}\right)^{-120}}{\dfrac{0.08}{12}} = \$110{,}304.67$$

meaning the amount paid on the loan $\$160{,}000 - \$110{,}304.67 = \$49{,}695.33$.
The couple's equity after 10 years is $\$40{,}000 + \$49{,}695.33 = \$89{,}695.33$

If the couple chooses to finance their home with a 25 year mortgage at 9% interest, they will have $n = 12 \cdot 25 = 300$ monthly payments, the interest rate per payment period is $i = \dfrac{0.09}{12} = 0.0075$, and their monthly payments will be

$$P = V\frac{i}{1-(1+i)^{-n}} = \$160{,}000\ \frac{0.0075}{1-(1+0.0075)^{-300}} = \$1342.71$$

The total interest paid on this loan is $(\$1342.71)(300) - \$160{,}000 = \$242{,}813$.

The couple's equity in the house is the sum of their down payment and the amount paid on the loan. After 10 years the couple still owes 180 payments. The amount still owed on the loan is the present value of these payments

$$V = P\frac{1-(1+i)^{-n}}{i} = \$1342.71\ \frac{1-(1+0.0075)^{-180}}{0.0075} = \$132{,}382.36$$

meaning the amount paid on the loan $\$160{,}000 - \$132{,}382.36 = \$27{,}617.64$.
The couple's equity after 10 years is $\$40{,}000 + \$27{,}617.64 = \$67{,}617.64$.

The 25 year loan at 9% interest has a larger monthly payment, and the total interest paid on the 25 year loan is $\$81{,}621$ more than that on the 20 year loan.
After 10 years the couple would have more equity in their purchase if they had the 8% loan.

19.   We first find how much John needs to accumulate prior to retirement. This is the present value V of $n = 12 \cdot 30 = 360$ payments of \$300 at $i = \dfrac{0.09}{12} = 0.0075$ per month.

$$V = P\frac{1-(1+i)^{-n}}{i} = \$300\,\frac{1-(1+0.0075)^{-360}}{0.0075} = \$37{,}284.56$$

We can now find the $n = 12 \cdot 20 = 240$ monthly payments necessary to meet this goal.

$$A = P\frac{(1+i)^{n}-1}{i}\,;\qquad \$37{,}284.56 = P\frac{(1+0.0075)^{240}-1}{0.0075}$$

$$\$37{,}284.56 = P(637.88687)$$
$$P = \$55.82$$

John should save \$55.82 each month to achieve his retirement goal.

21.   (a)   Dan's accumulated savings is actually an annuity with $P = \$100$, $n = 12$, and an interest rate $i = \dfrac{0.06}{52}$ per week. After 12 weeks Dan has

$$A = P\frac{(1+i)^{n}-1}{i} = \$100\,\frac{\left(1+\dfrac{0.06}{52}\right)^{12}-1}{\dfrac{0.06}{52}} = \$1207.64$$

(b)   Dan now amortizes his savings, he needs to find P if the present value $V = \$1207.64$, $n = 34$, and the interest per payment period is $i = \dfrac{0.06}{52}$.

$$P = V\frac{i}{1-(1+i)^{-n}} = \$1207.64\,\frac{\dfrac{0.06}{52}}{1-\left(1+\dfrac{0.06}{52}\right)^{-34}} = \$36.24$$

Dan can withdraw \$36.24 each week for 34 weeks.

23. (a)   Mike and Yola's down payment is 20% of $200,000.

$(0.20)(\$200,000) = \$40,000$

(b)   Their loan amount is the purchase price less the down payment.

$\$200,000 - \$40,000 = \$160,000$

(c)   The 30 year loan will have $n = 12 \cdot 30 = 360$ monthly payments at an interest rate $i = \dfrac{0.09}{12} = 0.0075$ per month. Their monthly payments will be

$$P = V\frac{i}{1-(1+i)^{-n}} = \$160,000\,\frac{0.0075}{1-(1+0.0075)^{-360}} = \$1287.40$$

(d)   The total interest paid is the difference between the total payments and the amount borrowed. Mike and Yola will pay $(\$1287.40)(360) - \$160,000 = \$303,464$ in interest.

(e)   If they pay an additional $100 each month, the payment $P = \$1387.40$. The present value V of the loan remains $160,000 and the interest rate $i = 0.0075$ per month. We solve for n.

$$V = P\frac{1-(1+i)^{-n}}{i};\qquad \$160,000 = \$1387.40\,\frac{1-(1+0.0075)^{-n}}{0.0075}$$

$$1 - \frac{160,000}{1387.40}\cdot(0.0075) = (1+0.0075)^{-n}$$

$$0.1350728 = 1.0075^{-n}$$

$$-n = \log_{1.0075}0.1350728$$

$$n = -\frac{\log 0.1350728}{\log 1.0075} = 267.93$$

If Mike and Yola increase their monthly payments by $100, they will pay off the loan in 268 months or 22 years and 4 months.

(f)   The total interest paid on the loan with the revised monthly payment of $1387.40 is

$(\$1387.40)(268) - \$160,000 = \$211,823.20$

25. The down payment on the car is 20% of $12,000 or $(0.20)(\$12,000) = \$2400$.
The amount of money borrowed is $\$12,000 - \$2400 = \$9600$.
The monthly payment on $V = \$9600$ when there are $n = 12 \cdot 3 = 36$ monthly payments at an interest rate $i = \frac{0.15}{12} = 0.0125$ per month is

$$P = V\frac{i}{1-(1+i)^{-n}} = \$9600\,\frac{0.0125}{1-(1+0.0125)^{-36}} = \$332.79$$

27. The down payment on the equipment is 10% of $20,000 or $(0.10)(\$20,000) = \$2000$.
The amount of money the restaurant owner borrowed is $\$20,000 - \$2000 = \$18,000$.
The monthly payment P, on $V = \$18,000$ when there are $n = 12 \cdot 4 = 48$ monthly payments at an interest rate $i = \frac{0.12}{12} = 0.01$ per month is

$$P = V\frac{i}{1-(1+i)^{-n}} = \$18,000\,\frac{0.01}{1-(1+0.01)^{-48}} = \$474.01$$

The restaurant owner will pay a total of $(\$474.01)(48) - \$18,000 = \$4752.43$ in interest on the loan.

29. The home buyer made a down payment of 20% of $140,000 or $(0.20)(\$140,000) = \$28,000$ on the house, and financed the remainder ($140,000 − $28,000 = $112,000). The terms of the mortgage were $n = 12 \cdot 30 = 360$ payments at an interest rate of $i = \dfrac{0.098}{12}$ per month. The monthly payment is

$$P = V\frac{i}{1-(1+i)^{-n}} = \$112,000\frac{\dfrac{0.098}{12}}{1-\left(1+\dfrac{0.098}{12}\right)^{-360}} = \$966.37$$

The total interest paid $I = (\$966.37)(360) - \$112,000 = \$235,892.50$

If the home buyer chose a 15 year mortgage, then $n = 12 \cdot 15 = 180$ monthly payments. The monthly payment would be

$$P = V\frac{i}{1-(1+i)^{-n}} = \$112,000\frac{\dfrac{0.098}{12}}{1-\left(1+\dfrac{0.098}{12}\right)^{-180}} = \$1189.89$$

and the total interest paid $I = (\$1189.89)(180) - \$112,000 = \$102,180.41$

Decreasing the term of the loan from 30 years to 15 years increases the monthly payment by $223.52, but saves $133,712.09 in interest payments.

31. We have $V = \$100,000$, $P = \$2000$ per month and $i = \dfrac{0.05}{12}$ per month. To determine how long the IRA payments will last need to find n.

$$V = P\frac{1-(1+i)^{-n}}{i}; \qquad \$100,000 = \$2000\frac{1-\left(1+\dfrac{0.05}{12}\right)^{-n}}{\dfrac{0.05}{12}}$$

$$1 - \left(\frac{0.05}{12}\right)50 = \left(1+\frac{0.05}{12}\right)^{-n}$$

$$0.79 = (1.0041667)^{-n}$$

$$-n = \log_{1.0041667}(0.79)$$

$$n = -\frac{\log 0.79}{\log 1.0041667} = 56.69$$

The payments will last 56.69 months, or about 4 years and 9 months.

33. The original loan had 360 payments at an interest rate $i = \dfrac{0.06825}{12} = 0.0056875$ per month.

The loan payments were

$$P = V\frac{i}{1-(1+i)^{-n}} = \$312{,}000\frac{0.0056875}{1-(1+0.0056875)^{-360}} = \$2039.20$$

After 60 months there are 300 payments still to be made, and the present value of the loan is

$$V = P\frac{1-(1+i)^{-n}}{i} = \$2039.20\frac{1-(1+0.0056875)^{-300}}{0.0056875} = \$293{,}133.88$$

Amount of the debt paid: $\$312{,}000 - \$293{,}133.88 = \$18{,}866.12$
The interest paid: $(\$2039.20)(60) - \$18{,}866.12 = \$122{,}352 - \$18{,}866.12 = \$103{,}485.88$

The new loan has a present value $V = \$293{,}133.88$, $n = 300$ monthly payments at an

interest rate $i = \dfrac{0.06125}{12}$ per month. The new monthly payment is

$$P = V\frac{i}{1-(1+i)^{-n}} = \$293{,}133.88\frac{\dfrac{0.06125}{12}}{1-\left(1+\dfrac{0.06125}{12}\right)^{-300}} = \$1911.13$$

With the refinancing the monthly payments are reduced by $\$2039.20 - \$1911.13 = \$128.07$ per month.

The interest that will be saved by refinancing at a lower interest rate is the difference between the interest that would have been paid without refinancing and the interest that will be paid with refinancing.

Interest without refinancing: $(\$2039.20)(360) - \$312{,}000 = \$422{,}112$
Interest with refinancing: interest paid on the old loan + interest on new loan
$$= \$103{,}486.18 + [(\$1911.13)(300) - \$293{,}133.81]$$
$$= \$383{,}691.37$$

The interest saved by refinancing: $\$422{,}112 - \$383{,}697.37 = \$38{,}420.63$

35.   The $V = \$235{,}000$ loan had $n = 12 \cdot 30 = 360$ payment at an interest rate $i = \dfrac{0.06125}{12}$ per month. The minimum monthly payment on the loan was

$$P = V\frac{i}{1-(1+i)^{-n}} = \$235{,}000\,\frac{\dfrac{0.06125}{12}}{1-\left(1+\dfrac{0.06125}{12}\right)^{-360}} = \$1427.88$$

After paying the loan for 4 years (48 payments) there are

$$V = P\frac{1-(1+i)^{-n}}{i} = \$1427.88\,\frac{1-\left(1+\dfrac{0.06125}{12}\right)^{-312}}{\dfrac{0.06125}{12}} = \$222{,}611.64 \text{ remaining on the loan.}$$

$\$12{,}388.36$ of the debt has been paid.

Interest paid: $(\$1427.88)(48) - \$12{,}388.36 = \$68{,}538.24 - \$12{,}388.36 = \$56{,}149.88$

To find the number of months that the term will be reduced, we find n with the new payment $P = \$1427.88 + \$150 = \$1577.88$

$$\$222{,}611.64 = \$1577.88\,\frac{1-\left(1+\dfrac{0.06125}{12}\right)^{-n}}{\dfrac{0.06125}{12}}$$

$$\left(1+\frac{0.06125}{12}\right)^{-n} = 1-\left(\frac{222{,}611.64}{1577.88}\right)\left(\frac{0.06125}{12}\right)$$

$$(1.005104167)^{-n} = 0.27989$$

$$-n = \log_{1.005104167}(0.27989)$$

$$n = -\frac{\log 0.27989}{\log 1.005104167} = 250.1$$

The term of the loan will be reduced by $312 - 250.1 = 61.9$ months.

The interest that will be saved by increasing the monthly payment is the difference between the interest that would have been paid if the couple continued to make minimum payments and the interest that will be paid with their revised payment.

Interest on the existing loan: $(\$1427.88)(360) - \$235{,}000 = \$279{,}036.80$
Interest with increased payments: interest paid before + interest paid with new payments
$$= \$56{,}149.88 + [(\$1577.88)(250.1) - \$222{,}611.64]$$
$$= \$228{,}166.03$$

The interest saved by increasing monthly payments:

$\$279{,}036.80 - \$228{,}166.03 = \$50{,}870.77$

5.5 Annuities and Amortization Using Recursive Sequences

1. If $B_0 = \$3000$ and $B_n = 1.01B_{n-1} - 100$, then
   (a) John's balance after 1 payment is $B_1 = 1.01(\$3000) - \$100 = \$2930$
   (b) Using the sequence mode, with $n\,\text{Min} = 0$
$$u(n) = 1.01u(n-1) - 100$$
$$u(n\,\text{Min}) = 3000$$
   and TABLE we find the balance is below \$2000 after 14 payments. The balance is \$1953.70.
   (c) John will pay off the debt on the 36th payment (of \$83.78).
   The total of all payments: $(35)(\$100) + \$83.78 = \$3583.78$
   (d) John's interest expense was $\$3583.78 - \$3000 = \$583.78$

3. If $p_0 = 2000$, and $p_n = 1.03p_{n-1} + 20$, then
   (a) After 2 months there are $p_2 = 1.03p_1 + 20$, where $p_1 = 1.03p_0 + 20 = 1.03(2000) + 20$
   So, $p_1 = 2080$, and $p_2 = 1.03(2080) + 20 = 2162$ trout in the pond.
   (b) Using the sequence mode with $n\text{Min} = 0$
$$u(n) = 1.03u(n-1) + 20$$
$$u(n\text{Min}) = \{2000\}$$
   and TABLE we find that the trout population reaches 5000 after 26 months (population $p_{26} = 5084.2$).

5. (a) If the payments are $P = \$500$ per quarter and the interest rate per payment period is
   $i = \dfrac{0.08}{4} = 0.02$, then a recursive formula that represents the balance at the end of
   each quarter is
   $A_0 = \$500$; $A_n = (1 + 0.02)A_{n-1} + P = 1.02A_{n-1} + \$500$
   (b) After 80 payments (80 quarters) the fund will have \$101,810.
   (c) The value of the account will be $A = \$159,738.48$.

7. (a) Since Bill and Laura borrowed $150,000 at an interest rate $i = \dfrac{0.06}{12} = 0.005$ per month and are repaying the loan with $n = 360$ monthly payments of $899.33, a recursive formula which represents their balance after each payment is

$$A_0 = \$150,000; \qquad A_n = (1 + 0.005)A_{n-1} - P = 1.005A_{n-1} - \$899.33$$

(b) After 1 payment, Bill and Laura's balance is

$$A_1 = (1.005)(150,000) - \$899.33 = \$149,850.67.$$

(c) To create a table showing Bill and Laura's balance after each payment, set the graphing utility in sequence mode. Then enter the recursive formula

$n\text{Min} = 0$
$u(n) = 1.005u(n-1) - 899.33$
$u(n\text{Min}) = \{150,000\}$

to obtain the TABLE reflecting the balance after each monthly payment.

(d) Using the recursive formula from part (c) and the TABLE, we find that after 58 payments the balance on the loan is $139,981.

(e) Bill and Laura pay off the balance with the 360th payment. It is a reduced payment of $890.65.

(f) Bill and Laura's interest expense is

$$I = [(359)(899.33) + 890.65] - 150,000 = \$173,750.12$$

(g) If Bill and Laura make payments of $999.33 instead of $899.33, the recursive formula representing their balance after each payment is

$$A_0 = \$150,000; \qquad A_n = (1 + 0.005)A_{n-1} - P = 1.005A_{n-1} - \$999.33$$

(b) After 1 payment, Bill and Laura's balance is

$$A_1 = (1.005)(150,000) - \$999.33 = \$149,750.67.$$

(c) To create a table showing Bill and Laura's balance after each payment, set the graphing utility in sequence mode. Then enter the recursive formula

$n\text{Min} = 0$
$u(n) = 1.005u(n-1) - 999.33$
$u(n\text{Min}) = \{150,000\}$

to obtain the TABLE reflecting the balance after each monthly payment.

(d) Using the recursive formula from part (c) and the TABLE, we find that after 37 payments the balance on the loan is $139,894.

(e) Bill and Laura pay off the balance with the 279th payment. It is a reduced payment of $353.69.

(f) Bill and Laura's interest expense is

$$I = [(278)(999.33) + 353.69] - 150,000 = \$128,167.43.$$

(h) Answers will vary.

5.6 Applications: Leasing; Capital Expenditure; Bonds

1. Suppose the corporation can invest money at 10% per year. The present value of an annuity of \$2000 (the cost of leasing) for 5 years is

$$V = P\frac{1-(1+i)^{-n}}{i} = \$2000 \cdot \frac{1-(1+0.10)^{-5}}{0.10} = \$7581.57 \text{ which is less than the purchase price}$$

of \$8100.00, so leasing is preferable.

3. We first find the equivalent annual cost of each machine.
Machine A: Machine A has an expected life of 8 years. We determine that at 10% interest an annuity for 8 years has a present value

$$V = P\frac{1-(1+i)^{-n}}{i} = P \cdot \frac{1-(1+0.10)^{-8}}{0.10} = 5.3349P$$

The equivalent annual cost of machine A is then $\dfrac{\$10,000}{5.33} = \$1874.44$.

Machine B: The present value of P for 6 years is

$$V = P\frac{1-(1+i)^{-n}}{i} = P \cdot \frac{1-(1+0.10)^{-6}}{0.10} = 4.355P, \text{ and the equivalent annual cost of machine}$$

B is $\dfrac{\$8000}{4.355} = \$1836.86$.

The net annual savings of each machine is

Machine	A	B
Labor Savings	\$2000.00	\$1800.00
Equivalent Annual Cost	1875.44	1836.86
Net Savings	\$ 125.56	(\$36.86)

Machine A has a greater net savings (Machine B results in a net loss), so Machine A is preferable.

5.   We have a $1000 bond with a nominal interest rate of 9% that matures in 15 years. We want to find the price that will yield an effective rate of 8%. To find the price we follow the steps outlined on pp. 307-308 in the text.

Step 1. We calculate the semi-annual interest payments,

$$I = Prt = (\$1000)(0.09)\left(\frac{1}{2}\right) = \$45$$

Step 2. We calculate the present value of the semi-annual payments; $P = \$45$, $n = 2\cdot15 = 30$, $i = \dfrac{0.08}{2} = 0.04$,

$$V = P\frac{1-(1+i)^{-n}}{i} = 45\cdot\frac{1-(1+0.04)^{-30}}{0.04} = \$778.14$$

Step 3. We calculate the present value of the bond at maturity.

$$P = A(1+i)^{-n} = \$1000(1.04)^{-30} = \$308.32$$

Step 4. We calculate the price of the bond by summing the results of steps 2 and 3.

$778.14 + $308.32 = $1086.46

7.   A 10 year treasury note with a 4.000% nominal rate of interest pays

Step 1. $I = Prt = (\$1000)(0.04)\left(\frac{1}{2}\right) = \$20$ every six months.

Step 2. The present value of the twenty $20 payments at a true rate of interest $i = \dfrac{0.04095}{2} = 0.020475$ every six months is

$$V = P\frac{1-(1+i)^{-n}}{i} = 20\cdot\frac{1-(1+0.020475)^{-20}}{0.020475} = \$325.53$$

Step 3. The present value of the bond at maturity at the true rate of interest is

$$P = A(1+i)^{-n} = \$1000(1.020475)^{-20} = \$666.73$$

Step 4. We calculate the price of the bond by summing the results of steps 2 and 3.

$325.53 + $666.73 = $992.26

9. The $1000 treasury note pays

$$I = Prt = (\$1000)(0.04375)\left(\frac{1}{2}\right) = \$21.875 \text{ in interest semiannually.}$$

The true interest rate is the rate of interest for which the present value of the note plus the present value of the 20 interest payments equals the price paid for the note.

$$\begin{pmatrix} \text{Price of} \\ \text{the Note} \end{pmatrix} = \begin{pmatrix} \text{Present Value} \\ \text{of } \$1000 \end{pmatrix} + \begin{pmatrix} \text{Present Value of an Annuity} \\ \text{Paying } \$21.875 \text{ Semiannually} \end{pmatrix}$$

$$\$998.80 = \$1000(1+i)^{-20} + \$21.875\left(\frac{1-(1+i)^{-20}}{i}\right)$$

We will use a graphing utility to solve for i by graphing

$$y_1 = 998.80 - 1000(1+x)^{-20}$$

$$y_2 = 21.875\left(\frac{1-(1+x)^{-20}}{x}\right)$$

and finding the intersection of the 2 graphs.

We know the true interest rate must be close to the nominal rate per payment period, or $i = \dfrac{0.04375}{2} = 0.021875$, so we set the window at Xmin = 0.0, Xmax = 0.04, Xscl = 0.1. We also evaluate y_1 at $x = 0.021875$, getting $y_1 = 350.1$ and set the y-values of the window at Ymin = 345, Ymax = 355, Yscl = 1.

Then using the intersect command on the graphing utility we find, $x = 0.02195$, making the true interest rate of the treasury note $2(0.02195) = 0.0439$ or 4.39%.

Chapter 5 Review

True-False Items

1. T   **3.** F

Fill in the Blanks

1. proceeds   **3.** annuity

Review Exercises

1. 3% of $500 = (0.03) \cdot (500) = 15$ **3.** 140% of $250 = (1.40) \cdot (250) = 350$

5. $x\%$ of $350 = 75$

$$\frac{x}{100} \cdot 350 = 75$$

$$350x = 7500$$

$$x = 21.4286$$

75 is 21.429% of 350.

7. $12 = 15\%$ of x

$$12 = 0.15x$$

$$x = \frac{12}{0.15} = 80$$

12 is 15% of 80.

9. $11 = 0.5\%$ of x
    $11 = 0.005x$

    $x = \dfrac{11}{0.005} = 2200$

    11 is 0.5% of 2200.

11. Dan will have to pay 6% of $330.00 in sales tax.

    $(0.06)(\$330.00) = \$19.80$

    Dan pays $19.80 in sales tax.

13. If Dan borrows $500 for 1 year and 2 months (14 months) at 9% simple interest, the interest charged is

    $I = Prt = (\$500)(0.09)\left(\dfrac{14}{12}\right) = \$52.50$

    and the amount due is $A = P + I = \$500.00 + \$52.50 = \$552.50$.

15. The proceeds of Warren's loan is $15,000. If the loan has an interest rate of 12%, Warren must repay L at the end of 2 years where

    $R = L(1 - rt)$
    $\$15,000 = L[1 - (0.12)(2)] = L(0.76)$
    $L = \$19,736.84$

17. If $P = \$100$ is invested for 2 years and 3 months ($n = 27$ months) at an interest rate $i = \dfrac{0.10}{12}$ per month, it will accumulate to

    $A = P(1 + i)^n = \$100\left(1 + \dfrac{0.10}{12}\right)^{27} = \$125.12$

19. Mike is choosing between two loans.
    Loan (a) is a $3000 loan for $t = 3$ years at $r = 12\%$ simple interest. This loan will cost Mike $A = P(1 + rt) = \$3000[1 + (0.12)(3)] = \$4080.00$.

    Loan (b) is a $3000 loan for $n = 12 \cdot 3 = 36$ months at an interest rate $i = \dfrac{0.10}{12}$ per month.

    This loan will cost $A = P(1 + i)^n = \$3000\left(1 + \dfrac{0.10}{12}\right)^{36} = \$4044.55$

    Loan (b) costs Mike less.

21. Katy wants the amount $A = \$75$ in $n = 6$ months. If her money can earn an interest rate of $i = \dfrac{0.10}{12}$ per month then Katy must invest $P = \$71.36$.

    $A = P(1 + i)^n \qquad \$75 = P\left(1 + \dfrac{0.10}{12}\right)^6$

    $P = \$75\left(1 + \dfrac{0.10}{12}\right)^{-6} = \$71.36$

23. Money doubles when $A = 2P$. The interest rate i, that allows money to double in $n = 12$ years, is given by $A = P(1+i)^n$ or $2P = P(1+i)^n$.

$$2 = (1+i)^{12}$$
$$i = \sqrt[12]{2} - 1 = 0.0595$$

25. The effective rate of interest is the simple interest rate that yields the same amount as the actual compounded rate. We want the interest rate compounded quarterly that is equivalent to the simple interest rate $r = 0.06$.

$$r_{\text{effective}} = (1+i)^4 - 1$$
$$0.06 = (1+i)^4 - 1$$
$$(1+i)^4 = 1.06$$
$$i = \sqrt[4]{1.06} - 1 = 0.0146738 \text{ per quarter}$$

The annual interest rate compounded quarterly equivalent to 6% simple interest is $4i = 4(0.0146738) = 0.05870 = 5.87\%$.

27. Mr. and Mrs. Corey are setting up an annuity to save the down payment for the house, where $A = \$40,000$, $n = 12 \cdot 2 = 24$ months, and the interest rate per payment period is $i = \dfrac{0.03}{12} = 0.0025$. We are looking for P, the Corey's monthly payment.

$$A = P\frac{(1+i)^n - 1}{i}; \quad \$40,000 = P\frac{(1+0.0025)^{24} - 1}{0.0025} = P(24.702818)$$
$$P = \$1619.25$$

Mr. and Mrs. Corey must save \$1619.25 per month in order to have the \$40,000 they need for the down payment.

29. Mr. and Mrs. Ostedt's new house cost $400,000. They made a $100,000 down payment, and are financing $300,000 at 10% for 25 years.

(a) The monthly payments are found by evaluating $P = V \dfrac{i}{1-(1+i)^{-n}}$. In this problem

$V = \$300,000$, $n = 12 \cdot 25 = 300$ months, and the interest rate per month is $i = \dfrac{0.10}{12}$.

$$P = \$300,000 \dfrac{\dfrac{0.10}{12}}{1-\left(1+\dfrac{0.10}{12}\right)^{-300}} = \$2726.10$$

The Ostedt's monthly payments are $2726.10.

(b) The Ostedt's will pay $(300)(\$2726.10) - \$300,000 = \$517,830$ in interest.

(c) The equity in a house is the sum of the down payment and the amount paid on the loan. After 5 years, the Ostedt's have made 60 payments. The present value of the remaining 240 payments is

$$V = P\dfrac{1-(1+i)^{-n}}{i} = \$2726.10 \dfrac{1-\left(1+\dfrac{0.10}{12}\right)^{-240}}{\dfrac{0.10}{12}} = \$282,491.07$$

which indicates that the Ostedt's have paid $\$300,000 - \$282,491.07 = \$17,508.93$ of the principal. Their equity is then

$$\$100,000 + \$17,508.93 = \$117,508.93$$

31. If $125,000 is to be amortized at a periodic interest rate $i = \dfrac{0.09}{12} = 0.0075$ per month over

$n = 12 \cdot 25 = 300$ months, the monthly payment will be

$$P = V \dfrac{i}{1-(1+i)^{-n}} = \$125,000 \dfrac{0.0075}{1-(1+0.0075)^{-300}} = \$1049.00$$

Equity is the down payment (if any) plus the amount of the loan already paid. After 10 years, 120 payments have been made and 180 remain to be made on the loan. The present value of the remaining payments is

$$V = P\dfrac{1-(1+i)^{-n}}{i} = \$1049.00 \dfrac{1-(1+0.0075)^{-180}}{0.0075} = \$103,424.49$$

meaning that $\$125,000 - \$103,424.49 = \$21,575.51$ of the loan has been paid. The equity in the property after 10 years is $21,576.

33. Let x denote the price Mr. Graf should pay for the gold mine. Then $0.15x$ represents a 15% annual Return on Investment (ROI). The annual Sinking Fund Contribution (SFC) needed to recover the purchase price x in 20 years can be calculated letting $A = x$, $n = 20$,

$i = 0.10$, and $P = $ SFC. $A = P\dfrac{(1+i)^n - 1}{i}$

$$x = \text{SFC}\dfrac{(1+0.10)^{20} - 1}{0.10}$$

$$x = \text{SFC}(57.274999)$$

$$\text{SFC} = 0.0174596x$$

The required annual ROI plus the annual SFC equals the annual net income of $20,000.

$$0.15x + 0.0174506x = \$20,000$$
$$0.1674596x = \$20,000$$
$$x = \$119,431.77$$

Mr. Graf should pay $119,431.77 for the gold mine.

35. Let x denote the price to be paid for the oil well. Then $0.20x$ represents a 20% annual Return on Investment (ROI). The annual Sinking Fund Contribution (SFC) needed to recover the purchase price x in 15 years can be calculated letting $A = x$, $n = 15$, $i = 0.10$,

and $P = $ SFC. $A = P\dfrac{(1+i)^n - 1}{i}$

$$x = \text{SFC}\dfrac{(1+0.10)^{15} - 1}{0.10}$$

$$x = \text{SFC}(31.77248169)$$

$$\text{SFC} = 0.03147377769x$$

The required annual ROI plus the annual SFC equals the annual net income of $25,000.

$$0.20x + 0.03147377769x = \$25,000$$
$$0.23147377769x = \$25,000$$
$$x = \$108,003.59$$

The investor should pay $108,003.59 for the oil well.

37. Mr. Jones is saving $500.00 every six months at an interest rate $i = \dfrac{0.06}{2} = 0.03$ per payment period. After 8 years, $n = 2 \cdot 8 = 16$ payments, he will have

$$A = P\dfrac{(1+i)^n - 1}{i} = \$500\dfrac{(1+0.03)^{16} - 1}{0.03} = \$10,078.44$$

39. If Bill is to pay his debt and interest in 7 years he will need

$$A = P(1+i)^n = \$1000(1+0.05)^7 = \$1407.10$$

at that time. Bill makes $n = 4 \cdot 7 = 28$ quarterly payments into the sinking fund that earns interest $i = \dfrac{0.08}{4} = 0.02$ per quarter. Each payment is

$$A = P\frac{(1+i)^n - 1}{i}; \quad \$1407.10 = P\frac{(1+0.02)^{28} - 1}{0.02}$$

$$\$1407.10 = P(37.0512103)$$
$$P = \$37.98$$

41. A loan of \$3000 at an interest rate per payment period of $i = \dfrac{0.12}{12} = 0.01$ is to be amortized in 2 years with $n = 12 \cdot 2 = 24$ equal monthly payments. Each payment is

$$P = V\frac{i}{1 - (1+i)^{-n}} = \$3000\frac{0.01}{1 - (1+0.01)^{-24}} = \$141.22$$

43. The effective rate of interest is the simple interest rate that yields the same amount as the actual compounded rate. The annual interest rate of 9% compounded monthly gives an interest rate $i = \dfrac{0.09}{12} = 0.0075$ per month, and

$$I_{effective} = (1 + 0.0075)^{12} - 1 = 0.0938$$

The effective rate of interest is 9.38%.

45. If John's trust fund is earning an interest rate $i = \dfrac{0.08}{2} = 0.04$ every 6 months it will be amortized in 15 years with $n = 2 \cdot 15 = 30$ equal payments. Each semiannual payment will be

$$P = V\frac{i}{1 - (1+i)^{-n}} = 20,000\frac{0.04}{1 - (1+0.04)^{-30}} = \$1156.60$$

47. At the end of $n = 30$ months, there will be

$$A = P\frac{(1+i)^n - 1}{i} = \$60\frac{(1+0.01)^{30} - 1}{0.01} = \$2087.09$$

in the employee's retirement fund.

49. The student is to amortize the \$4000 loan with $n = 4 \cdot 4 = 16$ quarterly payments. The interest rate per payment period of the loan is $i = \dfrac{0.14}{4} = 0.035$, and the quarterly payments are

$$P = V\frac{i}{1 - (1+i)^{-n}} = \$4000\frac{0.035}{1 - (1+0.035)^{-16}} = \$330.74$$

51. The present value of a 4 year annuity invested at 5% per annum with annual payments of $50,000 is

$$V = P\,\frac{1-(1+i)^{-n}}{i} = \$50{,}000\,\frac{1-(1+0.05)^{-4}}{0.05} = \$177{,}297.53$$

Since the present value of the annuity exceeds the purchase price of the trucks, purchasing the trucks is preferable.

53. We have a $10,000 bond with a nominal interest rate of 6.25% that matures in 8 years. We want to find the price that will yield a true rate of 6.5%. To find the price we follow the steps outlined on pp. 307–308 in the text.
Step 1. We calculate the semi-annual interest payments,

$$I = Prt = (\$10{,}000)(0.0625)\left(\frac{1}{2}\right) = \$312.50$$

Step 2. We calculate the present value of the semi-annual payments; $P = \$312.50$,

$$n = 2\cdot 8 = 16,\, i = \frac{0.065}{2} = 0.0325,$$

$$V = P\,\frac{1-(1+i)^{-n}}{i} = \$312.50\cdot\frac{1-(1+0.0325)^{-16}}{0.0325} = \$3851.36$$

Step 3. We calculate the present value of the bond at maturity.

$$P = A(1+i)^{-n} = \$10{,}000(1.0325)^{-16} = \$5994.58$$

Step 4. We calculate the price of the bond by summing the results of steps 2 and 3.

$$\$3851.36 + \$5994.58 = \$9845.94$$

Chapter 5 Project

1. The monthly payment for the 30 year mortgage with no points is

$$P = V\,\frac{i}{1-(1+i)^{-n}} = \$120{,}000\,\frac{\dfrac{0.0550}{12}}{1-\left(1+\dfrac{0.0550}{12}\right)^{-360}} = \$681.35$$

3. The difference in the monthly payments between the loan with no points and the mortgage with 0.5 point is $681.35 - $672.34 = $9.01

5. Answers will vary, but they should include justification.

7. If the loan is decreased by the amount that would have been spent on points and the larger interest rate of 5.50% is used, the monthly payments become:

Rate	Points	Fee Paid	Loan Amount	Monthly Payment
5.5%	0	$0.00	$120,000 - $3240 = $116,760$	$116,760\dfrac{\dfrac{0.055}{12}}{1-\left(1+\dfrac{0.055}{12}\right)^{-360}} = \$662.95$
5.38%	0.50	$600	$120,000 - $2640 = $117,360$	$117,360\dfrac{\dfrac{0.0538}{12}}{1-\left(1+\dfrac{0.0538}{12}\right)^{-360}} = \$657.55$
5.25%	1.10	$1320	$120,000 - $1920 = $118,080$	$118,080\dfrac{\dfrac{0.0525}{12}}{1-\left(1+\dfrac{0.0525}{12}\right)^{-360}} = \$652.04$
5.00%	2.70	$3240	$120,000$	$120,000\dfrac{\dfrac{0.500}{12}}{1-\left(1+\dfrac{0.5}{12}\right)^{-360}} = \$644.19$

Mathematical Questions from Professional Exams

1. (b)   **3.** (b)   **5.** (d)   **7.** (c)

Chapter 6

Sets; Counting Techniques

6.1 Sets

1. True   **3.** False   **5.** False   **7.** True

9. True   **11.** $\{1, 2, 3\} \cap \{2, 3, 4, 5\} = \{2, 3\}$

13. $\{1, 2, 3\} \cup \{2, 3, 4, 5\} = \{1, 2, 3, 4, 5\}$ **15.** $\{2, 4, 6, 8\} \cap \{1, 3, 5, 7\} = \varnothing$

17. $\{a, b, e\} \cup \{d, e, f, q\} = \{a, b, d, e, f, q\}$

19.
  (a) $A \cup B = \{0, 1, 2, 3, 5, 7, 8\}$ (b) $\overline{B \cap C} = \{5\}$
  (c) $A \cap B = \{5\}$ (d) $\overline{A \cap B} = \{0, 1, 2, 3, 4, 6, 7, 8, 9\}$
  (e) $\overline{A} \cap \overline{B} = \{2, 4, 6, 8, 9\} \cap \{0, 1, 4, 6, 7, 9\} = \{4, 6, 9\}$
  (f) $A \cup (B \cap A) = \{0, 1, 5, 7\} \cup \{5\} = \{0, 1, 5, 7\} = A$
  (g) $(C \cap A) \cap (\overline{A}) = \{5\} \cap \{2, 3, 4, 6, 8, 0\} = \varnothing$
  (h) $(A \cap B) \cup (B \cap C) = \{5\} \cup \varnothing = \{5\}$

21.
  (a) $A \cup B = \{b, c, d, e, f, g\}$
  (b) $A \cap B = \{c\}$
  (c) $\overline{A} \cap \overline{B} = \{a, h, i, j, k, l, m, n, o, p, q, r, s, t, u, v, w, x, y, z\}$
  (d) $\overline{A} \cup \overline{B} = \{a, b, d, e, f, g, h, i, j, k, l, m, n, o, p, q, r, s, t, u, v, w, x, y, z\}$

23. (a)

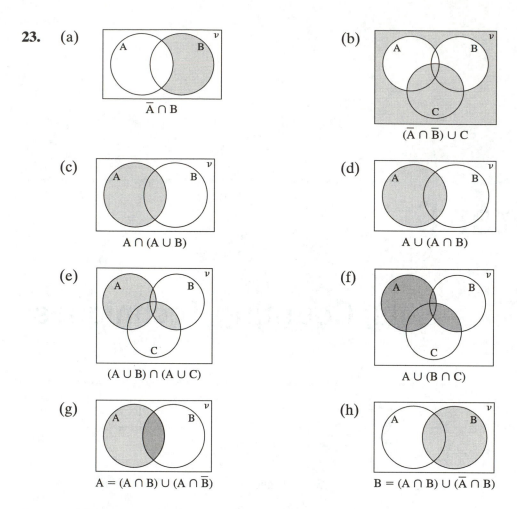

25. $A \cap E$ is the set of people who are both customers of IBM and are on the Board of Directors of IBM.

27. $A \cup D$ is the set of people who are customers of IBM or who are stockholders of IBM.

29. $\overline{A} \cap D$ is the set of people who are not customers of IBM, but who are stockholders of IBM.

31. $M \cap S$ is the set of all male students who smoke.

33. $\overline{M} \cup \overline{F}$ is the set of students who are female or who are not freshmen.

35. $F \cap S \cap M$ is the set of all male freshmen students who smoke.

37. The subsets of $\{a, b, c\}$ are $\varnothing$, $\{a\}$, $\{b\}$, $\{c\}$, $\{a, b\}$, $\{a, c\}$, $\{b, c\}$, and $\{a, b, c\}$.

6.2 The Number of Elements in a Set

1. $c(A) = 6$

3. $A \cap B = \{2, 4, 6\}$
$c(A \cap B) = 3$

5. $(A \cap B) \cup A = \{2, 4, 6\} \cup \{1, 2, 3, 4, 5, 6\}$
$ = \{1, 2, 3, 4, 5, 6\}$
$ = A$

$c[(A \cap B) \cup A] = c(A) = 6$

7. $c(A \cup B) = c(A) + c(B) - c(A \cap B)$
$ = 4 + 3 - 2$
$ = 5$

9. $c(A \cup B) = c(A) + c(B) - c(A \cap B)$
$c(A \cap B) = c(A) + c(B) - c(A \cup B)$
$ = 5 + 4 - 7$
$ = 2$

11. $c(A \cup B) = c(A) + c(B) - c(A \cap B)$
$c(A) = c(A \cup B) - c(B) + c(A \cap B)$
$ = 14 - 8 + 4$
$ = 10$

13. Here we are looking for the number of cars that have either automatic transmissions or power steering. It is the same as finding $c(A \cup B)$. So the number of cars we are looking for is:

$$\left(\begin{array}{c} \text{the number of cars with} \\ \text{automatic transmissions} \end{array} \right) + \left(\begin{array}{c} \text{the number of cars} \\ \text{with power steering} \end{array} \right) - \left(\begin{array}{c} \text{the number of cars} \\ \text{with both options} \end{array} \right)$$

$= 325 + 216 - 89$
$= 452$

The company manufactured 452 cars.

15. $c(A) = 10 + 6 + 3 + 5$
$ = 24$

17. Before determining the number of elements in A or B, we need to find the number of elements in $A \cap B$.

$c(A \cap B) = 6 + 3 = 9$

So,
$c(A \cup B) = c(A) + c(B) - c(A \cap B)$
$ = 24 + 19 - 9$
$ = 34$

19. To find the number of elements in A but not in B, subtract $c(A \cap B)$ from $c(A)$.

$c(A) - c(A \cap B) = 24 - 9$
$ = 15$

21. The number of elements in A or B or C is found by expanding the formula for finding the number of elements in A or C.

$c(A \cup B \cup C) = c(A) + c(B) + c(C) - c(A \cap B) - c(A \cap C) - c(B \cap C) + c(A \cap B \cap C)$
$ = 24 + 19 + 30 - 9 - 8 - 5 + 3$
$ = 54$

23.   The number of elements in A and B and C is the number of elements common to all three sets.

$$c(A \cap B \cap C) = 3$$

25.   The problem is made easier if we find some totals first. Adding the columns we get:

The number of voters under age 35:   195
The number of voters between ages 35 and 54:   265
The number of voters older than 54:   166

Adding the rows reveals how the voters of the two religions voted.

The number of Protestants who voted Republican: 345
The number of Protestants who voted Democrat: 90
The number of Catholics who voted Republican: 67
The number of Catholics who voted Democrat: 124

From these last four we can determine the number of Protestants polled and the number of Catholics polled.

The number of Protestants: 435
The number of Catholics: 191

(a)   To determine number of voters who are Catholic or Republican we need to find

$$\begin{pmatrix} \text{number of} \\ \text{Catholics} \end{pmatrix} + \begin{pmatrix} \text{number of} \\ \text{Republicans} \end{pmatrix} - \begin{pmatrix} \text{number of} \\ \text{Catholic Republicans} \end{pmatrix}$$

$$= 191 + 412 - 67$$
$$= 536$$

536 voters were either Catholic or Republican.

(b)   To find the number of voters who are Catholic or over 54 we determine

$$\begin{pmatrix} \text{number of} \\ \text{Catholics} \end{pmatrix} + \begin{pmatrix} \text{number of} \\ \text{voters older than 54} \end{pmatrix} - \begin{pmatrix} \text{number of Catholics} \\ \text{older than 54} \end{pmatrix}$$

$$= 191 + 166 - 40$$
$$= 317$$

Of the voters polled 317 were either Catholic or over 54 years of age.

(c)   Since a person cannot be both younger than 35 and older than 54, to find the number of persons who voted Democrat and are below 35 or over 54 simply add

$$\begin{pmatrix} \text{Protestants voting} \\ \text{Democrat under 35} \end{pmatrix} + \begin{pmatrix} \text{Catholics voting} \\ \text{Democrat under 35} \end{pmatrix} + \begin{pmatrix} \text{Protestants voting} \\ \text{Democrat older than 54} \end{pmatrix}$$

$$+ \begin{pmatrix} \text{Catholics voting} \\ \text{Democrat older than 54} \end{pmatrix}$$

$$= 42 + 44 + 15 + 33$$
$$= 134$$

134 voters under the age of 35 or over the age of 54 voted Democrat.

27. We denote the sets of females, seniors, and students on the dean's list by F, S, and D respectively. From the data given we determine that

$$c(F \cap S \cap D) = 31 \quad c(F \cap \overline{S} \cap D) = 62 \quad c(\overline{F} \cap S \cap D) = 45$$

$$c(F \cap S \cap \overline{D}) = 87 \quad c(\overline{F} \cap S \cap \overline{D}) = 96 \quad c(F \cap \overline{S} \cap \overline{D}) = 275$$

$$c(\overline{F} \cap \overline{S} \cap D) = 89 \quad c(\overline{F} \cap \overline{S} \cap \overline{D}) = 227$$

We next place the values on a Venn Diagram, so that we can answer the questions

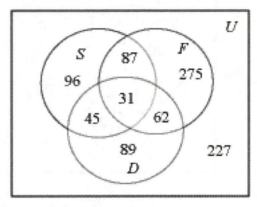

(a) There were $96 + 87 + 31 + 45 = 259$ seniors at the college.
(b) There were $275 + 87 + 31 + 62 = 455$ women at the college.
(c) There were $89 + 45 + 31 + 62 = 227$ students on the dean's list.
(d) There were $45 + 31 = 76$ seniors on the dean's list.
(e) There were $87 + 31 = 118$ female seniors.
(f) There were $31 + 62 = 93$ women on the dean's list.
(g) There were $96 + 87 + 275 + 45 + 31 + 62 + 89 + 227 = 912$ students at the college.

29.   We denote the sets of cars with air conditioning, automatic transmissions, and power
steering by A, T, and P respectively. From the data given we determine that

$$c(A)=90 \qquad c(T)=100 \qquad c(P)=75$$
$$c(A\cap T\cap P)=5 \quad c(\overline{A\cup T\cup P})=20 \quad c(A\cap\overline{T}\cap\overline{P})=20$$
$$c(\overline{A}\cap T\cap\overline{P})=60 \quad c(\overline{A}\cap\overline{T}\cap\overline{P})=30 \qquad c(T\cap P)=10$$

We next place the values on a Venn Diagram, so that we can answer the questions.

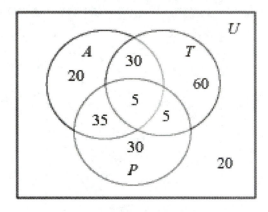

(a) $35 + 5 = 40$ cars had both power steering and air conditioning.
(b) $30 + 5 = 35$ cars had both automatic transmission and air conditioning.
(c) $20 + 20 = 40$ cars had neither power steering nor automatic transmissions.
(d)   There were $20 + 20 + 30 + 5 + 35 + 30 + 5 + 60 = 205$ cars sold in July.
(e) $100 + 90 - 35 = 155$ cars with automatic transmission or air conditioning or both were
sold during July.

31.   There are 8 different possible blood types, as shown in the Venn diagram below. They can
be listed as A^+, A^-, B^+, B^-, AB^+, AB^-, O^+, and O^-.

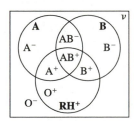

33. We denote the sets of users of IBM, Mac, and Dell by *I*, *M*, and *D* respectively. From the information given, we determine that

$$c(I)=27, \quad c(M)=35, \quad c(D)=35, \quad c(I\cap M)=10, \quad c(I\cap D)=10, \quad c(M\cap D)=10$$
$$c(I\cap M\cap D)=3, \quad \text{and} \quad c\left(\overline{I\cup M\cup D}\right)=30$$

Putting this information onto a Venn diagram will help answer the question.

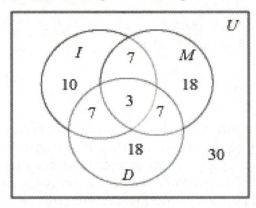

There are $10 + 18 + 18 = 46$ users who exclusively use only one brand of computer.

35. The subsets of $\{a, b, c, d\}$ are $\emptyset$, $\{a\}$, $\{b\}$, $\{c\}$, $\{d\}$, $\{a, b\}$, $\{a, c\}$, $\{a, d\}$, $\{b, c\}$, $\{b, d\}$, $\{c, d\}$, $\{a, b, c\}$, $\{a, b, d\}$, $\{a, c, d\}$, $\{b, c, d\}$, and $\{a, b, c, d\}$.

There are $2^4 = 16$ subsets, where 4 is the number of elements in the original set.

6.3 The Multiplication Principle

1. By the Multiplication Principle one can travel

$$2 \cdot 4 = 8$$

different routes from town A to town C through town B.

3. By the Multiplication Principle, the building company must show XYZ Company

$$3 \cdot 2 \cdot 4 = 24$$

different models to show every possible building complex.

5. By the Multiplication Principle, the man can wear

$$3 \cdot 8 \cdot 4 \cdot 9 = 864$$

different outfits.

7. There are 26 different letters and 10 different digits. Since there is no restriction on repetition of the letters or the digits, from the Multiplication Principle we can determine that

$$26 \cdot 26 \cdot 10 \cdot 10 = 67,600$$

different license plates are possible.

9.   There are five different mathematics books to be arranged. Once a book is placed on desk, it is no longer available to be used. So there are 5 books available for the first spot, 4 available for the second spot, 3 available for the third position, and so on.

According to the Multiplication Principle, the books can be arranged on the desk in

$5 \cdot 4 \cdot 3 \cdot 2 \cdot 1 = 120$ ways.

11.   Since there are 26 different letters and 10 different digits, by the Multiplication Principle, a user name formed by choosing 4 letters followed by four digits can be done

$26 \cdot 26 \cdot 26 \cdot 26 \cdot 10 \cdot 10 \cdot 10 \cdot 10$
$= 26^4 \cdot 10^4$
$= 4,569,760,000$

ways. So theoretically there are 4,569,760,000 different user names possible.

13.   In this problem letters and digits can be repeated, but no two adjacent symbols can be the same. In the first spot we can choose any one of 26 letters, but in the second position there are only 25 letters available. (We cannot repeat the letter to its left.) In the third position there are again 25 possible letters because the letter in the first position can be used, but the one in the second position cannot. The fourth position also has 25 available letters.

Similar reasoning is used for choosing the digits. There are 10 digits to choose from for the fifth position, but only 9 that can be used in the sixth position. The seventh and eighth positions each have 9 possible digits that can be used.

Using this reasoning and the Multiplication Principle, we find that there are

$26 \cdot 25 \cdot 25 \cdot 25 \cdot 10 \cdot 9 \cdot 9 \cdot 9 = 2,961,562,500$

user names for the system.

15.   (a)   If no letter can be repeated then according to the Multiplication Principle, there are

$6 \cdot 5 \cdot 4 \cdot 3 = 360$

4-letter code words possible when the first 6 letters of the alphabet are used.

(b)   If letters can be repeated then according to the Multiplication Principle, there are

$6 \cdot 6 \cdot 6 \cdot 6 = 6^4 = 1296$

different 4-letter code words possible.

17.   According to the Multiplication Principle, there are $7 \cdot 6 \cdot 5 \cdot 4 \cdot 3 \cdot 2 \cdot 1 = 5040$ ways to rank 7 candidates who apply for a job.

19.   The test has 10 questions with 4 choices, so each of these questions can be answered any of 4 ways. The test also contains 15 true-false questions. These can be answered either of 2 ways.

According to the Multiplication Principle, the 25 questions on the test can be answered $4^{10} \cdot 2^{15} = 34,359,738,368$ different ways.

21. The license plates have 2 letters (26 possibilities) followed by 4 digits (10 possibilities).

(a) If the letters and digits can be repeated, then using the Multiplication Principle we find that

$$26 \cdot 26 \cdot 10 \cdot 10 \cdot 10 \cdot 10 = 26^2 \cdot 10^4 = 6,760,000$$

different license plates can be made.

(b) If the letters can be repeated, but the digits cannot be repeated, then using the Multiplication Principle we find that

$$26 \cdot 26 \cdot 10 \cdot 9 \cdot 8 \cdot 7 = 3,407,040$$

different license plates can be made.

(c) If neither the letters nor the digits can be repeated, then using the Multiplication Principle we find that

$$26 \cdot 25 \cdot 10 \cdot 9 \cdot 8 \cdot 7 = 3,276,000$$

different license plates can be made.

23. This problem involves two steps since two models have 3 body styles and one model has only two body styles. It is easiest to see using a tree diagram.

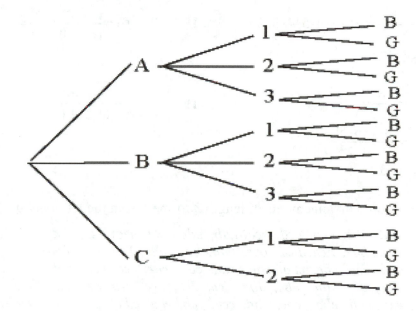

From the tree we see that there are 16 distinguishable car types.

25. By the Multiplication Principle, the contractor can build $5 \cdot 3 \cdot 4 = 60$ different types of homes.

27. By the Multiplication Principle, there can be $2^8 = 256$ different numbers (called bytes by computer scientists) possible.

29. Since there are 50 numbers on the lock and repetition is allowed, there are $50^3 = 125,000$ different lock combinations possible.

31. Think of the maze as if it were a tree diagram with each door of the maze as a branch on the tree. Then using the Multiplication Principle, we find that there are $4 \cdot 2 \cdot 1 = 8$ different paths from start to finish.

6.4 Permutations

1. $\dfrac{5!}{2!} = \dfrac{5 \cdot 4 \cdot 3 \cdot 2!}{2!} = 60$

3. $\dfrac{6!}{3!} = \dfrac{6 \cdot 5 \cdot 4 \cdot 3!}{3!} = 120$

5. $\dfrac{10!}{8!} = \dfrac{10 \cdot 9 \cdot 8!}{8!} = 90$

7. $\dfrac{9!}{8!} = \dfrac{9 \cdot 8!}{8!} = 9$

9. $\dfrac{8!}{2! \, 6!} = \dfrac{8 \cdot 7 \cdot 6!}{2 \cdot 1 \cdot 6!} = 28$

11. $P(7, 2) = 7 \cdot 6 = 42$

13. $P(8, 7) = \dfrac{8!}{(8-7)!} = 8 \cdot 7 \cdot 6 \cdot 5 \cdot 4 \cdot 3 \cdot 2$
$\qquad\quad = 40,320$

15. $P(6, 0) = \dfrac{6!}{(6-0)!} = \dfrac{6!}{6!} = 1$

17. $\dfrac{8!}{(8-3)! \, 3!} = \dfrac{8!}{5! \, 3!}$
$\qquad\quad = \dfrac{8 \cdot 7 \cdot 6 \cdot 5!}{5! \cdot 3 \cdot 2 \cdot 1}$
$\qquad\quad = 56$

19. $\dfrac{6!}{(6-6)! \, 6!} = \dfrac{6!}{0! \, 6!} = 1$

21. (a) The ordered arrangements of length 3 formed from the letters *a, b, c, d,* and *e* are:

abc, abd, abe, acb, acd, ace, adb, adc, ade, aeb, aec, aed,
bac, bad, bae, bca, bcd, bce, bda, bdc, bde, bea, bec, bed,
cab, cad, cae, cba, cbd, cbe, cda, cdb, cde, cea, ceb, ced,
dab, dac, dae, dba, dbc, dbe, dca, dcb, dce, dea, deb, dec,
eab, eac, ead, eba, ebc, ebd, eca, ecb, ecd, eda, edb, edc

(b) $P(5, 3) = 5 \cdot 4 \cdot 3 = 60$

23. (a) The ordered arrangements of length 3 formed from the objects 1, 2, 3, and 4 are:

123, 124, 132, 134, 142, 143, 213, 214, 231, 234, 241, 243
312, 314, 321, 324, 341, 342, 412, 413, 421, 423, 431, 432

(b) $P(4, 3) = 4 \cdot 3 \cdot 2 = 24$

25. Each letter of the code has 4 possible symbols, so there are $4^2 = 16$ possible two-letter codes using the letters *A, B, C,* and *D.*

27. Each digit of the number has two possible values, so there are $2^3 = 8$ possible three-digit numbers formed using the digits 0 and 1.

29. Four people can be lined up in $4! = 4 \cdot 3 \cdot 2 \cdot 1 = 24$ ways.

31. There are $5 \cdot 4 \cdot 3 = 60$ different three-letter codes that can be formed from the 5 letters *A, B, C, D,* and *E* if no letter can be used more than once.

This is the same as finding $P(5, 3) = 60$.

33. To find how many ways there are to seat 5 people in 8 chairs, find the number of ways to arrange 8 objects taken 5 at a time. This is $P(8, 5) = 8 \cdot 7 \cdot 6 \cdot 5 \cdot 4 = 6720$.

So there are 6720 ways to seat 5 people in 8 chairs.

35. Since the sets of symbols one-letter long, two-letters long, and three-letters long are disjoint, we can add the number of elements in each set.

$$\begin{pmatrix} \text{maximum number} \\ \text{of companies} \\ \text{in NYSE} \end{pmatrix} = c \begin{pmatrix} \text{companies with} \\ \text{1-letter symbol} \end{pmatrix} + c \begin{pmatrix} \text{companies with} \\ \text{2- letter symbol} \end{pmatrix} + c \begin{pmatrix} \text{companies with} \\ \text{3 -letter symbol} \end{pmatrix}$$

$$= 26 + 26^2 + 26^3$$
$$= 18,278$$

So, there can be 18,278 different companies listed on the New York Stock Exchange.

37. This is a permutation in which 2 different days are selected from a possible 365 days without a repetition. It is given by

$$P(365, \ 2) = \frac{365!}{(365-2)!} = \frac{365 \cdot 364 \cdot 363!}{363!} = 365 \cdot 364 = 132,860$$

So, 2 people can each have a different birthday 132,860 different ways.

39. (a) SUNDAY has 6 letters, and none of them are repeated. They can be arranged

$P(6, 6) = 6! = 720$ different ways.

(b) If the letter S must come first, then we are really only arranging 5 letters. This can be done $P(5, 5) = 5! = 120$ different ways.

(c) If the letter S must come first and the letter Y must come last, then only four letters are being arranged. It is the permutation of 4 objects.

$P(4, 4) = 4! = 24$

The letters of the word SUNDAY can be arranged 24 different ways if we insist S is the first letter and Y is the last letter.

41. Here we have to select 8 of the 12 children to whom we will distribute the books and then distribute the books among them. This is the permutation of $P(12, 8)$.

$$P(12, \ 8) = \frac{12!}{(12-8)!} = \frac{12 \cdot 11 \cdot 10 \cdot 9 \cdot 8 \cdot 7 \cdot 6 \cdot 5 \cdot 4!}{4!} = 19,958,400$$

The 8 books can be distributed to the 12 children 19,958,400 different ways.

43. We need to choose the three winning lottery tickets from the 1500 tickets sold. This can be done

$$P(1500,\ 3) = \frac{1500!}{(1500-3)!} = \frac{1500 \cdot 1499 \cdot 1498 \cdot 1497!}{1497!} = 3,368,253,000 \text{ different ways.}$$

45. The number of ways to choose the president, vice-president, secretary, and treasurer from a club with 15 members (assuming no one individual holds more than one position) is given by $P(15,4)$.

$$P(15,\ 4) = \frac{15!}{(15-4)!} = \frac{15 \cdot 14 \cdot 13 \cdot 12 \cdot 11!}{11!} = 32,760$$

So, the 4 officers can be chosen 32,760 different ways.

6.5 Combinations

1. $C(6,\ 4) = \dfrac{6!}{(6-4)!\ 4!} = \dfrac{6 \cdot 5 \cdot 4!}{2!\ 4!} = \dfrac{6 \cdot 5}{2} = 15$ **3.** $C(7,\ 2) = \dfrac{7!}{(7-2)!\ 2!} = \dfrac{7 \cdot 6 \cdot 5!}{5!\ 2!} = \dfrac{7 \cdot 6}{2 \cdot 1} = 21$

5. $C(5,\ 1) = \dfrac{5!}{(5-1)!\ 1!} = \dfrac{5 \cdot 4!}{4!\ 1!} = 5$ **7.** $C(8,\ 6) = \dfrac{8!}{(8-6)!\ 6!} = \dfrac{8 \cdot 7 \cdot 6!}{2!\ 6!} = \dfrac{8 \cdot 7}{2 \cdot 1} = 28$

9. (a) The combinations of 5 objects a, b, c, d, and e taken three at a time are:

   $abc, abd, abe, acd, ace, ade, bcd, bce, bde, cde$

   (b) $C(5,\ 3) = \dfrac{5!}{(5-3)!\ 3!} = \dfrac{5 \cdot 4 \cdot 3!}{2!\ 3!} = \dfrac{5 \cdot 4}{2 \cdot 1} = 10$

11. (a) The combinations of 4 objects 1, 2, 3, and 4 taken 3 at a time are:

   123, 124, 134, 234

   (b) $C(4,\ 3) = \dfrac{4!}{(4-3)!\ 3!} = \dfrac{4 \cdot 3!}{1!\ 3!} = 4$

13. A committee of 4 students chosen from a pool of 7 students can be formed $C(7,4) = 35$ ways.

$$C(7,\ 4) = \frac{7!}{(7-4)!\ 4!} = \frac{7 \cdot 6 \cdot 5 \cdot 4!}{3!\ 4!} = \frac{7 \cdot 6 \cdot 5}{3 \cdot 2 \cdot 1} = 35$$

15. The math department needs to select 4 professors without regard to order from a department of 17 eligible teachers. This can be done

$$C(17,\ 4) = \frac{17!}{(17-4)!\ 4!} = \frac{17 \cdot 16 \cdot 15 \cdot 14 \cdot 13!}{13!\ 4!} = \frac{17 \cdot 16 \cdot 15 \cdot 14}{4 \cdot 3 \cdot 2 \cdot 1} = 2380 \text{ ways.}$$

17. A committee is an unordered selection of people. We are interested in a committee of 3 members chosen from the 20 members of the Math Club. The committee can be selected

$$C(20, 3) = \frac{20!}{(20-3)! \; 3!} = \frac{20 \cdot 19 \cdot 18 \cdot 17!}{17! \; 3!} = \frac{20 \cdot 19 \cdot 18}{3 \cdot 2 \cdot 1} = 1140 \text{ ways.}$$

19. A bit is either a 0 or a 1. An 8-bit string is a list of 8 digits, all of which are either 0s or 1s. To determine how many 8-bit strings have exactly three 1s we need to select 3 positions to place the 1s. This can be done

$$C(8, 3) = \frac{8!}{(8-3)! \; 3!} = \frac{8 \cdot 7 \cdot 6 \cdot 5!}{5! \; 3!} = \frac{8 \cdot 7 \cdot 6}{3 \cdot 2 \cdot 1} = 56 \text{ ways.}$$

21. The historians are interested in choosing an unordered collection of 5 presidencies out of 43 presidencies. This can be done

$$C(43, 5) = \frac{43!}{(43-5)! \; 5!} = \frac{43 \cdot 42 \cdot 41 \cdot 40 \cdot 39 \cdot 38!}{38! \; 5!} = \frac{43 \cdot 42 \cdot 41 \cdot 40 \cdot 39}{5 \cdot 4 \cdot 3 \cdot 2 \cdot 1} = 962,598 \text{ ways.}$$

23. In the word ECONOMICS there are 9 letters, but they are not all distinct. There are 2 C's, 2 O's, and one each of the remaining 5 letters.

The number of 9-letter words that can be formed is given by

$$\frac{9!}{2! \; 2! \; 1! \; 1! \; 1! \; 1! \; 1!} = 90,720$$

25. There are 12 colored lights to be arranged in a string, but the colors are not all distinct. There are 3 reds, 4 yellows, and 5 blues. The number of different colored arrangements possible is

$$\frac{12!}{3! \; 4! \; 5!} = \frac{12 \cdot 11 \cdot 10 \cdot 9 \cdot 8 \cdot 7 \cdot 6 \cdot 5!}{3 \cdot 2 \cdot 1 \cdot 4 \cdot 3 \cdot 2 \cdot 1 \cdot 5!} = 27,720$$

27. This problem consists of two tasks: selecting the boys for the committee which can be done $C(4, 2)$ ways and selecting the girls for the committee which can be done $C(8, 3)$ ways. Then by the Multiplication Principle, we find that the committee can be formed in

$$C(4, 2) \cdot C(8, 3) = \frac{4!}{(4-2)! \; 2!} \cdot \frac{8!}{(8-3)! \; 3!} = \frac{4 \cdot 3 \cdot 2!}{2 \cdot 1 \cdot 2!} \cdot \frac{8 \cdot 7 \cdot 6 \cdot 5!}{5! \cdot 3 \cdot 2 \cdot 1} = 6 \cdot 56 = 336 \text{ ways.}$$

29. Once on a team the children are no longer distinct, so placing people on teams is much like forming words in which all the letters are not distinct.

The 12 children can be placed on 3 teams, a first having 3 players, a second having 5 players and a third having 4 players in

$$\frac{12!}{3! \cdot 5! \cdot 4!} = \frac{12 \cdot 11 \cdot 10 \cdot 9 \cdot 8 \cdot 7 \cdot 6 \cdot 5!}{3 \cdot 2 \cdot 1 \cdot 5! \cdot 4 \cdot 3 \cdot 2 \cdot 1} = 1 \cdot 11 \cdot 10 \cdot 9 \cdot 4 \cdot 7 \cdot 1 = 27,720 \text{ ways}$$

31.   The 100 senators can be placed on the committees in

$$\frac{100!}{22! \cdot 13! \cdot 10! \cdot 5! \cdot 16! \cdot 17! \cdot 17!} = 1.157 \times 10^{76} \text{ ways.}$$

33.   This problem consists of four separate tasks: choosing the winner in each of the three divisions which can be done $C(5, 1)$, $C(6, 1)$, and $C(5, 1)$ ways respectively, and then choosing the wild card team from the remaining 13 teams. This can be done $C(13, 1)$ ways. By the Multiplication Principle the playoff participants can be chosen

$$C(5, 1) \cdot C(6, 1) \cdot C(5, 1) \cdot C(13, 1) = 5 \cdot 6 \cdot 5 \cdot 13 = 1950 \text{ ways.}$$

35.   Since we do not care whether the sample includes smokers or non-smokers, we add the two groups together. Then we select the sample. This can be done $C(55, 8)$ ways.

$$C(55, 8) = \frac{55!}{(55-8)! \; 8!} = \frac{55!}{47! \; 8!} = 1,217,566,350 \text{ ways.}$$

37.   Three people from a group of 5 people can be chosen to participate in one of the distinct tests in

$$P(5, 3) = \frac{5!}{(5-3)!} = \frac{5!}{2!} = 60 \text{ ways.}$$

39.   In choosing a sample, order is not important. Four light bulbs can be selected from a box of 24 bulbs in

$$C(24, 4) = \frac{24!}{(24-4)! \cdot 4!} = \frac{24!}{20! \cdot 4!} = 10,626 \text{ ways.}$$

41.   Since the sportswriters rating is in order, the top 15 teams can be chosen from among 50 teams in

$$P(50, 15) = \frac{50!}{(50-15)!} = \frac{50!}{35!} = 2.94 \times 10^{24} \text{ different ways.}$$

43.   Here the order in which the cards are drawn is important, because different orders will result in different numbers. Also since the cards are not replaced, there is no repetition of the digits in the number formed. So there are $P(10, 4) = 10 \cdot 9 \cdot 8 \cdot 7 = 5040$ different numbers possible.

6.6 The Binomial Theorem

1.
$$(x+y)^5 = \binom{5}{0}x^5y^0 + \binom{5}{1}x^4y^1 + \binom{5}{2}x^3y^2 + \binom{5}{3}x^2y^3 + \binom{5}{4}x^1y^4 + \binom{5}{5}x^0y^5$$
$$= x^5 + 5x^4y + 10x^3y^2 + 10x^2y^3 + 5xy^4 + y^5$$

3.
$$(x+3y)^3 = \binom{3}{0}x^3(3y)^0 + \binom{3}{1}x^2(3y)^1 + \binom{3}{2}x^1(3y)^2 + \binom{3}{3}x^0(3y)^3$$
$$= x^3 + 3x^2 \cdot 3y + 3x \cdot 9y^2 + 27y^3$$
$$= x^3 + 9x^2y + 27xy^2 + 27y^3$$

5.
$$(2x-y)^4 = \binom{4}{0}(2x)^4(-y)^0 + \binom{4}{1}(2x)^3(-y)^1 + \binom{4}{2}(2x)^2(-y)^2 + \binom{4}{3}(2x)^1(-y)^3 + \binom{4}{4}(2x)^0(-y)^4$$
$$= 16x^4 - 4 \cdot 8x^3y + 6 \cdot 4x^2y^2 - 4 \cdot 2xy^3 + y^4$$
$$= 16x^4 - 32x^3y + 24x^2y^2 - 8xy^3 + y^4$$

7. The expansion of $(x+y)^5$ is

$$(x+y)^5 = \binom{5}{0}x^5 + \binom{5}{1}x^4y + \binom{5}{2}x^3y^2 + \binom{5}{3}x^2y^3 + \binom{5}{4}xy^4 + \binom{5}{5}y^5$$

The coefficient of x^2y^3 is $\binom{5}{3} = 10$.

9. The expansion of $(x+3)^{10}$ is

$$(x+3)^{10} = \binom{10}{0}x^{10} + \binom{10}{1}x^9 \cdot 3 + \binom{10}{2}x^8 \cdot 3^2 + \binom{10}{3}x^7 \cdot 3^3 + \binom{10}{4}x^6 \cdot 3^4$$
$$+ \binom{10}{5}x^5 \cdot 3^5 + \binom{10}{6}x^4 \cdot 3^6 + \binom{10}{7}x^3 \cdot 3^7 + \binom{10}{8}x^2 \cdot 3^8 + \binom{10}{9}x \cdot 3^9 + \binom{10}{10}3^{10}$$

The coefficient of x^8 is $\binom{10}{2} \cdot 3^2 = 405$.

11. A set with 8 elements has $2^8 = 256$ subsets.

13. A set with 10 elements has $2^{10} - 1 = 1023$ non-empty subsets.

15. For any set A half of the subsets of A have an even number of elements, and half of the subsets have an odd number of elements. So, a set with 10 elements has $\dfrac{2^{10}}{2} = \dfrac{1024}{2} = 512$ subsets with an odd number of elements.

17.   Show that $\dbinom{10}{7}=\dbinom{6}{6}+\dbinom{7}{6}+\dbinom{8}{6}+\dbinom{9}{6}$

We repeatedly use the identity $\dbinom{n}{k}=\dbinom{n-1}{k}+\dbinom{n-1}{k-1}$. So

$$\binom{10}{7}=\binom{9}{7}+\binom{9}{6}$$

$$=\left[\binom{8}{7}+\binom{8}{6}\right]+\binom{9}{6}$$

$$=\left[\binom{7}{7}+\binom{7}{6}\right]+\binom{8}{6}+\binom{9}{6}$$

Now since $\dbinom{7}{7}=\dbinom{6}{6}=1$, we substitute $\dbinom{6}{6}$ for $\dbinom{7}{7}$ in the line above and prove the statement.

$$\binom{10}{7}=\binom{6}{6}+\binom{7}{6}+\binom{8}{6}+\binom{9}{6}$$

19.   We use the identity $\dbinom{n}{k}=\dbinom{n-1}{k}+\dbinom{n-1}{k-1}$ to determine that

$$\binom{11}{6}+\binom{11}{5}=\binom{12}{6}$$

21.   Show that $k\dbinom{n}{k}=n\dbinom{n-1}{k-1}$

By definition,   and,

$$\binom{n}{k}=\frac{n!}{(n-k)!\,k!}\qquad\qquad\binom{n-1}{k-1}=\frac{(n-1)!}{(n-1-(k-1))!\,(k-1)!}=\frac{(n-1)!}{(n-k)!\,(k-1)!}$$

so,   so,

$$k\binom{n}{k}=k\cdot\frac{n!}{(n-k)!\,k!}\qquad\qquad n\binom{n-1}{k-1}=n\cdot\frac{(n-1)!}{(n-k)!\,(k-1)!}$$

$$=\frac{n!}{(n-k)!\,(k-1)!}\qquad\qquad\qquad =\frac{n!}{(n-k)!\,(k-1)!}$$

A comparison of the two expressions shows that they are equal and that $k\dbinom{n}{k}=n\dbinom{n-1}{k-1}$.

Chapter 6 Review

True–False

1.   True   **3.**   False It is 4! = 24   **5.**   True

7.   False The coefficient of 7 is $\binom{7}{3} = 35$.

Fill In The Blanks

1.   disjoint   **3.**   combination   **5.**   binomial coefficients

7. $\binom{5}{2} \cdot 2^2 = 40$

Review Exercises

1. $\subset, \subseteq$ **3.**   none of these   **5.**   none of these   **7.** $\subset, \subseteq$

9. $\subset, =$ **11.** $\subseteq, =$ **13.** $\subset, \subseteq$ **15.** $\subseteq, =$

17.   (a) $(A \cap B) \cup C = \{3, 6\} \cup \{6, 8, 9\} = \{3, 6, 8, 9\}$
       (b) $(A \cap B) \cap C = \{3, 6\} \cap \{6, 8, 9\} = \{6\}$
       (c) $(A \cup B) \cap B = \{1,2,3,5,6,7,8\} \cap \{2,3,6,7\} = \{2,3,6,7\} = B$
       (d) $B \cup \varnothing = B$
       (e) $A \cap \varnothing = \varnothing$
       (f) $(A \cup B) \cup C = \{1, 2, 3, 5, 6, 7, 8\} \cup \{6, 8, 9\} = \{1, 2, 3, 5, 6, 7, 8, 9\}$

19. (a)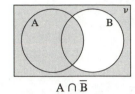

$$A \cap \overline{B}$$

(b)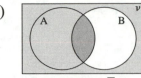

$$(A \cap B) \cup \overline{B}$$

(c)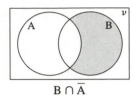

$$B \cap \overline{A}$$

(d)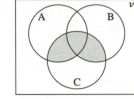

$$(A \cup B) \cap C$$

(e)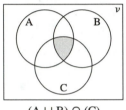

$$(A \cup B) \cap (C)$$

(f)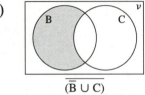

$$\overline{(\overline{B} \cup C)}$$

21. $A \cap B = \varnothing$

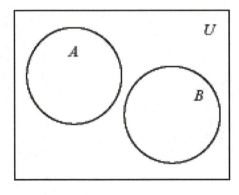

23. $A \cup V$ is the set of all states whose names begin with an A or end with a vowel.

25. $V \cap E$ is the set of all states whose names end with a vowel and which lie east of the Mississippi River.

27. $(A \cup V) \cap E$ is the set of all states whose names start with an A or end with a vowel and which lie east of the Mississippi.

29. $c(A \cup B) = c(A) + c(B) - c(A \cap B)$, so
$$
\begin{aligned}
c(A \cap B) &= c(A) + c(B) - c(A \cup B) \\
&= 24 + 12 - 33 \\
&= 3
\end{aligned}
$$

31.   (a) $c(A \cup B) = c(A) + c(B) - c(A \cap B)$
$$= 3 + 17 - 0$$
$$= 20$$

33. $A = \{2, 4, 6\}; \; B = \{1, 2, 3\}$
$A \cap B = \{2\}; \; c(A \cap B) = 1$
$A \cup B = \{1, 2, 3, 4, 6\}; \; c(A \cup B) = 5$

(b) Sets A and B are disjoint since their intersection has no elements.

35.   We denote the sets of cars with air conditioning, automatic transmissions, and power steering by A, T, and P respectively. From the data given we determine that

$$c(A) = 75 \qquad c(P) = 95 \qquad c(T) = 100$$
$$c(A \cap P \cap T) = 20 \quad c(\overline{A \cap P \cap T}) = 10 \quad c(A \cap \overline{P} \cap \overline{T}) = 10$$
$$c(P \cap T) = 50 \qquad c(A \cap T) = 60$$

We next place the values on a Venn Diagram, so that we can answer the questions.

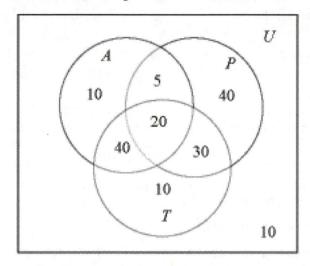

(a)   There were $10 + 5 + 20 + 40 + 10 + 30 + 40 + 10 = 165$ cars sold in June.
(b)   40 of the cars had only power steering.

37.   If $U = \{1, 2, 3, 4, 5\}; \; B = \{1, 4, 5\},$ and $A \cap B = \{1\},$ then A could equal any of the following sets:

$\{1\}, \{1, 2\}, \{1, 3\},$ or $\{1, 2, 3\}$

39. $0! = 1$

41. $\dfrac{7!}{4!} = \dfrac{7 \cdot 6 \cdot 5 \cdot 4!}{4!} = 210$

43. $\dfrac{12!}{11!} = \dfrac{12 \cdot 11!}{11!} = 12$

45. $P(5, 2) = 5 \cdot 4 = 20$

47. $P(12, 1) = 12$

49. $P(100, 2) = 100 \cdot 99 = 9900$

51. $C(10, 2) =$
$$\dfrac{10!}{(10-2)! \; 2!} = \dfrac{10!}{8! \; 2!} = \dfrac{10 \cdot 9 \cdot 8!}{8! \cdot 2 \cdot 1} = 45$$

53. $C(6, 6) = \dfrac{6!}{(6-6)! \; 6!} = \dfrac{6!}{0! \; 6!} = 1$

55. $\dbinom{7}{4} = \dfrac{7!}{(7-4)! \; 4!} = \dfrac{7!}{3! \; 4!}$

$\qquad = \dfrac{7 \cdot 6 \cdot 5 \cdot 4!}{3 \cdot 2 \cdot 1 \cdot 4!} = 35$

57. $\dbinom{9}{1} = \dfrac{9!}{(9-1)! \; 1!} = \dfrac{9!}{8! \; 1!} = \dfrac{9 \cdot 8!}{8! \cdot 1} = 9$

59. When choosing committees order is not important so a committee of 3 people can be chosen from 5 people in

$C(5, 3) = \dbinom{5}{3} = \dfrac{5!}{(5-3)! \; 3!} = 10$ ways.

61. When books are arranged on a shelf different orderings mean different arrangements. So 3 books can be arranged

$P(3, 3) = 3! = 6$ ways.

63. To count the number of different house styles that are available we use the Multiplication Principle. There are

$3 \cdot 4 \cdot 6 = 72$

different styles of houses possible.

65. Each question on a true-false test has 2 possible answers, and answers can be used more than once. So on a 10 question test, there are $2^{10} = 1024$ different ways the test can be answered.

67. (a) The number of 3-digit words that can be formed without repeating a digit from the symbols 1, 2, 3, 4, 5, 6 is given by

$P(6, 3) = 6 \cdot 5 \cdot 4 = 120$ words.

(b) If the same letters in different a order indicate the identical word, then there are only

$C(6, 3) = 20$ possible words.

69. (a) A committee of 3 boys and 4 girls can be formed from 7 boys and 6 girls in

$C(7, 3) \cdot C(6, 4) = 35 \cdot 15 = 525$ different ways.

(b) If the committee must have at least one person of each sex, the committee could consist of 1 boy and 6 girls or 2 boys and 5 girls or 3 boys and 4 girls or 4 boys and 3 girls or 5 boys and 2 girls or 6 boys and 1 girl. Since each configuration is mutually exclusive, we will determine the number of ways the committee can be formed by using the Multiplication Principle 6 times and adding the results. The number of different committees possible is

$$C(7, 1) \cdot C(6, 6) + C(7, 2) \cdot C(6, 5) + C(7, 3) \cdot C(6, 4) + C(7, 4) \cdot C(6, 3)$$
$$+ C(7, 5) \cdot C(6, 2) + C(7, 6) \cdot C(6, 1) = 7 \cdot 6 + 21 \cdot 15 + 35 \cdot 20 + 35 \cdot 15 + 21 \cdot 6 + 7 \cdot 1$$
$$= 42 + 315 + 700 + 525 + 126 + 7$$
$$= 1715$$

71. Since the books of the same subject must be kept together, there are 4! ways to arrange the history books, 5! ways to arrange the English books, and 6! ways to arrange the mathematics books. However, we must also decide the order the three subjects should be arranged on the shelf. This can be done 3! ways. Then by the Multiplication Principle there are $3! \cdot 4! \cdot 5! \cdot 6! = 12{,}441{,}600$ ways to arrange the books on the shelf.

73. In choosing a committee order is not important. So a committee of 8 boys and 5 girls can be formed form a group of 10 boys and 11 girls in

$$C(10, 8) \cdot C(11, 5) = 45 \cdot 462 = 20{,}790 \text{ ways.}$$

75. Three words, one each from five 3-letter words, six 4-letter words, and eight 5-letter words, can be chosen in

$$C(5, 1) \cdot C(6, 1) \cdot C(8, 1) = 5 \cdot 6 \cdot 8 = 240 \text{ ways.}$$

77. Although we are asked to order 5 speakers, we are told speaker B must come first. So in actuality we are only ordering 4 speakers which can be done $4! = 24$ ways.

79. Although we are told to order 5 speakers, we are told that B must go immediately after A. Think of tying them together. Then we need to order only 4 speakers which can be done

$$4! = 24 \text{ ways.}$$

81. We want to choose, without regard to order, 4 plums from a box of 25. Five of the plums are rotten, so we can partition the box into 20 good plums and 5 rotten plums.

(a) We can choose only good plums in $C(20, 4) = 4845$ ways.

(b) To determine how many ways there are to choose 3 good plums and 1 rotten plum, we use the Multiplication Principle. Three good plums and 1 rotten plum can be chosen in

$$C(20, 3) \cdot C(5, 1) = 1140 \cdot 5 = 5700 \text{ ways}$$

(c) The easiest way to find how many ways there are to get at least 1 rotten plum when choosing 4 plums is to subtract the number of ways we can get all good plums (part a) from the total number of ways we can choose 4 plums from the box of 25. This can be done

$$C(25, 4) - C(20, 4) = 12{,}650 - 4845 = 7805 \text{ ways.}$$

83. We need to find the number of permutations of 6 items where 2 of them are not distinct. The number of 6 letter words that can be made from the word FINITE is

$$\frac{6!}{1! \; 2! \; 1! \; 1! \; 1!} = \frac{720}{2} = 360$$

85. We need to find the number of permutations of 10 items where 5 of them are not distinct. The number of ways of arranging the 10 books on the shelf when there are 3 copies of one book and 2 copies of another is

$$\frac{10!}{3! \; 2! \; 1! \; 1! \; 1! \; 1! \; 1!} = \frac{3{,}628{,}800}{12} = 302{,}400$$

87. Using the binomial theorem, we get

$$(x+2)^4 = \binom{4}{0}x^4 + \binom{4}{1}x^3 \cdot 2^1 + \binom{4}{2}x^2 \cdot 2^2 + \binom{4}{3}x^1 \cdot 2^3 + \binom{4}{4}2^4$$

$$= x^4 + 8x^3 + 24x^2 + 32x + 16$$

89. To get the coefficient of x^3 in the expansion of $(x+2)^7$, first find the term of the expansion,

$$\binom{7}{4}x^3 \cdot 2^4 = 35 \cdot 2^4 x^3 = 560x^3$$

The coefficient of x^3 is 560.

Chapter 6 Project

1. Under the present system there are $7 \cdot 10^5 = 700{,}000$ different producer numbers available. The first digit can be chosen from the set $\{0, 2, 3, 4, 5, 6, 7\}$, but each of the other digits have 10 possible choices.

3. If all possible producer numbers are allowed there are 10^{11} possible correct UPC codes. The last digit is the check digit.

5. In Mauritius there are only 4 digits to be split between the publisher block and the title block. So the following 3 possibilities exist. There can be:

(1) digit for the publisher number and 3 digits for the title. This can be done $10 \cdot 10^3$ ways, which allows for $10^3 = 1000$ titles from each of 10 publishers or 10,000 titles.

(2) digits for the publisher number and 2 digits for the title. This can be done $10^2 \cdot 10^2$ ways, which allows for $10^2 = 100$ titles from each of 100 publishers or 10,000 titles.

(3) digits for the publisher number and 1 digit for the title. This can be done $10^3 \cdot 10$ ways, which allows for 10 titles from each of 1000 publishers or 10,000 titles.

Since only one ISBN number is assigned per title, we can use the counting formula.

$10{,}000 + 10{,}000 + 10{,}000 = 30{,}000$ titles can be published in Mauritius.

7. The restrictions placed on the publisher number limits the number of titles from English speaking areas. We consider the following 4 possibilities.

	Number of Country Codes	Number of Publisher Codes	Number of Titles	Total Number of Titles
1.	2	20	10^6	$2 \cdot 20 \cdot 10^6 = 40{,}000{,}000$
2.	2	500	10^5	$2 \cdot 500 \cdot 10^5 = 100{,}000{,}000$
3.	2	1500	10^4	$2 \cdot 1500 \cdot 10^4 = 30{,}000{,}000$
4.	2	500	10^3	$2 \cdot 500 \cdot 10^3 = 1{,}000{,}000$

Using the counting formula, we find that there are

$40{,}000{,}000 + 100{,}000{,}000 + 30{,}000{,}000 + 1{,}000{,}000 = 171{,}000{,}000$ possible titles from English speaking areas.

Chapter 7

Probability

7.1 Sample Spaces and the Assignment of Probabilities

1. The sample space is the set of possible outcomes when two coins are tossed one time.

$S = \{HH, HT, TH, TT\}$

3. The sample space is the set of possible outcomes when one coin is tossed three times.

$S = \{HHH, HHT, HTH, THH,$
$\qquad HTT, THT, TTH, TTT\}$

5. The sample space is the set of possible outcomes when a coin is tossed two times and then a die is thrown.

$S = \{HH1, HH2, HH3, HH4, HH5, HH6,$
HT1, HT2, HT3, HT4, HT5, HT6, TH1,
TH2, TH3, TH4, TH5, TH6, TT1, TT2,
TT3, TT4, TT5, TT6\}

7. Let G stand for green and R stand for red.

$S = \{GA, GB, GC, RA, RB, RC\}$

9. $S = \{AA, AB, AC, BA, BB, BC,$
$\qquad CA, CB, CC\}$

11. $S = \{$ AA1, AA2, AA3, AA4,
AB1, AB2, AB3, AB4,
BA1, BA2, BA3, BA4,
BB1, BB2, BB3, BB4,
AC1, AC2, AC3, AC4,
BC1, BC2, BC3, BC4\}

13. Let G stand for an outcome of green and R stand for an outcome of red on spinner 1.

$S = \{$GA1, GA2, GA3, GA4, GB1, GB2, GB3, GB4, GC1, GC2, GC3, GC4, RA1, RA2, RA3, RA4, RB1, RB2, RB3, RB4, RC1, RC2, RC3, RC4$\}$

15. Each time a coin is tossed there are 2 possible outcomes. So using the Multiplication Principle, when a coin is tossed 4 times there are

$2 \cdot 2 \cdot 2 \cdot 2 = 2^4 = 16$

outcomes in the sample space.

17. Each time a die is tossed there are 6 possible outcomes. Using the Multiplication Principle there are

$6 \cdot 6 \cdot 6 = 6^3 = 216$

outcomes in the sample space when 3 dice are tossed.

19. A regular deck contains 52 cards. There are 52 possibilities for the first outcome and 51 for the second outcome. So using the Multiplication Principle there are

$52 \cdot 51 = 2652$

outcomes in the sample space.

21. There are 26 possible outcomes for the first letter chosen and 26 possible outcomes for the second letter chosen. Using the Multiplication Principle, there are

$26 \cdot 26 = 676$

outcomes in the sample space.

23. Assignments 1, 2, 3, and 6 are valid. Valid assignments must be non-negative, and the sum of all probabilities in the sample space must equal one.

25. Assignment 2 should be used. If the coin always comes up tails HH, HT, and TH are impossible events and have probabilities of 0.

27. The sample space is $\{$H, T$\}$. Let x denote the probability tails occurs. Then

$P(\text{T}) = x \quad \text{and} \quad P(\text{H}) = 3x$

Since the sum of the probability assignments equals 1, we have

$$P(\text{T}) + P(\text{H}) = x + 3x = 1$$
$$4x = 1$$
$$x = \frac{1}{4}$$

So, $P(\text{T}) = \dfrac{1}{4}$ and $P(\text{H}) = \dfrac{3}{4}$

29. The sample space is $\{1, 2, 3, 4, 5, 6\}$. Let x denote the probability an even number occurs. Then

$$P(1) = 2x \qquad P(2) = x \qquad P(3) = 2x \qquad P(4) = x \qquad P(5) = 2x \qquad P(6) = x$$

Since the sum of the probability assignments of the elements in the sample space must equal 1, we have

$$P(1) + P(2) + P(3) + P(4) + P(5) + P(6) = 2x + x + 2x + x + 2x + x = 1$$
$$9x = 1$$
$$x = \frac{1}{9}$$

The probability model is:

$$P(1) = \frac{2}{9} \qquad P(2) = \frac{1}{9} \qquad P(3) = \frac{2}{9} \qquad P(4) = \frac{1}{9} \qquad P(5) = \frac{2}{9} \qquad P(6) = \frac{1}{9}$$

31. S is the sample space. $c(S) = 23$

Define event E: A white ball is picked. There are 3 ways E can occur, so $c(E) = 3$, and

$$P(E) = \frac{c(E)}{c(S)} = \frac{3}{23}$$

33. S is the sample space. $c(S) = 23$

Define event E: A green ball is picked. There are 7 ways E can occur, so $c(E) = 7$, and

$$P(E) = \frac{c(E)}{c(S)} = \frac{7}{23}$$

35. S is the sample space. $c(S) = 23$

Define event E: A white or a red ball is picked. There are 8 ways E can occur, so $c(E) = 8$, and

$$P(E) = \frac{c(E)}{c(S)} = \frac{8}{23}$$

37. S is the sample space. $c(S) = 23$

If neither red nor green is picked, then either white or blue is chosen.

Define event E: A white or a blue ball is picked. There are 11 ways E can occur, so $c(E) = 11$, and

$$P(E) = \frac{c(E)}{c(S)} = \frac{11}{23}$$

39. When 2 fair dice are thrown there are 36 possible outcomes. So $c(S) = 36$, and

$$P(A) = \frac{c(A)}{c(S)} = \frac{2}{36} = \frac{1}{18}$$

41. When 2 fair dice are thrown there are 36 possible outcomes. So $c(S) = 36$, and

$$P(C) = \frac{c(C)}{c(S)} = \frac{4}{36} = \frac{1}{9}$$

43. When 2 fair dice are thrown there are 36 possible outcomes. So $c(S) = 36$, and

$$P(E) = \frac{c(E)}{c(S)} = \frac{6}{36} = \frac{1}{6}$$

45. A regular deck of cards has 52 cards, so $c(S) = 52$.

Define event E: The ace of hearts is drawn.
There is one ace of hearts giving $c(E) = 1$, and

$$P(E) = \frac{c(E)}{c(S)} = \frac{1}{52}$$

47. A regular deck of cards has 52 cards, so $c(S) = 52$.

Define event E: A spade is drawn.
There are 13 spades, $c(E) = 13$, and

$$P(E) = \frac{c(E)}{c(S)} = \frac{13}{52} = \frac{1}{4}$$

49. A regular deck of cards has 52 cards, so $c(S) = 52$.

Define event E: A picture card is drawn.
There are 12 picture cards, making $c(E) = 12$, and

$$P(E) = \frac{c(E)}{c(S)} = \frac{12}{52} = \frac{3}{13}$$

51. A regular deck of cards has 52 cards, so $c(S) = 52$.

Define event E: A card with a number less than 6 is drawn.
There are 5 numbers less than 6 and 4 of each number, making $c(E) = 20$, and

$$P(E) = \frac{c(E)}{c(S)} = \frac{20}{52} = \frac{5}{13}$$

53. A regular deck of cards has 52 cards, so $c(S) = 52$.

Define event E: A card that is not an ace is drawn.
There are 48 cards that are not aces, making $c(E) = 48$, and

$$P(E) = \frac{c(E)}{c(S)} = \frac{48}{52} = \frac{12}{13}$$

55. All Americans are in the sample space, so

$$c(S) = 231,533,000 + 42,554,000 = 274,087,000$$

Define event E: An American who is covered by health insurance is chosen.

$$c(E) = 231,533,000$$

$$P(E) = \frac{c(E)}{c(S)} = \frac{231,533,000}{274,087,000} = 0.8447$$

The probability that a randomly selected American in the year 2000 was covered by health insurance was .845.

57. Simulation: Answers will vary.

Actual Probabilities: Since the coin is fair, we have

$$P(\text{H}) = \frac{1}{2} \qquad P(\text{T}) = \frac{1}{2}$$

59. Simulation: Answers will vary.

Actual Probabilities: Since the coin is loaded, and heads are 3 times as likely to occur than tails,

$$P(H) = \frac{3}{4} \qquad P(T) = \frac{1}{4}$$

61. Simulation: Answers will vary.

Actual Probabilities: $c(S) = 15$, $c(R) = 5$, $c(Y) = 2$, and $c(W) = 8$

$$P(R) = \frac{c(R)}{c(S)} = \frac{5}{15} = \frac{1}{3} \qquad P(Y) = \frac{c(Y)}{c(S)} = \frac{2}{15} \qquad P(W) = \frac{c(W)}{c(S)} = \frac{8}{15}$$

7.2 Properties of the Probability of an Event

1. Define event E: A person selected at random has Rh-positive blood.

$$\begin{aligned}
P(E) &= P(\text{O-Positive}) + P(\text{A-Positive}) + P(\text{B-Positive}) + P(\text{AB-Positive}) \\
&= 0.38 + 0.34 + 0.09 + 0.03 \\
&= 0.84
\end{aligned}$$

The probability that a person selected at random in the United States has Rh-positive blood is .84.

3. Define event E: A person selected at random has blood that contains the A antigen.

$$\begin{aligned}
P(E) &= P(\text{A-Negative}) + P(\text{A-Positive}) + P(\text{AB-Negative}) + P(\text{AB-Positive}) \\
&= 0.06 + 0.34 + 0.01 + 0.03 \\
&= 0.44
\end{aligned}$$

The probability that a person selected at random in the United States has blood containing the A antigen is .44.

5. The sample space S is defined as the ISP subscribers; $c(S) = 149$ million.

Define event E: A randomly selected ISP subscriber subscribes to America Online.
               F: A randomly selected ISP subscriber subscribes to MSN.

$$\begin{aligned}
P(E \cup F) &= P(E) + P(F) \\
&= \frac{26.7}{149} + \frac{9.0}{149} = \frac{35.7}{149} = 0.2396
\end{aligned}$$

The probability that a randomly selected ISP subscriber subscribes to America Online or MSN is 24.0%.

7.   The sample space S is defined as the ISP subscribers; $c(S) = 149$ million.

Define event E: A randomly selected ISP subscriber subscribes to an ISP other than one of the top 10.

$$P(E) = \frac{89.6}{149} = 0.6013$$

The probability that a randomly selected ISP subscriber subscribes to an ISP other than one of the top 10 is 60.1%.

9. $P(\overline{A}) = 1 - P(A)$
  $= 1 - 0.25$
  $= 0.75$

11. $P(A \cup B) = P(A) + P(B)$
  $= 0.25 + 0.40$
  $= 0.65$

13. $P(A \cup B) = P(A) + P(B) - P(A \cap B)$
  $= 0.25 + 0.40 - 0.15$
  $= 0.50$

15.   The sample space has 36 equally likely outcomes, so $c(S) = 36$
Let E and F be the events
E: Sum is 2. F: Sum is 12.
E and F are mutually exclusive, so the probability of obtaining either a 2 or a 12 is

$$P(E \cup F) = P(E) + P(F)$$
$$= \frac{1}{36} + \frac{1}{36} = \frac{2}{36} = \frac{1}{18}$$

17.   Let E and F be the events
E: The Bears win. F: The Bears tie.
E and F are mutually exclusive, so the probability that the Bears either win or tie is

$$P(E \cup F) = P(E) + P(F)$$
$$= 0.65 + 0.05$$
$$= 0.70$$

The probability that the Bears lose is the complement of their winning or tying.

$$P(\overline{E \cup F}) = 1 - P(E \cup F)$$
$$= 1 - 0.70$$
$$= 0.30$$

19.   Let M, and E be the events
M: Jenny passes mathematics. E: Jenny passes English.

We are given $P(M) = .4$; $P(E) = .6$, and $P(M \cup E) = .8$.

The probability that Jenny passes both courses is

$$P(M \cap E) = P(M) + P(E) - P(M \cup E)$$
$$= .4 + .6 - .8$$
$$= .2$$

21. $P(A) = .5$, $P(B) = .4$, and $P(A \cap B) = .2$

   (a)   *A or B*

   $$P(A \cup B) = P(A) + P(B) - P(A \cap B)$$
   $$= .5 + .4 - .2$$
   $$= .7$$

   (b)   *A but not B*

   $$P(A \text{ but not } B) = P(A) - P(A \cap B)$$
   $$= .5 - .2$$
   $$= .3$$

   (c) $P(B \text{ but not } A) = P(B) - P(A \cap B)$
   $$= .4 - .2$$
   $$= .2$$

   (d)   Neither *A* nor *B* is equivalent to finding the complement of $A \cup B$.

   $$P(\text{neither } A \text{ nor } B) = 1 - P(A \cup B)$$
   $$= 1 - .7$$
   $$= .3$$

23.   Define the events   *T*:   Car needs a tune-up.
   B: Car needs a brake job.

   $P(T) = .6$ $P(B) = .1$ $P(T \cap B) = .02$

   (a) $P(T \cup B) = P(T) + P(B) - P(T \cap B)$
   $$= .6 + .1 - .02$$
   $$= .68$$

   (b)   Probability a car needs a tune-up but not a brake job is

   $$P(T \cap \overline{B}) = P(T) - P(T \cap B)$$
   $$= 0.6 - 0.02 = .58$$

   (c) $P\left(\overline{T \cup B}\right) = 1 - P(T \cup B)$
   $$= 1 - .68$$
   $$= .32$$

25.   (a) $P(1 \text{ or } 2 \text{ TVs}) = P(1 \text{ TV}) + P(2 \text{ TVs})$
   $$= .24 + .33$$
   $$= .57$$

   (b) $P(1 \text{ or more TVs}) = 1 - P(0 \text{ TVs})$
   $$= 1 - 0.05$$
   $$= .95$$

   (c) $P(3 \text{ or fewer TVs}) = 1 - P(4 \text{ or more})$
   $$= 1 - 0.17$$
   $$= .83$$

   (d) $P(3 \text{ or more TVs})$
   $$= P(3 \text{ TVs}) + P(4 \text{ or more})$$
   $$= .21 + .17$$
   $$= .38$$

   (e) $P(\text{fewer than 2 TVs})$
   $$= P(0 \text{ TV}) + P(1 \text{ TV})$$
   $$= .05 + .24$$
   $$= .29$$

   (f) $P(\text{not even 1 TV}) = P(0 \text{ TV})$
   $$= .05$$

   (g) $P(1, 2, \text{ or } 3 \text{ TVs}) = .24 + .35 + .21$
   $$= .80$$

   (h) $P(2 \text{ or more TVs}) = .33 + .21 + .17$
   $$= .71$$

27. $P(E) = \dfrac{3}{3+1} = \dfrac{3}{4}$

29. $P(E) = \dfrac{5}{5+7} = \dfrac{5}{12}$

31. $P(E) = \dfrac{1}{1+1} = \dfrac{1}{2}$

33. $P(E) = .6$ $P(\overline{E}) = 1 - P(E)$
$\qquad\qquad\qquad\qquad\quad = 1 - .6$
$\qquad\qquad\qquad\qquad\quad = .4$

Odds for E: $\dfrac{P(E)}{P(\overline{E})} = \dfrac{6}{4}$ or 3 to 2

Odds against E: $\dfrac{P(\overline{E})}{P(E)} = \dfrac{4}{6}$ or 4 to 6

$\qquad\qquad\qquad\qquad\quad$ or 2 to 3

35. $P(F) = \dfrac{3}{4}$ $P(\overline{F}) = 1 - P(F)$

$\qquad\qquad\qquad\qquad\quad = 1 - \dfrac{3}{4} = \dfrac{1}{4}$

Odds for F: $\dfrac{P(F)}{P(\overline{F})} = \dfrac{\frac{3}{4}}{\frac{1}{4}} = \dfrac{3}{1}$ or 3 to 1

Odds against F: $\dfrac{P(\overline{F})}{P(F)} = \dfrac{\frac{1}{4}}{\frac{3}{4}} = \dfrac{1}{3}$

$\qquad\qquad\qquad\qquad\quad$ or 1 to 3

37. In the experiment of throwing two fair dice, define the events
E: A sum of 7 is thrown.
F: A sum of 11 is thrown.

(a) The odds for E are $\dfrac{P(E)}{P(\overline{E})} = \dfrac{P(E)}{1 - P(E)} = \dfrac{\frac{1}{6}}{\frac{5}{6}} = \dfrac{1}{5}$ or 1 to 5.

(b) The odds for F are $\dfrac{P(F)}{P(\overline{F})} = \dfrac{P(F)}{1 - P(F)} = \dfrac{\frac{2}{36}}{\frac{34}{36}} = \dfrac{2}{34} = \dfrac{1}{17}$ or 1 to 17.

(c) The odds for $E \cup F$ are $\dfrac{P(E \cup F)}{P(\overline{E \cup F})} = \dfrac{P(E \cup F)}{1 - P(E \cup F)} = \dfrac{\frac{8}{36}}{\frac{28}{36}} = \dfrac{8}{28} = \dfrac{2}{7}$ or 2 to 7.

39. Let E be the event
E: A person gets a job interview, and $P(E) = .54$

The odds against E are $\dfrac{P(\overline{E})}{P(E)} = \dfrac{46}{54} = \dfrac{23}{27}$ or   23 to 27.

41. If the odds that A will win are 1 to 2, then $P(A) = \dfrac{1}{3}$.

If the odds that B will win are 2 to 3, then $P(B) = \dfrac{2}{5}$.

If a tie is impossible, then A and B are mutually exclusive events, and

$$P(A \cup B) = P(A) + P(B) = \frac{1}{3} + \frac{2}{5} = \frac{11}{15}$$

The odds for $A \cup B = \dfrac{P(A \cup B)}{P(\overline{A \cup B})} = \dfrac{P(A \cup B)}{1 - P(A \cup B)} = \dfrac{\frac{11}{15}}{\frac{4}{15}} = \dfrac{11}{4}$ or 11 to 4.

43. Prove: $P(E \cup F) = P(E) + P(F) - P(E \cap F)$

Proof: $E \cup F$ can be written as the union of disjoint sets in the following way

$$E \cup F = (E \cap \overline{F}) \cup (E \cap F) \cup (\overline{E} \cap F)$$

and

$$
\begin{aligned}
P(E \cup F) &= P[(E \cap \overline{F}) \cup (E \cap F) \cup (\overline{E} \cap F)] \\
&= P(E \cap \overline{F}) + P(E \cap F) + P(\overline{E} \cap F) \qquad \text{(1) since the sets are disjoint.}
\end{aligned}
$$

Writing E and F as the union of disjoint sets

$$E = (E \cap \overline{F}) \cup (E \cap F) \qquad \text{and} \qquad F = (\overline{E} \cap F) \cup (E \cap F)$$

means

$$P(E) = P(E \cap \overline{F}) + P(E \cap F) \qquad \text{and} \qquad P(F) = P(\overline{E} \cap F) + P(E \cap F)$$

or,

$$P(E \cap \overline{F}) = P(E) - P(E \cap F) \qquad \text{and} \qquad P(\overline{E} \cap F) = P(F) - P(E \cap F)$$

Substituting into (1) we have

$$P(E \cup F) = [P(E) - P(E \cap F)] + P(E \cap F) + [P(F) - P(E \cap F)]$$

which simplifies to

$$P(E \cup F) = P(E) + P(F) - P(E \cap F)$$

45.   Show:
$$P(A \cup B \cup C) = P(A) + P(B) + P(C) - P(A \cap B) - P(A \cap C) - P(B \cap C) + P(A \cap B \cap C)$$

Proof:

Let $E = A \cup B$,

The addition rule states

$$P(E \cup C) = P(E) + P(C) - P(E \cap C)$$

Now substitute $A \cup B$ for E

$$P[(A \cup B) \cup C] = P(A \cup B) + P(C) - P[(A \cup B) \cap C)]$$

Apply the addition rule to $P(A \cup B)$

$$P[(A \cup B) \cup C] = [P(A) + P(B) - P(A \cap B)] + P(C) - P[(A \cup B) \cap C)]$$

Apply the distributive property to the sets in the last bracket

$$P[(A \cup B) \cup C] = [P(A) + P(B) - P(A \cap B)] + P(C) - P[(A \cap C) \cup (B \cap C)]$$

Apply the addition rule to the union of sets in the last bracket

$$P(A \cup B \cup C) = P(A) + P(B) - P(A \cap B) + P(C) - [P(A \cap C) + P(B \cap C) - P(A \cap B \cap C)]$$

Remove the brackets and rearrange terms

$$P(A \cup B \cup C) = P(A) + P(B) + P(C) - P(A \cap B) - P(A \cap C) - P(B \cap C) + P(A \cap B \cap C)$$

7.3 Probability Problems Using Counting Techniques

1.   The number of elements in the sample space S is equal to the number of combinations of 50 refrigerators taken 5 at a time,

$$C(50, 3) = \frac{50!}{5! \ 45!} = 2{,}118{,}760$$

(a)   Define E as the event, "5 refrigerators are defective."

$$P(E) = \frac{C(6, 5)}{C(50, 5)} = \frac{6}{2{,}118{,}760} = \frac{3}{1{,}059{,}380} = 2.83 \times 10^{-6}$$

(b)   Define F as the event, "at least 2 refrigerators are defective"

$P(F) = 1 - P(0 \text{ refrigerators are defective or 1 refrigerator is defective})$

$$= 1 - \left[\frac{C(6, 0)C(44, 5)}{2{,}118{,}760} + \frac{C(6, 1) \ C(44, 4)}{2{,}118{,}760} \right]$$

$$= 1 - \left[\frac{1{,}086{,}008}{2{,}118{,}760} + \frac{6 \cdot (135{,}751)}{2{,}118{,}760} \right]$$

$$= .103$$

3. The number of elements in the sample space S is found by using the Multiplication Principle. Since the coin is tossed 5 times, there are

$c(S) = 2^5 = 32$ elements.

(a) Define E as the event, "Exactly 3 heads appear."

$$P(E) = \frac{c(E)}{c(S)} = \frac{C(5, 3)}{32} = \frac{10}{32} = \frac{5}{16}$$

(b) Define F as the event, "No heads appear."

$$P(F) = \frac{c(F)}{c(S)} = \frac{C(5, 0)}{32} = \frac{1}{32}$$

5. The number of elements in the sample space S is found by using the Multiplication Principle. Since the dice are thrown 3 times,

$c(S) = 36^3 = 46,656$

(a) Define E as the event, "The sum of 7 appears 3 times."
When two dice are thrown, a sum of 7 can appear 6 ways, {(1, 6), (2, 5), (3, 4), (4, 3), (5, 2), (6, 1)}. Using the Multiplication Principle we find that on 3 throws a sum of 7 can appear 6^3 ways.

$$P(E) = \frac{c(E)}{c(S)} = \frac{6^3}{36^3} = \frac{1}{6^3} = \frac{1}{216}$$

(b) Define F as the event, "The sum of 7 or 11 appears at least twice."
When 2 dice are thrown a sum of 7 can appear 6 ways and a sum of 11 can appear 2 ways. Using the addition rule, the sum of 7 or 11 can appear 8 ways.

Event F is equivalent to the event: Exactly two throws result in a 7 or 11, or all three throws result in a 7 or 11. These are mutually exclusive, so we are interested in finding $P(F) = P(7 \text{ or } 11 \text{ appears twice}) + P(7 \text{ or } 11 \text{ appear } 3 \text{ times})$

Using the Multiplication Principle 7 or 11 can appear 3 times in $8^3 = 512$ ways.

7 or 11 can appear 2 times in $8^2 = 64$ ways, and the third number can appear $36 - 8 = 28$ ways. By the Multiplication Principle 2 throws of 7 or 11 and 1 throw of another number can appear $8^2 \cdot 28 = 1792$ ways. However, we still need to choose which of the two throws will result in the 7 or 11. We choose 2 out of 3 tries $C(3, 2) = 3$ ways.

$$P(F) = \frac{c(F)}{c(S)} = \frac{C(3, 2) \cdot 8^2 \cdot 28}{36^3} + \frac{8^3}{36^3}$$
$$= \frac{5376}{36^3} + \frac{512}{36^3} = \frac{5888}{46656} = .126$$

7. The number of elements in the sample space, S, is found by using the Multiplication Principle. Since there are 7 digits in the phone number, and there are no restrictions,

$c(S) = 10^7$

Define event E: A phone number has one or more repeated digits.

It is easier to do this problem looking at the complement of E.
$\overline{E}$: A phone number has no repeated digits. $c(\overline{E}) = P(10, 7)$
So we get $\quad P(E) = 1 - P(\overline{E})$

$$= 1 - \frac{c(\overline{E})}{c(S)}$$

$$= 1 - \frac{P(10, 7)}{10^7} = .940$$

9. The number of elements in the sample space, S, is found by using the Multiplication Principle. Since there are 26 letters in the alphabet and we are going to select 5,

$c(S) = 26^5$

Define event E: No letters are repeated. $c(E) = C(26, 5)$

$$P(E) = \frac{c(E)}{c(S)} = \frac{C(26, 5)}{26^5} = .0055$$

11. The number of elements in the sample space, S, is the number of orderings of the 4 movies.

$c(S) = 4! = 24$

Define event E: The movies are listed in order of revenue. $C(E) = 1$

$$P(E) = \frac{c(E)}{c(S)} = \frac{1}{24}$$

13. The number of elements in the sample space, S, is found by using the Multiplication Principle. Since there are 12 months in the year, and we are going to select 6,

$c(S) = 12^6$

Define event E: At least 2 were born in the same month.

It is easier to do this problem by looking at the complement of E,
$\overline{E}$: No two people were born in the same month.

$c(\overline{E}) = P(12, 6) = 12 \cdot 11 \cdot 10 \cdot 9 \cdot 8 \cdot 7 = 665280$

The probability that at least 2 persons were born in the same month is

$P(E) = 1 - P(\overline{E})$

$$= 1 - \frac{c(\overline{E})}{c(S)}$$

$$= 1 - \frac{P(12, 6)}{12^6} = .777$$

15.   The number of elements in the sample space, S, is found by using the Multiplication Principle. Since there are 365 days in a year, and 100 senators

$$c(S) = 365^{100}$$

Define event E: At least 2 senators have the same birthday.

It is easier to do this problem by looking at the complement of E,

$\overline{E}$: No two senators have the same birthday.

$$c(\overline{E}) = P(365,\ 100) = 365 \cdot 364 \cdot 363 \cdot \ ... \ \cdot 266$$

The probability that at least 2 senators have the same birthday is

$$P(E) = 1 - P(\overline{E})$$
$$= 1 - \frac{c(\overline{E})}{c(S)}$$
$$= 1 - \frac{P(365,\ 100)}{365^{100}} = .999$$

17.   The number of elements in the sample space, S, is the number of ways the 5 letters can be arranged.

$$c(S) = 5! = 120$$

Define event E: L will precede E in a scrambling of the letters.
The number of elements in E are determined as follows:

1.   Select 2 positions to arrange L and E with L preceding E. There are $P(5, 2)$ to arrange L and E in 5 positions, $\frac{1}{2}$ of which have L before E. Step 1 can be done in $\frac{P(5,\ 2)}{2}$ ways.

2.   Arrange the letters VOW in the remaining 3 positions. This can be done 3! Ways.
     Using the Multiplication Principle, $c(E) = \dfrac{P(5,\ 2)}{2} \cdot 3! = 60$

The probability that L will precede E in a scrambling of the letters of the word VOWEL is

$$P(E) = \frac{c(E)}{c(S)} = \frac{60}{120} = \frac{1}{2}$$

19.   The number of elements in the sample space, S, is the number of ways the 5 letters can be arranged.

$c(S) = 5! = 120$

Define event E: L comes first in the rearrangement.
The number of elements in E are determined as follows:
There is no choice for L, it must come first. The other 4 letters VOWE can be arranged $4! = 24$ ways.

$c(E) = 24$

The probability that L comes first in a rearrangement of the letters of the word VOWEL is

$$P(E) = \frac{c(E)}{c(S)} = \frac{24}{120} = \frac{1}{5}$$

21.   The number of elements in the sample space, S, is the number of ways the playoff teams can be chosen. Since the selection is made in several steps we use the Multiplication Principle.

$c(S) = C(5, 1) \cdot C(6, 1) \cdot C(5, 1) \cdot C(13, 1) = 1950$

Define event E: The Giants and Dodgers are in the playoffs.
The number of elements in E is determined as follows: One of the two teams is the division winner, this can happen 1 way; and the other team is the wild card, this can happen 1 way. Allowing for each team to be division winner doubles this number. The division winners from the East and Central Divisions have $C(5, 1)$ and $C(6, 1)$ ways of occurring. Using the Multiplication Principle, we find

$c(E) = 2(1 \cdot 5 \cdot 6 \cdot 1) = 60$

The probability that both the Giants and Dodgers are in the playoffs is

$$P(E) = \frac{c(E)}{c(S)} = \frac{60}{1950} = \frac{2}{65}$$

23.   The number of elements in the sample space, S, is the number of ways the playoff teams can be chosen from the three divisions. Since the selection is made in several steps we use the Multiplication Principle.

$c(S) = C(5, 1) \cdot C(6, 1) \cdot C(5, 1) \cdot C(13, 1) = 1950$

Define event E: The wild card team is from the Central Division.

The number of elements in E is determined using the Multiplication Principle.

$c(E) =$ East Winner $\cdot$ Central Winner $\cdot$ West Winner $\cdot$ Wild Card
$\quad = C(5, 1) \cdot C(6, 1) \cdot C(5, 1) \cdot C(5, 1)$
$\quad = 5 \cdot 6 \cdot 5 \cdot 5$
$\quad = 750$

The probability the wild card team is from the Central Division is

$$P(E) = \frac{750}{1950} = \frac{5}{13}$$

25. The sample space, S, consist of all the possible combinations of 13 cards,

$$c(S) = C(52, 13)$$

Define event E: A hand contains 5 spades, 4 hearts, 3 diamonds and 1 club.
The number of elements in E is determined using the Multiplication Principle,

$$c(E) = C(13, 5) \cdot C(13, 4) \cdot C(13, 3) \cdot C(13, 1)$$

$$P(E) = \frac{c(E)}{c(S)} = \frac{C(13, 5) \cdot C(13, 4) \cdot C(13, 3) \cdot C(13, 1)}{C(52, 13)} = .0054$$

27. The number of elements in the sample space, S, is the total number ways the 5 passengers can exit the elevator.

$$c(S) = 8^5 = 32{,}768$$

Define event E: No 2 passengers exit on the same floor.

$$c(E) = P(8, 5) = 8 \cdot 7 \cdot 6 \cdot 5 \cdot 4 = 6720$$

$$P(E) = \frac{c(E)}{c(S)} = \frac{6720}{32{,}768} = .2051$$

7.4 Conditional Probability

1. $P(E) = .2 + .3 = .5$

3. $P(E\,|\,F) = \dfrac{P(E \cap F)}{P(F)}$

$$= \frac{.3}{.7} = \frac{3}{7}$$

5. $P(E \cap F) = .3$

7. $P(\overline{E}) = 1 - P(E)$
$$= 1 - .5$$
$$= .5$$

9. $P(E\,|\,F) = \dfrac{P(E \cap F)}{P(F)}$

$$= \frac{1}{4}$$

$P(F\,|\,E) = \dfrac{P(E \cap F)}{P(E)}$

$$= \frac{1}{2}$$

11. $P(E\,|\,F) = \dfrac{P(E \cap F)}{P(F)}$

$P(F) = \dfrac{P(E \cap F)}{P(E\,|\,F)}$

$$= \frac{.2}{.4} = \frac{1}{2}$$

13. $P(E\,|\,F) = \dfrac{P(E \cap F)}{P(F)}$

$P(E \cap F) = P(F) \cdot P(E\,|\,F)$

$$= \frac{5}{13} \cdot \frac{4}{5} = \frac{4}{13}$$

15. (a) $P(F|E) = \dfrac{P(E \cap F)}{P(E)}$

$$P(E) = \dfrac{P(E \cap F)}{P(F|E)}$$

$$= \dfrac{\frac{1}{3}}{\frac{2}{3}} = \dfrac{1}{2}$$

(b) $P(E|F) = \dfrac{P(E \cap F)}{P(F)}$

$$P(F) = \dfrac{P(E \cap F)}{P(E|F)}$$

$$= \dfrac{\frac{1}{3}}{\frac{1}{2}} = \dfrac{2}{3}$$

17. $P(C) = (.7)(.9) + (.3)(.2)$
$$= .63 + .06$$
$$= .69$$

19. $P(C \mid A) = .9$

21. $P(C \mid B) = .2$

23. $P(E \cap F) = P(E) + P(F) - P(E \cup F)$
$$= .5 + .4 - .8$$
$$= .1$$

25. $P(F|E) = \dfrac{P(E \cap F)}{P(E)}$

$$= \dfrac{.1}{.5} = \dfrac{1}{5}$$

27. A Venn diagram helps to see this problem.

$$P\left(\overline{F}\right) = 1 - P(F) = 1 - .4 = .6$$

$$P\left(E \mid \overline{F}\right) = \dfrac{P\left(E \cap \overline{F}\right)}{P\left(\overline{F}\right)}$$

$$= \dfrac{.4}{.6} = \dfrac{2}{3}$$

29. $S = \{$ BBB, BBG, BGB, GBB, GGB, GBG, BGG, GGG$\}$; $c(S) = 8$

Let E be the event, "The family has 2 girls." $E = \{$GGB, GBG, BGG$\}$
Let F be the event, "The first child is a girl." $F = \{$ GBB, GGB, GBG, GGG$\}$

$$P(F) = \frac{4}{8} = \frac{1}{2}$$

$$E \cap F = \{\text{GGB, GBG}\} \qquad P(E \cap F) = \frac{1}{4}$$

So the probability a family with 3 children has exactly 2 girls, given the first child is a girl is

$$P(E \mid F) = \frac{P(E \cap F)}{P(F)}$$

$$= \frac{\dfrac{1}{4}}{\dfrac{1}{2}} = \frac{1}{2}$$

31.   The number of elements in the sample space, S, is $c(S) = 2^4 = 16$

Define event E: "4 heads occur." $c(E) = 1$

The probability 4 heads occur in 4 tosses of a coin is

$$P(E) = \frac{c(E)}{c(S)}$$

$$= \frac{1}{16}$$

If we are told the second toss is a head then we need to get 3 heads in 3 tosses, the probability of obtaining 4 heads is $\dfrac{1}{8}$.

33.   Define the event E: The first card is a heart.
Define the event F: The second card is red.

$$P(\text{E}) = \frac{13}{52} = \frac{1}{4} \qquad P(F|E) = \frac{25}{51}$$

The probability that when two cards are drawn without replacement, the first is a heart and the second is red is

$$P(E \cap F) = P(E) \cdot P(F|E)$$
$$= \frac{1}{4} \cdot \frac{25}{51}$$
$$= \frac{25}{204}$$

Define the event G: The first card is a red.
Define the event H: The second card is a heart.

$$P(G) = \frac{26}{52} = \frac{1}{2} \qquad P(H \mid G) = \frac{13}{51} \cdot \frac{1}{2} + \frac{12}{51} \cdot \frac{1}{2} = \frac{25}{102}$$

The probability that when two cards are drawn without replacement, the first is a red and the second is a heart is

$$P(G \cap H) = P(G) \cdot P(H|G)$$
$$= \frac{1}{2} \cdot \frac{25}{102} = \frac{25}{204}$$

35.   Define event E: When 2 balls are chosen without replacement a White and a Yellow result.

The probability of choosing a white and a yellow ball without replacement can be considered as the union of two mutually exclusive events.

$$P(E) = P(\text{W on first}) \cdot P(\text{Y} \mid \text{W on first}) + P(\text{Y on first}) \cdot P(\text{W} \mid \text{Y on first})$$
$$= \frac{3}{6} \cdot \frac{1}{5} + \frac{1}{6} \cdot \frac{3}{5}$$
$$= \frac{1}{10} + \frac{1}{10} = \frac{1}{5}$$

37. $c(S) = 52$

Define event E to be "A red ace is drawn." $\quad c(E) = 2$
Define event F to be "An ace is drawn." $\qquad c(F) = 4$
Define event G to be "A red card is picked." $\quad c(G) = 26$

(a) $P(E) = \dfrac{2}{52} = \dfrac{1}{26}$

(b) $P(E \mid F) = \dfrac{P(E \cap F)}{P(F)} = \dfrac{\frac{2}{52}}{\frac{4}{52}} = \dfrac{1}{2}$

(c) $P(E \mid G) = \dfrac{P(E \cap G)}{P(G)} = \dfrac{\frac{2}{52}}{\frac{26}{52}} = \dfrac{2}{26} = \dfrac{1}{13}$

39. Define event E: A family has more than 2 children.
Define event F: A family has at least 1 child.

$E \cap F = E$
$P(E \cap F) = 0.20 + 0.16 + 0.08 + 0.06 = 0.50$
$P(F) = 0.30 + 0.20 + 0.16 + 0.08 + 0.06 = 0.80$

The probability that a family has more than 2 children if it is known that it has at least 1 child is

$P(E \mid F) = \dfrac{P(E \cap F)}{P(F)}$

$= \dfrac{.5}{.8} = \dfrac{5}{8}$

41. $P(E) = .1 + .3 = .4$

43. $P(H) = .10 + .06 + .08 = .24$

45. $P(E \cap H) = .10$

47. $P(G \cap H) = .08$

49. $P(E \mid H) = \dfrac{P(E \cap H)}{P(H)}$

$= \dfrac{.10}{.24} = \dfrac{5}{12}$

51. $P(G \mid H) = \dfrac{P(G \cap H)}{P(H)}$

$= \dfrac{.08}{.24} = \dfrac{1}{3}$

53. Event $E \cap F$ is "A person has type B Positive blood."

$P(E) = 0.09 + 0.02 = 0.11$
$P(F) = 0.38 + 0.34 + 0.09 + 0.03 = 0.84$
$P(E \cap F) = 0.09$

$P(E \mid F) = \dfrac{P(E \cap F)}{P(F)}$

$= \dfrac{.09}{.84} = \dfrac{3}{28} = .107$

$P(F \mid E) = \dfrac{P(E \cap F)}{P(E)}$

$= \dfrac{.09}{.11} = \dfrac{9}{11} = .8181$

55. (a) $P(M) = \dfrac{1448}{2018} = .7175$

(b) $P(A) = \dfrac{666}{2018} = .3300$

(c) $P(F \cap B) = \dfrac{144}{2018} = .0714$

(d) $P(F \mid E) = \dfrac{P(F \cap E)}{P(E)}$

$\qquad\qquad = \dfrac{102}{526} = .1939$

(e) $P(A \mid M) = \dfrac{P(A \cap M)}{P(M)}$

$\qquad\quad = \dfrac{342}{1448} = .2362$

(f) $P(F \mid A \cup E) = \dfrac{P(F \cap (A \cup E))}{P(A \cup E)}$

$\qquad\qquad = \dfrac{324 + 102}{666 + 526}$

$\qquad\qquad = \dfrac{426}{1192} = .3574$

(g) $P(M \cap \overline{B}) = \dfrac{342 + 424}{2018}$

$\qquad\qquad = \dfrac{766}{2018} = .3796$

(h) $P(F \mid A \cup B) = \dfrac{P(F \cap (A \cup B))}{P(A \cup B)}$

$\qquad\qquad = \dfrac{324 + 144}{666 + 826}$

$\qquad\qquad = \dfrac{468}{1492} = .3137$

57. The number of elements in the sample space, S, is the total possible outcomes when a two 12 sided dice are thrown.

$c(S) = 12^2 = 144$

Event $(E \cap F) = \{(5, 9), (9, 5)\}$

$c(E \cap F) = 2$

$F = \{(1, 5), (2, 5), (3, 5), (4, 5), (5, 5), (6, 5), (7, 5), (8, 5), (9, 5), (10, 5), (11, 5), (12, 5), (5, 1),$
$\quad (5, 2), (5, 3), (5, 4), (5, 6), (5, 7), (5, 8), (5, 9). (5, 10), (5, 11), (5, 12)\}$

$c(F) = 23$

$P(E \mid F) = \dfrac{P(E \cap F)}{P(F)}$

$\qquad = \dfrac{\dfrac{2}{144}}{\dfrac{23}{144}} = \dfrac{2}{23}$

59. Define the event R: A person is a Republican.
Define the event D: A person is a Democrat.
Define the event V: A person voted for the Democratic Candidate.
We are given the odds that a person is a Republican are 3 to 1, so

$$P(R) = \frac{3}{3+1} = \frac{3}{4}$$

Since the Democratic candidate won by a ratio of 5 to 4, the winning candidate received $\frac{5}{9}$ of the vote. All the Democrats voted democratic, so they provided $\frac{1}{4}$ of the votes.

The Republicans made up the rest, $\frac{5}{9} - \frac{1}{4} = \frac{11}{36}$. So $P(R \cap V) = \frac{11}{36}$.

The probability that a voter selected at random is a Republican given that he or she voted for the Democratic candidate is

$$P(R \mid V) = \frac{P(R \cap V)}{P(V)}$$

$$= \frac{\frac{11}{36}}{\frac{5}{9}} = \frac{11}{20}$$

61. Define the events, M, F, and S as follows.
M: A person is male.
F: A person is female.
S: A person smokes.

We assume that half the population is male and half is female.

$$P(S) = P(S \mid M) \cdot P(M) + P(S \mid F) \cdot P(F)$$
$$= (.257)(.5) + (.210)(.5)$$
$$= .234$$

63. The sample space, S, is the set of possible outcomes when 2 dice are rolled. $c(S) = 36$

We define 3 events: W: The player wins. $c(W) = 5$

L: The player loses. $c(L) = 6$

R: The player rolls again. $c(R) = 25$

The player has already rolled an 8, so we can think of the next roll as the first.
We want $P(W)$. This can happen on the first roll or it can happen on 2nd roll or it can happen on the 3rd roll, and so on. The probability of each individual outcome is determined using the Multiplication Principle. See the table below.

Player wins on roll	Outcome	Probability
1	W	$\dfrac{5}{36}$
2	RW	$\dfrac{25}{36} \cdot \dfrac{5}{36}$
3	RRW	$\dfrac{25}{36} \cdot \dfrac{25}{36} \cdot \dfrac{5}{36} = \left(\dfrac{25}{36}\right)^2 \cdot \dfrac{5}{36}$
4	$RRRW$	$\left(\dfrac{25}{36}\right)^3 \cdot \dfrac{5}{36}$
$\vdots$	$\vdots$	$\vdots$

Each of these outcomes is mutually exclusive so we can use the addition rule.

$$P(W) = P(W) + P(R)\,P(W) + P(R)\,P(R)\,P(W) + P(R)\,P(R)\,P(R)\,P(W) + \ldots$$

This is a geometric series with $a = \dfrac{5}{36}$ and $r = \dfrac{25}{36}$. In Chapter 5 we learned that the sum of a geometric series is $S = a\left(\dfrac{1 - r^n}{1 - r}\right)$. If n gets larger $r^n = \left(\dfrac{25}{36}\right)^n$ becomes closer to 0. (Try it!) In an infinite series, one that can continue forever, as we have here, $S = a\left(\dfrac{1}{1 - r}\right)$. The probability of the player winning this game of craps is

$$S = \left(\dfrac{5}{36}\right)\left(\dfrac{1}{1 - \dfrac{25}{36}}\right)$$

$$= \left(\dfrac{5}{36}\right)\left(\dfrac{36}{11}\right)$$

$$= \dfrac{5}{11}$$

65. Given that $P(E) > 0$ and $P(F) > 0$,

$$P(F) \cdot P(E \mid F) = P(F) \cdot \frac{P(E \cap F)}{P(F)}$$

$$= P(E \cap F)$$

$$= \frac{P(E)}{P(E)} \cdot P(E \cap F)$$

$$= P(E) \cdot \frac{P(E \cap F)}{P(E)}$$

$$= P(E) \cdot P(F \mid E)$$

67. $P(E \mid F) + P(\overline{E} \mid F) = \dfrac{P(E \cap F)}{P(F)} + \dfrac{P(\overline{E} \cap F)}{P(F)}$

$$= \frac{P(E \cap F) + P(\overline{E} \cap F)}{P(F)}$$

Since $E \cap F$ and $\overline{E} \cap F$ are mutually exclusive events,

$$P(E \cap F) + P(\overline{E} \cap F) = P\big((E \cap F) \cup (\overline{E} \cap F)\big)$$

From set theory, we know that $(E \cap F) \cup (\overline{E} \cap F) = (E \cup \overline{E}) \cap F = S \cap F = F$.

So,

$$P(E \cap F) + P(\overline{E} \cap F) = P(F)$$

and

$$P(E \mid F) + P(\overline{E} \mid F) = \frac{P(F)}{P(F)} = 1$$

69. $P(E) > 0$ and $P(E \mid F) = P(E)$

$$P(E \mid F) = \frac{P(F \cap E)}{P(F)}$$

but we are given that $P(E \mid F) = P(E)$, so we have

$$\frac{P(F \cap E)}{P(F)} = P(E)$$

which means $P(F \cap E) = P(E) \cdot P(F)$
and so,

$$P(F \mid E) = \frac{P(F \cap E)}{P(E)}$$

$$= \frac{P(E) \cdot P(F)}{P(E)}$$

$$= P(F)$$

7.5 Independent Events

1. $P(F \cap E) = P(E) \cdot P(F)$
$\qquad = (0.4) \cdot (0.6)$
$\qquad = .24$

3. $P(E \cup F) = P(E) + P(F) - P(E \cap F)$
$\qquad\qquad = P(E) + P(F) - P(E)P(F)$
$\qquad\qquad = P(E) + P(F)(1 - P(E))$

$$P(F) = \frac{P(E \cup F) - P(E)}{1 - P(E)}$$

$$= \frac{.3 - .2}{1 - .2}$$

$$= \frac{.1}{.8} = .125$$

5. E and F are independent if

$$P(F \cap E) = P(E) \cdot P(F)$$

$$\text{Is } \frac{2}{9} = \left(\frac{4}{21}\right)\left(\frac{7}{12}\right)?$$

$$\frac{2}{9} \neq \frac{1}{9}$$

So, the events are not independent.

7. (a) $P(E \mid F) = P(E) = .2$
(c) $P(E \cap F) = P(E)\, P(F)$
$\qquad\qquad = (.2)\,(.4)$
$\qquad\qquad = .08$

(b) $P(F \mid E) = P(F) = .4$
(d) $P(E \cup F) = P(E) + P(F) - P(E \cap F)$
$\qquad\qquad = .2 + .4 - .08$
$\qquad\qquad = .52$

9. $P(E \cap F \cap G) = P(E) \cdot P(F) \cdot P(G)$

$$= \frac{2}{3} \cdot \frac{3}{7} \cdot \frac{2}{21}$$

$$= \frac{4}{147}$$

11. $P(E \cap F) = P(E) + P(F) - P(E \cup F)$
$\qquad\qquad = .3 + .2 - .4$
$\qquad\qquad = .1$

$$P(E \mid F) = \frac{P(F \cap E)}{P(F)}$$

$$= \frac{.1}{.2} = \frac{1}{2}$$

$P(E) \cdot P(F) = (.3)\,(.2) = .06 \neq P(E \cap F)$
E and F are not independent.

13. $E = \{1, 2, 3\} \qquad P(E) = \dfrac{1}{2} \qquad F = \{3, 4, 5\} \qquad P(F) = \dfrac{1}{2}$

$E \cap F = \{3\} \qquad P(E \cap F) = \dfrac{1}{6}$

Since $P(E) \cdot P(F) = \dfrac{1}{2} \cdot \dfrac{1}{2} = \dfrac{1}{4}$ is not equal to $P(E \cap F) = \dfrac{1}{6}$, the events E and F are not independent.

15. S is the set of outcomes on a 12-sided die. $c(S) = 12$

(a) $E = \{1, 2, 3, 4, 5, 6\}$ $P(E) = \dfrac{1}{2}$

$F = \{1, 3, 5, 7, 9, 11\}$ $P(F) = \dfrac{1}{2}$

$E \cap F = \{1, 3, 5\}$ $P(E \cap F) = \dfrac{1}{4}$

$P(E) \cdot P(F) = \dfrac{1}{2} \cdot \dfrac{1}{2} = \dfrac{1}{4} = P(E \cap F)$

Events E and F are independent.

(b) $E = \{8, 9, 10, 11, 12\}$ $P(E) = \dfrac{5}{12}$

$F = \{2, 4, 6, 8, 10, 12\}$ $P(F) = \dfrac{1}{2}$

$E \cap F = \{8, 10, 12\}$ $P(E \cap F) = \dfrac{1}{4}$

$P(E) \cdot P(F) = \dfrac{5}{12} \cdot \dfrac{1}{2} = \dfrac{5}{24} \neq P(E \cap F)$

Events E and F are not independent.

Probability Model for Problems 17–26.

Experiment: Toss a fair die and then a fair coin.

Let H stand for the outcome head, and T stand for the outcome tail. The sample space is

$S = \{1H, 1T, 2H, 2T, 3H, 3T, 4H, 4T, 5H, 5T, 6H, 6T\}$.

We next assign probabilities to each of the simple events in S. Since the die is fair we have

$P(1) = P(2) = P(3) = P(4) = P(5) = P(6) = \dfrac{1}{6}$. Since the coin is also fair, we have $P(H) = $

$P(T) = \dfrac{1}{2}$. Tossing a die and then tossing a coin are independent events, so the probability of

tossing a 1 followed by a H is the product $P(1)\,P(H) = \dfrac{1}{6} \cdot \dfrac{1}{2} = \dfrac{1}{12}$. Similarly, the probabilities

of the rest of the simple events in the sample space are also $\dfrac{1}{12}$.

This is a valid probability model since
1. each probability assignment is non-negative.
2. the sum of all the probability assignments is 1.

17. See the probability model above.

E: The coin comes up heads.

$P(E) = \dfrac{1}{2}$

19. See the probability model preceding Problem 17.

E: The die comes up 4.

$P(E) = \dfrac{2}{12} = \dfrac{1}{6}$

21. See the probability model preceding Problem 17.

E: The die comes up 4.

$P(\overline{E}) = 1 - P(E) = 1 - \dfrac{1}{6} = \dfrac{5}{6}$

23. See the probability model preceding Problem 17.

E: The die comes up 5.
F: The die comes up 6.

$P(E \cup F) = P(E) + P(F)$

$= \dfrac{1}{6} + \dfrac{1}{6}$

$= \dfrac{2}{6} = \dfrac{1}{3}$

25. See the probability model preceding Problem 17.

E: $3H \cup 4H \cup 5H$

$P(3H \cup 4H \cup 5H)$
$= P(3H) + P(4H) + P(5H)$

$= \dfrac{1}{12} + \dfrac{1}{12} + \dfrac{1}{12}$

$= \dfrac{3}{12} = \dfrac{1}{4}$

27. The sample space is the set of all possible outcomes. Let R stand for the mouse goes right and L stand for the mouse goes left.

$S = \{RRR, RRL, RLR, LRR, RLL, LRL, LLR, LLL\}$

We know $P(L) = P(R) = \dfrac{1}{2}$ on the first two runs, and we are told that the mouse is twice as

likely to choose L on the 3rd trial. From this we get $P(L) = \dfrac{2}{3}$ and $P(R) = \dfrac{1}{3}$.

A Venn diagram will help to assign the probabilities.

$P(RRR) = \dfrac{1}{12}$ $\qquad$ $P(RLL) = \dfrac{2}{12}$

$P(RRL) = \dfrac{2}{12}$ $\qquad$ $P(LRL) = \dfrac{2}{12}$

$P(RLR) = \dfrac{1}{12}$ $\qquad$ $P(LLR) = \dfrac{1}{12}$

$P(LRR) = \dfrac{1}{12}$ $\qquad$ $P(LLL) = \dfrac{2}{12}$

(a) $P(E) = P(RRR \cup RRL \cup LRR)$
$\qquad = P(RRR) + P(RRL) + P(LRR)$

$\qquad = \dfrac{1}{12} + \dfrac{2}{12} + \dfrac{1}{12}$

$\qquad = \dfrac{4}{12} = \dfrac{1}{3}$

(b) $P(F) = P(LLL)$

$\qquad = \dfrac{2}{12}$

(c)
$P(G) = P(LRR \cup LRL \cup LLR \cup LLL)$
$= P(LRR) + P(LRL) + P(LLR) + P(LLL)$

$= \dfrac{1}{12} + \dfrac{2}{12} + \dfrac{1}{12} + \dfrac{2}{12}$

$= \dfrac{6}{12} = \dfrac{1}{2}$

(d)
$P(H) = P(RRR \cup RRL \cup LRR \cup LRL)$
$= P(RRR) + P(RRL) + P(LRR) + P(LRL)$

$= \dfrac{1}{12} + \dfrac{2}{12} + \dfrac{1}{12} + \dfrac{2}{12}$

$= \dfrac{6}{12} = \dfrac{1}{2}$

29. $S = \{$ HH, HT, TH, TT$\}$

$E = \{$HH, HT$\}$ $\qquad P(E) = \dfrac{1}{2}$

$F = \{$HT, TT$\}$ $\qquad P(F) = \dfrac{1}{2}$

$E \cap F = \{$HT$\}$ $\qquad P(E \cap F) = \dfrac{1}{4}$

$P(E) \cdot P(F) = P(E \cap F)$

$\dfrac{1}{2} \cdot \dfrac{1}{2} = \dfrac{1}{4}$

S events E and F are independent.

31. Let H denote a child with heart disease, and N denote a child with no heart disease. The sample space is the set of all possible outcomes. The couple has two children.
$S = \{$ HH, HN, NH, NN$\}$

$P(H) = \dfrac{3}{4}$ $\qquad\qquad P(N) = \dfrac{1}{4}$

(a) $\begin{aligned} P(HH) &= P(H \cap H) \\ &= P(H) \cdot P(H) \\ &= \dfrac{3}{4} \cdot \dfrac{3}{4} \\ &= \dfrac{9}{16} \end{aligned}$

(b) $\begin{aligned} P(NN) &= P(N \cap N) \\ &= P(N) \cdot P(N) \\ &= \dfrac{1}{4} \cdot \dfrac{1}{4} \\ &= \dfrac{1}{16} \end{aligned}$

(c) $\begin{aligned} P(HN \cup NH) &= P(H \cap N) + P(N \cap H) \\ &= P(H) \cdot P(N) + P(N) \cdot P(H) \\ &= \dfrac{3}{4} \cdot \dfrac{1}{4} + \dfrac{1}{4} \cdot \dfrac{3}{4} \\ &= \dfrac{6}{16} = \dfrac{3}{8} \end{aligned}$

33.   Define the events H: Heads is thrown.

                            T: Tails is thrown.

The events are independent, but the coin is loaded with tails being 3 times more likely than heads.

$$P(H) = \frac{1}{4} \qquad P(T) = \frac{3}{4}$$

(a) $P(TTT) = P(T) \cdot P(T) \cdot P(T)$

$$= \frac{3}{4} \cdot \frac{3}{4} \cdot \frac{3}{4} = \frac{27}{64}$$

(b)   This problem involves two tasks.

    Task 1. Choose the throw that will result in a tail. This can be done $C(3, 1) = 3$ ways.

    Task 2. Find the probability of getting 2 heads and 1 tail.

$$P(HHT) = P(H) \cdot P(H) \cdot P(T)$$

$$= \frac{1}{4} \cdot \frac{1}{4} \cdot \frac{3}{4} = \frac{3}{64}$$

Using the Multiplication Principle, we find the probability of throwing 2 heads and 1 tail, in any order, is $3 \cdot \dfrac{3}{64} = \dfrac{9}{64}$.

35. Define the events R: Person will recover from the flu. $P(R) = .9$

$\quad\quad\quad\quad\quad\quad\quad\quad\quad D$: Person will not recover from the flu. $P(D) = .1$

The events are independent.

(a) $\quad P(RRRR) = P(R) \cdot P(R) \cdot P(R) \cdot P(R)$

$\quad\quad\quad\quad\quad\quad = .9^4$

$\quad\quad\quad\quad\quad\quad = .6561$

(b) Do (b) in 2 steps.

1. Choose the two people who will recover. This can be done $C(4, 2) = 6$ ways.
2. Determine the probability that a specific 2 recover, say $P(RRDD)$.

$\quad P(RRDD) = P(R) \cdot P(R) \cdot P(D) \cdot P(D)$

$\quad\quad\quad\quad\quad = 0.9 \cdot 0.9 \cdot 0.1 \cdot 0.1$

$\quad\quad\quad\quad\quad = 0.0081$

Using the Multiplication Principle, we find that the probability that exactly 2 will recover is $(6) \cdot (0.0081) = .0486$.

(c) The probability at least 2 will recover can be thought of as the probability exactly 2 will recover, or exactly 3 will recover or all four will recover. Since the three events are mutually exclusive we can add the probabilities. We found two of them parts (a) and (b). We will now find the probability exactly 3 people recover.

Probability exactly two people recover:
1. Choose the 3 people who will recover. This can be done $C(4, 3) = 4$ ways.
2. Determine the probability that a specific 3 people recover, say $P(RRRD)$.

$\quad P(RRRD) = P(R) \cdot P(R) \cdot P(R) \cdot P(D)$

$\quad\quad\quad\quad\quad = 0.9 \cdot 0.9 \cdot 0.9 \cdot 0.1$

$\quad\quad\quad\quad\quad = 0.0729$

Using the Multiplication Principle, we find that the probability that exactly 3 will recover is $(4) \cdot (0.0729) = .2916$.

The probability that at least 2 people recover from the flu is given the probability exactly 2 recover + probability exactly 3 recover + probability all 4 recover
$= .0486 + .2916 + .6561 = .9963$

37. Two marbles are chosen with replacement.

Define the events R: The marble chosen is red. $\quad\quad\quad P(R) = .60$

$\quad\quad\quad\quad\quad\quad\quad\quad W$: The marble chosen is white. $\quad\quad\quad P(W) = .40$

The events are independent.

(a) $\quad P(R \cap R) = P(R) \cdot P(R)$ (b) Probability exactly 1 of the marbles is red is

$\quad\quad\quad\quad\quad\quad = .60 \cdot .60$ $\quad\quad P(R \cap W) + P(W \cap R)$

$\quad\quad\quad\quad\quad\quad = .36$ $\quad\quad\quad\quad = P(R) \cdot P(W) + P(W) \cdot P(R)$

$\quad\quad\quad\quad\quad\quad\quad\quad\quad\quad\quad\quad\quad\quad = .60 \cdot .40 + .40 \cdot .60$

$\quad\quad\quad\quad\quad\quad\quad\quad\quad\quad\quad\quad\quad\quad = .48$

39.   (a) $P(A \mid U) = \dfrac{P(A \cap U)}{P(U)}$ (b) $P(A \mid \overline{U}) = \dfrac{P(A \cap \overline{U})}{P(\overline{U})}$

$= \dfrac{40}{325} = .123$ $= \dfrac{5}{515} = .010$

(c) A and U are not independent.   (d) $P(U) = \dfrac{325}{840} = .387$
      If they were $P(A \mid U)$ would
      equal $P(A) = .054$.

$P(\overline{A}) = \dfrac{795}{840} = .946$

$P(U \cap \overline{A}) = \dfrac{285}{840} = .339$

$P(U) \, P(\overline{A}) = .387 \cdot .946 = .366 \neq P(U \cap \overline{A})$

The events are not independent.

(e) $P(\overline{U}) = \dfrac{515}{840} = .613$ (f) $P(\overline{U}) = \dfrac{515}{840} = .613$

$P(A) = .054$

$P(\overline{A}) = \dfrac{795}{840} = .946$

$P(A \cap \overline{U}) = \dfrac{5}{840} = .006$

$P(\overline{A} \cap \overline{U}) = \dfrac{510}{840} = .607$

$P(\overline{U}) \cdot P(A) = .033 \neq P(A \cap \overline{U})$

$P(\overline{U}) \cdot P(\overline{A}) = .580 \neq P(\overline{A} \cap \overline{U})$

The events are not independent.   The events are not independent.

41.   Let V: The first voter votes for the candidate.
         W: The second voter votes for the candidate.

$P(V) = P(W) = \dfrac{2}{3}$

(a) $P(V \cap W) = P(V)P(W)$ (b) $P(\overline{V} \cap \overline{W}) = P(\overline{V}) \, P(\overline{W})$

$= \dfrac{2}{3} \cdot \dfrac{2}{3}$ $= \dfrac{1}{3} \cdot \dfrac{1}{3}$

$= \dfrac{4}{9}$ $= \dfrac{1}{9}$

(e) $P(V\overline{W} \cup \overline{V}W) = P(V\overline{W}) + P(\overline{V}W)$

$= \dfrac{2}{3} \cdot \dfrac{1}{3} + \dfrac{1}{3} \cdot \dfrac{2}{3}$

$= \dfrac{4}{9}$

43. (a) The probability of obtaining at least one 1 in four throws of a die is the complement of throwing no 1s.

$$P(\overline{E}) = \frac{5}{6}; \quad P(\overline{E}\ \overline{E}\ \overline{E}\ \overline{E}) = \left(\frac{5}{6}\right)^4$$

$$P(E) = 1 - P(\overline{E})$$

$$= 1 - \left(\frac{5}{6}\right)^4$$

$$= .518$$

(b) The probability of throwing a pair of 1s in a throw of 2 dice is $\frac{1}{36}$.

The probability of obtaining at least one pair of 1s in 24 throws of a pair of dice is the complement of obtaining no pairs.

$$P(\overline{E}) = \frac{35}{36}; \quad P(\overline{E}_1\ \overline{E}_2\ \overline{E}_3 \ldots \overline{E}_{24}) = \left(\frac{35}{36}\right)^{24}$$

$$P(E) = 1 - \left(\frac{35}{36}\right)^{24}$$

$$= .491$$

The first event is more likely.

45. If events E and F are mutually exclusive, then $P(E \cap F) = 0$.
If events E and F are independent, then $P(E \cap F) = P(E)\ P(F)$.

If the events are both mutually exclusive and independent then by substitution,

$$P(E)\ P(F) = 0$$

The zero property of multiplication indicates that either

$$P(E) = 0 \ \text{ or } \ P(F) = 0 \text{ or both.}$$

47. If E and F are independent events, meaning $P(E \cap F) = P(E)P(F)$, we can show that $\overline{E}$ and $\overline{F}$ are also independent, which would mean $P(\overline{E} \cap \overline{F}) = P(\overline{E})P(\overline{F})$.

We know $\overline{E} \cap \overline{F} = \overline{E \cup F}$, from De Morgan's properties, and $P(\overline{E \cup F}) = 1 - P(E \cup F)$. So

$$
\begin{aligned}
P(\overline{E} \cap \overline{F}) &= P(\overline{E \cup F}) \\
&= 1 - P(E \cup F) \\
&= 1 - [P(E) + P(F) - P(E \cap F)]
\end{aligned}
$$

Since E and F are independent, we can substitute $P(E)P(F)$ for $P(E \cap F)$.

$$
\begin{aligned}
P(\overline{E} \cap \overline{F}) &= 1 - [P(E) + P(F) - P(E)P(F)] \\
&= 1 - P(E) - P(F) + P(E)P(F)
\end{aligned}
$$

Factoring by grouping, we get

$$
\begin{aligned}
P(\overline{E} \cap \overline{F}) &= [1 - P(E)] - [1 - P(E)]\,P(F) \\
&= [1 - P(E)]\,[1 - P(F)] \\
&= P(\overline{E})P(\overline{F})
\end{aligned}
$$

So events $\overline{E}$ and $\overline{F}$ are also independent.

49. From the definition of conditional probability, we know that

$$
P(E \mid F) = \frac{P(E \cap F)}{P(F)} \quad \text{or} \quad P(E \cap F) = P(F) \cdot P(E \mid F)
$$

and

$$
P(F \mid E) = \frac{P(E \cap F)}{P(E)} \quad \text{or} \quad P(E \cap F) = P(E) \cdot P(F \mid E)
$$

So by transitivity we get

$$
P(E) \cdot P(F \mid E) = P(F) \cdot P(E \mid F)
$$

Since E is independent of F, then $P(E \mid F) = P(E)$, and

$$
\begin{aligned}
P(E) \cdot P(F \mid E) &= P(F) \cdot P(E) \\
P(F \mid E) &= P(F)
\end{aligned}
$$

So F is independent of E.

Chapter 7 Review

True-False Items

1. True   **3.** False   **5.** True   **7.** True

Fill in the Blanks

1. $\dfrac{1}{2}$ **3.** 1; 0

5. for   **7.** mutually exclusive

Review Exercises

1. $S = \{0, 1, 2, 3, 4, 5\}$

3. $S = \{BB, BG, GB, GG\}$

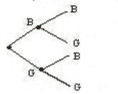

5. $S = \{P, Q, D\}$

$$P(P) = \frac{4}{15} \quad P(Q) = \frac{2}{5} \quad P(D) = \frac{1}{3}$$

7. $S = \{1, 2, 3, 4, 5, 6\}$

Let x denote the probability a 1 occurs.

$P(2) = P(5) = 2x$
$P(1) = P(3) = P(4) = P(6) = x$

$P(1) + P(2) + P(3) + P(4) + P(5) + P(6) = 1$
$x + 2x + x + x + 2x + x = 1$
$$8x = 1$$
$$x = \frac{1}{8}$$

Assign the probabilities

$$P(2) = P(5) = \frac{2}{8} = \frac{1}{4}$$

$$P(1) = P(3) = P(4) = P(6) = \frac{1}{8}$$

9.   (a) $S = \{0, 1, 2, 3, 4\}$

The simple event
0 occurs when there is no girl, BBBB.
1 occurs when there is 1 girl, GBBB, BGBB, BBGB, BBBG.
2 occurs when there are 2 girls, GGBB, GBGB, GBBG, BGGB, BGBG, BBGG.
3 occurs when there are 3 girls, GGGB, GGBG, GBGG, BGGG.
4 occurs when all 4 children are girls, GGGG.

We assign valid probabilities to each of these events, assuming it is equally likely for a child to be born G or B.

$$P(0) = \frac{1}{16}, \quad P(1) = \frac{4}{16} = \frac{1}{4}, \quad P(2) = \frac{6}{16} = \frac{3}{8}, \quad P(3) = \frac{4}{16} = \frac{1}{4}, \quad P(4) = \frac{1}{16}$$

(b)   *i.* $P(0) = \dfrac{1}{16}$ *ii.* $P(2) = \dfrac{3}{8}$

      iii. $P(1) = \dfrac{1}{4}$ *iv.* $1 - P(4) = 1 - \dfrac{1}{16} = \dfrac{15}{16}$

11.   (a) $S = \{0, 1, 2, 3\}$

Let H denote the coin shows heads and T denote the coin shows tails.
The simple event
0 occurs when there is no tail, HHH.
1 occurs when there is 1 tail, THH, HTH, HHT.
2 occurs when there are 2 tails, TTH, THT, HTT.
3 occurs when all 3 tosses are tails TTT.

We assign valid probabilities to each of these events, assuming it is equally likely for a coin to show H or T.

$$P(0) = \frac{1}{8}, \quad P(1) = \frac{3}{8}, \quad P(2) = \frac{3}{8}, \quad P(3) = \frac{1}{8}$$

(b)   *i.* $P(3) = \dfrac{1}{8}$ *ii.* $P(0) = \dfrac{1}{8}$

      iii. $P(2) = \dfrac{3}{8}$ *iv.* $P(2 \cup 3) = P(2) + P(3)$
$$= \frac{3}{8} + \frac{1}{8} = \frac{1}{2}$$

13. Let W denote a white marble is chosen, Y denote a yellow marble is chosen, R denote a red marble is chosen, and B denote a blue marble is chosen.

(a) $P(BB) = P(B \text{ on } 1^{ST})P(B \text{ on } 2^{ND})$

$$= \left(\frac{5}{14}\right)\left(\frac{4}{13}\right) = \frac{10}{91}$$

(b) The probability exactly one is blue is the union of two mutually exclusive events. The probability is

$$P(B\overline{B} \cup \overline{B}B) = \left(\frac{5}{14}\right)\left(\frac{9}{13}\right) + \left(\frac{9}{14}\right)\left(\frac{5}{13}\right)$$

$$= \frac{45}{91}$$

(c) The probability at least one is blue is the union of both are blue and exactly one is blue. The probability (using the results from parts a and b) is

$$\frac{10}{91} + \frac{45}{91} = \frac{55}{91}$$

15. (a) $P(A \cup B) = P(A) + P(B) - P(A \cap B)$
$$= .3 + .5 - .2$$
$$= .6$$

(b) $P(\overline{A}) = 1 - A$
$$= 1 - .3$$
$$= .7$$

(c) $P(\overline{A \cup B}) = 1 - P(A \cup B)$
$$= 1 - 0.6$$
$$= .4$$

(d) $P(\overline{A} \cup \overline{B}) = 1 - P(A \cap B)$
$$= 1 - .2$$
$$= .8$$

17. (a) $P(3) = \dfrac{c(3)}{c(S)} = \dfrac{84}{400} = \dfrac{21}{100}$

(b) $P(5) = \dfrac{c(5)}{c(S)} = \dfrac{92}{400} = \dfrac{23}{100}$

(c) $P(6) = \dfrac{c(6)}{c(S)} = \dfrac{73}{400}$

19. (a) $P(\overline{E}) = 1 - P(E)$
$$= 1 - .65$$
$$= .35$$

(b) $P(E \cup F) = P(E) + P(F) - P(E \cap F)$
$$= .65 + .40 - .30$$
$$= .75$$

(c) E and F are not mutually exclusive. If they were $P(E \cap F)$ would equal 0, but in this problem we are told that $P(E \cap F) = .30$.

21. (a) $P(\overline{E}) = 1 - P(E)$
$$= 1 - \frac{1}{2} = \frac{1}{2}$$

(b) $P(F) = P(E \cup F) + P(E \cap F) - P(E)$
$$= \frac{5}{8} + \frac{1}{3} - \frac{1}{2} = \frac{11}{24}$$

(c) $P(\overline{F}) = 1 - P(F)$
$$= 1 - \frac{11}{24} = \frac{13}{24}$$

23. (a) $P(\overline{E}) = 1 - P(E)$
   $= 1 - .30$
   $= .7$

(b) $P(\overline{F}) = 1 - P(F)$
   $= 1 - .45$
   $= .55$

(c) $P(E \cap F) = 0$

(d) $P(E \cup F) = P(E) + P(F)$
   $= .30 + .45 = .75$

(e) $P(\overline{E \cap F}) = 1 - P(E \cap F)$
   $= 1 - 0 = 1$

(f) $P(\overline{E \cup F}) = 1 - P(E \cup F)$
   $= 1 - .75$
   $= .25$

(g) $P(\overline{E} \cup \overline{F}) = P(\overline{E \cap F})$
   $= 1$

(h) $P(\overline{E} \cap \overline{F}) = P(\overline{E \cup F})$
   $= .25$

25. $E \cap F =$ The sum of the faces is 8

$P(E) = P(E \cap F) = \dfrac{5}{36}$

$P(F) = P(\text{sum} = 2 \text{ or sum} = 4 \text{ or sum} = 6 \text{ or sum} = 8 \text{ or sum} = 10 \text{ or sum} = 12)$

$= \dfrac{1}{36} + \dfrac{3}{36} + \dfrac{5}{36} + \dfrac{5}{36} + \dfrac{3}{36} + \dfrac{1}{36} = \dfrac{18}{36} = \dfrac{1}{2}$

$P(E \cup F) = P(E) + P(F) - P(E \cap F)$

$= \dfrac{5}{36} + \dfrac{1}{2} - \dfrac{5}{36} = \dfrac{1}{2}$

27. (a) The events are not equally likely.
(b) The outcome 0 has the highest probability.

(c) $P(F) = P(3, 3) \cdot \left(\dfrac{4}{8}\right)\left(\dfrac{3}{8}\right)\left(\dfrac{1}{8}\right)$

$= 6 \cdot \left(\dfrac{4}{8}\right)\left(\dfrac{3}{8}\right)\left(\dfrac{1}{8}\right) = \dfrac{9}{64}$

29. The sample space is the set of integers from 2 to 12, representing the sum when 2 dice are thrown.

Define the events: E: The sum is 5. $P(E) = \dfrac{4}{36}$

   F: The sum is 7. $P(F) = \dfrac{6}{36}$

   G: The sum is 9. $P(G) = \dfrac{4}{36}$

The events are mutually exclusive.
$P(E \cup F \cup G) = P(E) + P(F) + P(G)$

$= \dfrac{4}{36} + \dfrac{6}{36} + \dfrac{4}{36}$

$= \dfrac{14}{36} = \dfrac{7}{18}$

31. Define the event E: A 5 is obtained. $\qquad P(E) = \dfrac{1}{6}$

The odds in favor of E are defined as the ratio $\dfrac{P(E)}{P(\overline{E})}$.

The odds in favor of a 5 when a fair die is thrown is $\dfrac{1}{5}$ or 1 to 5.

33. Define the event E: A Bears win the title. The odds in favor of the Bears winning is 7 to 6.

$P(E) = \dfrac{7}{7+6} = \dfrac{7}{13}$

35. The sample space $S = \{HH, HT, TH, TT\}$. The probabilities assigned to the sample space

are $P(HH) = \dfrac{1}{16}$, $P(HT) = \dfrac{3}{16}$, $P(TH) = \dfrac{3}{16}$, and $P(TT) = \dfrac{9}{16}$.

$P(E \cap F) = \dfrac{3}{16}$; $\quad P(E) \cdot P(F) = \dfrac{1}{4} \cdot \dfrac{3}{4} = \dfrac{3}{16}$

Since $P(E \cap F) = P(E) \cdot P(F)$ the events are independent.

37. We define events $\quad E$: a student fails mathematics, $\quad P(E) = 0.38$; and

$\qquad\qquad\qquad\qquad F$: a student fails physics, $\qquad P(F) = 0.27$

We are given $P(E \cap F) = 0.09$.

(a) $\quad P(E \mid F) = \dfrac{P(E \cap F)}{P(F)} = \dfrac{0.09}{0.27} = \dfrac{1}{3}$ $\qquad$ (b) $\quad P(F \mid E) = \dfrac{P(E \cap F)}{P(E)} = \dfrac{0.09}{0.38} = \dfrac{9}{38}$

(c) $\quad P(E \cup F) = P(E) + P(F) - P(E \cap F) = 0.38 + 0.27 - 0.09 = 0.56$

$\qquad$ The probability a student fails either mathematics or physics is 0.56.

39. Define events E: an even sum occurs on the 1st toss, F: a sum less than 6 occurs on the 2nd toss, and G: a sum of 7 occurs on the 3rd toss. The events are independent so $P(E \cap F \cap G) = P(E) \cdot P(F) \cdot P(G)$.

$P(E) = P(\text{sum is 2 or 4 or 6 or 8 or 10 or 12}) = \dfrac{1}{36} + \dfrac{3}{36} + \dfrac{5}{36} + \dfrac{5}{36} + \dfrac{3}{36} + \dfrac{1}{36} = \dfrac{18}{36} = \dfrac{1}{2}$

$P(F) = P(\text{ the sum is 2 or 3 or 4 or 5}) = \dfrac{1}{36} + \dfrac{2}{36} + \dfrac{3}{36} + \dfrac{4}{36} = \dfrac{10}{36} = \dfrac{5}{18}$

$P(G) = \dfrac{1}{6}$

$P(E \cap F \cap G) = \dfrac{1}{2} \cdot \dfrac{5}{18} \cdot \dfrac{1}{6} = \dfrac{5}{216}$

41. Define events E: a person has blue eyes, $P(E) = 0.25$;
F: a person has brown eyes, $P(F) = 0.75$
G: a person is left handed.

We are told that $P(G \mid E) = 0.10$ and $P(G \mid F) = 0.05$.
Notice that $P(E) + P(F) = 0.25 + 0.75 = 1.00$, this indicates that no other eye-color is possible. This is important for part (b).

(a) $P(E \cap G) = P(G \mid E) \cdot P(E)$ (b) $P(G) = P(G \cap E) + P(G \cap F)$
    $= (0.10) \cdot (0.25)$ $= 0.025 + P(G \mid F) \cdot P(F)$
    $= 0.025$ $= 0.025 + (0.05) \cdot (0.75)$
The probability a person is $= 0.0625$
blue-eyed and left handed is 0.025.   The probability a person is left handed
                                           is 0.0625.

(c) $P(E \mid G) = \dfrac{P(E \cap G)}{P(G)} = \dfrac{0.025}{0.0625} = \dfrac{2}{5}$. The probability a person is blue-eyed given the

person is left handed is 0.40.

43. Define events E: a student took Form A,
F: a student took Form B,
G: a student scored over 80%,
H: a student scored under 80%.

(a) $P(E \mid G) = \dfrac{P(E \cap G)}{P(G)} = \dfrac{8}{20} = \dfrac{2}{5}$ (b) $P(G \mid E) = \dfrac{P(E \cap G)}{P(E)} = \dfrac{8}{40} = \dfrac{1}{5}$

The probability that a student who   The probability that a student who took
scored over 80% took form A is $\dfrac{2}{5}$.   form A scored over 80% is $\dfrac{1}{5}$.

(c) To show G and E are independent we need to show $P(E \cap G) = P(E) \cdot P(G)$.

$P(E \cap G) = \dfrac{8}{100}$; $P(E) = \dfrac{40}{100}$;   and $P(G) = \dfrac{20}{100}$

$P(E) \cdot P(G) = \dfrac{40}{100} \cdot \dfrac{20}{100} = \dfrac{8}{100} = P(E \cap G)$ So G and E are independent events.

(d) To determine if G and F are independent events we must check if
$P(F \cap G) = P(F) \cdot P(G)$.

$P(F \cap G) = \dfrac{12}{100}$; $P(F) = \dfrac{60}{100}$;   and $P(G) = \dfrac{20}{100}$

$P(F) \cdot P(G) = \dfrac{60}{100} \cdot \dfrac{20}{100} = \dfrac{12}{100} = P(F \cap G)$ So G and F are independent events.

45. The probability at least one person gets the correct letter is the complement of the probability everyone gets an incorrect letter.

Define event E: everyone get an incorrect letter.

$P(E) = \dfrac{c(E)}{c(S)} = \dfrac{2}{3!} = \dfrac{1}{3}$. So the probability at least one person gets the correct letter is

$1 - P(E) = 1 - \dfrac{1}{3} = \dfrac{2}{3}$.

47. The number of elements in the sample space S is equal to the number of ways 4 jars can be chosen from 72 jars of jam, or $C(72, 4) = 1{,}028{,}790$.

(a) Define E as the event, "All 4 jars are underweight." E can occur $C(10, 4)$ ways.

$$P(E) = \frac{C(10,\ 4)}{C(72,\ 4)} = \frac{210}{1{,}028{,}790} = 0.0002$$

(b) Define F as the event, "2 jars are underweight." F can occur $C(10, 2) \cdot C(62, 2)$ ways.

$$P(F) = \frac{C(10,\ 2) \cdot C(62,\ 2)}{C(72,\ 4)} = \frac{45 \cdot 1891}{1{,}028{,}790} = 0.0827$$

(c) Define G as the event, "At most 1 jar is underweight." The event G is equivalent to selecting either 0 or 1 underweight jars. Since the events are mutually exclusive, the sum of their probabilities will give the probability of G.

$P(G) = P(\text{No underweight jars}) + P(1 \text{ underweight jar})$

$$= \frac{C(10,\ 0) \cdot C(62,\ 4)}{C(72,\ 4)} + \frac{C(10,\ 1) \cdot C(62,\ 3)}{C(72,\ 4)}$$

$$= 0.5422 + 0.3676 = 0.9098$$

49. Define E as the event, " Each person has a different birth day of the month."

$c(S) = 31^{15}$ and $c(E) = P(31, 15)$, so $P(E) = \dfrac{P(31,\ 15)}{31^{15}} = 0.0167$

The probability that no two people in a room of 15 people have the birthdays on the same day of the month is 0.0167.

51. E and F are independent events,

$P(E \mid F) = \dfrac{P(E \cap F)}{P(F)} = \dfrac{0.2}{0.4} = \dfrac{1}{2}$

53. Define E: the basketball player makes a free throw. $P(E) = 0.7$
The probability the player misses a free throw, $P(\overline{E}) = 1 - P(E) = 1 - 0.7 = 0.3$.

(a) The probability the player misses a free throw and then makes 3 in a row is

$P(\overline{E} \cap E \cap E \cap E) = P(\overline{E}) \cdot P(E) \cdot P(E) \cdot P(E) = (0.3) \cdot (0.7)^3 = 0.1029$

(b) The probability of probability the player makes 10 free throws in a row is

$[P(E)]^{10} = (0.7)^{10} = 0.0282$

55. Define E: the car is black, F: the car is red, G: the interior is tan.
We make a table to organize the data.

	Black Car	Red Car	Total
Tan Interior	2	6	8
Other Interior	6	6	12
Total	8	12	20

$$P(E \mid G) = \frac{P(E \cap G)}{P(G)} = \frac{c(E \cap G)}{c(G)} = \frac{2}{8} = \frac{1}{4}$$

Chapter 7 Project

1. $P(E \mid H)$ is the probability a person who tossed a head is truly overweight. This probability will give the proportion of overweight students in the school.

3. $P(E \mid T) = 1$ Since respondent was instructed to answer "Yes" if a coin toss resulted in a tail. (No one who flipped a tail answered, "No.")

$$P(E) = P(H) \cdot P(E \mid H) + P(T) \cdot P(E \mid T)$$

$$P(E \mid H) = \frac{P(E) - P(T) \cdot P(E \mid T)}{P(H)}$$

5. $P(A \mid H) = \dfrac{P(A \cap H)}{P(H)}$

7. $P(C \mid H) = \dfrac{P(C \cap H)}{P(H)}$

Mathematical Questions from Professional Exams

1. (b) $P \cap Q = \emptyset$ means $P(P \cap Q) = 0$, but we were told $P(P) > 0$ and $P(Q) > 0$ so $P(P) \cdot P(Q) > 0$ which indicates P and Q are not independent.

3. (b) $\left(\dfrac{26}{52}\right)\left(\dfrac{25}{51}\right)\left(\dfrac{24}{50}\right) = \dfrac{2}{17}$

5. (b) The probability that all are not the same is the complement of the probability that the same face shows each time.

$$P(\text{ at least one different}) = 1 - P(\text{all faces the same}) > 0.999$$
$$P(\text{all faces the same}) < 0.001$$
$$\left(\frac{1}{6}\right)^n < 0.001$$
$$n \log\left(\frac{1}{6}\right) < \log(.001)$$
$$n > \frac{\log(0.001)}{\log\left(\dfrac{1}{6}\right)} > 3.9$$

7. (b) Define E: the first roll is an even number
   F: the sum of the rolls is 8

$$P(F \mid E) = \frac{P(E \cap F)}{P(E)} = \frac{\dfrac{3}{36}}{\dfrac{1}{2}} = \frac{1}{6}$$

9. (c) $P(S \cap T) = P\left(S \cap \overline{T}\right) = P\left(\overline{S} \cap T\right) = p$
   $P(S \cup T) = P(S) + P(T) - P(S \cap T)$
   $\qquad\qquad = [P\left(S \cap \overline{T}\right) + P(S \cap T)] + [P\left(\overline{S} \cap T\right) + P(S \cap T)] - P(S \cap T)$
   $\qquad\qquad = p + p + p + p - p = 3p$

Chapter 8

Additional Probability Topics

8.1 Bayes' Formula

1. $P(E \mid A) = 0.4$ **3.** $P(E \mid B) = 0.2$

5. $P(E \mid C) = 0.7$

7. $P(E) = P(E \cap A) + P(E \cap B) + P(E \cap C)$
$= (0.3)(0.4) + (0.6)(0.2) + (0.1)(0.7)$
$= 0.31$

9. $P(A \mid E) = \dfrac{P(A \cap E)}{P(E)} = \dfrac{P(E \mid A) \cdot P(A)}{P(E)} = \dfrac{(0.4)(0.3)}{0.31} = \dfrac{12}{31} = 0.387$

11. $P(C \mid E) = \dfrac{P(C \cap E)}{P(E)} = \dfrac{P(E \mid C) \cdot P(C)}{P(E)} = \dfrac{(0.7)(0.1)}{0.31} = \dfrac{7}{31} = 0.226$

13. $P(B \mid E) = \dfrac{P(B \cap E)}{P(E)} = \dfrac{P(E \mid B) \cdot P(B)}{P(E)} = \dfrac{(0.2)(0.6)}{0.31} = \dfrac{12}{31} = 0.387$

15.

$$P(E) = P(E \cap A_1) + P(E \cap A_2)$$
$$= P(A_1) \cdot P(E \mid A_1) + P(A_2) \cdot P(E \mid A_2)$$
$$= (0.4) \cdot (0.03) + (0.6) \cdot (0.02)$$
$$= 0.024$$

17.
$$P(E) = P(E \cap A_1) + P(E \cap A_2) + P(E \cap A_3)$$
$$= P(A_1) \cdot P(E \mid A_1) + P(A_2) \cdot P(E \mid A_2) + P(A_3) \cdot P(E \mid A_3)$$
$$= (0.6) \cdot (0.01) + (0.2) \cdot (0.03) + (0.2) \cdot (0.02)$$
$$= 0.016$$

19.
$$P(A_1 \mid E) = \frac{P(A_1 \cap E)}{P(E)} = \frac{P(A_1) \cdot P(E \mid A_1)}{P(E)} = \frac{(0.4)(0.03)}{0.024} = \frac{12}{24} = \frac{1}{2} = 0.5$$
$$P(A_2 \mid E) = \frac{P(A_2 \cap E)}{P(E)} = \frac{P(A_2) \cdot P(E \mid A_2)}{P(E)} = \frac{(0.6)(0.02)}{0.024} = \frac{12}{24} = \frac{1}{2} = 0.5$$

21.
$$P(A_1 \mid E) = \frac{P(A_1 \cap E)}{P(E)} = \frac{P(A_1) \cdot P(E \mid A_1)}{P(E)} = \frac{(0.6)(0.01)}{0.016} = \frac{6}{16} = \frac{3}{8} = 0.375$$
$$P(A_2 \mid E) = \frac{P(A_2 \cap E)}{P(E)} = \frac{P(A_2) \cdot P(E \mid A_2)}{P(E)} = \frac{(0.2)(0.03)}{0.016} = \frac{6}{16} = \frac{3}{8} = 0.375$$
$$P(A_3 \mid E) = \frac{P(A_3 \cap E)}{P(E)} = \frac{P(A_3) \cdot P(E \mid A_3)}{P(E)} = \frac{(0.2)(0.02)}{0.016} = \frac{4}{16} = \frac{1}{4} = 0.250$$

23.
$$P(A_1 \mid E) = \frac{P(A_1 \cap E)}{P(E)} = \frac{P(A_1) \cdot P(E \mid A_1)}{P(E)} = \frac{(0.4)(0.02)}{0.029} = \frac{8}{29} = 0.276$$
$$P(A_2 \mid E) = \frac{P(A_2 \cap E)}{P(E)} = \frac{P(A_2) \cdot P(E \mid A_2)}{P(E)} = \frac{(0.5)(0.04)}{0.029} = \frac{20}{29} = 0.690$$
$$P(A_3 \mid E) = \frac{P(A_3 \cap E)}{P(E)} = \frac{P(A_3) \cdot P(E \mid A_3)}{P(E)} = \frac{(0.1)(0.01)}{0.029} = \frac{1}{29} = 0.034$$

25.
$$P(A_2 \mid E) = \frac{P(A_2 \cap E)}{P(E)} = \frac{P(A_2) \cdot P(E \mid A_2)}{P(E)} = \frac{(0.2)(0)}{0.31} = 0$$
$$P(A_3 \mid E) = \frac{P(A_3 \cap E)}{P(E)} = \frac{P(A_3) \cdot P(E \mid A_3)}{P(E)} = \frac{(0.1)(0.2)}{0.31} = \frac{2}{31} = 0.065$$
$$P(A_4 \mid E) = \frac{P(A_4 \cap E)}{P(E)} = \frac{P(A_4) \cdot P(E \mid A_4)}{P(E)} = \frac{(0.3)(0)}{0.31} = 0$$
$$P(A_5 \mid E) = \frac{P(A_5 \cap E)}{P(E)} = \frac{P(A_5) \cdot P(E \mid A_5)}{P(E)} = \frac{(0.1)(0.2)}{0.31} = \frac{2}{31} = 0.065$$

27. Define the events: U_1: Jar 1 is selected.

U_2: Jar 2 is selected.

U_3: Jar 3 is selected.

E: The ball is red.

Since the jar is selected at random, $P(U_1) = P(U_2) = P(U_3) = \dfrac{1}{3}$.

To solve this using Bayes' Formula, we need to determine $P(E)$.

$$P(E) = P(E \cap U_1) + P(E \cap U_2) + P(E \cap U_3)$$
$$= P(U_1) \cdot P(E \mid U_1) + P(U_2) \cdot P(E \mid U_2) + P(U_3) \cdot P(E \mid U_3)$$
$$= \left(\frac{1}{3}\right) \cdot \left(\frac{5}{16}\right) + \left(\frac{1}{3}\right) \cdot \left(\frac{3}{16}\right) + \left(\frac{1}{3}\right) \cdot \left(\frac{7}{16}\right) = \frac{15}{48} = \frac{5}{16} = 0.3125$$

$$P(U_1 \mid E) = \frac{P(U_1 \cap E)}{P(E)} = \frac{P(U_1) \cdot P(E \mid U_1)}{P(E)} = \frac{\left(\frac{1}{3}\right)\left(\frac{5}{16}\right)}{\left(\frac{5}{16}\right)} = \frac{1}{3} = 0.3333$$

$$P(U_2 \mid E) = \frac{P(U_2 \cap E)}{P(E)} = \frac{P(U_2) \cdot P(E \mid U_2)}{P(E)} = \frac{\left(\frac{1}{3}\right)\left(\frac{3}{16}\right)}{\left(\frac{5}{16}\right)} = \frac{1}{5} = 0.20$$

$$P(U_3 \mid E) = \frac{P(U_3 \cap E)}{P(E)} = \frac{P(U_3) \cdot P(E \mid U_3)}{P(E)} = \frac{\left(\frac{1}{3}\right)\left(\frac{7}{16}\right)}{\left(\frac{5}{16}\right)} = \frac{7}{15} = 0.4667$$

29. Define events: A_1: The person chosen is male.

A_2: The person chosen is female.

E: The person is colorblind.

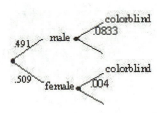

$$P(A_1) = 0.491 \quad P(A_2) = 0.509$$

$$P(E \mid A_1) = \frac{1}{12} = 0.0833 \quad P(E \mid A_2) = 0.004$$

$$P(A_1 \mid E) = \frac{P(A_1 \cap E)}{P(E)} = \frac{P(A_1) \cdot P(E \mid A_1)}{P(E)}$$

$$P(E) = P(E \cap A_1) + P(E \cap A_2)$$
$$= P(A_1) \cdot P(E \mid A_1) + P(A_2) \cdot P(E \mid A_2)$$
$$= 0.491 \cdot (0.0833) + 0.509 \cdot (0.004)$$
$$= 0.043$$

$$P(A_1 \mid E) = \frac{P(A_1 \cap E)}{P(E)} = \frac{P(A_1) \cdot P(E \mid A_1)}{P(E)} = \frac{(.491)(0.08333)}{.04294} = 0.953$$

The probability that a colorblind person is male is 0.953.

31. Define events: D: The person is a Democrat.
R: The person is a Republican.
I: The person is an Independent.
E: The person voted.

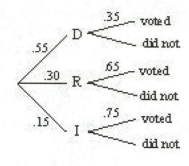

$$P(E) = P(E \cap D) + P(E \cap R) + P(E \cap I)$$
$$= P(D) \cdot P(E \mid D) + P(R) \cdot P(E \mid R) + P(I) \cdot P(E \mid I)$$
$$= (0.55) \cdot (0.35) + (0.30) \cdot (0.65) + (0.15) \cdot (0.75)$$
$$= 0.5$$

$$P(D \mid E) = \frac{P(D \cap E)}{P(E)} = \frac{P(D) \cdot P(E \mid D)}{P(E)} = \frac{(0.55)(0.35)}{0.5} = 0.385$$

The probability that a voter was a Democrat was 0.385.

$$P(R \mid E) = \frac{P(R \cap E)}{P(E)} = \frac{P(R) \cdot P(E \mid R)}{P(E)} = \frac{(0.30)(0.65)}{0.5} = 0.39$$

The probability that a voter was a Republican was 0.390.

$$P(I \mid E) = \frac{P(I \cap E)}{P(E)} = \frac{P(I) \cdot P(E \mid I)}{P(E)} = \frac{(0.15)(0.75)}{0.5} = 0.225$$

The probability that a voter was an Independent was 0.225.

33. Define events: A_1: The soil is rock.
A_2: The soil is clay.
A_3: The is soil is sand.
E: The geological test is positive.

$$P(A_1) = 0.53 \qquad P(A_2) = 0.21 \qquad P(A_3) = 0.26$$
$$P(E \mid A_1) = 0.35 \qquad P(E \mid A_2) = 0.48 \qquad P(E \mid A_3) = 0.75$$
$$P(E) = P(A_1) \cdot P(E \mid A_1) + P(A_2) \cdot P(E \mid A_2) + P(A_3) \cdot P(E \mid A_3)$$
$$= (0.53) \cdot (0.35) + (0.21) \cdot (0.48) + (0.26) \cdot (0.75)$$
$$= 0.4813$$

$$P(A_1 \mid E) = \frac{P(A_1 \cap E)}{P(E)} = \frac{P(A_1) \cdot P(E \mid A_1)}{P(E)} = \frac{(0.53)(0.35)}{0.4813} = 0.385$$

The probability that the soil is rock given the test is positive is 0.385.

$$P(A_2 \mid E) = \frac{P(A_2 \cap E)}{P(E)} = \frac{P(A_2) \cdot P(E \mid A_2)}{P(E)} = \frac{(0.21)(0.48)}{0.4813} = 0.209$$

There is a 0.209 probability that the soil is clay given that the test is positive.

$$P(A_3 \mid E) = \frac{P(A_3 \cap E)}{P(E)} = \frac{P(A_3) \cdot P(E \mid A_3)}{P(E)} = \frac{(0.26)(0.75)}{0.4813} = 0.405$$

The probability is 0.405 that the soil is sand given that the test came out positive.

35. Define events N: The person is from the Northeast.
$\quad\quad\quad\quad\quad\quad$ S: The person is from the South.
$\quad\quad\quad\quad\quad\quad$ M: The person is from the Midwest.
$\quad\quad\quad\quad\quad\quad$ W: The person is from the West.
$\quad\quad\quad\quad\quad\quad$ E: The person votes Republican.

$$P(N) = 0.40 \quad\quad P(S) = 0.10 \quad\quad P(M) = 0.25 \quad\quad P(W) = 0.25$$
$$P(E \mid N) = 0.40 \quad\quad P(E \mid S) = 0.56 \quad\quad P(E \mid M) = 0.48 \quad\quad P(E \mid W) = 0.52$$

$$P(E) = P(E \cap N) + P(E \cap S) + P(E \cap M) + P(E \cap W)$$
$$= P(N) \cdot P(E \mid N) + P(S) \cdot P(E \mid S) + P(M) \cdot P(E \mid M) + P(W) \cdot P(E \mid W)$$
$$= (0.40) \cdot (0.40) + (0.10) \cdot (0.56) + (0.25) \cdot (0.48) + (0.25) \cdot (0.52) = 0.466$$

The probability that a person chosen at random votes Republican is 0.466.

$$P(N \mid E) = \frac{P(N \cap E)}{P(E)} = \frac{P(N) \cdot P(E \mid N)}{P(E)} = \frac{(0.40)(0.40)}{0.466} = 0.343$$

If a person has voted Republican, there is a 34.3% probability the person is from the Northeast.

37. Define events A_1: The nurse forgets to give Mr. Brown his pill.
$\quad\quad\quad\quad\quad\quad$ A_2: The nurse remembers to give Mr. Brown his pill.
$\quad\quad\quad\quad\quad\quad$ E: Mr. Brown dies.

$$P(A_1) = \frac{2}{3} \quad\quad P(A_2) = \frac{1}{3} \quad\quad P(E \mid A_1) = \frac{3}{4} \quad\quad P(E \mid A_2) = \frac{1}{3}$$

$$P(E) = P(A_1) \cdot P(E \mid A_1) + P(A_2) \cdot P(E \mid A_2)$$
$$= \frac{2}{3} \cdot \frac{3}{4} + \frac{1}{3} \cdot \frac{1}{3} = \frac{11}{18}$$

$$P(A_1 \mid E) = \frac{P(A_1 \cap E)}{P(E)} = \frac{P(A_1) \cdot P(E \mid A_1)}{P(E)} = \frac{\frac{2}{3} \cdot \frac{3}{4}}{\frac{11}{18}} = \frac{9}{11} = 0.8182$$

The probability the nurse forgot to give Mr. Brown his pill given that he died is $\frac{9}{11}$.

39. Define events: A_1: The student is majoring in engineering. $P(A_1) = 0.26$
A_2: The student is majoring in business. $P(A_2) = 0.30$
A_3: The student is majoring in education. $P(A_3) = 0.09$
A_4: The student is majoring in social science. $P(A_4) = 0.12$
A_5: The student is majoring in natural science. $P(A_5) = 0.12$
A_6: The student is majoring in humanities. $P(A_6) = 0.09$
A_7: The customer is student is majoring in something else. $P(A_7) = 0.02$
E: The student is female.

$P(E \mid A_1) = 0.40$ $P(E \mid A_2) = 0.35$ $P(E \mid A_3) = 0.80$ $P(E \mid A_4) = 0.52$
$P(E \mid A_5) = 0.56$ $P(E \mid A_6) = 0.65$ $P(E \mid A_7) = 0.51$

$$P(E) = P(A_1){\cdot}P(E \mid A_1) + P(A_2){\cdot}P(E \mid A_2) + P(A_3){\cdot}P(E \mid A_3) + P(A_4){\cdot}P(E \mid A_4) +$$
$$P(A_5){\cdot}P(E \mid A_5) + P(A_6){\cdot}P(E \mid A_6) + P(A_7){\cdot}P(E \mid A_7)$$
$$= (0.26){\cdot}(0.40) + (0.30){\cdot}(0.35) + (0.09){\cdot}(0.80) + (0.12){\cdot}(0.52) +$$
$$(0.12){\cdot}(0.56) + (0.09){\cdot}(0.65) + (0.02){\cdot}(0.51)$$
$$= 0.4793$$

$$P(A_1 \mid E) = \frac{P(A_1 \cap E)}{P(E)} = \frac{P(A_1) \cdot P(E \mid A_1)}{P(E)} = \frac{(0.26) \cdot (0.40)}{0.4793} = 0.217$$

The probability that a female student is majoring in engineering is 0.217

41. Define events A_1: A patient has the disease.

A_2: A patient does not have the diseases.

E: The test is positive.

(a) If 20% of the population has the disease, we have

$$P(A_1) = 0.20 \qquad P(A_2) = 0.80 \qquad P(E \mid A_1) = 0.97 \qquad P(E \mid A_2) = 0.04$$

$$\begin{aligned} P(E) &= P(A_1) \cdot P(E \mid A_1) + P(A_2) \cdot P(E \mid A_2) \\ &= (0.20) \cdot (0.97) + (0.80) \cdot (0.04) = 0.226 \end{aligned}$$

$$P(A_1 \mid E) = \frac{P(A_1 \cap E)}{P(E)} = \frac{P(A_1) \cdot P(E \mid A_1)}{P(E)} = \frac{(0.20)(0.97)}{0.226} = 0.858$$

The probability that a patient with a positive test actually has the disease is 0.858.

(b) If 4% of the population has the disease, we have

$$P(A_1) = 0.04 \qquad P(A_2) = 0.96 \qquad P(E \mid A_1) = 0.97 \qquad P(E \mid A_2) = 0.04$$

$$\begin{aligned} P(E) &= P(A_1) \cdot P(E \mid A_1) + P(A_2) \cdot P(E \mid A_2) \\ &= (0.04) \cdot (0.97) + (0.96) \cdot (0.04) = 0.0772 \end{aligned}$$

$$P(A_1 \mid E) = \frac{P(A_1 \cap E)}{P(E)} = \frac{P(A_1) \cdot P(E \mid A_1)}{P(E)} = \frac{(0.04)(0.97)}{0.0772} = 0.503$$

In this population a patient with a positive test has a 0.503 probability of having the disease.

(c) Define event F: A person has 2 positive tests.

Assuming the tests are independent, we get

$$P(F \mid A_1) = P(E \mid A_1) \cdot P(E \mid A_1) = 0.97^2 = 0.9409 \quad \text{and}$$
$$P(F \mid A_2) = P(E \mid A_2) \cdot P(E \mid A_2) = 0.04^2 = 0.0016$$

$$\begin{aligned} P(F) &= P(A_1) \cdot P(F \mid A_1) + P(A_2) \cdot P(F \mid A_2) \\ &= (0.04) \cdot (0.9409) + (0.96) \cdot (0.0016) = 0.0392 \end{aligned}$$

$$P(A_1 \mid F) = \frac{P(A_1 \cap F)}{P(F)} = \frac{P(A_1) \cdot P(F \mid A_1)}{P(F)} = \frac{(0.04)(0.9409)}{0.0392} = 0.961$$

A patient with 2 positive tests has a 96.1% chance of having the disease.

43. When F is a subset of E, $E \cap F = F$, and $P(E \cap F) = P(F)$.

By the definition of conditional probability, $P(E \mid F) = \dfrac{P(E \cap F)}{P(F)}$ provided $P(F) \neq 0$.

So $P(E \mid F) = \dfrac{P(F)}{P(F)} = 1$.

8.2 The Binomial Probability Model

1. $b(7, 4; .20) = \dbinom{7}{4}(.2)^4(.8)^3 = (35)(.0016)(.512) = 0.0287$

3. $b(15, 8; .80) = \dbinom{15}{8}(.80)^8(.2)^7 = (6435)(.16777)(1.28 \times 10^{-5}) = 0.01382$

5. $b(15, 10; \frac{1}{2}) = \dbinom{15}{10}\left(\frac{1}{2}\right)^{10}\left(\frac{1}{2}\right)^5 = 0.09164$

7. $b(15, 3; .3) + b(15, 2; .3) + b(15, 1; .3) + b(15, 0; .3)$

$$= \dbinom{15}{3}(.3)^3(.7)^{12} + \dbinom{15}{2}(.3)^2(.7)^{13} + \dbinom{15}{1}(.3)^1(.7)^{14} + \dbinom{15}{0}(.3)^0(.7)^{15}$$

$$= 0.17004 + 0.09156 + 0.03052 + 0.00475 = 0.29687$$

9. $b\left(3, 2; \frac{1}{3}\right) = \dbinom{3}{2}\left(\frac{1}{3}\right)^2\left(\frac{2}{3}\right)^1 = \frac{2}{9}$

11. $b\left(3, 0; \frac{1}{6}\right) = \dbinom{3}{0}\left(\frac{1}{6}\right)^0\left(\frac{5}{6}\right)^3 = \frac{125}{216}$

13. $b\left(5, 3; \frac{2}{3}\right) = \dbinom{5}{3}\left(\frac{2}{3}\right)^3\left(\frac{1}{3}\right)^2 = \frac{80}{243}$

15. $n = 10, k = 6, p = .3$

$$b(10, 6; .3) = \dbinom{10}{6}(.3)^6(.7)^4 = 0.0368$$

17. $n = 12, k = 9, p = .8$

$$b(12, 9; .8) = \dbinom{12}{9}(.8)^9(.2)^3 = 0.2362$$

19. $n = 8, \; p = .30$

The probability P of at least 5 successes is the probability of 5 or 6 or 7 or 8 successes. Since the events are mutually exclusive we can add the probabilities.

$$P = b(8, 5; .30) + b(8, 6; .30) + b(8, 7; .30) + b(8, 8; .30)$$

$$= \dbinom{8}{5}(.30)^5(.70)^3 + \dbinom{8}{6}(.30)^6(.70)^2 + \dbinom{8}{7}(.30)^7(.70)^1 + \dbinom{8}{8}(.30)^8(.70)^0$$

$$= 0.0467 + 0.0100 + 0.0012 + 0.0001 = 0.058$$

21. $n = 8, k = 1; p = .5$

$P(\text{exactly 1 head}) = b(8, 1; .5) = \binom{8}{1}(.5)^1(.5)^7 = 0.03125$

23. $n = 8, p = .5$

$P(\text{at least 5 tails}) = P(\text{exactly 5 tails}) + P(\text{exactly 6 tails}) + P(\text{exactly 7 tails}) +$
$\qquad\qquad P(\text{exactly 8 tails})$
$\qquad\qquad = b(8, 5; .5) + b(8, 6; .5) + b(8, 7; .5) + b(8, 8; .5) = 0.3633$

25. $n = 8, p = .5$

$P(\text{at least 1 head}) = P(\text{exactly 1 head}) + P(\text{exactly 2 heads}) + \ldots P(\text{exactly 8 heads})$
$\qquad\qquad = 1 - P(0 \text{ heads})$

The intersection of "at least 1 head" and "exactly 2 heads" is "exactly 2 heads"

$P(\text{at exactly 2 heads} \mid \text{at least 1 head}) = \dfrac{P(\text{exactly 2 heads})}{1 - P(\text{no heads})} = \dfrac{b(8, \ 2; \ .5)}{1 - b(8, \ 0; \ .5)} = 0.1098$

27. The probability of rolling a sum of 7 with two dice is $\dfrac{1}{6}$.

So we have $n = 5, k = 2, p = \dfrac{1}{6}$.

$P(\text{exactly 2 sums of 7}) = b(5, 2; \dfrac{1}{6}) = 0.1608$

29. $n = 8, p = 0.05$
    (a) $P(\text{exactly 1 is defective}) = b(8, 1; 0.05) = 0.2793$
    (b) $P(\text{exactly 2 are defective}) = b(8, 2; 0.05) = 0.0515$
    (c) $P(\text{at least 1 is defective}) = 1 - P(\text{none are defective})$
$\qquad\qquad\qquad\qquad\qquad = 1 - b(8, 0; 0.05)$
$\qquad\qquad\qquad\qquad\qquad = 1 - 0.6634 = 0.3366$

    (d) $P(\text{fewer than 3 defective}) = P(\text{no defective}) + P(\text{exactly 1 defective})$
$\qquad\qquad\qquad\qquad + P(\text{exactly 2 defective})$
$\qquad\qquad\qquad\qquad = b(8, 0; .05) + b(8, 1; .05) + b(8, 2; .05) = 0.9942$

31. $n = 6, k = 3, p = .5$

$b(6, 3; .5) = 0.3125$

The probability that a family with 6 children has exactly 3 boys and 3 girls is 0.3125.

33.   (a)

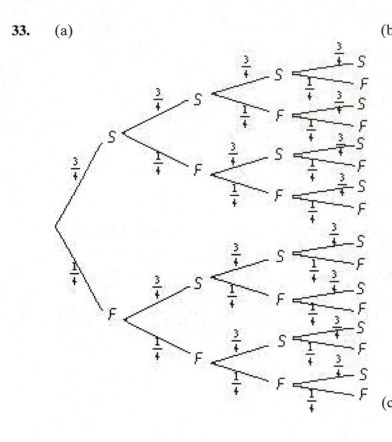

(b) P(exactly 2 success, 2 failures)
$P(SSFF) + P(SFSF) + P(SFFS)$
$+ P(FSSF) + P(FSFS) +$
$P(FFSS)$

$$= \left(\frac{3}{4}\right)\left(\frac{3}{4}\right)\left(\frac{1}{4}\right)\left(\frac{1}{4}\right)+$$

$$\left(\frac{3}{4}\right)\left(\frac{1}{4}\right)\left(\frac{3}{4}\right)\left(\frac{1}{4}\right)+$$

$$\left(\frac{3}{4}\right)\left(\frac{1}{4}\right)\left(\frac{1}{4}\right)\left(\frac{3}{4}\right)+$$

$$\left(\frac{1}{4}\right)\left(\frac{3}{4}\right)\left(\frac{3}{4}\right)\left(\frac{1}{4}\right)+$$

$$\left(\frac{1}{4}\right)\left(\frac{3}{4}\right)\left(\frac{1}{4}\right)\left(\frac{3}{4}\right)+$$

$$\left(\frac{1}{4}\right)\left(\frac{1}{4}\right)\left(\frac{3}{4}\right)\left(\frac{3}{4}\right)$$

$$= 6 \cdot \left(\frac{9}{256}\right) = \frac{27}{128}$$

(c) P(exactly 2 success, 2 failures)

$$= b\left(4, 2; \frac{3}{4}\right) = \binom{4}{2}\left(\frac{3}{4}\right)^2\left(\frac{1}{4}\right)^2 = \frac{27}{128}$$

35. $n = 10, p = \dfrac{2}{3}$

$$P(\text{at least 2 hits}) = 1 - [P(\text{no hits}) + P(\text{exactly 1 hit})]$$

$$= 1 - [b(10, 0; \frac{2}{3}) + b(10, 1; \frac{2}{3})]$$

$$= 1 - \left(\frac{1}{3}\right)^{10} - 20 \cdot \left(\frac{1}{3}\right)^{10} = 0.9996$$

The probability the target is hit at least twice is 0.9996.

37. (a) $n = 15, p = .5$

$P(\text{at least 10 correct}) = P(\text{exactly 10}) + P(\text{exactly 11}) + P(\text{exactly 12}) + P(\text{exactly 13}) +$
$\qquad P(\text{exactly 14}) + P(\text{exactly 15})$
$\qquad = b(15, 10; .5) + b(15, 11, .5) + b(15, 12; .5) + b(15, 13; .5)$
$\qquad + b(15, 14; 5) + b(15, 15; .5)$
$\qquad = 0.0916 + 0.0417 + 0.0139 + 0.0032 + 0.0005 + 0.0000$
$\qquad = 0.1509$

A student who guesses on a 15 question true/false test has a probability of .1509 of getting at least 10 questions correct.

(b) $n = 15, p = .8$

$P(\text{at least 12 correct}) = P(\text{exactly 12}) + P(\text{exactly 13}) + P(\text{exactly 14}) + P(\text{exactly 15})$
$\qquad = b(15, 12; .8) + b(15, 13; .8) + b(15, 14; .8) + b(15, 15; .8)$
$\qquad = 0.2501 + 0.2309 + 0.1319 + 0.0352$
$\qquad = 0.6481$

A student who has an 80% chance of getting an answer correct will get at least 12 questions correct on a 15 question true/false test .6481 of the time.

39. $n = 8, k = 8; p = .40$

$P(\text{all 8 voters prefer Ms. Moran}) = b(8, 8; .40) = 0.0007.$

The probability that all 8 voters prefer Ms. Moran is 0.07%.

41. $n = 10, k = 4, p = .23$

$P(\text{exactly 4 deaths are due to heart attack}) = b(10, 4; .23) = 0.1225$

There is a .1225 probability that 4 of the next 10 unexpected deaths are due to heart attacks.

43. (a) $n = 6, p = .5$

$P(\text{identifies at least 5 cups}) = P(\text{identifies exactly 5 cups}) + P(\text{identifies exactly 6 cups})$
$\qquad = b(6, 5; .5) + b(6, 6; .5)$
$\qquad = 0.09375 + 0.015625 = 0.109375$

The probability of correctly identifying at least 5 out of 6 cups of coffee if merely guessing is 0.1094.

(b) $n = 6, p = .80$

$P(\text{identifies fewer than 5 cups}) = 1 - P(\text{identifies at least 5 cups})$
$\qquad = 1 - [P(\text{identifies exactly 5 cups}) + P(\text{identifies exactly 6 cups})]$
$\qquad = 1 - [b(6, 5; .80) + b(6, 6; .80)]$
$\qquad = 1 - 0.393216 - 0.262144 = 0.34464$

The probability her claim is rejected when she really has the ability to identify the coffee is .3446.

45. $n = 10, p = .124$

(a) $P(\text{exactly 4 are 65 or older}) = b(10, 4; .124) = 0.0224$

There is a probability of .0224 that exactly 4 of the 10 people are 65 or older.

(b) $P(\text{no one is 65 or older}) = b(10, 0; .124) = 0.2661$

There is a probability of .2661 that none of the 10 people are 65 or older.

(c) $P(\text{at most 5 are} \geq 65) = P(\text{no one} \geq 65) + P(\text{exactly 1} \geq 65) + P(\text{exactly 2} \geq 65) +$
$$P(\text{exactly 3} \geq 65) + P(\text{exactly 4} \geq 65) + P(\text{exactly 5} \geq 65)$$
$$= b(10, 0; .124) + b(10, 1; .124) + b(10, 2; .124) +$$
$$b(10, 3; .124) + b(10, 4; .124) + b(10, 5; .124)$$
$$= 0.2661 + 0.3767 + 0.2399 + 0.0906 + 0.0224 + 0.0038 = 0.9995$$

The probability that no more than 5 people selected are 65 years of age or older is .9995.

47.   Estimates will vary, but all should be close to the actual theoretical probabilities.

The actual values of $P(k)$ are obtained from $b(4, k; .5)$ where k denotes the number of heads obtained in 4 tosses of the coin.

k	Estimate of $P(k)$	Actual $P(k)$
0		0.0625
1		0.2500
2		0.3750
3		0.2500
4		0.0625

49.   Estimates will vary, but they should be close to the actual theoretical probability. The actual value of $P(\text{exactly 3 heads}) = b(8, 3; .5) = 0.2188$

8.3 Expected Value

1. $E = (2)(.4) + (3)(.2) + (-2)(.1) + (0)(.3) = 1.2$

3. $E = (30,000)(.08) + (40,000)(.42) + (60,000)(.42) + (80,000)(.08) = 50,800$

5. $E = (8)(.1) + (0)(.90) = 0.8$

Mary should pay $0.80 for one draw.

7. $P(\text{double when throwing 2 dice}) = \dfrac{1}{6}$

$$E = (10)\left(\frac{1}{6}\right) + (0)\left(\frac{5}{6}\right) = \frac{10}{6} = 1.6667$$

David should pay $1.67 for a throw.

9. $E = (100)(0.001) + (50)(0.003) + (0)(0.996) = 0.25$

The price of a ticket exceeds the expected value by $0.75.

11. $P(3 \text{ tails}) = \dfrac{1}{8}$ $\qquad P(2 \text{ tails}) = \dfrac{3}{8}$ $\qquad P(1 \text{ tail}) = \dfrac{3}{8}$ $\qquad P(0 \text{ tails}) = \dfrac{1}{8}$

(a) $E = (3)\left(\dfrac{1}{8}\right) + (2)\left(\dfrac{3}{8}\right) + (0)\left(\dfrac{3}{8}\right) + (-3)\left(\dfrac{1}{8}\right) = \dfrac{6}{8} = 0.75$

The expected value of the game is $0.75.

(b) The game is not fair. Fair games have an expected value of 0.

(c) Let x represent the payoff for tossing 1 tail.

$x\left(\dfrac{3}{8}\right) = -\dfrac{6}{8}$ or $x = -2.$

To make the game fair the player should lose $2.00 if one tail is thrown.

13. $P(\text{team A wins}) = \dfrac{9}{14}$ $\qquad P(\text{team B wins}) = \dfrac{5}{14}$

If team A wins you lose $4; if team B wins you win $6. The expected value of the game

is $E = (-4)\left(\dfrac{9}{14}\right) + (6)\left(\dfrac{5}{14}\right) = -\dfrac{6}{14} = -0.4286$

The bet is not fair to you. You should expect to lose 42 cents.

15. $P(\text{selecting a heart other than the ace}) = \dfrac{12}{52}$; $P(\text{selecting an ace other than the heart}) = \dfrac{3}{52}$

$P(\text{selecting the ace of hearts}) = \dfrac{1}{52}$

$E = (40)\left(\dfrac{12}{52}\right) + (50)\left(\dfrac{3}{52}\right) + (90)\left(\dfrac{1}{52}\right) = \dfrac{720}{52} = 13.8462$

Sarah's expected winnings are $13.846 - 15 = -1.15$ cents.

17. $P(\text{horse wins}) = \dfrac{7}{12}$ $\qquad P(\text{horse loses}) = \dfrac{5}{12}$. Let x denote the amount the bettor should

wager to make the game fair. A fair game has an expected value of 0.

$E = (5)\left(\dfrac{7}{12}\right) + x\left(\dfrac{5}{12}\right) = 0$

$35 + 5x = 0$ or $x = -7$. A bettor should bet $7.00 to make the game fair.

19. We compare the two expected profits. $P(\text{success}) = \dfrac{1}{2}$

Location 1: $\qquad\qquad\qquad\qquad\qquad$ Location 2:

$E = (15,000)\left(\dfrac{1}{2}\right) + (-3000)\left(\dfrac{1}{2}\right) \qquad E = (20,000)\left(\dfrac{1}{2}\right) + (-6000)\left(\dfrac{1}{2}\right)$

$\quad = 6000 \qquad\qquad\qquad\qquad\qquad\qquad = 7000$

The first location will provide an $\qquad$ The second location will provide an expected
expected profit of $6000. $\qquad\qquad\qquad$ profit of $7000.

The management should choose the Location 2. It has a higher expected profit.

21. This is an example of 2000 Bernoulli trials, where $p = \dfrac{1}{6}$ is the probability of success.

$$E = np = 2000\left(\dfrac{1}{6}\right) = 333.333$$

We expect 333.333 fives in 2000 rolls of a fair die.

23. This can be considered an example of 500 Bernoulli trials, where $p = .02$ is the probability a light bulb is defective.

$$E = np = (500)(.02) = 10$$

We expect 10 defective light bulbs in a shipment of 500.

25. This is an example of 500 Bernoulli trials, where $p = .002$ is the probability of having an unfavorable reaction to the drug.

$$E = np = (500)(.002) = 1$$

The doctor can expect 1 patient to have an unfavorable reaction to the drug.

27. If $P(H) = \dfrac{1}{4}$ and $P(T) = \dfrac{3}{4}$ then the expected number of tosses needed to obtain either 1 head or 4 tails is

$$E = (1)P(H) + 2P(TH) + 3P(TTH) + 4P(TTTH) + 4P(TTTT)$$

$$= (1)\left(\dfrac{1}{4}\right) + 2\left(\dfrac{3}{4}\right)\left(\dfrac{1}{4}\right) + 3\left(\dfrac{3}{4}\right)\left(\dfrac{3}{4}\right)\left(\dfrac{1}{4}\right) + 4\left(\dfrac{3}{4}\right)\left(\dfrac{3}{4}\right)\left(\dfrac{3}{4}\right)\left(\dfrac{1}{4}\right) + 2\left(\dfrac{3}{4}\right)\left(\dfrac{3}{4}\right)\left(\dfrac{3}{4}\right)\left(\dfrac{3}{4}\right)$$

$$= \dfrac{1}{4} + \dfrac{6}{16} + \dfrac{27}{64} + \dfrac{27}{64} + \dfrac{81}{64} = \dfrac{175}{64} = 2.73$$

We expect to have to throw the coin 2.73 times before getting either a head or 4 tails.

29. Profit = Revenue – Cost. We will compare the profit from each aircraft.

Aircraft A
Expected number of passengers:

$$E = (150)(.2) + (180)(.3) + (200)(.2 + .2 + .1) = 30 + 54 + 100 = 184$$

Expected revenue: ticket price times the number of tickets sold = (500)(184) = 92,000
Expected cost: fixed cost plus passenger cost = 16,000 + 200(184) = 52,800
Expected profit: Revenue – Cost = 92,000 – 52,800 = 39,200

The company can expect a profit of $39,200 if it uses aircraft A.

Aircraft B
Expected number of passengers:

$$E = (150)(.2) + (180)(.3) + (200)(.2) + (250)(.2) + (300)(.1) = 30 + 54 + 40 + 50 + 30 = 204$$

Expected revenue: ticket price times the number of tickets sold = (500)(204) = 102,000
Expected cost: fixed cost plus passenger cost = 18,000 + 230(204) = 64,920
Expected profit: Revenue – Cost = 102,000 – 64,920 = 37,080

The company can expect a profit of $37,080 if it uses aircraft B.
The company should use aircraft A; it generates a larger expected profit.

8.4 Applications

1. The expected number of customers is 9.
$$E = 7p_7 + 8p_8 + 9p_9 + 10p_{10} + 11p_{11}$$
$$= (7)(.10) + (8)(.20) + (9)(.40) + (10)(.20) + (11)(.10) = 9$$
To decide the optimal number of cars to have on hand, we need to find the expected profit for each possible number of cars.
Profit = Revenue − Cost; Expected profit is the product of the profit and the probability of obtaining it.

1. If 7 cars are on hand, all will be rented, and the expected profit will be
    $$E(\text{profit}) = (7)(30) - (7)(10) = 210 - 70 = \$140$$
2. If 8 cars are on hand, 7 will be rented with probability .10 and 8 will be rented with probability .90   (.90 = .20 + .40 + .20 + .10)
    $$E(\text{profit}) = [(30)(7) - (10)(8)](.10) + [(30)(8) - (10)(8)](.90) = \$157$$
3. If 9 cars are on hand,

Number of cars rented	7	8	9
Probability of being rented	.10	.20	.40 + .20 + .10 = .70
Cost of holding cars	$90	$90	$90

and the expected profit will be

$$E(\text{profit}) = [(30)(7) - (90)](.10) + [(30)(8) - (90)](.20) + [(30)(9) - (90)](.70) = \$168$$

3. If 10 cars are on hand,

Number of cars rented	7	8	9	10
Probability of being rented	.10	.20	.40	.20 + .10 = .30
Cost of holding cars	$100	$100	$100	$100

and the expected profit will be

$$E(\text{profit}) = [(30)(7) - (100)](.10) + [(30)(8) - (100)](.20) + [(30)(9) - (100)](.40) +$$
$$[(30)(10) - (100)](.30) = \$167$$

4. Although not truly needed, we look at the expected profit if 11 cars are on hand,

Number of cars rented	7	8	9	10	11
Probability of being rented	.10	.20	.40	.20	.10
Cost of holding cars	$110	$110	$110	$110	$110

and the expected profit will be

$$E(\text{profit}) = [(30)(7) - (110)](.10) + [(30)(8) - (110)](.20) + [(30)(9) - (110)](.40) +$$
$$[(30)(10) - (110)](.20) + [(30)(11) - (110)](.10) = \$160$$

The largest expected profit is $168, and the rental agency should keep 9 cars on hand.

3.   We are asked to show that if the probability a component is good, $p = .95$, in Example 2, (p. 453 of the text), the optimal group size is 5.

For the explanation of how we derive the formula, see the solution to Problem 2.

The optimal group size is that which saves the most tests per component.

Group Size	Expected Tests Saved per Component $p = .95$	Percent Saving
2	$.95^2 - \dfrac{1}{2} = 0.4025$	40.25%
3	$.95^3 - \dfrac{1}{3} = 0.5240$	52.4%
4	$.95^4 - \dfrac{1}{4} = 0.5645$	56.45%
5	$.95^5 - \dfrac{1}{5} = 0.5738$	57.38%
6	$.95^6 - \dfrac{1}{6} = 0.5684$	56.84%
7	$.95^7 - \dfrac{1}{7} = 0.5555$	55.55%

57.38% is the largest percent savings, making $n = 5$ the optimal group size.

5.   We are told the probability a diver finds the instrument is .95, which means the probability a diver fails to find the instrument is $1 - .95 = .05$.

(a)   Let x denote the number of divers hired. The probability x divers cannot find the instrument is the same as the probability they all fail or $.05^x$. So with x divers looking, the probability the instrument is found is given by $1 - .05^x$.

The expected net, $E(x)$, gain is the is expected value of the equipment if it is returned less the cost of the search.

$$E(x) = [75{,}000(1 - .05^x) + 0(.05^x)] - 500x$$
$$= 75{,}000 - 75{,}000(.05^x) - 500x$$

(b)   To find the number of divers needed to maximize the net gain, we evaluate $E(x)$ for various values of x, and choose the largest.

Number of Divers	Expected Gain
1	$75{,}000 - 75{,}000(.05)^1 - 500x = \$70{,}750$
2	$75{,}000 - 75{,}000(.05)^2 - 500x = \$73{,}813$
3	$75{,}000 - 75{,}000(.05)^3 - 500x = \$73{,}491$
4	$75{,}000 - 75{,}000(.05)^4 - 500x = \$73{,}000$
5	$75{,}000 - 75{,}000(.05)^5 - 500x = \$72{,}500$
6	$75{,}000 - 75{,}000(.05)^6 - 500x = \$72{,}000$

The largest expected gain, $73,813 occurs when 2 divers are hired.

7. The code is of length 15, and corrects 1 error. The probability a digit is transmitted correctly is .98. So we have $n = 15$, $p = .98$.
The probability a message is received correctly will be the probability it was received without error plus the probability if was received with one corrected error.

$b(15, 15; .98) + b(15, 14; .98) = .7386 + .2261 = .9647$

The message will be received correctly 96.47% of the time using this Hamming code.

8.5 Random Variables

1. Let X denote the number of heads that appear when a fair coin is tossed twice.

X	$x = 0$	$x = 1$	$x = 2$
$P(X = x)$	$p(0) = b(2, 0; .5) = \dfrac{1}{4}$	$p(1) = b(2, 1; .5) = \dfrac{1}{2}$	$p(2) = b(2, 2; .5) = \dfrac{1}{4}$

3. Let X denote the number of female children in 3-child family assuming probability that a child is female is $\dfrac{1}{2}$.

X	$P(X = x)$
$x = 0$	$0) = b\left(3,\ 0;\ \dfrac{1}{2}\right) = \dfrac{1}{8}$
$x = 1$	$p(1) = b\left(3,\ 1;\ \dfrac{1}{2}\right) = \dfrac{3}{8}$
$x = 2$	$p(2) = b\left(3,\ 2;\ \dfrac{1}{2}\right) = \dfrac{3}{8}$
$x = 3$	$p(3) = b\left(3,\ 3;\ \dfrac{1}{2}\right) = \dfrac{1}{8}$

5. Let X denote the number of red balls that are drawn when 3 balls are drawn with replacement from an urn with 10 balls.

$P(\text{red ball chosen}) = .4$;
$P(\text{white ball chosen}) = .6$

X	$P(X = x)$
0	$p(0) = b(3, 0; .4) = 0.216$
1	$p(1) = b(3, 1; .4) = 0.432$
2	$p(2) = b(3, 2; .4) = 0.288$
3	$p(3) = b(3, 3; .4) = 0.064$

7. $E(X) = x_1 p_1 + x_2 p_2 + x_3 p_3 + x_4 p_4$
$= (2)(.4) + (3)(.2) + (-2)(.1) + (0)(.3) = 1.2$

9. Estimates will vary, but all should be close to the actual theoretical probabilities.

k	1	2	3	4	5	6
estimate of $P(X = k)$						
actual value of $P(X = k)$	$\dfrac{1}{6}$	$\dfrac{1}{6}$	$\dfrac{1}{6}$	$\dfrac{1}{6}$	$\dfrac{1}{6}$	$\dfrac{1}{6}$

11. Estimates will vary, but all should be close to the actual theoretical probability.
     The actual probability $P(0.1 \leq X < 0.3) = \dfrac{0.3 - 0.1}{1.0} = 0.2$

13. Estimates will vary, but all should be close to the actual theoretical probability.
     The actual probability $P(X = 2) = \dfrac{1}{12}$

Chapter 8 Review

True-False Items

1. T   3. F   5. F

Fill in the Blanks

1. Bayes' formula   3. expected value of the experiment

5. expected value

Review Exercises

1. $P(E \mid A) = .82$ 3. $P(E \mid B) = .10$

5. $P(E) = P(E \cap A) + P(E \cap B) = P(A) \cdot P(E \mid A) + P(B) \cdot P(E \mid B)$
    $$= (.90)(.82) + (.10)(.10) = .748$$

    $$P(A \mid E) = \frac{P(E \cap A)}{P(E)} = \frac{P(A) \cdot P(E \mid A)}{P(E)} = \frac{(.90)(.82)}{.748} = .9866$$

7. $P(E) = P(E \cap A) + P(E \cap B) = P(A) \cdot P(E \mid A) + P(B) \cdot P(E \mid B)$
    $$= (.90)(.82) + (.10)(.10) = .748$$

    $$P(B \mid E) = \frac{P(E \cap B)}{P(E)} = \frac{P(B) \cdot P(E \mid B)}{P(E)} = \frac{(.10)(.10)}{.748} = .0134$$

9. $P(E \mid A) = .5$ 11. $P(E \mid B) = .4$

13. $P(E \mid C) = .3$

15. $P(E) = P(E \cap A) + P(E \cap B) + P(E \cap C) = P(A) \cdot P(E \mid A) + P(B) \cdot P(E \mid B) + P(C) \cdot P(E \mid C)$
     $$= (.4)(.5) + (.5)(.4) + (.1)(.3) = .43$$

17. $P(A \mid E) = \dfrac{P(E \cap A)}{P(E)} = \dfrac{P(A) \cdot P(E \mid A)}{P(E)} = \dfrac{(.4)(.5)}{.43} = .4651$

19. $P(B \mid E) = \dfrac{P(E \cap B)}{P(E)} = \dfrac{P(B) \cdot P(E \mid B)}{P(E)} = \dfrac{(.5)(.4)}{.43} = .4651$

21. $P(C \mid E) = \dfrac{P(E \cap C)}{P(E)} = \dfrac{P(C) \cdot P(E \mid C)}{P(E)} = \dfrac{(.1)(.3)}{.43} = .0698$

23. Using the chart given in the Problem, we first get the totals

$c(E) = 345 \qquad c(F) = 210 \qquad c(G) = 260 \qquad c(H) = 207 \qquad c(K) = 127 \qquad n = 594$

(a) $P(E \mid G) = \dfrac{P(E \cap G)}{P(G)} = \dfrac{180}{260} = \dfrac{9}{13}$ (b) $P(G \mid E) = \dfrac{P(G \cap E)}{P(E)} = \dfrac{180}{345} = \dfrac{12}{23}$

(c) $P(H \mid E) = \dfrac{P(H \cap E)}{P(E)} = \dfrac{110}{345} = \dfrac{22}{69}$ (d) $P(K \mid E) = \dfrac{P(K \cap E)}{P(E)} = \dfrac{55}{345} = \dfrac{11}{69}$

(e) $P(F \mid G) = \dfrac{P(F \cap G)}{P(G)} = \dfrac{60}{260} = \dfrac{3}{13}$ (f) $P(G \mid F) = \dfrac{P(G \cap F)}{P(F)} = \dfrac{60}{210} = \dfrac{2}{7}$

(g) $P(H \mid F) = \dfrac{P(H \cap F)}{P(F)} = \dfrac{85}{210} = \dfrac{17}{42}$ (h) $P(K \mid F) = \dfrac{P(K \cap F)}{P(F)} = \dfrac{65}{210} = \dfrac{13}{42}$

25. We make a tree diagram to illustrate the problem.

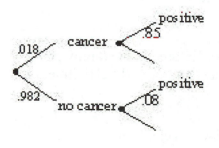

Define the events:
E: The patient has lung cancer.
F: The patient does not have lung cancer
G: The test is positive.

$$P(E \mid G) = \dfrac{P(E \cap G)}{P(G)}$$

$$= \dfrac{P(E) \cdot P(G \mid E)}{P(E) \cdot P(G \mid E) + P(F) \cdot P(G \mid F)}$$

$$= \dfrac{(.018)(.85)}{(.018)(.85) + (.982)(.08)} = .1630$$

There is a probability of .163 that a patient with a positive test actually has lung cancer.

27. This is a Bernoulli experiment. We have $n = 5$ and $p = .2$.
The probability that none of the people chosen will purchase the product is

$P(\text{exactly 0 purchases}) = b(5, 0; .2) = C(5, 0)(.2^0)(.8^5) = (1)(1)(.32768) = .32768$

The probability that exactly 3 will purchase the product is

$P(\text{exactly 3 purchases}) = b(5, 3; .2) = C(5, 3)(.2^3)(.8^2) = (10)(.008)(.64) = .0512$

29. This is a Bernoulli experiment with $n = 12$ and $p = .5$

    (a) The probability the student will get all answers correct is

$$P(\text{exactly 12 correct}) = b(12, 12; .5) = C(12, 12)(.5^{12}) = .0002$$

    (b) The probability the student passes the test is

$P(\text{at least 7 correct})$
$$\begin{aligned}
&= P(\text{exactly 7 correct}) + P(\text{exactly 8 correct}) + P(\text{exactly 9 correct}) + \\
&\quad P(\text{exactly 10 correct}) + P(\text{exactly 11 correct}) + P(\text{exactly 12 correct}) \\
&= b(12, 7; .50) + b(12, 8; .50) + b(12, 9; .50) + b(12, 10; .50) + b(12, 11; .50) \\
&\quad + b(12, 12; .50) \\
&= .1934 + .1208 + .0537 + .0161 + .0029 + .0002 = .3871
\end{aligned}$$

    (c) The odds in favor of passing are 387 to 613.

31. This is a Bernoulli experiment. The probability of throwing a sum of 11 with 2 dice is $\dfrac{1}{18}$.

So in this experiment we have $n = 5$ and $p = \dfrac{1}{18}$.

The probability of throwing at least 3 sums of 11 is

$$P(\text{at least 3 11's}) = P(\text{exactly 3 11's}) + P(\text{exactly 4 11's}) + P(\text{exactly 5 11's})$$

$$= b\left(5, 3; \frac{1}{18}\right) + b\left(5, 4; \frac{1}{18}\right) + b\left(5, 5; \frac{1}{18}\right)$$
$$= .00153 + .00004 + .00000 = .0016$$

33. This is a Bernoulli experiment with $n = 3$, the number of children in the family and $p = .5$, the probability of having a girl child. The expected value of a Bernoulli experiment is given by $E = np$.

The expected number of girls in a 3 child family is

$E = np = (3)(.5) = 1.5$. So we expect 1.5 girls in a family with 3 children.

35. The expected return of the game, or the payoff, is the expected value of the experiment.

In this problem $\dfrac{1}{7}$ of the time the player wins \$90 and $\dfrac{1}{3}$ or the time the player wins \$50,

the rest of the time $1 - \left(\dfrac{1}{7} + \dfrac{1}{3}\right) = \dfrac{11}{21}$ of the time the player wins nothing.

If m denotes the winnings, and p represents the probability of winning, then

$$E = m_1 p_1 + m_2 p_2 + m_3 p_3 = (90)\left(\frac{1}{7}\right) + (50)\left(\frac{1}{3}\right) + (0)\left(\frac{11}{21}\right) = \frac{620}{21} = \$29.52$$

The expected return of the game is \$29.52.

37. A game is fair if the expected value of the game and the price to play are equal.
    (a)   The expected value of this game is

$$E = (100)(.001) + (50)(.001) + (30)(.001) = .18$$

        A fair price to pay for the game is 18 cents.
    (b)   Alice paid \$0.25 to play the game. She paid $0.25 - 0.18 = \$0.07$ extra per ticket, or \$0.56 extra for 8 tickets.

39. In this game $\frac{1}{3}$ the time you go to box 1 and $\frac{2}{3}$ the time you go to box 2.

The expected value of the game is

$$E = \frac{1}{3}\left[(0)\left(\frac{1}{8}\right) + (5)\left(\frac{1}{4}\right) + (8)\left(\frac{1}{4}\right) + (10)\left(\frac{1}{8}\right) + (70)\left(\frac{1}{8}\right) + (80)\left(\frac{1}{8}\right)\right] + \frac{2}{3}\left[(1)\left(\frac{1}{2}\right) + (5)\left(\frac{1}{2}\right)\right]$$
$$= 9.75$$

41. (a)   The expected profit from the event is $E = (750{,}000)(.95) + (20{,}000)(.05) = \$713{,}500$
    (b)   The insurance costs \$50,000, and purchasing it will decrease profit. If the expected profit adjusted to reflect the premium is more than the expected profit without insurance, then the buying the insurance policy is a wise idea.

        Adjusted $E = (750{,}000 - 50{,}000)(.95) + (500{,}000 - 50{,}000)(.05) = \$687{,}500$

        Buying insurance is not a wise idea.

43. This is a Bernoulli experiment with $n = 500$ and $p = \frac{1}{6}$. The expected value gives the

number of doubles you expect to roll. $E = np = (500)\left(\frac{1}{6}\right) = \frac{500}{6} = 83.3333$ doubles.

45. The code is of length 31, and corrects up to 2 errors. The probability a digit is transmitted correctly is .97. So we have $n = 31$, $p = .97$.
The probability a message is received correctly will be the probability it was received without error or with one or two or corrected errors.

$$b(31, 31; .97) + b(31, 30; .97) + b(31, 29; .97)$$
$$= .3890 + .3729 + .1730 = .9349$$

The message will be received correctly 93.49% of the time using this BCH code.

47. If the probability a test is positive is p, the probability the test is negative is $(1 - p)$.
    (a)   The test for a pooled sample will be positive if at least one person tests positive. So,

$$P(\text{pooled test is positive}) = 1 - P(\text{all tests are negative})$$
$$= 1 - (1 - p)^{30}$$

    (b)   If the pooled test is negative then 1 test suffices for 30 people, but if the pooled test is positive then 31 tests must be done. The expected number of tests necessary using the pooled method is

$$E = (1)(1 - p)^{30} + 31\left[1 - (1 - p)^{30}\right] = (1 - p)^{30} + 31 - 31(1 - p)^{30} = 31 - 30(1 - p)^{30}$$

49. (a) The random variable X has the values 0, 1, 2, 3, 4, 5.

    (b) The probability distribution of X, the number of defective pens in the 5-pen sample is

$$p(0) = \frac{\binom{10}{0}\binom{90}{5}}{\binom{100}{5}} = .5838 \qquad p(1) = \frac{\binom{10}{1}\binom{90}{4}}{\binom{100}{5}} = .3394 \qquad p(2) = \frac{\binom{10}{2}\binom{90}{3}}{\binom{100}{5}} = .0702$$

$$p(3) = \frac{\binom{10}{3}\binom{90}{2}}{\binom{100}{5}} = .0064 \qquad p(4) = \frac{\binom{10}{4}\binom{90}{1}}{\binom{100}{5}} = .0003 \qquad p(5) = \frac{\binom{10}{5}\binom{90}{0}}{\binom{100}{5}} = .0000$$

    (c) The expected value of X is $E(X) = x_1 p(x_1) + x_2 p(x_2) + x_3 p(x_3) + x_4 p(x_4) + x_5 p(x_5)$.

        Here $E(X) = (0)(.5838) + (1)(.3394) + (2)(.0702) + (3)(.0064) + (4)(.0003) + (5)(.0000)$

                $= .5002$

51. (a) The random variable X has the values 0, 1, 2, 3 where X is the number of yellow flowers that grow.

    (b) The probability distribution of X, the number of yellow flowers in the 3-bulb sample is

$$p(0) = \frac{\binom{3}{0}\binom{7}{3}}{\binom{10}{3}} = .2917 \qquad p(1) = \frac{\binom{3}{1}\binom{7}{2}}{\binom{10}{3}} = .525 \qquad p(2) = \frac{\binom{3}{2}\binom{7}{1}}{\binom{10}{3}} = .175$$

$$p(3) = \frac{\binom{3}{3}\binom{7}{0}}{\binom{10}{3}} = .0083$$

    (c) The expected value of X is $E(X) = x_1 p(x_1) + x_2 p(x_2) + x_3 p(x_3)$.

        Here $E(X) = (0)(.2917) + (1)(.525) + (2)(.175) + (3)(.0083) = .8999$. We expect .8999 yellow flowers to grow.

53. The random variable X is the number of persons in the sample who are 65 or older. X can take on the values, 0, 1, 2, 3, 4, and 5. X has a binomial distribution with $n = 5$ and $p = .124$. The distribution is

$p(0) = b(5, 0; .124) = .5158$ $p(1) = b(5, 1; .124) = .3651$

$p(2) = b(5, 2; .124) = .1034$ $p(3) = b(5, 3; .124) = .0146$

$p(4) = b(5, 4; .124) = .0010$ $p(5) = b(5, 5; .124) = .00002$

The expected value of this binomial distribution is $E(X) = np = (5)(.124) = .62$. We expect .62 of the sample to be at least 65 years of age.

Chapter 8 Project

1. The probability that a defective spacecraft is detected using 4 or fewer tests means that a defect was found on the 1st test, or it passed the 1st test but failed the 2nd test, or it passed the tests 1 and 2 but failed the 3rd test, or it passed tests 1 through 3 but failed test 4. Since the tests are independent, $P(E \cap F) = P(E)P(F)$.

    The probability a defect is found in 4 or fewer tests is

    $$P(F) + P(E)P(F) + P(E)P(E)P(F) + P(E)P(E)P(E)P(F)$$
    $$= (.9) + (.1)(.9) + (.1)(.1)(.9) + (.1)(.1)(.1)(.9) = .9999$$

3. We use G: The spacecraft is good, and D: The spacecraft is defective.

    (a) $P(G \mid E) = \dfrac{P(G) \cdot P(E \mid G)}{P(E)} = \dfrac{P(G) \cdot P(E \mid G)}{P(G) \cdot P(E \mid G) + P(D) \cdot P(E \mid D)} = \dfrac{(.05)(1)}{(.05)(1) + (.95)(.1)}$

    $\qquad = .3448$

    (b) $P(D \mid E) = \dfrac{P(D) \cdot P(E \mid D)}{P(E)} = \dfrac{P(D) \cdot P(E \mid D)}{P(G) \cdot P(E \mid G) + P(D) \cdot P(E \mid D)} = \dfrac{(.95)(.1)}{(.05)(1) + (.95)(.1)}$

    $\qquad = .6552$

5. Define event K: The spacecraft passes 3 tests.
    Assuming the tests are independent, we get

    $$P(K \mid G) = P(E \mid G) \cdot P(E \mid G) \cdot P(E \mid G) = 1 \quad \text{and}$$
    $$P(K \mid D) = P(E \mid D) \cdot P(E \mid D) \cdot P(E \mid D) = 0.1^3 = 0.001$$

    $$P(K) = P(G) \cdot P(K \mid G) + P(D) \cdot P(K \mid D)$$
    $$= (0.05) \cdot (1) + (0.95) \cdot (0.001) = 0.05095$$

    $$P(G \mid K) = \dfrac{P(G) \cdot P(K \mid G)}{P(K)} = \dfrac{(0.05)(1)}{0.05095} = 0.9814$$

7.

Number of Tests	0	1	2	3	4
$P(G)$	.05	.3448	.8403	.9814	.9981

Mathematical Questions from the Professional Exams

1. (d) $P(\text{no fewer than 1 and no more than 9}) = 1 - [P(\text{no heads}) + P(10 \text{ heads})]$

$$= 1 - \frac{1}{2^{10}} - \frac{1}{2^{10}} = 1 - 2\left(\frac{1}{2^{10}}\right) = 1 - \frac{1}{2^9}$$

3. (a) First we find the expected value of the investment if the investor holds it, and then we compare it to $10,000, the return if the investor sells now.

$E = (5000)(.4) + (8000)(.2) + (12,000)(.3) + (30,000)(.1) = \$10,200$

5. (b) $E = (6000)(.2) + (8000)(.2) + (10,000)(.2) + (12,000)(.2) + (14,000)(.1) +$
$(16,000)(.1) = 10,200$

7. (b) or (c)

$E(\#1) = (100,000)(.7) + (70,000)(.3)$
$E(\#2) = (170,000 - 40,000)(.8) + (80,000 - 40,000)(.2)$
$\qquad = (130,000)(.8) + (40,000)(.2)$
$\qquad = (170,000)(.8) + (80,000)(.2) - (40,000)(.8 + .2)$
$\qquad = (170,000)(.8) + (80,000)(.2) - 40,000$

Chapter 9

Statistics

9.1 Introduction to Statistics: Data and Sampling

1. The variable is the number of heads thrown, and it is discrete.

3. The variable is the miles per gallon, and it is continuous.

5. The variable is the time of waiting in line, and it is continuous.

7. The variable is the number of flights, and it is discrete

9. The variable is the number of people crossing the intersection, and it is discrete.

11. The variable is length of time, and it is continuous.

13. Answers will vary. All answers should include a method to choose a group of viewers for which each viewer of the program has an equal chance of being chosen.

15. Answers will vary. All answers should include a method to choose a sample in which each member of the population has an equal chance of being selected.

17. Answers will vary. All answers should include a method to choose a sample in which each member of the population has an equal chance of being selected.

18–22. Answers will vary. All answers should give examples of possible bias.

9.2 Representing Data Graphically: Bar Graphs; Pie Charts

1. **(a)**

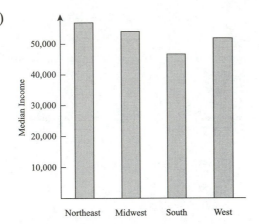

(b) The Northeast has the highest median income.

(c) The South has the lowest median income.

(d) Answers will vary.

3. **(a)**

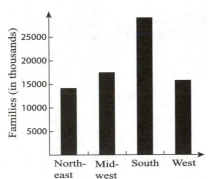

(b)

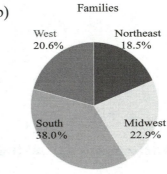

(c) Answers will vary.

(d) The South has the most families.

(e) The Northeast has the fewest families.

(f) Answers will vary.

5. **(a)**

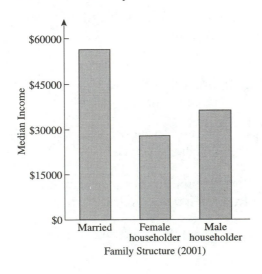

(b) Married – couple families have the highest median income.

(c) Female householder – no spouse families have the lowest median income.

(d) Answers will vary.

7. (a)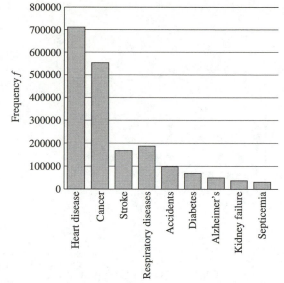

Causes of Death for Americans in 2000

(b)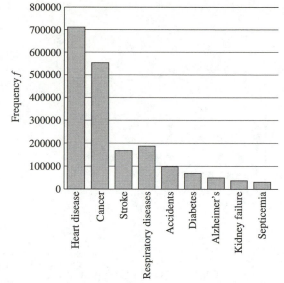

(c) Answers will vary.
(d) The leading cause of death among Americans in 2000 was heart disease.

9. (a) Sky West had the highest percentage of on-time flights with 85.4% of its flights on time.
(b) Atlantic Coast had the lowest percentage of on-time flights. Only 56.9% of its flights were on time.
(c) 84.3% of United Airlines' flights were on time in February 2003.

11. (a) The largest component of the CPI is housing, fuel, and utilities. It comprises 42% of the Consumer Price Index.
(b) The smallest component of the CPI is miscellaneous goods and services. It accounts for only 4% of the CPI.
(c) Answers will vary.

9.3 Organization of Data

1. (a)

Score	Frequency	Score	Frequency	Score	Frequency
25	1	36	2	47	1
26	1	37	4	48	3
27	0	38	1	49	1
28	1	39	1	50	1
29	1	40	1	51	1
30	3	41	5	52	3
31	2	42	3	53	2
32	1	43	1	54	2
33	2	44	2	55	1
34	2	45	1		
35	1	46	2		

(b)

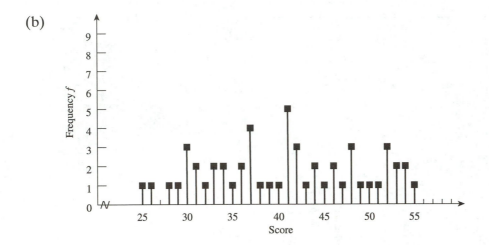

(c) and
(f)

Class Interval	Frequency	Cumulative Frequency	Class Interval	Frequency	Cumulative Frequency
24 – 25.9	1	1	40 – 41.9	6	29
26 – 27.9	1	2	42 – 43.9	4	33
28 – 29.9	2	4	44 – 45.9	3	36
30 – 31.9	5	9	46 – 47.9	3	39
32 – 33.9	3	12	48 – 49.9	4	43
34 – 35.9	3	15	50 – 51.9	2	45
36 – 37.9	6	21	52 – 53.9	5	50
38 – 39.9	2	23	54 – 55.9	3	53

(d)

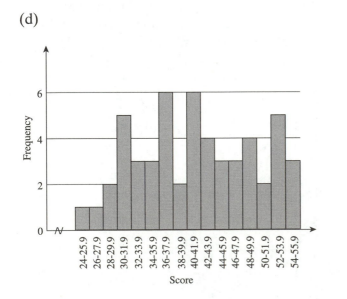

(e)

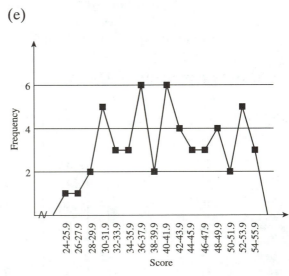

(g)

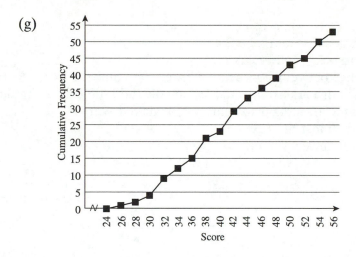

3.   (a) and (d)

Class Interval	Frequency	Cumulative Frequency
50 – 54.9	1	1
55 – 59.9	6	7
60 – 64.9	3	10
65 – 69.9	6	16
70 – 74.9	8	24
75 – 79.9	11	35
80 – 84.9	2	37
85 – 89.9	12	49
90 – 94.9	12	61
95 – 99.9	2	63
100 – 104.9	2	65
105 – 109.9	4	69
110 – 114.9	0	69
115 – 119.9	2	71

(b)

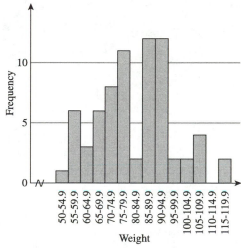

(c)

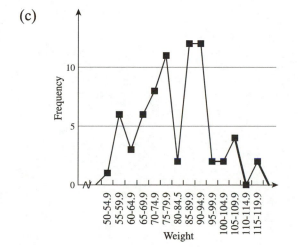

(e)

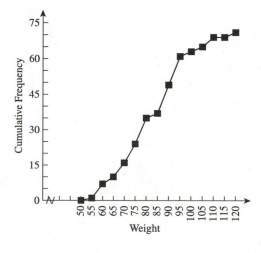

5. (a) There are 13 class intervals.
   (b) The lower class limit of the first class interval is 20.
       The upper class limit of the first class interval is 24.
   (c) The class width is the difference between consecutive lower class limits. It is 5.
   (d) To find the number of licensed drivers from 70 to 84 years old, add the drivers in the
       class intervals 70–74, 75–79, and 80–84, approximately 1,300,000 drivers.
   (e) The 30–34 class interval has the most drivers.
   (f) The 80–84 class interval has the fewest drivers.

   (g)

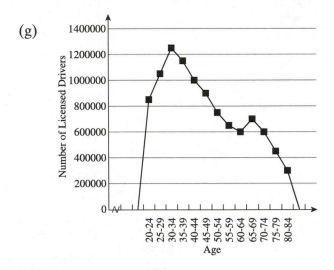

7. (a) There are 7 class intervals.
(b) The lower class limit of the last class interval is 80.
The upper class limit of the last class interval is 89.

(c) and (e)

Class Interval	Frequency	Cumulative Frequency
20 – 29	2,000,000	2,000,000
30 – 39	2,400,000	4,400,000
40 – 49	1,870,000	6,270,000
50 – 59	1,420,000	7,690,000
60 – 69	1,230,000	8,920,000
70 – 79	1,030,000	9,950,000
80 – 89	250,000	10,200,000

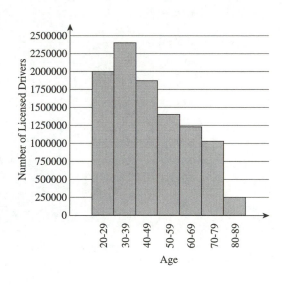

(d)
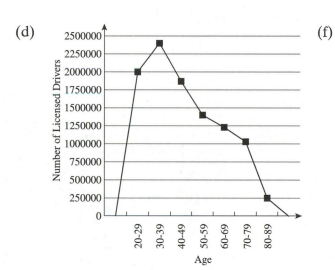

(f)
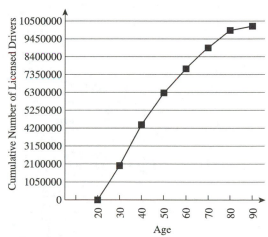

(g) Answers will vary.

9. (a) There are 13 class intervals.
    (b) The lower class limit of the first class interval is 20.
         The upper class limit of the first class interval is 24.
    (c) The class width is difference between consecutive lower class limits. Here the class
         width is 5.

    (d) 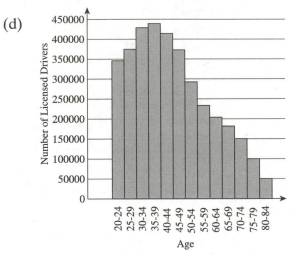   (e)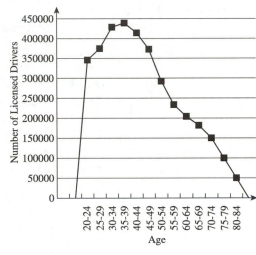

    (f) Most licensed drivers are in the 35–39 year-old age group.
    (g) The fewest licensed drivers are in the 80–84 year-old age group.

11. (a) There are 15 class intervals.
    (b) The lower class limit of the first class interval is 0.
         The upper class limit of the first class interval is $999.
    (c) The class width is difference between consecutive lower class limits. Here the class
         width is $1000.

    (d) 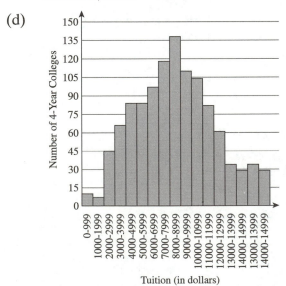   (e)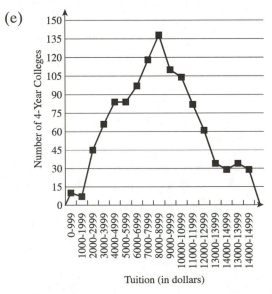

    (f) Tuition in 1992–1993 was most frequently in the $8000–$8999 range.

13. (a)

	Class Interval	Frequency
1	11.0–11.9	0
2	12.0–12.9	3
3	13.0–13.9	5
4	14.0–14.9	6
5	15.0–15.9	2
6	16.0–16.9	3
7	17.0–17.9	1
8	18.0–18.9	0

(b)

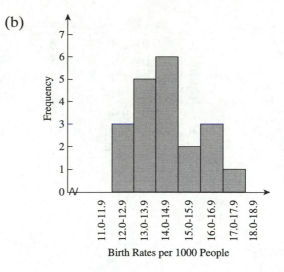

(c)

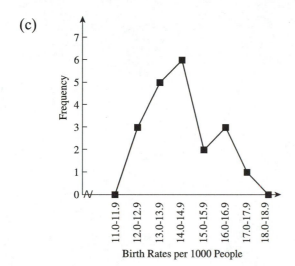

15.   (a)

	Class Interval	Frequency
1	0.0 – 1.9	6
2	2.0 – 3.9	5
3	4.0 – 5.9	4
4	6.0 – 7.9	1
5	8.0 – 9.9	0
6	10.0 – 11.9	3
7	12.0 – 13.9	1

(b)

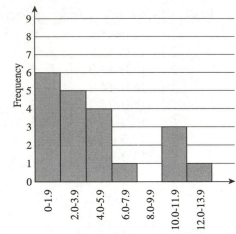

Death Rates (per 1000) from HIV Related Illness

(c)

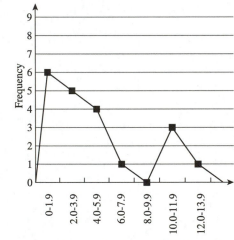

Death Rates (per 1000) from HIV Related Illness

9.4 Measures of Central Tendency

1.   21, 25, 43, 36

(a) Mean:

$$\overline{x} = \frac{21 + 25 + 43 + 36}{4} = 31.25$$

(b) To find the median the data must be ranked from smallest to largest.

21, 25, 36, 43

Then since there is an even number of data points, the median is given by:

$$\frac{25 + 36}{2} = 30.5$$

(c) Since no point is repeated more than once, there is no mode.

3.   55, 55, 80, 92, 70

(a) Mean:

$$\overline{x} = \frac{55 + 55 + 80 + 92 + 70}{5} = 70.4$$

(b) To find the median, the data must be ranked from smallest to largest.

55, 55, 70, 80, 92

Now since there is an odd number of data points, the median is given by the middle point, 70.

(c) The mode is 55.

5.   65, 82, 82, 95, 70
   (a) Mean:

$$\bar{x} = \frac{65 + 82 + 82 + 95 + 70}{5} = 78.8$$

   (b) To find the median, the data must be ranked from smallest to largest.

65, 70, 82, 82, 95

Now since there is an odd number of data points, the median is given by the middle point, 82.

   (c) The mode is 82.

7.   48, 65, 80, 92, 80, 75
   (a) Mean:

$$\bar{x} = \frac{48 + 65 + 80 + 92 + 80 + 75}{6} = 73.333$$

   (b) To find the median, the data must be ranked from smallest to largest.

48, 65, 75, 80, 80, 92

Then since there is an even number of data points, the median is given by

$$\frac{75 + 80}{2} = 77.5$$

   (c) The mode is 80.

9.   (a) To find the mean age of the players, add all their ages and divide by 31, the number of players on the team.

$$\bar{x} = 31.29 \text{ years of age.}$$

   (b) To find the median, the data must be ranked from smallest to largest.

24, 24, 24, 24, 24, 27, 28, 29, 29, 30, 30, 30, 31, 31, 32, 32,
32, 32, 32, 32, 32, 32, 33, 34, 35, 35, 36, 37, 38, 40, 41

Since there is an odd number of players on the team, the median age is the middle value on the ordered list. The median is 32.

   (c) The mode is the value that occurs most often in the list. The mode is 32.

11.   To find the mean cost per share, we use a method similar to finding the mean of grouped data. The result is often called the weighted average, or the weighted mean.

$$\bar{x} = \frac{\sum(\text{cost per share})(\text{number of shares purchased})}{\text{total number of shares purchased}}$$

$$\bar{x} = \frac{(\$85)(50) + (\$105)(90) + (\$110)(120) + (\$130)(75)}{50 + 90 + 120 + 75}$$

$$= \frac{36650}{335}$$

$$= 109.40$$

The investor paid a mean price of $109.40 per share of the IBM stock.

13. (a) To find the mean of grouped data we use the following steps:

Step 1: Multiply the midpoint, m_i, of each class interval by the frequency, f_i, for that class interval.

Step 2: Sum the products found in step 1.

Step 3: Compute the mean by dividing the sum of $f_i m_i$ by the number n of entries.

Class Interval	Midpoint, m_i	Frequency, f_i	$f_i m_i$
10–14	12.5	9	112.5
15–19	17.5	469	8207.5
20–24	22.5	1018	22905
25–29	27.5	1088	29920
30–34	32.5	929	30192.5
35–39	37.5	452	16950
40–44	42.5	90	3825
45–49	47.5	4	190
		n = 4059	112302.5 = Sum of $f_i m_i$

The mean is $\bar{x} = \dfrac{112302.5}{4059} = 27.668$ years.

(b) To find the median of grouped data we use the following steps.

Step 1: Find the interval containing the median. The median is the middle value when the 4059 items are in ascending order. So the median in this example is the 2030th entry. The 2030th entry is in the interval 25–29.

Step 2: In the interval, count the number p of items remaining to reach the median. The first three intervals account for 1496 data points, there are $2030 - 1496 = 534$ points left, so $p = 534$.

Step 3: Calculate the interpolation factor.

q is the frequency for the interval containing the median; $q = 1088$

i is the size of the interval; $i = 5$

$$\text{interpolation factor} = \frac{p}{q} \cdot i$$

$$= \frac{534}{1088} \cdot (5) = 2.45$$

Step 4: The median M is

$$M = \begin{bmatrix} \text{lower limit of interval} \\ \text{containing the median} \end{bmatrix} + \begin{bmatrix} \text{interpolation factor} \end{bmatrix}$$

$$M = 25 + 2.45 = 27.45$$

15. (a) To find the mean of grouped data we use the following steps:

Step 1: Multiply the average number of employees for each size company, (2), by the sales per employee, (1), for that size company.

Step 2: Sum the products found in step 1.

Step 3: Compute the mean by dividing the sum of sales by the number n of employees.

Number of Employees	Sales per Employee, (thousands of dollars) (1)	Average number of Employees (2)	Sales per company (1)·(2)
1–4	112	2.5	280
5–19	128	12	1536
20–99	127	59.5	7556.5
100–499	118	299.5	35,341.0
500–4999	120	2749.5	329,940.0
		$n = 3123$	$374,653.5 =$ Sum of Sales

The mean is 119.966 thousand dollars ($119,966) in sales per employee.

(b) To find the median of grouped data we use the following steps

Step 1: Find the interval containing the median. The median is the middle value when the 3123 items are in ascending order. So the median in this example is the 1562nd entry. The 1562nd entry is in the interval 500–4999.

Step 2: We need no interpolation factor here, since we are looking for the median sales per employee. The sales per employee for companies employing between 500 and 4999 people, are $120,000.

The median sales per employee for firms having fewer than 5000 employee is:

$M = \$120,000.$

17. (a) To find the mean of grouped data we use the following steps:

Step 1: Multiply the midpoint, m_i, of each class interval by the frequency, f_i for that class interval.

Step 2: Sum the products found in step 1.

Step 3: Compute the mean by dividing the sum of $f_i m_i$ by the number n of entries.

Class Interval	Frequency, f	Midpoint, m_i	$f_i m_i$
20 – 24	345941	22.5	7783672.5
25 – 29	374629	27.5	10302297.5
30 – 34	428748	32.5	13934310.0
35 – 39	439137	37.5	16467637.5
40 – 44	414344	42.5	17609620.0
45 – 49	372814	47.5	17708665.0
50 – 54	292460	52.5	15354150.0
55 – 59	233615	57.5	13432862.5
60 – 64	204235	62.5	12764687.5
65 – 69	181977	67.5	12283447.5
70 – 74	150347	72.5	10900157.5
75 – 79	100068	77.5	7755270.0
80 – 84	50190	82.5	4140675.0
	$n = 3588505$		160437452.5

The mean is $\bar{x} = \dfrac{\Sigma f_i m_i}{n} = \dfrac{160437452.5}{3388505} = 44.709$ years old.

(b) To find the median of grouped data we use the following steps

Step 1: Find the interval containing the median. The median is the middle value when the 3588505 items are in ascending order. So the median in this example is the 1,794,253[rd] entry. This entry is in the interval 40 – 44.

Step 2: In the interval, count the number p of items remaining to reach the median. The first four intervals account for 1,588,455 data points, there are left $1,794,253 - 1,588,455 = 205,798$ items, so $p = 205,798$.

Step 3: Calculate the interpolation factor.

q is the frequency for the interval containing the median; $q = 414,344$

i is the size of the interval; $i = 5$

$$\text{interpolation factor} = \frac{p}{q} \cdot i$$

$$= \frac{205,798}{414,344} \cdot (5) = 2.483$$

Step 4: The median M is

$$M = \begin{bmatrix} \text{lower limit of interval} \\ \text{containing the median} \end{bmatrix} + \begin{bmatrix} \text{interpolation factor} \end{bmatrix}$$

$$M = 40 + 2.48 = 42.48 \text{ years of age.}$$

19. To find the mean of grouped data we use the following steps:
    Step 1: Multiply the midpoint, m_i, of each class interval by the frequency, f_i, for that class interval.
    Step 2: Sum the products found in step 1.
    Step 3: Compute the mean by dividing the sum of fm_i by the number n of entries.

Class Interval	Frequency, f	Midpoint, m_i	$f_i m_i$
0 – 999	10	500	5000
1000 – 1999	7	1500	10500
2000 – 2999	45	2500	112500
3000 – 3999	66	3500	231000
4000 – 4999	84	4500	378000
5000 – 5999	84	5500	462000
6000 – 6999	97	6500	630500
7000 – 7999	118	7500	885000
8000 – 8999	138	8500	1173000
9000 – 9999	110	9500	1045000
10000 – 10999	104	10500	1092000
11000 – 11999	82	11500	943000
12000 – 12999	61	12500	762500
13000 – 13999	34	13500	459000
14000 – 14999	29	14500	420500
	$n = 1069$		8609500

The mean tuition is $\bar{x} = \dfrac{\Sigma f_i m_i}{n} = \dfrac{8,609,500}{1069} = 8053.789 = \$8053.79$.

21. (a) Mean:

$$\bar{x} = \frac{34,000 + 35,000 + 36,000 + 36,500 + 65,000}{5} = \frac{206,500}{5} = 41,300$$

The mean faculty salary is $41,300.
Median: The median is the middle entry once the items are placed in ascending order.

$$M = \$36000$$

(b) The median describes the situation more realistically because the data is skewed to the right. This means that 4 salaries are clustered, while the salary of $65,000 is much higher.

(c) Answers will vary.

9.5 Measures of Dispersion

1.

Score, x_i	Deviation from the Mean, $x - \bar{x}$	Deviation Squared $\left(x - \bar{x}\right)^2$
4	−6.85714	47.02041
5	−5.85714	34.30612
9	−1.85714	3.44898
9	−1.85714	3.44898
10	−0.85714	0.73469
14	3.142857	9.87755
25	14.14286	200.0204
Mean $= 10.86$ $n = 7$	Sum $= 0$	Sum $= 298.8571$

The standard deviation is

$$S = \sqrt{\frac{\sum (x - \bar{x})^2}{n - 1}} = \sqrt{\frac{298.8571}{7 - 1}} = 7.058$$

3.

Score, x_i	Deviation from the Mean, $x - \bar{x}$	Deviation Squared $\left(x - \bar{x}\right)^2$
62	−3	9
58	−7	49
70	5	25
70	5	25
$n = 4$ $\bar{x} = 65$		sum $= 108$

The standard deviation is

$$S = \sqrt{\frac{\sum (x - \bar{x})^2}{n - 1}} = \sqrt{\frac{108}{3}} = 6$$

5.

Score, x_i	Deviation from the Mean, $x - \bar{x}$	Deviation Squared $(x - \bar{x})^2$
85	5	25
75	−5	25
62	−18	324
78	−2	4
100	20	400
$\bar{x} = 80$		sum = 778

The standard deviation is

$$S = \sqrt{\frac{\sum(x - \bar{x})^2}{n-1}} = \sqrt{\frac{778}{4}} = 13.946$$

7.

Class Interval	Class Midpoint m_i	Frequency f_i	$m_i \cdot f_i$	$m_i - \bar{x}$	$(m_i - \bar{x})^2 i$	$(m_i - \bar{x})^2 \cdot f_i$
10–16	13	1	13	−18.879	356.409	356.409
17–23	20	3	60	−11.879	141.106	423.317
24–30	27	10	270	−4.879	23.803	238.026
31–37	34	12	408	2.121	4.500	53.994
38–44	41	5	205	9.121	83.197	415.983
45–51	48	2	96	16.121	259.893	519.787
Sum		33	1052			2007.515

The mean is

$$\bar{x} = \frac{\sum m_i \cdot f_i}{n} = \frac{1052}{33} = 31.879$$

The standard deviation is

$$S = \sqrt{\frac{\sum(x - \bar{x})^2}{n-1}} = \sqrt{\frac{2007.515}{32}} = 7.921$$

9. The mean is

$$\bar{x} = \frac{968 + 893 + 769 + 845 + 922 + 915}{6} = \frac{5312}{6} = 885.333$$

Time x	Deviation from the Mean $x - \bar{x} = x - 885.333$	Deviation Squared $\left(x - \bar{x}\right)^2 = \left(x - 885.333\right)^2$
968	82.667	6833.778
893	7.667	58.778
769	−116.333	13533.444
845	− 40.333	1626.778
922	36.667	1344.444
915	29.667	880.111
Sum = 5312		24277.333

The standard deviation is

$$S = \sqrt{\frac{\sum (x - \bar{x})^2}{n-1}} = \sqrt{\frac{24277.333}{5}} = 69.681$$

11. (a) The range is the difference between the largest value and the smallest value. The range for the Yankees' ages is $41 - 24 = 17$ years.

For (b) and (c):

The mean (from 9.4 Problem 9) is 31.29 years of age.

Age x_i	Frequency f_i	$f_i \cdot x_i$	$x_i - \overline{x}$	$\left(x_i - \overline{x}\right)^2$	$\left(x_i - \overline{x}\right)^2 \cdot f_i$
24	5	120	−7.290	53.149	265.744
27	1	27	−4.290	18.407	18.407
28	1	28	−3.290	10.826	10.826
29	2	58	−2.290	5.246	10.491
30	3	90	−1.290	1.665	4.995
31	2	62	−0.290	0.084	0.169
32	8	256	0.710	0.504	4.029
33	1	33	1.710	2.923	2.923
34	1	34	2.710	7.342	7.342
35	2	70	3.710	13.762	27.523
36	1	36	4.710	22.181	22.181
37	1	37	5.710	32.600	32.600
38	1	38	6.710	45.020	45.020
40	1	40	8.710	75.858	75.858
41	1	41	9.710	94.278	94.278
Sum	31	970			622.387

(b) The standard deviation assuming sample data is

$$S = \sqrt{\frac{\sum(x - \overline{x})^2 \cdot f_i}{n-1}} = \sqrt{\frac{622.387}{30}} = 4.555 \text{ years.}$$

(c) The standard deviation assuming population data is

$$\sigma = \sqrt{\frac{\sum(x - \mu)^2 \cdot f_i}{n}} = \sqrt{\frac{622.387}{31}} = 4.481 \text{ years.}$$

(d) Answers will vary.

13. (a) These are population data. We have all the mothers in the United States represented.

   (b) The mean age of the mother (from 9.4 Problem 13) is 27.668 years. This is $\bar{x}$ in the following table.

Class Midpoint m_i	Frequency f_i	$m_i - \mu$	$(m_i - \mu)^2$	$(m_i - \mu)^2 \cdot f_i$
12.5	9	−15.168	230.054	2,070.485
17.5	469	−10.168	103.379	48,484.584
22.5	1018	−5.168	26.703	27,184.016
27.5	1088	−0.168	0.028	30.536
32.5	929	4.832	23.353	21,694.729
37.5	452	9.832	96.677	43,698.224
42.5	90	14.832	220.002	19,800.198
47.5	4	19.832	393.327	1,573.308
Sum	4059			164,536.080

The standard deviation, assuming population data, is

$$\sigma = \sqrt{\frac{\sum (x - \mu)^2 \cdot f_i}{n}} = \sqrt{\frac{164536.080}{4059}} = 6.367$$

15. (a) These are population data since all earthquakes are included.
To compute the mean and standard deviation, set up the table below. Use the first three columns to determine the mean. Then use the mean to complete the table and to calculate the standard deviation.

Class Midpoint m_i	Frequency f_i	$m_i \cdot f_i$	$m_i - \mu$	$(m_i - \mu)^2$	$(m_i - \mu)^2 \cdot f$
0.45	2,389	1,075.05	−2.828	7.996	19,102.929
1.45	752	1,090.40	−1.828	3.341	2,512.200
2.45	3,851	9,434.95	−0.828	0.685	2,638.627
3.45	5,639	19,454.55	0.172	0.030	167.298
4.45	6,943	30,896.35	1.172	1.374	9,540.767
5.45	832	4,534.40	2.172	4.719	3,925.912
6.45	113	728.85	3.172	10.063	1,137.134
7.45	10	74.50	4.172	17.408	174.076
Sum	20,529	67,289.05			39,198.945

   (b) The mean magnitude of the earthquakes worldwide in 1998 is

$$\mu = \frac{\sum m_i \cdot f_i}{n} = \frac{67,289.05}{20,529} = 3.278 \text{ in magnitude.}$$

   (c) The population standard deviation is

$$\sigma = \sqrt{\frac{\sum (x - \mu)^2 \cdot f_i}{n}} = \sqrt{\frac{39,198.945}{20,529}} = 1.382$$

17. In Problem 17 in 9.4 we found the mean age of the drivers in Tennessee to be 44.709 years of age. We will use the mean in the table below:

Class Midpoint m_i	Frequency f_i	$m_i - \bar{x}$	$\left(m_i - \bar{x}\right)^2$	$\left(m_i - \bar{x}\right)^2 \cdot f_i$
22.5	345941	−22.209	493.227	170,627,471.3
27.5	374629	−17.209	296.140	110,942,602.7
32.5	428748	−12.209	149.053	63,906,071.5
37.5	439137	−7.209	51.966	22,820,014.5
42.5	414344	−2.209	4.878	2,021,347.5
47.5	372814	2.791	7.791	2,904,692.3
52.5	292460	7.791	60.704	17,753,520.9
57.5	232615	12.791	163.617	38,223,370.3
62.5	204235	17.791	316.530	64,646,457.7
67.5	181977	22.791	519.443	94,526,607.2
72.5	150347	27.791	772.355	116,121,323.6
77.5	100068	32.791	1075.268	107,599,946.0
82.5	50190	37.791	1428.181	71,680,410.1
Sum	3588505			883,773,835.6

(a) The standard deviation of the age of drivers in Tennessee, assuming that the data are from a sample is

$$S = \sqrt{\frac{\sum (x - \bar{x})^2 \cdot f_i}{n-1}} = \sqrt{\frac{883,773,835.6}{3588504}} = 15.69328467 \approx 15.693$$

(b) The standard deviation of the age of drivers in Tennessee, assuming that the data are the population is

$$\sigma = \sqrt{\frac{\sum (x - \mu)^2 \cdot f_i}{n}} = \sqrt{\frac{883,773,835.6}{3588505}} = 15.69328248 \approx 15.693$$

(c) Answers will vary.

19. (a)   These are population data; all 4-year colleges are represented.
  (b)   In Problem 19 of Section 9.4 we found that the mean tuition at 4-year colleges in 1992–93 was \$8053.79. We will use the mean in the table that follows.

Class Midpoint m_i	Frequency f_i	$m_i - \mu$	$(m_i - \mu)^2$	$(m_i - \mu)^2 \cdot f_i$
500	10	−7553.789	57,059,722.02	570,597,220.2
1500	7	−6553.789	42,952,144.85	300,665,013.9
2500	45	−5553.789	30,844,567.67	1,388,005,545.0
3500	66	−4553.789	20,736,990.50	1,368,641,373.0
4500	84	−3553.789	12,629,413.32	1,060,870,719.0
5500	84	−2553.789	6,521,836.15	547,834,236.4
6500	97	−1553.789	2,414,258.97	234,183,120.4
7500	118	− 553.789	306,681.80	36,188,452.3
8500	138	446.211	199,104.63	27,476,438.3
9500	110	1446.211	2,091,527.45	230,068,019.6
10500	104	2446.211	5,983,950.28	622,330,828.8
11500	82	3446.211	11,876,373.10	973,862,594.5
12500	61	4446.211	19,768,795.93	1,205,896,552.0
13500	34	5446.211	29,661,218.76	1,008,481,438.0
14500	29	6446.211	41,553,641.58	1,205,055,606.0
Sum	1069			10,780,157,156.2

The standard deviation is

$$\sigma = \sqrt{\frac{\sum (x - \mu)^2 \cdot f_i}{n}} = \sqrt{\frac{10,780,157,156.2}{1069}} = 3175.584 = \$3175.58$$

21. We are told that $\mu = 25$ and $\sigma = 3$

(a) We want the outcome to be between 19 and 31, so $k = \mu - 19 = 25 - 19 = 6$, and according to Chebychev's theorem, the probability is at least

$$1 - \frac{\sigma^2}{k^2} = 1 - \frac{3^2}{6^2} = 1 - \frac{9}{36} = \frac{1}{4} = .75$$

At least 75% of the outcomes are between 19 and 31.

(b) We want the outcome to be between 20 and 30, so $k = \mu - 20 = 25 - 20 = 5$, and according to Chebychev's theorem, the probability is at least

$$1 - \frac{\sigma^2}{k^2} = 1 - \frac{3^2}{5^2} = 1 - \frac{9}{25} = .64$$

At least 64% of the outcomes are between 20 and 30.

(c) We want the outcome to be between 16 and 34, so $k = \mu - 16 = 25 - 16 = 9$, and according to Chebychev's theorem, the probability is at least

$$1 - \frac{\sigma^2}{k^2} = 1 - \frac{3^2}{9^2} = 1 - \frac{9}{81} = \frac{8}{9} = .889$$

At least 88.9% of the outcomes are between 16 and 34.

(d) We want the outcome to be less than 19 or more than 31. This is the opposite event from part (a), so the probability is at most $1 - .75 = .25$.

At most 25% of the outcomes are less than 19 or greater than 31.

(e) We want the outcome to be less than 16 or more than 34. This is the opposite event from part (c) so the probability is at most $1 - .889 = .111$.

At most 11.1% of the outcomes are less than 16 or more than 34.

23. Chebychev's theorem can be used to solve the problem.
We are told that $\mu = 6$ and $\sigma = 2$. We want to estimate the number of boxes that have between 0 and 12 defective watches.

$$k = \mu - 0 = 6$$

According to Chebychev's theorem the probability is at least $1 - \frac{2^2}{6^2} = 0.8889$ that there are

between 0 and 12 defective watches in a box.

So we expect at least 66.7% of the 1000 boxes or at least 889 boxes to have between 0 and 12 defective watches.

25. (a) These are population data because they record all live births in the United States for the years listed.

(b) The mean number of live births is

$$\mu = \frac{4,051,814 + 3,959,417 + 3,941,553 + 3,880,894 + 3,891,494 + 3,899,689}{6}$$

$$= 3,937,476.83$$

(c) To find the standard deviation, first complete the table below.

Year	Births, f_i	$f_i - \mu$	$(f_i - \mu)^2$
2000	4,051,814	114,337.167	13,072,987,758
1999	3,959,417	21,940.167	481,370,928
1998	3,941,553	4,076.167	16,615,137
1997	3,880,894	−56,582.833	3,201,616,990
1996	3,891,494	−45,982.833	2,114,420,931
1995	3,899,689	−37,787.833	1,427,920,323
Sum	23,624,861		20,314,932,067

The standard deviation of births over the 6-year period is

$$\sigma = \sqrt{\frac{(f_i - \mu)^2}{n}} = \sqrt{\frac{20,314,932,067}{6}} = 58,187.817$$

(d) The mean and standard deviations are exact because the data is not grouped.

(e) Answers will vary.

9.6 The Normal Distribution

1. The mean is always at the center of the normal curve.

$\mu = 8$

34.135% of the area under the curve lies between μ and σ.

$\sigma = 1$

3. The mean is always at the center of the normal curve.

$\mu = 18$

68.27% of the area under the normal curve lies between $\mu - \sigma$ and $\mu + \sigma$.

$\sigma = 1$

5. $Z = \dfrac{x - \mu}{\sigma}$ Here we are told $\mu = 13.1$ and $\sigma = 9.3$.

(a) $x = 7$

$Z = \dfrac{x - \mu}{\sigma} = \dfrac{7 - 13.1}{9.3} = -0.66$

(b) $x = 9$

$Z = \dfrac{x - \mu}{\sigma} = \dfrac{9 - 13.1}{9.3} = -0.44$

(c) $x = 13$

$Z = \dfrac{x - \mu}{\sigma} = \dfrac{13 - 13.1}{9.3} = -0.01$

(d) $x = 29$

$Z = \dfrac{x - \mu}{\sigma} = \dfrac{29 - 13.1}{9.3} = 1.71$

(e) $x = 37$

$Z = \dfrac{x - \mu}{\sigma} = \dfrac{37 - 13.1}{9.3} = 2.57$

(f) $x = 41$

$Z = \dfrac{x - \mu}{\sigma} = \dfrac{41 - 13.1}{9.3} = 3$

7.   Using the Standard Normal Curve Table on the inside back cover of the text, we can find the area under the standard normal curve between the standard score, Z, and the mean

(a) $Z = 0.89$

Read down the table under Z until you reach the row beginning 0.8. Then read across the row until you reach the entry under the column marked 0.09. The area under the standard normal curve between 0 and $Z = 0.89$ is 0.3133.

(b) $Z = 1.10$

Read down the table under Z until you reach the row beginning 1.1. The next entry in the row, 0.3642, represents the area under the standard normal curve between the mean and $Z = 1.10$.

(c) $Z = 3.06$

Read down the table under Z until you reach the row beginning 3.0. Then read across the row until you reach the entry under the column marked 0.06. The area under the standard normal curve between 0 and $Z = 3.06$ is 0.4989.

(d) $Z = -1.22$

There are no negative Z-scores on this table, but we use the symmetry of the normal curve to find the area under the curve between $Z = -1.22$ and the mean.
Read down the table under Z until you reach the row beginning 1.2. Then read across the row until you reach the entry under the column marked 0.02. Because of symmetry, 0.3888 is area under the standard normal curve between $Z = -1.22$ and 0, as well as between 0 and $Z = 1.22$.

(e) $Z = 2.30$

Read down the table under Z until you reach the row beginning 2.3. The next entry in the row, 0.4893, represents the area under the standard normal curve between the mean and $Z = 2.30$.

(f) $Z = -0.75$

There are no negative Z-scores on this table, but we use the symmetry of the normal curve to find the area under the curve between $Z = -0.75$ and the mean.
Read down the table under Z until you reach the row beginning 0.7. Then read across the row until you reach the entry under the column marked 0.05. Because of symmetry, 0.2734 is area under the standard normal curve between $Z = -0.75$ and 0, as well as between 0 and $Z = 0.75$.

9. We use the Standard Normal Curve Table and the interpretation of a Z-score as the number of standard deviations the original score is from its mean to solve this problem.

A – a score exceeds $\mu + 1.6\sigma$.

  Here $Z = 1.6$; the area under the standard normal curve between 0 and 1.6 is 0.4452. We need the area to the right of $Z = 1.6$. So we subtract 0.4452 from 0.5, the area under the curve to the right of 0.

$0.5000 - 0.4452 = 0.0548$

So, 5.48% of the class will get a grade of A.

B – a score is between $\mu + 0.6\,\sigma$ and $\mu + 1.6\sigma$.

  We know that the area under the curve from 0 to $Z = 1.6$ is 0.4452.
We find the area under the curve from 0 to $Z = 0.6$ is 0.2257. The area under the curve between the two Z-scores is the difference between the two areas, $0.4452 - 0.2257$, which is 0.2195.
So, 21.95% of the class will get a grade of B.

C – a score is between $\mu - 0.3\sigma$ and $\mu + 0.6\sigma$.

  We use symmetry and find the area under the curve between 0 and $Z = -0.3$ is 0.1179.
We found in part (b) that the area under the curve from 0 to $Z = 0.6$ is 0.2257.
Here since the Z-scores have opposite signs we add the two areas, $0.1179 + 0.2257$, and get 0.3436.
So, 34.36% of the class will get a grade of C.

D – a score is between $\mu - 1.4\sigma$ and $\mu - 0.3\sigma$.

  We use the symmetry of the normal curve to determine the areas between both 0 and $Z = -1.4$, which is 0.4192, and 0 and $Z = -0.3$, which is 0.1179. Now since both Z-scores are negative, the area between them is the difference between 0.4192 and 0.1179, which is 0.3013.
So, 30.13% of the class will get a grade of D.

F – a score is below $\mu - 1.4\sigma$.

  Use the symmetry of the curve to determine that the area between 0 and $Z = -1.4$ is 0.4192. But we need the area to the left of Z, so we subtract 0.4192 from 0.5, the total area under the curve to the left of 0.

$0.5000 - 0.4192 = 0.0808$

So, 8.08% of the class will get a grade of F.

(Note: When you add all the percents you should get 100%, the entire class.)

11. $Z = -0.5; A = 0.1915$
Since we want the area to the left of Z, subtract A from 0.5000.

Area $= 0.5000 - 0.1915$
Area $= 0.3085$

13. $Z_1 = -1.2; A_1 = 0.3849$
$Z_2 = 1.5; A_2 = 0.4332$
Since the Z-scores are on opposite sides of the mean, add the areas.

Area $= A_1 + A_2$
Area $= 0.3849 + 0.4332$
Area $= 0.8181$

15. We are told that $\overline{x} = 64$ and $S = 2$. We will convert the given heights to Z-scores to determine the percent of women in the required intervals. Then we will calculate how many of the 2000 women are in the interval.

(a) between 62 and 66 inches: These women are within 1 standard deviation of the mean, $64 - 2 = 62$ and $64 + 2 = 66$. This gives

$Z_1 = -1.0; A_1 = 0.3413$ and $Z_2 = 1.0; A_2 = 0.3413$

Since the two Z-scores are on opposite sides of the mean, add the corresponding areas

$A = A_1 + A_2 = 0.3413 + 0.3413 = 0.6826$

So, approximately 68.26% of the women or 1365 of the 2000 women sampled will be between 62 and 66 inches tall.

(b) between 60 and 68 inches tall: These women are within 2 standard deviations of the mean, $64 - 2(2) = 60$ and $64 + 2(2) = 68$. This gives

$Z_1 = -2.0; A_1 = 0.4772$ and $Z_2 = 2.0; A_2 = 0.4772$

Since the two Z-scores are on opposite sides of the mean, add the corresponding areas

$A = A_1 + A_2 = 0.4772 + 0.4772 = 0.9544$

So, approximately 95.44% of the women or 1909 of the 2000 women sampled will be between 60 and 68 inches tall.

(c) between 58 and 70 inches tall: These women are within 3 standard deviations of the mean, $64 - 3(2) = 58$ and $64 + 3(2) = 70$. This gives

$Z_1 = -3.0; A_1 = 0.4987$ and $Z_2 = 3.0; A_2 = 0.4987$

Since the two Z-scores are on opposite sides of the mean, add the corresponding areas

$A = A_1 + A_2 = 0.4987 + 0.4987 = 0.9974$

So, approximately 99.74% of the women or 1995 of the 2000 women sampled will be between 58 and 70 inches tall.

(d) more than 70 inches tall: These women are more than 3 standard deviations from the mean. We need to subtract $A = 0.4987$ from 0.5000 to find the percent of the population more than 3 standard deviations from the mean.

$0.5000 - 0.4987 = 0.0013$

So, approximately 0.13% of the women or 3 of the 2000 women sampled will be more than 70 inches tall.

(e) shorter than 58 inches: These women are more than 3 standard deviations below the mean. This gives $Z = -3$ and $A = 0.4987$. Since we want the area to the left of Z, we need to subtract A from 0.5000.

$0.5000 - 0.4987 = 0.0013$

So, approximately 0.13% of the women or 3 of the 2000 women sampled will be shorter than 58 inches.

17. We are given that $\mu = 130$ and $\sigma = 5.2$ pounds.
   (a)   Convert 142 pounds to a Z-score.

$$Z = \frac{x - \mu}{\sigma} = \frac{142 - 130}{5.2} = 2.308 \, ; \, A = 0.4896$$

We are interested in the area under the normal curve that to the right of Z, so we subtract A from 0.5000.

$$0.5000 - 0.4896 = 0.0104$$

Approximately 1.04% of the students or 1 student weighs at least 142 pounds.
   (b)   To find the range of weights that includes the middle 70% of the students, we need to find the Z-score corresponding to A = 0.35. (Because of symmetry 35% of the students will weigh more than the mean, and 35% will weigh less than the mean.)
   Looking in the body of the Standard Normal Curve Table, we find that 0.3508 is closest to A = 0.3500, and 0.3508 corresponds to $Z = 1.04$ and $Z = -1.04$.
   Using $Z = \pm 1.04$, $\mu = 130$ and $\sigma = 5.2$, we solve the equations

$$Z = \frac{x - \mu}{\sigma} \qquad\qquad Z = \frac{x - \mu}{\sigma}$$

$$1.04 = \frac{x - 130}{5.2} \qquad\qquad -1.04 = \frac{x - 130}{5.2}$$

$$5.408 = x - 130 \qquad\qquad -5.408 = x - 130$$

$$x = 135.408 \qquad\qquad x = 124.592$$

So, we would expect 70% of the students to weight between 124.59 and 135.41 pounds.

19. We are given that $\mu = 40$ months and $\sigma = 7$ months.
   Standardize 28 months and 42 months by finding the Z-scores corresponding to 28 and 40.

$$Z_1 = \frac{x_1 - \mu}{\sigma} = \frac{28 - 40}{7} = -1.71 \, ; \, A_1 = 0.4564$$

$$Z_2 = \frac{x_2 - \mu}{\sigma} = \frac{42 - 40}{7} = 0.29 \, ; \, A_2 = 0.1141$$

Now find the area under the standard normal curve between the two Z-scores. Since the Z-scores are on opposite sides of the mean, add the areas.

$$A = A_1 + A_2 = 0.4564 + 0.1141 = 0.5705$$

57.05% of the clothing can be expected to last between 28 and 42 months.

21. We are given that $\mu = 10,000$ and $\sigma = 1000$ persons.

(a) The lowest 70% of the attendance figures includes the 50% that are less than the mean and the 20% that are between the mean and some positive value of Z. We look in the body of the Standard Normal Curve Table for the number closest to 0.2000. We find 0.2000 is almost half way between 0.1985 and 0.2019, which correspond to $Z = 0.52$ and $Z = 0.53$ respectively. We will use an approximate $Z = 0.525$.

Using $Z = 0.525$, $\mu = 10,000$ and $\sigma = 1000$, we will solve the equation:

$$Z = \frac{x - \mu}{\sigma}$$

$$0.525 = \frac{x - 10000}{1000}$$

$$525 = x - 10000$$

$$x = 10,525$$

Attendance lower than 10,525 will be in the lowest 70% of the figures.

(b) To find the percent of attendance figures that falls between 8500 and 11,000 persons, find the Z-score for each and determine the area under the standard normal curve between the two Z-scores.

$$Z_1 = \frac{x_1 - \mu}{\sigma} = \frac{8500 - 10000}{1000} = -1.5 \,;\, A_1 = 0.4332$$

$$Z_2 = \frac{x_2 - \mu}{\sigma} = \frac{11000 - 10000}{1000} = 1.0 \,;\, A_2 = 0.3413$$

Since the Z-scores are on opposite sides of the mean, add the areas.

$$A = A_1 + A_2 = 0.4332 + 0.3413 = 0.7745$$

Approximately 77.45% of the attendance figures are between 8500 and 11,000 persons.

(c) Here we are looking for the percent of attendance figures that are more than 11,500 or less than 8500. First find the Z-scores and areas corresponding to $x = 11,500$ and $x = 8500$.

$$Z_1 = \frac{x_1 - \mu}{\sigma} = \frac{11500 - 10000}{1000} = 1.5 \,;\, A_1 = 0.4332$$

$Z_2 = -1.5$; $A_2 = 0.4332$ (from part (b))

Now find the area under the standard normal curve outside the two Z-scores. Since the Z-scores are on opposite sides of the mean, add these areas.

$$A = (0.5 - A_1) + (0.5 - A_2) = (0.5 - 0.4332) + (0.5 - 0.4332) = 0.1336$$

Approximately 13.36% of the attendance figures differ from the mean by 1500 persons or more.

23.   Transform each score to a standard score and then compare.

Colleen's score was 76; Colleen's standard score is $Z = \dfrac{x - \mu}{\sigma} = \dfrac{76 - 82}{7} = -0.857$

Mary's score was 89; Mary's standard score is $Z = \dfrac{x - \mu}{\sigma} = \dfrac{89 - 93}{2} = -2.0$

Kathleen's score was 21; Kathleen's standard score is $Z = \dfrac{x - \mu}{\sigma} = \dfrac{21 - 24}{9} = -0.33$

Kathleen has the highest relative standing.

25.   (a)   To find the line chart and the frequency curve, we need to first find the frequency distribution for the experiment. It is binomial with $n = 15$ and $p = .3$. The distribution is

Number of Heads, k	Probability $b(15,\ k,\ 0.3)$
0	.005
1	.031
2	.092
3	.170
4	.219
5	.206
6	.147
7	.081
8	.035
9	.012
10	.003
11	.0006
12	.0001
13	< 0.0001
14	< 0.0001
15	< 0.0001

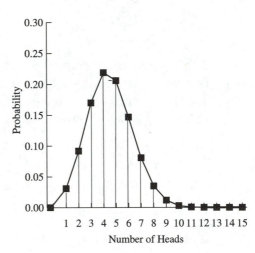

(b)   Answers will vary.

(c)   Mean: $\mu = n \cdot p = 15 \cdot (0.3) = 4.5$

Standard Deviation: $\sigma = \sqrt{np(1-p)} = \sqrt{15 \cdot (0.3)(.7)} = 1.775$

27.   To approximate the probability of obtaining between 285 and 315 successes in the 750 trials, we find the area under a normal curve from $x = 285$ to $x = 315$. We convert to Z-scores,

$x = 285:\ Z_1 = \dfrac{x - \mu}{\sigma} = \dfrac{285 - 300}{13.4} = -1.12 \quad A_1 = 0.3686$

$x = 315:\ Z_2 = \dfrac{x - \mu}{\sigma} = \dfrac{315 - 300}{13.4} = 1.12 \quad A_2 = 0.3686$

Since the values are on opposite sides of the mean, we add the areas.

$A = A_1 + A_2 = 0.3686 + 0.3686 = 0.7372$

The approximate probability that there are between 285 and 315 successes is .7372.

29. To approximate the probability of obtaining 300 or more successes in the 750 trials, we find the area under a normal curve to the right of $x = 300$. But 300 is the mean and we know the area to the right of the mean is 0.5.

So, the approximate probability of obtaining 300 or more successes is .5.

31. To approximate the probability of obtaining 325 or more successes in the 750 trials, we find the area under a normal curve to the right of $x = 325$. We first convert 325 to a Z-score.

$$x = 325: \quad Z = \frac{x - \mu}{\sigma} = \frac{325 - 300}{13.4} = 1.87 \quad A = 0.4693$$

We need the area to the right of $Z = 1.87$, so subtract A from 0.5000.

Area $= 0.5000 - 0.4693 = 0.0307$

The approximate probability of obtaining 325 or more successes is .0307.

33. This is a binomial distribution which we will approximate with a normal curve. We are told that the player's lifetime batting average is .250; this is his probability of success, and that he will bat 300 times; this is n.

We first find the mean and the standard deviation of the distribution.
mean: $\mu = np = (300)(.250) = 75$ hits
standard deviation: $\sigma = \sqrt{np(1-p)} = \sqrt{300 \cdot (.250)(.750)} = 7.5$

   (a) To approximate the probability that he gets at least 80 but no more than 90 hits, we find the area under a normal curve from $x = 80$ to $x = 90$. Convert 80 and 90 to Z-scores.

$$x = 80: \quad Z_1 = \frac{x - \mu}{\sigma} = \frac{80 - 75}{7.5} = 0.67 \quad A_1 = 0.2486$$

$$x = 90 \quad Z_2 = \frac{x - \mu}{\sigma} = \frac{90 - 75}{7.5} = 2 \quad A_2 = 0.4772$$

Since both 80 and 90 are greater than the mean, the area between the points is the difference between A_1 and A_2.

Area $= A_2 - A_1 = 0.4772 - 0.2486 = 0.2286$

The approximate probability of having at least 80, but no more than 90 hits is .2286.

   (b) To approximate the probability that 85 or more hits occur, we find the area under a normal curve to the right of $x = 85$. We first convert 85 to a Z-score.

$$x = 85: \quad Z = \frac{x - \mu}{\sigma} = \frac{85 - 75}{7.5} = 1.33 \quad A = 0.4082$$

Since the area to the right of the mean is 0.5, we subtract A from 0.5 to obtain the area to the right of Z.

Area $= 0.5000 - 0.4082 = 0.0918$

The approximate probability of 85 or more hit occurring is .0918.

35. This is a binomial distribution which we will approximate with a normal curve. We are told that the 1% of the packages do not seal properly; we will consider this p. 500 packages will be selected; 500 will be n.

We first find the mean and the standard deviation of the distribution.

mean: $\mu = np = (500)(.01) = 5$ packages

standard deviation: $\sigma = \sqrt{np(1-p)} = \sqrt{500 \cdot (.01)(.99)} = 2.2$

To approximate the probability that at least 10 packages are not properly sealed, we find the area under a normal curve to the right of $x = 10$. Converting 10 to a Z-score gives

$x = 10$: $Z = \dfrac{x-\mu}{\sigma} = \dfrac{10-5}{2.2} = 2.27$ $A = 0.4884$

Since we want the area to the right of $x = 10$, we subtract A from 0.5000.

Area $= 0.5000 - 0.4884 = 0.0116$

The approximate probability of selecting at least 10 unsealed packages is .0116.

37. $y = \dfrac{1}{\sqrt{2\pi}} e^{-(1/2)x^2}$

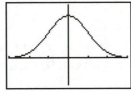

window: Xmin $= -3$; Xmax $= 3$
            Ymin $= -.25$; Ymax $= .5$

The function assumes its maximum at
$x = 0$.

39. From Problem 35, we have a binomial distribution with $n = 500$ and $p = .01$.
We want the probability that at least 10 bags of jelly beans are not sealed properly.
The TI-83 Plus has a function, called the cumulative binomial distribution, that calculates the probability of being less than or equal to x.

    The probability at least 10 bags are improperly sealed will be given by 1 minus the opposite event, the probability that at most 9 bags are improperly sealed.

```
1-binomcdf(500,.
01,9)
        .0311021071
```

The exact probability of selecting at least 10 unsealed packages is .0311.

Chapter 9 Review

True-False Items

1. False – The range is the difference between the largest number and the smallest number.

3. False – A small standard deviation indicates that the measures are cluster around the mean.

5. False – Approximately 68.27% of the area under the curve is within 1 standard deviation of the mean.

Fill in the Blanks

1. (a) mean
    (b) median
    (c) mode

3. bell

5. $\mu - k$ and $\mu + k$

Review Exercises

1. Variable: Circumference of head;
    Continuous

3. Variable: Number of people;
    Discrete

5. Variable: Number of defective products;
    Discrete

7. Answers will vary. All answers should include a method to choose a sample of 100 students from the population for which each student has an equal chance of being chosen.

9. Answers will vary. All answers should give examples of possible bias.

11. (a)

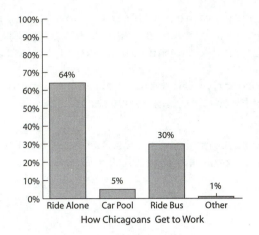

(b)

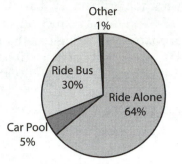

(c)   Answers will vary.

13. (a)

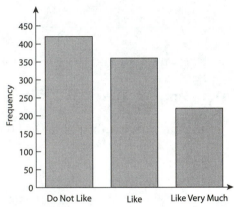

(b)

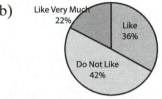

(c)   Answers will vary.

15. (a) American Indians made up the smallest percentage of college enrollment with 1%.

(b) Asian-Americans were overrepresented in four-year colleges. They were 4% of the general population, but 6% of the college student population.

(c) Approximately

$$(0.06)(8,897,000) = 533,820$$

Hispanics were enrolled in four-year colleges in 1997.

17. (a) Most Americans' highest level of educational attainment is a high school diploma.

(b) About 45 million Americans have at least a bachelor's degree. (Add the numbers of people with bachelors' and graduate/professional degrees.)

$$30,000 + 15,000 = 45,000 \text{ (thousands)}$$
$$= 45,000,000$$

(c) About 32 million Americans do not have a high school diploma.

$$12,000 + 20,000 = 32,000 \text{ (thousands)}$$
$$= 32,000,000$$

(d) About 48 million Americans went to college but do not have a bachelor's degree. (Add the numbers with some college and with associates' degrees.)

$$36,000 + 12,000 = 48,000 \text{ (thousands)}$$
$$= 48,000,000$$

19. (a)

Score	Frequency	Score	Frequency	Score	Frequency	Score	Frequency
21	2	62	1	74	1	87	2
33	1	63	2	75	1	89	1
41	2	66	2	77	1	90	2
42	1	68	1	78	2	91	1
44	1	69	1	80	4	92	1
48	1	70	2	82	1	95	1
52	2	71	1	83	1	99	1
55	1	72	2	85	2	100	2
60	1	73	2				

The range of scores is 79.

(b)
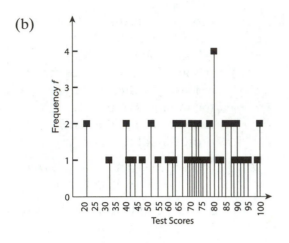

(c)
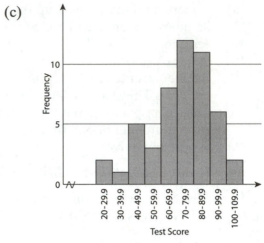

(d)
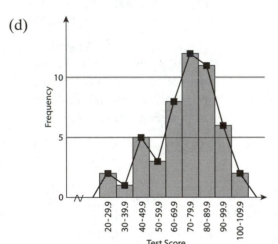

(f)
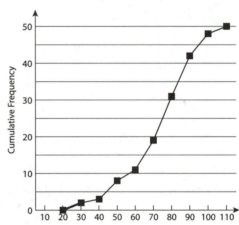

(e)

Score	Cumulative Frequency	Score	Cumulative Frequency	Score	Cumulative Frequency	Score	Cumulative Frequency
21	2	62	13	74	27	87	41
33	3	63	15	75	28	89	42
41	5	66	17	77	29	90	44
42	6	68	18	78	31	91	45
44	7	69	19	80	35	92	46
48	8	70	21	82	36	95	47
52	10	71	22	83	37	99	48
55	11	72	24	85	39	100	50
60	12	73	26				

21. (a)

Time	Freq.	Time	Freq.	Time	Freq.	Time	Freq.	Time	Freq
4'12"	1	4'46"	2	5'08"	1	5'43"	1	6'12"	1
4'15"	1	4'50"	1	5'12"	2	5'48"	1	6'30"	1
4'22"	1	4'52"	1	5'18"	1	5'50"	1	6'32"	1
4'30"	2	4'56"	1	5'20"	3	5'55"	1	6'40"	1
4'36"	1	5'01"	1	5'31"	2	6'01"	1	7'05"	1
4'39"	1	5'02"	1	5'37"	1	6'02"	1	7'15"	1
4'40"	2	5'06"	2	5'40"	2	6'10"	1		

The range is 3 minutes, 3 seconds.

(b)

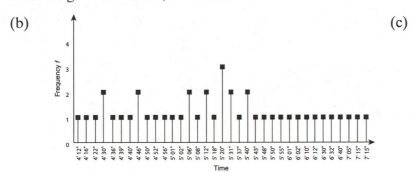

(c)

Class Interval	Frequency f_i
4'00" – 4'29"	3
4'30" – 4'59"	10
5'00" – 5'29"	11
5'30" – 5'59"	9
6'00" – 6'29"	4
6'30" – 6'59"	3
7'00" – 7'29"	2

(d)

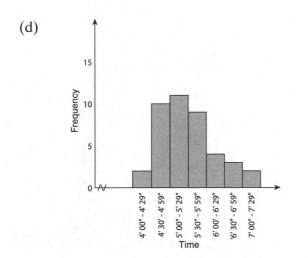

(e)

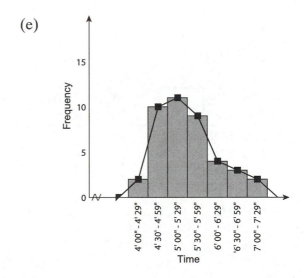

(f)

Class Interval	Cumulative Frequency
4'00"–4'29"	3
4'30"–4'59"	14
5'00"–5'29"	25
5'30"–5'59"	33
6'00"–6'29"	37
6'30"–6'59"	40
7'00"–7'29"	42

(g)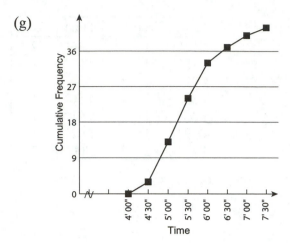

23. (a)

Age	Frequency f_i	Age	Frequency f_i
24	5	34	1
27	1	35	2
28	1	36	1
29	2	37	1
30	3	38	1
31	2	40	1
32	8	41	1
33	1		

(b)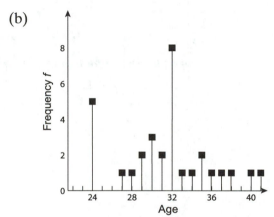

(c) and (f)

Class Interval	Frequency	Cumulative Frequency
20.0–24.9	5	5
25.0–29.9	4	9
30.0–34.9	15	24
35.0–39.9	5	29
40.0–44.9	2	31

There are 5 class intervals.

(d)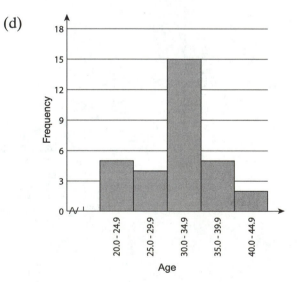

(e)

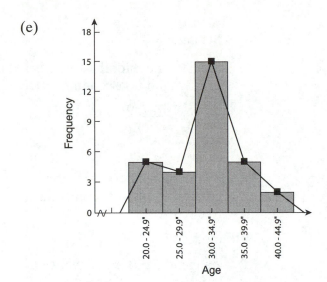

(g)

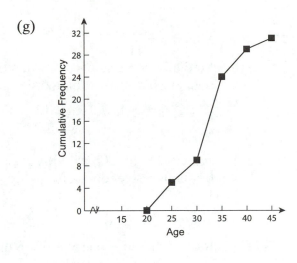

25. (a)

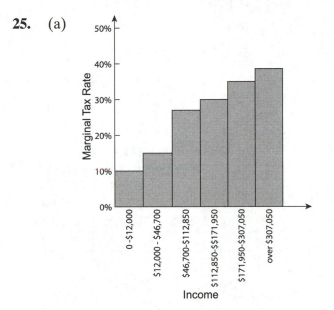

(b)

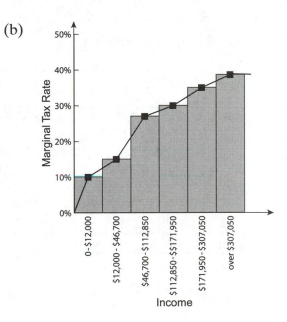

27. (a) Mean: $\overline{x} = \dfrac{63}{11} = 5.727$

    (b) Median: the middle value when the data are put in ascending order.

       0, 0, 2, 4, 4, 5, 8, 8, 10, 10, 12

       $M = 5$

    (c) Mode: the value(s) with the greatest frequency exceeding a frequently of 1.

       modes: 0, 4, 8, 10

    (d) Range: the difference between the largest and smallest value.

       Range $= 12 - 0 = 12$

    (e) Standard Deviation:

value x	freq f	$x - \overline{x}$	$\left(x - \overline{x}\right)^2$	$\left(x - \overline{x}\right)^2 f$
0	2	−5.727	32.799	65.597
2	1	−3.727	13.891	13.891
4	2	−1.727	2.983	5.965
5	1	−0.727	0.529	0.529
8	2	2.273	5.167	10.333
10	2	4.273	18.259	36.517
12	1	6.273	39.351	39.351
Sum	11			172.182

$$S = \sqrt{\dfrac{\sum (x_i - \overline{x})^2 f}{n-1}} = \sqrt{\dfrac{172.182}{10}} = 4.149$$

29. (a) Mean: $\overline{x} = \dfrac{162}{10} = 16.2$

    (b) Median: the middle value when the data are put in ascending order.

       2, 5, 5, 7, 7, 7, 9, 9, 11, 100

       $M = \dfrac{7+7}{2} = 7$

    (c) Mode: the value(s) with the greatest frequency exceeding a frequently of 1.

       mode $= 7$

    (d) Range: the difference between the largest and smallest value.

       Range $= 100 - 2 = 98$

    (e) Standard Deviation

value x	freq f	$x - \overline{x}$	$\left(x - \overline{x}\right)^2$	$\left(x - \overline{x}\right)^2 f$
2	1	−14.2	201.64	201.64
5	2	−11.2	125.44	250.88
7	3	−9.2	84.64	253.92
9	2	−7.2	51.84	103.68
11	1	−5.2	27.04	27.04
100	1	83.8	7022.44	7022.44
162	10			7859.6

$$S = \sqrt{\dfrac{\sum (x_i - \overline{x})^2 f}{n-1}} = \sqrt{\dfrac{7859.6}{9}} = 29.55$$

31. (a) Mean: $\bar{x} = \dfrac{70}{10} = 7$

(b) Median: the middle value when the data are put in ascending order.

1, 2, 5, 6, <u>7, 7</u>, 9, 10, 11, 12

$M = \dfrac{7+7}{2} = 7$

(c) Mode: the value(s) with the greatest frequency exceeding a frequently of 1.

mode = 7

(d) Range: the difference between the largest and smallest value.

Range = 12 − 1 = 11

(e) Standard Deviation.

value x	$x - \bar{x}$	$\left(x - \bar{x}\right)^2$
5	−2	4
7	0	0
7	0	0
9	2	4
10	3	9
11	4	16
1	−6	36
6	−1	1
2	−5	25
12	5	25
Sum = 70		120

$$S = \sqrt{\dfrac{\sum (x_i - \bar{x})^2}{n-1}} = \sqrt{\dfrac{120}{9}} = 3.65$$

33. The mean is a poor measure of central tendency in Problems 28 and 29. The mean is influenced by extreme values called outliers.

35. Answers will vary.

37. (a) Answers may vary. For this problem, we will assume we have sample data; it is implied that Joe has played more than 7 rounds of golf.

(b) Mean:

$$\bar{x} = \frac{529}{7} = 75.57$$

(c) Standard deviation:

value x	$x - \bar{x}$	$\left(x - \bar{x}\right)^2$
74	−1.571	2.469
72	−3.571	12.755
76	0.429	0.184
81	5.429	29.469
77	1.429	2.041
76	0.429	0.184
73	−2.571	6.612
Sum = 529		53.714

$$S = \sqrt{\frac{\sum (x_i - \bar{x})^2}{n-1}} = \sqrt{\frac{53.714}{6}} = 2.99$$

39.

Midpoint m	Frequency f	$f \cdot m$	$(m - \mu)$	$(m - \mu)^2$	$(m - \mu)^2 f$
2.5	9263	23157.5	−35.34	1249.50	11,574,147.6
7.5	9611	72082.5	−30.34	921.02	8,851,923.47
12.5	9765	122062.5	−25.34	642.54	6,274,372.91
17.5	9668	169190.0	−20.34	414.05	4,003,072.05
22.5	9162	206145.0	−15.34	235.57	2,158,298.51
27.5	8855	243512.5	−10.34	107.09	948,260.31
32.5	9890	321425.0	−5.4	28.60	262,897.90
37.5	11087	415762.5	−0.34	0.12	1,345.09
42.5	11473	487602.5	4.65	21.64	248,255.11
47.5	10202	484595.0	9.65	93.16	950,368.19
52.5	9049	475072.5	14.65	214.67	1,942,566.65
57.5	6992	402040.0	19.65	386.19	2,700,232.45
62.5	5670	354375.0	24.65	607.71	3,445,691.51
67.5	5087	343372.5	29.65	879.22	4,472,605.45
72.5	4972	360470.0	34.65	1200.74	5,970,076.79
77.5	4316	334490.0	39.65	1572.26	6,785,858.54
82.5	3072	253440.0	44.65	1993.77	6,124,871.47
87.5	1834	160475.0	49.65	2465.29	4,521,342.13
92.5	871	80567.5	54.65	2986.81	2,601,508.92
97.5	286	27885.0	59.65	3558.32	1,017,680.64
102.5	56	5740.0	64.65	4179.84	234,071.09
Sum	141,181	5,343,463			75,109,446.8

(a)　The approximate mean age of a female in the year 2000 was

$$\mu = \frac{5,343,463}{141,181} = 37.8 \text{ years of age.}$$

(b)　To find the median, we need to determine where the middle value is located.

　　　STEP 1: There are 141,181 total entries, making the middle entry the 70,591[st] one on the list. It is found in the class interval from 35 – 39. The first 7 intervals account for 66,214 entries.

　　　STEP 2: In the 8 interval there are 70,591 – 66,214 = 4377 more entries to reach the median. We let $p = 4377$.

　　　STEP 3: The interpolation factor will be

$$\frac{p}{q} i = \frac{4377}{11087} \cdot 5 = 1.974 \quad \text{where } q \text{ is the number of entries in the interval}$$

　　　　　　and i is the interval width.

　　　STEP 4: The approximate median age of a female in the year 2000 was
　　　　　　 M = 35 + 1.974 = 36.974 years of age.

(c)　The approximate standard deviation of the age of females in the year 2000 was

$$\sigma = \sqrt{\frac{\sum (x_i - \mu)^2}{n}} = \sqrt{\frac{75,109,446.8}{141,181}} = 23.07 \text{ years.}$$

41. Here $\mu = 12$ and $\sigma = 0.05$. We want to estimate the probability a jar contains between 11.9 and 12.1 ounces, so

$k = 12.1 - \mu = 12.1 - 12 = 0.1$ or
$k = \mu - 11.9 = 12 - 11.9 = 0.1$

By Chebychev's theorem, the probability a jar has between 11.9 and 12.1 ounces is at least

$$1 - \frac{\sigma^2}{k^2} = 1 - \frac{0.05^2}{0.1^2} = .75$$

43. Here $\mu = 10$ and $\sigma = 0.25$ pounds. We want to estimate the probability a bag weighs less than 9.5 or more than 10.5 pounds.

$k = 10.5 - \mu = 10.5 - 10 = 0.5$, or
$k = \mu - 9.5 = 10 - 9.5 = 0.5$

By Chebychev's theorem, the probability a bag weighs between 9.5 and 10.5 pounds is at least

$$1 - \frac{\sigma^2}{k^2} = 1 - \frac{0.25^2}{0.5^2} = .75$$

So the probability that a bag weighs less than 9.5 or more than 10.5 pounds is less than
$1 - 0.75 = 0.25$.

45.
$$Z = \frac{x - \mu}{\sigma} = \frac{8 - 10}{3}$$
$$Z = -\frac{2}{3} = -0.67$$

47.
$$Z = \frac{x - \mu}{\sigma} = \frac{8 - 1}{5}$$
$$Z = \frac{7}{5} = 1.40$$

49.
$$Z = \frac{x - \mu}{\sigma} = \frac{60 - 55}{3}$$
$$Z = \frac{5}{3} = 1.67$$

51.
$Z_1 = -1.35 \qquad A_1 = 0.4115$
$Z_2 = -2.75 \qquad A_2 = 0.4970$

Since both Z-scores are negative, they are both less than the mean. The area under the normal curve is the difference in A_1 and A_2.

Area $= A_2 - A_1 = 0.4970 - 0.4115$
$\qquad = 0.0855$

53.
$Z_1 = -0.75 \qquad A_1 = 0.2734$
$Z_2 = 2.1 \qquad A_2 = 0.4821$

Since the Z-scores have opposite signs, one is less than and the other is greater than the mean. The area under the normal curve is the sum of A_1 and A_2.

Area $= A_1 + A_2 = 0.2734 + 0.4821$
$\qquad = 0.7555$

55. We are told that $\mu = 25$ and $\sigma = 5$.

(a) We need to find the area under a normal curve between $x = 20$ and $x = 30$.

$$x = 20: \quad Z_1 = \frac{x - \mu}{\sigma} = \frac{20 - 25}{5} = -1 \quad A_1 = 0.3413$$

$$x = 30: \quad Z_2 = \frac{x - \mu}{\sigma} = \frac{30 - 25}{5} = 1 \quad A_2 = 0.3413$$

Since the x-values are on opposite sides of the mean, the area under the normal curve is the sum of A_1 and A_2.

Area $= A_1 + A_2 = 0.3413 + 0.3413 = 0.6826$

So, 68.26% of the scores fall between 20 and 30.

(b) We need to find the area under a normal curve to the right of $x = 35$.

$$x = 35: \quad Z = \frac{x - \mu}{\sigma} = \frac{35 - 25}{5} = 2 \quad A = 0.4772$$

Since the area to the right of the mean is 0.5, the area to the right of $x = 35$ will be

Area $= 0.5000 - 0.4772 = 0.0228$

So, 2.28% of the scores are above 35.

57. Here $\mu = 14$ and $\sigma = 1.25$ years. We need the area under a normal curve to the left of 10 years, 4 months which is equivalent to $10\frac{1}{3} = \frac{31}{3}$ years. We convert $x = \frac{31}{3}$ to a Z-score.

$$x = \frac{31}{3}: \quad Z = \frac{x - \mu}{\sigma} = \frac{\frac{31}{3} - 14}{1.25} = -2.93 \quad A = 0.4983$$

The area to the left of x is

Area $= 0.500 - 0.4983 = 0.0017$

Approximately 0.17% of dogs will die before reaching the age of 10 years 4 months.

59. We need to convert both test grades to standard scores, and then compare the grades.

Mathematics: score: 89

$\mu = 79; \sigma = 5$

$$Z = \frac{x - \mu}{\sigma} = \frac{89 - 79}{5} = 2.0$$

Sociology: score: 79

$\mu = 72; \sigma = 3.5$

$$Z = \frac{x - \mu}{\sigma} = \frac{79 - 72}{3.5} = 2.0$$

Bob scored equally well on both exams.

61. We need to find the area under a normal curve between $x = 30$ and $x = 50$. We convert these values to Z-scores, using $\mu = 40$ and $\sigma = \sqrt{25} = 5$ units.

$x = 30$: $\quad Z_1 = \dfrac{x - \mu}{\sigma} = \dfrac{30 - 40}{5} = -2 \qquad A_1 = 0.4772$

$x = 50$: $\quad Z_2 = \dfrac{x - \mu}{\sigma} = \dfrac{50 - 40}{5} = 2 \qquad A_2 = 0.4772$

The area between the x-values is the sum of A_1 and A_2 which equals 0.9544. So, the probability that the week's production will be between 30 and 50 is .9544.

63. This is a binomial distribution which we will approximate using a normal curve. We obtain the area under a normal curve to the right of $x = 160$ by converting x to a Z-score. We have $p = .7$, the probability of a positive result, and
$\qquad n = 200$, the number of blood samples.

We can calculate

$\mu = np = (200)(0.7) = 140$, and
$\sigma = \sqrt{np(1-p)} = \sqrt{200(0.7)(0.3)} = 6.48$

$x = 160$: $Z = \dfrac{x - \mu}{\sigma} = \dfrac{160 - 140}{6.48} = 3.09 \quad A = 0.4990$

The probability of obtaining more than 160 positive results is $0.5000 - 0.4990 = .001$.

65. This is a binomial distribution which we will approximate using a normal curve. We obtain the area under a normal curve between $x = 20$ and $x = 30$ by converting both to Z-scores.
We have $p = 0.05$, the probability the gate does not open, and
$\qquad n = 500$, the number of times the gate is opened.

We can calculate

$\mu = np = (500)(0.05) = 25$, and
$\sigma = \sqrt{np(1-p)} = \sqrt{500(0.05)(0.95)} = 4.87$

$x = 20$: $\quad Z_1 = \dfrac{x - \mu}{\sigma} = \dfrac{20 - 25}{4.87} = -1.03 \qquad A_1 = 0.3485$

$x = 30$: $\quad Z_2 = \dfrac{x - \mu}{\sigma} = \dfrac{30 - 25}{4.87} = 1.03 \qquad A_2 = 0.3485$

The area between the x-values is the sum of A_1 and A_2 which equals 0.697. the probability that the toll gate fails to work between 20 and 30 times is .697.

Chapter 9 Project

1. Mean high temperature: $\dfrac{2394}{30} = 79.8°$ Fahrenheit.

    Mean low temperature: $\dfrac{1893}{30} = 63.1°$ Fahrenheit.

    Total rainfall: 4.5 inches

3. Mean monthly temperature = $\dfrac{\text{sum of 30 high and 30 low temperatures}}{60} = \dfrac{4287}{60} = 71.45°$

    Mean of mean high and mean low temperatures:

    $$\dfrac{1}{2}\left[\left(\dfrac{\text{sum of 30 daily high temperatures}}{30}\right) + \left(\dfrac{\text{sum of 30 daily low temperatures}}{30}\right)\right]$$

    $$= \dfrac{1}{2}\left[\dfrac{\text{sum of 60 daily high and low temperatures}}{30}\right]$$

    $$= \left[\dfrac{\text{sum of 60 daily high and low temperatures}}{60}\right]$$

    = mean monthly temperature.

5. Modal high temperature was 85°.
    Modal low temperatures were 56° and 70°.

7. High temperatures: Low temperatures:
    Standard deviation: Standard deviation:

    $$\sigma = \sqrt{\dfrac{\sum (x - \mu)^2}{n}} = \sqrt{\dfrac{1788.8}{30}} = 7.72 \qquad \sigma = \sqrt{\dfrac{\sum (x - \mu)^2}{n}} = \sqrt{\dfrac{1168.7}{30}} = 6.24$$

    Since for the high temperatures, the mean < median < mode, the distribution is not symmetric, but is skewed to the left. This means that there are a few extremely low temperatures that are influencing the mean.

    One might underestimate the probability of obtaining a temperature of between 90° and 92° using a normal distribution.

9.

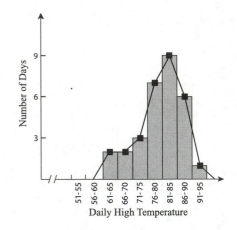

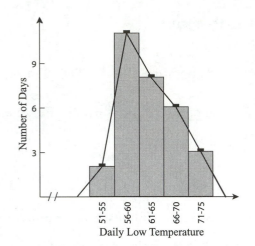

11.

Midpoint	f	$f \cdot m$	$m - \mu$	$(m - \mu)^2$	$f \cdot (m - \mu)^2$
62.5	2	125	−16.833	283.361	566.722
67.5	2	135	−11.833	140.028	280.056
72.5	3	217.5	−6.833	46.694	140.083
77.5	7	542.5	−1.833	3.361	23.528
82.5	9	742.5	3.167	10.028	90.25
87.5	6	525	8.167	66.694	400.167
92.5	1	92.5	13.167	173.361	173.361
Sum	30	2380			1674.167

Approximate Mean: $\mu = \dfrac{\sum f \cdot m}{n} = \dfrac{2380}{30} = 79.33\,°$

Approximate Standard Deviation: $\sigma = \sqrt{\dfrac{\sum f \cdot (m - \mu)^2}{n}} = \sqrt{\dfrac{1674.167}{30}} = 7.47°$

Answers and reasons will vary.

Mathematical Questions from Professional Exams

1.   Answer: (e) 10
In a normal curve, approximately 95% of the area under the curve lies within 2 standard deviations of the mean, so $K = 2\sigma$.
The experiment is binomial, meaning $\sigma = \sqrt{np(1-p)}$.
$p = \dfrac{1}{6}$, the probability of throwing a sum of 7
$n = 180$ throws
$K = 2\sqrt{(180)\left(\dfrac{1}{6}\right)\left(\dfrac{5}{6}\right)} = 10$

3.   Answer (c) .68
We are looking for the probability the balls are within 1 standard deviation of the mean weight of the 100 balls. The area under a normal curve between $\mu - \sigma$ and $\mu + \sigma$ is approximately .68.

Chapter 10
Functions and Their Graphs

10.1 Graphs of Equations

1. The point: $(3, 4)$
   (a) The point symmetric with respect to the x-axis: $(3, -4)$.

   (b) The point symmetric with respect to the y-axis: $(-3, 4)$.

   (c) The point symmetric with respect to the origin: $(-3, -4)$.

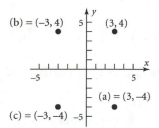

3. The point: $(-2, 1)$
   (a) The point symmetric with respect to the x-axis: $(-2, -1)$.

   (b) The point symmetric with respect to the y-axis: $(2, 1)$.

   (c) The point symmetric with respect to the origin: $(2, -1)$.

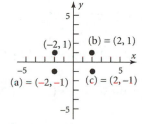

5. The point: $(1, 1)$
   (a) The point symmetric with respect to the x-axis: $(1, -1)$.

   (b) The point symmetric with respect to the y-axis: $(-1, 1)$.

   (c) The point symmetric with respect to the origin: $(-1, -1)$.

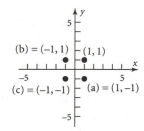

7. The point: $(-3, -4)$
   (a) The point symmetric with respect to the x-axis: $(-3, 4)$.

   (b) The point symmetric with respect to the y-axis: $(3, -4)$.

   (c) The point symmetric with respect to the origin: $(3, 4)$.

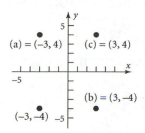

9. The point: $(0, -3)$
    (a) The point symmetric with respect to the x-axis: $(0, 3)$.

    (b) The point symmetric with respect to the y-axis: $(0, -3)$.

    (c) The point symmetric with respect to the origin: $(0, 3)$.

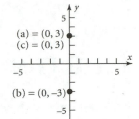

11. (a) The x-intercepts are $(1, 0)$ and $(-1, 0)$. There is no y-intercept.

    (b) The graph is symmetric with respect to the x-axis, the y-axis, and the origin.

13. (a) The x-intercepts are $\left(-\dfrac{\pi}{2}, 0\right)$ and $\left(\dfrac{\pi}{2}, 0\right)$. The y-intercept is $(0, 1)$.

    (b) The graph is symmetric only with respect to the y-axis.

15. (a) The x-intercept is $(0, 0)$. The y-intercept is also $(0, 0)$.

    (b) The graph is symmetric only with respect to the x-axis.

17. (a) The x-intercept is $(1, 0)$; there is no y-intercept.

    (b) The graph has no symmetry with respect to the x-axis, the y-axis, or the origin.

19. (a) The x-intercepts are $(1, 0)$ and $(-1, 0)$. The y-intercept is $(0, -1)$.

    (b) The graph is symmetric only with respect to the y-axis.

21. (a) There is no x-intercept, and there is no y-intercept.

    (b) The graph is symmetric only with respect to the origin.

23. For each point we check to see if the point satisfies the equation $y = x^4 - \sqrt{x}$.
    (a) $(0, 0)$: $0^4 - \sqrt{0} = 0$
    The equation is satisfied so the point $(0, 0)$ is on the graph.

    (b) $(1, 1)$: $1^4 - \sqrt{1} = 0 \neq 1$
    The equation is not satisfied so the point $(1, 1)$ is not on the graph.

    (c) $(-1, 0)$: $y = (-1)^4 - \sqrt{(-1)}$ is not a real number, so the point $(-1, 0)$ is not on the graph.

25. For each point we check to see if the point satisfies the equation $y^2 = x^2 + 9$ or $y^2 - x^2 = 9$.

(a) $(0, 3)$: $3^2 - 0^2 = 9$
The equation is satisfied so the point $(0, 3)$ is on the graph.

(b) $(3, 0)$: $0^2 - 3^2 = -9 \neq 9$
The equation is not satisfied so the point $(0, 3)$ is not on the graph.

(c) $(-3, 0)$: $0^2 - (-3)^2 = -9 \neq 9$
The equation is not satisfied so the point $(0, -3)$ is not on the graph.

27. For each point we check to see if the point satisfies the equation $x^2 + y^2 = 4$.

(a) $(0, 2)$: $0^2 + 2^2 = 4$
The equation is satisfied so the point $(0, 2)$ is on the graph.

(b) $(-2, 2)$: $(-2)^2 + 2^2 = 8 \neq 4$
The equation is not satisfied so the point $(-2, 2)$ is not on the graph.

(c) $\left(\sqrt{2}, \sqrt{2}\right)$: $\left(\sqrt{2}\right)^2 + \left(\sqrt{2}\right)^2 = 2 + 2 = 4$
The equation is satisfied so the point $\left(\sqrt{2}, \sqrt{2}\right)$ is on the graph.

29. To find the x-intercept(s) we let $y = 0$ and solve the equation $x^2 = 0$ or $x = 0$. So the
x-intercept is $(0, 0)$.

To find the y-intercept(s) we let $x = 0$ and get $0 = y$. So the y-intercept is also $(0, 0)$.

To test the graph of the equation $x^2 = y$ for symmetry with respect to the x-axis, we
replace y by $-y$ in the equation $x^2 = y$. Since the resulting equation $x^2 = -y$ is not
equivalent to the original equation, the graph is not symmetric with respect to the x-axis.

To test the graph of the equation $x^2 = y$ for symmetry with respect to the y-axis, we
replace x by $-x$ in the equation and simplify.
$$x^2 = y$$
$$(-x)^2 = y \text{ or } x^2 = y$$
Since the resulting equation is equivalent to the original equation, the graph is symmetric
with respect to the y-axis.

To test the graph of the equation $x^2 = y$ for symmetry with respect to the origin, we
replace x by $-x$ and y by $-y$ in the equation and simplify.
$$x^2 = y$$
$$(-x)^2 = -y \text{ or } x^2 = -y$$
Since the resulting equation is not equivalent to the original equation, the graph is not
symmetric with respect to the origin.

31. To find the x-intercept(s) we let $y = 0$ and solve the equation $0 = 3x$ or $x = 0$. So the
x-intercept is $(0, 0)$.

To find the y-intercept(s) we let $x = 0$ and get $y = 3 \cdot 0 = 0$. So the y-intercept is also
$(0, 0)$.

To test the graph of the equation $y = 3x$ for symmetry with respect to the x-axis, we replace y by $-y$ in the equation. Since the resulting equation $-y = 3x$ is not equivalent to the original equation, the graph is not symmetric with respect to the x-axis.

To test the graph of the equation $y = 3x$ for symmetry with respect to the y-axis, we replace x by $-x$ in the equation and simplify.
$$y = 3x$$
$$y = 3 \cdot (-x) = -3x$$
Since the resulting equation is not equivalent to the original equation, the graph is not symmetric with respect to the y-axis.

To test the graph of the equation $y = 3x$ for symmetry with respect to the origin, we replace x by $-x$ and y by $-y$ in the equation and simplify.
$$y = 3x$$
$$-y = 3 \cdot (-x)$$
$$-y = -3x$$
$$y = 3x$$
Since the resulting equation is equivalent to the original equation, the graph is symmetric with respect to the origin.

33. To find the x-intercept(s) we let $y = 0$ and solve the equation
$$x^2 + 0 - 9 = 0$$
$$x^2 = 9$$
$$x = -3 \text{ or } x = 3$$
So the x-intercepts are $(-3, 0)$ and $(3, 0)$.

To find the y-intercept(s) we let $x = 0$ and solve the equation
$$0^2 + y - 9 = 0$$
$$y = 9$$
So the y-intercept is $(0, 9)$.

To test the graph of the equation $x^2 + y - 9 = 0$ for symmetry with respect to the x-axis, we replace y by $-y$ in the equation.
$$x^2 + (-y) - 9 = 0$$
$$x^2 - y - 9 = 0$$
Since the resulting equation is not equivalent to the original equation, the graph is not symmetric with respect to the x-axis.

To test the graph of the equation $x^2 + y - 9 = 0$ for symmetry with respect to the y-axis, we replace x by $-x$ in the equation and simplify.
$$x^2 + y - 9 = 0$$
$$(-x)^2 + y - 9 = 0$$
$$x^2 + y - 9 = 0$$
Since the resulting equation is equivalent to the original equation, the graph is symmetric with respect to the y-axis.

To test the graph of the equation $x^2 + y - 9 = 0$ for symmetry with respect to the origin, we replace x by $-x$ and y by $-y$ in the equation and simplify.

$$x^2 + y - 9 = 0$$
$$(-x)^2 + (-y) - 9 = 0$$
$$x^2 - y - 9 = 0$$

Since the resulting equation is not equivalent to the original equation, the graph is not symmetric with respect to the origin.

35. To find the x-intercept(s) we let $y = 0$ and solve the equation
$$9x^2 + 4 \cdot 0^2 = 36$$
$$9x^2 = 36$$
$$x^2 = 4$$
$$x = -2 \text{ or } x = 2$$
So the x-intercepts are $(-2, 0)$ and $(2, 0)$.

To find the y-intercept(s) we let $x = 0$ and solve the equation
$$9 \cdot 0^2 + 4y^2 = 36$$
$$4y^2 = 36$$
$$y^2 = 9$$
$$y = -3 \text{ or } y = 3$$
So the y-intercepts are $(0, -3)$ and $(0, 3)$.

To test the graph of the equation $9x^2 + 4y^2 = 36$ for symmetry with respect to the x-axis, we replace y by $-y$ in the equation.
$$9x^2 + 4y^2 = 36$$
$$9x^2 + 4(-y)^2 = 36$$
$$9x^2 + 4y^2 = 36$$
Since the resulting equation is equivalent to the original equation, the graph is symmetric with respect to the x-axis.

To test the graph of the equation $9x^2 + 4y^2 = 36$ for symmetry with respect to the y-axis, we replace x by $-x$ in the equation and simplify.
$$9x^2 + 4y^2 = 36$$
$$9(-x)^2 + 4y^2 = 36$$
$$9x^2 + 4y^2 = 36$$
Since the resulting equation is equivalent to the original equation, the graph is symmetric with respect to the y-axis.

To test the graph of the equation $9x^2 + 4y^2 = 36$ for symmetry with respect to the origin, we replace x by $-x$ and y by $-y$ in the equation and simplify.
$$9x^2 + 4y^2 = 36$$
$$9(-x)^2 + 4(-y)^2 = 36$$
$$9x^2 + 4y^2 = 36$$
Since the resulting equation is equivalent to the original equation, the graph is symmetric with respect to the origin.

37. To find the x-intercept(s) we let $y = 0$ and solve the equation
$$0 = x^3 - 27$$
$$x^3 = 27$$
$$x = 3$$
So the x-intercept is $(3, 0)$.

To find the y-intercept(s) we let $x = 0$ and solve the equation
$$y = 0^3 - 27$$
$$y = -27$$
So the y-intercept is $(0, -27)$.

To test the graph of the equation $y = x^3 - 27$ for symmetry with respect to the x-axis, we replace y by $-y$ in the equation.
$$y = x^3 - 27$$
$$-y = x^3 - 27$$
Since the resulting equation is not equivalent to the original equation, the graph is not symmetric with respect to the x-axis.

To test the graph of the equation $y = x^3 - 27$ for symmetry with respect to the y-axis, we replace x by $-x$ in the equation and simplify.
$$y = x^3 - 27$$
$$y = (-x)^3 - 27$$
$$y = -x^3 - 27$$
Since the resulting equation is not equivalent to the original equation, the graph is not symmetric with respect to the y-axis.

To test the graph of the equation $y = x^3 - 27$ for symmetry with respect to the origin, we replace x by $-x$ and y by $-y$ in the equation and simplify.
$$y = x^3 - 27$$
$$(-y) = (-x)^3 - 27$$
$$-y = -x^3 - 27$$
$$y = x^3 + 27$$
Since the resulting equation is not equivalent to the original equation, the graph is not symmetric with respect to the origin.

39. To find the x-intercept(s) we let $y = 0$ and solve the equation
$$0 = x^2 - 3x - 4$$
$$0 = (x - 4)(x + 1)$$
$$x = 4 \text{ or } x = -1$$
So the x-intercepts are $(4, 0)$ and $(-1, 0)$.

To find the y-intercept(s) we let $x = 0$ and solve the equation
$$y = 0^2 - 3 \cdot 0 - 4$$
$$y = -4$$
So the y-intercept is $(0, -4)$.

To test the graph of the equation $y = x^2 - 3x - 4$ for symmetry with respect to the x-axis, we replace y by $-y$ in the equation.
$$y = x^2 - 3x - 4$$
$$-y = x^2 - 3x - 4$$
Since the resulting equation is not equivalent to the original equation, the graph is not symmetric with respect to the x-axis.

To test the graph of the equation $y = x^2 - 3x - 4$ for symmetry with respect to the y-axis, we replace x by $-x$ in the equation and simplify.

$$y = x^2 - 3x - 4$$
$$y = (-x)^2 - 3 \cdot (-x) - 4$$
$$y = x^2 + 3x - 4$$

Since the resulting equation is not equivalent to the original equation, the graph is not symmetric with respect to the y-axis.

To test the graph of the equation $y = x^2 - 3x - 4$ for symmetry with respect to the origin, we replace x by $-x$ and y by $-y$ in the equation and simplify.

$$y = x^2 - 3x - 4$$
$$(-y) = (-x)^2 - 3 \cdot (-x) - 4$$
$$-y = x^2 + 3x - 4$$

Since the resulting equation is not equivalent to the original equation, the graph is not symmetric with respect to the origin.

41. To find the x-intercept(s) we let $y = 0$ and solve the equation

$$0 = \frac{3x}{x^2 + 9}$$

$$0 = 3x \qquad\qquad \text{Since } x^2 + 9 \neq 0, \text{ multiply both sides by } x^2 + 9.$$

$$x = 0$$

So the x-intercept is $(0, 0)$.

To find the y-intercept(s) we let $x = 0$ and solve the equation

$$y = \frac{3 \cdot 0}{0^2 + 9} = 0$$

So the y-intercept is also $(0, 0)$.

To test the graph of the equation $y = \dfrac{3x}{x^2 + 9}$ for symmetry with respect to the x-axis, we replace y by $-y$ in the equation.

$$y = \frac{3x}{x^2 + 9}$$

$$-y = \frac{3x}{x^2 + 9}$$

Since the resulting equation is not equivalent to the original equation, the graph is not symmetric with respect to the x-axis.

To test the graph of the equation $y = \dfrac{3x}{x^2 + 9}$ for symmetry with respect to the y-axis, we replace x by $-x$ in the equation and simplify.

$$y = \frac{3x}{x^2 + 9}$$

$$y = \frac{3(-x)}{(-x)^2 + 9} = -\frac{3x}{x^2 + 9}$$

Since the resulting equation is not equivalent to the original equation, the graph is not symmetric with respect to the y-axis.

To test the graph of the equation $y = \dfrac{3x}{x^2+9}$ for symmetry with respect to the origin, we replace x by $-x$ and y by $-y$ in the equation and simplify.

$$y = \frac{3x}{x^2+9}$$

$$-y = \frac{3(-x)}{(-x)^2+9}$$

$$-y = -\frac{3x}{x^2+9}$$

$$y = \frac{3x}{x^2+9}$$

Since the resulting equation is equivalent to the original equation, the graph is symmetric with respect to the origin.

43.   To find the x-intercept(s) we let $y = 0$ and solve the equation

$$0 = |x|$$
$$0 = x$$

So the x-intercept is $(0, 0)$.

To find the y-intercept(s) we let $x = 0$ and solve the equation

$$y = |0| = 0$$

So the y-intercept is also $(0, 0)$.

To test the graph of the equation $y = |x|$ for symmetry with respect to the x-axis, we replace y by $-y$ in the equation.

$$y = |x|$$
$$-y = |x|$$

Since the resulting equation is not equivalent to the original equation, the graph is not symmetric with respect to the x-axis.

To test the graph of the equation $y = |x|$ for symmetry with respect to the y-axis, we replace x by $-x$ in the equation and simplify.

$$y = |x|$$
$$y = |-x| = |x|$$

Since the resulting equation is equivalent to the original equation, the graph is symmetric with respect to the y-axis.

To test the graph of the equation $y = |x|$ for symmetry with respect to the origin, we replace x by $-x$ and y by $-y$ in the equation and simplify.

$$y = |x|$$
$$-y = |-x| = |x|$$

Since the resulting equation is not equivalent to the original equation, the graph is not symmetric with respect to the origin.

45.

x	$x^3 - 1 = y$	(x, y)
-2	$(-2)^3 - 1 = -9$	$(-2, -9)$
-1	$(-1)^3 - 1 = -2$	$(-1, -2)$
0	$0^3 - 1 = -1$	$(0, -1)$
1	$1^3 - 1 = 0$	$(1, 0)$
2	$2^3 - 1 = 7$	$(2, 7)$

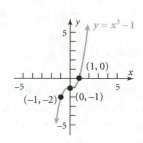
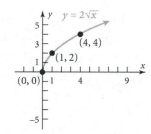

47.

x	$2\sqrt{x} = y$	(x, y)
0	$2\sqrt{0} = 0$	$(0, 0)$
1	$2\sqrt{1} = 2$	$(1, 2)$
4	$2\sqrt{4} = 4$	$(4, 4)$
9	$2\sqrt{9} = 6$	$(9, 6)$

49. If $(a, 2)$ is a point on the graph, then $x = a$ and $y = 2$ must satisfy the equation
$$y = 3x + 5$$
$$2 = 3a + 5$$
$$3a = -3$$
$$a = -1$$

51. If (a, b) is a point on the graph, then $x = a$ and $y = b$ must satisfy the equation
$$2x + 3y = 6$$
$$2a + 3b = 6$$
$$2a = 6 - 3b$$
$$a = \frac{6 - 3b}{2} = 3 - \frac{3}{2}b$$

53. (a)

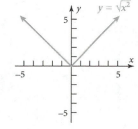
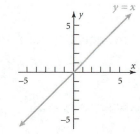

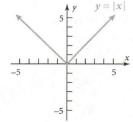
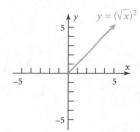

The graphs of $y = \sqrt{x^2}$ and $y = |x|$ are the same.

(b) - (d) Answers will vary.

55.   Let (x, y) be a point on the graph of the equation.
(1) Assume the graph is symmetric with respect to both axes. Then because of symmetry with respect to the y-axis, the point $(-x, y)$ is on the graph. Similarly, because of symmetry with respect to the x-axis, the point $(-x, -y)$ is also on the graph. So the graph is symmetric with respect to the origin.

(2) Assume the graph is symmetric with respect to the x-axis and the origin. Then because of symmetry with respect to the x-axis, the point $(x, -y)$ is on the graph. Similarly, because of symmetry with respect to the origin, the point $(-x, y)$ is also on the graph. So the graph of the equation is symmetric with respect to the y-axis.

(3) Assume the graph is symmetric with respect to the y-axis and the origin. Then because of symmetry with respect to the y-axis, the point $(-x, y)$ is on the graph. Similarly, because of symmetry with respect to the origin, the point $(x, -y)$ is also on the graph. So the graph of the equation is symmetric with respect to the x-axis.

10.2 Functions

1.   (a) We substitute 0 for x in the equation for f to get
$$f(0) = 3 \cdot 0^2 + 2 \cdot 0 - 4 = -4$$

(b) We substitute 1 for x in the equation for f to get
$$f(1) = 3 \cdot 1^2 + 2 \cdot 1 - 4 = 1$$

(c) We substitute (-1) for x in the equation for f to get
$$f(-1) = 3 \cdot (-1)^2 + 2 \cdot (-1) - 4 = -3$$

(d) We substitute $(-x)$ for x in the equation for f to get
$$f(-x) = 3 \cdot (-x)^2 + 2 \cdot (-x) - 4 = 3x^2 - 2x - 4$$

(e) $-f(x) = -(3x^2 + 2x - 4) = -3x^2 - 2x + 4$

(f) We substitute $(x + 1)$ for x in the equation for f to get
$$f(x+1) = 3(x+1)^2 + 2(x+1) - 4 = (3x^2 + 6x + 3) + (2x + 2) - 4 = 3x^2 + 8x + 1$$

(g) We substitute $2x$ for x in the equation for f to get
$$f(2x) = 3 \cdot (2x)^2 + 2 \cdot 2x - 4 = 12x^2 + 4x - 4$$

(h) We substitute $(x + h)$ for x in the equation for f to get
$$f(x+h) = 3(x+h)^2 + 2(x+h) - 4 = (3x^2 + 6xh + 3h^2) + (2x + 2h) - 4$$

$$= 3x^2 + 2x + 2h + 6xh + 3h^2 - 4$$

3. (a) We substitute 0 for x in the equation for f to get
$$f(0) = \frac{0}{0^2 + 1} = 0$$

(b) We substitute 1 for x in the equation for f to get
$$f(1) = \frac{1}{1^2 + 1} = \frac{1}{2}$$

(c) We substitute -1 for x in the equation for f to get
$$f(-1) = \frac{-1}{(-1)^2 + 1} = -\frac{1}{2}$$

(d) We substitute $-x$ for x in the equation for f to get
$$f(-x) = \frac{-x}{(-x)^2 + 1} = -\frac{x}{x^2 + 1}$$

(e) $-f(x) = -\dfrac{x}{x^2 + 1}$

(f) We substitute $(x + 1)$ for x in the equation for f to get
$$f(x + 1) = \frac{x + 1}{(x + 1)^2 + 1} = \frac{x + 1}{x^2 + 2x + 2}$$

(g) We substitute $2x$ for x in the equation for f to get
$$f(2x) = \frac{2x}{(2x)^2 + 1} = \frac{2x}{4x^2 + 1}$$

(h) We substitute $(x + h)$ for x in the equation for f to get
$$f(x + h) = \frac{x + h}{(x + h)^2 + 1} = \frac{x + h}{x^2 + 2xh + h^2 + 1}$$

5. (a) We substitute 0 for x in the equation for f to get
$$f(0) = |0| + 4 = 4$$

(b) We substitute 1 for x in the equation for f to get
$$f(1) = |1| + 4 = 5$$

(c) We substitute -1 for x in the equation for f to get
$$f(-1) = |-1| + 4 = 5$$

(d) We substitute $-x$ for x in the equation for f to get
$$f(-x) = |-x| + 4 = |x| + 4 = f(x)$$

(e) $-f(x) = -[|x|+4] = -|x|-4$

(f) We substitute $(x + 1)$ for x in the equation for f to get
$$f(x+1) = |x+1|+4$$

(g) We substitute $2x$ for x in the equation for f to get
$$f(2x) = |2x|+4 = 2|x|+4$$

(h) We substitute $(x + h)$ for x in the equation for f to get
$$f(x+h) = |x+h|+4$$

7. (a) We substitute 0 for x in the equation for f to get
$$f(0) = \frac{2 \cdot 0 + 1}{3 \cdot 0 - 5} = \frac{1}{-5} = -\frac{1}{5}$$

(b) We substitute 1 for x in the equation for f to get
$$f(1) = \frac{2 \cdot 1 + 1}{3 \cdot 1 - 5} = \frac{3}{-2} = -\frac{3}{2}$$

(c) We substitute -1 for x in the equation for f to get
$$f(-1) = \frac{2 \cdot (-1) + 1}{3 \cdot (-1) - 5} = \frac{-1}{-8} = \frac{1}{8}$$

(d) We substitute $-x$ for x in the equation for f to get
$$f(-x) = \frac{2(-x) + 1}{3(-x) - 5} = \frac{-2x + 1}{-3x - 5} = \frac{2x - 1}{3x + 5}$$

(e) $-f(x) = -\dfrac{2x+1}{3x-5}$

(f) We substitute $(x + 1)$ for x in the equation for f to get
$$f(x+1) = \frac{2(x+1) + 1}{3(x+1) - 5} = \frac{2x+3}{3x-2}$$

(g) We substitute $2x$ for x in the equation for f to get
$$f(2x) = \frac{2(2x) + 1}{3(2x) - 5} = \frac{4x+1}{6x-5}$$

(h) We substitute $(x + h)$ for x in the equation for f to get
$$f(x+h) = \frac{2(x+h) + 1}{3(x+h) - 5} = \frac{2x+2h+1}{3x+3h-5}$$

9. If $f(x) = 4x + 3$ then $f(x + h) = 4(x + h) + 3 = 4x + 4h + 3$, and the difference quotient is
$$\frac{f(x+h)-f(x)}{h} = \frac{4x+4h+3-(4x+3)}{h} = \frac{4h}{h} = 4$$

11. If $f(x) = x^2 - x + 4$ then $f(x + h) = (x + h)^2 - (x + h) + 4 = x^2 + 2xh + h^2 - x - h + 4$, and the difference quotient is
$$\frac{f(x+h)-f(x)}{h} = \frac{x^2+2xh+h^2-x-h+4-(x^2-x+4)}{h} = \frac{2xh+h^2-h}{h}$$
$$= \frac{h(2x+h-1)}{h} = 2x+h-1$$

13. If $f(x) = x^3$ then $f(x + h) = (x + h)^3 = x^3 + 3x^2h + 3xh^2 + h^3$, and the difference quotient is
$$\frac{f(x+h)-f(x)}{h} = \frac{x^3+3x^2h+3xh^2+h^3-x^3}{h} = \frac{3x^2h+3xh^2+h^3}{h}$$
$$= \frac{h(3x^2+3xh+h^2)}{h} = 3x^2+3xh+h^2$$

15. If $f(x) = x^4$ then $f(x + h) = (x + h)^4 = x^4 + 4x^3h + 6x^2h^2 + 4xh^3 + h^4$, and the difference quotient is
$$\frac{f(x+h)-f(x)}{h} = \frac{(x^4+4x^3h+6x^2h^2+4xh^3+h^4)-x^4}{h}$$
$$= \frac{4x^3h+6x^2h^2+4xh^3+h^4}{h} = \frac{h(4x^3+6x^2h+4xh^2+h^3)}{h}$$
$$= 4x^3+6x^2h+4xh^2+h^3$$

17. Since there is only one y-value for each x-value, $y = x^2$ is a function.

19. Since there is only one y-value for each x-value, $y = \dfrac{1}{x}$ is a function.

21. To determine whether the equation $y^2 = 4 - x^2$ is a function, we need to solve the equation for y.
$$y = \pm\sqrt{4-x^2}$$
For values of x between -2 and 2, two values of y result. This means the equation is not a function.

23. To determine whether the equation $x = y^2$ is a function, we need to solve the equation for y.
$$y = \pm\sqrt{x}$$
For values of $x > 0$, two values of y result. This means the equation is not a function.

25. Since there is only one y-value for each x-value, $y = 2x^2 - 3x + 4$ is a function.

27. To determine whether the equation $2x^2 + 3y^2 = 1$ is a function, we need to solve the equation for y.

$$3y^2 = 1 - 2x^2$$

$$y = \pm\sqrt{\frac{1 - 2x^2}{3}}$$

For values of x between $-\frac{\sqrt{2}}{2}$ and $\frac{\sqrt{2}}{2}$, two values of y result. This means the equation is not a function.

29. Since the operations of multiplication and addition can be performed on any real number, the domain of $f(x) = -5x + 4$ is all real numbers. (The domain of a polynomial function is always all real numbers.)

31. The function $f(x) = \dfrac{x}{x^2 + 1}$ is defined provided the denominator is not equal to zero. Since $x^2 + 1$ never equals zero, the domain of f is all real numbers.

33. The function $g(x) = \dfrac{x}{x^2 - 16}$ is defined provided the denominator is not equal to zero. $x^2 - 16 = 0$ when $x = -4$ or when $x = 4$, so the domain of function g is the set $\{x \mid x \neq -4, x \neq 4\}$.

35. The function $F(x) = \dfrac{x - 2}{x^3 + x}$ is defined provided the denominator is not equal to zero. $x^3 + x = x(x^2 + 1) = 0$ when $x = 0$, so the domain of function F is the set $\{x \mid x \neq 0\}$.

37. The function $h(x) = \sqrt{3x - 12}$ is defined provided the radicand is nonnegative. $3x - 12 \geq 0$ when $3x \geq 12$ or when $x \geq 4$. So the domain of function h is the set $\{x \mid x \geq 4\}$ or the interval $[4, \infty)$.

39. The function $f(x) = \dfrac{4}{\sqrt{x - 9}}$ is defined provided the radicand is positive. Zero must be eliminated to avoid dividing by zero. $x - 9 > 0$ when $x > 9$. So the domain of function f is the set $\{x \mid x > 9\}$ or the interval $(9, \infty)$.

41. The function $p(x) = \sqrt{\dfrac{2}{x - 1}}$ is defined provided the denominator is positive. (Radicals must be nonnegative and division by zero is not defined.) $x - 1 > 0$ when $x > 1$. So the domain of function p is the set $\{x \mid x > 1\}$ or the interval $(1, \infty)$.

43. If $f(2) = 5$ then $x = 2$, $y = 5$ must satisfy the equation $y = 2x^3 + Ax^2 + 4x - 5$. We substitute for x and y and solve for A.

$$5 = 2 \cdot 2^3 + A \cdot 2^2 + 4 \cdot 2 - 5$$

$$5 = 16 + 4A + 8 - 5$$
$$5 = 4A + 19$$
$$-14 = 4A$$
$$A = -\frac{7}{2}$$

45. If $f(0) = 2$ then $x = 0$, $y = 2$ must satisfy the equation $y = \dfrac{3x + 8}{2x - A}$. We substitute for x and y and solve for A.

$$2 = \frac{3 \cdot 0 + 8}{2 \cdot 0 - A} = \frac{8}{-A}$$
$$A = -4$$

The function f is undefined when $2x - A = 0$ because division by zero is undefined. Since $A = -4$, f is undefined when $2x + 4 = 0$ or when $x = -2$.

47. If $f(4) = 0$ then $x = 4$, $y = 0$ must satisfy the equation $y = \dfrac{2x - A}{x - 3}$. We substitute for x and y and solve for A.

$$0 = \frac{2 \cdot 4 - A}{4 - 3} = 8 - A$$
$$A = 8$$

The function f is undefined when $x - 3 = 0$ or when $x = 3$ because division by zero is undefined.

49. $G(x) = 10x$ dollars where \$10 is the hourly wage and x is the number of hours worked. The domain of G is $\{x \mid x \geq 0\}$ since a person cannot work a negative number of hours.

51. Revenue $R = R(x) = xp$.

$$R(x) = x\left(-\frac{1}{5}x + 100\right)$$
$$= -\frac{1}{5}x^2 + 100x \qquad 0 \leq x \leq 500$$

53. Revenue $R = R(x) = xp$. Here the demand equation is expressed in terms of p, so we first solve for p.

$$x = -20p + 100 \qquad\qquad 0 \leq p \leq 5$$
$$x - 100 = -20p$$
$$p = \frac{x - 100}{-20} = -\frac{1}{20}x + 5 \qquad 0 \leq x \leq 100$$

We then construct the revenue function R.

$$R(x) = x\left(-\frac{1}{20}x + 5\right) = -\frac{1}{20}x^2 + 5x \qquad 0 \leq x \leq 100$$

55. If 1990 is represented by $t = 0$, then 2010, $(1990 + 20 = 2010)$, will be represented by $t = 20$, $(0 + 20 = 20)$. To project the number (in thousands) of acres of wheat that will be planted in 2010 we evaluate the function A at $t = 20$.

$$A(20) = -119(20^2) + 113(20) + 73,367 = 28,027$$

In 2010, it is projected that 28,027,000 acres of wheat will be planted.

57. 2010 is 16 years after 1994. The expected SAT mathematics score will be

$$A(16) = -0.04(16^3) + 0.43(16^2) + 0.24(16) + 506 = 456$$

59. (a) With no wind, $x = 500$ miles per hour, the cost per passenger is

$$C(500) = 100 + \frac{500}{10} + \frac{36,000}{500} = 222 \text{ dollars.}$$

(b) With a head wind of 50 miles per hour, $x = 500 - 50 = 450$ miles per hour, and the cost per passenger is

$$C(450) = 100 + \frac{450}{10} + \frac{36,000}{450} = 225 \text{ dollars.}$$

(c) With a tail wind of 100 miles per hour, $x = 500 + 100 = 600$ miles per hour and the cost per passenger is

$$C(600) = 100 + \frac{600}{10} + \frac{36,000}{600} = 220 \text{ dollars.}$$

(d) With a head wind of 100 miles per hour, $x = 500 - 100 = 400$ miles per hour, and the cost per passenger is

$$C(400) = 100 + \frac{400}{10} + \frac{36,000}{400} = 230 \text{ dollars.}$$

61. (a) $h(x) = 2x$

$$\begin{aligned} h(a+b) &= 2(a+b) \\ &= 2a + 2b \\ &= h(a) + h(b) \end{aligned}$$

(b) $g(x) = x^2$

$$\begin{aligned} g(a+b) &= (a+b)^2 \\ &= a^2 + 2ab + b^2 \\ &= g(a) + g(b) + 2ab \\ &\neq g(a) + g(b) \end{aligned}$$

(c) $F(x) = 5x - 2$

$$\begin{aligned} F(a+b) &= 5(a+b) - 2 \\ &= 5a + 5b - 2 \\ &= 5a - 2 + 2 + 5b - 2 \\ &= F(a) + F(b) + 2 \\ &\neq F(a) + F(b) \end{aligned}$$

(d) $G(x) = \dfrac{1}{x}$

$$G(a+b) = \frac{1}{a+b}$$

$$\neq G(a) + G(B) = \frac{1}{a} + \frac{1}{b} = \frac{b+a}{ab}$$

Only (a) has the property, $f(a + b) = f(a) + f(b)$.

63. Research sources and explanations may vary.

10.3 Graphs of Functions; Properties of Functions

1. The graph fails the vertical line test. It is not the graph of a function.

3. This is the graph of a function.
(a) The domain is the set $\{x \mid -\pi \leq x \leq \pi\}$ or the interval $[-\pi, \pi]$; the range is the set $\{y \mid -1 \leq y \leq 1\}$ or the interval $[-1, 1]$.

(b) The x-intercepts are $\left(-\dfrac{\pi}{2}, 0\right)$ and $\left(\dfrac{\pi}{2}, 0\right)$; the y-intercept is the point $(0, 1)$.

(c) The graph is symmetric with respect to the y-axis.

5. The graph fails the vertical line test. It is not the graph of a function.

7. This is the graph of a function.
(a) The domain is the set $\{x \mid x > 0\}$ or the interval $(0, \infty)$; the range is all real numbers.

(b) There is no y-intercept; the x-intercept is the point $(1, 0)$.

(c) The graph has no symmetries with respect to the x-axis, the y-axis, or the origin.

9. This is the graph of a function.
(a) The domain is all real numbers; the range is the set $\{y \mid y \leq 2\}$ or the interval $(-\infty, 2]$.

(b) The x-intercepts are the points $(-3, 0)$ and $(3, 0)$; the y-intercept is the point $(0, 2)$.

(c) The graph is symmetric with respect to the y-axis.

11. This is the graph of a function.
(a) The domain is the set of all real numbers; the range is the set $\{y \mid y \geq -3\}$ or the interval $[-3, \infty)$.

(b) The x-intercepts are the points $(1, 0)$ and $(3, 0)$; the y-intercept is the point $(0, 9)$.

(c) The graph has no symmetries with respect to the x-axis, the y-axis, or the origin.

13. (a) $f(0) = 3; f(-6) = -3$

(b) $f(6) = 0; f(11) = 1$

(c) $f(3)$ is above the x-axis so it is positive.

(d) $f(-4) = -1$ is below the x-axis so it is negative.

(e) $f(x) = 0$ whenever the graph crosses or touches the x-axis. So $f(x) = 0$ when $x = -3$, when $x = 6$, and when $x = 10$.

(f) $f(x) > 0$ whenever the graph of f is above the x-axis. $f(x) > 0$ on the intervals $[-3, 6]$ and $[10, 11]$.

(g) The domain of f is the set $\{x \mid -6 \leq x \leq 11\}$ or the interval $[-6, 11]$.

(h) The range of f is the set $\{y \mid -3 \leq y \leq 4\}$ or the interval $[-3, 4]$.

(i) The x-intercepts are the points $(-3, 0)$, $(6, 0)$, and $(10, 0)$.

(j) The y-intercept is the point $(0, 3)$.

(k) Draw the horizontal line $y = \dfrac{1}{2}$ on the same axes as the graph. Count the number of times the two graphs intersect. They intersect 3 times.

(l) Draw the vertical line $x = 5$ on the same axes as the graph of f. Count the number of times the two graphs intersect. They intersect once.

(m) $f(x) = 3$ whenever the y-value equals 3. This occurs when $x = 0$ and when $x = 4$.

(n) $f(x) = -2$ whenever the y-value equals -2. This occurs when $x = -5$ and when $x = 8$.

15. (a) The point $(-1, 2)$ is on the graph of f if $x = -1$, $y = 2$ satisfies the equation.
$$f(x) = 2x^2 - x - 1$$
$$2(-1)^2 - (-1) - 1 = 2$$
So the point $(-1, 2)$ is on the graph of f.

(b) If $x = -2, f(-2) = 2(-2)^2 - (-2) - 1 = 9$. The point $(-2, 9)$ is on the graph of f.

(c) If $f(x) = -1$, then
$$2x^2 - x - 1 = -1$$
$$2x^2 - x = 0$$
$$x(2x - 1) = 0$$
$$x = 0 \text{ or } 2x - 1 = 0$$
$$x = 0 \quad \text{or} \quad x = \frac{1}{2}$$

The points $(0, -1)$ and $\left(\dfrac{1}{2}, -1\right)$ are on the graph of f.

(d) The domain of f is all real numbers.

(e) To find the x-intercepts, we let $f(x) = y = 0$ and solve for x.
$$2x^2 - x - 1 = 0$$
$$(2x + 1)(x - 1) = 0$$

$$x = -\frac{1}{2} \quad \text{or} \quad x = 1$$

The x-intercepts are $\left(-\frac{1}{2}, 0\right)$ and $(1, 0)$.

(f) To find the y-intercept, we let $x = 0$ and solve for y.
$$f(0) = 2(0^2) - 0 - 1 = -1$$
The y-intercept is $(0, -1)$.

17. (a) The point $(3, 14)$ is on the graph of f if $x = 3$, $y = 14$ satisfies the equation.
$$f(x) = \frac{x+2}{x-6}$$
$$\frac{3+2}{3-6} = \frac{5}{-3} = -\frac{5}{3}$$

The point $\left(3, -\frac{5}{3}\right)$ is on the graph of f, but the point $(3, 14)$ is not.

(b) If $x = 4$, $f(4) = \frac{4+2}{4-6} = \frac{6}{-2} = -3$. The point $(4, -3)$ is on the graph of f.

(c) If $f(x) = 2$, then
$$\frac{x+2}{x-6} = 2$$
$$x + 2 = 2(x-6)$$
$$x + 2 = 2x - 12$$
$$x = 14$$
The point $(14, 2)$ is on the graph of f.

(d) The function f is a rational function. Rational functions are not defined at values of x that would cause the denominator to equal zero. So the domain of f is the set $\{x \mid x \neq 6\}$.

(e) To find the x-intercepts, we let $f(x) = y = 0$ and solve for x.
$$\frac{x+2}{x-6} = 0$$
$$x + 2 = 0$$
$$x = -2$$
The x-intercept is $(-2, 0)$.

(f) To find the y-intercept, we let $x = 0$ and solve for y.
$$f(0) = \frac{0+2}{0-6} = -\frac{1}{3}$$
The y-intercept is $\left(0, -\frac{1}{3}\right)$.

19. (a) The point $(-1, 1)$ is on the graph of f if $x = -1, y = 1$ satisfies the equation.

$$f(x) = \frac{2x^2}{x^4 + 1}$$

$$\frac{2(-1)^2}{(-1)^4 + 1} = \frac{2}{2} = 1$$

The point $(-1, 1)$ is on the graph of f.

(b) If $x = 2, f(2) = \frac{2 \cdot 2^2}{2^4 + 1} = \frac{8}{17}$. The point $\left(2, \frac{8}{17}\right)$ is on the graph of f.

(c) If $f(x) = 1$, then

$$\frac{2x^2}{x^4 + 1} = 1$$
$$2x^2 = x^4 + 1$$
$$x^4 - 2x^2 + 1 = 0$$
$$(x^2 - 1)(x^2 - 1) = 0$$
$$(x - 1)(x + 1) = 0$$
$$x - 1 = 0 \quad \text{or} \quad x + 1 = 0$$
$$x = 1 \quad \text{or} \quad x = -1$$

The points $(1, 1)$ and $(-1, 1)$ are on the graph of f.

(d) The function f is a rational function. Rational functions are not defined at values of x that would cause the denominator to equal zero, but this denominator can never equal zero. So the domain of f is the set of all real numbers.

(e) To find the x-intercepts, we let $f(x) = y = 0$ and solve for x. The x-intercept is $(0, 0)$.

(f) To find the y-intercept, we let $x = 0$ and solve for y.

$$f(0) = \frac{2 \cdot 0^2}{0^4 + 1} = 0$$

The y-intercept is also $(0, 0)$.

21. Yes, the function is increasing when $-8 < x < -2$.

23. No, the function is decreasing on the interval $(2, 5)$ and then it increases on the interval $(5, 10)$.

25. The function f is increasing on the intervals $(-8, -2)$; $(0, 2)$; and $(5, \infty)$ or for $-8 < x < -2$; $0 < x < 2$; and $x > 5$.

27. There is a local maximum at $x = 2$. The maximum value is $f(2) = 10$.

29. The function f has local maxima at $x = -2$ and at $x = 2$. The local maxima are $f(-2) = 6$ and $f(2) = 10$.

31. (a) The x-intercepts of the graph are $(-2, 0)$ and $(2, 0)$. The y-intercept is $(0, 3)$.

(b) The domain of f is the set $\{x \mid -4 \le x \le 4\}$ or the interval $[-4, 4]$.
The range of f is the set $\{y \mid 0 \le y \le 3\}$ or the interval $[0, 3]$.

(c) The function is increasing on the intervals $(-2, 0)$ and $(2, 4)$ or when $-2 < x < 0$ and $2 < x < 4$. The function is decreasing on the intervals $(-4, -2)$ and $(0, 2)$ or when $-4 < x < -2$ and $0 < x < 2$.

(d) The graph of the function is symmetric with respect to the y-axis, so the function is even.

33. (a) There is no x-intercept of the graph. The y-intercept is $(0, 1)$.

(b) The domain of f is the set of all real numbers.
The range of f is the set $\{y \mid y > 0\}$ or the interval $(0, \infty)$.
(c) The function is always increasing.

(d) The graph of the function is not symmetric with respect to the y-axis or the origin, so the function is neither even nor odd.

35. (a) The x-intercepts of the graph are $(-\pi, 0)$, $(0, 0)$, and $(\pi, 0)$. The y-intercept is $(0, 0)$.

(b) The domain of f is the set $\{x \mid -\pi \le x \le \pi\}$ or the interval $[-\pi, \pi]$. The range of f is the set $\{y \mid -1 \le y \le 1\}$ or the interval $[-1, 1]$.

(c) The function is increasing on the interval $\left(-\dfrac{\pi}{2}, \dfrac{\pi}{2}\right)$ or for $-\dfrac{\pi}{2} < x < \dfrac{\pi}{2}$. The function is decreasing on the intervals $\left(-\pi, -\dfrac{\pi}{2}\right)$ and $\left(\dfrac{\pi}{2}, \pi\right)$ or for $-\pi < x < -\dfrac{\pi}{2}$ and $\dfrac{\pi}{2} < x < \pi$.

(d) The graph of the function is symmetric with respect to the origin, so the function is odd.

37. (a) The x-intercepts of the graph are $\left(\dfrac{1}{2}, 0\right)$ and $\left(\dfrac{5}{2}, 0\right)$. The y-intercept is $\left(0, \dfrac{1}{2}\right)$.

(b) The domain of f is the set $\{x \mid -3 \le x \le 3\}$ or the interval $[-3, 3]$. The range of f is the set $\{y \mid -1 \le y \le 2\}$ or the interval $[-1, 2]$.

(c) The function is increasing on the interval $(2, 3)$ or $2 < x < 3$; it is decreasing on the interval $(-1, 1)$ or $-1 < x < 1$; and it is constant on the intervals $(-3, -1)$ and $(1, 2)$ or when $-3 < x < -1$ and $1 < x < 2$.

(d) The graph of the function is not symmetric with respect to the y-axis or the origin, so the function is neither even nor odd.

39. (a) f has a local maximum at 0 since for all x close to 0, $x \neq 0, f(x) < f(0)$. The local maximum is $f(0) = 3$.

    (b) f has local minima at -2 and at 2. The local minima are $f(-2) = 0$ and $f(2) = 0$.

41. (a) f has a local maximum at $\dfrac{\pi}{2}$ since for all x close to $\dfrac{\pi}{2}, x \neq \dfrac{\pi}{2}, f(x) < f\left(\dfrac{\pi}{2}\right)$. The local maximum is $f\left(\dfrac{\pi}{2}\right) = 1$.

    (b) f has local minimum at $-\dfrac{\pi}{2}$ since for all x close to $-\dfrac{\pi}{2}, x \neq -\dfrac{\pi}{2}, f(x) > f\left(-\dfrac{\pi}{2}\right)$.
    The local minimum is $f\left(-\dfrac{\pi}{2}\right) = -1$.

43. (a) The average rate of change from 0 to 2 is
    $$\frac{\Delta y}{\Delta x} = \frac{f(2) - f(0)}{2 - 0} = \frac{\left[-2 \cdot 2^2 + 4\right] - \left[-2 \cdot 0^2 + 4\right]}{2} = \frac{-4 - 4}{2} = -4$$

    (b) The average rate of change from 1 to 3 is
    $$\frac{\Delta y}{\Delta x} = \frac{f(3) - f(1)}{3 - 1} = \frac{\left[-2 \cdot 3^2 + 4\right] - \left[-2 \cdot 1^2 + 4\right]}{2} = \frac{-16}{2} = -8$$

    (c) The average rate of change from 1 to 4 is
    $$\frac{\Delta y}{\Delta x} = \frac{f(4) - f(1)}{4 - 1} = \frac{\left[-2 \cdot 4^2 + 4\right] - \left[-2 \cdot 1^2 + 4\right]}{3} = \frac{-30}{3} = -10$$

45. (a) $\dfrac{f(x) - f(1)}{x - 1} = \dfrac{5x - 5(1)}{x - 1} = \dfrac{5(x - 1)}{x - 1} = 5$

    (b) Using $x = 2$ in part (a), we get 5. This is the slope of the secant line containing the points $(1, f(1))$ and $(2, f(2))$.

    (c) $m_{\text{sec}} = 5$. Using the point-slope form of the line and the point $(1, f(1)) = (1, 5)$, we get
    $$\begin{aligned} y - y_1 &= m_{\text{sec}}(x - x_1) \\ y - 5 &= 5(x - 1) \\ y &= 5x - 5 + 5 \\ y &= 5x \end{aligned}$$

47. (a) $\dfrac{f(x) - f(1)}{x - 1} = \dfrac{[1 - 3x] - [1 - 3]}{x - 1} = \dfrac{3 - 3x}{x - 1} = \dfrac{3(1 - x)}{x - 1} = -3$

(b) Using $x = 2$ in part (a), we get -3. This is the slope of the secant line containing the points $(1, f(1))$ and $(2, f(2))$.

(c) $m_{sec} = -3$. Using the point-slope form of the line and the point $(1, f(1)) = (1, -2)$, we get

$$y - y_1 = m_{sec}(x - x_1)$$
$$y - (-2) = -3(x - 1)$$
$$y = -3x + 3 - 2$$
$$y = -3x + 1$$

49. (a) $\dfrac{f(x) - f(1)}{x - 1} = \dfrac{[x^2 - 2x] - [1^2 - 2(1)]}{x - 1} = \dfrac{[x^2 - 2x] - [-1]}{x - 1} = \dfrac{x^2 - 2x + 1}{x - 1}$

$$= \dfrac{(x - 1)^2}{x - 1} = x - 1$$

(b) Using $x = 2$ in part (a), we get 1. This is the slope of the secant line containing the points $(1, f(1))$ and $(2, f(2))$.

(c) $m_{sec} = 1$. Using the point-slope form of the line and the point $(1, f(1)) = (1, -1)$, we get

$$y - y_1 = m_{sec}(x - x_1)$$
$$y - (-1) = 1(x - 1)$$
$$y = x - 1 - 1$$
$$y = x - 2$$

51. (a) $\dfrac{f(x) - f(1)}{x - 1} = \dfrac{[x^3 - x] - [1^3 - 1]}{x - 1} = \dfrac{[x^3 - x] - [0]}{x - 1} = \dfrac{x(x^2 - 1)}{x - 1}$

$$= \dfrac{x\cancel{(x-1)}(x + 1)}{\cancel{x-1}} = x^2 + x$$

(b) Using $x = 2$ in part (a), we get $2^2 + 2 = 6$. This is the slope of the secant line containing the points $(1, f(1))$ and $(2, f(2))$.

(c) $m_{sec} = 6$. Using the point-slope form of the line and the point $(1, f(1)) = (1, 0)$, we get

$$y - y_1 = m_{sec}(x - x_1)$$
$$y - 0 = 6(x - 1)$$
$$y = 6x - 6$$

53. (a) $\dfrac{f(x) - f(1)}{x - 1} = \dfrac{\left[\dfrac{2}{x+1}\right] - \left[\dfrac{2}{1+1}\right]}{x - 1} = \dfrac{\left[\dfrac{2}{x+1}\right] - [1]}{x - 1} = \dfrac{2 - (1)(x+1)}{(x-1)(x+1)} = \dfrac{2 - x - 1}{(x-1)(x+1)}$

$$= \dfrac{\cancel{-x+1}^{-1}}{\cancel{(x-1)}(x+1)} = -\dfrac{1}{x+1}$$

(b) Using $x = 2$ in part (a), we get $-\dfrac{1}{2+1} = -\dfrac{1}{3}$. This is the slope of the secant line containing the points $(1, f(1))$ and $(2, f(2))$.

(c) $m_{\text{sec}} = -\dfrac{1}{3}$. Using the point-slope form of the line and the point $(1, f(1)) = (1, 1)$, we get

$$y - y_1 = m_{\text{sec}}(x - x_1)$$
$$y - 1 = -\frac{1}{3}(x - 1)$$
$$y = -\frac{1}{3}x + \frac{1}{3} + 1$$
$$y = -\frac{1}{3}x + \frac{4}{3}$$

55. (a) $\dfrac{f(x) - f(1)}{x - 1} = \dfrac{\sqrt{x} - \sqrt{1}}{x - 1} = \dfrac{\sqrt{x} - 1}{(\sqrt{x} - 1)(\sqrt{x} + 1)} = \dfrac{1}{\sqrt{x} + 1}$

(b) Using $x = 2$ in part (a), we get $\dfrac{1}{\sqrt{2} + 1}$. This is the slope of the secant line containing the points $(1, f(1))$ and $(2, f(2))$.

(c) $m_{\text{sec}} = \dfrac{1}{\sqrt{2} + 1}$

Using the point-slope form of the line and the point $(1, f(1)) = (1, 1)$, we get
$$y - y_1 = m_{\text{sec}}(x - x_1)$$
$$y - 1 = \frac{1}{\sqrt{2} + 1}(x - 1)$$
$$y \approx 0.414x - 0.414 + 1$$
$$y \approx 0.414x + 0.586$$

57. To determine algebraically whether a function is even, odd, or neither we replace x with $-x$ and simplify.
$$f(-x) = 4(-x)^3 = 4 \cdot (-x^3) = -4x^3 = -f(x)$$
Since $f(-x) = -f(x)$, the function is odd.

59. To determine algebraically whether a function is even, odd, or neither we replace x with $-x$ and simplify.
$$g(-x) = -3(-x)^2 - 5 = -3x^2 - 5 = g(x)$$
Since $g(-x) = g(x)$, the function is even.

61. To determine algebraically whether a function is even, odd, or neither we replace x with $-x$ and simplify.

$$F(-x) = \sqrt[3]{-x} = -\sqrt[3]{x} = -F(x)$$

Since $F(-x) = -F(x)$, the function is odd.

63. To determine algebraically whether a function is even, odd, or neither we replace x with $-x$ and simplify.

$$f(-x) = -x + |-x| = -x + |x|$$

Since $f(-x)$ equals neither $f(x)$ nor $-f(x)$, the function is neither even nor odd.

65. To determine algebraically whether a function is even, odd, or neither we replace x with $-x$ and simplify.

$$g(-x) = \frac{1}{(-x)^2} = \frac{1}{x^2} = g(x)$$

Since $g(-x) = g(x)$, the function is even.

67. To determine algebraically whether a function is even, odd, or neither we replace x with $-x$ and simplify.

$$h(-x) = \frac{-(-x)^3}{3(-x)^2 - 9} = \frac{x^3}{3x^2 - 9} = -\left(\frac{-x^3}{3x^2 - 9}\right) = -h(x)$$

Since $h(-x) = -h(x)$, the function is odd.

69. Using 2nd, calculate, we find there is a local maximum at -1. The local maximum is $f(-1) = 4$.
There is a local minimum at 1. The local minimum is $f(1) = 0$.

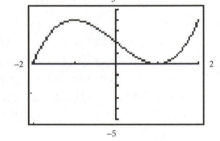

The function is increasing on the intervals $(-2, -1)$ and $(1, 2)$.

The function is decreasing on the interval $(-1, 1)$.

71. Using 2nd, CALCULATE, we find there is a local maximum at -0.77. The local maximum is $f(-0,77) = 0.19$.
There is a local minimum at 0.77. The local minimum is $f(0.77) = -0.19$.

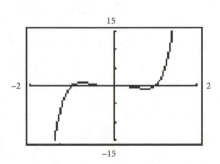

The function is increasing on the intervals $(-2, -0.77)$ and $(0.77, 2)$.

The function is decreasing on the interval $(-0.77, 0.77)$.

73. Using 2nd, calculate, we find there is a local maximum at 1.77. The local maximum is $f(1.77) = -1.91$.
There is a local minimum at -3.77. The local minimum is $f(-3.77) = -18.89$.

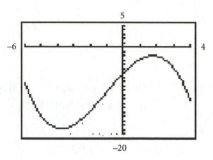

The function is increasing on the interval $(-3.77, 1.77)$.

The function is decreasing on the intervals $(-6, -3.77)$ and $(1.77, 4)$.

75. Using 2nd, calculate, we find there is a local maximum at 0. The local maximum is $f(0) = 3$.
There is a local minimum at -1.87. The local minimum is $f(-1.87) = 0.95$.
There is another local minimum at 0.97. The local minimum is $f(0.97) = 2.65$.

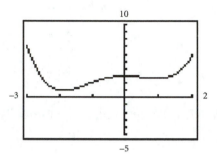

The function is increasing on the intervals $(-1.87, 0)$ and $(0.97, 2)$.

The function is decreasing on the intervals $(-3, -1.87)$ and $(0, 0.97)$.

77. The average rate of change of f from 0 to x is

$$\frac{\Delta y}{\Delta x} = \frac{f(x) - f(0)}{x - 0} = \frac{x^2 - 0}{x} = x \qquad x \neq 0$$

(a) The average rate of change from 0 to 1 is 1.
(b) The average rate of change from 0 to 0.5 is 0.5.
(c) The average rate of change from 0 to 0.1 is 0.1.
(d) The average rate of change from 0 to 0.01 is 0.01.
(e) The average rate of change from 0 to 0.001 is 0.001.

(g) and (h) Answers will vary.

(f)

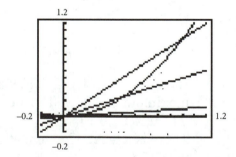

79.

(a) $h(100) = \dfrac{-32(100)^2}{130^2} + 100 = 81.07$ feet high

(b) $h(300) = \dfrac{-32(300)^2}{130^2} + 300 = 129.59$ feet high

(c) $h(500) = \dfrac{-32(500)^2}{130^2} + 500 = 26.63$ feet high

(d) To determine how far the ball was hit, we find $h(x) = 0$.

$$\dfrac{-32x^2}{130^2} + x = 0$$
$$-32x^2 + 130^2 x = 0$$
$$x(-32x + 130^2) = 0$$
$$x = 0 \quad \text{or} \quad -32x + 130^2 = 0$$
$$x = 0 \quad \text{or} \quad x = 528.125$$

The ball traveled 528.125 feet.

(e)

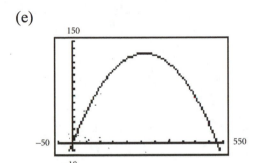

(f)

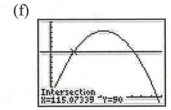 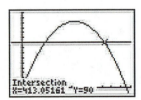

The golf ball is 90 feet high after it has traveled about 115 feet and again when it has traveled about 413 feet.

(g)

X	Y₁
0	0
25	23.817
50	45.266
75	64.349
100	81.065
125	95.414
150	107.4

X=0

X	Y₁	Y₂
200	124.26	90
225	129.14	90
250	131.66	90
275	131.8	90
300	129.59	90
325	125	90
350	118.05	90

X=275

(h)   The ball travels about 275 feet when it reaches its maximum height of about 131.8 feet.

(i) The ball actually travels only 264 feet before reaching its maximum height of 132.03 feet.

X	Y₁	Y₂
260	132	50
261	132.01	50
262	132.02	50
263	132.03	50
264	132.03	50
265	132.03	50
266	132.02	50

X=264

81. (a) Volume of a box is the product of its length, width, and height. The box has height x inches and length and width equal to $24 - 2x$ inches.
$$V = V(x) = x(24 - 2x)^2 \text{ cubic inches}$$

(b) If $x = 3$ inches, then $V = 3(24 - 2 \cdot 3)^2 = 972$ cubic inches.

(c) If $x = 10$ inches, then $V = 10(24 - 2 \cdot 10)^2 = 10(4)^2 = 160$ cubic inches.

(d)

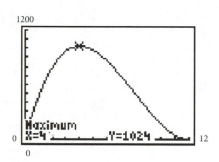

83. (a)

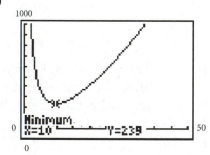

(b) The average cost is minimized when 10 riding mowers are produced.
(c) The average cost of producing each of the 10 mowers is $239.

85. Best matches are are (a) graph (II), (b) graph (V), (c) graph (IV), (d) graph (III), (e) graph (I). Reasons may vary.

87. Graphs will vary.

89. Explanations and graphs will vary.

91. Descriptions will vary.

93. Functions will vary.

10.4 Library of Functions; Piecewise-defined Functions

1. (C) Graphs of square functions are parabolas.

3. (E) Graphs of square root functions are defined for $x \geq 0$ and are increasing.

5. (B) Graphs of linear functions are straight lines.

7. (F) Graph of a reciprocal function is not defined at zero.

9.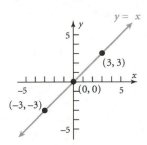

11.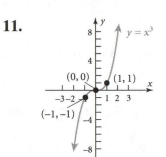

13.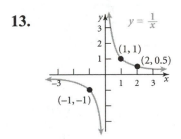

15.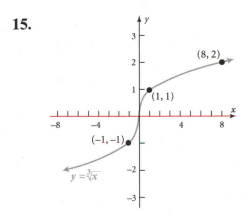

17. (a) When $x = -2$, the equation for f is $f(x) = x^2$, so $f(-2) = (-2)^2 = 4$

  (b) When $x = 0$, the equation for f is $f(x) = 2$, so $f(0) = 2$

  (c) When $x = 2$, the equation for f is $f(x) = 2x + 1$, so $f(2) = 2(2) + 1 = 5$

19. (a) $f(1.2) = \text{int}(2 \cdot 1.2) = \text{int}(2.4) = 2$

  (b) $f(1.6) = \text{int}(2 \cdot 1.6) = \text{int}(3.2) = 3$

  (c) $f(-1.8) = \text{int}(2 \cdot (-1.8)) = \text{int}(-3.6) = -4$

21. (a) The domain of f is all real numbers or the interval $(-\infty, \infty)$.

  (b) To find the y-intercept we let $x = 0$ and solve. When $x = 0$, the equation for f is $f(0) = 1$. So the y-intercept is $(0, 1)$.

  To find the x-intercept, we let $y = 0$ and solve for x. $f(x)$ never equals zero, so there is no x-intercept.

(c)

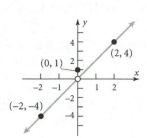

(d) The range of f is the set $\{y \mid y \neq 0\}$.

23. (a) The domain of f is all real numbers or the interval $(-\infty, \infty)$.

(b) To find the y-intercept we let $x = 0$ and solve. When $x = 0$, the equation for f is $f(x) = -2x + 3$ and $f(0) = 3$. So the y-intercept is $(0, 3)$.

To find the x-intercept, we let $y = 0$ and solve. If $x < 1$, then $y = -2x + 3 = 0$ or
$x = \dfrac{3}{2} = 1.5$.

Since $x = 1.5$ is not less than 1, we ignore this solution. If $x \geq 1$, then $y = 3x - 2 = 0$
or $x = \dfrac{2}{3}$.

Since $x = \dfrac{2}{3} < 1$, we ignore this solution and conclude $f(x)$ never equals zero and there

is no x-intercept.

(c)

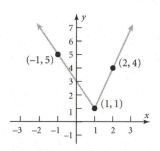

(d) The range of f is the set $\{y \mid y \geq 1\}$ or $[1, \infty)$.

25. (a) The domain of f is the set $\{x \mid x \geq -2\}$ or the interval $[-2, \infty)$.

(b) To find the y-intercept we let $x = 0$ and solve. When $x = 0$, the equation for f is $f(x) = x + 3$, and $f(0) = 3$. So the y-intercept is $(0, 3)$.

To find the x-intercept(s), we let $y = 0$ and solve.
If $-2 \leq x < 1$, $y = f(x) = x + 3 = 0$ or $x = -3$. Since $x = -3$ is not in the interval $[-2, 1)$, we ignore this solution.
If $x > 1$, $y = -x + 2 = 0$ or $x = 2$. So there is an x-intercept at $(2, 0)$.

(c)

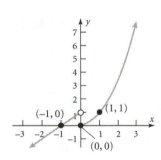

(d) The range of f is the set $\{y \mid y = 5; y < 4\}$.

27. (a) The domain of f is all real numbers.

(b) To find the y-intercept we let $x = 0$ and solve. When $x = 0$, the equation for f is $f(x) = x^2$, and $f(0) = 0$. So the y-intercept is $(0, 0)$.

To find the x-intercept(s), we let $y = 0$ and solve.
If $x < 0$, $y = f(x) = 1 + x = 0$ or $x = -1$. So there is an x-intercept at $(-1, 0)$.
If $x > 0$, $y = f(x) = x^2 = 0$ or $x = 0$. So there is an x-intercept at $(0, 0)$.

(c)

(d) The range of f is the set of all real numbers.

29. (a) The domain of f is the set $\{x \mid x \geq -2\}$ or the interval $[-2, \infty)$.

(b) To find the y-intercept we let $x = 0$ and solve. When $x = 0$, the equation for f is $f(x) = 1$, and $f(0) = 1$. So the y-intercept is $(0, 1)$.

To find the x-intercept(s), we let $y = 0$ and solve. If $-2 \leq x < 0$, $y = f(x) = |x| = 0$ or $x = 0$.
Since $x = 0$ is not in the interval $[-2, 0)$, we ignore this solution.
If $x > 0$, $y = x^3 = 0$ or $x = 0$. Since $x = 0$ is not in the interval $(0, \infty)$, we also ignore this solution, and conclude there is no x-intercept.

(c)

(d) The range of f is the set $\{y \mid y > 0\}$ or the interval $(0, \infty)$.

31. (a) The domain of *f* is the set of all real numbers.

(b) To find the *y*-intercept we let $x = 0$ and solve. When $x = 0$, the equation for *f* is $f(x) = 2$ int (x), and $f(0) = 0$. So the *y*-intercept is $(0, 0)$.

To find the *x*-intercept(s), we let $y = 0$ and solve. $y = f(x) = 2$ int $(x) = 0$ for all *x* in the set $\{x \mid 0 \le x < 1\}$ or the interval $[0, 1)$. So there are *x*-intercepts at all points in the interval.

(c)   (d) The range of *f* is the set of even integers.

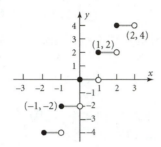

33. This graph is made of two line segments. We find each.
For $-1 \le x \le 0$, we have two points $(-1, 1)$ and $(0, 0)$. The slope of the line between the points is

$$m = \frac{y_2 - y_1}{x_2 - x_1} = \frac{1 - 0}{-1 - 0} = -1$$

Using the point-slope form of an equation of a line and the point $(0, 0)$, we get

$$y - y_1 = m(x - x_1)$$
$$y - 0 = -1(x - 0)$$
$$y = -x$$

For $0 < x \le 2$, we have two points $(0, 0)$ and $(2, 1)$. The slope of the line between the points is

$$m = \frac{y_2 - y_1}{x_2 - x_1} = \frac{1 - 0}{2 - 0} = \frac{1}{2}$$

Using the point-slope form of an equation of a line and the point $(0, 0)$, we get

$$y - y_1 = m(x - x_1)$$
$$y - 0 = \frac{1}{2}(x - 0)$$
$$y = \frac{1}{2}x$$

The piecewise function is

$$f(x) = \begin{cases} -x & -1 \le x \le 0 \\ \dfrac{1}{2}x & 0 < x \le 2 \end{cases}$$

35. This graph is made of two line segments. We find each.
For $x \le 0$, we have two points $(-1, 1)$ and $(0, 0)$. The slope of the line between the points is

$$m = \frac{y_2 - y_1}{x_2 - x_1} = \frac{1 - 0}{-1 - 0} = -1$$

Using the point-slope form of an equation of a line and the point $(0, 0)$, we get

$$y - y_1 = m(x - x_1)$$
$$y - 0 = -1(x - 0)$$
$$y = -x$$

For $0 < x \le 2$, we have two points $(0, 2)$ and $(2, 0)$. The slope of the line between the points is

$$m = \frac{y_2 - y_1}{x_2 - x_1} = \frac{2 - 0}{0 - 2} = -1$$

Using the point-slope form of an equation of a line and the point $(0, 2)$, we get

$$y - y_1 = m(x - x_1)$$
$$y - 2 = -1(x - 0)$$
$$y = -x + 2$$

The piecewise function is

$$f(x) = \begin{cases} -x & x \le 0 \\ -x + 2 & 0 < x \le 2 \end{cases}$$

37. (a) If $x = 200$ minutes are used, the equation for C is $C(x) = 39.99$. So $C(200) = \$39.99$.

(b) If $x = 365$ minutes are used, the equation for C is $C(x) = 0.25x - 47.51$. So $C(365) = 0.25(365) - 47.51 = \$43.74$.

(c) If $x = 351$ minutes are used, the equation for C is $C(x) = 0.25x - 47.51$. So $C(351) = 0.25(351) - 47.51 = \$40.24$.

39. (a) If 50 therms are used, the charge C will be
$$C(50) = 9.45 + 0.36375(50) + 0.6338(50) = \$59.33$$

(b) If 500 therms are used the charge will be
$$C(500) = 9.45 + 0.36375(50) + 0.11445(500 - 50) + 0.6338(500) = \$396.04$$

(c) When $0 \le x \le 50$ the equation that relates the monthly charge for using x therms of gas is
$$C(x) = 9.45 + 0.36375x + 0.6338x = 0.99755x + 9.45$$
When $x > 50$ the equation that relates the monthly charge for using x therms of gas is
$$C(x) = 9.45 + 0.36375(50) + 0.11445(x - 50) + 0.6338x$$
$$= 21.915 + 0.74825x$$
The function that relates the monthly charge C for x therms of gas is
$$C(x) = \begin{cases} 9.45 + 0.99755x & 0 \le x \le 50 \\ 21.915 + 0.74825x & x > 50 \end{cases}$$

(d)

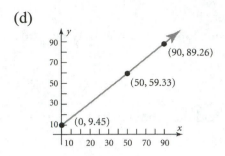

41. (a) When $v = 1$ meter per second and $t = 10°$, the equation representing the wind chill is
$W = t = 10°$.

(b) When $v = 5$ m/sec. and $t = 10°$, the equation representing the wind chill is
$$W = 33 - \frac{(10.45 + 10\sqrt{v} - v)(33 - t)}{22.04} = 33 - \frac{(10.45 + 10\sqrt{5} - 5)(33 - 10)}{22.04}$$
$$= 33 - \frac{(5.45 + 10\sqrt{5})(23)}{22.04} = 3.98°$$

(c) When $v = 15$ m/sec. and $t = 10°$, the equation representing the wind chill is
$$W = 33 - \frac{(10.45 + 10\sqrt{v} - v)(33 - t)}{22.04} = 33 - \frac{(10.45 + 10\sqrt{15} - 15)(33 - 10)}{22.04}$$
$$= 33 - \frac{(10\sqrt{15} - 4.55)(23)}{22.04} = -2.67°$$

(d) When $v = 25$ m/sec. and $t = 10°$, the equation representing the wind chill is
$W = 33 - 1.5958(33 - t) = 33 - 1.5958(33 - 10) = 33 - 1.5958(23) = -3.70°$

(e) Answers may vary.

(f) Answers may vary.

43.
$$y = f(x) = \begin{cases} 0.10x & 0 < x < 7000 \\ 700 + 0.15(x - 7000) & 7000 \leq x < 28,400 \\ 3910 + 0.25(x - 28,400) & 28,400 \leq x < 68,800 \\ 14,010 + 0.28(x - 68,800) & 68,800 \leq x < 143,500 \\ 34,926 + 0.33(x - 143,500) & 143,500 \leq x < 311,950 \\ 90,514 + 0.35(x - 311,950) & x \geq 311,950 \end{cases}$$

45.

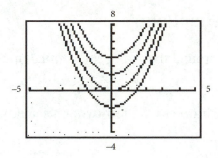

Answers may vary.

47.

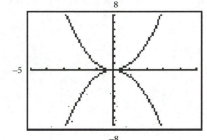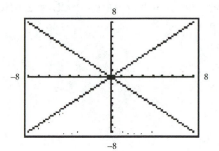

Answers may vary.

49.

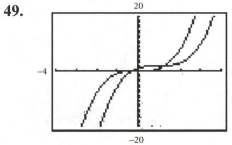

Answers may vary.

51.

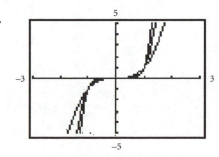

Answers may vary.

10.5 Graphing Techniques: Shifts and Reflections

1. (*B*) This is the graph of a square function reflected around the *x*-axis and shifted up two units.

3. (*H*) This is the graph of an absolute value function shifted to the left two units and then reflected over the *x*-axis.

5. (*A*) This is the graph of a square function shifted up two units.

7. (*F*) This is the graph of a square function shifted to the left two units and then reflected over the *x*-axis.

9. $y = (x - 4)^3$ 11. $y = x^3 + 4$

13. $y = (-x)^3 = -x^3$

15.

(1) Shift up 2 units.	Add 2.	$y = \sqrt{x} + 2$
(2) Reflect about the *x*-axis.	Multiply *y* by -1.	$y = -\left(\sqrt{x} + 2\right) = -\sqrt{x} - 2$
(3) Reflect about the *y*-axis.	Replace *x* by $-x$.	$y = -\sqrt{-x} - 2$

17.

(1) Reflect about the *x*-axis.	Multiply *y* by -1.	$y = -\sqrt{x}$
(2) Shift up 2 units.	Add 2.	$y = -\sqrt{x} + 2$
(3) Shift left 3 units.	Replace *x* by $x + 3$.	$y = -\sqrt{x+3} + 2$

19. Ans: (c) $-f(x)$ changes the sign of *y*. So if (3, 0) is a point on *f*, (3, 0) is a point on $-f(x)$.

21.

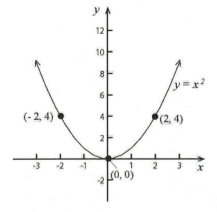

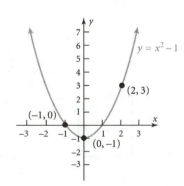

(a) $y = x^2$ Subtract 1; vertical   (b) $y = x^2 - 1$
                        shift down 1 unit.

23.

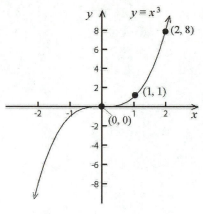

(a) $y = x^3$

$\rightarrow$

Add 1; vertical
shift up 1 unit.

(b) $y = x^3 + 1$

25.

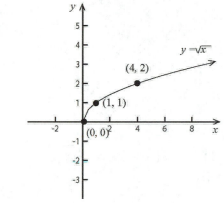

(a) $y = \sqrt{x}$

$\rightarrow$

Replace x by $x - 2$;
horizontal shift right 2 units.

(b) $y = \sqrt{x - 2}$

27.

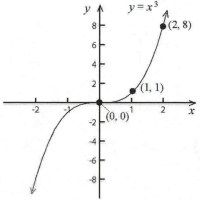

(a) $y = x^3$

$\rightarrow$

Replace x by $x - 1$;
horizontal shift right
1 unit.

(b) $y = (x - 1)^3$

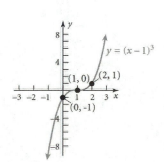

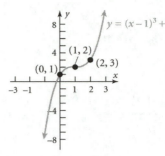

$\rightarrow$

Add 2: vertical shift
up 2 units.

(c) $y = (x-1)^3 + 2$

29.

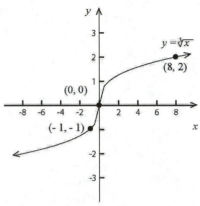

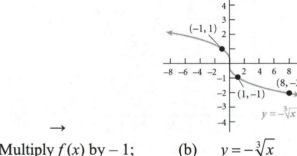

(a) $y = \sqrt[3]{x}$

$\rightarrow$

Multiply $f(x)$ by -1;
reflect about the x-axis.

(b) $y = -\sqrt[3]{x}$

31.

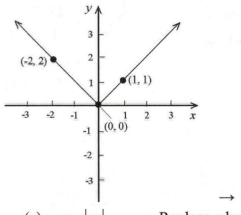

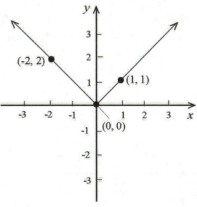

(a) $y = |x|$

$\rightarrow$

Replace x by $-x$;
reflect about y-axis.

(b) $y = |-x|$

33.

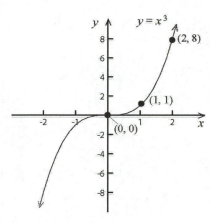

(a) $y = x^3$

$\xrightarrow{\text{Multiply } f(x) \text{ by } -x;}$
reflect about y-axis.

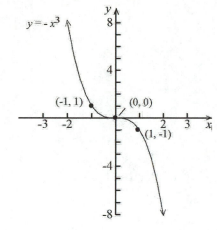

(b) $y = -x^3$

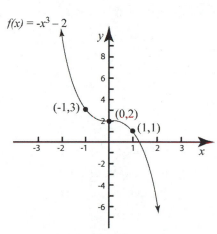

$\xrightarrow{\text{Add 2; vertical}}$
shift 2 units.

(c) $y = -x^3 + 2$

35.

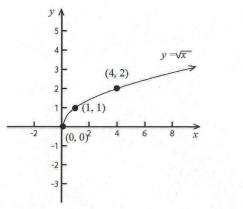

(a) $y = \sqrt{x}$

$\xrightarrow{\text{Replace } x \text{ by } x - 2;}$
horizontal shift right 2 units.

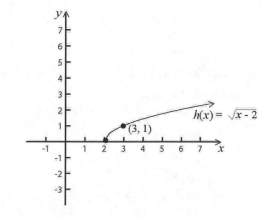

(b) $y = \sqrt{x - 2}$

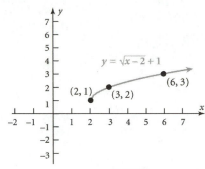

→

Add 1; vertical shift
up 1 unit.

(c) $y = \sqrt{x-2} + 1$

37.

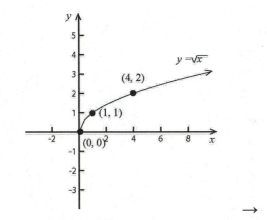

(a) $y = \sqrt{x}$

→

Replace x by $-x$;
reflect about y-axis.

(b) $y = \sqrt{-x}$

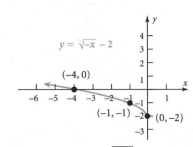

→

Subtract 2; vertical
down 2 units.

(c) $y = \sqrt{-x} - 2$

39.

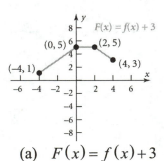

(a) $F(x) = f(x) + 3$

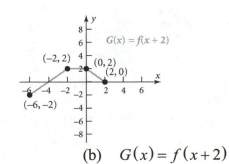

(b) $G(x) = f(x+2)$

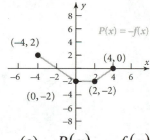

(c) $P(x) = -f(x)$

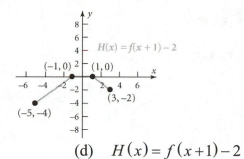

(d) $H(x) = f(x+1) - 2$

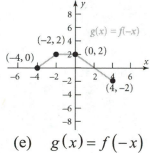

(e) $g(x) = f(-x)$

41. (a)

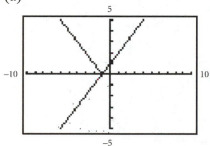

(b)

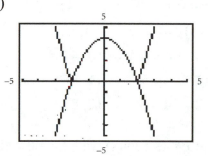

(c)

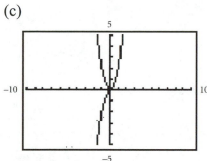

(d) Answers may vary.

43.

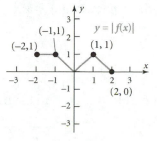

(a) $y = |f(x)|$

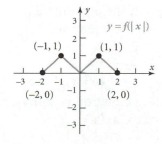

(b) $y = f(|x|)$

Chapter 10 Review

TRUE-FALSE ITEMS

1. False

3. False

5. True

FILL IN THE BLANKS

1. independent; dependent

3. $5;\ -3$

5. $(-5, 0), (-2, 0), (2, 0)$

REVIEW EXERCISES

1.

x	$x^2 + 4$	(x, y)
-2	$(-2)^2 + 4 = 8$	$(-2, 8)$
-1	$(-1)^2 + 4 = 5$	$(-1, 5)$
0	$0^2 + 4 = 4$	$(0, 4)$
1	$1^2 + 4 = 5$	$(1, 5)$
2	$2^2 + 4 = 8$	$(2, 8)$

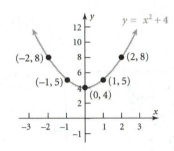

3. To find the x-intercept(s) we let $y = 0$ and solve the equation
$$2x = 3 \cdot 0^2$$
$$x = 0$$
So the x-intercept is $(0, 0)$.

To find the y-intercept(s) we let $x = 0$ and solve the equation
$$2 \cdot 0 = 3y^2$$
$$y = 0$$
So the y-intercept is $(0, 0)$.

To test the graph of the equation $2x = 3y^2$ for symmetry with respect to the x-axis, we replace y by $-y$ in the equation.
$$2x = 3y^2$$
$$2x = 3(-y)^2$$
$$2x = 3y^2$$
Since the resulting equation is equivalent to the original equation, the graph is symmetric with respect to the x-axis.

To test the graph of the equation $2x = 3y^2$ for symmetry with respect to the y-axis, we replace x by $-x$ in the equation and simplify.

$$2x = 3y^2$$
$$2(-x) = 3y^2$$
$$-2x = 3y^2$$

Since the resulting equation is not equivalent to the original equation, the graph is not symmetric with respect to the y-axis.

To test the graph of the equation $2x = 3y^2$ for symmetry with respect to the origin, we replace x by $-x$ and y by $-y$ in the equation and simplify.
$$2(-x) = 3(-y)^2$$
$$-2x = 3y^2$$

Since the resulting equation is not equivalent to the original equation, the graph is not symmetric with respect to the origin.

5. To find the x-intercept(s) we let $y = 0$ and solve the equation
$$x^2 + 4 \cdot 0^2 = 16$$
$$x^2 = 16$$
$$x = -4 \text{ or } x = 4$$
So the x-intercepts are $(-4, 0)$ and $(4, 0)$.

To find the y-intercept(s) we let $x = 0$ and solve the equation
$$0^2 + 4y^2 = 16$$
$$4y^2 = 16$$
$$y^2 = 4$$
$$y = -2 \text{ or } y = 2$$
So the y-intercepts are $(0, -2)$ and $(0, 2)$.

To test the graph of the equation $x^2 + 4y^2 = 16$ for symmetry with respect to the x-axis, we replace y by $-y$ in the equation.
$$x^2 + 4y^2 = 16$$
$$x^2 + 4(-y)^2 = 16$$
$$x^2 + 4y^2 = 16$$

Since the resulting equation is equivalent to the original equation, the graph is symmetric with respect to the x-axis.

To test the graph of the equation $x^2 + 4y^2 = 16$ for symmetry with respect to the y-axis, we replace x by $-x$ in the equation and simplify.
$$x^2 + 4y^2 = 16$$
$$(-x)^2 + 4y^2 = 16$$
$$x^2 + 4y^2 = 16$$

Since the resulting equation is equivalent to the original equation, the graph is symmetric with respect to the y-axis.

To test the graph of the equation $x^2 + 4y^2 = 16$ for symmetry with respect to the origin, we replace x by $-x$ and y by $-y$ in the equation and simplify.
$$x^2 + 4y^2 = 16$$
$$(-x)^2 + 4(-y)^2 = 16$$
$$x^2 + 4y^2 = 16$$

Since the resulting equation is equivalent to the original equation, the graph is symmetric with respect to the origin.

7. To find the x-intercept(s) we let $y = 0$ and solve the equation
$$x^4 + 2x^2 + 1 = 0$$
$$(x^2 + 1)(x^2 + 1) = 0$$
$$x^2 + 1 = 0$$
has no real solution. So there is no x-intercept.

To find the y-intercept(s) we let $x = 0$ and solve the equation
$$0^4 + 2 \cdot 0^2 + 1 = y$$
$$y = 1$$
So the y-intercept is $(0, 1)$.

To test the graph of the equation $x^4 + 2x^2 + 1 = y$ for symmetry with respect to the x-axis, we replace y by $-y$ in the equation.
$$x^4 + 2x^2 + 1 = -y$$
Since the resulting equation is not equivalent to the original equation, the graph is not symmetric with respect to the x-axis.

To test the graph of the equation $x^4 + 2x^2 + 1 = y$ for symmetry with respect to the y-axis, we replace x by $-x$ in the equation and simplify.
$$(-x)^4 + 2(-x)^2 + 1 = y$$
$$x^4 + 2x^2 + 1 = y$$
Since the resulting equation is equivalent to the original equation, the graph is symmetric with respect to the y-axis.

To test the graph of the equation $x^4 + 2x^2 + 1 = y$ for symmetry with respect to the origin, we replace x by $-x$ and y by $-y$ in the equation and simplify.
$$(-x)^4 + 2(-x)^2 + 1 = -y$$
$$x^4 + 2x^2 + 1 = -y$$
Since the resulting equation is not equivalent to the original equation, the graph is not symmetric with respect to the origin.

9. To find the x-intercept(s) we let $y = 0$ and solve the equation
$$x^2 + x + 0^2 + 2 \cdot 0 = 0$$
$$x^2 + x = x(x + 1) = 0$$
$$x = 0 \quad \text{or} \quad x + 1 = 0$$
$$x = -1$$
So the x-intercepts are $(-1, 0)$ and $(0, 0)$.

To find the y-intercept(s) we let $x = 0$ and solve the equation
$$0^2 + 0 + y^2 + 2y = 0$$
$$y^2 + 2y = y(y + 2) = 0$$

$$y = 0 \quad \text{or} \quad y + 2 = 0$$
$$y = -2$$

So the y-intercepts are $(0, -2)$ and $(0, 0)$.

To test the graph of the equation $x^2 + x + y^2 + 2y = 0$ for symmetry with respect to the x-axis, we replace y by $-y$ in the equation.

$$x^2 + x + (-y)^2 + 2(-y) = 0$$
$$x^2 + x + y^2 - 2y = 0$$

Since the resulting equation is not equivalent to the original equation, the graph is not symmetric with respect to the x-axis.

To test the graph of the equation $x^2 + x + y^2 + 2y = 0$ for symmetry with respect to the y-axis, we replace x by $-x$ in the equation and simplify.

$$(-x)^2 + (-x) + y^2 + 2y = 0$$
$$x^2 - x + y^2 + 2y = 0$$

Since the resulting equation is not equivalent to the original equation, the graph is not symmetric with respect to the y-axis.

To test the graph of the equation $x^2 + x + y^2 + 2y = 0$ for symmetry with respect to the origin, we replace x by $-x$ and y by $-y$ in the equation and simplify.

$$(-x)^2 + (-x) + (-y)^2 + 2(-y) = 0$$
$$x^2 - x + y^2 - 2y = 0$$

Since the resulting equation is not equivalent to the original equation, the graph is not symmetric with respect to the origin.

11. (a) We substitute 2 for x in the equation for f to get
$$f(2) = \frac{3(2)}{2^2 - 1} = \frac{6}{3} = 2$$

(b) We substitute -2 for x in the equation for f to get
$$f(-2) = \frac{3(-2)}{(-2)^2 - 1} = \frac{-6}{3} = -2$$

(c) We substitute $(-x)$ for x in the equation for f to get
$$f(-x) = \frac{3(-x)}{(-x)^2 - 1} = \frac{-3x}{x^2 - 1}$$

(d) $-f(x) = -\dfrac{3x}{x^2 - 1}$

(e) We substitute $(x - 2)$ for x in the equation for f to get
$$f(x-2) = \frac{3(x-2)}{(x-2)^2 - 1} = \frac{3x - 6}{x^2 - 4x + 4 - 1} = \frac{3x - 6}{x^2 - 4x + 3}$$

(f) We substitute $2x$ for x in the equation for f to get

$$f(2x) = \frac{3(2x)}{(2x)^2 - 1} = \frac{6x}{4x^2 - 1}$$

13. (a) We substitute 2 for x in the equation for f to get

$$f(2) = \sqrt{2^2 - 4} = 0$$

(b) We substitute -2 for x in the equation for f to get

$$f(-2) = \sqrt{(-2)^2 - 4} = 0$$

(c) We substitute $(-x)$ for x in the equation for f to get

$$f(-x) = \sqrt{(-x)^2 - 4} = \sqrt{x^2 - 4} = f(x)$$

(d) $-f(x) = -\sqrt{x^2 - 4}$

(e) We substitute $(x - 2)$ for x in the equation for f to get

$$f(x-2) = \sqrt{(x-2)^2 - 4} = \sqrt{x^2 - 4x + 4 - 4} = \sqrt{x^2 - 4x}$$

(f) We substitute $2x$ for x in the equation for f to get

$$f(2x) = \sqrt{(2x)^2 - 4} = \sqrt{4x^2 - 4} = 2\sqrt{x^2 - 1}$$

15. (a) We substitute 2 for x in the equation for f to get

$$f(2) = \frac{2^2 - 4}{2^2} = 0$$

(b) We substitute -2 for x in the equation for f to get

$$f(-2) = \frac{(-2)^2 - 4}{(-2)^2} = 0$$

(c) We substitute $(-x)$ for x in the equation for f to get

$$f(-x) = \frac{(-x)^2 - 4}{(-x)^2} = \frac{x^2 - 4}{x^2} = f(x)$$

(d) $-f(x) = -\dfrac{x^2 - 4}{x^2}$

(e) We substitute $(x - 2)$ for x in the equation for f to get

$$f(x-2) = \frac{(x-2)^2 - 4}{(x-2)^2} = \frac{x^2 - 4x + 4 - 4}{x^2 - 4x + 4} = \frac{x^2 - 4x}{x^2 - 4x + 4}$$

(f) We substitute $2x$ for x in the equation for f to get

$$f(2x) = \frac{(2x)^2 - 4}{(2x)^2} = \frac{4x^2 - 4}{4x^2} = \frac{x^2 - 1}{x^2}$$

17. The denominator of the f cannot equal 0, so $x^2 - 9 \neq 0$ or $x \neq 3$ and $x \neq -3$.
    The domain is the set $\{x \mid x \neq 3 \text{ and } x \neq -3\}$.

19. The radicand must be nonnegative, so $2 - x \geq 0$ or $x \leq 2$.
    The domain is the set $\{x \mid x \leq 2\}$ or the interval $(-\infty, -2]$.

21. The radicand must be nonnegative and the denominator cannot equal 0.
    The domain is the set $\{x \mid x > 0\}$ or the interval $(0, \infty)$.

23. The denominator of the function f cannot equal 0, so $x^2 + 2x - 3 \neq 0$.
    $$x^2 + 2x - 3 = 0$$
    $$(x - 1)(x + 3) = 0$$
    $$x - 1 = 0 \text{ or } x + 3 = 0$$
    $$x = 1 \text{ or } x = -3$$
    The domain is the set $\{x \mid x \neq 1 \text{ and } x \neq -3\}$.

25. $\dfrac{f(x+h) - f(x)}{h} = \dfrac{-2(x+h)^2 + (x+h) + 1 - [-2x^2 + x + 1]}{h}$

    $$= \frac{-2x^2 - 4xh - 2h^2 + x + h + 1 + 2x^2 - x - 1}{h}$$

    $$= \frac{-4xh - 2h^2 + h}{h}$$

    $$= \frac{\cancel{h}(-4x - 2h + 1)}{\cancel{h}}$$

    $$= -4x - 2h + 1$$

27. (a) The domain of f is the set $\{x \mid -4 \leq x \leq 3\}$ or the interval $[-4, 3]$.
    The range of f is the set $\{y \mid -3 \leq y \leq 3\}$ or the interval $[-3, 3]$.

    (b) The x-intercept is $(0, 0)$, and the y-intercept is $(0, 0)$.

    (c) $f(-2) = -1$

    (d) $f(x) = -3$ when $x = -4$.

    (e) $f(x) > 0$ on the interval $(0, 3]$.

29. (a) The domain of f is the set of all real numbers or the interval $(-\infty, \infty)$.
    The range of f is the set $\{y \mid y \leq 1\}$ or the interval $(-\infty, 1]$.

    (b) f is increasing on the intervals $(-\infty, -1)$ and $(3, 4)$; f is decreasing on the intervals $(-1, 3)$ and $(4, \infty)$.

(c) The local maxima are 1 at $f(-1)=1$ and 0 at $f(4)=0$. There is a local minimum of -3 at $f(3)=-3$.

(d) The graph is not symmetric with respect to the x-axis, y-axis, or the origin.

(e) Since the graph of the function has no symmetry, the function is neither even nor odd.

(f) The x-intercepts are $(-2, 0)$, $(0, 0)$ and $(4, 0)$; the y-intercept is $(0, 0)$.

31. $f(-x)=(-x)^3-4(-x)=-x^3+4x=-f(x)$ Since $f(-x)=-f(x)$, the function is odd.

33. $h(-x)=\dfrac{1}{(-x)^4}+\dfrac{1}{(-x)^2}+1=\dfrac{1}{x^4}+\dfrac{1}{x^2}+1=h(x)$ Since $h(-x)=h(x)$, the function is even.

35. $G(-x)=1-(-x)+(-x)^3=1+x-x^3$ Since $G(-x)$ is not equal to $G(x)$ or to $-G(x)$, the function is neither even nor odd.

37. $f(-x)=\dfrac{(-x)}{1+(-x)^2}=\dfrac{-x}{1+x^2}=-f(x)$

Since $f(-x)=-f(x)$, the function is odd.

39. There is a local maximum of 4.043 at $x=-0.913$, and a local minimum of -2.043 at $x=0.913$.

The function is increasing on the intervals $(-3, -0.913)$ and $(0.913, 3)$. The function is decreasing on the interval $(-0.913, 0.913)$.

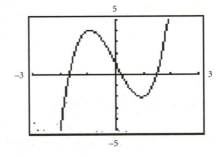

41. There is a local maximum of 1.532 at $x=0.414$, and local minima of 0.543 at $x=-0.336$ and of -3.565 at $x=1.798$.

The function is increasing on the intervals $(-0.336, 0.414)$ and $(1.798, 3)$. The function is decreasing on the intervals $(-2, -0.336)$ and $(0.414, 1.798)$.

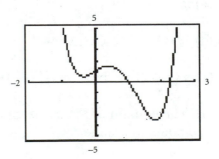

43.

(a) $\dfrac{\Delta y}{\Delta x}=\dfrac{f(2)-f(1)}{2-1}=\dfrac{[8\cdot 2^2-2]-[8\cdot 1^2-1]}{1}=30-7=23$

(b) $\dfrac{\Delta y}{\Delta x} = \dfrac{f(1) - f(0)}{1 - 0} = \dfrac{[8 \cdot 1^2 - 1] - \left[8 \cdot 0^2 - 0\right]}{1} = 7 - 0 = 7$

(c) $\dfrac{\Delta y}{\Delta x} = \dfrac{f(4) - f(2)}{4 - 2} = \dfrac{[8 \cdot 4^2 - 4] - \left[8 \cdot 2^2 - 2\right]}{2} = \dfrac{124 - 30}{2} = 47$

45.
$\dfrac{\Delta y}{\Delta x} = \dfrac{f(x) - f(2)}{x - 2} = \dfrac{[2 - 5x] - [2 - 5 \cdot 2]}{x - 2} = \dfrac{2 - 5x - 2 + 10}{x - 2}$

$= \dfrac{-5x + 10}{x - 2} = \dfrac{-5(x - 2)}{x - 2} = -5$

47.
$\dfrac{\Delta y}{\Delta x} = \dfrac{f(x) - f(2)}{x - 2} = \dfrac{[3x - 4x^2] - \left[3 \cdot 2 - 4 \cdot 2^2\right]}{x - 2} = \dfrac{3x - 4x^2 - 6 + 16}{x - 2}$

$= \dfrac{-4x^2 + 3x + 10}{x - 2} = \dfrac{-(4x + 5)(x - 2)}{x - 2} = -4x - 5$

49. (b), (c), (d), and (e) are graphs of functions because they pass the vertical line test.

51.

53. (a) The domain of f is the set $\{x \mid x > -2\}$ or the interval $(-2, \infty)$.

(b) To find the y-intercept we let $x = 0$ and solve. When $x = 0$, the equation for f is $f(x) = 3x$, and $f(0) = 0$. So the y-intercept is $(0, 0)$.

To find the x-intercept(s), we let $y = 0$ and solve.
If $-2 \le x < 1$, $y = f(x) = 3x = 0$ or $x = 0$. So there is an x-intercept at $(0, 0)$
If $x > 1$, $y = x + 1 = 0$ or $x = -1$. Since $x = -1$ is not in the interval $(1, \infty)$, we ignore this solution.

(c)

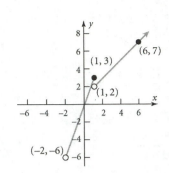

(d) The range of f is the set $\{y \mid y > -6\}$.

55. (a) The domain of f is the set $\{x \mid x \geq -4\}$ or the interval $[-4, \infty)$.

(b) To find the y-intercept we let $x = 0$ and solve. When $x = 0$, the equation for f is $f(x) = 1$. So the y-intercept is $(0, 1)$.

To find the x-intercept(s), we let $y = 0$ and solve.
If $-4 \leq x < 0$, $y = f(x) = x = 0$. Since $x = 0$ is not in the interval $[-4, 0)$, we ignore this solution.
If $x > 0$, $y = 3x = 0$ or $x = 0$. Since $x = 0$ is not in the interval $(0, \infty)$, we ignore this solution and we conclude that there is no x-intercept on this graph.

(c)

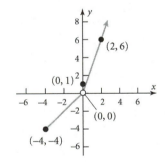

(d) The range of f is the set $\{y \mid y \geq -4, \text{ but } y \neq 0\}$ or the interval $[-4, 0)$ and $(0, \infty)$.

57.

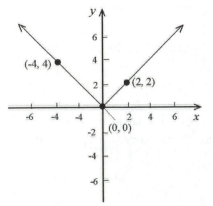

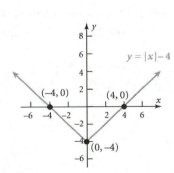

(a) $F(y) = |x|$

Subtract 4; vertical shift down 4 units.

(b) $F(x) = |x| - 4$

The x-intercepts of F are $(-4, 0)$ and $(4, 0)$; the y-intercept is $(0, -4)$. The domain is all real numbers and the range is $\{y \mid y \geq -4\}$.

59.

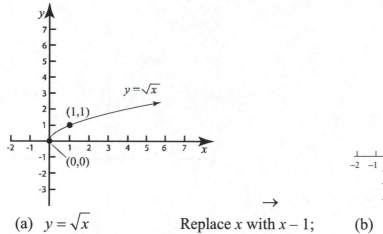

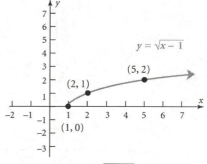

      (a) $y = \sqrt{x}$ $\rightarrow$ Replace x with $x - 1$;   (b) $F(x) = \sqrt{x - 1}$

horizontal shift right 1 unit.

The x-intercept of F is $(1, 0)$; there is no y-intercept. The domain of F is the interval $[1, \infty)$ and the range is $[0, \infty)$.

61.

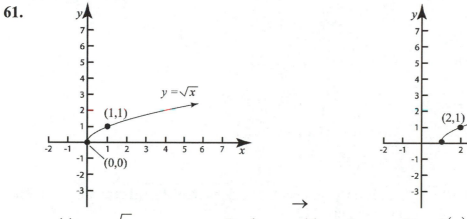

      (a) $y = \sqrt{x}$ $\rightarrow$ Replace x with $x - 1$;   (b) $f(x) = \sqrt{x - 1}$

horizontal shift right 1 unit.

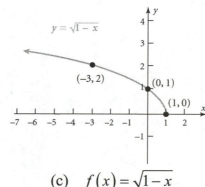

     $\rightarrow$ Replace $x - 1$ by $-(x - 1)$;   (c) $f(x) = \sqrt{1 - x}$

reflect about the y-axis.

The x-intercept of f is $(1, 0)$; the y-intercept is $(0, 1)$. The domain of f is the set $\{x \mid x \leq 1\}$ or the interval $[-\infty, 1)$. The range is the set $\{y \mid y \geq 0\}$ or the interval $[0, \infty)$.

63.

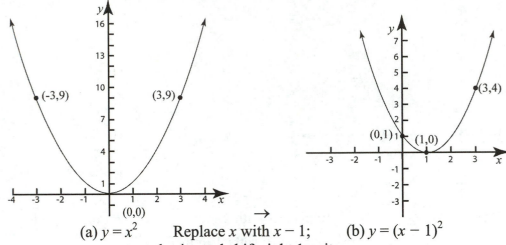

(a) $y = x^2$ Replace x with $x - 1$; $\rightarrow$ (b) $y = (x - 1)^2$
horizontal shift right 1 unit.

$\rightarrow$

Add -2; vertical   (c) $f(x) = (x - 1)^2 + 2$
shift up 2 units.

The y-intercept is $(0, 3)$; there is no x-intercept. The domain of f is all real numbers. The range is the set $\{y \,|\, y \geq 2\}$ or the interval $[2, \infty)$.

65.   (a) $y = f(-x)$ (b) $y = -f(x)$

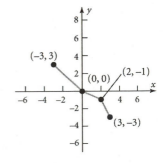

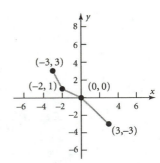

(c) $y = f(x + 2)$ (d) $y = f(x) + 2$

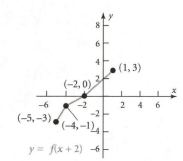   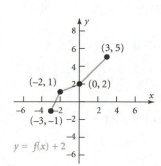

67.   Since f is linear, we first find the slope of the function using the 2 given points $(4, -5)$ and $(0, 3)$.

$$m = \frac{y_2 - y_1}{x_2 - x_1} = \frac{-5 - 3}{4 - 0} = \frac{-8}{4} = -2$$

Then we get the point-slope form of a line by using m and the point $(0, 3)$.

$$y - y_1 = m(x - x_1)$$
$$y - 3 = -2(x - 0)$$
$$y = -2x + 3$$

So the linear function is $f(x) = -2x + 3$.

69. $f(1) = 4$ means the point $(1, 4)$ satisfies the equation for f. So

$$4 = \frac{A(1) + 5}{6(1) - 2} = \frac{A + 5}{4}$$
$$16 = A + 5$$
$$A = 11$$

71.   Since the height is twice the radius, we can write $h = 2r$. Then the volume of the cylinder can be expressed as $V = \pi r^2 (2r) = 2\pi r^3$.

73.   (a) $R = R(x) = xp$

$$R(x) = x\left(-\frac{1}{6}x + 100\right)$$
$$= -\frac{1}{6}x^2 + 100x \qquad 0 \le x \le 600$$

(b) If 200 units are sold, $x = 200$ and the revenue is

$$R(200) = -\frac{1}{6}(200)^2 + 100(200) = \$13,333.33$$

75.   (a) To find the revenue function $R = R(x)$, we first solve the demand equation for p.

$$x = -5p + 100 \qquad (1)$$
$$5p = 100 - x$$
$$p = 20 - \frac{1}{5}x \qquad 0 \le x \le 100$$

We find the domain by using equation (1) and solving

$$\text{when } p = 0, \quad x = -5(0) + 100 = 100$$
$$\text{when } p = 20, \quad x = -5(20) + 100 = 0$$

So the revenue R can be expressed as

$$R(x) = x\left(20 - \frac{1}{5}x\right)$$

$$R(x) = 20x - \frac{1}{5}x^2 \qquad 0 \le x \le 100$$

(b) If 15 units are sold, $x = 15$, and the revenue is

$$R(15) = 20(15) - \frac{1}{5}(15)^2 = \$255.00$$

77. (a) The cost of making the drum is the sum of the costs of making the top and bottom and the side.

The amount of material used in the top and bottom is the area of the two circles,
$$A_{\text{top}} + A_{\text{bottom}} = \pi r^2 + \pi r^2 = 2\pi r^2 \text{ square centimeters.}$$
At \$0.06 per square centimeter, the cost of the top and the bottom of the drum is
$$C = 0.06(2\pi r^2) = 0.12\pi r^2 \text{ dollars.}$$

The amount of material used in the side of the drum is the area of the rectangle of material measured by the circumference of the top and the height of the drum.
$$A_{\text{side}} = 2\pi r h \text{ square centimeters.}$$
To express the area of the side as a function of r, we use the fact that we are told the volume of the drum is 500 cubic centimeters.
$$V = \pi r^2 h = 500$$
$$h = \frac{500}{\pi r^2}$$

So $A_{\text{side}} = 2\pi r h = 2\pi r\left(\dfrac{500}{\pi r^2}\right) = \dfrac{1000}{r}$ square centimeters.

At \$0.04 per square centimeter, the cost of making the side of the drum is
$$C = 0.04\left(\frac{1000}{r}\right) = \frac{40}{r} \text{ dollars.}$$

The total cost of making the drum is $C = C(r) = 0.12\pi r^2 + \dfrac{40}{r}$ dollars.

(b) If the radius is 4 cm, the cost of making the drum is
$$C(4) = 0.12\pi(4)^2 + \frac{40}{4} = \$16.03$$

(c) If the radius is 8 cm, the cost of making the drum is
$$C(8) = 0.12\pi(8)^2 + \frac{40}{8} = \$29.13$$

(d)

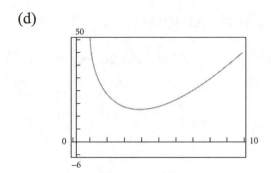

Making the can with a radius of 3.758 centimeters minimizes the cost of making the drum. The minimum cost is $15.97.

Chapter 10 Project

1. Since Avis has unlimited mileage, the cost of driving an Avis car x miles is
$$A = A(x) = 64.99$$
A is a constant function.

3.

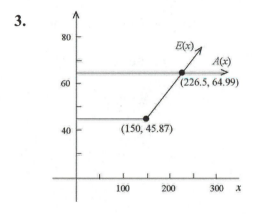

   If you drive more than 226 miles, Avis becomes the better choice.

5. If driving fewer than 130 miles, SaveALot Car Rental costs the least.
   If driving between 130 and 227 miles Enterprise is the most economical.
   If driving more than 227 miles Avis is the best buy.

7. If driving fewer than 53 miles, USave Car Rental is the cheapest.
   If driving between 53 and 130 miles, SaveALot Car Rental costs the least.
   If driving between 130 and 227 miles Enterprise is the most economical.
   If driving more than 227 miles Avis is the best buy.

Mathematical Questions from Professional Exams

1. (d) On the open interval $(-1, 2)$, The minimum value of f is 0, and the maximum value of f approaches 4. So the range of f is $0 \le y < 4$.

3. (c) The domain of f is nonnegative. That is,

    $$x^3 - x \ge 0$$
    $$x(x^2 - 1) \ge 0$$
    $$x(x-1)(x+1) \ge 0$$

    Solving the related equation for 0 and testing a point in each interval gives

    $$x(x-1)(x+1) = 0$$
    $$x = 0 \quad \text{or} \quad x = 1 \quad \text{or} \quad x = -1$$

    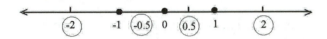

    When $x = -2$, $(-2)^3 - (-2) = -6 < 0$, so the interval $(-\infty, -1)$ is not part of the domain.
    When $x = -0.5$, $(-0.5)^3 - (-0.5) = 0.375 \ge 0$, so the interval $[-1, 0]$ is part of the domain.
    When $x = 0.5$, $(0.5)^3 - (0.5) = -0.375 < 0$, so the interval $(0, 1)$ is not part of the domain.
    When $x = 2$, $(2)^3 - (2) = 6 \ge 0$, so the interval $[1, \infty)$ is part of the domain.

Chapter 11
Classes of Functions

11.1 Quadratic Functions $614/1-33(ODD)$

1. **(C)** f is a quadratic function whose graph opens up, and whose vertex is
$$\left(-\frac{b}{2a}, f\left(-\frac{b}{2a}\right)\right) = (0, -1).$$
$$-\frac{b}{2a} = -\frac{0}{2} = 0; \qquad f(0) = 0^2 - 1 = -1$$

3. **(F)** f is a quadratic function whose graph opens up, and whose vertex is
$$\left(-\frac{b}{2a}, f\left(-\frac{b}{2a}\right)\right) = (1, 0).$$
$$-\frac{b}{2a} = \frac{2}{2} = 1; \qquad f(1) = 1^2 - 2 \cdot 1 + 1 = 0$$

5. **(G)** f is a quadratic function whose graph opens up, and whose vertex is
$$\left(-\frac{b}{2a}, f\left(-\frac{b}{2a}\right)\right) = (1, 1).$$
$$-\frac{b}{2a} = \frac{2}{2} = 1; \qquad f(1) = 1^2 - 2 \cdot 1 + 2 = 1$$

7. **(H)** f is a quadratic function whose graph opens up, and whose vertex is
$$\left(-\frac{b}{2a}, f\left(-\frac{b}{2a}\right)\right) = (1, -1).$$
$$-\frac{b}{2a} = \frac{2}{2} = 1; \qquad f(1) = 1^2 - 2 \cdot 1 = -1$$

9. $a = 1$, $b = 2$, $c = 0$. Since $a > 0$, the parabola opens up.

The x-coordinate of the vertex is $-\frac{b}{2a} = -\frac{2}{2(1)} = -1$.

The y-coordinate of the vertex is $f(-1) = (-1)^2 + 2(-1) = -1$.

So the vertex is $(-1, -1)$ and the axis of symmetry is the line $x = -1$.
Since $f(0) = c = 0$, the y-intercept is $(0, 0)$.
The x-intercepts are found by solving $f(x) = 0$.
$$x^2 + 2x = 0$$
$$x(x+2) = 0$$
$$x = 0 \ \text{ or } \ x = -2$$
The x-intercepts are $(0, 0)$ and $(-2, 0)$.
The domain is the set of all real numbers or the interval $(-\infty, \infty)$; the range is the set $\{y \mid$

$y \geq -1\}$ or the interval $[-1, \infty)$.

The function is increasing to the right of the axis or on the interval $(-1, \infty)$, and it is decreasing to the left of the axis or on the interval $(-\infty, -1)$.

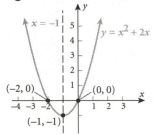

11. $a = -1$, $b = -6$, $c = 0$. Since $a < 0$, the parabola opens down.

The x-coordinate of the vertex is $-\dfrac{b}{2a} = -\dfrac{-6}{2(-1)} = -3$.

The y-coordinate of the vertex is $f(-3) = -(-3)^2 - 6(-3) = 9$.

So the vertex is $(-3, 9)$ and the axis of symmetry is the line $x = -3$.

Since $f(0) = c = 0$, the y-intercept is $(0, 0)$.

The x-intercepts are found by solving $f(x) = 0$.

$$-x^2 - 6x = 0$$
$$-x(x + 6) = 0$$
$$x = 0 \ \text{ or } \ x = -6$$

The x-intercepts are $(0, 0)$ and $(-6, 0)$.

The domain is the set of all real numbers or the interval $(-\infty, \infty)$; the range is the set $\{y \mid y \leq 9\}$ or the interval $(-\infty, 9]$.

The function is increasing to the left of the axis or on the interval $(-\infty, -3)$, and it is decreasing to the right of the axis or on the interval $(-3, \infty)$.

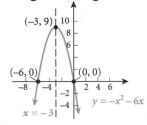

13. $a = 2$, $b = -8$, $c = 0$. Since $a > 0$, the parabola opens up.

The x-coordinate of the vertex is $-\dfrac{b}{2a} = -\dfrac{-8}{2(2)} = 2$.

The y-coordinate of the vertex is $f(2) = 2(2)^2 - 8(2) = -8$.

So the vertex is $(2, -8)$ and the axis of symmetry is the line $x = 2$.

Since $f(0) = c = 0$, the y-intercept is $(0, 0)$.

The x-intercepts are found by solving $f(x) = 0$.

$$2x^2 - 8x = 0$$
$$2x(x - 4) = 0$$
$$x = 0 \ \text{ or } \ x = 4$$

The x-intercepts are $(0, 0)$ and $(4, 0)$.

The domain is the set of all real numbers or the interval $(-\infty, \infty)$; the range is the set $\{y \mid y \geq -8\}$ or the interval $[-8, \infty)$.

The function is increasing to the right of the axis or on the interval $(2, \infty)$, and it is decreasing to the left of the axis or on the interval $(-\infty, 2)$.

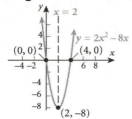

15. $a = 1$, $b = 2$, $c = -8$. Since $a > 0$, the parabola opens up.

The x-coordinate of the vertex is $-\dfrac{b}{2a} = -\dfrac{2}{2(1)} = -1$.

The y-coordinate of the vertex is $f(-1) = (-1)^2 + 2(-1) - 8 = -9$.

So the vertex is $(-1, -9)$ and the axis of symmetry is the line $x = -1$.

Since $f(0) = c = -8$, the y-intercept is $(0, -8)$.

The x-intercepts are found by solving $f(x) = 0$.
$$x^2 + 2x - 8 = 0$$
$$(x - 2)(x + 4) = 0$$
$$x = 2 \quad \text{or} \quad x = -4$$

The x-intercepts are $(2, 0)$ and $(-4, 0)$.

The domain is the set of all real numbers or the interval $(-\infty, \infty)$; the range is the set $\{y \mid y \geq -9\}$ or the interval $[-9, \infty)$.

The function is increasing to the right of the axis or on the interval $(-1, \infty)$, and it is decreasing to the left of the axis or on the interval $(-\infty, -1)$.

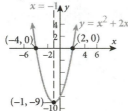

17. $a = 1$, $b = 2$, $c = 1$. Since $a > 0$, the parabola opens up.

The x-coordinate of the vertex is $-\dfrac{b}{2a} = -\dfrac{2}{2(1)} = -1$.

The y-coordinate of the vertex is $f(-1) = (-1)^2 + 2(-1) + 1 = 0$.

So the vertex is $(-1, 0)$ and the axis of symmetry is the line $x = -1$.

Since $f(0) = c = 1$, the y-intercept is $(0, 1)$.

The x-intercepts are found by solving $f(x) = 0$.
$$x^2 + 2x + 1 = 0$$
$$(x + 1)(x + 1) = 0$$
$$x = -1$$

The x-intercept is $(-1, 0)$.

The vertex and the x-intercept are the same, so we use symmetry and the y-intercept to obtain a third point $(-2, 1)$ on the graph.

The domain is the set of all real numbers or the interval $(-\infty, \infty)$; the range is the set

$\{y \mid y \geq 0\}$ or the interval $[0, \infty)$.
The function is increasing to the right of the axis or on the interval $(-1, \infty)$, and it is
decreasing to the left of the axis or on the interval $(-\infty, -1)$.

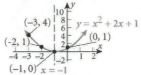

19. $a = 2, b = -1, c = 2$. Since $a > 0$, the parabola opens up.

The x-coordinate of the vertex is $-\dfrac{b}{2a} = -\dfrac{-1}{2(2)} = \dfrac{1}{4}$.

The y-coordinate of the vertex is $f\left(\dfrac{1}{4}\right) = 2\left(\dfrac{1}{4}\right)^2 - \left(\dfrac{1}{4}\right) + 2 = \dfrac{15}{8}$.

So the vertex is $\left(\dfrac{1}{4}, \dfrac{15}{8}\right) = (0.25, 1.875)$ and the axis of symmetry is the line $x = 0.25$.

Since $f(0) = c = 2$, the y-intercept is $(0, 2)$.

The x-intercepts are found by solving $f(x) = 0$. Since the discriminant
$b^2 - 4ac = (-1)^2 - 4(2)(2) = -15$ is negative, the equation $f(x) = 0$ has no real solution,
and therefore, the parabola has no x-intercept.

To graph the function, we choose a point and use symmetry. If we choose $x = 1$,
$f(1) = 2(1)^2 - (1) + 2 = 3$. Using symmetry, we obtain the point $(-0.5, 3)$.

The domain is the set of all real numbers or the interval $(-\infty, \infty)$; the range is the set
$\{y \mid y \geq 1.875\}$ or the interval $[1.875, \infty)$.
The function is increasing to the right of the axis or on the interval $(0.25, \infty)$, and it is
decreasing to the left of the axis or on the interval $(-\infty, 0.25)$.

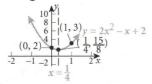

21. $a = -2, b = 2, c = -3$. Since $a < 0$, the parabola opens down.

The x-coordinate of the vertex is $-\dfrac{b}{2a} = -\dfrac{2}{2(-2)} = \dfrac{1}{2} = 0.5$.

The y-coordinate of the vertex is $f\left(\dfrac{1}{2}\right) = -2\left(\dfrac{1}{2}\right)^2 + 2\left(\dfrac{1}{2}\right) - 3 = -\dfrac{5}{2} = -2.5$.

So the vertex is $\left(\dfrac{1}{2}, -\dfrac{5}{2}\right) = (0.5, -2.5)$, and the axis of symmetry is the line $x = 0.5$.

Since $f(0) = c = -3$, the y-intercept is $(0, -3)$.

The x-intercepts are found by solving $f(x) = 0$. Since the discriminant

$$b^2 - 4ac = 2^2 - 4(-2)(-3) = -20$$

is less than zero, the equation $f(x) = 0$ has no real solution, and therefore, the parabola has no x-intercept.

To graph the function, we choose an additional point and use symmetry. If we choose $x = 2$, $f(2) = -2(2)^2 + 2(2) - 3 = -7$. Using symmetry, we obtain the point $(-1, -7)$.

The domain is the set of all real numbers or the interval $(-\infty, \infty)$; the range is the set $\{y \mid y \le -2.5\}$ or the interval $(-\infty, -2.5]$.

The function is increasing to the left of the axis or on the interval $(-\infty, 0.5)$, and it is decreasing to the right of the axis or on the interval $(0.5, \infty)$.

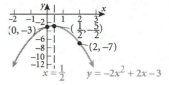

23. $a = 3$, $b = 6$, $c = 2$. Since $a > 0$, the parabola opens up.

The x-coordinate of the vertex is $-\dfrac{b}{2a} = -\dfrac{6}{2(3)} = -1$.

The y-coordinate of the vertex is $f(-1) = 3(-1)^2 + 6(-1) + 2 = -1$.

So the vertex is $(-1, -1)$ and the axis of symmetry is the line $x = -1$.

Since $f(0) = c = 2$, the y-intercept is $(0, 2)$.

The x-intercepts are found by solving $f(x) = 0$. Using the quadratic formula, we obtain

$$x = \frac{-6 \pm \sqrt{6^2 - 4(3)(2)}}{2(3)} = \frac{-6 \pm \sqrt{36 - 24}}{6} = \frac{-6 \pm \sqrt{12}}{6} = \frac{-3 \pm \sqrt{3}}{3}$$

$$x \approx -1.58 \text{ or } x \approx -0.42$$

The x-intercepts are approximately $(-1.58, 0)$ and $(-0.42, 0)$.

The domain is the set of all real numbers or the interval $(-\infty, \infty)$; the range is the set $\{y \mid y \ge -1\}$ or the interval $[-1, \infty)$.

The function is increasing to the right of the vertex or on the interval $(-1, \infty)$, and it is decreasing to the left of the vertex or on the interval $(-\infty, -1)$.

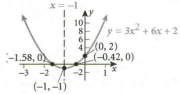

25. $a = -4$, $b = -6$, $c = 2$. Since $a < 0$, the parabola opens down.

The x-coordinate of the vertex is $-\dfrac{b}{2a} = -\dfrac{-6}{2(-4)} = -\dfrac{3}{4} = -0.75$.

The y-coordinate of the vertex is $f\left(-\dfrac{3}{4}\right)=-4\left(-\dfrac{3}{4}\right)^{2}-6\left(-\dfrac{3}{4}\right)+2=\dfrac{17}{4}$.

So the vertex is $\left(-\dfrac{3}{4},\dfrac{17}{4}\right)=(-0.75,\ 4.25)$, and the axis of symmetry is the line $x=-0.75$.

Since $f(0)=c=2$, the y-intercept is $(0,\ 2)$.

The x-intercepts are found by solving $f(x)=0$. Using the quadratic formula, we obtain

$$x=\dfrac{6\pm\sqrt{(-6)^{2}-4(-4)(2)}}{2(-4)}=\dfrac{6\pm\sqrt{36+32}}{-8}=\dfrac{6\pm\sqrt{68}}{-8}=\dfrac{3\pm\sqrt{17}}{-4}$$

$$x\approx-1.78\ \ \text{or}\ \ x\approx0.28$$

The x-intercepts are approximately $(-1.78,\ 0)$ and $(0.28,\ 0)$.

The domain is the set of all real numbers or the interval $(-\infty,\ \infty)$; the range is the set $\{y\,|\,y\le4.25\}$ or the interval $(-\infty,\ 4.25]$.

The function is increasing to the left of the axis or on the interval $(-\infty,\ -0.75)$, and it is decreasing to the right of the axis or on the interval $(-0.75,\ \infty)$.

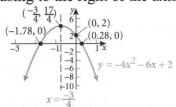

27. $a=2,\ b=12,\ c=0$. Since $a>0$, the parabola opens up, and the function has a minimum value. The minimum value occurs at

$$x=-\dfrac{b}{2a}=-\dfrac{12}{2(2)}=-3$$

The minimum value is $f(-3)=2(-3)^{2}+12(-3)=-18$.

29. $a=2,\ b=12,\ c=-3$. Since $a>0$, the parabola opens up, and the function has a minimum value. The minimum value occurs at

$$x=-\dfrac{b}{2a}=-\dfrac{12}{2(2)}=-3$$

The minimum value is $f(-3)=2(-3)^{2}+12(-3)-3=-21$.

31. $a=-1,\ b=10,\ c=-4$. Since $a<0$, the parabola opens down, and the function has a maximum value. The maximum value occurs at

$$x=-\dfrac{b}{2a}=-\dfrac{10}{2(-1)}=5$$

The maximum value is $f(5)=-(5)^{2}+10(5)-4=21$.

33. $a = -3, b = 12, c = 1$. Since $a < 0$, the parabola opens down and the function has a maximum value. The maximum value occurs at

$$x = -\frac{b}{2a} = -\frac{12}{2(-3)} = 2$$

The maximum value is $f(2) = -3(2)^2 + 12(2) + 1 = 13$.

35.   If $r_1 = -3$, and $r_2 = 1$, and $f(x) = a(x - r_1)(x - r_2)$
(a) Then if $a = 1, f(x) = 1(x - (-3))(x - 1) = (x + 3)(x - 1)$.
If $a = 2, f(x) = 2(x - (-3))(x - 1) = 2(x + 3)(x - 1)$.
If $a = -2, f(x) = -2(x - (-3))(x - 1) = -2(x + 3)(x - 1)$.
If $a = 5, f(x) = 5(x - (-3))(x - 1) = 5(x + 3)(x - 1)$.

(b) The x-intercepts are found by solving $f(x) = 0$.
$$a(x - r_1)(x - r_2) = 0$$
$$(x - r_1)(x - r_2) = 0$$
So the value of a, $a \neq 0$, has no affect on the x-intercept.

The y-intercept is found by letting $x = 0$ and simplifying.
$$f(x) = a(x - r_1)(x - r_2)$$
$$f(0) = a(0 - r_1)(0 - r_2)$$
$$f(0) = ar_1r_2$$
So the y-intercept is $(0, ar_1r_2)$ is the product of the a and the x-intercepts.

(c) The axis of symmetry is the line $x = -\dfrac{b}{2a}$. To determine b, we multiply out the factors of f.

$$\begin{aligned} f(x) &= a(x - r_1)(x - r_2) \\ &= a(x^2 - r_2 x - r_1 x + r_1 r_2) \\ &= a[x^2 - (r_1 + r_2)x + r_1 r_2] \\ &= ax^2 - a(r_1 + r_2)x + ar_1 r_2 \end{aligned}$$

We find $b = -a(r_1 + r_2)$. The line of symmetry is

$$x = -\frac{b}{2a} = -\frac{-a(r_1 + r_2)}{2a} = \frac{r_1 + r_2}{2}$$

which does not involve a. So the value of a does not affect the axis of symmetry.

(d) The vertex is the point $\left(\dfrac{r_1 + r_2}{2}, f\left(\dfrac{r_1 + r_2}{2}\right)\right)$. If we evaluate f at $\dfrac{r_1 + r_2}{2}$, we get

$$\begin{aligned} f\left(\frac{r_1 + r_2}{2}\right) &= a\left(\frac{r_1 + r_2}{2} - r_1\right)\left(\frac{r_1 + r_2}{2} - r_2\right) \\ &= a\left[\left(\frac{r_1 + r_2}{2} - r_1\right)\left(\frac{r_1 + r_2}{2} - r_2\right)\right] \end{aligned}$$

We see the y-value of the vertex is changed by a factor of a.

(e) The x-coordinate of the vertex is $x = \dfrac{r_1 + r_2}{2}$. The midpoint of the x-intercepts is

$$\text{midpoint} = \left(\frac{x_1 + x_2}{2}, \frac{y_1 + y_2}{2}\right)$$

$$= \left(\frac{r_1 + r_2}{2}, \frac{0+0}{2}\right) = \left(\frac{r_1 + r_2}{2}, 0\right)$$

The x-coordinate of the vertex and the midpoint of the x-intercepts is the same.

37. Since R is a quadratic function with $a = -4 < 0$, the vertex will give the maximum revenue.

The unit price to be charged should be $p = -\dfrac{b}{2a} = -\dfrac{4000}{2(-4)} = 500$ dollars.

If the dryers cost \$500, the revenue R will be maximized. The maximum revenue will be
$$R(500) = -4(500)^2 + 4000(500) = -1,000,000 + 2,000,000 = \$1,000,000$$

39. (a) $R(x) = xp$

$$R(x) = x\left(-\frac{1}{6}x + 100\right) = -\frac{1}{6}x^2 + 100x \qquad 0 \le x \le 600$$

(b) If 200 units are sold, $x = 200$, and the revenue R is
$$R(200) = -\frac{1}{6}(200)^2 + 100(200) = \frac{-40,000 + 120,000}{6} = \frac{80,000}{6} = \$13,333.33$$

(c) Since R is a quadratic function with $a = -\dfrac{1}{6} < 0$, the vertex will give the maximum

revenue. Revenue is maximized when $x = -\dfrac{b}{2a} = -\dfrac{100}{2\left(-\dfrac{1}{6}\right)} = 300$ units are sold. The

maximum revenue is $R(300) = -\dfrac{1}{6}(300)^2 + 100(300) = -15,000 + 30,000 = \$15,000$

(d) The company should charge $p = -\dfrac{1}{6}(300) + 100 = -50 + 100 = \$50.00$ per unit to

maximize revenue.

41. (a) We first solve the demand equation for x.
$$x = -5p + 100 \qquad\qquad 0 \le p \le 20,$$
$$5p = 100 - x$$
$$p = \frac{100 - x}{5} = 20 - \frac{1}{5}x$$

Since when $p = 0$, $x = 100$ and when $p = 20$, $x = 0$, the domain of the function p is $0 \le x \le 100$.

The revenue function R is
$$R(x) = x\left(20 - \frac{1}{5}x\right) = 20x - \frac{1}{5}x^2 \qquad 0 \le x \le 100$$

(b) If 15 units are sold, $x = 15$, and the revenue R is

$$R(15) = 20(15) - \frac{1}{5}(15)^2 = 300 - 45 = \$255$$

(c) Since R is a quadratic function with $a = -\frac{1}{5} < 0$, the vertex will give the maximum revenue. Revenue is maximized when $x = -\frac{b}{2a} = -\frac{20}{2\left(-\frac{1}{5}\right)} = 50$ units are sold. The

maximum revenue is $R(50) = 20(50) - \frac{1}{5}(50)^2 = 1000 - 500 = \$500$

(d) The company should charge $p = 20 - \frac{1}{5}(50) = 20 - 10 = \$10$ per unit to maximize revenue.

43. If the rectangle is shown at right and we are told the width is x and the perimeter, or the distance around the edge of the rectangle, is 400 yards, then we can find the length of the rectangle.

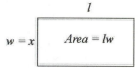

$$P = 2l + 2w$$
$$400 = 2l + 2x$$
$$l = \frac{400 - 2x}{2} = 200 - x \text{ yards}$$

(a) $A(x) = lw = (200 - x)x = 200x - x^2$ square yards.

(b) A is a quadratic function with $a = -1$, $b = 200$, and $c = 0$. Since $a < 0$, the vertex gives the maximum area. The area is maximum when $x = -\frac{b}{2a} = -\frac{200}{2(-1)} = 100$ yards.

(c) The maximum area is $A(100) = 200(100) - 100^2 = 10,000$ square yards.

45. From the figure we see that the width of the plot measures x meters and the length of the plot measures $4000 - 2x$ meters. The area of the plot is

$$A(x) = lw = (4000 - 2x)x = 4000x - 2x^2 \text{ meters squared.}$$

A is a quadratic function with $a = -2$, $b = 4000$, and $c = 0$. Since $a < 0$, the vertex gives the maximum area. The area is maximum when $x = -\frac{b}{2a} = -\frac{4000}{2(-2)} = 1000$ meters. The

maximum area is $A(1000) = 4000(1000) - 2(1000)^2 = 2,000,000$ meters squared.

47.

(a) h is a quadratic function with $a = -\frac{32}{50^2}$, $b = 1$, and $c = 200$. Since $a < 0$, the vertex gives the maximum height.

The height is maximum when $x = -\dfrac{b}{2a} = -\dfrac{1}{2\left(\dfrac{-32}{50^2}\right)} = \dfrac{2500}{64} = 39.06$ feet from the cliff.

(b) The maximum height of the projectile is

$$h(39.06) = \dfrac{-32(39.06)^2}{50^2} + 39.06 + 200 = 219.53 \text{ feet above the water.}$$

(c) The projectile will hit the water when $h(x) = 0$.

$$-\dfrac{32}{50^2}x^2 + x + 200 = 0$$

$$x = 170.02 \text{ feet from the cliff.}$$

(d)

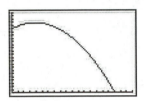

(e)

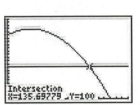

When the projectile is 100 feet above the water, it is 135.70 feet from the base of the cliff.

49. If we denote the depth of the rain gutter by x, then the area A of the cross-section is given by

$$A = lw$$
$$A(x) = (12 - 2x)x$$
$$= 12x - 2x^2$$

The function A is quadratic, with $a = -2 < 0$, so the vertex is the maximum point. The cross-sectional area is maximum when

$$x = -\dfrac{b}{2a} = -\dfrac{12}{2(-2)} = 3 \text{ inches.}$$

51. If x denotes the width of the rectangle (and the diameter of the circle), then the perimeter of the track is

$$P = 2l + 2\left(\dfrac{1}{2}\pi x\right) = 2l + \pi x = 400$$

We can express l in terms of x by solving the equation for l.

$$l = \frac{400 - \pi x}{2}$$

The area of the rectangle is

$$A = \left(\frac{400 - \pi x}{2}\right)x = 200x - \frac{\pi}{2}x^2$$

A is a maximum when

$$x = -\frac{b}{2a} = -\frac{200}{2 \cdot \left(-\frac{\pi}{2}\right)} = \frac{200}{\pi} \approx 63.66 \text{ meters, and}$$

$$l = \frac{1}{2}\left[400 - \pi\left(\frac{200}{\pi}\right)\right] = \frac{1}{2}(200) = 100 \text{ meters.}$$

53. (a) Since $H(x)$ is a quadratic function, the income level for which there are the most hunters is given by x-value of the vertex. For this function $a = -1.01$, $b = 114.3$, and $c = 451$.

$$x = -\frac{b}{2a} = -\frac{114.3}{2(-1.01)} = 56.584$$

The most hunters have an income level of approximately \$56,584.
There are about

$$H(56.584) = -1.01(56.584)^2 + 114.3(56.584) + 451 = 3684.785$$

About 3685 hunters have an annual income of \$56,584.

(b)

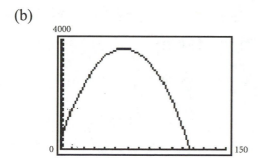

The number of hunters earning between \$20,000 and \$40,000 is increasing.

55. (a) $M(23) = 0.76(23)^2 - 107.00(23) + 3854.18 = 1795.2$

Approximately 1795 males who are 23 years old are murdered.

(b) Using a graphing utility, and finding the intersection of $M(x)$ and the line $y = 1456$, we find that 1456 males age 28 are murdered.

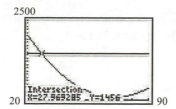

0

(c)

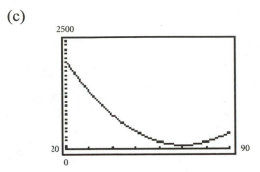

(d) The number of male murder victims decreases with age until age 70, and then it begins to increase.

57. The reaction rate is modeled by the quadratic function V.
$$V = V(x) = akx - kx^2$$
where $a = -k$, $b = ak$, and $c = 0$. The reaction rate is maximum at the vertex of V.
$$x = -\frac{b}{2a} = -\frac{ak}{2(-k)} = \frac{a}{2}$$
The reaction rate is maximum when half the initial amount of the compound is present.

59. $y = f(x) = -5x^2 + 8$

$y_0 = f(-1) = -5(-1)^2 + 8 = 3$

$y_1 = f(0) = -5(0)^2 + 8 = 8$

$y_2 = f(1) = -5(1)^2 + 8 = 3$

$$\text{Area} = \frac{h}{3}(y_0 + 4y_1 + y_2)$$

$$= \frac{1}{3}(3 + 4(8) + 3)$$

$$= \frac{38}{3} \text{ square units}$$

61. $y = f(x) = x^2 + 3x + 5$

$y_0 = f(-4) = (-4)^2 + 3(-4) + 5 = 9$

$y_1 = f(0) = (0)^2 + 3(0) + 5 = 5$

$y_2 = f(4) = (4)^2 + 3(4) + 5 = 33$

$$\text{Area} = \frac{h}{3}(y_0 + 4y_1 + y_2)$$

$$= \frac{4}{3}(9 + 4(5) + 33)$$

$$= \frac{248}{3} \text{ square units}$$

63. The rectangle is drawn on the right. The width of the rectangle is x units, and the length is $y = 10 - x$ units.

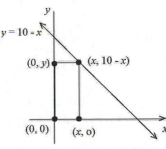

The area A of the rectangle is
$$A = lw$$
$$A(x) = (10 - x)x$$
$$= 10x - x^2$$
A is a quadratic function with $a = -1$, $b = 10$, and $c = 0$. Since $a < 0$, the maximum area occurs at the vertex of A. That is, when the width of the rectangle is
$$x = -\frac{b}{2a} = -\frac{10}{2(-1)} = 5$$

The largest area enclosed by the rectangle is
$$A = 10(5) - 5^2 = 25 \text{ square units.}$$

65. Functions will vary. All answers should have $a < 0$ and be perfect squares.

67.

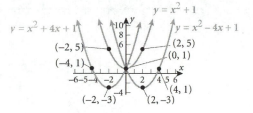

Descriptions may vary.

69. Answers will vary.

$625 / 1 - 39 \text{ (ODD)}$

11.2 Power Functions; Polynomial Functions; Rational Functions

1. Answers will vary.

3. origin

5.

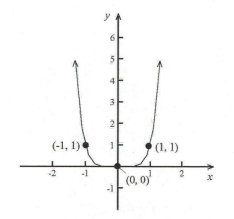

$y = x$ Add 2; vertical $f(x) = y + 2 = x^6 + 2$
                 shift up 2 units.

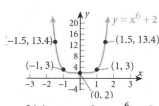

7.

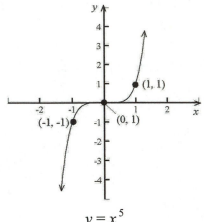

$$y = x^5$$

Replace y by $-y$;

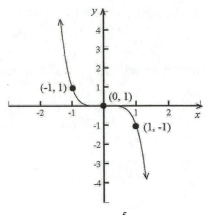

$$-y = -x^5$$

→
Add 2; vertical
shift up 2 units.

$$f(x) = -y + 2 = -x^5 + 2$$

9.

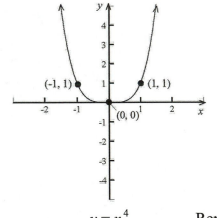

$$y = x^4$$

→
Replace x with $x - 2$;
horizontal shift to the
right 2 units.

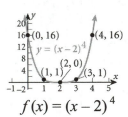

$$f(x) = (x - 2)^4$$

11. f is a polynomial. Its degree is 3.

13. g is a polynomial. Its degree is 2.

15. f is not a polynomial. The exponent of x is -1.

17. g is not a polynomial. One of the exponents is not an integer.

19. F is a polynomial. Its degree is 4.

21. G is a polynomial. Its degree is 4.

23. $y = 3x^4$

25. $y = -2x^5$

27.   Multiply the function out. $5(x+1)^2(x-2) = 5(x^2+2x+1)(x-2)$
$$= (5x^2+10x+5)(x-2)$$
$$= 5x^3-10x^2+10x^2-20x+5x-10$$
$$= 5x^3-15x-10$$
So the power function is $y = 5x^3$.

29. R is a rational function. The domain is all real numbers except those for which the denominator is 0.
$$x-3 \ne 0 \text{ or } x \ne 3$$
The domain of R is the set $\{x \mid x \ne 3\}$.

31. H is a rational function. The domain is all real numbers except those for which the denominator is 0.
$$(x-2)(x+4) \ne 0$$
$$x-2 \ne 0 \quad \text{or} \quad x+4 \ne 0$$
$$x \ne 2 \quad \text{or} \quad x \ne -4$$

The domain of H is the set $\{x \mid x \ne 2 \text{ and } x \ne -4\}$.

33. F is a rational function. The domain is all real numbers except those for which the denominator is 0.
$$2x^2 - 5x - 3 \ne 0$$
$$(2x+1)(x-3) \ne 0$$
$$2x+1 \ne 0 \quad \text{or} \quad x-3 \ne 0$$
$$x \ne -\frac{1}{2} \quad \text{or} \quad x \ne 3$$

The domain of F is the set $\left\{x \mid x \ne -\dfrac{1}{2} \text{ and } x \ne 3\right\}$.

35. R is a rational function. The domain is all real numbers except those for which the denominator is 0.
$$x^3 - 8 \ne 0$$
$$(x-2)(x^2+2x+4) \ne 0$$
$$x-2 \ne 0 \qquad\qquad x^2+2x+4 = 0 \text{ has a negative discriminant,}$$
$$\text{and so has no real solutions.}$$
$$x \ne 2$$

The domain of R is the set $\{x \mid x \ne 2\}$.

37. H is a rational function. The domain is all real numbers except those for which the denominator is 0.
$$x^2 + 4 \ne 0$$

$x^2 + 4 = 0$ has a negative discriminant, and so has no real solutions. The denominator never equals 0.

The domain of H is the set of all real numbers or the interval $(-\infty, \infty)$.

39. R is a rational function. The domain is all real numbers except those for which the denominator is 0.
$$4(x^2 - 9) \neq 0$$
$$4(x - 3)(x + 3) \neq 0$$
$$x - 3 \neq 0 \quad \text{or} \quad x + 3 \neq 0$$
$$x \neq 3 \quad \text{or} \quad x \neq -3$$

The domain of R is the set $\{x \mid x \neq 3 \text{ and } x \neq -3\}$.

41. (a) The year 2000 is 70 years since the year 1930. $(2000 - 1930 = 70)$
The percentage of union membership in 2000 is $u(70)$.
$$u(70) = 11.93 + 1.9(70) - 0.052(70^2) + 0.00037(70^3)$$
$$= 17.04\%$$
(b) $u(75) = 11.93 + 1.9(75) - 0.052(75^2) + 0.00037(75^3) = 18.02\%$
Answers will vary.

637/ 19-26,27 -55 (ODD)

11.3 Exponential Functions 640/57, 59, 61, 67

1. (a) $3^{2.2} = 11.2116$ (b) $3^{2.23} = 11.5873$
(c) $3^{2.236} = 11.6639$ (d) $3^{\sqrt{5}} = 11.6648$

3. (a) $2^{3.14} = 8.8152$ (b) $2^{3.141} = 8.8214$
(c) $2^{3.1415} = 8.8244$ (d) $2^{\pi} = 8.8250$

5. (a) $3.1^{2.7} = 21.2166$ (b) $3.14^{2.71} = 22.2167$
(c) $3.141^{2.718} = 22.4404$ (d) $\pi^e = 22.4592$

7. $e^{1.2} = 3.3201$

9. $e^{-0.85} = 0.4274$

11.

x	$f(x)$	$\dfrac{f(x)}{f(x-1)}$
-1	3	
0	6	$\dfrac{6}{3} = 2$
1	12	$\dfrac{12}{6} = 2$
2	18	$\dfrac{18}{12} = \dfrac{3}{2}$
3	30	$\dfrac{30}{18} = \dfrac{5}{3}$

The ratio of consecutive outputs is not constant for unit increases in inputs. So the function f is not exponential.

13.

x	$H(x)$	$\dfrac{H(x)}{H(x-1)}$
-1	$\dfrac{1}{4}$	
0	1	$\dfrac{1}{\frac{1}{4}} = 4$
1	4	$\dfrac{4}{1} = 4$
2	16	$\dfrac{16}{4} = 4$
3	64	$\dfrac{64}{16} = 4$

The ratio of consecutive outputs is constant for unit increases in inputs. So the function H is exponential. The base $a = 4$.

15.

x	$f(x)$	$\dfrac{f(x)}{f(x-1)}$
-1	$\dfrac{3}{2}$	
0	3	$\dfrac{3}{\frac{3}{2}} = 2$
1	6	$\dfrac{6}{3} = 2$
2	12	$\dfrac{12}{6} = 2$
3	24	$\dfrac{24}{12} = 2$

The ratio of consecutive outputs is constant for unit increases in inputs. So the function f is exponential. The base $a = 2$.

17.

x	$H(x)$	$\dfrac{H(x)}{H(x-1)}$
-1	2	
0	4	$\dfrac{4}{2}=2$
1	6	$\dfrac{6}{4}=\dfrac{3}{2}$
2	8	$\dfrac{8}{6}=\dfrac{4}{3}$
3	10	$\dfrac{10}{8}=\dfrac{5}{4}$

The ratio of consecutive outputs is not constant for unit increases in inputs. So the function H is not exponential.

19. B. This is the graph of $y=3^x$ reflected about the y-axis. It is the graph of $y=3^{-x}$.

21. D. This is the graph of $y=3^x$ reflected over both the x- and the y-axes. It is the graph of $y=-3^{-x}$.

23. A. This is the graph of $y=3^x$.

25. E. This is the graph of $y=3^x$ vertically shifted down one unit. It is the graph of $y=3^x-1$.

27.

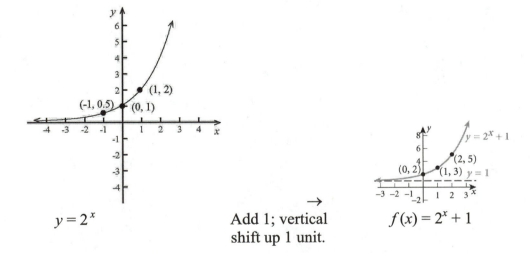

$y=2^x$ Add 1; vertical shift up 1 unit. $f(x)=2^x+1$

The domain of f is all real numbers or the interval $(-\infty, \infty)$; the range is the set $\{y \mid y>1\}$ or the interval $(1, \infty)$. The horizontal asymptote is the line $y=1$.

29.

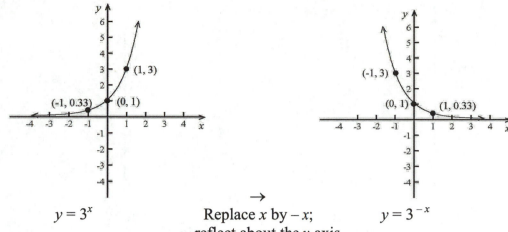

$$y = 3^x \qquad \overset{\rightarrow}{\underset{\text{reflect about the } y\text{-axis.}}{\text{Replace } x \text{ by } -x;}} \qquad y = 3^{-x}$$

$$\overset{\rightarrow}{\underset{\text{shift down 2 units.}}{\text{Subtract 2; vertical}}} \qquad f(x) = 3^{-x} - 2$$

The domain of f is all real numbers or the interval $(-\infty, \infty)$; the range is the set $\{y \mid y > -2\}$ or the interval $(-2, \infty)$. The horizontal asymptote is the line $y = -2$.

31.

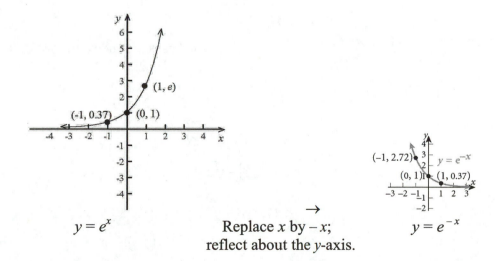

$$y = e^x \qquad \overset{\rightarrow}{\underset{\text{reflect about the } y\text{-axis.}}{\text{Replace } x \text{ by } -x;}} \qquad y = e^{-x}$$

The domain of f is all real numbers or the interval $(-\infty, \infty)$; the range is the set $\{y \mid y > 0\}$ or the interval $(0, \infty)$. The horizontal asymptote is the line $y = 0$.

33.

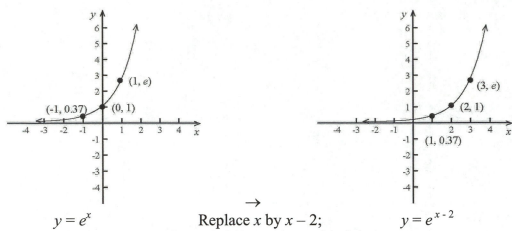

$$y = e^x$$ $\rightarrow$ Replace x by $x - 2$; $$y = e^{x-2}$$
horizontal shift 2 units to the right.

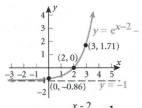

$\rightarrow$ Subtract 1; vertical $$y = e^{x-2} - 1$$
shift 1 unit down.

The domain of f is all real numbers or the interval $(-\infty, \infty)$; the range is the set $\{y \,|\, y > -1\}$ or the interval $(-1, \infty)$. The horizontal asymptote is the line $y = -1$.

35. $2^{2x+1} = 4 = 2^2$
Since the bases are the same, we obtain
$$2x + 1 = 2$$
$$2x = 1$$
$$x = \frac{1}{2}$$
$$\left\{\frac{1}{2}\right\}$$

37. $3^{x^3} = 9^x = \left(3^2\right)^x = 3^{2x}$
Since the bases are the same, we obtain
$$x^3 = 2x$$
$$x^3 - 2x = 0$$
$$x(x^2 - 2) = 0$$
$$x = 0 \quad \text{or} \quad x^2 - 2 = 0$$
$$x = \pm\sqrt{2}$$
$$\left\{-\sqrt{2}, \, 0, \, \sqrt{2}\right\}$$

39.
$$8^{x^2-2x} = \frac{1}{2}$$
$$\left(2^3\right)^{x^2-2x} = 2^{-1}$$
$$2^{3\left(x^2-2x\right)} = 2^{-1}$$

Since the bases are the same, we obtain
$$3(x^2 - 2x) = -1$$
$$3x^2 - 6x + 1 = 0, \text{ where } a = 3, b = -6, c = 1$$

41. $2^x \cdot 8^{-x} = 4^x$
$$2^x \cdot \left(2^3\right)^{-x} = \left(2^2\right)^x$$
$$2^x \cdot 2^{-3x} = 2^{2x}$$
$$2^{x-3x} = 2^{2x}$$

Since the bases are the same, we obtain
$$x - 3x = 2x$$
$$-4x = 0$$
$$x = 0$$
$$\{0\}$$

Using the quadratic formula we find

$$x = \frac{6 \pm \sqrt{(-6)^2 - 4(3)(1)}}{2(3)} = \frac{6 \pm \sqrt{36 - 12}}{6}$$

$$= \frac{6 \pm \sqrt{24}}{6} = \frac{6 \pm 2\sqrt{6}}{6} = \frac{3 \pm \sqrt{6}}{3}$$

$$\left\{ \frac{3 - \sqrt{6}}{3}, \frac{3 + \sqrt{6}}{3} \right\}$$

43.
$$\left(\frac{1}{5}\right)^{2-x} = 25$$
$$5^{-(2-x)} = 5^2$$

Since the bases are the same, we obtain
$$-(2 - x) = 2$$
$$x = 4$$

$\{4\}$

45.
$$4^x = 8$$
$$2^{2x} = 2^3$$

Since the bases are the same, we obtain
$$2x = 3$$

$$x = \frac{3}{2}$$

$$\left\{ \frac{3}{2} \right\}$$

47.
$$e^{x^2} = \left(e^{3x}\right) \cdot \frac{1}{e^2}$$
$$e^{x^2} = e^{3x-2}$$

Since the bases are the same, we obtain
$$x^2 = 3x - 2$$
$$x^2 - 3x + 2 = 0$$
$$(x - 2)(x - 1) = 0$$
$$x - 2 = 0 \quad \text{or} \quad x - 1 = 0$$
$$x = 2 \quad \text{or} \quad x = 1$$

$\{2, 1\}$

49. $4^{-2x} = (4^x)^{-2}$

So if $4^x = 7$, $4^{-2x} = 7^{-2} = \frac{1}{7^2} = \frac{1}{49}$

51. If $3^{-x} = 2$, then $3^x = 2^{-1}$, and
$$3^{2x} = (2^{-1})^2$$
$$= 2^{-2} = \frac{1}{4}$$

53. The graph is increasing and $\left(-1, \frac{1}{3}\right)$, $(0, 1)$, and $(1, 3)$ are points on the graph. So the function $f(x) = 3^x$.

55. The graph is decreasing, negative, and $\left(-1, -\dfrac{1}{6}\right)$, $(0, -1)$, and $(1, -6)$ are points on the graph. So the function $f(x) = -6^x$.

57. (a) The percent of light passing through 10 panes of glass is
$$p(10) = 100e^{-0.03(10)} = 74.08\%$$

   (b) The percent of light passing through 25 panes of glass is
$$p(25) = 100e^{-0.03(25)} = 47.24\%$$

59. (a) After 30 days there will be
$$w(30) = 50e^{-0.004(30)} = 44.35 \text{ watts of power.}$$

   (b) After 365 days there will be
$$w(365) = 50e^{-0.004(365)} = 11.61 \text{ watts of power.}$$

61. After 1 hour, there will be
$$D(1) = 5e^{-0.4(1)} = 3.35 \text{ milligrams of drug will be in the patient's bloodstream.}$$

After 6 hours, there will be
$$D(6) = 5e^{-0.4(6)} = 0.45 \text{ milligrams of drug will be in the patient's bloodstream.}$$

63. (a) The probability that a car will arrive within 10 minutes of 12:00 PM is
$$F(10) = 1 - e^{-0.1(10)} = 0.632$$

   (b) The probability that a car will arrive within 40 minutes of 12:00 PM is
$$F(40) = 1 - e^{-0.1(40)} = 0.982$$

   (c) As t becomes unbounded in the positive direction, $e^{-0.1t} = \dfrac{1}{e^{-0.1t}}$ approaches 0. So F approaches 1.

   (d)

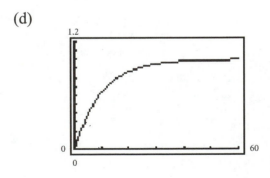

(e)

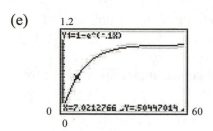

It takes 6.9 minutes for the probability to reach 50%.

65. (a) If 15 cars arrive, then $x = 15$ and
$$P(x) = P(15) = \frac{20^{15}e^{-20}}{15!} = 0.0516$$
The probability that 15 cars arrive between 5:00 PM and 6:00 PM is 0.0516.

(b) The probability that 20 cars arrive between 5:00 PM and 6:00 PM is given by
$$P(x) = P(20) = \frac{20^{20}e^{-20}}{20!} = 0.0888$$

67. (a) If the Civic is 3 years old $x = 3$, and its cost is
$$p(x) = p(3) = 16,630(0.90)^3 = \$12,123.27$$

(b) If the Civic is 9 years old $x = 9$, and its cost is
$$p(x) = p(9) = 16,630(0.90)^9 = \$6442.80$$

69. (a) If $E = 120$ volts, $R = 10$ ohms, and $L = 5$ henrys, then the amperage is given by the function
$$I(t) = \frac{120}{10}\left[1 - e^{-(10/5)t}\right] = 12\left[1 - e^{-2t}\right]$$

After 0.3 second $t = 0.3$ and $I(t) = I(0.3) = 12\left[1 - e^{-2(0.3)}\right] = 5.414$ amps.

After 0.5 second $t = 0.5$ and $I(t) = I(0.5) = 12\left[1 - e^{-2(0.5)}\right] = 7.585$ amps.

After 1 second $t = 1$ and $I(t) = I(1) = 12\left[1 - e^{-2}\right] = 10.375$ amps.

(b)

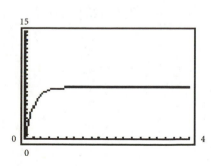

(c) The maximum current of $I_1(t)$ approaches 12 amps as t becomes unbounded in the positive direction.

(d) If $E = 120$ volts, $R = 5$ ohms, and $L = 10$ henrys, then the amperage is given by the function

$$I(t) = \frac{120}{5}\left[1 - e^{-(5/10)t}\right] = 24\left[1 - e^{-(1/2)t}\right]$$

After 0.3 second $t = 0.3$ and $I(t) = I(0.3) = 24\left[1 - e^{-(1/2)(0.3)}\right] = 3.343$ amps.

After 0.5 second $t = 0.5$ and $I(t) = I(0.5) = 24\left[1 - e^{-(1/2)(0.5)}\right] = 5.309$ amps.

After 1 second $t = 1$ and $I(t) = I(1) = 24\left[1 - e^{-1/2}\right] = 9.443$ amps.

(e)

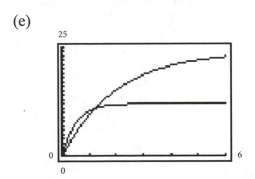

(f) The maximum current of $I_2(t)$ approaches 24 amps as t becomes unbounded in the positive direction.

71. Using a calculator, we get

$$n = 4, \qquad 2 + \frac{1}{2!} + \frac{1}{3!} + \frac{1}{4!} = 2.70833333$$

$$n = 6, \qquad 2 + \frac{1}{2!} + \frac{1}{3!} + \frac{1}{4!} + \frac{1}{5!} + \frac{1}{6!} = 2.71805556$$

$$n = 8, \qquad 2 + \frac{1}{2!} + \frac{1}{3!} + \frac{1}{4!} + \frac{1}{5!} + \frac{1}{6!} + \frac{1}{7!} + \frac{1}{8!} = 2.71827877$$

$$n = 10 \qquad 2 + \frac{1}{2!} + \frac{1}{3!} + \frac{1}{4!} + \frac{1}{5!} + \frac{1}{6!} + \frac{1}{7!} + \frac{1}{8!} + \frac{1}{9!} + \frac{1}{10!} = 2.718281801$$

and $e = 2.718281828$

Comparisons might differ.

73.
$$\frac{f(x+h)-f(x)}{h}=\frac{a^{(x+h)}-a^x}{h}$$

$$=\frac{a^x a^h-a^x}{h}$$ Use the exponential property $a^r \cdot a^s = a^{r+s}$.

$$=\frac{a^x(a^h-1)}{h}=a^x\left(\frac{a^h-1}{h}\right)$$ Factor.

75.   If $f(x)=a^x$, then $f(-x)=a^{-x}$

$$=\frac{1}{a^x}$$ Use the exponential property $a^{-r}=\dfrac{1}{a^r}$.

$$=\frac{1}{f(x)}$$ Substitute.

77.   (a) If F = 50° and D = 41°, then
$$R=10^{\frac{4221}{T+459.4}-\frac{4221}{D+459.4}+2}=10^{\frac{4221}{50+459.4}-\frac{4221}{41+459.4}+2}$$
$$=10^{\frac{4221}{509.4}-\frac{4221}{500.4}+2}$$
$$=70.95$$
The relative humidity is 70.95%.

(b) If F = 68° and D = 59°, then
$$R=10^{\frac{4221}{T+459.4}-\frac{4221}{D+459.4}+2}=10^{\frac{4221}{68+459.4}-\frac{4221}{59+459.4}+2}$$
$$=10^{\frac{4221}{527.4}-\frac{4221}{518.4}+2}$$
$$=72.62$$
The relative humidity is 72.62%.

(c) If F = D, then $\dfrac{4221}{T+459.4}=\dfrac{4221}{D+459.4}$. So $\dfrac{4221}{T+459.4}-\dfrac{4221}{D+459.4}=0$, and

$$R=10^{0+2}=100$$
When the temperature and the dew point are equal, the relative humidity is 100%.

79.   (a) To show $f(x)=\sinh x$ is an odd function, we evaluate $f(-x)$ and simplify.
$$f(-x)=\frac{1}{2}\left(e^{-x}-e^{-(-x)}\right)$$
$$=\frac{1}{2}\left(e^{-x}-e^x\right)$$ Simplify.
$$=\frac{1}{2}\left(-e^x+e^{-x}\right)$$ Rearrange the terms.
$$=-\frac{1}{2}\left(e^x-e^{-x}\right)$$ Factor out -1.
$$=-f(x)$$

So f is an odd function.

(b)

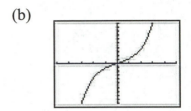

81. It takes 59 minutes. Explanations will vary.

83. There is no power function that increases more rapidly than an exponential function whose base is greater than 1. Explanations will vary.

85. Answers will vary.

11.4 Logarithmic Functions   650 (1-69) ODD ; 652 (91-109) ODD

1. We use the fact that $y = \log_a x$ and $x = a^y$, $a > 0$, $a \neq 1$ are equivalent.
If $9 = 3^2$, then $\log_3 9 = 2$.

3. We use the fact that $y = \log_a x$ and $x = a^y$, $a > 0$, $a \neq 1$ are equivalent.
If $a^2 = 1.6$, then $\log_a 1.6 = 2$.

5. We use the fact that $y = \log_a x$ and $x = a^y$, $a > 0$, $a \neq 1$ are equivalent.
If $1.1^2 = M$, then $\log_{1.1} M = 2$.

7. We use the fact that $y = \log_a x$ and $x = a^y$, $a > 0$, $a \neq 1$ are equivalent.
If $2^x = 7.2$, then $\log_2 7.2 = x$.

9. We use the fact that $y = \log_a x$ and $x = a^y$, $a > 0$, $a \neq 1$ are equivalent.
If $x^{\sqrt{2}} = \pi$, then $\log_x \pi = \sqrt{2}$.

11. We use the fact that $y = \log_a x$ and $x = a^y$, $a > 0$, $a \neq 1$ are equivalent.
If $e^x = 8$, then $\log_e 8 = x$ or $\ln 8 = x$.

13. We use the fact that $x = a^y$, $a > 0$, $a \neq 1$ and $y = \log_a x$ are equivalent.
    If $\log_2 8 = 3$, then $2^3 = 8$.

15. We use the fact that $x = a^y$, $a > 0$, $a \neq 1$ and $y = \log_a x$ are equivalent.
    If $\log_a 3 = 6$, then $a^6 = 3$.

17. We use the fact that $x = a^y$, $a > 0$, $a \neq 1$ and $y = \log_a x$ are equivalent.
    If $\log_3 2 = x$, then $3^x = 2$.

19. We use the fact that $x = a^y$, $a > 0$, $a \neq 1$ and $y = \log_a x$ are equivalent.
    If $\log_2 M = 1.3$, then $2^{1.3} = M$.

21. We use the fact that $x = a^y$, $a > 0$, $a \neq 1$ and $y = \log_a x$ are equivalent.
    If $\log_{\sqrt{2}} \pi = x$, then $\left(\sqrt{2}\right)^x = \pi$.

23. We use the fact that $x = a^y$, $a > 0$, $a \neq 1$ and $y = \log_a x$ are equivalent.
    If $\ln 4 = x$, then $e^x = 4$.

25. To find the exact value of the logarithm, we change the expression to its equivalent exponential expression and simplify.

    $$y = \log_2 1$$

$2^y = 1$	Write in exponential form.
$2^y = 2^0$	$1 = 2^0$
$y = 0$	Equate exponents.

    Therefore, $\log_2 1 = 0$

27. To find the exact value of the logarithm, we change the expression to its equivalent exponential expression and simplify.

    $$y = \log_5 25$$

$5^y = 25$	Write in exponential form.
$5^y = 5^2$	$25 = 5^2$
$y = 2$	Equate exponents.

    Therefore, $\log_5 25 = 2$.

29. To find the exact value of the logarithm, we change the expression to its equivalent exponential expression and simplify.

    $$y = \log_{1/2} 16$$

$\left(\dfrac{1}{2}\right)^y = 16$	Write in exponential form.
$\left(\dfrac{1}{2}\right)^y = \left(\dfrac{1}{2}\right)^{-4}$	$16 = \left(\dfrac{1}{2}\right)^{-4}$
$y = -4$	Equate exponents.

    Therefore, $\log_{1/2} 16 = -4$.

31. To find the exact value of the logarithm, we change the expression to its equivalent exponential expression and simplify.

$$y = \log_{10} \sqrt{10}$$

$$10^y = \sqrt{10}$$ Write in exponential form.

$$10^y = 10^{1/2}$$ $\sqrt{10} = 10^{1/2}$

$$y = \frac{1}{2}$$ Equate exponents.

Therefore, $\log_{10} \sqrt{10} = \frac{1}{2}$.

33. To find the exact value of the logarithm, we change the expression to its equivalent exponential expression and simplify.

$$y = \log_{\sqrt{2}} 4$$

$$\left(\sqrt{2}\right)^y = 4$$ Write in exponential form.

$$\left(\sqrt{2}\right)^y = \left(\sqrt{2}\right)^4$$ $4 = 2^2 = \left[\left(\sqrt{2}\right)^2\right]^2 = \left(\sqrt{2}\right)^4$

$$y = 4$$ Equate exponents.

Therefore, $\log_{\sqrt{2}} 4 = 4$.

35. To find the exact value of the logarithm, we change the expression to its equivalent exponential expression and simplify.

$$y = \ln \sqrt{e}$$

$$e^y = \sqrt{e}$$ Write in exponential form.

$$e^y = e^{1/2}$$ $\sqrt{e} = e^{1/2}$

$$y = \frac{1}{2}$$ Equate exponents.

Therefore, $\ln \sqrt{e} = \frac{1}{2}$.

37. The domain of a logarithmic function is limited to all positive real numbers, so for $f(x) = \ln(x-3)$, $x - 3 > 0$. The domain is all $x > 3$, or using interval notation $(3, \infty)$.

39. The domain of a logarithmic function is limited to all positive real numbers, so for $F(x) = \log_2 x^2$, $x^2 > 0$. x^2 is positive except when $x = 0$, meaning the domain is the set $\{x \mid x \neq 0\}$.

41. The domain of $f(x) = 3 - 2\log_4 \frac{x}{2}$ is restricted by $\log_4 \frac{x}{2}$ which is defined only when $\frac{x}{2}$ is positive. So the domain of f is all $x > 0$, or using interval notation $(0, \infty)$.

43. There are two restrictions on the domain of $f(x) = \sqrt{\ln x}$.

First, $y = \sqrt{x}$ is defined only when x is nonnegative, that is when $x \geq 0$.
Second, $y = \ln x$ is defined only when x is positive, that is when $x > 0$. However, on the interval $(0, 1)$, $\ln x < 0$, so $f(x) = \sqrt{\ln x}$ is not defined.
So the domain of f is all $x \geq 1$, or using interval notation $[1, \infty)$.

45. $\ln \dfrac{5}{3} = 0.511$

47. $\dfrac{\ln \dfrac{10}{3}}{0.04} = 30.099$

49. If the graph of f contains the point $(2, 2)$, then $x = 2$ and $y = 2$ must satisfy the equation

$$y = f(x) = \log_a x$$
$$2 = \log_a 2$$
$$a^2 = 2$$
$$a = \sqrt{2}$$

51. The graph of $y = \log_3 x$ has the properties:

The domain is $(0, \infty)$. The range is $(-\infty, \infty)$.
The x-intercept is $(1, 0)$.
The y-axis is a vertical asymptote.
The graph is increasing since $a = 3 > 1$.

The graph contains the points $(3, 1)$ and $\left(\dfrac{1}{3}, -1\right)$.

The graph is continuous.

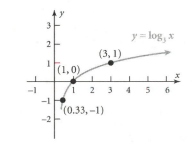

53. The graph of $y = \log_{1/5} x$ has the properties:

The domain is $(0, \infty)$. The range is $(-\infty, \infty)$.
The x-intercept is $(1, 0)$.
The y-axis is a vertical asymptote.

The graph is decreasing since $a = \dfrac{1}{5} < 1$.

The graph contains the points $(5, -1)$ and $\left(\dfrac{1}{5}, 1\right)$.

The graph is continuous.

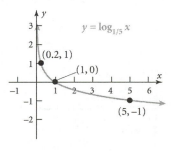

55. B. This graph is reflected over the y-axis.

57. D. This graph was reflected about both the x- and y-axes.

59. A. This graph has not been shifted or reflected.

61. E. This graph was shifted vertically down 1 unit.

63.

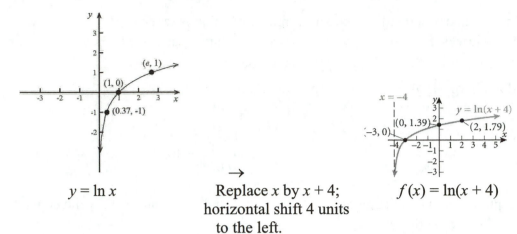

$y = \ln x$ Replace x by $x + 4$; $f(x) = \ln(x + 4)$
horizontal shift 4 units
to the left.

The domain of f is all $x > -4$, or in interval notation, $(-4, \infty)$. The range is all real numbers or $(-\infty, \infty)$. The vertical asymptote is the line $x = -4$.

65.

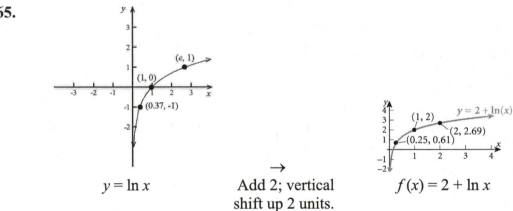

$y = \ln x$ Add 2; vertical $f(x) = 2 + \ln x$
shift up 2 units.

The domain of f is all $x > 0$, or in interval notation, $(0, \infty)$. The range is all real numbers or $(-\infty, \infty)$. The vertical asymptote is the y-axis, that is the line $x = 0$.

67.

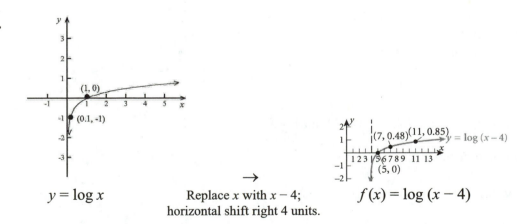

$y = \log x$ Replace x with $x - 4$; $f(x) = \log(x - 4)$
horizontal shift right 4 units.

The domain of f is all $x > 4$, or in interval notation, $(4, \infty)$. The range is all real numbers or $(-\infty, \infty)$. The vertical asymptote is the line $x = 4$.

69.

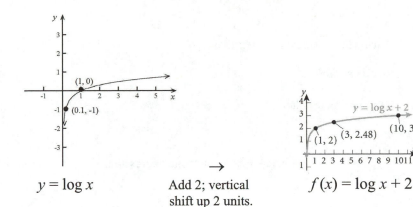

$y = \log x$ Add 2; vertical $f(x) = \log x + 2$
shift up 2 units.

The domain of f is all $x > 0$, or in interval notation, $(0, \infty)$. The range is all real numbers or $(-\infty, \infty)$. The vertical asymptote is the y-axis, that is the line $x = 0$.

71. $\log_3 x = 2$

     $x = 3^2 = 9$ Change to exponential form and simplify.

73. $\log_2(2x+1) = 3$

     $2x + 1 = 2^3$ Change to exponential form.

     $2x + 1 = 8$ Solve the linear equation.

       $2x = 7$

         $x = \dfrac{7}{2}$

75. $\log_x 4 = 2$

     $4 = x^2$ Change to exponential form.

     $x = 2$ Solve using the Square Root Method. ($x \neq -2$, the base is positive.)

77. $\ln e^x = 5$

     $e^x = e^5$ Change to exponential form.

     $x = 5$ Since the bases are equal, the exponents are equal.

79. $\log_4 64 = x$

       $64 = 4^x$ Change to exponential form.

       $4^3 = 4^x$

       $3 = x$ Since the bases are equal, the exponents are equal.

81. $\log_3 243 = 2x + 1$

     $3^{2x+1} = 243$ Change to exponential form.

     $3^{2x+1} = 3^5$

     $2x + 1 = 5$ Since the bases are equal, the exponents are equal.

       $2x = 4$ Solve the linear equation.

        $x = 2$

83.

$$e^{3x} = 10$$

$\ln 10 = 3x$ Change to a logarithmic expression.

$$x = \frac{\ln 10}{3}$$ Exact solution.

$$\approx 0.768$$ Approximate solution.

85.

$$e^{2x+5} = 8$$

$\ln 8 = 2x + 5$ Change to a logarithmic expression.

$2x = -5 + \ln 8$

$$x = \frac{-5 + \ln 8}{2}$$ Exact solution.

$$\approx -1.460$$ Approximate solution.

87. $\log_3\left(x^2 + 1\right) = 2$

$x^2 + 1 = 3^2$ Change to an exponential expression.

$x^2 + 1 = 9$

$x^2 = 8$

$x = \pm\sqrt{8}$ Use the Square Root Method.

$= \pm 2\sqrt{2}$ Simplify.

89. $\log_2 8^x = -3$

$8^x = 2^{-3}$ Change to an exponential expression.

$(2^3)^x = 2^{-3}$

$2^{3x} = 2^{-3}$

$3x = -3$ Since the bases are equal, the exponents are equal.

$x = -1$

91.   (a) When $H^+ = 0.1$, $pH = -\log(0.1) = 1$.

        (b) When $H^+ = 0.01$, $pH = -\log(0.01) = 2$.

        (c) When $H^+ = 0.001$, $pH = -\log(0.001) = 3$.

        (d) As the hydrogen ion concentration decreases, pH increases.

        (e) If $pH = 3.5$, then $3.5 = -\log x$

$$\log x = -3.5$$
$$10^{-3.5} = x$$
$$x = 0.000316$$

        (f) If $pH = 7.4$, then $7.4 = -\log x$

$$\log x = -7.4$$
$$10^{-7.4} = x$$
$$x = 3.981 \times 10^{-8}$$

93.   (a) If $p = 320$ mm,

$$320 = 760e^{-0.145h}$$

$$e^{-0.145h} = \frac{320}{760} = \frac{8}{19} \qquad \text{Divide both sides by 760.}$$

$$-0.145h = \ln\left(\frac{8}{19}\right) \qquad \text{Change to a logarithmic expression.}$$

$$h = -\frac{1000}{145}\ln\left(\frac{8}{19}\right) \qquad 0.145 = \frac{145}{1000}$$

$$\approx 5.965$$

The aircraft is at an altitude of approximately 5.965 kilometers above sea level.

(b) If $p = 667$ mm,
$$667 = 760\,e^{-0.145h}$$

$$e^{-0.145h} = \frac{667}{760} \qquad \text{Divide both sides by 760.}$$

$$-0.145h = \ln\left(\frac{667}{760}\right) \qquad \text{Change to a logarithmic expression.}$$

$$h = -\frac{1000}{145}\ln\left(\frac{667}{760}\right) \qquad 0.145 = \frac{145}{1000}$$

$$\approx 0.900$$

The mountain is approximately 0.9 kilometer above sea level.

95. (a) We want $F(t) = 0.50$.
$$0.50 = 1 - e^{-0.1t}$$
$$e^{-0.1t} = 0.50$$
$$-0.1t = \ln 0.5$$
$$t = -10\ln 0.5$$
$$\approx 6.93$$

After approximately 6.9 minutes the probability that a car arrives at the drive-thru reaches 50%.

(b) We want $F(t) = 0.80$.
$$0.80 = 1 - e^{-0.1t}$$
$$e^{-0.1t} = 0.20$$
$$-0.1t = \ln 0.2$$
$$t = -10\ln 0.2$$
$$\approx 16.09$$

After approximately 16.1 minutes the probability that a car arrives at the drive-thru reaches 80%.

97. We want $D = 2$.
$$2 = 5\,e^{-0.4h}$$

$$e^{-0.4h} = \frac{2}{5} = 0.4$$

$$-0.4h = \ln 0.4$$

$$h = -\frac{5}{2} \ln 0.4$$

$$\approx 2.29$$

The drug should be administered every 2.3 hours (about 2 hours 17 minutes).

99. We want $I = 0.5$

$$0.5 = \frac{12}{10}\left[1 - e^{-(10/5)t}\right] = 1.2\left(1 - e^{-2t}\right)$$

$$1 - e^{-2t} = \frac{0.5}{1.2} = \frac{5}{12}$$

$$e^{-2t} = \frac{7}{12}$$

$$-2t = \ln\left(\frac{7}{12}\right)$$

$$t = -\frac{1}{2}\ln\left(\frac{7}{12}\right)$$

$$\approx 0.269$$

It takes about 0.27 seconds to obtain a current of 0.5 ampere.

Next we want $I = 1.0$

$$1.0 = 1.2\left(1 - e^{-2t}\right)$$

$$1 - e^{-2t} = \frac{1.0}{1.2} = \frac{5}{6}$$

$$e^{-2t} = \frac{1}{6}$$

$$-2t = \ln\left(\frac{1}{6}\right)$$

$$t = -\frac{1}{2}\ln\left(\frac{1}{6}\right)$$

$$\approx 0.896$$

It takes about 0.9 seconds to obtain a current of 1 ampere.

101. We are interested in the population when $t = 11$, so we want $P(11)$.

$$P(10) = 298,710,000 + 10,000,000 \log 11$$

$$= 309,123,926.9$$

On January 1, 2020 the population (to the nearest thousand) will be 309,124,000.

103. Normal conversation: $x = 10^{-7}$ watt per square meter,

$$L\left(10^{-7}\right) = 10 \log \frac{10^{-7}}{10^{-12}} = 50 \text{ decibels}$$

105. Amplified rock music: $x = 10^{-1}$ watt per square meter,

$$L\left(10^{-1}\right) = 10 \log \frac{10^{-1}}{10^{-12}} = 110 \text{ decibels}$$

107. Mexico City, 1985: $x = 125,892$ mm.

$$M(125,892) = \log\left(\frac{125,892}{10^{-3}}\right) = 8.1 = 8.1$$

109. (a) Since $R = 10$ when $x = 0.06$, we can find k.
$$10 = 3\,e^{0.06k}$$

$$0.06k = \ln\left(\frac{10}{3}\right)$$

$$k = \frac{100}{6}\ln\left(\frac{10}{3}\right) = \frac{50}{3}\ln\left(\frac{10}{3}\right) \approx 20.066$$

(b) If $x = 0.17$, then the risk of having a car accident is
$$R = 3e^{\left(\frac{50}{3}\ln\left(\frac{10}{3}\right)\right)(0.17)} = 90.91\%$$

(c) If the risk is 100%, then $R = 100$.
$$3e^{\left(\frac{50}{3}\ln\left(\frac{10}{3}\right)\right)x} = 100$$

$$e^{\left(\frac{50}{3}\ln\left(\frac{10}{3}\right)\right)x} = \frac{100}{3}$$

$$\frac{50}{3}\ln\left(\frac{10}{3}\right)x = \ln\frac{100}{3}$$

$$x \approx 0.1747$$

(d) $$R = 15 = 3e^{\left(\frac{50}{3}\ln\left(\frac{10}{3}\right)\right)x}$$

$$e^{\left(\frac{50}{3}\ln\left(\frac{10}{3}\right)\right)x} = 5$$

$$\frac{50}{3}\ln\left(\frac{10}{3}\right)x = \ln 5$$

$$x \approx 0.080$$

A driver with a blood alcohol concentration greater than or equal to 0.08 should be arrested and charged with DUI.

(e) Answers will vary.

111. Answers may vary.

11.5 Properties of Logarithms 661/1-89 (ODD)

1. $\log_3 3^{71} = 71$ $\log_a a^r = r$ **3.** $\ln e^{-4} = -4$ $\log_a a^r = r$

5. $2^{\log_2 7} = 7$ $a^{\log_a M} = M$

7. $\log_8 2 + \log_8 4 = \log_8 (2 \cdot 4) = \log_8 8 = 1$

9.
$$\log_6 18 - \log_6 3 = \log_6 \left(\frac{18}{3}\right) = \log_6 6 = 1$$

11. Use a change of base formula,
$$\log_2 6 \cdot \log_6 4 = \frac{\ln 6}{\ln 2} \cdot \frac{\ln 4}{\ln 6} = \frac{\ln 4}{\ln 2}$$

Next simplify $\ln 4$,
$$\frac{\ln 4}{\ln 2} = \frac{\ln 2^2}{\ln 2} = \frac{2 \ln 2}{\ln 2} = 2$$

So, $\log_2 6 \cdot \log_6 4 = 2$

13.
$$3^{\log_3 5 - \log_3 4} = 3^{\log_3 \left(\frac{5}{4}\right)} = \frac{5}{4}$$

15. First we write the exponent as $y = \log_{e^2} 16$, and express it as an exponential.
$$y = \log_{e^2} 16$$
$$e^{2y} = 16 \quad \text{or} \quad e^y = 4. \quad \text{So} \quad y = \ln 4.$$
Then, $e^{\log_{e^2} 16} = e^y = e^{\ln 4} = 4.$

17. $\ln 6 = \ln (2 \cdot 3) = \ln 2 + \ln 3 = a + b$

19. $\ln 1.5 = \ln \dfrac{3}{2} = \ln 3 - \ln 2 = b - a$

21. $\ln 8 = \ln 2^3 = 3 \ln 2 = 3a$

23. $\ln \sqrt[5]{6} = \dfrac{1}{5} \ln 6 = \dfrac{1}{5} \ln (2 \cdot 3) = \dfrac{1}{5} (\ln 2 + \ln 3) = \dfrac{1}{5} (a + b)$

25. $\log_5 (25x) = \log_5 \left(5^2 x\right) = \log_5 5^2 + \log_5 x = 2 \log_5 5 + \log_5 x$

27. $\log_2 z^3 = 3 \log_2 z$

29. $\ln (ex) = \ln e + \ln x = 1 + \ln x$

31. $\ln (xe^x) = \ln x + \ln e^x = \ln x + x \ln e = \ln x + x$

33. $\log_a \left(u^2 v^3\right) = \log_a u^2 + \log_a v^3 = 2 \log_a u + 3 \log_a v$

35. $\ln \left(x^2 \sqrt{1-x}\right) = \ln x^2 + \ln (1-x)^{1/2} = 2 \ln x + \dfrac{1}{2} \ln (1-x)$

37.

$$\log_2\left(\frac{x^3}{x-3}\right) = \log_2 x^3 - \log_2(x-3) = 3\log_2 x - \log_2(x-3)$$

39.

$$\log\left[\frac{x(x+2)}{(x+3)^2}\right] = \log\left[x(x+2)\right] - \log(x+3)^2 = \log x + \log(x+2) - 2\log(x+3)$$

41.

$$\ln\left[\frac{x^2-x-2}{(x+4)^2}\right]^{1/3} = \frac{1}{3}\ln\left[\frac{x^2-x-2}{(x+4)^2}\right] = \frac{1}{3}\left[\ln(x^2-x-2) - \ln(x+4)^2\right]$$

$$= \frac{1}{3}\left[\ln\left[(x-2)(x+1)\right] - \ln(x+4)^2\right]$$

$$= \frac{1}{3}\left[\ln(x-2) + \ln(x+1) - 2\ln(x+4)\right]$$

$$= \frac{1}{3}\ln(x-2) + \frac{1}{3}\ln(x+1) - \frac{2}{3}\ln(x+4)$$

43.

$$\ln\frac{5x\sqrt{1+3x}}{(x-4)^3} = \ln\left[5x\sqrt{1+3x}\right] - \ln(x-4)^3 = \ln 5 + \ln x + \frac{1}{2}\ln(1+3x) - 3\ln(x-4)$$

45. $3\log_5 u + 4\log_5 v = \log_5 u^3 + \log_5 v^4 = \log_5\left(u^3 v^4\right)$

47.

$$\log_3\sqrt{x} - \log_3 x^3 = \log_3\left(\frac{\sqrt{x}}{x^3}\right) = \log_3 x^{-5/2} = -\log_3 x^{5/2} \text{ or } -\frac{5}{2}\log_3 x$$

49.

$$\log_4\left(x^2-1\right) - 5\log_4(x+1) = \log_4\left(x^2-1\right) - \log_4(x+1)^5 = \log_4\left[\frac{x^2-1}{(x+1)^5}\right]$$

$$= \log_4\left[\frac{(x-1)\cancel{(x+1)}}{(x+1)^{\cancel{5}4}}\right] = \log_4\left[\frac{x-1}{(x+1)^4}\right]$$

51.

$$\ln\left(\frac{x}{x-1}\right) + \ln\left(\frac{x+1}{x}\right) - \ln(x^2-1) = \ln\left[\frac{\cancel{x}}{x-1} \cdot \frac{x+1}{\cancel{x}} \cdot \frac{1}{(x-1)\cancel{(x+1)}}\right]$$

$$= \ln\left[\frac{1}{(x-1)^2}\right] = \ln(x-1)^{-2} = -2\ln(x-1)$$

53. $8\log_2\sqrt{3x-2} - \log_2\left(\frac{4}{x}\right) + \log_2 4 = \log_2\left(\sqrt{3x-2}\right)^8 - \log_2\left(\frac{4}{x}\right) + \log_2 4$

$$= \log_2(3x-2)^4 - \log_2\left(\frac{4}{x}\right) + \log_2 4$$

$$= \log_2 \left[\frac{(3x-2)^4 \cdot \cancel{A}}{\dfrac{\cancel{A}}{x}} \right] = \log_2 \left[x(3x-2)^4 \right]$$

55.

$$2\log_a\left(5x^3\right) - \frac{1}{2}\log_a\left(2x+3\right) = \log_a\left(5x^3\right)^2 - \log_a\sqrt{2x+3}$$

$$= \log_a \frac{\left(5x^3\right)^2}{\sqrt{2x+3}} = \log_a \frac{25x^6}{\sqrt{2x+3}}$$

57.

$$2\log_2(x+1) - \log_2(x+3) - \log_2(x-1) = \log_2(x+1)^2 - \log_2(x+3) - \log_2(x-1)$$

$$= \log_2 \frac{(x+1)^2}{(x+3)(x-1)} = \log_2 \frac{(x+1)^2}{x^2+2x-3}$$

59.

$$\log_3 21 = \frac{\ln 21}{\ln 3} = 2.771$$

61.

$$\log_{1/3} 71 = \frac{\ln 71}{\ln \dfrac{1}{3}} = -3.880$$

63.

$$\log_{\sqrt{2}} 7 = \frac{\ln 7}{\ln \sqrt{2}} = 5.615$$

65.

$$\log_\pi e = \frac{\ln e}{\ln \pi} = 0.874$$

67.

$$y = \log_4 x = \frac{\ln x}{\ln 4}$$

69.

$$y = \log_2(x+2) = \frac{\ln(x+2)}{\ln 2}$$

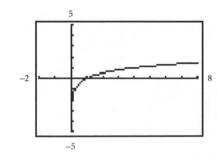

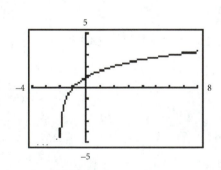

71.

$$y = \log_{x-1}(x+1) = \frac{\ln(x+1)}{\ln(x-1)}$$

73.

$$\ln y = \ln x + \ln C$$
$$\ln y = \ln(Cx)$$
$$y = Cx$$

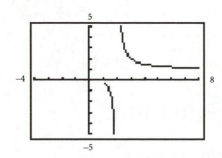

75. $\ln y = \ln x + \ln(x+1) + \ln C$

$\ln y = \ln \left[x(x+1)C \right]$

$y = x(x+1)C$

77. $\ln y = 3x + \ln C$

$\ln y = \ln e^{3x} + \ln C$

$\ln y = \ln Ce^{3x}$

$y = Ce^{3x}$

79. $\ln(y-3) = -4x + \ln C$

$\ln(y-3) = \ln e^{-4x} + \ln C$

$\ln(y-3) = \ln\left(Ce^{-4x}\right)$

$y - 3 = Ce^{-4x}$

$y = Ce^{-4x} + 3$

81.

$3 \ln y = \dfrac{1}{2} \ln(2x+1) - \dfrac{1}{3}\ln(x+4) + \ln C$

$\ln y^3 = \ln(2x+1)^{1/2} - \ln(x+4)^{1/3} + \ln C$

$\ln y^3 = \ln\left[C(2x+1)^{1/2} \right] - \ln(x+4)^{1/3}$

$\ln y^3 = \ln\left[\dfrac{C(2x+1)^{1/2}}{(x+4)^{1/3}} \right]$

$y^3 = \dfrac{C(2x+1)^{1/2}}{(x+4)^{1/3}}$

$y = \left[\dfrac{C(2x+1)^{1/2}}{(x+4)^{1/3}} \right]^{1/3} = \dfrac{C(2x+1)^{1/6}}{(x+4)^{1/9}}$ $C^{1/3}$ is still a positive constant. We write C.

83. We use the change of base formula and simplify.

$\log_2 3 \cdot \log_3 4 \cdot \log_4 5 \cdot \log_5 6 \cdot \log_6 7 \cdot \log_7 8$

$= \dfrac{\log 3}{\log 2} \cdot \dfrac{\log 4}{\log 3} \cdot \dfrac{\log 5}{\log 4} \cdot \dfrac{\log 6}{\log 5} \cdot \dfrac{\log 7}{\log 6} \cdot \dfrac{\log 8}{\log 7}$

$= \dfrac{\log 8}{\log 2} = \dfrac{\log 2^3}{\log 2}$

$$= \frac{3 \log 2}{\log 2} = 3$$

85. We use the change of base formula and simplify.

$$\log_2 3 \cdot \log_3 4 \cdot \log_4 5 \cdot \log_5 6 \cdot \log_6 7 \cdot \log_7 8 \cdot \ldots \cdot \log_n (n+1) \cdot \log_{n+1} 2$$

Noticing that the first $n-1$ factors follow the pattern of problem 83, we get

$$\left[\log_2 3 \cdot \log_3 4 \cdot \log_4 5 \cdot \log_5 6 \cdot \log_6 7 \cdot \log_7 8 \cdot \ldots \cdot \log_n (n+1) \right] \cdot \log_{n+1} 2$$

$$= \left[\frac{\log 3}{\log 2} \cdot \frac{\log 4}{\log 3} \cdot \frac{\log 5}{\log 4} \cdot \ldots \cdot \frac{\log (n+1)}{\log n} \right] \cdot \frac{\log 2}{\log(n+1)}$$

$$= 1$$

87.

$$\log_a \left(x + \sqrt{x^2 - 1} \right) + \log_a \left(x - \sqrt{x^2 - 1} \right) = \log_a \left[\left(x + \sqrt{x^2 - 1} \right)\left(x - \sqrt{x^2 - 1} \right) \right]$$

$$= \log_a \left(x^2 - \left(\sqrt{x^2 - 1} \right)^2 \right)$$

$$= \log_a \left(x^2 - \left| x^2 - 1 \right| \right)$$

$$= \log_a \left(x^2 - x^2 + 1 \right)$$

$$= \log_a 1 = 0$$

89.

$$\ln\left(1 + e^{2x} \right) = 2x + \ln\left(1 + e^{-2x} \right)$$

We work from the complicated side (the right) and simplify to get to the left side of the equation.

$$2x + \ln\left(1 + e^{-2x} \right) = \ln e^{2x} + \ln\left(1 + e^{-2x} \right)$$

$$= \ln\left[e^{2x} \cdot \left(1 + e^{-2x} \right) \right]$$

$$= \ln\left[e^{2x} + e^{2x} \cdot e^{-2x} \right]$$

$$= \ln\left[e^{2x} + 1 \right] = \ln\left(1 + e^{2x} \right)$$

91. We show that $-f(x) = \log_{1/a} x$ by using the Change of Base Formula.

$$f(x) = \log_a x \quad \text{so} \quad -f(x) = -\log_a x = -\frac{\ln x}{\ln a}$$

$$= \frac{\ln x}{-\ln a}$$

$$= \frac{\ln x}{\ln a^{-1}} = \frac{\ln x}{\ln \dfrac{1}{a}}$$

Using the Change of base formula in reverse, the last expression becomes $\ln_{1/a} x$. So
$$-f(x) = \log_{1/a} x$$

93. If $f(x) = \log_a x$, then
$$f\left(\frac{1}{x}\right) = \log_a\left(\frac{1}{x}\right) = \log_a x^{-1} = -\log_a x = -f(x)$$

95. If $A = \log_a M$ and $B = \log_a N$, then (writing each expression as an exponential) we get
$$a^A = M \quad \text{and} \quad a^B = N.$$
$$\begin{aligned}
\log_a\left(\frac{M}{N}\right) &= \log_a\left(\frac{a^A}{a^B}\right)\\
&= \log_a a^{A-B}\\
&= A - B\\
&= \log_a M - \log_a N
\end{aligned}$$

97.

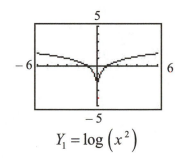

$$Y_1 = \log\left(x^2\right) \qquad Y_2 = 2\log\left(x\right)$$

Explanations may vary.

11.6 Continuously Compounded Interest

1. If $\$1000$ is invested at 4% compounded continuously, the amount A after 3 years is
$$A = Pe^{rt} = 1000e^{(0.04)(3)} = 1000e^{0.12} = \$1127.50$$

3. If $\$500$ is invested at 5% compounded continuously, the amount A after 3 years is
$$A = Pe^{rt} = 500e^{(0.05)(3)} = 500e^{0.15} = \$580.92$$

5. The present value of $\$100$ that will be invested at 4% compounded continuously for 6 months is
$$P = Ae^{-rt} = 100e^{-(0.04)(0.5)} = 100e^{-0.02} = \$98.02$$

7. The present value of $\$500$ that will be invested at 7% compounded continuously for 1 year is
$$P = Ae^{-rt} = 500e^{-(0.07)(1)} = 500e^{-0.07} = \$466.20$$

9. If $1000 is invested at 2% compounded continuously, the amount A after 1 year is
$$A = Pe^{rt} = 1000e^{(0.02)(1)} = 1000e^{0.02} = \$1020.20$$

$A - P = \$20.20$ interest was earned.

11. We need to deposit the present value of $5000 that will be invested at 3% compounded continuously for 4 years.
$$P = Ae^{-rt} = 5000e^{-(0.03)(4)} = 5000e^{-0.12} = \$4434.60$$

We need to deposit the present value of $5000 that will be invested at 3% compounded continuously for 8 years.
$$P = Ae^{-rt} = 5000e^{-(0.03)(8)} = 5000e^{-0.24} = \$3933.14$$

13. If P is invested, it will double when amount $A = 2P$.
$$A = 2P = Pe^{rt}$$
$$2 = e^{3r}$$
$$3r = \ln 2$$
$$r = \frac{\ln 2}{3} \approx 0.2310$$
It will require an interest rate of about 23.1% compounded continuously to double an investment in 3 years.

15. If P is invested, it will triple when amount $A = 3P$.
$$A = 3P = Pe^{rt}$$
$$3 = e^{0.10t}$$
$$0.10t = \ln 3$$
$$t = \frac{\ln 3}{0.10} \approx 10.986$$
It will take about 11 years for an investment to triple at 10 % compounded continuously.

17. We need the present value of $1000 that will be invested at 9% compounded continuously for 1 year.
$$P = Ae^{-rt} = 1000e^{-(0.09)(1)} = 1000e^{-0.09} = \$913.93$$

We need the present value of $1000 that will be invested at 9% compounded continuously for 2 years.
$$P = Ae^{-rt} = 1000e^{-(0.09)(2)} = 1000e^{-0.18} = \$835.27$$

19. Tami and Todd need to invest the present value of $40,000 that will earn 3% compounded continuously for 4 years.
$$P = Ae^{-rt} = 40,000e^{-(0.03)(4)} = 40,000e^{-0.12} = \$35,476.82$$

21. If P is invested, it will triple when amount $A = 3P$.
$$A = 3P = Pe^{rt}$$
$$3 = e^{5r}$$

$$5r = \ln 3$$

$$r = \frac{\ln 3}{5} \approx 0.2197$$

It will require an interest rate of about 22% compounded continuously to triple an investment in 5 years.

23. The rule of 70:
   (a) The actual time it takes to double an investment if $r = 1\%$ is

   $$t = \frac{\ln 2}{0.01} = 69.3147 \text{ years.}$$

   The estimated time it takes to double an investment using the Rule of 70 if $r = 1\%$ is

   $$t = \frac{0.70}{0.01} = 70.0 \text{ years.}$$

   The Rule of 70 overestimates the time to double money by 0.6853 year, about 8 months.

   (b) The actual time it takes to double an investment if $r = 5\%$ is

   $$t = \frac{\ln 2}{0.05} = 13.8629 \text{ years.}$$

   The estimated time it takes to double an investment using the Rule of 70 if $r = 5\%$ is

   $$t = \frac{0.70}{0.05} = 14 \text{ years.}$$

   The Rule of 70 overestimates the time to double money by 0.1371 year, about 1.6 months.

   (c) The actual time it takes to double an investment if $r = 10\%$ is

   $$t = \frac{\ln 2}{0.10} = 6.9315 \text{ years.}$$

   The estimated time it takes to double an investment using the Rule of 70 if $r = 10\%$ is

   $$t = \frac{0.70}{0.10} = 7.0 \text{ years.}$$

   The Rule of 70 overestimates the time to double money by 0.0685 year, almost 1 month.

Chapter 11 Review

TRUE-FALSE ITEMS

1. True   3. True   5. False

7. False   9. False

FILL IN THE BLANKS

1. parabola

3. $x = -\dfrac{b}{2a}$

5. one

7. $(0, \infty)$

9. one

REVIEW EXERCISES

1. First we expand the function.
$$f(x) = (x-2)^2 + 2 = x^2 - 4x + 4 + 2 = x^2 - 4x + 6$$
and find that $a = 1$, $b = -4$, $c = 6$.

$a > 0$; the parabola opens up.

The x-coordinate of the vertex is $-\dfrac{b}{2a} = -\dfrac{(-4)}{2(1)} = 2$, and the y-coordinate is
$$f(2) = (2)^2 - 4(2) + 6 = 2.$$
So the vertex is (2, 2) and the axis of symmetry is the line $x = 2$.

Since $f(0) = c = 6$, the y-intercept is (0, 6).
The x-intercepts are found by solving $f(x) = 0$. Since the discriminant of f
$$b^2 - 4ac = (-4)^2 - 4(1)(6) = -8$$
is negative, the equation $f(x) = 0$ has no real solution, and therefore, the parabola has no x-intercept.

To graph the function, we use symmetry. If we choose the y-intercept, we obtain its symmetric point (4, 6).

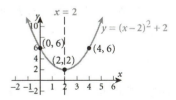

3. $a = \dfrac{1}{4}$, $b = 0$, $c = -16$

$a > 0$, the parabola opens up.

The x-coordinate of the vertex is $-\dfrac{b}{2a} = 0$, and the y-coordinate is
$$f(0) = \frac{1}{4}(0)^2 - 16 = -16.$$

So the vertex is (0, -16), and the axis of symmetry is the line $x = 0$.

Since $f(0) = c = -16$, the y-intercept is also (0, -16)
The x-intercepts are found by solving $f(x) = 0$.

$$\frac{1}{4}x^2 - 16 = 0$$
$$x^2 - 64 = 0$$
$$x^2 = 64$$
$$x = \pm 8$$

The x-intercepts are $(8, 0)$ and $(-8, 0)$.

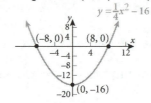

5. $a = -4$, $b = 4$, $c = 0$

$a < 0$, the parabola opens down.

The x-coordinate of the vertex is $-\dfrac{b}{2a} = -\dfrac{4}{2(-4)} = \dfrac{1}{2}$, and the y-coordinate is

$$f\left(\frac{1}{2}\right) = -4\left(\frac{1}{2}\right)^2 + 4x = 1$$

So the vertex is $\left(\dfrac{1}{2}, 1\right)$, and the axis of symmetry is the line $x = \dfrac{1}{2}$.

Since $f(0) = c = 0$, the y-intercept is $(0, 0)$.
The x-intercepts are found by solving $f(x) = 0$.

$$-4x^2 + 4x = 0$$
$$-4x(x - 1) = 0$$
$$x = 0 \quad \text{or} \quad x = 1$$

The x-intercepts are $(0, 0)$ and $(1, 0)$.

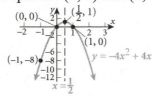

7. $a = \dfrac{9}{2}$, $b = 3$, $c = 1$

$a > 0$; the parabola opens up.

The x-coordinate of the vertex is $-\dfrac{b}{2a} = -\dfrac{3}{2\left(\dfrac{9}{2}\right)} = -\dfrac{1}{3}$, and the y-coordinate is

$$f\left(-\frac{1}{3}\right) = \frac{9}{2}\left(-\frac{1}{3}\right)^2 + 3\left(-\frac{1}{3}\right) + 1 = \frac{1}{2}$$

So the vertex is $\left(-\dfrac{1}{3}, \dfrac{1}{2}\right)$, and the axis of symmetry is the line $x = -\dfrac{1}{3}$.

Since $f(0) = c = 1$, the y-intercept is $(0, 1)$.
The x-intercepts are found by solving $f(x) = 0$. Since the discriminant of f

$$b^2 - 4ac = 3^2 - (4)\left(\frac{9}{2}\right)(1) = -9$$

is negative, the equation $f(x) = 0$ has no real solution, and therefore, the parabola has no x-intercept.

To graph the function, we use symmetry. If we choose the y-intercept, we obtain its symmetric point $\left(-\frac{2}{3}, 1\right)$. Because this point is so close to the vertex, we also use the point $\left(-1, \frac{5}{2}\right)$.

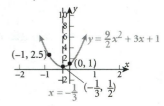

9. $a = 3, b = 4, c = -1$

$a > 0$; the parabola opens up.

The x-coordinate of the vertex is $-\dfrac{b}{2a} = -\dfrac{4}{2(3)} = -\dfrac{2}{3}$, and the y-coordinate is

$$f\left(-\frac{2}{3}\right) = 3\left(-\frac{2}{3}\right)^2 + 4\left(-\frac{2}{3}\right) - 1 = -\frac{7}{3}$$

So the vertex is $\left(-\dfrac{2}{3}, -\dfrac{7}{3}\right)$ and the axis of symmetry is the line $x = -\dfrac{2}{3}$.

Since $f(0) = c = -1$, the y-intercept is $(0, -1)$.
The x-intercepts are found by solving $f(x) = 0$.

$$3x^2 + 4x - 1 = 0$$

$$x = \frac{-4 \pm \sqrt{4^2 - 4(3)(-1)}}{2(3)} = \frac{-4 \pm \sqrt{28}}{6}$$

$$x = \frac{-4 \pm 2\sqrt{7}}{6} = \frac{-2 \pm \sqrt{7}}{3}$$

The x-intercepts are difficult to graph, so we choose another point and use symmetry.
We chose $(-2, 3)$ and by symmetry $\left(\dfrac{2}{3}, 3\right)$.

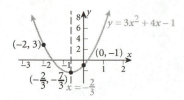

11. $a = 3$, $b = -6$, $c = 4$

The quadratic function has a minimum since $a = 3$ is greater than zero.

$$x = -\frac{b}{2a} = -\frac{-6}{2(3)} = 1$$

$$f(1) = 3(1)^2 - 6(1) + 4 = 1$$

The minimum value is 1, and it occurs at $x = 1$.

13. $a = -1$, $b = 8$, $c = -4$

The quadratic function has a maximum since $a = -1$ is less than zero.

$$x = -\frac{b}{2a} = -\frac{8}{2(-1)} = 4$$

$$f(4) = -(4)^2 + 8(4) - 4 = 12$$

The maximum value is 12, and it occurs at $x = 4$.

15. $a = -3$, $b = 12$, $c = 4$

The quadratic function has a maximum since $a = -3$ is less than zero.

$$x = -\frac{b}{2a} = -\frac{12}{2(-3)} = 2$$

$$f(2) = -3(2)^2 + 12(2) + 4 = 16$$

The maximum value is 16, and it occurs at $x = 2$.

17. Answers will vary.

19.

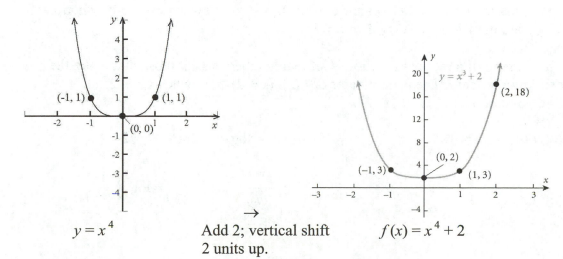

$y = x^4$ Add 2; vertical shift $f(x) = x^4 + 2$
                 2 units up.

21.

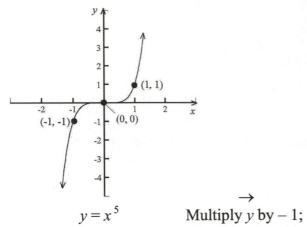

$$y = x^5$$

$\overrightarrow{}$ Multiply y by -1; reflect about x-axis.

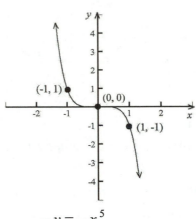

$$-y = -x^5$$

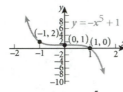

$\overrightarrow{}$ Add 1; vertical shift up 1 unit.

$$f(x) = -x^5 + 1$$

23. f is a polynomial function. Its degree is 5.

25. f is not a polynomial function. The exponent on the middle term is not a positive integer.

27. The power function that models the end behavior of the function f is $p(x) = -2x^4$.

29. The domain of a rational function is all real numbers except those that make the denominator zero. The denominator of $R(x)$, $x^2 - 9 = 0$ when $x = -3$ or when $x = 3$. The domain of R is $\{x \mid x \neq -3 \text{ or } x \neq 3\}$.

31. The domain of a rational function is all real numbers except those that make the denominator zero. The denominator of $R(x)$, $(x + 2)^2 = 0$, when $x = -2$. The domain of R is $\{x \mid x \neq -2\}$.

33. (a) $f(4) = 3^4 = 81$

    (b) $g(9) = \log_3 9 = \log_3 3^2 = 2$

    (c) $f(-2) = 3^{-2} = \dfrac{1}{9}$

    (d) $g\left(\dfrac{1}{27}\right) = \log_3\left(\dfrac{1}{27}\right) = \log_3\left(\dfrac{1}{3^3}\right)$
$$= \log_3 3^{-3} = -3$$

35. $5^2 = z$
    $\log_5 z = 2$

37. $\log_5 u = 13$
       $5^{13} = u$

39.

The domain of f consists of all x for which $3x - 2 > 0$, that is for all $x > \dfrac{2}{3}$, or using interval notation, $\left(\dfrac{2}{3}, \infty \right)$.

41.

The domain of H consists of all x for which $-3x + 2 > 0$, that is for all $x < \dfrac{2}{3}$, or using interval notation, $\left(-\infty, \dfrac{2}{3} \right)$.

43.
$$\log_2 \left(\frac{1}{8} \right) = \log_2 \left(2^{-3} \right) = -3 \log_2 2 = -3$$

45. $\log_3 81 = \log_3 \left(3^4 \right) = 4 \log_3 3 = 4$

47. $\ln e^2 = 2 \ln e = 2$

49. $\ln e^{\sqrt{2}} = \sqrt{2} \ln e = \sqrt{2}$

51. $2^{\log_2 0.4} = 0.4$

53.
$$\log_3 \left(\frac{uv^2}{w} \right) = \log_3 \left(uv^2 \right) - \log_3 w$$
$$= \log_3 u + \log_3 v^2 - \log_3 w$$
$$= \log_3 u + 2 \log_3 v - \log_3 w$$

55.
$$\log \left(x^2 \sqrt{x^3 + 1} \right) = \log x^2 + \log \sqrt{x^3 + 1}$$
$$= 2 \log x + \frac{1}{2} \log \left(x^3 + 1 \right)$$

57.
$$\ln \left(\frac{x \sqrt[3]{x^2 + 1}}{x - 3} \right) = \ln \left(x \sqrt[3]{x^2 + 1} \right) - \ln (x - 3)$$
$$= \ln x + \ln \sqrt[3]{x^2 + 1} - \ln (x - 3)$$
$$= \ln x + \frac{1}{3} \ln \left(x^2 + 1 \right) - \ln (x - 3)$$

59.
$$3 \log_4 x^2 + \frac{1}{2} \log_4 \sqrt{x} = \log_4 \left(x^2 \right)^3 + \log_4 \left(\sqrt{x} \right)^{1/2}$$
$$= \log_4 x^6 + \log_4 x^{1/4}$$
$$= \log_4 \left(x^6 \cdot x^{1/4} \right)$$
$$= \log_4 x^{25/4}$$

61.

$$\ln\left(\frac{x-1}{x}\right)+\ln\left(\frac{x}{x+1}\right)-\ln\left(x^2-1\right)=\ln\left[\left(\frac{x-1}{\cancel{x}}\right)\left(\frac{\cancel{x}}{x+1}\right)\right]-\ln\left(x^2-1\right)$$

$$=\ln\left[\frac{\left(\frac{x-1}{x+1}\right)}{\left(x^2-1\right)}\right]$$

$$=\ln\left[\left(\frac{x-1}{x+1}\right)\left(\frac{1}{x^2-1}\right)\right]$$

$$=\ln\left[\left(\frac{\cancel{x-1}}{x+1}\right)\left(\frac{1}{(\cancel{x-1})(x+1)}\right)\right]$$

$$=\ln\left(\frac{1}{(x+1)^2}\right)=\ln(x+1)^{-2}$$

$$=-2\ln(x+1)$$

63.

$$2\log 2+3\log x-\frac{1}{2}\left[\log(x+3)+\log(x-2)\right]=2\log 2+3\log x-\frac{1}{2}\left[\log(x+3)(x-2)\right]$$

$$=\log 2^2+\log x^3-\log\left[(x+3)(x-2)\right]^{1/2}$$

$$=\log\left(4x^3\right)-\log\left[(x+3)(x-2)\right]^{1/2}$$

$$=\log\left(\frac{4x^3}{\left[(x+3)(x-2)\right]^{1/2}}\right)$$

$$=\log\left(\frac{4x^3}{\sqrt{x^2+x-6}}\right)$$

65.

$$\log_4 19=\frac{\ln 19}{\ln 4}=2.124$$

67.

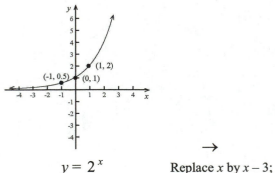

$$y=2^x$$

→ Replace x by $x-3$; horizontal shift right 3 units.

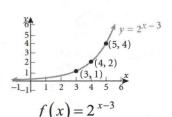

$$f(x)=2^{x-3}$$

The domain of the function f is all real numbers, or in interval notation $(-\infty, \infty)$; the range is all $\{y \mid y > 0\}$. The x-axis is a horizontal asymptote as f becomes unbounded in the negative direction.

69.

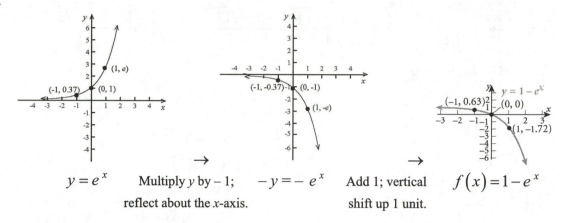

$$y = e^x \qquad \text{Multiply } y \text{ by } -1; \qquad -y = -e^x \qquad \text{Add 1; vertical} \qquad f(x) = 1 - e^x$$
$$\text{reflect about the } x\text{-axis.} \qquad\qquad \text{shift up 1 unit.}$$

The domain of the function f is all real numbers, or in interval notation $(-\infty, \infty)$; the range is all $\{y \mid y < 1\}$. The line $y = 1$ is a horizontal asymptote as f becomes unbounded in the negative direction.

71.

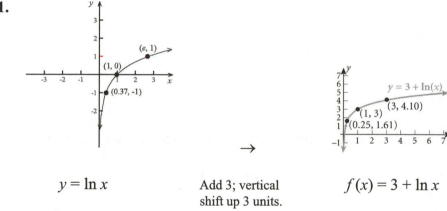

$$y = \ln x \qquad\qquad \text{Add 3; vertical} \qquad f(x) = 3 + \ln x$$
$$\text{shift up 3 units.}$$

The domain of the function f is $\{x \mid x > 0\}$, or in interval notation $(0, \infty)$; the range is all real numbers. The y-axis is a vertical asymptote.

73.
$$4^{1-2x} = 2$$
$$\left(2^2\right)^{1-2x} = 2$$
$$2^{2(1-2x)} = 2^1$$
$$2 - 4x = 1 \qquad \text{Set exponents equal.}$$
$$4x = 1$$
$$x = \frac{1}{4}$$

75.
$$3^{x^2+x} = \sqrt{3} = 3^{1/2}$$
$$x^2 + x = \frac{1}{2}$$

$$x^2 + x - \frac{1}{2} = 0$$
$$2x^2 + 2x - 1 = 0$$

$$x = \frac{-2 \pm \sqrt{2^2 - 4(2)(-1)}}{2(2)} = \frac{-2 \pm \sqrt{4+8}}{4} = \frac{-2 \pm \sqrt{12}}{4} = \frac{-\cancel{2} \pm \cancel{2}\sqrt{3}}{\cancel{4}2} = \frac{-1 \pm \sqrt{3}}{2}$$

$$x = \left\{ \frac{-1+\sqrt{3}}{2}, \frac{-1-\sqrt{3}}{2} \right\}$$

77. $\log_x 64 = -3$

$$x^{-3} = 64 = 4^3$$

$$x^{-3} = \left[\left(\frac{1}{4} \right)^{-1} \right]^3 = \left(\frac{1}{4} \right)^{-3}$$

$$x = \frac{1}{4}$$

79. $9^{2x} = 27^{3x-4}$

$$\left(3^2 \right)^{2x} = \left(3^3 \right)^{3x-4}$$

$$3^{4x} = 3^{3(3x-4)}$$

$$4x = 3(3x-4)$$

$$4x = 9x - 12$$

$$5x = 12$$

$$x = \frac{12}{5}$$

81. $\log_3(x-2) = 2$

$$3^2 = x - 2$$
$$9 = x - 2$$
$$x = 11$$

83. If $100 is invested at 10% compounded continuously, the amount A after 2.25 years (2 years and 3 months) is

$$A = Pe^{rt} = 100e^{(0.1)(2.25)} = 100e^{0.225} = \$125.23$$

85. I need to invest the present value of $1000 at 4% compounded continuously for 2 years.

$$P = Ae^{-rt} = 1000e^{-(0.04)(2)} = 1000e^{-0.08} = \$923.12$$

87. If P is invested, it will double when amount $A = 2P$.

$$A = 2P = Pe^{rt}$$
$$2(220,000) = 220,000e^{0.06t}$$
$$2 = e^{0.06t}$$
$$0.06t = \ln 2$$
$$t = \frac{\ln 2}{0.06} \approx 11.55$$

It will take about 11 and one half years for an investment to double at 6 % compounded continuously.

89. When $T = 0°$ and $P_0 = 760$ mm. of mercury,

$$h(x) = 8000 \log\left(\frac{760}{x}\right)$$

$$h(300) = 8000 \log\left(\frac{760}{300}\right) = 3229.54$$

The Piper Cub is flying at an altitude of approximately 3230 meters above sea level.

91. The perimeter of the pond is $P = 200 = 2l + 2w$

$$l = \frac{200 - 2w}{2} = 100 - w$$

If x denotes the width of the pond, then
$$A = lw = (100 - x)x = 100x - x^2$$

$A = A(x)$ is a quadratic function with $a = -1$, so the maximum value of A is found at the vertex of A.

$$x = -\frac{b}{2a} = -\frac{100}{2(-1)} = 50$$

$$A(50) = 100(50) - 50^2 = 2500 \text{ square feet.}$$

The dimensions should be 50 feet by 50 feet.

93. The perimeter of the window is 100 feet and can be expressed as
$$P = 2l + \pi d = 2l + \pi x = 100$$

Solving for l in terms of x, we find
$$l = \frac{100 - \pi x}{2} = 50 - \frac{\pi x}{2}$$
The area A of the rectangle is a quadratic function given by
$$A(x) = \left(50 - \frac{\pi x}{2}\right)x = 50x - \frac{\pi x^2}{2}$$

The maximum area is found at the vertex of A. The width of the rectangle with maximum area is

$$x = -\frac{b}{2a} = -\frac{50}{2\left(-\frac{\pi}{2}\right)} = \frac{50}{\pi} \text{ feet}$$

The length of the rectangle is

$$l = 50 - \frac{\pi x}{2} = 50 - \frac{\not\pi\left(\frac{50}{\not\pi}\right)}{2} = 50 - 25 = 25 \text{ feet}$$

95. (a) A 3.5 inch telescope has a limiting magnitude of
$$L = 9 + 5.1 \log 3.5 = 11.77$$

(b) If the star's magnitude is 14, the telescope must have a lens with a diameter of

$$9 + 5.1 \log d = 14$$
$$5.1 \log d = 5$$
$$\log d = \frac{5}{5.1}$$
$$d = 10^{5/5.1} = 9.56 \text{ inches.}$$

97. (a) If $620.17 grows to $5000 in 20 years when interest is compounded continuously, the interest rate is 10.436%

$$A = Pe^{rt}$$
$$5000 = 620.17 e^{20r}$$
$$e^{20r} = \frac{5000}{620.17}$$
$$20r = \ln\left(\frac{5000}{620.17}\right)$$
$$r = \frac{1}{20} \ln\left(\frac{5000}{620.17}\right) = 0.10436$$

(b) An investment of $4000 will have a value A in 20 years if it is invested at 10.436% compounded continuously.

$$A = Pe^{rt} = 4000 e^{(0.10436)(20)} = \$32,249.24$$

99. (a) The Calloway Company will minimize marginal cost if it produces

$$x = -\frac{b}{2a} = -\frac{(-617.4)}{2(4.9)} = 63 \text{ golf clubs.}$$

(b) The marginal cost of making the 64 golf club is

$$C(63) = 4.9(63)^2 - 617.4(63) + 19,600 = \$151.90$$

CHAPTER 11 PROJECT

1.

3. (a) $t = 150$ represents the year 2000. The projected population at $t = 150$ is
$$P(150) = 2.4 \cdot 1.59^{150/10} = 2.4 \cdot 1.59^{15} = 2518.7 \text{ thousand persons.}$$

(b) The projected population overestimates the actual population by 565.1 thousand persons, or by 28.9%.

(c) Explanations will vary.

(d) $t = 160$ represents the year 2010. Using the exponential growth function, the predicted population is
$$P(160) = 2.4 \cdot 1.59^{160/10} = 2.4 \cdot 1.59^{16} = 4004.7 \text{ thousand persons.}$$

(e) $t = 200$ represents the year 2050. Using the exponential growth function, the predicted population is
$$P(200) = 2.4 \cdot 1.59^{200/10} = 2.4 \cdot 1.59^{20} = 25,594$$
The predicted population is 25.594 million people.

5. The growth rate for Houston is $P(t) = 2.4 \, e^{\ln 1.59(t/10)}$

7. Answers will vary.

9. Answers will vary.

11. 3. To do this problem with the result from above, use TBLSET Indpnt: Ask, TABLE.
(a) The population of Houston in 2000 ($t = 50$) is predicted to be 2159.1 thousand.

(b) The prediction overestimates the actual population by 205.5 thousand, or 10.5%.

(c) Answers will vary.

(d) The population of Houston in 2010 ($t = 60$) is predicted to be 2702.0 thousand.

(e) The population of Houston in 2050 ($t = 100$) is predicted to be 6627.0 thousand.

4. If $P = P_0 a^{t/10} = P_0 e^{kt/10}$, where $P_0 = 703.4597$ and $a = 1.0227$ then

$$a^{t/10} = e^{kt/10}$$
$$k = \ln a = \ln 1.0227 = 0.022446$$

5. The growth rate for Houston is $P(t) = 703.4597\, e^{\ln 1.0227(t/10)}$

Explanations will vary.

MATHEMATICAL QUESTIONS FROM PROFESSIONAL EXAMS

1.

(a) $4 \cdot 27 = 2^2 \cdot 3^3 = 6^2 \cdot 3 = \dfrac{6^2 \cdot 3 \cdot 2}{2} = \dfrac{6^2 \cdot 6}{2} = \dfrac{6^3}{2}$

$\log_6(4 \cdot 27) = \log_6\left(\dfrac{6^3}{2}\right) = \log_6(6^3) - \log_6(2) = 3 - b$

3.

(c)
$$y = \frac{e^x - e^{-x}}{2}$$
$$2y = e^x - e^{-x}$$
$$2ye^x = e^x \cdot e^x - e^{-x} \cdot e^x \qquad \text{Multiply both sides by } e^x.$$
$$2ye^x = e^{2x} - 1$$
$$e^{2x} - 2ye^x - 1 = 0 \qquad\qquad a = 1, b = -2y, c = -1$$
$$e^x = \frac{2y \pm \sqrt{4y^2 + 4}}{2}$$
$$= \frac{2y \pm 2\sqrt{y^2 + 1}}{2}$$
$$= y \pm \sqrt{y^2 + 1}$$

Since $\dfrac{e^x - e^{-x}}{2} > 0$, $y > 0$, so $e^x = y + \sqrt{y^2 + 1}$.

$$x = \ln\left(y + \sqrt{y^2 + 1}\right)$$

5.

(b) $\left(\log_a b\right)\left(\log_b a\right) = \dfrac{\log b}{\log a} \cdot \dfrac{\log a}{\log b} = 1$

7. (a) $e^{2\ln(x-1)} = 4$

$e^{\ln(x-1)^2} = 4$

$(x-1)^2 = 4$

$x - 1 = \pm 2$

$x = 3$

Chapter 12
The Limit of a Function

12.1 Finding Limits Using Tables and Graphs 681/1 – 39 (ODD)

1. Here $f(x) = 2x$, and $c = 1$. We complete the table by evaluating the function f at each value of x.

x	0.9	0.99	0.999
$f(x) = 2x$	1.8	1.98	1.998
x	1.1	1.01	1.001
$f(x) = 2x$	2.2	2.02	2.002

We infer from the table that $\displaystyle\lim_{x \to 1} f(x) = \lim_{x \to 1} 2x = 2$.

3. Here $f(x) = x^2 + 2$, and $c = 0$. We complete the table by evaluating the function f at each value of x.

x	−0.1	−0.01	−0.001
$f(x) = x^2 + 2$	2.01	2.0001	2.0000
x	0.1	0.01	0.001
$f(x) = x^2 + 2$	2.01	2.0001	2.0000

We infer from the table that $\displaystyle\lim_{x \to 0} f(x) = \lim_{x \to 0}\left(x^2 + 2\right) = 2$.

5. Here $f(x) = \dfrac{x^2 - 4}{x + 2}$, and $c = -2$. We complete the table by evaluating the function f at each value of x.

x	−2.1	−2.01	−2.001
$f(x) = \dfrac{x^2 - 4}{x + 2}$	−4.1	−4.01	−4.001
x	−1.9	−1.99	−1.999
$f(x) = \dfrac{x^2 - 4}{x + 2}$	−3.9	−3.99	−3.999

We infer from the table that $\displaystyle\lim_{x \to -2} f(x) = \lim_{x \to -2}\left(\dfrac{x^2 - 4}{x + 2}\right) = -4$.

7. Here $f(x) = \dfrac{x^3 + 1}{x + 1}$, and $c = -1$. We complete the table by evaluating the function f at each value of x.

x		-1.1	-1.01	-1.001
$f(x) = \dfrac{x^3 + 1}{x + 1}$		3.31	3.0301	3.0030
x		-0.9	-0.99	-0.999
$f(x) = \dfrac{x^3 + 1}{x + 1}$		2.71	2.9701	2.9970

We infer from the table that $\displaystyle\lim_{x \to -1} f(x) = \lim_{x \to -1}\left(\dfrac{x^3 + 1}{x + 1}\right) = 3$.

9. Here $f(x) = 4x^3$, and $c = 2$. We choose values of x close to 2, starting at 1.99. Then we select additional numbers that get closer to 2, but remain less than 2. Next we choose values of x greater than 2, starting with 2.01, that get closer to 2. Finally we evaluate the function f at each choice to obtain the table:

x	1.99	1.999	1.9999	$\to$ $\leftarrow$	2.0001	2.001	2.01
$f(x) = 4x^3$	31.522	31.952	31.995	$\to$ $\leftarrow$	32.005	32.048	32.482

We infer that as x gets closer to 2, f gets closer to 32. That is,

$$\lim_{x \to 2} f(x) = \lim_{x \to 2}\left(4x^3\right) = 32$$

11. Here $f(x) = \dfrac{x + 1}{x^2 + 1}$, and $c = 0$. We choose values of x close to 0, starting at -0.01. Then we select additional numbers that get closer to 0, but remain less than 0. Next we choose values of x greater than 0, starting with 0.01, that get closer to 0. Finally we evaluate the function f at each choice to obtain the table:

x	-0.01	-0.001	-0.0001	$\to$ $\leftarrow$	0.0001	0.001	0.01
$f(x) = \dfrac{x + 1}{x^2 + 1}$	0.9899	0.9990	0.9999	$\to$ $\leftarrow$	1.0001	1.001	1.0099

We infer that as x gets closer to 0, f gets closer to 1. That is,

$$\lim_{x \to 0} f(x) = \lim_{x \to 0}\left(\dfrac{x + 1}{x^2 + 1}\right) = 1$$

13. Here $f(x) = \dfrac{x^2 - 4x}{x - 4}$, and $c = 4$. We choose values of x close to 4, starting at 3.99. Then we select additional numbers that get closer to 4, but remain less than 4. Next we choose values of x greater than 4, starting with 4.01, that get closer to 4. Finally we evaluate the function f at each choice to obtain the table:

x	3.99	3.999	3.9999	$\to$ $\leftarrow$	4.0001	4.001	4.01
$f(x) = \dfrac{x^2 - 4x}{x - 4}$	3.99	3.999	3.9999	$\to$ $\leftarrow$	4.0001	4.001	4.01

We infer that as x gets closer to 4, f gets closer to 4. That is,

$$\lim_{x \to 4} f(x) = \lim_{x \to 4} \left(\frac{x^2 - 4x}{x - 4} \right) = 4$$

15. Here $f(x) = e^x + 1$, and $c = 0$. We choose values of x close to 0, starting at -0.01. Then we select additional numbers that get closer to 0, but remain less than 0. Next we choose values of x greater than 0, starting with 0.01, that get closer to 0. Finally we evaluate the function f at each choice to obtain the table:

x	-0.01	-0.001	-0.0001	$\to$	$\leftarrow$	0.0001	0.001	0.01
$f(x) = e^x + 1$	1.9900	1.9990	1.9999	$\to$	$\leftarrow$	2.0001	2.0010	2.0101

We infer that as x gets closer to 0, f gets closer to 2. That is,

$$\lim_{x \to 0} f(x) = \lim_{x \to 0} \left(e^x + 1 \right) = 2$$

17. To determine the $\lim\limits_{x \to 2} f(x)$ we observe that as x gets closer to 2, $f(x)$ gets closer to 3. So We conclude that

$$\lim_{x \to 2} f(x) = 3$$

19. To determine the $\lim\limits_{x \to 2} f(x)$ we observe that as x gets closer to 2, $f(x)$ gets closer to 4. We conclude that

$$\lim_{x \to 2} f(x) = 4$$

21. To determine the $\lim\limits_{x \to 3} f(x)$ we observe that as x gets closer to 3, but remains less than 3, the value of f gets closer to 3. However, we see that as x gets closer to 3, but remains greater than 3, the value of f gets closer to 6.   Since there is no single number that the values of f are close to when x is close to 3, we conclude that $\lim\limits_{x \to 3} f(x)$ does not exist.

23. We conclude from the graph that
$$\lim_{x \to 4} f(x) = \lim_{x \to 4} (3x + 1) = 13$$

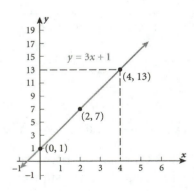

25. We conclude from the graph that
$$\lim_{x \to 2} f(x) = \lim_{x \to 2}\left(1 - x^2\right) = -3$$

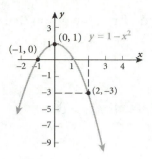

27. We conclude from the graph that

$$\lim_{x \to -3} f(x) = \lim_{x \to -3}\left(|x| - 2\right) = 1$$

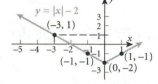

29. We conclude from the graph that

$$\lim_{x \to 0} f(x) = \lim_{x \to 0} e^x = 1$$

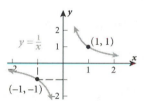

31. We conclude from the graph that
$$\lim_{x \to -1} f(x) = \lim_{x \to -1} \frac{1}{x} = -1$$

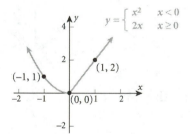

33. We conclude from the graph that

$$\lim_{x \to 0} f(x) = 0$$

Notice that
$$\lim_{x \to 0^-} f(x) = \lim_{x \to 0^-} x^2 = 0$$
and
$$\lim_{x \to 0^+} f(x) = \lim_{x \to 0^+}\left(2x\right) = 0$$

35. We conclude from the graph that
$$\lim_{x \to 1} f(x) \text{ does not exist.}$$

Notice that
$$\lim_{x \to 1^-} f(x) = \lim_{x \to 1^-} 3x = 3$$
but
$$\lim_{x \to 1^+} f(x) = \lim_{x \to 1^+}\left(x + 1\right) = 2$$

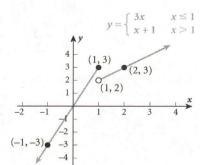

37. We conclude from the graph that
$$\lim_{x \to 0} f(x) = 0$$

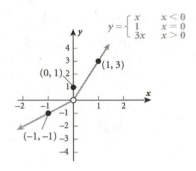

Notice that
$$\lim_{x \to 0^-} f(x) = \lim_{x \to 0^-} x = 0$$
and
$$\lim_{x \to 0^+} f(x) = \lim_{x \to 0^+} (3x) = 0$$

39. We conclude from the graph that
$$\lim_{x \to 0} f(x) = 0$$

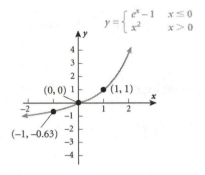

Notice that
$$\lim_{x \to 0^-} f(x) = \lim_{x \to 0^-} (e^x - 1) = 1 - 1 = 0$$
and
$$\lim_{x \to 0^+} f(x) = \lim_{x \to 0^+} x^2 = 0$$

41. To find the limit we create the tables shown below and conclude
$$\lim_{x \to 1} \frac{x^3 - x^2 + x - 1}{x^4 - x^3 + 2x - 2} = \frac{2}{3}$$

X	Y1
0	.5
.5	.58824
.8	.65287
.9	.66325
.99	.66663
.999	.66667
.9999	.66667

X=.9999

X	Y1
2	.5
1.5	.60465
1.2	.65451
1.1	.66346
1.01	.66663
1.001	.66667
1.0001	.66667

X=1.0001

43. To find the limit we create the tables shown below and conclude
$$\lim_{x \to -2} \frac{x^3 - 2x^2 + 4x - 8}{x^2 + x - 6} = 1.6$$

X	Y1
1	1.25
1.5	1.3889
1.8	1.5083
1.9	1.5531
1.99	1.5952
1.999	1.5995
1.9999	1.6

X=1.9999

X	Y1
3	2.1667
2.5	1.8636
2.2	1.7
2.1	1.649
2.01	1.6048
2.001	1.6005
2.0001	1.6

X=2.0001

45. To find the limit we create the tables shown below and conclude
$$\lim_{x \to -1} \frac{x^3 + 2x^2 + x}{x^4 + x^3 + 2x + 2} = 0$$

12.2 Techniques for Finding Limits of Functions 690/1-51 (ODD))

1. Using formula (1) (p. 243), we find
$$\lim_{x \to 1} 5 = 5$$

3. Using formula (2) (p. 243), we find
$$\lim_{x \to 4} x = 4$$

5.
$$\lim_{x \to 2}(3x + 2) = \lim_{x \to 2}(3x) + \lim_{x \to 2} 2 = \left[\lim_{x \to 2} 3\right]\left[\lim_{x \to 2} x\right] + \lim_{x \to 2} 2$$
$$= (3) \cdot (2) + 2 = 6 + 2 = 8$$

7.
$$\lim_{x \to -1}(3x^2 - 5x) = 3 \cdot (-1)^2 - 5 \cdot (-1) = 3 \cdot 1 + 5 = 3 + 5 = 8$$

9.
$$\lim_{x \to 1}(5x^4 - 3x^2 + 6x - 9) = 5 \cdot 1^4 - 3 \cdot 1^2 + 6 \cdot 1 - 9$$
$$= 5 \cdot 1 - 3 \cdot 1 + 6 - 9$$
$$= 5 - 3 - 3 = -1$$

11.
$$\lim_{x \to 1}(x^2 + 1)^3 = \left[\lim_{x \to 1}(x^2 + 1)\right]^3 = (2)^3 = 8$$

13.
$$\lim_{x \to 1}\sqrt{5x + 4} = \sqrt{\lim_{x \to 1}(5x + 4)} = \sqrt{9} = 3$$

15. The limit we seek is the limit of a rational function whose domain is $\{x \mid x \neq -2, x \neq 2\}$. Since 0 is in the domain, we use formula (12).
$$\lim_{x \to 0}\frac{x^2 - 4}{x^2 + 4} = \frac{0^2 - 4}{0^2 + 4} = \frac{-4}{4} = -1$$

17.
$$\lim_{x \to 2}(3x - 2)^{\frac{5}{2}} = \left[\lim_{x \to 2}(3x - 2)\right]^{\frac{5}{2}} = (4)^{\frac{5}{2}} = (2)^5 = 32$$

19. The domain of the rational function $R(x) = \dfrac{x^2 - 4}{x^2 - 2x}$ is $\{x \mid x \neq 0, x \neq 2\}$. Since 2 is not in the domain, we cannot evaluate $R(2)$, but we notice that the function can be factored as
$$\frac{x^2 - 4}{x^2 - 2x} = \frac{(x + 2)(x - 2)}{x(x - 2)}$$

Since x is near 2, but $x \neq 2$, we can cancel the $(x - 2)$'s. Formula (11), can then be used to find the limit of the function as x approaches 2.

$$\lim_{x \to 2} \frac{x^2 - 4}{x^2 - 2x} = \lim_{x \to 2} \frac{(x+2)\cancel{(x-2)}}{x\cancel{(x-2)}} = \frac{\lim\limits_{x \to 2}(x+2)}{\lim\limits_{x \to 2}(x)} = \frac{4}{2} = 2$$

21. The domain of the rational function $R(x) = \dfrac{x^2 - x - 12}{x^2 - 9}$ is $\{x \mid x \neq -3, x \neq 3\}$. Since -3 is not in the domain, we cannot evaluate $R(-3)$, but we notice that the function can be factored as

$$\frac{x^2 - x - 12}{x^2 - 9} = \frac{(x-4)(x+3)}{(x-3)(x+3)}$$

Since x is near -3, but $x \neq -3$, we can cancel the $(x + 3)$'s. Formula (11), can then be used to find the limit of the function as x approaches -3.

$$\lim_{x \to -3} \frac{x^2 - x - 12}{x^2 - 9} = \lim_{x \to -3} \frac{(x-4)\cancel{(x+3)}}{(x-3)\cancel{(x+3)}} = \frac{\lim\limits_{x \to -3}(x-4)}{\lim\limits_{x \to -3}(x-3)} = \frac{(-3)-4}{(-3)-3} = \frac{-7}{-6} = \frac{7}{6}$$

23. The domain of the rational function $R(x) = \dfrac{x^3 - 1}{x - 1}$ is $\{x \mid x \neq 1\}$. Since 1 is not in the domain, we cannot evaluate $R(1)$, but we notice that the function can be factored as

$$\frac{x^3 - 1}{x - 1} = \frac{(x-1)(x^2 + x + 1)}{x - 1}$$

Since x is near 1, but $x \neq 1$, we can cancel the $(x - 1)$'s. Formula (11), can then be used to find the limit of the function as x approaches 1.

$$\lim_{x \to 1} \frac{x^3 - 1}{x - 1} = \lim_{x \to 1} \frac{\cancel{(x-1)}(x^2 + x + 1)}{\cancel{x-1}} = \frac{\lim\limits_{x \to 1}(x^2 + x + 1)}{\lim\limits_{x \to 1}(1)} = \frac{1^2 + 1 + 1}{1} = 3$$

25. The domain of the rational function $R(x) = \dfrac{(x+1)^2}{x^2 - 1}$ is $\{x \mid x \neq -1, x \neq 1\}$. Since -1 is not in the domain, we cannot evaluate $R(-1)$, but we notice that the function can be factored as

$$\frac{(x+1)^2}{x^2 - 1} = \frac{(x+1)^2}{(x-1)(x+1)}$$

Since x is near -1, but $x \neq -1$, we can cancel $(x + 1)$'s. Formula (11), can then be used to find the limit of the function as x approaches -1.

$$\lim_{x \to -1} \frac{(x+1)^2}{x^2 - 1} = \lim_{x \to -1} \frac{(x+1)^{\cancel{2}}}{(x-1)\cancel{(x+1)}} = \lim_{x \to -1} \frac{(x+1)}{(x-1)} = \frac{(-1)+1}{(-1)-1} = \frac{0}{-2} = 0$$

27. The limit of the denominator of this function as x approaches 1 is zero, so formula (11), cannot be used directly. We first factor the function by grouping.

$$\frac{x^3 - x^2 + x - 1}{x^4 - x^3 + 2x - 2} = \frac{x^2(x-1) + 1\cdot(x-1)}{x^3(x-1) + 2\cdot(x-1)} = \frac{(x-1)(x^2+1)}{(x-1)(x^3+2)}$$

Since x is near 1, but $x \neq 1$, we can cancel $(x-1)$'s. Then using formula (11), we get

$$\lim_{x\to 1}\frac{x^3 - x^2 + x - 1}{x^4 - x^3 + 2x - 2} = \lim_{x\to 1}\frac{\cancel{(x-1)}(x^2+1)}{\cancel{(x-1)}(x^3+2)} = \frac{\lim_{x\to 1}(x^2+1)}{\lim_{x\to 1}(x^3+2)} = \frac{2}{3}$$

29. The limit of the denominator of this function as x approaches 2 is zero, so formula (11), cannot be used directly. We first factor the function, using grouping to factor the numerator.

$$\frac{x^3 - 2x^2 + 4x - 8}{x^2 + x - 6} = \frac{x^2(x-2) + 4\cdot(x-2)}{(x+3)(x-2)} = \frac{(x^2+4)(x-2)}{(x+3)(x-2)}$$

Since x is near 2, but $x \neq 2$, we can cancel $(x-2)$'s. Then using formula (11), we get

$$\lim_{x\to 2}\frac{x^3 - 2x^2 + 4x - 8}{x^2 + x - 6} = \lim_{x\to 2}\frac{(x^2+4)\cancel{(x-2)}}{(x+3)\cancel{(x-2)}} = \frac{\lim_{x\to 2}(x^2+4)}{\lim_{x\to 2}(x+3)} = \frac{2^2+4}{2+3} = \frac{8}{5}$$

31. The limit of the denominator of this function as x approaches -1 is zero, so formula (11), cannot be used directly. We first factor the function, using grouping to factor the denominator.

$$\frac{x^3 + 2x^2 + x}{x^4 + x^3 + 2x + 2} = \frac{x(x^2 + 2x + 1)}{x^3(x+1) + 2(x+1)} = \frac{x(x+1)(x+1)}{(x^3+2)(x+1)}$$

Since x is near -1, but $x \neq -1$, we can cancel $(x+1)$'s. Then using formula (11), we get

$$\lim_{x\to -1}\frac{x^3 + 2x^2 + x}{x^4 + x^3 + 2x + 2} = \lim_{x\to -1}\frac{x(x+1)\cancel{(x+1)}}{(x^3+2)\cancel{(x+1)}} = \frac{\lim_{x\to -1}x(x+1)}{\lim_{x\to -1}(x^3+1)} = \frac{(-1)\cdot[(-1)+1]}{(-1)^3+2} = \frac{(-1)\cdot(0)}{1} = 0$$

33. The average rate of change of f from 2 to x is

$$\frac{\Delta y}{\Delta x} = \frac{f(x) - f(2)}{x - 2} = \frac{(5x-3) - (5\cdot 2 - 3)}{x - 2} = \frac{5x - 3 - 7}{x - 2} = \frac{5x - 10}{x - 2} = \frac{5(x-2)}{x - 2}$$

The limit of the average rate of change as x approaches 2 is

$$\lim_{x\to 2}\frac{f(x) - f(2)}{x - 2} = \lim_{x\to 2}\frac{(5x-3) - 7}{x - 2} = \lim_{x\to 2}\frac{5x - 10}{x - 2} = \lim_{x\to 2}\frac{5\cancel{(x-2)}}{\cancel{x-2}} = 5$$

35. The average rate of change of f from 3 to x is

$$\frac{\Delta y}{\Delta x} = \frac{f(x) - f(3)}{x - 3} = \frac{x^2 - 3^2}{x - 3} = \frac{(x+3)(x-3)}{x - 3}$$

The limit of the average rate of change as x approaches 3 is

$$\lim_{x \to 3} \frac{f(x)-f(3)}{x-3} = \lim_{x \to 3} \frac{x^2-3^2}{x-3} = \lim_{x \to 3} \frac{(x+3)\,(x\cancel{-3})}{\cancel{x-3}} = 6$$

37. The average rate of change of f from -1 to x is

$$\frac{\Delta y}{\Delta x} = \frac{f(x)-f(-1)}{x-(-1)} = \frac{\left(x^2+2x\right)-\left[(-1)^2+2\cdot(-1)\right]}{x+1} = \frac{\left(x^2+2x\right)-(-1)}{x+1} = \frac{x^2+2x+1}{x+1} = \frac{(x+1)^2}{x+1}$$

The limit of the average rate of change as x approaches -1 is

$$\lim_{x \to -1} \frac{f(x)-f(-1)}{x-(-1)} = \lim_{x \to -1} \frac{\left(x^2+2x\right)-(-1)}{x+1} = \lim_{x \to -1} \frac{x^2+2x+1}{x+1} = \lim_{x \to -1} \frac{(x+1)^{\cancel{2}}}{\cancel{x+1}} = (-1)+1 = 0$$

39. The average rate of change of f from 0 to x is

$$\frac{\Delta y}{\Delta x} = \frac{f(x)-f(0)}{x-0} = \frac{\left(3x^3-2x^2+4\right)-(4)}{x} = \frac{3x^3-2x^2}{x} = \frac{x^2(3x-2)}{x}$$

The limit of the average rate of change as x approaches 0 is

$$\lim_{x \to 0} \frac{f(x)-f(0)}{x-0} = \lim_{x \to 0} \frac{3x^3-2x^2}{x} = \lim_{x \to 0} \frac{x^{\cancel{2}}(3x-2)}{\cancel{x}} = 0$$

41. The average rate of change of f from 1 to x is

$$\frac{\Delta y}{\Delta x} = \frac{f(x)-f(1)}{x-1} = \frac{\dfrac{1}{x}-\dfrac{1}{1}}{x-1} = \frac{\dfrac{1}{x}-\dfrac{x}{x}}{x-1} = \frac{\dfrac{1-x}{x}}{x-1} = \frac{1-x}{x(x-1)} = \frac{(-1)\cdot(x-1)}{x(x-1)}$$

$\qquad\qquad\qquad\qquad\quad\uparrow\qquad\qquad\qquad\uparrow\qquad\qquad\uparrow$

$\qquad\qquad$ Find a common denominator Simplify Factor out (-1)

The limit of the average rate of change as x approaches 1 is

$$\lim_{x \to 1} \frac{f(x)-f(1)}{x-1} = \lim_{x \to 1} \frac{1-x}{x(x-1)} = \lim_{x \to 1} \frac{(-1)\cdot(\cancel{x-1})}{x(\cancel{x-1})} = -1$$

43. The average rate of change of f from 4 to x is

$$\frac{\Delta y}{\Delta x} = \frac{f(x)-f(4)}{x-4} = \frac{\sqrt{x}-\sqrt{4}}{x-4} = \frac{\sqrt{x}-2}{\left(\sqrt{x}+2\right)\left(\sqrt{x}-2\right)}$$

$\qquad\qquad\qquad\qquad\qquad\qquad\uparrow$

$\qquad\qquad\qquad\qquad$ Factor the denominator

The limit of the average rate of change as x approaches 4 is

$$\lim_{x \to 4} \frac{f(x)-f(4)}{x-4} = \lim_{x \to 4} \frac{\sqrt{x}-\sqrt{4}}{x-4} = \lim_{x \to 4} \frac{\cancel{\sqrt{x}-2}}{\left(\sqrt{x}+2\right)\left(\cancel{\sqrt{x}-2}\right)} = \frac{1}{\sqrt{4}+2} = \frac{1}{4}$$

45. Since $\lim\limits_{x \to c} f(x) = 5$ and $\lim\limits_{x \to c} g(x) = 2$, $\lim\limits_{x \to c} \left[2f(x) \right] = \left(\lim\limits_{x \to c} 2 \right)\left(\lim\limits_{x \to c} f(x) \right) = 2 \cdot 5 = 10$

47. Since $\lim\limits_{x \to c} f(x) = 5$ and $\lim\limits_{x \to c} g(x) = 2$, $\lim\limits_{x \to c} \left[g(x) \right]^3 = \left[\lim\limits_{x \to c} g(x) \right]^3 = 2^3 = 8$

49. Since $\lim\limits_{x \to c} f(x) = 5$ and $\lim\limits_{x \to c} g(x) = 2$, $\lim\limits_{x \to c} \dfrac{4}{f(x)} = \dfrac{\lim\limits_{x \to c} 4}{\lim\limits_{x \to c} f(x)} = \dfrac{4}{5}$

51. Since $\lim\limits_{x \to c} f(x) = 5$ and $\lim\limits_{x \to c} g(x) = 2$,

$$\lim\limits_{x \to c} \left[4f(x) - 5g(x) \right] = \lim\limits_{x \to c} \left[4f(x) \right] - \lim\limits_{x \to c} \left[5g(x) \right]$$

$$= \left(\lim\limits_{x \to c} 4 \right) \cdot \left(\lim\limits_{x \to c} f(x) \right) - \left(\lim\limits_{x \to c} 5 \right) \cdot \left(\lim\limits_{x \to c} g(x) \right)$$

$$= 4 \cdot 5 - 5 \cdot 2 = 20 - 10 = 10$$

12.3 One-Sided Limits; Continuous Functions $\quad$ 636 / 1–61 (ODD)

1. The domain of f is $\{x \mid -8 \le x < -3 \text{ or } -3 < x < 4 \text{ or } 4 < x \le 6\}$ or the intervals $[8, -3)$ or $(-3, 4)$ or $(4, 6]$.

3. The x-intercepts of the graph of f are $(-8, 0)$ and $(-5, 0)$. At these points the graph of f either crosses or touches the x-axis.

5. $f(-8) = 0$ and $f(-4) = 2$

7. To find $\lim\limits_{x \to -6^-} f(x)$, we look at the values of f when x is close to -6, but less than -6. Since the graph of f is approaching $y = 3$ for these values, we have $\lim\limits_{x \to -6^-} f(x) = 3$.

9. To find $\lim\limits_{x \to -4^-} f(x)$, we look at the values of f when x is close to -4, but less than -4. Since the graph of f is approaching $y = 2$ for these values, we have $\lim\limits_{x \to -4^-} f(x) = 2$.

11. To find $\lim\limits_{x \to 2^-} f(x)$, we look at the values of f when x is close to 2, but less than 2. Since the graph of f is approaching $y = 1$ for these values, we have $\lim\limits_{x \to 2^-} f(x) = 1$.

13. The $\lim_{x \to 4} f(x)$ exists because both $\lim_{x \to 4^-} f(x) = 0$ and $\lim_{x \to 4^+} f(x) = 0$. Since both one-sided limits exist and are equal, the limit of f as x approaches 4 exists and is equal to the one-sided limits. That is, $\lim_{x \to 4} f(x) = 0$.

15. The function f is not continuous at $x = -6$, because $\lim_{x \to -6} f(x)$ does not exist. (The one-sided limits are not equal. See Problems 7 and 8.)

17. The function f is continuous at $x = 0$. The function is defined at zero, $f(0) = 3$, and $\lim_{x \to 0} f(x) = f(0) = 3$. (See Problem 14.)

19. The function f is not continuous at $x = 4$. The function is not defined at 4. That is, 4 is not part of the domain of f.

21. To find the one-sided limit we look at values of x close to 1, but greater than 1. Since $f(x) = 2x + 3$ for such numbers, we conclude that
$$\lim_{x \to 1^+} f(x) = \lim_{x \to 1^+} (2x + 3) = 5$$

23. To find the one-sided limit we look at values of x close to 1, but less than 1. Since $f(x) = 2x^3 + 5x$ for such numbers, we conclude that
$$\lim_{x \to 1^-} f(x) = \lim_{x \to 1^-} (2x^3 + 5x) = 7$$

25. To find the one-sided limit we look at values of x close to 0, but less than 0. Since $f(x) = e^x$ for such numbers, we conclude that
$$\lim_{x \to 0^-} f(x) = \lim_{x \to 0^-} (e^x) = 1$$

27. To find the one-sided limit we look at values of x close to 2, but greater than 2. Since $f(x) = \dfrac{x^2 - 4}{x - 2}$ for such numbers, we first factor the function.
$$f(x) = \frac{x^2 - 4}{x - 2} = \frac{(x + 2)(x - 2)}{x - 2}$$
We can then conclude that
$$\lim_{x \to 2^+} f(x) = \lim_{x \to 2^+} \frac{x^2 - 4}{x - 2} = \lim_{x \to 2^+} \frac{(x + 2)\cancel{(x - 2)}}{\cancel{x - 2}} = 4$$

29. To find the one-sided limit we look at values of x close to -1, but less than -1. Since $f(x) = \dfrac{x^2 - 1}{x^3 + 1}$ for such numbers, we first factor the function.
$$f(x) = \frac{x^2 - 1}{x^3 + 1} = \frac{(x + 1)(x - 1)}{(x + 1)(x^2 - x + 1)}$$

We can then conclude that

$$\lim_{x \to -1^-} f(x) = \lim_{x \to -1^-} \frac{x^2-1}{x^3+1} = \lim_{x \to -1^-} \frac{(x+1)(x-1)}{(x+1)(x^2-x+1)} = \frac{-2}{3} = -\frac{2}{3}$$

31. To find the one-sided limit we look at values of x close to -2, but greater than -2. Since $f(x) = \dfrac{x^2+x-2}{x^2+2x}$ for such numbers, we first factor the function.

$$f(x) = \frac{x^2+x-2}{x^2+2x} = \frac{(x+2)(x-1)}{x(x+2)}$$

We can then conclude that

$$\lim_{x \to 2^+} f(x) = \lim_{x \to 2^+} \frac{x^2+x-2}{x^2+2x} = \lim_{x \to 2^+} \frac{(x+2)(x-1)}{x(x+2)} = \lim_{x \to 2^+} \frac{x-1}{x} = \frac{1}{2}$$

33. $f(x) = x^3 - 3x^2 + 2x - 6$ is continuous at $c = 2$ because f is a polynomial, and polynomials are continuous at every number.

35. $f(x) = \dfrac{x^2+5}{x-6}$ is a rational function whose domain is $\{x \mid x \neq 6\}$. f is continuous at $c = 3$ since f is defined at 3.

37. $f(x) = \dfrac{x+3}{x-3}$ is a rational function whose domain is $\{x \mid x \neq 3\}$. f is not continuous at $c = 3$ since f is not defined at 3.

39. $f(x) = \dfrac{x^3+3x}{x^2-3x}$ is a rational function whose domain is $\{x \mid x \neq 0, x \neq 3\}$. f is not continuous at $c = 0$ since f is not defined at 0.

41. To determine whether $f(x) = \begin{cases} \dfrac{x^3+3x}{x^2-3x} & \text{if } x \neq 0 \\ 1 & \text{if } x = 0 \end{cases}$ is continuous at $c = 0$, we investigate f when $x = 0$.

$$f(0) = 1$$

$$\lim_{x \to 0^-} f(x) = \lim_{x \to 0^-} \frac{x^3+3x}{x^2-3x} = \lim_{x \to 0^-} \frac{x(x^2+3)}{x(x-3)} = \frac{3}{-3} = -1$$

$$\lim_{x \to 0^+} f(x) = \lim_{x \to 0^+} \frac{x^3+3x}{x^2-3x} = \lim_{x \to 0^+} \frac{x(x^2+3)}{x(x-3)} = \frac{3}{-3} = -1$$

Since $\lim\limits_{x \to 0} f(x) = -1 \neq f(0) = 1$, the function f is not continuous at $c = 0$.

43. To determine whether $f(x) = \begin{cases} \dfrac{x^3 + 3x}{x^2 - 3x} & \text{if } x \neq 0 \\ -1 & \text{if } x = 0 \end{cases}$ is continuous at $c = 0$, we investigate f when $x = 0$.

$$f(0) = -1$$

$$\lim_{x \to 0^-} f(x) = \lim_{x \to 0^-} \frac{x^3 + 3x}{x^2 - 3x} = \lim_{x \to 0^-} \frac{\cancel{x}(x^2 + 3)}{\cancel{x}(x - 3)} = \frac{3}{-3} = -1$$

$$\lim_{x \to 0^+} f(x) = \lim_{x \to 0^+} \frac{x^3 + 3x}{x^2 - 3x} = \lim_{x \to 0^+} \frac{\cancel{x}(x^2 + 3)}{\cancel{x}(x - 3)} = \frac{3}{-3} = -1$$

Since $\lim_{x \to 0} f(x) = -1 = f(0)$, the function f is continuous at $c = 0$.

45. To determine whether $f(x) = \begin{cases} \dfrac{x^3 - 1}{x^2 - 1} & \text{if } x < 1 \\ 2 & \text{if } x = 1 \\ \dfrac{3}{x + 1} & \text{if } x > 1 \end{cases}$ is continuous at $c = 1$, we investigate f when $x = 1$.

$$f(1) = 2$$

$$\lim_{x \to 1^-} f(x) = \lim_{x \to 1^-} \frac{x^3 - 1}{x^2 - 1} = \lim_{x \to 1^-} \frac{\cancel{(x-1)}(x^2 + x + 1)}{\cancel{(x-1)}(x + 1)} = \frac{1 + 1 + 1}{1 + 1} = \frac{3}{2}$$

$$\lim_{x \to 1^+} f(x) = \lim_{x \to 1^+} \frac{3}{x + 1} = \frac{3}{2}$$

Since $\lim_{x \to 1} f(x) = \dfrac{3}{2} \neq f(1)$, the function f is not continuous at $c = 1$.

47. To determine whether $f(x) = \begin{cases} 2e^x & \text{if } x < 0 \\ 2 & \text{if } x = 0 \\ \dfrac{x^3 + 2x^2}{x^2} & \text{if } x > 0 \end{cases}$ is continuous at $c = 0$, we investigate f when $x = 0$.

$$f(0) = 2$$

$$\lim_{x \to 0^-} f(x) = \lim_{x \to 0^-} 2e^x = 2$$

$$\lim_{x \to 0^+} f(x) = \lim_{x \to 0^+} \frac{x^3 + 2x^2}{x^2} = \lim_{x \to 0^+} \frac{\cancel{x^2}(x + 2)}{\cancel{x^2}} = 2$$

The $\lim_{x \to 0} f(x)$ exists, and $\lim_{x \to 0} f(x) = f(0) = 2$. So we conclude that the function f is continuous at $c = 0$.

49. $f(x) = 2x + 3$ is a first degree polynomial function. Polynomial functions are continuous at all real numbers.

51. $f(x) = 3x^2 + x$ is a second degree polynomial function. Polynomial functions are continuous at all real numbers.

53. $f(x) = 4 \ln x$ is the product of a constant function $h(x) = 4$, which is continuous at every number, and the logarithmic function $g(x) = \ln x$, which is continuous for every number in the domain $(0, \infty)$. So $f(x) = 4 \ln x$ is continuous for all values $x > 0$.

55. $f(x) = 3e^x$ is the product of a constant function $h(x) = 3$, which is continuous at every number, and the exponential function $g(x) = e^x$, which is continuous for every number in the domain $(-\infty, \infty)$. So $f(x) = 3e^x$ is continuous for all real numbers.

57. $f(x) = \dfrac{2x+5}{x^2 - 4}$ is a rational function. Rational functions are continuous at every number in the domain. The domain of f is $\{x \mid x \neq -2, \ x \neq 2\}$, and f is continuous at all those numbers. f is discontinuous at $x = -2$ and $x = 2$.

59. $f(x) = \dfrac{x-3}{\ln x}$ is the quotient of a polynomial function, which is continuous at all real numbers and the logarithmic function, which is continuous at all numbers in the domain $(0, \infty)$. So f is continuous at all positive numbers or for $x > 0$.

61. The "pieces" of f, that is, $y = 3x + 1$, $y = -x^2$, and $y = \dfrac{1}{2}x - 5$, are each continuous for every number since they are polynomials. So we only need to investigate $x = 0$ and $x = 2$, the two points at which the pieces change.

For $x = 0$:
$$f(0) = 3(0) + 1 = 1$$
$$\lim_{x \to 0^-} f(x) = \lim_{x \to 0^-} (3x + 1) = 1$$
$$\lim_{x \to 0^+} f(x) = \lim_{x \to 0^+} (-x^2) = 0$$

Since $\lim_{x \to 0^+} f(x) \neq f(0)$, we conclude that the function f is discontinuous at $x = 0$.

For $x = 2$:
$$f(2) = -2^2 = -4$$
$$\lim_{x \to 2^-} f(x) = \lim_{x \to 2^-} (-x^2) = -4$$
$$\lim_{x \to 2^+} f(x) = \lim_{x \to 2^+} \left(\frac{1}{2}x - 5\right) = -4$$

Since $\lim_{x \to 2} f(x) = f(2) = -4$, we conclude that f is continuous at $x = 2$.

63. The cost function is $C(x) = \begin{cases} 39.99 & \text{if } 0 < x \le 350 \\ 0.25x - 47.51 & \text{if } x > 350 \end{cases}$.

(a) $\lim\limits_{x \to 350^-} C(x) = \lim\limits_{x \to 350^-} (39.99) = 39.99$

(b) $\lim\limits_{x \to 350^+} C(x) = \lim\limits_{x \to 350^+} (0.25x - 47.51) = 39.99$

(c) The left limit equals the right limit, so $\lim\limits_{x \to 350} C(x)$ exists. C is continuous at $x = 350$

since $\lim\limits_{x \to 350} C(x) = C(350) = 39.99$.

(d) Answers will vary.

65. (a) If $t = 10°$ then $W(v) = \begin{cases} 10 & 0 \le v < 1.79 \\ 33 - \dfrac{23(10.45 + 10\sqrt{v} - v)}{22.04} & 1.79 \le v \le 20 \\ -3.7034 & v > 20 \end{cases}$.

(b) $\lim\limits_{v \to 0^+} W(v) = \lim\limits_{v \to 0^+} (10) = 10$

(c) $\lim\limits_{v \to 1.79^-} W(v) = \lim\limits_{v \to 1.79^-} (10) = 10$

(d) $\lim\limits_{v \to 1.79^+} W(v) = \lim\limits_{v \to 1.79^+} \left[33 - \dfrac{23(10.45 + 10\sqrt{1.79} - 1.79)}{22.04} \right] = 10.00095$

(e) $W(1.79) = 10.00095$

(f) W is not continuous at $v = 1.79$ since $\lim\limits_{v \to 1.79^-} W(v) \ne \lim\limits_{v \to 1.79^+} W(v)$, and therefore,

$\lim\limits_{v \to 1.79} W(v)$ does not exist. In order to be continuous at 1.79, $\lim\limits_{v \to 1.79} W(v)$ must exist.

(g) Rounded to two decimal places,
$$\lim\limits_{v \to 1.79^-} W(v) = 10.00 \qquad \lim\limits_{v \to 1.79^+} W(v) = 10.00 \qquad W(1.79) = 10.00$$
Now the function W is continuous at 1.79.

(h) Answers will vary.

(i) $\lim\limits_{v \to 20^-} W(v) = \lim\limits_{v \to 20^-} \left[33 - \dfrac{23(10.45 + 10\sqrt{20} - 20)}{22.04} \right] = -3.7033$

(j) $\lim\limits_{v \to 20^+} W(v) = \lim\limits_{v \to 20^+} (-3.7034) = -3.7034$

(k) $W(20) = -3.7033$

(l) W is not continuous at 20. The right limit is not equal to $W(20)$.

(m) Rounded to two decimal places,

$$\lim_{v \to 20^-} W(v) = -3.70 \qquad \lim_{v \to 20^+} W(v) = -3.70 \qquad W(20) = -3.70$$

Now the function W is continuous at 20.

(n) Answers will vary.

$705 \mid 1-41 \; (ODD)$

12.4 Limits at Infinity; Infinite Limits; End Behavior; Asymptotes

1. As $x \to \infty$, $x^3 + x^2 + 2x - 1 = x^3$, and $x^3 + x + 1 = x^3$, so

$$\lim_{x \to \infty} \frac{x^3 + x^2 + 2x - 1}{x^3 + x + 1} = \lim_{x \to \infty} \frac{x^3}{x^3} = 1$$

3. As $x \to \infty$, $2x + 4 = 2x$, and $x - 1 = x$, so

$$\lim_{x \to \infty} \frac{2x + 4}{x - 1} = \lim_{x \to \infty} \frac{2x}{x} = 2$$

5. As $x \to \infty$, $3x^2 - 1 = 3x^2$, and $x^2 + 4 = x^2$, so

$$\lim_{x \to \infty} \frac{3x^2 - 1}{x^2 + 4} = \lim_{x \to \infty} \frac{3x^2}{x^2} = 3$$

7. As $x \to -\infty$, $5x^3 - 1 = 5x^3$, and $x^4 + 1 = x^4$, so

$$\lim_{x \to -\infty} \frac{5x^3 - 1}{x^4 + 1} = \lim_{x \to -\infty} \frac{5x^3}{x^4} = \lim_{x \to -\infty} \frac{5}{x} = 5 \lim_{x \to -\infty} \frac{1}{x} = 0$$

9. As $x \to \infty$, $5x^3 + 3 = 5x^3$, and $x^2 + 1 = x^2$, so

$$\lim_{x \to \infty} \frac{5x^3 + 3}{x^2 + 1} = \lim_{x \to \infty} \frac{5x^3}{x^2} = \lim_{x \to \infty} 5x = \infty$$

11. As $x \to -\infty$, $4x^5 = 4x^5$, and $x^2 + 1 = x^2$, so

$$\lim_{x \to -\infty} \frac{4x^5}{x^2 + 1} = \lim_{x \to -\infty} \frac{4x^5{}^3}{x^2} = \lim_{x \to -\infty} 4x^3 = 4 \cdot \lim_{x \to -\infty} x^3 = -\infty$$

13. Here $f(x) = \dfrac{1}{x - 2}$, $x \neq 2$. To determine $\displaystyle\lim_{x \to 2^+} \frac{1}{x - 2}$, we examine the values of f that are close to 2, but remain greater than 2.

x	2.1	2.01	2.001	2.0001	2.00001
$f(x) = \dfrac{1}{x-2}$	10	100	1000	10,000	100,000

We see that as x gets closer to 2 from the right, the value of $f(x) = \dfrac{1}{x-2}$ becomes unbounded in the positive direction, and we write

$$\lim_{x \to 2^+} \frac{1}{x-2} = \infty$$

15. Here $f(x) = \dfrac{x}{(x-1)^2}, x \neq 1$. To determine $\displaystyle\lim_{x \to 1^-} \dfrac{x}{(x-1)^2}$, we examine the values of f that are close to 1, but remain smaller than 1.

x	0.9	0.99	0.999	0.9999
$f(x) = \dfrac{x}{(x-1)^2}$	90	9900	999,000	99,990,000

We see that as x gets closer to 1 from the left, the value of $f(x) = \dfrac{x}{(x-1)^2}$ becomes unbounded in the positive direction, and we write

$$\lim_{x \to 1^-} \frac{x}{(x-1)^2} = \infty$$

17. Here $f(x) = \dfrac{x^2+1}{x^3-1}, x \neq 1$. To determine $\displaystyle\lim_{x \to 1^+} \dfrac{x^2+1}{x^3-1}$, we examine the values of f that are close to 1, but remain larger than 1.

x	1.1	1.01	1.001	1.0001	1.00001
$f(x) = \dfrac{x^2+1}{x^3-1}$	6.6767	66.66777	666.66678	6666.667	66,666.667

We see that as x gets closer to 1 from the right, the value of $f(x) = \dfrac{x^2+1}{x^3-1}$ becomes unbounded in the positive direction, and we write

$$\lim_{x \to 1^+} \frac{x^2+1}{x^3-1} = \infty$$

19. Here $f(x) = \dfrac{1-x}{3x-6}, x \neq 2$. To determine $\displaystyle\lim_{x \to 2^-} \dfrac{1-x}{3x-6}$, we examine the values of f that are close to 2, but remain smaller than 2.

x	1.9	1.99	1.999	1.9999	1.99999
$f(x) = \dfrac{1-x}{3x-6}$	3	33	333	3333	33,333

We see that as x gets closer to 2 from the left, the value of $f(x) = \dfrac{1-x}{3x-6}$ becomes unbounded in the positive direction, and we write

$$\lim_{x \to 2^-} \frac{1-x}{3x-6} = \infty$$

21. To find horizontal asymptotes, we need to find two limits, $\lim_{x \to \infty} f(x)$ and $\lim_{x \to -\infty} f(x)$.

$$\lim_{x \to \infty} f(x) = \lim_{x \to \infty}\left(3 + \frac{1}{x^2}\right) = \lim_{x \to \infty} 3 + \lim_{x \to \infty} \frac{1}{x^2} = 3 + 0 = 3$$

We conclude that the line $y = 3$ is a horizontal asymptote of the graph when x becomes unbounded in the positive direction.

$$\lim_{x \to -\infty} f(x) = \lim_{x \to -\infty}\left(3 + \frac{1}{x^2}\right) = \lim_{x \to -\infty} 3 + \lim_{x \to -\infty} \frac{1}{x^2} = 3 + 0 = 3$$

We conclude that the line $y = 3$ is a horizontal asymptote of the graph when x becomes unbounded in the negative direction.

To find vertical asymptotes, we need to examine the behavior of the graph of f when x is near 0, the point where f is not defined. This will require looking at the one-sided limits of f at 0.

$\lim_{x \to 0^-} f(x)$: Since $x \to 0^-$, we know $x < 0$, but $x^2 > 0$. It follows that the expression $\dfrac{1}{x^2}$ is positive and becomes unbounded as $x \to 0^-$.

$$\lim_{x \to 0^-} f(x) = \lim_{x \to 0^-}\left(3 + \frac{1}{x^2}\right) = \infty$$

$\lim_{x \to 0^+} f(x)$: Since $x \to 0^+$, we know $x > 0$, and $x^2 > 0$. It follows that the expression $\dfrac{1}{x^2}$ is positive and becomes unbounded as $x \to 0^+$.

$$\lim_{x \to 0^+} f(x) = \lim_{x \to 0^+}\left(3 + \frac{1}{x^2}\right) = \infty$$

We conclude that the graph of f has a vertical asymptote at $x = 0$.

23. To find horizontal asymptotes, we need to find two limits, $\lim_{x \to \infty} f(x)$ and $\lim_{x \to -\infty} f(x)$.

$$\lim_{x \to \infty} f(x) = \lim_{x \to \infty} \frac{2x^2}{(x-1)^2} = \lim_{x \to \infty} \frac{2x^2}{x^2 - 2x + 1} = \lim_{x \to \infty} \frac{2x^2}{x^2} = 2$$

We conclude that the line $y = 2$ is a horizontal asymptote of the graph when x becomes unbounded in the positive direction.

$$\lim_{x \to -\infty} f(x) = \lim_{x \to -\infty} \frac{2x^2}{(x-1)^2} = \lim_{x \to -\infty} \frac{2x^2}{x^2 - 2x + 1} = \lim_{x \to -\infty} \frac{2x^2}{x^2} = 2$$

We conclude that the line $y = 2$ is a horizontal asymptote of the graph when x becomes unbounded in the negative direction.

To find vertical asymptotes, we need to examine the behavior of the graph of f when x is near 1, the point where f is not defined. This will require looking at the one-sided limits of f at 1.

$\lim\limits_{x \to 1^-} f(x)$: Since $x \to 1^-$, we know $x < 1$, so $x - 1 < 0$, but $(x - 1)^2 > 0$. It follows

that the expression $\dfrac{2x^2}{(x-1)^2}$ is positive and becomes unbounded as $x \to 1^-$.

$$\lim\limits_{x \to 1^-} f(x) = \lim\limits_{x \to 1^-} \frac{2x^2}{(x-1)^2} = \infty$$

$\lim\limits_{x \to 1^+} f(x)$: Since $x \to 1^+$, we know $x > 1$, so both $x - 1 > 0$ and $(x-1)^2 > 0$. It

follows that the expression $\dfrac{2x^2}{(x-1)^2}$ is positive and becomes unbounded as $x \to 1^+$.

$$\lim\limits_{x \to 1^+} f(x) = \lim\limits_{x \to 1^+} \frac{2x^2}{(x-1)^2} = \infty$$

We conclude that the graph of f has a vertical asymptote at $x = 1$.

25. To find horizontal asymptotes, we need to find two limits, $\lim\limits_{x \to \infty} f(x)$ and $\lim\limits_{x \to -\infty} f(x)$.

$$\lim\limits_{x \to \infty} f(x) = \lim\limits_{x \to \infty} \frac{x^2}{x^2 - 4} = \lim\limits_{x \to \infty} \frac{x^2}{x^2} = \lim\limits_{x \to \infty} 1 = 1$$

We conclude that the line $y = 1$ is a horizontal asymptote of the graph when x becomes unbounded in the positive direction.

$$\lim\limits_{x \to -\infty} f(x) = \lim\limits_{x \to -\infty} \frac{x^2}{x^2 - 4} = \lim\limits_{x \to -\infty} \frac{x^2}{x^2} = \lim\limits_{x \to -\infty} 1 = 1$$

We conclude that the line $y = 1$ is a horizontal asymptote of the graph when x becomes unbounded in the negative direction.

To find vertical asymptotes, we need to examine the behavior of the graph of f when x is near -2 and 2, the points where f is not defined. This will require looking at the one-sided limits of f.

$\lim\limits_{x \to -2^-} f(x)$: Since $x \to -2^-$, we know $x < -2$ and $x^2 > 4$, so $x^2 - 4 > 0$. It follows

that the expression $\dfrac{x^2}{x^2 - 4}$ is positive and becomes unbounded as $x \to -2^-$.

$$\lim\limits_{x \to -2^-} f(x) = \lim\limits_{x \to -2^-} \frac{x^2}{x^2 - 4} = \infty$$

$\lim\limits_{x \to -2^+} f(x)$: Since $x \to -2^+$, we know $x > -2$ and $x^2 < 4$, so $x^2 - 4 < 0$. It

follows that the expression $\dfrac{x^2}{x^2 - 4}$ is negative and becomes unbounded as $x \to -2^+$.

$$\lim_{x \to -2^+} f(x) = \lim_{x \to -2^+} \frac{x^2}{x^2 - 4} = -\infty$$

We now examine the limits as $x \to 2$.

$\lim_{x \to 2^-} f(x)$: Since $x \to 2^-$, we know $x < 2$ and $x^2 < 4$, so $x^2 - 4 < 0$. It follows that

the expression $\dfrac{x^2}{x^2 - 4}$ is negative and becomes unbounded as $x \to 2^-$.

$$\lim_{x \to 2^-} f(x) = \lim_{x \to 2^-} \frac{x^2}{x^2 - 4} = -\infty$$

$\lim_{x \to 2^+} f(x)$: Since $x \to 2^+$, we know $x > 2$ and $x^2 > 4$, so $x^2 - 4 > 0$. It follows that

the expression $\dfrac{x^2}{x^2 - 4}$ is positive and becomes unbounded as $x \to 2^+$.

$$\lim_{x \to 2^+} f(x) = \lim_{x \to 2^+} \frac{x^2}{x^2 - 4} = \infty$$

We conclude that the graph of f has a vertical asymptotes at $x = -2$ and at $x = 2$.

27. (a) We observe from the graph that the domain continues indefinitely toward the infinities, since there are arrows on both ends. We also observe a vertical asymptote at $x = 6$. We conclude that the domain of f is $\{x \mid x \neq 6\}$ or all real numbers except 6.

(b) The arrows pointing upward as x approaches 6 (the vertical asymptote) indicate that the range of f is the set of positive numbers or $\{y \mid y \geq 0\}$ or the interval $[0, \infty)$.

(c) (-4, 0) and (0, 0) are the x-intercepts; (0, 0) is also the y-intercept.

(d) Since $f(x) = y$, $f(-2) = 2$.

(e) If $f(x) = 4$, then $x = 8$ or $x = 4$.

(f) f is discontinuous at $x = 6$; 6 is not in the domain of f.

(g) The vertical asymptote is $x = 6$.

(h) $y = 4$ is a horizontal asymptote of the graph when x becomes unbounded in the negative direction.

(i) There is only one local maximum. It occurs at $(-2, 2)$ where the local maximum is $y = 2$.

(j) There are 3 local minima. They occur at $(-4, 0)$ where the local minimum is $y = 0$; at $(0, 0)$ where the local minimum is $y = 0$; and at $(8, 4)$ where the local minimum is $y = 4$.

(k) The function f is increasing on the intervals $(-4, -2)$, $(0, 6)$, and $(8, \infty)$.

(l) The function f is decreasing on the intervals $(-\infty, -4)$, $(-2, 0)$, and $(6, 8)$.

(m) As x approaches $-\infty$, y approaches 4, so $\lim\limits_{x \to -\infty} f(x) = 4$.

(n) As x approaches ∞, y becomes unbounded in the positive direction, so $\lim\limits_{x \to \infty} f(x) = \infty$.

(o) As x approaches 6 from the left, we see that y becomes unbounded in the positive direction, so $\lim\limits_{x \to 6^-} f(x) = \infty$.

(p) As x approaches 6 from the right, we see that y becomes unbounded in the positive direction, so $\lim\limits_{x \to 6^+} f(x) = \infty$.

29. $R(x) = \dfrac{x-1}{x^2 - 1} = \dfrac{x-1}{(x-1)(x+1)}$. To determine the behavior of the graph near -1 and 1, we look at $\lim\limits_{x \to -1} R(x)$ and $\lim\limits_{x \to 1} R(x)$.

For $\lim\limits_{x \to -1} R(x)$, we have

$$\lim\limits_{x \to -1} R(x) = \lim\limits_{x \to -1} \frac{x-1}{x^2 - 1} = \lim\limits_{x \to -1} \frac{\cancel{x-1}}{\cancel{(x-1)}(x+1)} = \lim\limits_{x \to -1} \frac{1}{x+1}$$

If $x < -1$ and x is getting closer to -1, the value of $\dfrac{1}{x+1} < 0$ and is becoming unbounded; that is, $\lim\limits_{x \to -1^-} R(x) = -\infty$.

If $x > -1$ and x is getting closer to -1, the value of $\dfrac{1}{x+1} > 0$ and is becoming unbounded; that is, $\lim\limits_{x \to -1^+} R(x) = \infty$.

The graph of R will have a vertical asymptote at $x = -1$.

For $\lim\limits_{x \to 1} R(x)$, we have

$$\lim\limits_{x \to 1} R(x) = \lim\limits_{x \to 1} \frac{x-1}{x^2 - 1} = \lim\limits_{x \to 1} \frac{\cancel{x-1}}{\cancel{(x-1)}(x+1)} = \lim\limits_{x \to 1} \frac{1}{x+1} = \frac{1}{2}$$

As x gets closer to 1, the graph of R gets closer to $\dfrac{1}{2}$. Since R is not defined at 1, the graph will have a hole at $\left(1, \dfrac{1}{2}\right)$.

31. $R(x) = \dfrac{x^2 + x}{x^2 - 1} = \dfrac{x(x+1)}{(x-1)(x+1)}$. To determine the behavior of the graph near -1 and 1,

look at $\lim\limits_{x \to -1} R(x)$ and $\lim\limits_{x \to 1} R(x)$.

For $\lim\limits_{x \to -1} R(x)$, we have

$$\lim_{x \to -1} R(x) = \lim_{x \to -1} \frac{x^2 + x}{x^2 - 1} = \lim_{x \to -1} \frac{x(x+1)}{(x-1)(x+1)} = \lim_{x \to -1} \frac{x}{x-1} = \frac{-1}{-2} = \frac{1}{2}$$

As x gets closer to -1, the graph of R gets closer to $\dfrac{1}{2}$. Since R is not defined at -1, the

graph will have a hole at $\left(-1, \dfrac{1}{2}\right)$.

For $\lim\limits_{x \to 1} R(x)$, we have

$$\lim_{x \to 1} R(x) = \lim_{x \to 1} \frac{x^2 + x}{x^2 - 1} = \lim_{x \to 1} \frac{x(x+1)}{(x-1)(x+1)} = \lim_{x \to 1} \frac{x}{x-1}$$

If $x < 1$ and x is getting closer to 1, the value of $\dfrac{x}{x-1} < 0$ and is becoming unbounded;

that is, $\lim\limits_{x \to 1^-} R(x) = -\infty$.

If $x > 1$ and x is getting closer to 1, the value of $\dfrac{x}{x-1} > 0$ and is becoming unbounded;

that is, $\lim\limits_{x \to 1^+} R(x) = \infty$.

The graph of R will have a vertical asymptote at $x = 1$.

33. A rational function is undefined at every number that makes the denominator zero. So we solve

$$\begin{aligned}
x^4 - x^3 + 8x - 8 &= 0 && \text{Set the denominator} = 0. \\
x^3(x-1) + 8(x-1) &= 0 && \text{Factor by grouping.} \\
(x^3 + 8)(x-1) &= 0 \\
x^3 + 8 = 0 \quad \text{or} \quad x - 1 &= 0 && \text{Apply the Zero-Product Property.} \\
x = -2 \quad \text{or} \quad x &= 1 && \text{Solve for } x.
\end{aligned}$$

To determine the behavior of the graph near -2 and near 1, we look at $\lim\limits_{x \to -2} R(x)$ and

$\lim\limits_{x \to 1} R(x)$.

For $\lim\limits_{x \to -2} R(x)$, we have

$$\lim_{x \to -2} R(x) = \lim_{x \to -2} \frac{x^3 - x^2 + x - 1}{x^4 - x^3 + 8x - 8} = \lim_{x \to -2} \frac{x^2(x-1) + (x-1)}{x^3(x-1) + 8(x-1)}$$

$$= \lim_{x \to -2} \frac{\left(x^2+1\right)\left(x-1\right)}{\left(x^3+8\right)\left(x-1\right)} = \lim_{x \to -2} \frac{x^2+1}{x^3+8}$$

Since the limit of the denominator is 0, we use one-sided limits. If $x < -2$ and x is getting closer to -2, the value of $x^3+8 < 0$, so the quotient $\dfrac{x^2+1}{x^3+8} < 0$ and is becoming unbounded; that is, $\lim\limits_{x \to -2^-} R(x) = -\infty$.

If $x > -2$ and x is getting closer to -2, the value of $x^3+8 > 0$, so the quotient $\dfrac{x^2+1}{x^3+8} > 0$ and is becoming unbounded; that is, $\lim\limits_{x \to -2^+} R(x) = \infty$.

The graph of R will have a vertical asymptote at $x = -2$.

For $\lim\limits_{x \to 1} R(x)$, we have

$$\lim_{x \to 1} R(x) = \lim_{x \to 1} \frac{x^3-x^2+x-1}{x^4-x^3+8x-8} = \lim_{x \to 1} \frac{x^2+1}{x^3+8} = \frac{2}{9}$$

As x gets closer to 1, the graph of R gets closer to $\dfrac{2}{9}$. Since R is not defined at 1, the graph will have a hole at $\left(1, \dfrac{2}{9}\right)$.

35. A rational function is undefined at every number that makes the denominator zero. So we solve

$$
\begin{array}{ll}
x^2+x-6 = 0 & \text{Set the denominator} = 0. \\
(x+3)(x-2) = 0 & \text{Factor.} \\
x+3 = 0 \quad \text{or} \quad x-2 = 0 & \text{Apply the Zero-Product Property.} \\
x = -3 \quad \text{or} \quad x = 2 & \text{Solve for } x.
\end{array}
$$

To determine the behavior of the graph near -3 and near 2, we look at $\lim\limits_{x \to -3} R(x)$ and $\lim\limits_{x \to 2} R(x)$.

For $\lim\limits_{x \to -3} R(x)$, we have

$$\lim_{x \to -3} R(x) = \lim_{x \to -3} \frac{x^3-2x^2+4x-8}{x^2+x-6} = \lim_{x \to -3} \frac{x^2(x-2)+4(x-2)}{(x+3)(x-2)}$$

$$= \lim_{x \to -3} \frac{\left(x^2+4\right)\left(x-2\right)}{(x+3)\left(x-2\right)} = \lim_{x \to -3} \frac{x^2+4}{x+3}$$

Since the limit of the denominator is 0, we use one-sided limits. If $x < -3$ and x is getting closer to -3, the value of $x+3 < 0$, so the quotient $\dfrac{x^2+4}{x+3} < 0$ and is becoming unbounded; that is, $\lim\limits_{x \to -3^-} R(x) = -\infty$.

If $x > -3$ and x is getting closer to -3, the value of $x+3 > 0$, so the quotient

$\dfrac{x^2+4}{x+3} > 0$ and is becoming unbounded; that is, $\lim\limits_{x\to -3^+} R(x) = \infty$.

The graph of R will have a vertical asymptote at $x = -3$.

For $\lim\limits_{x\to 2} R(x)$, we have

$$\lim\limits_{x\to 2} R(x) = \lim\limits_{x\to 2} \frac{x^3 - 2x^2 + 4x - 8}{x^2 + x - 6} = \lim\limits_{x\to 2} \frac{x^2 + 4}{x + 3} = \frac{8}{5}$$

As x gets closer to 2, the graph of R gets closer to $\dfrac{8}{5}$. Since R is not defined at 2, the

graph will have a hole at $\left(2, \dfrac{8}{5}\right)$.

37. A rational function is undefined at every number that makes the denominator zero. So we solve

$$
\begin{array}{ll}
x^4 + x^3 + x + 1 = 0 & \text{Set the denominator} = 0. \\
x^3(x+1) + (x+1) = 0 & \text{Factor by grouping.} \\
(x^3 + 1)(x + 1) = 0 & \\
x^3 + 1 = 0 \quad \text{or} \quad x + 1 = 0 & \text{Apply the Zero-Product Property.} \\
x = -1 \quad \text{or} \quad x = -1 & \text{Solve for } x.
\end{array}
$$

To determine the behavior of the graph near -1, we look at $\lim\limits_{x\to -1} R(x)$.

$$\lim\limits_{x\to -1} R(x) = \lim\limits_{x\to -1} \frac{x^3 + 2x^2 + x}{x^4 + x^3 + x + 1} = \lim\limits_{x\to -1} \frac{x(x^2 + 2x + 1)}{(x^3 + 1)(x + 1)}$$

$$= \lim\limits_{x\to -1} \frac{x\,\cancel{(x+1)}\,\cancel{(x+1)}}{\cancel{(x+1)}\,(x^2 - x + 1)\,\cancel{(x+1)}} = \lim\limits_{x\to -1} \frac{x}{x^2 - x + 1} = -\frac{1}{3}$$

As x gets closer to -1, the graph of R gets closer to $-\dfrac{1}{3}$. Since R is not defined at 1,

the graph will have a hole at $\left(-1, -\dfrac{1}{3}\right)$.

39. (a) Production costs are the sum of fixed costs and variable costs. So the cost function C of producing x calculators is
$$C = C(x) = 10x + 79{,}000$$

(b) The domain of C is $\{x \mid x \geq 0\}$.

(c) The average cost per calculator, when x calculators are produced is given by the

function $\overline{C}(x) = \dfrac{C(x)}{x} = \dfrac{10x + 79{,}000}{x} = 10 + \dfrac{79{,}000}{x}$.

(d) The domain of $\overline{C}$ is $\{x \mid x > 0\}$.

(e) $\lim\limits_{x\to 0^+} \overline{C} = \lim\limits_{x\to 0^+} \left(10 + \dfrac{79{,}000}{x}\right) = \infty$

The average cost of making nearly 0 calculators becomes unbounded.

(f) $\lim\limits_{x \to \infty} \overline{C}(x) = \lim\limits_{x \to \infty} \dfrac{10x + 79{,}000}{x} = \lim\limits_{x \to \infty} \dfrac{10\not{x}}{\not{x}} = 10$

The average cost of producing a calculator when a very large number of calculators are produced is $10.

41. (a) $\lim\limits_{x \to 100^-} C(x) = \lim\limits_{x \to 100^-} \dfrac{5x}{100 - x}$ Since x is approaching 100, but is remaining less than

100, $100 - x > 0$, and the quotient $\dfrac{5x}{100 - x} > 0$ and is becoming unbounded; so

$\lim\limits_{x \to 100^-} C(x) = \infty$.

(b) It is not possible to remove 100% of the pollutant. Explanations will vary.

43. Graphs will vary.

Chapter 12 Review

TRUE-FALSE ITEMS

1. True   **3.** True

5. True   **7.** True

FILL-IN-THE-BLANKS

1. $\lim\limits_{x \to c} f(x) = N$ **3.** not exist

5. is not equal to   **7.** $y = 2$ ... horizontal

REVIEW EXERCISES

1. Here $f(x) = \dfrac{x^3 - 8}{x - 2}$, and $c = 2$. We find the limit by evaluating the function f at values of x close to 2.

x	1.9	1.99	1.999	1.9999
$f(x) = \dfrac{x^3 - 8}{x - 2}$	11.41	11.94	11.9940	11.9994
x	2.1	2.01	2.001	2.0001
$f(x) = \dfrac{x^3 - 8}{x - 2}$	12.61	12.0601	2.0060	12.0006

We infer from the table that $\lim\limits_{x \to 2} f(x) = \lim\limits_{x \to 2} \dfrac{x^3 - 8}{x - 2} = 12$.

3.

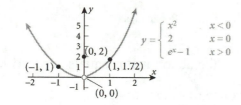

$$y = \begin{cases} x^2 & x < 0 \\ 2 & x = 0 \\ e^x - 1 & x > 0 \end{cases}$$

$$\lim\limits_{x \to 0} f(x) = 0$$

5. $f(x) = 3x^2 - 2x + 1$ is a polynomial. We know that for polynomials, $\lim\limits_{x \to c} f(x) = f(c)$.

$$\lim\limits_{x \to 2} \left(3x^2 - 2x + 1\right) = 3 \cdot 2^2 - 2 \cdot 2 + 1 = 12 - 4 + 1 = 9 \qquad \text{Limit of a polynomial.}$$

7. $f(x) = x^2 + 1$ is a polynomial. We know that for polynomials, $\lim\limits_{x \to c} f(x) = f(c)$.

$$\lim\limits_{x \to -2} \left(x^2 + 1\right)^2 = \left(\lim\limits_{x \to -2} \left(x^2 + 1\right)\right)^2 = \left[(-2)^2 + 1\right]^2 = (5)^2 = 25$$

$\uparrow$ $\uparrow$

Limit of a Power   Limit of a polynomial

9. $f(x) = \sqrt{x^2 + 7}$; its domain is all real numbers.

$$\lim\limits_{x \to 3} \sqrt{x^2 + 7} = \sqrt{\lim\limits_{x \to 3}\left(x^2 + 7\right)} = \sqrt{\left(3^2 + 7\right)} = \sqrt{16} = 4$$

$\uparrow$ $\uparrow$

Limit of a Root   Limit of a polynomial

11. $f(x) = \sqrt{1 - x^2}$. Its domain is the set of numbers that keeps $1 - x^2 \geq 0$.

$$1 - x^2 \geq 0 \text{ or } x^2 \leq 1 \text{ or } x \geq -1 \text{ and } x \leq 1$$

So the domain of f is $\{x \mid -1 \leq x \leq 1\}$ or x in the interval $[-1, 1]$.

As $x \to 1^-$, x gets closer to 1, but remains less than 1; x is in the domain of f. So we need only to consider x as it approaches 1.

$$\lim\limits_{x \to 1^-} \sqrt{1 - x^2} = \lim\limits_{x \to 1} \sqrt{1 - x^2} = \sqrt{\lim\limits_{x \to 1^-}\left(1 - x^2\right)} = \sqrt{1 - 1} = 0$$

$\uparrow$ $\uparrow$

Limit of a Root   Limit of a Polynomial

13. $f(x) = 5x + 6$ is a polynomial, so as x approaches c, $f(x)$ approaches $f(c)$.

$$\lim\limits_{x \to 2}(5x + 6)^{3/2} = \left[\lim\limits_{x \to 2}(5x + 6)\right]^{3/2} = (5 \cdot 2 + 6)^{3/2} = 16^{3/2} = 64$$

$\uparrow$ $\uparrow$

Limit of a Power   Limit of a Polynomial

15. Here $f(x) = x^2 + x + 2$ and $g(x) = x^2 - 9$ are both polynomials. So,

$$\lim_{x \to -1} \left(x^2 + x + 2\right)\left(x^2 - 9\right) = \left[\lim_{x \to -1}\left(x^2 + x + 2\right)\right]\left[\lim_{x \to -1}\left(x^2 - 9\right)\right] = \left[(-1)^2 + (-1) + 2\right]\left[(-1)^2 - 9\right] = -16$$

$\qquad\qquad\uparrow\qquad\qquad\qquad\qquad\qquad\qquad\qquad\qquad\qquad\uparrow$

$\qquad$ Limit of a Product $\qquad\qquad\qquad\qquad\qquad\qquad$ Limits of Polynomials

17. Here $f(x) = \dfrac{x-1}{x^3 - 1}$. As x approaches 1, the limit of the denominator equals zero, so

Formula (11) cannot be used directly. We factor the expression first.

$$\lim_{x \to 1} \frac{x-1}{x^3 - 1} = \lim_{x \to 1} \frac{x-1}{(x-1)\left(x^2 + x + 1\right)} = \lim_{x \to 1} \frac{1}{x^2 + x + 1} = \frac{\displaystyle\lim_{x \to 1} 1}{\displaystyle\lim_{x \to 1} x^2 + x + 1} = \frac{1}{3}$$

$\qquad\qquad\quad\uparrow\qquad\qquad\qquad\qquad\qquad\qquad\qquad\qquad\quad\uparrow$

$\qquad\qquad$ Factor. $\qquad\qquad\qquad\qquad\qquad\qquad\qquad$ Limit of a Quotient

19. Here $f(x) = \dfrac{x^2 - 9}{x^2 - x - 12}$. As x approaches -3, the limit of the denominator equals zero,

so Formula (11) cannot be used directly. We factor the expression first.

$$\lim_{x \to -3} \frac{x^2 - 9}{x^2 - x - 12} = \lim_{x \to -3} \frac{(x-3)(x+3)}{(x-4)(x+3)} = \lim_{x \to -3} \frac{x-3}{x-4} = \frac{\displaystyle\lim_{x \to -3}(x-3)}{\displaystyle\lim_{x \to -3}(x-4)} = \frac{-6}{-7} = \frac{6}{7}$$

$\qquad\qquad\qquad\uparrow\qquad\qquad\qquad\qquad\qquad\qquad\qquad\qquad\qquad\uparrow$

$\qquad\qquad\quad$ Factor. $\qquad\qquad\qquad\qquad\qquad\qquad\qquad$ Limit of a Quotient

21.

Here $f(x) = \dfrac{x^2 - 1}{x^3 - 1}$. As x approaches -1 from the left, the limit of the denominator

equals zero, so Formula (11) cannot be used directly. We factor the expression first.

$$\lim_{x \to -1^-} \frac{x^2 - 1}{x^3 - 1} = \lim_{x \to -1^-} \frac{(x-1)(x+1)}{(x-1)\left(x^2 + x + 1\right)} = \lim_{x \to -1^-} \frac{x+1}{x^2 + x + 1} = \frac{\displaystyle\lim_{x \to -1^-}(x+1)}{\displaystyle\lim_{x \to -1^-}\left(x^2 + x + 1\right)} = \frac{0}{1} = 0$$

$\qquad\qquad\qquad\uparrow\qquad\qquad\qquad\qquad\qquad\qquad\qquad\qquad\qquad\qquad\uparrow$

$\qquad\qquad\quad$ Factor. $\qquad\qquad\qquad\qquad\qquad\qquad\qquad\qquad$ Limit of a Quotient

23. Here $f(x) = \dfrac{x^3 - 8}{x^3 - 2x^2 + 4x - 8}$. As x approaches 2, the limit of the denominator equals

zero, so Formula (11) cannot be used directly. We factor the expression first.

$$\lim_{x \to 2} \frac{x^3 - 8}{x^3 - 2x^2 + 4x - 8} = \lim_{x \to 2} \frac{(x-2)\left(x^2 + 2x + 4\right)}{x^2(x-2) + 4(x-2)} = \lim_{x \to 2} \frac{(x-2)\left(x^2 + 2x + 4\right)}{\left(x^2 + 4\right)(x-2)}$$

$$= \lim_{x \to 2} \frac{x^2 + 2x + 4}{x^2 + 4} = \frac{\displaystyle\lim_{x \to 2}\left(x^2 + 2x + 4\right)}{\displaystyle\lim_{x \to 2}\left(x^2 + 4\right)} = \frac{2^2 + 2 \cdot 2 + 4}{2^2 + 4} = \frac{12}{8} = \frac{3}{2}$$

$\qquad\qquad\qquad\qquad\qquad\qquad\qquad\qquad\qquad\qquad\uparrow$

$\qquad\qquad\qquad\qquad\qquad\qquad\qquad\qquad$ Limit of a Quotient

25. Here $f(x) = \dfrac{x^4 - 3x^3 + x - 3}{x^3 - 3x^2 + 2x - 6}$. As x approaches 3, the limit of the denominator equals zero, so Formula (11) cannot be used directly. We factor the expression first.

$$\frac{x^4 - 3x^3 + x - 3}{x^3 - 3x^2 + 2x - 6} = \frac{x^3(x-3) + 1(x-3)}{x^2(x-3) + 2(x-3)} = \frac{(x^3 + 1)(x-3)}{(x^2 + 2)(x-3)}$$

Factor both the numerator and the denominator by grouping.

$$\lim_{x \to 3} \frac{x^4 - 3x^3 + x - 3}{x^3 - 3x^2 + 2x - 6} = \lim_{x \to 3} \frac{(x^3 + 1)(\cancel{x-3})}{(x^2 + 2)(\cancel{x-3})} = \frac{\lim_{x \to 3}(x^3 + 1)}{\lim_{x \to 3}(x^2 + 2)} = \frac{28}{11}$$

↑
Limit of a Quotient

27. $\displaystyle\lim_{x \to \infty} \frac{5x^4 - 8x^3 + x}{3x^4 + x^2 + 5} = \lim_{x \to \infty} \frac{5x^4}{3x^4}$

As $x \to \infty$, $5x^4 - 8x^3 + x = 5x^4$ and $3x^4 + x^2 + 5 = 3x^4$.

$$= \lim_{x \to \infty} \frac{5}{3} = \frac{5}{3}$$

29. $f(x) = \dfrac{x^2}{x-3}$ is not defined at $x = 3$. When $x \to 3^-$, $x - 3 < 0$. Since $x^2 \geq 0$, it follows

that the expression $\dfrac{x^2}{x-3}$ is negative and becomes unbounded as $x \to 3^-$.

$$\lim_{x \to 3^-} \frac{x^2}{x-3} = -\infty$$

31. $\displaystyle\lim_{x \to \infty} \frac{8x^4 - x^2 + 2}{-4x^3 + 1} = \lim_{x \to \infty} \frac{8x^4}{-4x^3}$

As $x \to \infty$, $8x^4 - x^2 + 2 = 8x^4$ and $-4x^3 + 1 = -4x^3$.

$$= \lim_{x \to \infty} \frac{8x}{-4} = -\infty$$

33. $f(x) = \dfrac{1 - 9x^2}{x^2 - 9}$ is not defined at $x = -3$. When $x \to -3^+$, $x > -3$ and $x^2 - 9 < 0$. Since

$1 - 9x^2 < 0$, it follows that $\dfrac{1 - 9x^2}{x^2 - 9}$ is positive and as becomes unbounded $x \to -3^+$.

$$\lim_{x \to -3^+} \frac{1 - 9x^2}{x^2 - 9} = \infty$$

35. $f(x) = 3x^4 - x^2 + 2$ is a polynomial function, and polynomial functions are continuous at all values of x. So $f(x)$ is continuous at $c = 5$.

37. $f(x) = \dfrac{x^4 - 4}{x + 2}$ is a rational function which is continuous at all values of x in its domain. Since $x = -2$ is not in the domain of f, the function f is not continuous at $c = -2$.

39. The function f is defined at $c = -2$; $f(-2) = 4$.

The $\displaystyle\lim_{x \to -2} f(x) = \lim_{x \to -2} \frac{x^2 - 4}{x + 2} = \lim_{x \to -2} (x - 2) = -4$

Since the limit as x approaches -2 does not equal $f(-2)$, the function is not continuous at $c = -2$.

41. The function f is defined at $c = -2$; $f(-2) = -4$.

The $\displaystyle\lim_{x \to -2} f(x) = \lim_{x \to -2} \frac{x^2 - 4}{x + 2} = \lim_{x \to -2} (x - 2) = -4$

Since the limit as x approaches -2 equals $f(-2)$, the function is continuous at $c = -2$

43. To find any horizontal asymptotes we need to find $\displaystyle\lim_{x \to \infty} f(x)$ and $\displaystyle\lim_{x \to -\infty} f(x)$.

$$\lim_{x \to \infty} f(x) = \lim_{x \to \infty} \frac{3x}{x^2 - 1} = \lim_{x \to \infty} \frac{3x}{x^2} = \lim_{x \to \infty} \frac{3}{x} = 0$$

The line $y = 0$ is a horizontal asymptote of the graph when x is sufficiently positive.

$$\lim_{x \to -\infty} f(x) = \lim_{x \to -\infty} \frac{3x}{x^2 - 1} = \lim_{x \to -\infty} \frac{3x}{x^2} = \lim_{x \to -\infty} \frac{3}{x} = 0$$

The line $y = 0$ is a horizontal asymptote of the graph when x is sufficiently negative.

The domain of f is $\{x \mid x \neq -1, x \neq 1\}$. To locate any vertical asymptotes we look at $\displaystyle\lim_{x \to -1} f(x)$ and $\displaystyle\lim_{x \to 1} f(x)$.

Looking at one-sided limits of f at -1, we find
$\displaystyle\lim_{x \to -1^-} f(x)$: When $x \to -1$ from the left, $x < -1$ and $x^2 > 1$ or $x^2 - 1 > 0$. So, the expression $\dfrac{3x}{x^2 - 1}$ is negative and becomes unbounded.

$$\lim_{x \to -1^-} f(x) = \lim_{x \to -1^-} \frac{3x}{x^2 - 1} = -\infty$$

$\displaystyle\lim_{x \to -1^+} f(x)$: When $x \to -1$ from the right, $x > -1$ and $x^2 < 1$ or $x^2 - 1 < 0$. So, the expression $\dfrac{3x}{x^2 - 1}$ is positive and becomes unbounded.

$$\lim_{x \to -1^+} f(x) = \lim_{x \to -1^+} \frac{3x}{x^2 - 1} = \infty$$

We conclude f has a vertical asymptote at $x = -1$.

$\displaystyle\lim_{x \to 1^-} f(x)$: When $x \to 1$ from the left, $x < 1$ and $x^2 < 1$ or $x^2 - 1 < 0$. So, the expression $\dfrac{3x}{x^2-1}$ is negative and becomes unbounded.

$$\lim_{x\to 1^-} f(x) = \lim_{x\to 1^-} \frac{3x}{x^2-1} = -\infty$$

$\lim\limits_{x\to 1^+} f(x)$: When $x \to 1$ from the right, $x > 1$ and $x^2 - 1 > 0$. So, the expression

$\dfrac{3x}{x^2-1}$ is positive and becomes unbounded.

$$\lim_{x\to 1^+} f(x) = \lim_{x\to 1^+} \frac{3x}{x^2-1} = \infty$$

We conclude f has a vertical asymptote at $x = 1$.

45.   To find any horizontal asymptotes we need to find $\lim\limits_{x\to\infty} f(x)$ and $\lim\limits_{x\to -\infty} f(x)$.

$$\lim_{x\to\infty} f(x) = \lim_{x\to\infty} \frac{5x}{x+2} = \lim_{x\to\infty} \frac{5x}{x} = \lim_{x\to\infty} \frac{5}{1} = 5$$

The line $y = 5$ is a horizontal asymptote of the graph when x is sufficiently positive.

$$\lim_{x\to -\infty} f(x) = \lim_{x\to -\infty} \frac{5x}{x+2} = \lim_{x\to -\infty} \frac{5x}{x} = \lim_{x\to -\infty} \frac{5}{1} = 5$$

The line $y = 5$ is a horizontal asymptote of the graph when x is sufficiently negative.

The domain of f is $\{x \mid x \neq -2\}$. To locate any vertical asymptotes we look at $\lim\limits_{x\to -2} f(x)$.

Looking at one-sided limits of f at -2, we find

$\lim\limits_{x\to -2^-} f(x)$: When $x \to -2$ from the left, $x < -2$ and $x + 2 < 0$. So, the expression

$\dfrac{5x}{x+2}$ is positive and becomes unbounded.

$$\lim_{x\to -2^-} f(x) = \lim_{x\to -2^-} \frac{5x}{x+2} = \infty$$

$\lim\limits_{x\to -2^+} f(x)$: When $x \to -2$ from the right, $x > -2$ and $x + 2 > 0$. So, the expression

$\dfrac{5x}{x+2}$ is negative and becomes unbounded.

$$\lim_{x\to -2^+} f(x) = \lim_{x\to -2^+} \frac{5x}{x+2} = -\infty$$

We conclude f has a vertical asymptote at $x = -2$.

47. (a) There is a vertical asymptote at $x = 2$ and f is not defined at 2, so the domain of f is the intervals $(-\infty, 2)$ or $(2, 5)$ or $(5, \infty)$.

(b) The range of f is the set of all real numbers, that is all y in the interval $(-\infty, \infty)$.

(c) The x-intercepts are the points at which the graph crosses or touches the x-axis. The x-intercepts are $(-2, 0)$, $(0, 0)$, $(1, 0)$, and $(6, 0)$.

(d) The y-intercept is $(0, 0)$.

(e) $f(-6) = 2$ and $f(-4) = 1$

(f) $f(-2) = 0$ and $f(6) = 0$

(g) $\lim_{x \to -4^-} f(x) = 4$; $\lim_{x \to -4^+} f(x) = -2$

(h) $\lim_{x \to -2^-} f(x) = -2$; $\lim_{x \to -2^+} f(x) = 2$

(i) $\lim_{x \to 5^-} f(x) = 2$; $\lim_{x \to 5^+} f(x) = 2$

(j) The $\lim_{x \to 0} f(x)$ does not exist since $\lim_{x \to 0^-} f(x) = 4$ and $\lim_{x \to 0^+} f(x) = 1$ are not equal.

(k) The $\lim_{x \to 2} f(x)$ does not exist since $\lim_{x \to 2^-} f(x) = -\infty$ and $\lim_{x \to 2^+} f(x) = \infty$.

(l) f is not continuous at -2 since $\lim_{x \to -2} f(x)$ does not exist.

(m) f is not continuous at -4 since $\lim_{x \to -4} f(x)$ does not exist.

(n) f is not continuous at 0 since $\lim_{x \to 0} f(x)$ does not exist.

(o) f is not continuous at 2; there is a vertical asymptote at 2.

(p) f is continuous at 4.

(q) f is not continuous at 5 since f is not defined at $x = 5$.

(r) f is increasing on the open intervals $(-6, -4)$, $(-2, 0)$, and $(6, \infty)$.

(s) f is decreasing on the open intervals $(-\infty, -6)$, $(0, 2)$, $(2, 5)$, and $(5, 6)$.

(t) $\lim_{x \to -\infty} f(x) = \infty$ and $\lim_{x \to \infty} f(x) = 2$

(u) There are no local maxima. There is a local minimum of 2 at $x = -6$, a local minimum of 0 at $x = 0$, and a local minimum of 0 at $x = 6$.

(v) There is a horizontal asymptote of $y = 2$ as x becomes unbounded in the positive direction, and a vertical asymptote at $x = 2$.

49. The average rate of change of $f(x)$ from -2 to x is

$$\frac{\Delta y}{\Delta x} = \frac{f(x) - f(-2)}{x - (-2)} = \frac{\left(2x^2 - 3x\right) - \left(2(-2)^2 - 3(-2)\right)}{x - (-2)}$$

$$= \frac{2x^2 - 3x - 8 - 6}{x + 2} \qquad \text{Remove parentheses.}$$

$$= \frac{(2x - 7)(x + 2)}{x + 2} \qquad \text{Factor.}$$

The limit as $x \to -2$ is

$$\lim_{x \to -2} \frac{(2x - 7)\cancel{(x + 2)}}{\cancel{x + 2}} = \lim_{x \to -2} (2x - 7) = -11$$

51. The average rate of change of $f(x)$ from 3 to x is

$$\frac{\Delta y}{\Delta x} = \frac{f(x) - f(3)}{x - 3} = \frac{\dfrac{x}{x-1} - \dfrac{3}{3-1}}{x - 3} = \frac{\dfrac{x}{x-1} - \dfrac{3}{2}}{x - 3}$$

$$= \frac{2x - 3(x - 1)}{2(x - 1)(x - 3)} \qquad \text{Write as a single fraction.}$$

$$= \frac{-x + 3}{2(x - 1)(x - 3)}$$

The limit as $x \to 3$ is

$$\lim_{x \to 3} \frac{\cancel{-x + 3}^{-1}}{2(x - 1)\cancel{(x - 3)}} = \lim_{x \to 3} \frac{-1}{2(x - 1)} = -\frac{1}{4}$$

53. $R(x) = \dfrac{x + 4}{x^2 - 16}$. To determine the behavior of the graph near -4 and 4, we look at $\displaystyle\lim_{x \to -4} R(x)$ and $\displaystyle\lim_{x \to 4} R(x)$.

For $\displaystyle\lim_{x \to -4} R(x)$, we have

$$\lim_{x \to -4} R(x) = \lim_{x \to -4} \frac{x + 4}{x^2 - 16} = \lim_{x \to -4} \frac{\cancel{x + 4}}{\cancel{(x + 4)}(x - 4)} = \lim_{x \to -4} \frac{1}{x - 4} = -\frac{1}{8}$$

As x gets closer to -4, the graph of R gets closer to $-\dfrac{1}{8}$. Since R is not defined at -4, the graph will have a hole at $\left(-4, -\dfrac{1}{8}\right)$.

For $\lim\limits_{x \to 4} R(x)$, we have

$$\lim_{x \to 4} R(x) = \lim_{x \to 4} \frac{x+4}{x^2-16} = \lim_{x \to 4} \frac{\cancel{x+4}}{\cancel{(x+4)}(x-4)} = \lim_{x \to 4} \frac{1}{x-4}$$

Since the limit of the denominator is 0, we use one-sided limits to investigate $\lim\limits_{x \to 4} \dfrac{1}{x-4}$.

If $x < 4$ and x is getting closer to 4, the value of $\dfrac{1}{x-4} < 0$ and is becoming unbounded;

that is, $\lim\limits_{x \to 4^-} \dfrac{1}{x-4} = -\infty$.

If $x > 4$ and x is getting closer to 4, the value of $\dfrac{1}{x-4} > 0$ and is becoming unbounded;

that is, $\lim\limits_{x \to 4^+} \dfrac{1}{x-4} = \infty$.

The graph of R will have a vertical asymptote at $x = 4$.

55.   Rational functions are undefined at values of x that would make the denominator of the function equal zero. Solving $x^2 - 11x + 18 = 0$ or $(x-9)(x-2) = 0$ we get $x = 9$ or $x = 2$. So R is undefined at $x = 2$ and $x = 9$.

To analyze the behavior of the graph near 2 and 9, we look at $\lim\limits_{x \to 2} R(x)$ and $\lim\limits_{x \to 9} R(x)$.

For $\lim\limits_{x \to 2} R(x)$, we have

$$\lim_{x \to 2} R(x) = \lim_{x \to 2} \frac{x^3 - 2x^2 + 4x - 8}{x^2 - 11x + 18} = \lim_{x \to 2} \frac{x^2(x-2) + 4(x-2)}{(x-2)(x-9)}$$

$$= \lim_{x \to 2} \frac{(x^2+4)\cancel{(x-2)}}{\cancel{(x-2)}(x-9)} = \lim_{x \to 2} \frac{x^2+4}{x-9} = -\frac{8}{7}$$

As x gets closer to 4, the graph of R gets closer to $-\dfrac{8}{7}$. Since R is not defined at 2, the graph will have a hole at $\left(2, -\dfrac{8}{7}\right)$.

For $\lim\limits_{x \to 9} R(x)$, we have

$$\lim_{x \to 9} R(x) = \lim_{x \to 9} \frac{x^3 - 2x^2 + 4x - 8}{x^2 - 11x + 18} = \lim_{x \to 9} \frac{(x^2+4)\cancel{(x-2)}}{\cancel{(x-2)}(x-9)} = \lim_{x \to 9} \frac{x^2+4}{x-9}$$

Since the limit of the denominator is 0, we use one-sided limits to investigate $\lim\limits_{x \to 9} \dfrac{x^2+4}{x-9}$.

If $x < 9$ and x is getting closer to 9, the value of $\dfrac{x^2+4}{x-9} < 0$ and is becoming unbounded;

that is, $\displaystyle\lim_{x\to 9^-}\dfrac{x^2+4}{x-9}=-\infty$.

If $x > 9$ and x is getting closer to 9, the value of $\dfrac{x^2+4}{x-9} > 0$ and is becoming unbounded;

that is, $\displaystyle\lim_{x\to 9^+}\dfrac{x^2+4}{x-9}=\infty$.

The graph of R will have a vertical asymptote at $x=9$.

57. Answers will vary.

59. (a) $\displaystyle\lim_{x\to\infty}S(x)=\lim_{x\to\infty}\dfrac{2000x^2}{3.5x^2+1000}=\lim_{x\to\infty}\dfrac{2000x^2}{3.5x^2}=\lim_{x\to\infty}\dfrac{2000}{3.5}=571.43$

CHAPTER 12 PROJECT

1.
$$R(x)=\begin{cases}0.10 & 0 < x \le 7000 \\ 0.15 & 7000 < x \le 28,400 \\ 0.25 & 28,400 < x \le 68,800 \\ 0.28 & 68,800 < x \le 143,500 \\ 0.33 & 143,500 < x \le 311,950 \\ 0.35 & x > 311,950\end{cases}$$

3. The function R is not continuous. It is discontinuous at the endpoints of each tax bracket.

5.
$$A(x)=\begin{cases}0.10x & 1 < x \le 7000 \\ 700+0.15(x-7000) & 7000 < x \le 28,400 \\ 3910+0.25(x-28,400) & 28,400 < x \le 68,800 \\ 14,010+0.28(x-68,800) & 68,800 < x \le 143,500 \\ 34,926+0.33(x-143,500) & 143,500 < x \le 311,950 \\ 90,514+0.35(x-311,950) & x > 311,950\end{cases}$$

7. The function A is not continuous if your income is \$311,950.

9. To compute column 3, we find the amount of tax paid if a person earns the highest dollar amount allowable in the previous row. That is

Row 2, Column 3: A couple earning \$14,000 pays
$$0.10(14,000)=\$1400$$
So the entry will be \$1400.

Row 3, Column 3: We calculate the taxes paid by a couple earning $56,800.

$$\$1400 + 0.15(\$56,800 - \$14,000) = \$9710$$

So the entry will be $9710.

Row 4, Column 3: We calculate the taxes paid by a couple earning $114,650.

$$\$9710 + 0.25(\$114,650 - \$56,800) = \$24,172.50$$

So the entry will be $24,172.50.

Row 5, Column 3: We calculate the taxes paid by a couple earning $174,700.

$$\$24,172.50 + 0.28(\$174,700 - \$114,650) = \$40,986.50$$

So the entry will be $40,986.50.

Row 6, Column 3: We calculate the taxes paid by a couple earning $311,950.

$$\$40,986.50 + 0.33(\$311,950 - \$174,700) = \$86,279.00$$

So the entry will be $86,279.00.

MATHEMATICAL QUESTIONS FROM PROFESSIONAL EXAMS

1. (b) $\dfrac{5}{6}$ $\quad\quad \lim\limits_{x \to 3} \dfrac{x^2 - x - 6}{x^2 - 9} = \lim\limits_{x \to 3} \dfrac{(x-3)(x+2)}{(x-3)(x+3)} = \lim\limits_{x \to 3} \dfrac{(x+2)}{(x+3)} = \dfrac{5}{6}$

3. (d) $\dfrac{\sqrt{2}}{4}$ $\quad\quad \lim\limits_{h \to 0} \dfrac{\sqrt{2+h} - \sqrt{2}}{h} = \lim\limits_{h \to 0} \left(\dfrac{\sqrt{2+h} - \sqrt{2}}{h} \cdot \dfrac{\sqrt{2+h} + \sqrt{2}}{\sqrt{2+h} + \sqrt{2}} \right)$

$$= \lim\limits_{h \to 0} \left(\dfrac{\left(\sqrt{2+h}\right)^2 - \left(\sqrt{2}\right)^2}{h\left(\sqrt{2+h} + \sqrt{2}\right)} \right) = \lim\limits_{h \to 0} \left(\dfrac{2 + h - 2}{h\left(\sqrt{2+h} + \sqrt{2}\right)} \right)$$

$$= \lim\limits_{h \to 0} \dfrac{h}{h\left(\sqrt{2+h} + \sqrt{2}\right)} = \lim\limits_{h \to 0} \dfrac{1}{\left(\sqrt{2+h} + \sqrt{2}\right)}$$

$$= \dfrac{1}{2\sqrt{2}} = \dfrac{\sqrt{2}}{4}$$

Chapter 13
The Derivative of a Function

13.1 The Definition of a Derivative 723/1-69 (ODD)

1. The slope of the tangent line to the graph of $f(x) = 3x + 5$ at the point (1, 8) is

$$m_{\tan} = \lim_{x \to 1} \frac{f(x) - f(1)}{x - 1} = \lim_{x \to 1} \frac{(3x + 5) - 8}{x - 1} = \lim_{x \to 1} \frac{3x - 3}{x - 1} = \lim_{x \to 1} \frac{3(x - 1)}{x - 1} = \lim_{x \to 1} 3 = 3$$

An equation of the tangent line is

$$y - 8 = 3(x - 1) \qquad\qquad y - f(c) = m_{\tan}(x - c)$$
$$y = 3x + 5 \qquad\qquad \text{Simplify.}$$

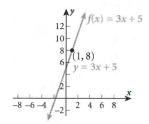

3. The slope of the tangent line to the graph of $f(x) = x^2 + 2$ at the point (− 1, 3) is

$$m_{\tan} = \lim_{x \to -1} \frac{f(x) - f(-1)}{x - (-1)} = \lim_{x \to -1} \frac{(x^2 + 2) - (3)}{x - (-1)} = \lim_{x \to -1} \frac{x^2 - 1}{x + 1} = \lim_{x \to -1} \frac{(x - 1)(x + 1)}{(x + 1)}$$

$$= \lim_{x \to -1} (x - 1) = -2$$

An equation of the tangent line is

$$y - 3 = (-2)[x - (-1)] \qquad\qquad y - f(c) = m_{\tan}(x - c)$$
$$y - 3 = -2x - 2 \qquad\qquad \text{Simplify.}$$
$$y = -2x + 1 \qquad\qquad \text{Add 3 to both sides.}$$

5. The slope of the tangent line to the graph of $f(x) = 3x^2$ at the point (2, 12) is

$$m_{\tan} = \lim_{x \to 2} \frac{f(x) - f(2)}{x - 2} = \lim_{x \to 2} \frac{3x^2 - 12}{x - 2} = \lim_{x \to 2} \frac{3(x^2 - 4)}{x - 2} = \lim_{x \to 2} \frac{3(x - 2)(x + 2)}{x - 2}$$

$$= \lim_{x \to 2} [3(x + 2)] = 12$$

An equation of the tangent line is

$$y - 12 = 12(x - 2) \qquad\qquad y - f(c) = m_{\tan}(x - c)$$

$$y - 12 = 12x - 24 \qquad \text{Simplify.}$$
$$y = 12x - 12 \qquad \text{Add 12 to both sides.}$$

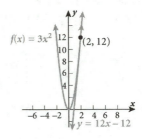

7. The slope of the tangent line to the graph of $f(x) = 2x^2 + x$ at the point $(1, 3)$ is

$$m_{\tan} = \lim_{x \to 1} \frac{f(x) - f(1)}{x - 1} = \lim_{x \to 1} \frac{(2x^2 + x) - 3}{x - 1} = \lim_{x \to 1} \frac{(2x + 3)(x - 1)}{x - 1} = \lim_{x \to 1} (2x + 3) = 5$$

An equation of the tangent line is

$$y - 3 = 5(x - 1) \qquad y - f(c) = m_{\tan}(x - c)$$
$$y = 5x - 2 \qquad \text{Simplify.}$$

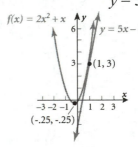

9. The slope of the tangent line to the graph of $f(x) = x^2 - 2x + 3$ at the point $(-1, 6)$ is

$$m_{\tan} = \lim_{x \to -1} \frac{f(x) - f(-1)}{x - (-1)} = \lim_{x \to -1} \frac{(x^2 - 2x + 3) - 6}{x + 1} = \lim_{x \to -1} \frac{x^2 - 2x - 3}{x + 1}$$

$$= \lim_{x \to -1} \frac{(x - 3)(x + 1)}{x + 1} = \lim_{x \to -1} (x - 3) = -4$$

An equation of the tangent line is

$$y - 6 = (-4)[x - (-1)] \qquad y - f(c) = m_{\tan}(x - c)$$
$$y - 6 = -4x - 4 \qquad \text{Simplify.}$$
$$y = -4x + 2 \qquad \text{Add 6 to both sides.}$$

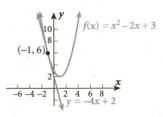

11. The slope of the tangent line to the graph of $f(x) = x^3 + x^2$ at the point $(-1, 0)$ is

$$m_{\tan} = \lim_{x \to -1} \frac{(x^3 + x^2) - 0}{x - (-1)} = \lim_{x \to -1} \frac{x^2(x + 1)}{(x + 1)} = \lim_{x \to -1} x^2 = 1$$

An equation of the tangent line is
$$y - 0 = 1[x - (-1)] \qquad y - f(c) = m_{\tan}(x - c)$$
$$y = x + 1 \qquad\qquad \text{Simplify.}$$

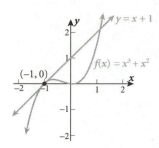

13.   To find $f'(3)$, we follow the three steps outlined in the text.
   Step 1: $f(3) = -4(3) + 5 = -12 + 5 = -7$
   Step 2: $\dfrac{f(x) - f(3)}{x - 3} = \dfrac{(-4x + 5) - (-7)}{x - 3} = \dfrac{-4x + 12}{x - 3} = \dfrac{(-4)(x - 3)}{x - 3}$
   Step 3: The derivative of f at 3 is
$$f'(3) = \lim_{x \to 3} \frac{f(x) - f(3)}{x - 3} = \lim_{x \to 3} \frac{(-4)(x-3)}{x-3} = -4$$

15.   To find $f'(0)$, we follow the three steps outlined in the text.
   Step 1: $f(0) = (0)^2 - 3 = -3$
   Step 2: $\dfrac{f(x) - f(0)}{x - 0} = \dfrac{(x^2 - 3) - (-3)}{x} = \dfrac{x^2}{x}$
   Step 3: The derivative of f at 0 is
$$f'(0) = \lim_{x \to 0} \frac{f(x) - f(0)}{x - 0} = \lim_{x \to 0} \frac{x^2}{x} = \lim_{x \to 0} x = 0$$

17.   To find $f'(1)$, we follow the three steps outlined in the text.
   Step 1: $f(1) = 2 \cdot 1^2 + 3 \cdot 1 = 5$
   Step 2: $\dfrac{f(x) - f(1)}{x - 1} = \dfrac{(2x^2 + 3x) - (5)}{x - 1} = \dfrac{2x^2 + 3x - 5}{x - 1} = \dfrac{(2x + 5)(x - 1)}{x - 1}$
   Step 3: The derivative of f at 1 is
$$f'(1) = \lim_{x \to 1} \frac{f(x) - f(1)}{x - 1} = \lim_{x \to 1} \frac{(2x+5)(x-1)}{x-1} = \lim_{x \to 1} (2x + 5) = 7$$

19.   To find $f'(0)$, we follow the three steps outlined in the text.
   Step 1: $f(0) = 0^3 + 4 \cdot 0 = 0$
   Step 2: $\dfrac{f(x) - f(0)}{x - 0} = \dfrac{(x^3 + 4x) - (0)}{x} = \dfrac{x(x^2 + 4)}{x}$
   Step 3: The derivative of f at 0 is
$$f'(0) = \lim_{x \to 0} \frac{f(x) - f(0)}{x - 0} = \lim_{x \to 0} \frac{x(x^2+4)}{x} = \lim_{x \to 0} (x^2 + 4) = 4$$

21. To find $f'(1)$, we follow the three steps outlined in the text.

Step 1: $f(1) = 1^3 + 1^2 - 2 \cdot 1 = 0$

Step 2: $\dfrac{f(x) - f(1)}{x - 1} = \dfrac{(x^3 + x^2 - 2x) - 0}{x - 1} = \dfrac{x(x^2 + x - 2)}{x - 1} = \dfrac{x(x + 2)(x - 1)}{x - 1}$

Step 3: The derivative of f at 1 is

$$f'(1) = \lim_{x \to 1} \frac{f(x) - f(1)}{x - 1} = \lim_{x \to 1} \frac{x(x + 2)\cancel{(x - 1)}}{\cancel{x - 1}}$$

$$= \lim_{x \to 1} [x(x + 2)] = \lim_{x \to 1} x \cdot \lim_{x \to 1} (x + 2) = 1 \cdot 3 = 3$$

23. To find $f'(1)$, we follow the three steps outlined in the text.

Step 1: $f(1) = \dfrac{1}{1} = 1$

Step 2: $\dfrac{f(x) - f(1)}{x - 1} = \dfrac{\left(\dfrac{1}{x}\right) - (1)}{x - 1} = \dfrac{\dfrac{1 - x}{x}}{x - 1} = \dfrac{(-1)(x - 1)}{x(x - 1)}$

Step 3: The derivative of f at 1 is

$$f'(1) = \lim_{x \to 1} \frac{f(x) - f(1)}{x - 1} = \lim_{x \to 1} \frac{(-1)\cancel{(x - 1)}}{x\cancel{(x - 1)}} = \lim_{x \to 1} \frac{-1}{x} = -1$$

25. First we find the difference quotient of $f(x) = 2x$.

$$\frac{f(x + h) - f(x)}{h} = \frac{2(x + h) - 2x}{h} = \frac{2x + 2h - 2x}{h} = \frac{2h}{h} = 2$$

↑   ↑
Simplify   Cancel the h's.

The derivative of f is the limit of the difference quotient as $h \to 0$, that is,

$$f'(x) = \lim_{h \to 0} \frac{f(x + h) - f(x)}{h} = \lim_{h \to 0} 2 = 2$$

27. First we find the difference quotient of $f(x) = 1 - 2x$.

$$\frac{f(x + h) - f(x)}{h} = \frac{[1 - 2(x + h)] - [1 - 2x]}{h} = \frac{1 - 2x - 2h - 1 + 2x}{h} = \frac{-2h}{h} = -2$$

↑   ↑
Simplify   Cancel the h's.

The derivative of f is the limit of the difference quotient as $h \to 0$, that is,

$$f'(x) = \lim_{h \to 0} \frac{f(x + h) - f(x)}{h} = \lim_{h \to 0} (-2) = -2$$

29. First we find the difference quotient of $f(x) = x^2 + 2$.

$$\frac{f(x + h) - f(x)}{h} = \frac{[(x + h)^2 + 2] - [x^2 + 2]}{h}$$

$$= \frac{x^2 + 2xh + h^2 + 2 - x^2 - 2}{h}$$

$$= \frac{2xh + h^2}{h} \qquad \text{Simplify.}$$

$$= \frac{h(2x + h)}{h} \qquad \text{Factor out } h.$$

$$= 2x + h \qquad \text{Cancel the } h\text{'s.}$$

The derivative of f is the limit of the difference quotient as $h \to 0$, that is,

$$f'(x) = \lim_{h \to 0} \frac{f(x+h) - f(x)}{h} = \lim_{h \to 0} (2x + h) = 2x$$

31.   First we find the difference quotient of $f(x) = 3x^2 - 2x + 1$.

$$\frac{f(x+h) - f(x)}{h} = \frac{\left[3(x+h)^2 - 2(x+h) + 1\right] - \left[3x^2 - 2x + 1\right]}{h}$$

$$= \frac{3x^2 + 6xh + 3h^2 - 2x - 2h + 1 - 3x^2 + 2x - 1}{h}$$

$$= \frac{6xh + 3h^2 - 2h}{h} \qquad \text{Simplify.}$$

$$= \frac{h(6x + 3h - 2)}{h} \qquad \text{Factor out } h.$$

$$= 6x + 3h - 2 \qquad \text{Cancel the } h\text{'s.}$$

The derivative of f is the limit of the difference quotient as $h \to 0$, that is,

$$f'(x) = \lim_{h \to 0} \frac{f(x+h) - f(x)}{h} = \lim_{h \to 0} (6x + 3h - 2) = 6x - 2$$

33.   First we find the difference quotient of $f(x) = x^3$.

$$\frac{f(x+h) - f(x)}{h} = \frac{(x+h)^3 - x^3}{h}$$

$$= \frac{x^3 + 3x^2h + 3xh^2 + h^3 - x^3}{h}$$

$$= \frac{3x^2h + 3xh^2 + h^3}{h} \qquad \text{Simplify.}$$

$$= \frac{h(3x^2 + 3xh + h^2)}{h} \qquad \text{Factor out } h.$$

$$= 3x^2 + 3xh + h^2 \qquad \text{Cancel the } h\text{'s.}$$

The derivative of f is the limit of the difference quotient as $h \to 0$, that is,

$$f'(x) = \lim_{h \to 0} \frac{f(x+h) - f(x)}{h} = \lim_{h \to 0} (3x^2 + 3xh + h^2) = 3x^2$$

35.   First we find the difference quotient of $f(x) = mx + b$.

$$\frac{f(x+h) - f(x)}{h} = \frac{\left[m(x+h) + b\right] - \left[mx + b\right]}{h}$$

$$= \frac{mx + mh + b - mx - b}{h}$$

$$= \frac{mh}{h} \qquad \text{Simplify}$$

$$= m \qquad \text{Cancel the } h\text{'s.}$$

The derivative of f is the limit of the difference quotient as $h \to 0$, that is,

$$f'(x) = \lim_{h \to 0} \frac{f(x+h) - f(x)}{h} = \lim_{h \to 0} m = m$$

37. (a) The average rate of change of $f(x) = 3x + 4$ as x changes from 1 to 3 is

$$\frac{\Delta f}{\Delta x} = \frac{f(3) - f(1)}{3 - 1} = \frac{(3 \cdot 3 + 4) - (3 \cdot 1 + 4)}{3 - 1} = \frac{13 - 7}{2} = \frac{6}{2} = 3$$

(b) The instantaneous rate of change at $x = 1$ is the derivative of f at 1.

$$f'(1) = \lim_{x \to 1} \frac{f(x) - f(1)}{x - 1} = \lim_{x \to 1} \frac{(3x + 4) - 7}{x - 1} = \lim_{x \to 1} \frac{3x - 3}{x - 1} = \lim_{x \to 1} \frac{3(x - 1)}{x - 1} = 3$$

The instantaneous rate of change of f at 1 is 3.

39. (a) The average rate of change of $f(x) = 3x^2 + 1$ as x changes from 1 to 3 is

$$\frac{\Delta f}{\Delta x} = \frac{f(3) - f(1)}{3 - 1} = \frac{(3 \cdot 3^2 + 1) - (3 \cdot 1^2 + 1)}{2} = \frac{28 - 4}{2} = \frac{24}{2} = 12$$

(b) The instantaneous rate of change at $x = 1$ is the derivative of f at 1.

$$f'(1) = \lim_{x \to 1} \frac{f(x) - f(1)}{x - 1} = \lim_{x \to 1} \frac{(3x^2 + 1) - (4)}{x - 1} = \lim_{x \to 1} \frac{3x^2 - 3}{x - 1}$$

$$= \lim_{x \to 1} \frac{3(x^2 - 1)}{x - 1} = \lim_{x \to 1} \frac{3(x - 1)(x + 1)}{x - 1} = \lim_{x \to 1} 3(x + 1) = 6$$

The instantaneous rate of change of f at 1 is 6.

41. (a) The average rate of change of $f(x) = x^2 + 2x$ as x changes from 1 to 3 is

$$\frac{\Delta f}{\Delta x} = \frac{f(3) - f(1)}{3 - 1} = \frac{(3^2 + 2 \cdot 3) - (1^2 + 2 \cdot 1)}{2} = \frac{15 - 3}{2} = \frac{12}{2} = 6$$

(b) The instantaneous rate of change at $x = 1$ is the derivative of f at 1.

$$f'(1) = \lim_{x \to 1} \frac{f(x) - f(1)}{x - 1} = \lim_{x \to 1} \frac{(x^2 + 2x) - (3)}{x - 1} = \lim_{x \to 1} \frac{x^2 + 2x - 3}{x - 1}$$

$$= \lim_{x \to 1} \frac{(x + 3)(x - 1)}{x - 1} = \lim_{x \to 1} (x + 3) = 4$$

The instantaneous rate of change of f at 1 is 4.

43. (a) The average rate of change of $f(x) = 2x^2 - x + 1$ as x changes from 1 to 3 is

$$\frac{\Delta f}{\Delta x} = \frac{f(3) - f(1)}{3 - 1} = \frac{(2 \cdot 3^2 - 3 + 1) - (2 \cdot 1^2 - 1 + 1)}{2} = \frac{16 - 2}{2} = \frac{14}{2} = 7$$

(b) The instantaneous rate of change at $x = 1$ is the derivative of f at 1.

$$f'(1) = \lim_{x \to 1} \frac{f(x) - f(1)}{x - 1} = \lim_{x \to 1} \frac{(2x^2 - x + 1) - (2)}{x - 1} = \lim_{x \to 1} \frac{2x^2 - x - 1}{x - 1}$$

$$= \lim_{x \to 1} \frac{(2x + 1)(x - 1)}{x - 1} = \lim_{x \to 1} (2x + 1) = 3$$

The instantaneous rate of change of f at 1 is 3.

45. The display below is from a TI-83 Plus graphing calculator.

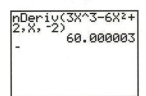

So $f'(-2) = 60$.

47. The display below is from a TI-83 Plus graphing calculator.

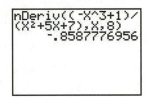

So $f'(8) = -0.85878$.

49. The display below is from a TI-83 Plus graphing calculator.

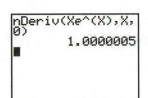

So $f'(0) = 1$.

51. The display below is from a TI-83 Plus graphing calculator.

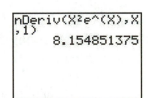

So $f'(1) = 8.15485$.

53. The display below is from a TI-83 Plus graphing calculator.

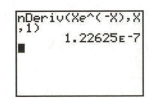

So $f'(1) = 0$.

55. We first find an equation of the tangent line to the graph of $y = x^2$ at $(1, 1)$.

The slope of the tangent line at $(1, 1)$ is

$$m_{\tan} = \lim_{x \to 1} \frac{f(x) - f(1)}{x - 1} = \lim_{x \to 1} \frac{x^2 - 1}{x - 1} = \lim_{x \to 1} \frac{(x+1)\cancel{(x-1)}}{\cancel{(x-1)}} = \lim_{x \to 1}(x+1) = 2$$

An equation of the tangent line is

$$y - 1 = 2(x - 1) \qquad \qquad y - f(c) = m_{\tan}(x - c)$$
$$y - 1 = 2x - 2 \qquad \qquad \text{Simplify.}$$
$$y = 2x - 1 \qquad \qquad \text{Add 1 to both sides.}$$

Now we see if the point $(2, 5)$ satisfies the equation of the tangent line.

$$2 \cdot 2 - 1 = 3 \qquad \qquad y = 2x - 1;\ x = 2,\ y = 5.$$
$$5 \neq 3$$

So the graph of the tangent line does not pass through the point $(2, 5)$.

57. For the rocket bomb to hit its target, the point $(1, 0)$ must be on the graph of the tangent line to the graph of $y = x^2$ at some point (c, c^2).

The slope of the tangent line at (c, c^2) is

$$\lim_{x \to c} \frac{x^2 - c^2}{x - c} = \lim_{x \to c} \frac{\cancel{(x-c)}(x+c)}{\cancel{x-c}} = \lim_{x \to c}(x+c) = 2c$$

An equation of the tangent line is

$$y - c^2 = 2c(x - c) \qquad \qquad y - f(c) = m_{\tan}(x - c)$$
$$y - c^2 = 2cx - 2c^2 \qquad \qquad \text{Simplify.}$$
$$y = 2cx - c^2 \qquad \qquad \text{Add } c^2 \text{ to both sides.}$$

The point $(1, 0)$ satisfies the equation of the tangent line, so

$$0 = 2c(1) - c^2 \qquad \qquad y = 2cx - c^2;\ x = 1;\ y = 0$$
$$c^2 - 2c = 0 \qquad \qquad \text{Simplify.}$$
$$c(c - 2) = 0 \qquad \qquad \text{Factor.}$$
$$c = 0 \qquad c = 2 \qquad \qquad \text{Apply the Zero-Product Property.}$$

Since the dive bomber is flying from right to left, the bomber reaches $c = 2$ first and should release the bomb at point $(2, 4)$.

59. (a) The average rate of change in sales S from day $x = 1$ to day $x = 5$ is

$$\frac{\Delta S}{\Delta x} = \frac{S(5) - S(1)}{5 - 1} = \frac{\left(4(5)^2 + 50(5) + 5000\right) - \left(4(1)^2 + 50(1) + 5000\right)}{4}$$
$$= \frac{5350 - 5054}{4} = \frac{296}{4} = 74 \text{ tickets per day.}$$

(b) The average rate of change in sales S from day $x = 1$ to day $x = 10$ is

$$\frac{\Delta S}{\Delta x} = \frac{S(10) - S(1)}{10 - 1} = \frac{\left(4(10)^2 + 50(10) + 5000\right) - \left(4(1)^2 + 50(1) + 5000\right)}{9}$$
$$= \frac{5900 - 5054}{9} = \frac{846}{9} = 94 \text{ tickets per day.}$$

(c) The average rate of change in sales S from day $x = 5$ to day $x = 10$ is

$$\frac{\Delta S}{\Delta x} = \frac{S(10) - S(5)}{10 - 5} = \frac{\left(4(10)^2 + 50(10) + 5000\right) - \left(4(5)^2 + 50(5) + 5000\right)}{5}$$

$$= \frac{5900 - 5350}{5} = \frac{550}{5} = 110 \text{ tickets per day.}$$

(d) The instantaneous rate of change in sales on day 5 is the derivative of S at $x = 5$.

$$S'(5) = \lim_{x \to 5} \frac{S(x) - S(5)}{x - 5} = \lim_{x \to 5} \frac{(4x^2 + 50x + 5000) - (5350)}{x - 5} = \lim_{x \to 5} \frac{4x^2 + 50x - 350}{x - 5}$$

$$= \lim_{x \to 5} \frac{2(2x + 35)(x - 5)}{x - 5} = \lim_{x \to 5} [2(2x + 35)] = 2 \lim_{x \to 5} (2x + 35) = 2 \cdot (10 + 35) = 90$$

The instantaneous rate of change of S on day 5 is 90 ticket sales per day.

(e) The instantaneous rate of change in sales on day 10 is the derivative of S at $x = 10$.

$$S'(10) = \lim_{x \to 10} \frac{S(x) - S(10)}{x - 10} = \lim_{x \to 10} \frac{(4x^2 + 50x + 5000) - (5900)}{x - 10} = \lim_{x \to 10} \frac{4x^2 + 50x - 900}{x - 10}$$

$$= \lim_{x \to 10} \frac{2(2x + 45)(x - 10)}{x - 10} = \lim_{x \to 10} [2(2x + 45)] = 2 \lim_{x \to 10} (2x + 45) = 2 \cdot 65 = 130$$

The instantaneous rate of change of S on day 10 is 130 ticket sales per day.

61. (a) At $x = \$10$ per crate, the farmer is willing to supply
$$S(10) = 50 \cdot 10^2 - 50 \cdot 10 = 4500 \text{ crates of grapefruits.}$$
$$\uparrow$$
$$S(x) = 50x^2 - 50x$$

(b) At $x = \$13$ per crate, the farmer is willing to supply
$$S(13) = 50 \cdot 13^2 - 50 \cdot 13 = 7800 \text{ crates of grapefruits.}$$

(c) The average rate of change in supply from \$10 to \$13 is
$$\frac{\Delta S}{\Delta x} = \frac{S(13) - S(10)}{13 - 10} = \frac{7800 - 4500}{3} = \frac{3300}{3} = 1100$$

The average rate of change in crates of grapefruit supplied is 1100 crates per dollar increase in price.

(d) The instantaneous rate of change in supply at $x = 10$ is the derivative $S'(10)$.

$$S'(10) = \lim_{x \to 10} \frac{S(x) - S(10)}{x - 10} = \lim_{x \to 10} \frac{[50x^2 - 50x] - [4500]}{x - 10} = \lim_{x \to 10} \frac{50(x^2 - x - 90)}{x - 10}$$

$$= \lim_{x \to 10} \frac{50(x - 10)(x + 9)}{x - 10} = \lim_{x \to 10} [50(x + 9)] = 50 \lim_{x \to 10} (x + 9) = 50 \cdot 19 = 950$$

The instantaneous rate of change in supply at $x = \$10$ is 950 crates.

(e) The average rate of change in supply over the price interval from \$10 to \$13 is 1100 crates of grapefruit per \$1.00 change in price.

The instantaneous rate of change in supply of 950 crates is the increase in supply of grapefruit as the price changes from \$10 to \$11.

63. (a) The marginal revenue is the derivative $R'(x)$.

$$R'(x) = \lim_{h \to 0} \frac{R(x+h) - R(x)}{h} = \lim_{h \to 0} \frac{\left[8(x+h) - (x+h)^2\right] - \left[8x - x^2\right]}{h}$$

$$= \lim_{h \to 0} \frac{8x + 8h - x^2 - 2xh - h^2 - 8x + x^2}{h} \qquad \text{Simplify.}$$

$$= \lim_{h \to 0} \frac{8h - 2xh - h^2}{h} \qquad \text{Simplify.}$$

$$= \lim_{h \to 0} \frac{\cancel{h}(8 - 2x - h)}{\cancel{h}} \qquad \text{Factor out the } h.$$

$$= \lim_{h \to 0} (8 - 2x - h) = 8 - 2x \qquad \text{Cancel the } h\text{'s. Go to the limit.}$$

The marginal revenue is $R'(x) = 8 - 2x$.

(b) The marginal cost is the derivative $C'(x)$.

$$C'(x) = \lim_{h \to 0} \frac{C(x+h) - C(x)}{h} = \lim_{h \to 0} \frac{\left[2(x+h) + 5\right] - \left[2x + 5\right]}{h}$$

$$= \lim_{h \to 0} \frac{2x + 2h + 5 - 2x - 5}{h} = \lim_{h \to 0} \frac{2\cancel{h}}{\cancel{h}} = \lim_{h \to 0} 2 = 2$$

(c) To find the break-even point we solve the equation $R(x) = C(x)$.

$$8x - x^2 = 2x + 5$$
$$x^2 - 6x + 5 = 0 \qquad \text{Put the quadratic equation in standard form.}$$
$$(x - 5)(x - 1) = 0 \qquad \text{Factor.}$$
$$x - 5 = 0 \quad x - 1 = 0 \qquad \text{Apply the Zero-Product Property.}$$
$$x = 5 \qquad x = 1 \qquad \text{Solve.}$$

There are two break-even points. One is when 1000 units are produced, and the other is when 5000 units are produced.

(d) To find the number x for which marginal revenue equals marginal cost, we solve the equation $R'(x) = C'(x)$.

$$8 - 2x = 2$$
$$2x = 6$$
$$x = 3$$

Marginal revenue equals marginal cost when 3000 units are produced and sold.

(e)

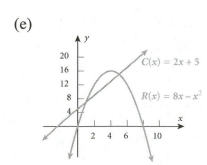

65. (a) The revenue function $R(x) = xp = x(-10x + 2000) = -10x^2 + 2000x$.

(b) The marginal revenue is the derivative $R'(x)$.

$$R'(x) = \lim_{h \to 0} \frac{R(x+h) - R(x)}{h} = \lim_{h \to 0} \frac{\left[-10(x+h)^2 + 2000(x+h)\right] - \left[-10x^2 + 2000x\right]}{h}$$

$$= \lim_{h \to 0} \frac{-10x^2 - 20xh - 10h^2 + 2000x + 2000h + 10x^2 - 2000x}{h} \qquad \text{Simplify.}$$

$$= \lim_{h \to 0} \frac{-20xh - 10h^2 + 2000h}{h} \qquad \text{Simplify.}$$

$$= \lim_{h \to 0} \frac{h(-20x - 10h + 2000)}{h} \qquad \text{Factor out an } h.$$

$$= \lim_{h \to 0} (-20x - 10h + 2000) \qquad \text{Cancel out the } h\text{'s.}$$

$$= -20x + 2000 \qquad \text{Go to the limit.}$$

(c) The marginal revenue at $x = 100$ tons is $R'(100) = (-20) \cdot 100 + 2000 = 0$ dollars.

(d) The average rate in change in revenue from $x = 100$ to $x = 101$ tons is

$$\frac{\Delta R}{\Delta x} = \frac{R(101) - R(100)}{101 - 100} = \frac{\left[(-10)(101^2) + 2000(101)\right] - \left[(-10)(100^2) + 2000(1000)\right]}{1}$$

$$= 99,990 - 100,000 = -10$$

(e) $R'(100) = 0$ indicates that there is no additional revenue gained by selling the 101st ton of cement.

The average rate of change in revenue from selling the 101st ton of cement represents a decrease in revenue of $10.

67. (a) The revenue function $R(x) = xp$, where p is the unit price and x is the number of units sold. $R(x) = xp = x(90 - 0.02x) = 90x - 0.02x^2$

(b) The marginal revenue is the derivative $R'(x)$.

$$R'(x) = \lim_{h \to 0} \frac{R(x+h) - R(x)}{h} = \lim_{h \to 0} \frac{\left[90(x+h) - 0.02(x+h)^2\right] - \left[90x - 0.02x^2\right]}{h}$$

$$= \lim_{h \to 0} \frac{90x + 90h - 0.02x^2 - 0.04xh - 0.02h^2 - 90x + 0.02x^2}{h} \qquad \text{Simplify.}$$

$$= \lim_{h \to 0} \frac{90h - 0.04xh - 0.02h^2}{h} \qquad \text{Simplify.}$$

$$= \lim_{h \to 0} \frac{h(90 - 0.04x - 0.02h)}{h} \qquad \text{Factor out an } h.$$

$$= \lim_{h \to 0} (90 - 0.04x - 0.02h) = 90 - 0.04x \qquad \text{Cancel the } h\text{'s; take the limit.}$$

(c) It costs $10 per unit to produce the product, so the cost function $C = C(x) = 10x$. The marginal cost is the derivative $C'(x)$.

$$C'(x) = \lim_{h \to 0} \frac{C(x+h) - C(x)}{h} = \lim_{h \to 0} \frac{\left[10(x+h)\right] - \left[10x\right]}{h}$$

$$= \lim_{h \to 0} \frac{10x + 10h - 10x}{h} = \lim_{h \to 0} \frac{10\cancel{h}}{\cancel{h}} = \lim_{h \to 0} 10 = 10$$

(d) A break-even point is a number x for which $R(x) = C(x)$. We solve the equation

$R(x) = C(x)$	
$90x - 0.02x^2 = 10x$	
$0.02x^2 - 80x = 0$	Put the quadratic equation in standard form.
$x^2 - 4000x = 0$	Multiply both sides by 50.
$x(x - 4000) = 0$	Factor.
$x = 0 \qquad x - 4000 = 0$	Apply the Zero-Product Property.
$x = 0 \qquad\quad x = 4000$	Solve for x.

There are two break-even points. One is when no units are produced and sold; the other is when $x = 4000$ units are produced and sold.

(e) The marginal revenue equals marginal cost when $R'(x) = C'(x)$.

$$R'(x) = C'(x)$$
$$90 - 0.04x = 10$$
$$80 = 0.04x$$
$$x = 2000$$

The marginal revenue equals marginal cost when 2000 units are produced and sold.

69. The instantaneous rate of change of the volume of the cylinder with respect to the radius r at $r = 3$ is the derivative $V'(3)$.

$$V'(3) = \lim_{r \to 3} \frac{V(r) - V(3)}{r - 3} = \lim_{r \to 3} \frac{3\pi r^2 - 3\pi(3^2)}{r - 3} = \lim_{r \to 3} \frac{3\pi r^2 - 27\pi}{r - 3} = \lim_{r \to 3} \frac{3\pi(r^2 - 9)}{r - 3}$$

$$= \lim_{r \to 3} \frac{3\pi(\cancel{r - 3})(r + 3)}{\cancel{r - 3}} = \lim_{r \to 3} [3\pi(r + 3)] = 18\pi \approx 56.55$$

$\qquad\qquad\qquad\qquad\qquad\qquad 732/1 - 69 \text{ (odd)}$

13.2 The Derivative of a Power Function; Sum and Difference Formulas

1. The function $f(x) = 4$ is a constant. $f'(x) = 0$

3. The function $f(x) = x^5$ is a power function. $f'(x) = 5x^{5-1} = 5x^4$

5. $f'(x) = \dfrac{d}{dx}(6x^2) = 6\dfrac{d}{dx}x^2 = 6 \cdot 2x = 12x$

7. $f'(t) = \dfrac{d}{dt}\left(\dfrac{t^4}{4}\right) = \dfrac{1}{4}\dfrac{d}{dt}t^4 = \dfrac{1}{4} \cdot 4t^{4-1} = t^3$

9. $f'(x) = \dfrac{d}{dx}(x^2 + x) = \dfrac{d}{dx}x^2 + \dfrac{d}{dx}x$ Use the derivative of a sum formula (Formula (5)).

$$= 2x + 1 \qquad\qquad\qquad\qquad \dfrac{d}{dx}x = 1$$

11. $f'(x) = \dfrac{d}{dx}(x^3 - x^2 + 1) = \dfrac{d}{dx}x^3 - \dfrac{d}{dx}x^2 + \dfrac{d}{dx}1$ Use Formulas (6) and (5).

$$= 3x^2 - 2x \qquad\qquad \dfrac{d}{dx}1 = 0$$

13. $f'(t) = \dfrac{d}{dt}(2t^2 - t + 4) = \dfrac{d}{dt}(2t^2) - \dfrac{d}{dt}t + \dfrac{d}{dt}4$ Use Formulas (5) and (6).

$$= 2\dfrac{d}{dt}t^2 - \dfrac{d}{dt}t + 0 \qquad \text{Use Formulas (4) and (2).}$$

$$= 4t - 1 \qquad\qquad\qquad \text{Differentiate.}$$

15. $f'(x) = \dfrac{d}{dx}\left(\dfrac{1}{2}x^8 + 3x + \dfrac{2}{3}\right) = \dfrac{d}{dx}\left(\dfrac{1}{2}x^8\right) + \dfrac{d}{dx}(3x) + \dfrac{d}{dx}\dfrac{2}{3}$ Use Formula (5).

$$= \dfrac{1}{2}\dfrac{d}{dx}x^8 + 3\dfrac{d}{dx}x + 0 \qquad \text{Use Formula (4); } \dfrac{d}{dx}\dfrac{2}{3} = 0.$$

$$= \dfrac{1}{2}\cdot 8x^7 + 3 = 4x^7 + 3 \qquad \text{Differentiate and simplify.}$$

17. $f'(x) = \dfrac{d}{dx}\left[\dfrac{1}{3}(x^5 - 8)\right] = \dfrac{1}{3}\dfrac{d}{dx}(x^5 - 8) = \dfrac{1}{3}\left[\dfrac{d}{dx}x^5 - \dfrac{d}{dx}8\right] = \dfrac{1}{3}[5x^4 - 0] = \dfrac{5}{3}x^4$

$$\qquad\qquad\qquad\quad \uparrow \qquad\qquad \uparrow \qquad\qquad\qquad \uparrow \qquad\qquad \uparrow$$
$$\qquad\quad \text{Use Formula (4).} \quad \text{Use Formula (6).} \quad \text{Differentiate.} \quad \text{Simplify.}$$

19. $f'(x) = \dfrac{d}{dx}[ax^2 + bx + c] = \dfrac{d}{dx}(ax^2) + \dfrac{d}{dx}(bx) + \dfrac{d}{dx}c = a\dfrac{d}{dx}x^2 + b\dfrac{d}{dx}x + 0$

$$\qquad\qquad\qquad\qquad \uparrow \qquad\qquad\qquad\qquad \uparrow$$
$$\qquad\qquad \text{Use Formula (5).} \qquad\qquad \text{Use Formula (4); } \dfrac{d}{dx}c = 0$$

$$= a\cdot 2x + b\cdot 1 = 2ax + b$$
$$\qquad\quad \uparrow \qquad\qquad \uparrow$$
$$\qquad \text{Differentiate.} \quad \text{Simplify.}$$

21. $\dfrac{d}{dx}(-6x^2 + x + 4) = \dfrac{d}{dx}(-6x^2) + \dfrac{d}{dx}x + \dfrac{d}{dx}4 = (-6)\dfrac{d}{dx}x^2 + 1 + 0 = (-6)\cdot 2x + 1 = -12x + 1$

23. $\dfrac{d}{dt}(-16t^2 + 80t) = \dfrac{d}{dt}(-16t^2) + \dfrac{d}{dt}(80t) = (-16)\dfrac{d}{dt}t^2 + 80\dfrac{d}{dt}t = (-16)\cdot 2t + 80\cdot 1 = -32t + 80$

25. $\dfrac{dA}{dr} = \dfrac{d}{dr}(\pi r^2) = \pi\dfrac{d}{dr}r^2 = \pi\cdot 2r = 2\pi r$

27. $\dfrac{dV}{dr} = \dfrac{d}{dr}\left(\dfrac{4}{3}\pi r^3\right) = \dfrac{4}{3}\pi\dfrac{d}{dr}r^3 = \dfrac{4}{3}\pi\cdot 3r^2 = 4\pi r^2$

29. To find $f'(-3)$, first we find the derivative of the function f.

$$f'(x) = \frac{d}{dx}(4x^2) = 8x$$

Then we substitute -3 for x.

$$f'(-3) = 8 \cdot (-3) = -24$$

31. To find $f'(4)$, first we find the derivative of the function f.

$$f'(x) = \frac{d}{dx}(2x^2 - x) = \frac{d}{dx}(2x^2) - \frac{d}{dx}x = 2\frac{d}{dx}x^2 - 1 = 4x - 1$$

Then we substitute 4 for x.

$$f'(4) = 4 \cdot 4 - 1 = 15$$

33. To find $f'(3)$, first we find the derivative of the function f.

$$f'(x) = \frac{d}{dt}\left(-\frac{1}{3}t^3 + 5t\right) = \frac{d}{dt}\left(-\frac{1}{3}t^3\right) + \frac{d}{dt}(5t) = -\frac{1}{3}\frac{d}{dt}t^3 + 5\frac{d}{dt}t = -\frac{1}{3} \cdot 3t^2 + 5 = -t^2 + 5$$

Then we substitute 3 for t.

$$f'(3) = -3^2 + 5 = -4$$

35. To find $f'(1)$, first we find the derivative of the function f.

$$f'(x) = \frac{d}{dx}\left[\frac{1}{2}(x^6 - x^4)\right] = \frac{1}{2}\frac{d}{dx}(x^6 - x^4) = \frac{1}{2}\left[\frac{d}{dx}x^6 - \frac{d}{dx}x^4\right]$$

$$= \frac{1}{2}(6x^5 - 4x^3) = 3x^5 - 2x^3$$

Then we substitute 1 for x.

$$f'(1) = 3 \cdot 1^5 - 2 \cdot 1^3 = 1$$

37. First we find the derivative of the function f. In Problem 19 we found $f'(x) = 2ax + b$.

So we now substitute $-\dfrac{b}{2a}$ for x.

$$f'\left(-\frac{b}{2a}\right) = 2a\left(-\frac{b}{2a}\right) + b = -b + b = 0$$

39. First we find the derivative $\dfrac{dy}{dx} = 4x^3$

Then we evaluate the derivative at the point $(1, 1)$ by substituting 1 for x.

$$\frac{dy}{dx} = 4 \cdot 1^3 = 4$$

41. First we find the derivative $\dfrac{dy}{dx} = 2x - 0 = 2x$

Then we evaluate the derivative at the point $(4, 2)$ by substituting 4 for x.

$$\frac{dy}{dx} = 2 \cdot 4 = 8$$

43. First we find the derivative $\dfrac{dy}{dx} = 6x - 1$

Then we evaluate the derivative at the point $(-1, 4)$ by substituting -1 for x.

$$\frac{dy}{dx} = 6 \cdot (-1) - 1 = -7$$

45. First we find the derivative $\dfrac{dy}{dx} = \dfrac{1}{2} \cdot 2x = x$

Then we evaluate the derivative at the point $\left(1, \dfrac{1}{2}\right)$ by substituting 1 for x.

$$\frac{dy}{dx} = 1$$

47. First we find the derivative $\dfrac{dy}{dx} = 0 - 2 + 3x^2 = -2 + 3x^2$

Then we evaluate the derivative at the point $(2, 6)$ by substituting 2 for x.

$$\frac{dy}{dx} = -2 + 3 \cdot 2^2 = 10$$

49. The slope of the tangent line to the graph of $f(x) = x^3 + 3x - 1$ at the point $(0, -1)$ is the derivative of the function f evaluated at the point $(0, -1)$. The derivative of f is

$$f'(x) = 3x^2 + 3$$
$$m_{\tan} = f'(0) = 3 \cdot (0)^2 + 3 = 3$$

An equation of the tangent line is

$$\begin{aligned} y - (-1) &= 3(x - 0) & y - f(c) &= m_{\tan}(x - c) \\ y + 1 &= 3x & &\text{Simplify.} \\ y &= 3x - 1 & &\text{Subtract 1 from both sides.} \end{aligned}$$

51. We first find the derivative $f'(x)$.

$$f'(x) = 6x - 12 + 0 = 6x - 12$$

We then solve the equation $f'(x) = 0$.

$$\begin{aligned} 6x - 12 &= 0 \\ 6x &= 12 \\ x &= 2 \end{aligned}$$

53. We first find the derivative $f'(x)$.

$$f'(x) = 3x^2 - 3 + 0 = 3x^2 - 3$$

We then solve the equation $f'(x) = 0$.

$$\begin{aligned} 3x^2 - 3 &= 0 & f'(x) &= 3x^2 - 3 \\ 3(x^2 - 1) &= 0 & &\text{Factor out the 3.} \\ x^2 - 1 &= 0 & &\text{Divide both sides by 3.} \\ x = 1 \qquad x &= -1 & &\text{Solve using the Square Root Method.} \end{aligned}$$

55. We first find the derivative $f'(x)$.
$$f'(x) = 3x^2 + 1$$

We then solve the equation $f'(x) = 0$.
$$3x^2 + 1 = 0 \qquad\qquad f'(x) = 3x^2 + 1$$
has no solutions.

57. The slope of the tangent line to the function $f(x) = 9x^3$ is the derivative of f.
$$f'(x) = 27x^2$$
To find the slope of the line $3x - y + 2 = 0$, we put the equation in slope-intercept form.
$$y = 3x + 2$$
The slope of the line is $m = 3$. For the tangent line to the graph of f to be parallel to the line y, $m_{\tan} = 3$. We solve the equation $f'(x) = 3$.
$$27x^2 = 3 \qquad\qquad f'(x) = 27x^2$$
$$x^2 = \frac{1}{9} \qquad\qquad \text{Divide both sides by 27.}$$
$$x = \frac{1}{3} \qquad x = -\frac{1}{3} \qquad \text{Solve using the Square Root Method.}$$
There are two points on the graph of the function $y = 9x^3$ for which the slope of the tangent line is parallel to the graph of the line $3x - y + 2 = 0$. They are
$$\left(\frac{1}{3},\, f\left(\frac{1}{3}\right)\right) = \left(\frac{1}{3}, \frac{1}{3}\right) \quad \text{and} \quad \left(-\frac{1}{3},\, f\left(-\frac{1}{3}\right)\right) = \left(-\frac{1}{3}, -\frac{1}{3}\right)$$

59. If $(x, y) = (x, 2x^2 - 4x + 1)$ is a point on the graph of the function $y = 2x^2 - 4x + 1$ for which the tangent line to the graph of y passes through the point $(1, -3)$, then the slope of the tangent line is given by
$$m_{\tan} = \frac{\Delta y}{\Delta x} = \frac{y_2 - y_1}{x_2 - x_1} = \frac{(2x^2 - 4x + 1) - (-3)}{x - 1} = \frac{2x^2 - 4x + 4}{x - 1}$$
The slope of the tangent line to the graph of the function y is also given by the derivative of y.
$$m_{\tan} = \frac{d}{dx}(2x^2 - 4x + 1) = 4x - 4$$
Since both these expressions for the slope must be same, we will set them equal to each other and solve for x.

$$\frac{2x^2 - 4x + 4}{x - 1} = 4x - 4 \qquad \text{Set the slope equal to the derivative.}$$
$$2x^2 - 4x + 4 = (4x - 4)(x - 1) \qquad \text{Multiply both sides by } x - 1.$$
$$2x^2 - 4x + 4 = 4x^2 - 8x + 4 \qquad \text{Multiply out the right side.}$$
$$2x^2 - 4x = 0 \qquad \text{Put the quadratic equation in standard form.}$$
$$2x(x - 2) = 0 \qquad \text{Factor.}$$
$$2x = 0 \quad x - 2 = 0 \qquad \text{Apply the Zero-Product Property.}$$
$$x = 0 \qquad x = 2 \qquad \text{Solve for } x.$$

We evaluate the tangent lines to the graph of y using each of these two values of x.

If $x = 0$ then $y = f(0) = 1$. The point on the graph of y is $(0, 1)$. The slope of the tangent line to the graph of y at $(0, 1)$ is $f'(0) = 4 \cdot 0 - 4 = -4$. The equation of the tangent line through $(0, 1)$ and $(1, -3)$ is

$$y - 1 = -4(x - 0)$$
$$y = -4x + 1$$

If $x = 2$ then $y = f(2) = 2 \cdot 2^2 - 4 \cdot 2 + 1 = 1$. The point on the graph of y is $(2, 1)$. The slope of the tangent line to the graph of y at $(2, 1)$ is $f'(2) = 4 \cdot 2 - 4 = 4$. The equation of the tangent line through $(2, 1)$ and $(1, -3)$ is

$$y - 1 = 4(x - 2)$$

$$y - 1 = 4x - 8$$
$$y = 4x - 7$$

61. (a) The average cost of producing 10 additional pairs of eyeglasses is

$$\frac{\Delta y}{\Delta x} = \frac{C(110) - C(100)}{110 - 100} = \frac{\left[0.2 \cdot 110^2 + 3 \cdot 110 + 1000\right] - \left[0.2 \cdot 100^2 + 3 \cdot 100 + 1000\right]}{10}$$

$$= \frac{3750 - 3300}{10} = \frac{450}{10} = 45 \text{ dollars per pair.}$$

(b) The marginal cost of producing an additional pair of eyeglasses is
$$C'(x) = 0.4x + 3$$

(c) The marginal cost at $x = 100$ is $C'(100) = 0.4(100) + 3 = 43$ dollars.

(d) The marginal cost at $x = 100$ is the cost of producing one additional pair of eyeglasses when 100 pairs have already been produced.

63. (a) The derivative $V'(R) = 4kR^3$.

(b) $V'(0.3) = 4k \cdot 0.3^3 = 0.108k$ centimeters cubed

(c) $V'(0.4) = 4k \cdot 0.4^3 = 0.256k$ centimeters cubed

65. (a) It costs $C(40) = 2000 + 50 \cdot 40 - 0.05 \cdot 40^2 = 3920$ dollars to produce 40 microwave ovens.

(b) The marginal cost function is the derivative of $C(x)$.
$$C'(x) = 50 - 0.10x$$

(c) $C'(40) = 50 - 0.10 \cdot 40 = 46$. The marginal cost at $x = 40$ indicates that the cost of producing the 41st microwave oven is \$46.00.

(d) An estimate of the cost of producing 41 microwave ovens can be obtained by adding $C(40)$ and $C'(40)$.
$$C(40) + C'(40) = 3920 + 46 = 3966 \text{ dollars to produce 41 microwaves.}$$

(e) The actual cost of producing 41 microwave ovens is
$$C(41) = 2000 + 50 \cdot 41 - 0.05 \cdot 41^2 = 3965.95 \text{ dollars.}$$
The actual cost of producing 41 microwave ovens is \$0.05 less than the estimated cost.

(f) The actual cost of producing the 41st microwave oven is $45.95.
$$C(41) - C(40) = 3965.95 - 3920 = \$45.95$$

(g) The average cost function for producing x microwave ovens is
$$\overline{C}(x) = \frac{2000 + 50x - 0.05x^2}{x} = \frac{2000}{x} + 50 - 0.05x$$

(h) The average cost of producing 41 microwave ovens is $\overline{C}(41) = \$96.73$
$$\overline{C}(41) = \frac{2000 + 50 \cdot 41 - 0.05 \cdot 41^2}{41} = 96.73$$

67. The marginal price of beans in year t is the derivative of $p(t)$.
$$p'(t) = 0.021t^2 - 1.26t + 0.005$$
(a) The marginal price of beans in 1995 is $p'(2)$, since $t = 0$ represents 1993, $t = 2$ represents 1995.
$$p'(2) = 0.021 \cdot 2^2 - 1.26 \cdot 2 + 0.005 = -2.431$$

(b) The marginal price of beans in 2002 is $p'(9)$, since $t = 0$ represents 1993, $t = 9$ represents 2002.
$$p'(9) = 0.021 \cdot 9^2 - 1.26 \cdot 9 + 0.005 = -9.634$$

(c) Answers will vary.

69. The instantaneous rate of change of the volume V of a sphere with respect to its radius r when $r = 2$ feet is $V'(2)$.
$$V(r) = \frac{4}{3}\pi r^3$$
$$V'(r) = 4\pi r^2$$
$$V'(2) = 4\pi \cdot 2^2 = 16\pi \text{ feet cubed}$$

71. The instantaneous rate of change of work output at time t is the derivative of $A(t)$.
$$A'(t) = 3a_3 t^2 + 2a_2 t + a_1$$

73. Formula (3) from Section 2 of the text is $\dfrac{d}{dx}x^n = nx^{n-1}$

To prove this formula we begin with the difference quotient.
$$\frac{d}{dx}x^n = f'(x) = \lim_{h \to 0} \frac{f(x+h) - f(x)}{h}$$

$$= \lim_{h \to 0} \frac{(x+h)^n - x^n}{h} \qquad \text{Use the difference quotient.}$$

$$= \lim_{h \to 0} \frac{\left[x^n + nx^{n-1}h + \dfrac{n(n-1)}{2}x^{n-2}h^2 + \ldots + h^n \right] - \left[x^n \right]}{h} \qquad \text{Use the hint provided.}$$

$$= \lim_{h \to 0} \frac{nx^{n-1}h + \frac{n(n-1)}{2}x^{n-2}h^2 + \ldots + h^n}{h}$$ Simplify.

$$= \lim_{h \to 0} \frac{h\left(nx^{n-1} + \frac{n(n-1)}{2}x^{n-2}h + \ldots + h^{n-1}\right)}{h}$$ Factor out an h.

$$= \lim_{h \to 0} \left(nx^{n-1} + \frac{n(n-1)}{2}x^{n-2}h + \ldots + h^{n-1}\right)$$ Cancel the h's.

$$= nx^{n-1}$$ Find the limit.

13.3 Product and Quotient Formulas 741/11-25, 741/27-51 (ODD)

1. The function f is the product of two functions $g(x) = 2x + 1$ and $h(x) = 4x - 3$ so that using the formula for the derivative of a product, we have

$$f'(x) = (2x+1)\left[\frac{d}{dx}(4x-3)\right] + (4x-3)\left[\frac{d}{dx}(2x+1)\right]$$ Derivative of a product formula.

$$= (2x + 1)(4) + (4x - 3)(2)$$ Differentiate.
$$= 8x + 4 + 8x - 6$$ Simplify.
$$= 16x - 2$$ Simplify.

3. The function f is the product of two functions $g(t) = t^2 + 1$ and $h(t) = t^2 - 4$ so that using the formula for the derivative of a product, we have

$$f'(t) = (t^2 + 1)\left[\frac{d}{dt}(t^2 - 4)\right] + (t^2 - 4)\left[\frac{d}{dt}(t^2 + 1)\right]$$ Derivative of a product formula.

$$= (t^2 + 1)(2t) + (t^2 - 4)(2t)$$ Differentiate.
$$= 2t^3 + 2t + 2t^3 - 8t$$ Simplify.
$$= 4t^3 - 6t$$ Simplify.

5. The function f is the product of two functions $g(x) = 3x - 5$ and $h(x) = 2x^2 + 1$ so that using the formula for the derivative of a product, we have

$$f'(x) = (3x-5)\left[\frac{d}{dx}(2x^2 + 1)\right] + (2x^2 + 1)\left[\frac{d}{dx}(3x - 5)\right]$$ Derivative of a product formula.

$$= (3x - 5)(4x) + (2x^2 + 1)(3)$$ Differentiate.
$$= 12x^2 - 20x + 6x^2 + 3$$ Simplify.
$$= 18x^2 - 20x + 3$$ Simplify.

7. The function f is the product of two functions $g(x) = x^5 + 1$ and $h(x) = 3x^3 + 8$ so that using the formula for the derivative of a product, we have

$$f'(x) = (x^5 + 1)\left[\frac{d}{dx}(3x^3 + 8)\right] + (3x^3 + 8)\left[\frac{d}{dx}(x^5 + 1)\right]$$ Derivative of a product formula.

$$= (x^5 + 1)(9x^2) + (3x^3 + 8)(5x^4)$$ Differentiate.
$$= 9x^7 + 9x^2 + 15x^7 + 40x^4$$ Simplify.
$$= 24x^7 + 40x^4 + 9x^2$$ Simplify.

9. The function f is the quotient of two functions $g(x) = x$ and $h(x) = x + 1$. We use the formula for the derivative of a quotient to get

$$\frac{d}{dx}\left(\frac{x}{x+1}\right) = \frac{(x+1)\dfrac{d}{dx}x - x\dfrac{d}{dx}(x+1)}{(x+1)^2}$$ Derivative of a quotient formula.

$$= \frac{(x+1)(1)-x(1)}{(x+1)^2}$$ Differentiate.

$$= \frac{x+1-x}{(x+1)^2}$$ Simplify.

$$= \frac{1}{(x+1)^2}$$ Simplify.

11. The function f is the quotient of two functions $g(x) = 3x + 4$ and $h(x) = 2x - 1$. We use the formula for the derivative of a quotient to get

$$\frac{d}{dx}\left(\frac{3x+4}{2x-1}\right) = \frac{(2x-1)\dfrac{d}{dx}(3x+4)-(3x+4)\dfrac{d}{dx}(2x-1)}{(2x-1)^2}$$ Derivative of a quotient formula.

$$= \frac{(2x-1)(3)-(3x+4)(2)}{(2x-1)^2}$$ Differentiate.

$$= \frac{6x-3-6x-8}{(2x-1)^2}$$ Simplify.

$$= -\frac{11}{(2x-1)^2}$$ Simplify.

13. The function f is the quotient of two functions $g(x) = x^2$ and $h(x) = x - 4$. We use the formula for the derivative of a quotient to get

$$\frac{d}{dx}\left(\frac{x^2}{x-4}\right) = \frac{(x-4)\dfrac{d}{dx}x^2 - x^2\dfrac{d}{dx}(x-4)}{(x-4)^2}$$ Derivative of a quotient formula.

$$= \frac{(x-4)(2x)-x^2(1)}{(x-4)^2}$$ Differentiate.

$$= \frac{2x^2-8x-x^2}{(x-4)^2}$$ Simplify.

$$= \frac{x^2-8x}{(x-4)^2}$$ Simplify.

15. The function f is the quotient of two functions $g(x) = 2x + 1$ and $h(x) = 3x^2 + 4$. We use the formula for the derivative of a quotient to get

$$\frac{d}{dx}\left(\frac{2x+1}{3x^2+4}\right) = \frac{(3x^2+4)\dfrac{d}{dx}(2x+1)-(2x+1)\dfrac{d}{dx}(3x^2+4)}{(3x^2+4)^2}$$ Derivative of a quotient formula.

$$= \frac{(3x^2+4)(2)-(2x+1)(6x)}{(3x^2+4)^2} \qquad \text{Differentiate.}$$

$$= \frac{6x^2+8-12x^2-6x}{(3x^2+4)^2} \qquad \text{Simplify.}$$

$$= -\frac{6x^2+6x-8}{(3x^2+4)^2} \qquad \text{Simplify.}$$

17. $\dfrac{d}{dx}\dfrac{-2}{t^2} = \dfrac{d}{dx}(-2t^{-2}) = -2\dfrac{d}{dx}t^{-2} = -2(-2t^{-3}) = \dfrac{4}{t^3}$

19. $\dfrac{d}{dx}\left(1+\dfrac{1}{x}+\dfrac{1}{x^2}\right) = \dfrac{d}{dx}(1+x^{-1}+x^{-2}) = \dfrac{d}{dx}1 + \dfrac{d}{dx}x^{-1} + \dfrac{d}{dx}x^{-2} = 0 - 1x^{-2} - 2x^{-3} = -\dfrac{1}{x^2} - \dfrac{2}{x^3}$

21. The slope of the tangent line to the function f at $(1, 2)$ is the derivative of f at $x = 1$.

$$f'(x) = \frac{d}{dx}(x^3-2x+2)(x+1)$$

$$= (x^3-2x+2)\frac{d}{dx}(x+1)+(x+1)\frac{d}{dx}(x^3-2x+2) \qquad \text{Derivative of a product.}$$

$$= (x^3-2x+2)(1)+(x+1)(3x^2-2) \qquad \text{Differentiate.}$$

$$= x^3-2x+2+3x^3+3x^2-2x-2 \qquad \text{Simplify.}$$

$$= 4x^3+3x^2-4x \qquad \text{Simplify.}$$

$$m_{\tan} = f'(1) = 3$$

An equation of the tangent line is

$$y - 2 = 3(x - 1) \qquad y - y_1 = m(x - x_1)$$
$$y - 2 = 3x - 3 \qquad \text{Simplify.}$$
$$y = 3x - 1 \qquad \text{Add 2 to both sides.}$$

23. The slope of the tangent line to the function f at $\left(1, \dfrac{1}{2}\right)$ is the derivative of f at $x = 1$.

$$f'(x) = \frac{d}{dx}\frac{x^3}{x+1} = \frac{(x+1)\dfrac{d}{dx}x^3 - x^3\dfrac{d}{dx}(x+1)}{(x+1)^2} \qquad \text{Derivative of a quotient.}$$

$$= \frac{(x+1)(3x^2)-x^3(1)}{(x+1)^2} \qquad \text{Differentiate.}$$

$$= \frac{3x^3+3x^2-x^3}{(x+1)^2} \qquad \text{Simplify.}$$

$$= \frac{2x^3+3x^2}{(x+1)^2} \qquad \text{Simplify.}$$

$$m_{\tan} = f'(1) = \frac{2(1^3)+3(1^2)}{(1+1)^2} = \frac{5}{4}$$

An equation of the tangent line is

$$y - \frac{1}{2} = \frac{5}{4}(x-1) \qquad\qquad y - y_1 = m(x - x_1)$$

$$y - \frac{1}{2} = \frac{5}{4}x - \frac{5}{4} \qquad\qquad \text{Simplify.}$$

$$y = \frac{5}{4}x - \frac{3}{4} \qquad\qquad \text{Add } \frac{1}{2} \text{ to both sides.}$$

25. We first find the derivative $f'(x)$.

$$f'(x) = \frac{d}{dx}\left[(x^2 - 2)(2x-1)\right] = (x^2 - 2)\frac{d}{dx}(2x-1) + (2x-1)\frac{d}{dx}(x^2 - 2)$$

$$= (x^2 - 2)(2) + (2x-1)(2x)$$

$$= 2x^2 - 4 + 4x^2 - 2x = 6x^2 - 2x - 4$$

We then solve the equation $f'(x) = 0$.

$$6x^2 - 2x - 4 = 0 \qquad\qquad f'(x) = 6x^2 - 2x - 4$$

$$3x^2 - 1x - 2 = 0$$

$$(3x+2)(x-1) = 0 \qquad\qquad \text{Factor.}$$

$$x = \frac{2}{3} \quad x = 1 \qquad\qquad \text{Apply the Zero-Product Property.}$$

27. We first find the derivative $f'(x)$.

$$f'(x) = \frac{d}{dx}\frac{x^2}{x+1} = \frac{(x+1)\frac{d}{dx}x^2 - x^2\frac{d}{dx}(x+1)}{(x+1)^2}$$

$$= \frac{(x+1)(2x) - x^2(1)}{(x+1)^2}$$

$$= \frac{2x^2 + 2x - x^2}{(x+1)^2} = \frac{x^2 + 2x}{(x+1)^2}$$

We then solve the equation $f'(x) = 0$.

$$\frac{x^2 + 2x}{(x+1)^2} = 0 \qquad\qquad f'(x) = \frac{x^2 + 2x}{(x+1)^2}$$

$$x^2 + 2x = 0 \qquad\qquad \text{Multiply both sides by } (x+2)^2;\ x \neq -2.$$

$$x(x+2) = 0 \qquad\qquad \text{Factor.}$$

$$x = 0 \quad x + 2 = 0 \qquad\qquad \text{Apply the Zero-Product Property.}$$

$$x = -2 \qquad\qquad \text{Solve for } x.$$

29. y is the product of two functions, we will use the formula for the derivative of a product.

$$y' = \frac{d}{dx}\left[x^2(3x-2)\right] = x^2\frac{d}{dx}(3x-2) + (3x-2)\frac{d}{dx}x^2$$

$$= x^2(3) + (3x-2)(2x) \qquad\qquad \text{Differentiate.}$$

$$= 3x^2 + 6x^2 - 4x \qquad\qquad \text{Simplify.}$$

$$= 9x^2 - 4x \qquad\qquad \text{Simplify.}$$

31. y is the product of two functions, we will use the formula for the derivative of a product.

$$y' = \frac{d}{dx}\left[(x^2+4)(4x^2+3)\right] = (x^2+4)\frac{d}{dx}(4x^2+3) + (4x^2+3)\frac{d}{dx}(x^2+4)$$

$$= (x^2+4)(8x) + (4x^2+3)(2x) \qquad \text{Differentiate.}$$
$$= 8x^3 + 32x + 8x^3 + 6x \qquad \text{Simplify.}$$
$$= 16x^3 + 38x \qquad \text{Simplify.}$$

33. y is the quotient of two functions, we will use the formula for the derivative of a quotient.

$$y' = \frac{d}{dx}\frac{2x+3}{3x+5} = \frac{(3x+5)\frac{d}{dx}(2x+3) - (2x+3)\frac{d}{dx}(3x+5)}{(3x+5)^2}$$

$$= \frac{(3x+5)(2) - (2x+3)(3)}{(3x+5)^2} \qquad \text{Differentiate.}$$

$$= \frac{6x+10-6x-9}{(3x+5)^2} \qquad \text{Simplify.}$$

$$= \frac{1}{(3x+5)^2} \qquad \text{Simplify.}$$

35. y is the quotient of two functions, so we will use the formula for the derivative of a quotient.

$$y' = \frac{d}{dx}\frac{x^2}{x^2-4} = \frac{(x^2-4)\frac{d}{dx}x^2 - x^2\frac{d}{dx}(x^2-4)}{(x^2-4)^2}$$

$$= \frac{(x^2-4)(2x) - x^2(2x)}{(x^2-4)^2} \qquad \text{Differentiate.}$$

$$= \frac{2x^3 - 8x - 2x^3}{(x^2-4)^2} \qquad \text{Simplify.}$$

$$= -\frac{8x}{(x^2-4)^2} \qquad \text{Simplify.}$$

37. y is the quotient of two functions, but its numerator is the product of two more functions. So we will use the formula for the derivative of a quotient and when differentiating the numerator the formula for the derivative of a product.

$$y' = \frac{d}{dx}\frac{(3x+4)(2x-3)}{2x+1} = \frac{(2x+1)\frac{d}{dx}\left[(3x+4)(2x-3)\right] - \left[(3x+4)(2x-3)\right]\frac{d}{dx}(2x+1)}{(2x+1)^2}$$

$$= \frac{(2x+1)\left[(3x+4)\frac{d}{dx}(2x-3) + (2x-3)\frac{d}{dx}(3x+4)\right] - (3x+4)(2x-3)\frac{d}{dx}(2x+1)}{(2x+1)^2}$$

$$= \frac{(2x+1)\left[(3x+4)(2)+(2x-3)(3)\right]-(3x+4)(2x-3)(2)}{(2x+1)^2} \qquad \text{Differentiate.}$$

$$= \frac{(2x+1)\left[6x+8+6x-9\right]-\left(6x^2+8x-9x-12\right)(2)}{(2x+1)^2} \qquad \text{Simplify.}$$

$$= \frac{(2x+1)(12x-1)-\left(6x^2-x-12\right)(2)}{(2x+1)^2} \qquad \text{Add like terms.}$$

$$= \frac{24x^2-2x+12x-1-12x^2+2x+24}{(2x+1)^2} \qquad \text{Multiply.}$$

$$= \frac{12x^2+12x+23}{(2x+1)^2} \qquad \text{Simplify.}$$

39. y is the quotient of two functions, so we will use the formula for the derivative of a quotient.

$$y' = \frac{d}{dx}\frac{4x^3}{x^2+4} = \frac{\left(x^2+4\right)\dfrac{d}{dx}\left(4x^3\right)-4x^3\dfrac{d}{dx}\left(x^2+4\right)}{\left(x^2+4\right)^2}$$

$$= \frac{\left(x^2+4\right)\left(12x^2\right)-4x^3(2x)}{\left(x^2+4\right)^2} \qquad \text{Differentiate.}$$

$$= \frac{12x^4+48x^2-8x^4}{\left(x^2+4\right)^2} \qquad \text{Simplify.}$$

$$= \frac{4x^4+48x^2}{\left(x^2+4\right)^2} \qquad \text{Simplify.}$$

41. (a) The average change in value from $t=2$ to $t=5$ is

$$\frac{\Delta V}{\Delta t} = \frac{V(5)-V(2)}{5-2} = \frac{\left[\dfrac{10,000}{5}+6000\right]-\left[\dfrac{10,000}{2}+6000\right]}{3}$$

$$= = \frac{2000-5000}{3} = \frac{-3000}{3} = -1000$$

(b) The instantaneous rate of change in value is the derivative of function V.

$$V(t) = \frac{10,000}{t}+6000 = 10,000t^{-1}+6000$$

$$V'(t) = -10,000t^{-2} = -\frac{10,000}{t^2}$$

(c) The instantaneous rate of change after 2 years is $V'(2)$.

$$V'(2) = -\frac{10,000}{2^2} = -\frac{10,000}{4} = -2500$$

(d) The instantaneous rate of change after 5 years is $V'(5)$.
$$V'(5) = -\frac{10,000}{5^2} = -\frac{10,000}{25} = -400$$

(e) Answers will vary.

43.   (a) The revenue function R is the product of the unit price and the number of units sold.
$$R = R(x) = px = \left(10 + \frac{40}{x}\right)x = 10x + 40$$

(b) The marginal revenue is the derivative of the revenue function R.
$$R'(x) = 10$$

(c) The marginal revenue when $x = 4$, is $R'(4) = 10$.

(d) The marginal revenue when $x = 6$, is $R'(6) = 10$

45.   (a) $D'(p) = \dfrac{d}{dp}\dfrac{100,000}{p^2 + 10p + 50} = \dfrac{(p^2 + 10p + 50)\dfrac{d}{dp}(100,000) - 100,000\dfrac{d}{dp}(p^2 + 10p + 50)}{(p^2 + 10p + 50)^2}$

$$= \frac{(p^2 + 10p + 50)(0) - 100,000(2p + 10)}{(p^2 + 10p + 50)^2} \qquad \frac{d}{dp}100,000 = 0$$

$$= \frac{-200,000p - 1,000,000}{(p^2 + 10p + 50)^2}$$

(b) $D'(5) = \dfrac{-200,000(5) - 1,000,000}{(5^2 + 10(5) + 50)^2} = -128$

$D'(10) = \dfrac{-200,000(10) - 1,000,000}{(10^2 + 10(10) + 50)^2} = -48$

$D'(15) = \dfrac{-200,000(15) - 1,000,000}{(15^2 + 10(15) + 50)^2} = -22.145$

47.   The rate at which the population is growing is given by the derivative of the population function P.

$$P'(t) = \frac{d}{dt}1000\left(1 + \frac{4t}{100 + t^2}\right) = 1000\left[\frac{d}{dt}1 + \frac{d}{dt}\frac{4t}{100 + t^2}\right]$$

$\dfrac{d}{dx}(cf(x)) = c\dfrac{d}{dx}f(x)$;
The derivative of a sum is the sum of the derivatives

$$= 1000\left[\frac{(100 + t^2)\dfrac{d}{dt}(4t) - 4t\dfrac{d}{dt}(100 + t^2)}{(100 + t^2)^2}\right]$$

Derivative of a quotient formula.

$$= 1000\left[\frac{(100 + t^2)(4) - 4t(2t)}{(100 + t^2)^2}\right]$$

Differentiate.

$$= 1000 \left[\frac{400 + 4t^2 - 8t^2}{\left(100 + t^2\right)^2} \right] = \frac{4000\left(100 - t^2\right)}{\left(100 + t^2\right)^2} \qquad \text{Simplify.}$$

In Parts (a) – (d), we evaluate the derivative $P'(t)$ at the indicated time.

(a) $t = 1$ hour, $P'(1) = \dfrac{4000\left(100 - 1^2\right)}{\left(100 + 1^2\right)^2} = 38.820$

(b) $t = 2$ hours, $P'(2) = \dfrac{4000\left(100 - 2^2\right)}{\left(100 + 2^2\right)^2} = 35.503$

(c) $t = 3$ hours, $P'(3) = \dfrac{4000\left(100 - 3^2\right)}{\left(100 + 3^2\right)^2} = 30.637$

(d) $t = 4$ hours, $P'(4) = \dfrac{4000\left(100 - 4^2\right)}{\left(100 + 4^2\right)^2} = 24.970$

49.   First we must find the function I that describes the intensity of light with respect to the distance r of the object from the source of the light. Since I is inversely related to the square of distance, and we are told that $I = 1000$ units when $r = 1$ meter, we solve for the constant of proportionality k.

$$I(1) = \frac{k}{1^2} = 1000 \qquad\qquad I(r) = \frac{k}{r^2}$$

$$k = 1000 \qquad\qquad \text{Solve for } k.$$

So, we have

$$I = I(r) = \frac{1000}{r^2}$$

The rate of change of intensity with respect to distance r is the derivative $I'(r)$ of the function I.

$$I'(r) = \frac{dI}{dr} = \frac{d}{dr} \frac{1000}{r^2} = \frac{d}{dr}\left(1000 r^{-2}\right) = 1000 \frac{d}{dr} r^{-2} = 1000\left(-2r^{-3}\right) = -\frac{2000}{r^3}$$

When $r = 10$ meters the rate of change of the intensity of the light is

$$I'(10) = -\frac{2000}{10^3} = -\frac{2000}{1000} = -2 \text{ units per meter.}$$

51.   (a) The marginal cost is the derivative of the cost function C.

$$C'(x) = \frac{d}{dx}\left(100 + \frac{x}{10} + \frac{36,000}{x}\right) = \frac{d}{dx}\left(100 + \frac{x}{10} + 36,000 x^{-1}\right)$$

$$= \frac{d}{dx} 100 + \frac{1}{10} \cdot \frac{d}{dx} x + 36,000 \frac{d}{dx} x^{-1}$$

The derivative of a sum is the sum of the derivatives;

$$\frac{d}{dx}\left(cf(x)\right) = c\frac{d}{dx} f(x)$$

$$= 0 + \frac{1}{10} + 36,000\left(-1x^{-2}\right) \qquad \text{Differentiate.}$$

$$= \frac{1}{10} - \frac{36,000}{x^2}$$

In Parts (b) – (d), the derivative $C'(x)$ is evaluated at the indicated ground speeds.

(b) ground speed $x = 500$ mph $\qquad C'(500) = \dfrac{1}{10} - \dfrac{36,000}{500^2} = -0.044$

(c) ground speed $x = 550$ mph $\qquad C'(550) = \dfrac{1}{10} - \dfrac{36,000}{550^2} = -0.019$

(d) ground speed $x = 450$ mph $\qquad C'(450) = \dfrac{1}{10} - \dfrac{36,000}{450^2} = -0.078$

53. (a) First we find the derivative of the function S with respect to reward r.

$$S'(r) = \frac{d}{dr}\frac{ar}{g-r} = \frac{(g-r)\dfrac{d}{dr}(ar) - ar\dfrac{d}{dr}(g-r)}{(g-r)^2} \qquad \text{Use the derivative of a quotient formula.}$$

$$= \frac{(g-r)(a) - ar(-1)}{(g-r)^2} \qquad \text{Differentiate.}$$

$$= \frac{ag - ar + ar}{(g-r)^2} = \frac{ag}{(g-r)^2} \qquad \text{Simplify.}$$

Since both a and g are constants for a given individual $k = ag$, and $S'(r)$ is of the form,

$S'(r) = \dfrac{k}{(g-r)^2}$ which is inversely proportional to the square of the difference between

the personal goal of the individual and the amount of reward received.

(b) Answers may vary.

13.4 The Power Rule   747/1 – 27, 29 – 33 (ODD)

1. $f'(x) = \dfrac{d}{dx}f(x) = \dfrac{d}{dx}(2x-3)^4 = 4(2x-3)^3 \dfrac{d}{dx}(2x-3) = 4(2x-3)^3(2) = 8(2x-3)^3$

$\qquad\qquad\qquad\qquad$ ↑ $\qquad\qquad\qquad$ ↑ $\qquad\qquad$ ↑
$\qquad\qquad\qquad$ Use the Power Rule. $\qquad$ Differentiate. $\qquad$ Simplify.

3. $f'(x) = \dfrac{d}{dx}f(x) = \dfrac{d}{dx}(x^2+4)^3 = 3(x^2+4)^2 \dfrac{d}{dx}(x^2+4) = 3(x^2+4)^2(2x) = 6x(x^2+4)^2$

$\qquad\qquad\qquad\qquad$ ↑ $\qquad\qquad\qquad$ ↑ $\qquad\qquad$ ↑
$\qquad\qquad\qquad$ Use the Power Rule. $\qquad$ Differentiate. $\qquad$ Simplify.

5. $f'(x) = \dfrac{d}{dx}f(x) = \dfrac{d}{dx}(3x^2+4)^2 = 2(3x^2+4)\dfrac{d}{dx}(3x^2+4) = 2(3x^2+4)(6x)$

$\qquad\qquad\qquad\qquad$ ↑ $\qquad\qquad\qquad$ ↑
$\qquad\qquad\qquad$ Use the Power Rule. $\qquad$ Differentiate.

$$= 12x(3x^2 + 4) = 36x^3 + 48x$$

↑ Simplify. ↑ Simplify.

7. The function f is the product of x and $(x+1)^3$. We begin by using the formula for the derivative of a product. That is,

$$f'(x) = \frac{d}{dx} f(x) = x\frac{d}{dx}(x+1)^3 + (x+1)^3 \frac{d}{dx}x$$

We continue by using the Power Rule:

$$f'(x) = x\left[3(x+1)^2 \frac{d}{dx}(x+1)\right] + (x+1)^3 \frac{d}{dx}x$$

$$= x\left[3(x+1)^2 \cdot 1\right] + (x+1)^3 \cdot 1 \qquad \text{Differentiate.}$$

$$= 3x(x+1)^2 + (x+1)^3 \qquad \text{Simplify.}$$

$$= (x+1)^2 \left[3x + (x+1)\right] \qquad \text{Factor.}$$

$$= (x+1)^2 (4x+1) \qquad \text{Simplify.}$$

9. The function f is the product of $4x^2$ and $(2x+1)^4$. We begin by using the formula for the derivative of a product. That is,

$$f'(x) = 4x^2 \frac{d}{dx}(2x+1)^4 + (2x+1)^4 \frac{d}{dx}(4x^2) = 4x^2\frac{d}{dx}(2x+1)^4 + (2x+1)^4 \cdot 4 \cdot \frac{d}{dx}x^2$$

We continue by using the Power Rule:

$$f'(x) = 4x^2 \left[4(2x+1)^3 \frac{d}{dx}(2x+1)\right] + (2x+1)^4 \cdot 4 \cdot \frac{d}{dx}x^2$$

$$= 4x^2 \left[4(2x+1)^3 (2)\right] + (2x+1)^4 \cdot 4 \cdot 2x \qquad \text{Differentiate.}$$

$$= 32x^2 (2x+1)^3 + 8x(2x+1)^4 \qquad \text{Simplify.}$$

$$= 8x(2x+1)^3 \left[4x + (2x+1)\right] \qquad \text{Factor.}$$

$$= 8x(2x+1)^3 (6x+1) \qquad \text{Simplify.}$$

11. Before differentiating the function f, we simplify it, then we use the Power Rule.

$$f(x) = [x(x-1)]^3 = (x^2 - x)^3$$

$$f'(x) = \frac{d}{dx}(x^2 - x)^3 = 3(x^2 - x)^2 \frac{d}{dx}(x^2 - x) \qquad \text{Use the Power Rule.}$$

$$= 3(x^2 - x)^2 (2x - 1) \qquad \text{Differentiate.}$$

$$= 3x^2 (x-1)^2 (2x-1) \qquad \text{Factor.}$$

13. $$f'(x) = \frac{d}{dx}(3x-1)^{-2} = -2(3x-1)^{-3}\frac{d}{dx}(3x-1) \qquad \text{Use the Power Rule.}$$

$$= -2(3x-1)^{-3}(3) \qquad \text{Differentiate.}$$

$$= -\frac{6}{(3x-1)^3} \qquad \text{Simplify.}$$

15. We rewrite $f(x)$ as $f(x) = 4(x^2 + 4)^{-1}$. Then we use the Power Rule.

$$f'(x) = \frac{d}{dx}\left[4(x^2 + 4)^{-1}\right] = 4\frac{d}{dx}(x^2 + 4)^{-1} = 4 \cdot (-1)(x^2 + 4)^{-2}\frac{d}{dx}(x^2 + 4)$$

$$\uparrow \qquad\qquad\qquad \uparrow$$
$$f'(cx) = cf'(x) \qquad \text{Use the Power Rule.}$$

$$= -4(x^2 + 4)^{-2}(2x) = -\frac{8x}{(x^2 + 4)^2}$$

$$\uparrow \qquad\qquad \uparrow$$
$$\text{Differentiate.} \qquad \text{Simplify.}$$

17. We rewrite $f(x)$ as $f(x) = -4(x^2 - 9)^{-3}$. Then we use the Power Rule.

$$f'(x) = \frac{d}{dx}\left[-4(x^2 - 9)^{-3}\right] = -4 \cdot \frac{d}{dx}(x^2 - 9)^{-3} = -4 \cdot (-3)(x^2 - 9)^{-4}\frac{d}{dx}(x^2 - 9)$$

$$\uparrow \qquad\qquad\qquad \uparrow$$
$$f'(cx) = cf'(x) \qquad \text{Use the Power Rule.}$$

$$= 12(x^2 - 9)^{-4}(2x) = \frac{24x}{(x^2 - 9)^4}$$

$$\uparrow \qquad\qquad \uparrow$$
$$\text{Differentiate.} \qquad \text{Simplify.}$$

19. In this problem the function f is a quotient raised to the power 3. We begin with the Power Rule and then use the formula for the derivative of a quotient.

$$f'(x) = \frac{d}{dx}\left(\frac{x}{x+1}\right)^3 = 3\left(\frac{x}{x+1}\right)^2\left[\frac{d}{dx}\left(\frac{x}{x+1}\right)\right] \qquad \text{The Power Rule.}$$

$$= 3\left(\frac{x}{x+1}\right)^2\left[\frac{(x+1)\dfrac{d}{dx}(x) - x\dfrac{dy}{dx}(x+1)}{(x+1)^2}\right] \qquad \text{The derivative of a quotient.}$$

$$= 3\left(\frac{x}{x+1}\right)^2\left[\frac{(x+1)(1) - x(1)}{(x+1)^2}\right] \qquad \text{Differentiate.}$$

$$= 3\left(\frac{x}{x+1}\right)^2\left[\frac{1}{(x+1)^2}\right] = \frac{3x^2}{(x+1)^4} \qquad \text{Simplify.}$$

21. Here the function f is the quotient of two functions, the numerator of which is raised to a power. We will first use the formula for the derivative of a quotient and then use the power rule when differentiating the numerator.

$$f'(x) = \frac{d}{dx}\left[\frac{(2x+1)^4}{3x^2}\right] = \frac{3x^2\dfrac{d}{dx}(2x+1)^4 - (2x+1)^4\dfrac{d}{dx}(3x^2)}{(3x^2)^2}$$

$$= \frac{3x^2\left[4(2x+1)^3\dfrac{d}{dx}(2x+1)\right] - (2x+1)^4\dfrac{d}{dx}(3x^2)}{(3x^2)^2} \qquad \text{The Power Rule.}$$

$$= \frac{3x^2 \left[4(2x+1)^3 (2) \right] - (2x+1)^4 (6x)}{\left(3x^2 \right)^2}$$ Differentiate.

$$= \frac{24x^2 (2x+1)^3 - 6x(2x+1)^4}{9x^4}$$ Simplify.

$$= \frac{6x(2x+1)^3 \left[4x - (2x+1) \right]}{9x^4}$$ Factor.

$$= \frac{\cancel{6}^2 \cancel{x} (2x+1)^3 (2x-1)}{\cancel{9}^3 x^{\cancel{4}3}}$$ Simplify.

$$= \frac{2(2x+1)^3 (2x-1)}{3x^3}$$ Cancel.

23. Here the function f is the quotient of two functions, the numerator of which is raised to a power. We will first use the formula for the derivative of a quotient and then use the power rule when differentiating the numerator.

$$f'(x) = \frac{d}{dx} \left[\frac{(x^2+1)^3}{x} \right] = \frac{x \frac{d}{dx}(x^2+1)^3 - (x^2+1)^3 \frac{d}{dx} x}{x^2}$$

$$= \frac{x \left[3(x^2+1)^2 \frac{d}{dx}(x^2+1) \right] - (x^2+1)^3 \frac{d}{dx} x}{x^2}$$ The Power Rule.

$$= \frac{x \left[3(x^2+1)^2 (2x) \right] - (x^2+1)^3 (1)}{x^2}$$ Differentiate.

$$= \frac{6x^2 (x^2+1)^2 - (x^2+1)^3}{x^2}$$ Simplify.

$$= \frac{(x^2+1)^2 \left[6x^2 - (x^2+1) \right]}{x^2}$$ Factor.

$$= \frac{(x^2+1)^2 (5x^2-1)}{x^2}$$ Simplify.

25. We rewrite f as $f(x) = (x + x^{-1})^3$ and use the Power Rule.

$$f'(x) = \frac{d}{dx} f(x) = \frac{d}{dx}(x+x^{-1})^3 = 3(x+x^{-1})^2 \frac{d}{dx}(x+x^{-1}) = 3(x+x^{-1})^2 (1-x^{-2})$$

$$\underset{\text{Use the Power Rule.}}{\uparrow} \qquad\qquad \underset{\text{Differentiate.}}{\uparrow}$$

$$= 3\left(x+\frac{1}{x} \right)^2 \left(1-\frac{1}{x^2} \right) = 3\left(\frac{x^2+1}{x} \right)^2 \left(\frac{x^2-1}{x^2} \right) = \frac{3(x^2+1)^2 (x^2-1)}{x^4}$$

$$\underset{\substack{\text{Rewrite with positive} \\ \text{exponents.}}}{\uparrow} \qquad \underset{\substack{\text{Write with a single} \\ \text{denominator.}}}{\uparrow} \qquad \underset{\text{Simplify.}}{\uparrow}$$

27. Here the function f is the quotient of two functions, the denominator of which is raised to a power. We will first use the formula for the derivative of a quotient and then use the Power Rule when differentiating the denominator.

$$f'(x) = \frac{d}{dx}\frac{3x^2}{(x^2+1)^2} = \frac{(x^2+1)^2\frac{d}{dx}(3x^2) - 3x^2\frac{d}{dx}(x^2+1)^2}{\left[(x^2+1)^2\right]^2}$$ The derivative of a quotient.

$$= \frac{(x^2+1)^2\frac{d}{dx}(3x^2) - 3x^2 \cdot 2(x^2+1)\frac{d}{dx}(x^2+1)}{(x^2+1)^4}$$ Use the Power Rule.

$$= \frac{(x^2+1)^2(6x) - 3x^2 \cdot 2(x^2+1)(2x)}{(x^2+1)^4}$$ Differentiate.

$$= \frac{(x^2+1)^2(6x) - 12x^3(x^2+1)}{(x^2+1)^4}$$ Simplify.

$$= \frac{6x(x^2+1)\left[(x^2+1) - 2x^2\right]}{(x^2+1)^3}$$ Factor.

$$= \frac{6x(1-x^2)}{(x^2+1)^3}$$ Simplify.

29. The rate at which the car is depreciating is the derivative $V'(t)$.

$$V'(t) = \frac{d}{dt}\frac{29,000}{1+0.4t+0.1t^2} = \frac{(1+0.4t+0.1t^2)\frac{d}{dt}29,000 - 29,000\frac{d}{dt}(1+0.4t+0.1t^2)}{(1+0.4t+0.1t^2)^2}$$

$$= \frac{-29,000(0.4+0.2t)}{(1+0.4t+0.1t^2)^2}$$

(a) The rate of depreciation 1 year after purchase is $V'(1)$

$$V'(1) = \frac{-29,000(0.4+0.2)}{(1+0.4+0.1)^2} = \frac{-29,000(0.6)}{1.5^2} = -7733.33$$

The car is depreciating at a rate of $7733.33 per year when it is one year old.

(b) The rate of depreciation 2 years after purchase is $V'(2)$

$$V'(2) = \frac{-29,000(0.4+0.2\cdot2)}{(1+0.4\cdot2+0.1\cdot2^2)^2} = \frac{-29,000(0.8)}{2.2^2} = -4793.39$$

The car is depreciating at a rate of $4793.39 per year when it is two years old.

(c) The rate of depreciation 3 years after purchase is $V'(3)$

$$V'(3) = \frac{-29,000(0.4+0.2 \cdot 3)}{\left(1+0.4 \cdot 3+0.1 \cdot 3^2\right)^2} = \frac{-29,000(1.0)}{3.1^2} = -3017.69$$

The car is depreciating at a rate of $3017.69 per year when it is three years old.

(d) The rate of depreciation 4 years after purchase is $V'(4)$

$$V'(4) = \frac{-29,000(0.4+0.2 \cdot 4)}{\left(1+0.4 \cdot 4+0.1 \cdot 4^2\right)^2} = \frac{-29,000(1.2)}{4.2^2} = -1972.79$$

The car is depreciating at a rate of $1972.79 per year when it is four years old.

31. (a) The rate of change is given by the derivative.

$$\frac{dp}{dx} = \frac{d}{dx}\left(\frac{10,000}{5x+100}-5\right) = \frac{d}{dx}\frac{10,000}{5x+100} - \frac{d}{dx}5 = \frac{d}{dx}\left[10,000\,(5x+100)^{-1}\right]$$

⟨↑⟩   ↑

Derivative of a difference. $\frac{d}{dx}5 = 0$

$$= 10,000 \cdot \frac{d}{dx}(5x+100)^{-1} = 10,000 \cdot (-1)(5x+100)^{-2}(5) = -\frac{50,000}{(5x+100)^2} = -\frac{2000}{(x+20)^2}$$

↑   ↑   ↑

$f'(cx) = cf'(x)$ Differentiate.   Simplify.

(b) The revenue function $R = px$.

$$R = R(x) = \left(\frac{10,000}{5x+100}-5\right)x = \frac{10,000x}{5x+100} - 5x$$

(c) The marginal revenue is the derivative $R'(x)$.

$$R'(x) = \frac{d}{dx}\left(\frac{10,000x}{5x+100}-5x\right) = \frac{d}{dx}\frac{10,000x}{5x+100} - \frac{d}{dx}(5x)$$

$$= \frac{(5x+100)(10,000)-10,000x(5)}{(5x+100)^2} - 5 = \frac{1,000,000}{(5x+100)^2} - 5 = \frac{40,000}{(x+20)^2} - 5$$

(d) $R'(10) = \dfrac{1,000,000}{(5 \cdot 10 +100)^2} - 5 = \dfrac{1,000,000}{150^2} - 5 = 39.44$ dollars

$$R'(40) = \frac{1,000,000}{(5 \cdot 40 +100)^2} - 5 = \frac{1,000,000}{300^2} - 5 = 6.11 \text{ dollars}$$

33. (a) The average rate of change in the mass of the protein is

$$\frac{\Delta M}{\Delta t} = \frac{M(2)-M(0)}{2-0} = \frac{\dfrac{28}{2+2}-\dfrac{28}{0+2}}{2} = \frac{7-14}{2} = -\frac{7}{2}$$

grams per hour.

(b) $M'(t) = \dfrac{d}{dt}\dfrac{28}{t+2} = 28\dfrac{d}{dt}(t+2)^{-1} = 28\cdot(-1)(t+2)^{-2}\cdot 1 = -\dfrac{28}{(t+2)^2}$

$M'(0) = -\dfrac{28}{(0+2)^2} = -\dfrac{28}{4} = -7$

13.5 The Derivatives of the Exponential and Logarithmic Functions; The Chain Rule $757 / 1-61$ (ODD)

1. $f'(x) = \dfrac{d}{dx}(x^3 - e^x) = \dfrac{d}{dx}x^3 - \dfrac{d}{dx}e^x = 3x^2 - e^x$

3. Using the formula for the derivative of a product,

$$f'(x) = \dfrac{d}{dx}(x^2 e^x) = x^2\dfrac{d}{dx}e^x + e^x\dfrac{d}{dx}x^2 = x^2 e^x + 2xe^x = x(x+2)e^x$$

5. Using the formula for the derivative of a quotient,

$$f'(x) = \dfrac{d}{dx}\dfrac{e^x}{x^2} = \dfrac{x^2\dfrac{d}{dx}e^x - e^x\dfrac{d}{dx}x^2}{(x^2)^2} = \dfrac{x^2 e^x - e^x\cdot 2x}{x^4} = \dfrac{x(x-2)e^x}{x^4} = \dfrac{(x-2)e^x}{x^3}$$

$\qquad\qquad\qquad\qquad\quad \uparrow \qquad\qquad\qquad \uparrow \qquad\qquad \uparrow$
$\qquad\qquad\qquad\qquad$ Differentiate. $\qquad$ Factor. $\qquad$ Simplify.

7. Using the formula for the derivative of a quotient,

$$f'(x) = \dfrac{d}{dx}\dfrac{4x^2}{e^x} = \dfrac{e^x\dfrac{d}{dx}(4x^2) - 4x^2\dfrac{d}{dx}e^x}{(e^x)^2} = \dfrac{e^x\cdot 8x - 4x^2 e^x}{e^{2x}} = \dfrac{4x(2-x)e^x}{e^{2x}} = \dfrac{4x(2-x)}{e^x}$$

$\qquad\qquad\qquad\qquad\quad \uparrow \qquad\qquad\qquad \uparrow \qquad\qquad \uparrow$
$\qquad\qquad\qquad\qquad$ Differentiate. $\qquad$ Factor. $\qquad$ Simplify.

$$= \dfrac{8x - 4x^2}{e^x}$$

9. $y = f(u) = u^5 = (x^3 + 1)^5 = f(x)$

$$\dfrac{dy}{dx} = \dfrac{d}{dx}(x^3 + 1)^5 = 5(x^3 + 1)^4\dfrac{d}{dx}(x^3 + 1) = 5(x^3 + 1)^4(3x^2) = 15x^2(x^3 + 1)^4$$

$\qquad\qquad\qquad \uparrow \qquad\qquad\qquad\qquad \uparrow \qquad\qquad\qquad \uparrow$
$\qquad\qquad\quad$ Chain Rule. $\qquad\qquad$ Differentiate. $\qquad$ Simplify.

11. $y = f(u) = \dfrac{u}{u+1} = \dfrac{x^2+1}{(x^2+1)+1} = \dfrac{x^2+1}{x^2+2} = f(x)$

$$\frac{dy}{dx} = \frac{d}{dx} \frac{x^2+1}{x^2+2} = \frac{(x^2+2)\frac{d}{dx}(x^2+1)-(x^2+1)\frac{d}{dx}(x^2+2)}{(x^2+2)^2}$$

↑
Derivative of a quotient

$$= \frac{(x^2+2)(2x)-(x^2+1)(2x)}{(x^2+2)^2} = \frac{2x\left[(x^2+2)-(x^2+1)\right]}{(x^2+2)^2} = \frac{2x}{(x^2+2)^2}$$

↑   ↑   ↑
Differentiate.   Factor.   Simplify.

13. $y = f(u) = (u+1)^2 = \left(\frac{1}{x}+1\right)^2 = (x^{-1}+1)^2 = f(x)$

$$\frac{dy}{dx} = \frac{d}{dx}(x^{-1}+1)^2 = 2(x^{-1}+1)\frac{d}{dx}(x^{-1}+1) = 2(x^{-1}+1)\cdot(-1)x^{-2} = \frac{-2\left(\frac{1}{x}+1\right)}{x^2} = -\frac{2(1+x)}{x^3}$$

↑   ↑   ↑   ↑
Use the Chain Rule.   Differentiate.   Write with positive   Simplify.
                                                 exponents.

15. $y = f(u) = (u^3-1)^5 = \left[(x^{-2})^3-1\right]^5 = (x^{-6}-1)^5 = f(x)$

$$\frac{dy}{dx} = \frac{d}{dx}(x^{-6}-1)^5 = 5(x^{-6}-1)^4 \frac{d}{dx}(x^{-6}-1) = 5(x^{-6}-1)^4\cdot(-6)x^{-7}$$

↑   ↑
Use the Chain Rule.   Differentiate.

$$= \frac{-30(x^{-6}-1)^4}{x^7} = -\frac{30\left(\frac{1}{x^6}-1\right)^4}{x^7} = -\frac{30\left(\frac{1-x^6}{x^6}\right)^4}{x^7} = -\frac{30(1-x^6)^4}{(x^6)^4 x^7} = -\frac{30(1-x^6)^4}{x^{31}}$$

↑   ↑   ↑   ↑
Simplify.   Write with positive Write the numerator   Simplify.
                   exponent.   as a single quotient.

17. $y = f(u) = u^3 = (e^x)^3 = e^{3x} = f(x)$

$$\frac{dy}{dx} = \frac{d}{dx}e^{3x} = e^{3x}\cdot\frac{d}{dx}(3x) = e^{3x}\cdot 3 = 3e^{3x}$$

↑   ↑   ↑
Chain Rule   Differentiate.   Simplify.

19. $y = f(u) = e^u = e^{x^3} = f(x)$

$$\frac{dy}{dx} = \frac{d}{dx}e^{x^3} = e^{x^3}\cdot\frac{d}{dx}x^3 = e^{x^3}\cdot 3x^2 = 3x^2 e^{x^3}$$

↑   ↑
Chain Rule   Differentiate.

21. (a) Using the Chain Rule, $y = (x^3 + 1)^2$ is thought of as $y = u^2$ and $u = x^3 + 1$.

$$\frac{dy}{du} = 2u \quad \text{and} \quad \frac{du}{dx} = 3x^2$$

$$\frac{dy}{dx} = \frac{dy}{du} \cdot \frac{du}{dx} = 2u \cdot 3x^2 = 2(x^3 + 1) \cdot 3x^2 = 6x^2(x^3 + 1) = 6x^5 + 6x^2$$

$\qquad\qquad\qquad\qquad\qquad\qquad \uparrow \qquad\qquad\qquad \uparrow$

$\qquad\qquad\qquad\qquad\quad$ Substitute $u = x^3 + 1 \qquad$ Simplify.

(b) Using the Power Rule,

$$\frac{dy}{dx} = \frac{d}{dx}(x^3 + 1)^2 = 2(x^3 + 1) \cdot \frac{d}{dx}(x^3 + 1) = 2(x^3 + 1) \cdot (3x^2) = 6x^2(x^3 + 1) = 6x^5 + 6x^2$$

$\qquad\qquad\qquad\quad \uparrow \qquad\qquad\qquad\qquad\qquad \uparrow \qquad\qquad\qquad\qquad \uparrow$

$\qquad\qquad\qquad$ Power Rule $\qquad\qquad\qquad$ Differentiate. $\qquad\qquad$ Simplify.

(c) Expanding, $y = (x^3 + 1)^2 = x^6 + 2x^3 + 1$, and

$$\frac{dy}{dx} = \frac{d}{dx}(x^6 + 2x^3 + 1) = \frac{d}{dx}x^6 + 2\frac{d}{dx}x^3 + \frac{d}{dx}1 = 6x^5 + 6x^2 = 6x^2(x^3 + 1)$$

$\qquad\qquad\qquad\qquad\qquad\qquad\qquad\qquad\qquad\qquad\qquad \uparrow \qquad\qquad \uparrow$

$\qquad\qquad\qquad\qquad\qquad\qquad\qquad\qquad\qquad\quad$ Differentiate. $\quad$ Factor.

23. $f'(x) = \dfrac{d}{dx}e^{5x} = e^{5x} \cdot \dfrac{d}{dx}(5x) = e^{5x} \cdot 5 = 5e^{5x}$

25. $f'(x) = \dfrac{d}{dx}8e^{-x^2} = 8 \cdot \dfrac{d}{dx}e^{-x^2} = 8e^{-x^2}\dfrac{d}{dx}(-x^2) = 8e^{-x^2} \cdot (-2x) = -16xe^{-x^2}$

27. $f'(x) = \dfrac{d}{dx}x^2 e^{x^2} = x^2\dfrac{d}{dx}e^{x^2} + e^{x^2}\dfrac{d}{dx}x^2 = x^2\left(e^{x^2}\dfrac{d}{dx}x^2\right) + e^{x^2} \cdot 2x$

$\qquad\qquad\qquad\qquad \uparrow \qquad\qquad\qquad\qquad\qquad\qquad \uparrow$

$\qquad\qquad$ The derivative of a product $\qquad\quad$ The Chain Rule

$$= x^2 e^{x^2} \cdot 2x + e^{x^2} \cdot 2x = e^{x^2}(2x^3 + 2x) = 2xe^{x^2}(x^2 + 1)$$

$\quad \uparrow \qquad\qquad\qquad\qquad\qquad \uparrow$

Differentiate. $\qquad\qquad$ Factor.

29. $f'(x) = \dfrac{d}{dx}\left[5(e^x)^3\right] = 5\dfrac{d}{dx}(e^x)^3 = 5 \cdot 3(e^x)^2 \cdot \dfrac{d}{dx}e^x = 15(e^x)^2 e^x = 15e^{3x}$

$\qquad\qquad\qquad\qquad\qquad\qquad\qquad \uparrow \qquad\qquad\qquad \uparrow \qquad\qquad\qquad \uparrow$

$\qquad\qquad\qquad\qquad\qquad$ Use the Power Rule. $\quad$ Differentiate. $\quad$ Simplify.

31. $f(x) = \dfrac{x^2}{e^x} = x^2 e^{-x}$

$$f'(x) = \frac{d}{dx}(x^2 e^{-x}) = x^2\frac{d}{dx}e^{-x} + e^{-x}\frac{d}{dx}x^2 = x^2\left(e^{-x} \cdot \frac{d}{dx}(-x)\right) + e^{-x} \cdot 2x$$

$\qquad\qquad\qquad \uparrow \qquad\qquad\qquad\qquad\qquad\qquad \uparrow$

$\qquad\qquad$ Derivative of a product $\qquad\qquad$ Chain Rule

$$= x^2 \cdot e^{-x} \cdot (-1) + 2x e^{-x} = -x^2 e^{-x} + 2x e^{-x} = x e^{-x}(2-x) = \frac{2x - x^2}{e^x}$$

$\uparrow$ Differentiate. $\uparrow$ Simplify. $\uparrow$ Factor.

33. $f(x) = \dfrac{(e^x)^2}{x} = x^{-1} e^{2x}$

$$f'(x) = \frac{d}{dx}\left(x^{-1} e^{2x}\right) = x^{-1} \frac{d}{dx} e^{2x} + e^{2x} \frac{d}{dx} x^{-1} = x^{-1}\left(e^{2x} \frac{d}{dx}(2x)\right) + e^{2x} \cdot (-1) x^{-2}$$

$\uparrow$ The derivative of a product $\uparrow$ Apply the Chain Rule.

$$= 2x^{-1} e^{2x} - x^{-2} e^{2x} = e^{2x}\left(2x^{-1} - x^{-2}\right) = \frac{e^{2x}(2x-1)}{x^2}$$

$\uparrow$ Differentiate. $\uparrow$ Factor. $\uparrow$ Simplify.

35. $f'(x) = \dfrac{d}{dx}\left(x^2 - 3\ln x\right) = \dfrac{d}{dx} x^2 - 3\dfrac{d}{dx}\ln x = 2x - 3 \cdot \dfrac{1}{x} = 2x - \dfrac{3}{x}$

$\uparrow$ The derivative of a difference $\uparrow$ Differentiate. $\uparrow$ Simplify.

37. $f'(x) = \dfrac{d}{dx}\left(x^2 \ln x\right) = x^2 \dfrac{d}{dx}\ln x + \ln x \cdot \dfrac{d}{dx} x^2 = x^2 \cdot \dfrac{1}{x} + \ln x \cdot 2x$

$\uparrow$ The derivative of a product $\uparrow$ Differentiate.

$$= x + 2x \ln x = x(1 + 2\ln x) = x\left(1 + \ln x^2\right)$$

$\uparrow$ Simplify. $\uparrow$ Factor. $\uparrow$ Alternate form of the answer

39. $f'(x) = \dfrac{d}{dx}\left[3\ln(5x)\right] = 3\dfrac{d}{dx}\ln(5x) = 3 \cdot \dfrac{\dfrac{d}{dx}(5x)}{5x} = 3 \cdot \dfrac{5}{5x} = \dfrac{3}{x}$

$\uparrow$ Derivative of $\ln g(x)$

41. Here the f is the product of two functions, we use the product rule first and then the Chain Rule when differentiating $\ln(x^2 + 1)$.

$$f'(x) = \frac{d}{dx}\left[x\ln(x^2+1)\right] = x \cdot \frac{d}{dx}\ln(x^2+1) + \ln(x^2+1) \cdot \frac{d}{dx} x \qquad \text{The derivative of a product.}$$

$$= x \cdot \frac{\dfrac{d}{dx}(x^2+1)}{x^2+1} + \ln(x^2+1) \cdot 1 \qquad \text{Use the Chain Rule; } \frac{d}{dx} x = 1.$$

$$= x \cdot \frac{2x}{x^2+1} + \ln(x^2+1) \qquad \text{Differentiate.}$$

$$= \frac{2x^2}{x^2+1} + \ln(x^2+1) \qquad \text{Simplify.}$$

43. $f'(x) = \dfrac{d}{dx}\left[x + 8\ln(3x)\right] = \dfrac{d}{dx}x + 8\dfrac{d}{dx}\ln(3x) = \dfrac{d}{dx}x + 8 \cdot \dfrac{\dfrac{d}{dx}(3x)}{3x}$

$\uparrow$ The derivative of a sum $\qquad\qquad$ $\uparrow$ The Chain Rule

$$= 1 + 8 \cdot \frac{3}{3x} = 1 + \frac{8}{x} = \frac{x+8}{x}$$

$\uparrow$ Differentiate. $\qquad$ $\uparrow$ Simplify. $\quad$ $\uparrow$ Alternate form of the answer.

45. $f'(x) = \dfrac{d}{dx}\left[8(\ln x)^3\right] = 8\dfrac{d}{dx}(\ln x)^3 = 8 \cdot 3(\ln x)^2 \cdot \dfrac{d}{dx}\ln x = 24(\ln x)^2 \cdot \dfrac{1}{x} = \dfrac{24(\ln x)^2}{x}$

$\qquad\qquad$ $\uparrow$ Use the Power Rule. $\quad$ $\uparrow$ Differentiate. $\qquad$ $\uparrow$ Simplify.

47. $f'(x) = \dfrac{d}{dx}\log_3 x = \dfrac{1}{x\ln 3} \qquad\qquad \dfrac{d}{dx}\log_a x = \dfrac{1}{x\ln a}$

49. $f'(x) = \dfrac{d}{dx}(x^2\log_2 x) = x^2 \cdot \dfrac{d}{dx}\log_2 x + \log_2 x \cdot \dfrac{d}{dx}x^2$

$$= x^2 \cdot \frac{1}{x\ln 2} + \log_2 x \cdot 2x$$

$$= \frac{x}{\ln 2} + 2x\log_2 x = \frac{x}{\ln 2} + 2x\frac{\ln x}{\ln 2} \qquad \text{Use the Change of Base Formula.}$$

$$= \frac{x + 2x\ln x}{\ln 2}$$

51. $f'(x) = \dfrac{d}{dx}3^x = 3^x\ln 3 \qquad\qquad \dfrac{d}{dx}a^x = a^x\ln a$

53. $f'(x) = \dfrac{d}{dx}(x^2 \cdot 2^x) = x^2 \cdot \dfrac{d}{dx}2^x + 2^x \cdot \dfrac{d}{dx}x^2 = x^2 \cdot 2^x\ln 2 + 2^x \cdot 2x = 2^x(x^2\ln 2 + 2x)$

$\qquad$ $\uparrow$ The derivative of a product $\qquad$ $\uparrow$ Differentiate; $\dfrac{d}{dx}a^x = a^x\ln a$. $\quad$ $\uparrow$ Factor.

55. The slope of the tangent line to the graph of $f(x) = e^{3x}$ at the point (0, 1) is the derivative of the function f evaluated at the point (0, 1). The derivative of f is
$$f'(x) = 3e^{3x}$$
The slope of the tangent line is $m_{\tan} = f'(0) = 3 \cdot e^0 = 3$.

An equation of the tangent line is $y - 1 = 3(x - 0)$ or $y = 3x + 1$.

57. The slope of the tangent line to the graph of $f(x) = \ln x$ at the point $(1, 0)$ is the derivative of the function f evaluated at the point $(1, 0)$. The derivative of f is

$$f'(x) = \frac{1}{x}$$

The slope of the tangent line is $m_{\tan} = f'(1) = 1$.

An equation of the tangent line is $y - 0 = 1(x - 1)$ or $y = x - 1$.

59. The slope of the tangent line to the graph of $f(x) = e^{3x-2}$ at the point $\left(\frac{2}{3}, 1\right)$ is the derivative of the function f evaluated at the point $\left(\frac{2}{3}, 1\right)$. The derivative of f is

$$f'(x) = 3e^{3x-2}$$

The slope of the tangent line is $m_{\tan} = 3e^{3 \cdot 2/3 - 2} = 3e^0 = 3$

An equation of the tangent line is $y - 1 = 3\left(x - \frac{2}{3}\right)$ or $y = 3x - 1$.

61. The slope of the tangent line to the graph of $f(x) = x \ln x$ at the point $(1, 0)$ is the derivative of the function f evaluated at the point $(1, 0)$. The derivative of f is

$$f'(x) = x \cdot \frac{d}{dx} \ln x + \ln x \cdot \frac{d}{dx} x = x \cdot \frac{1}{x} + \ln x \cdot 1 = 1 + \ln x$$

| ↑ | ↑ | ↑ |
| The derivative of a product | Differentiate. | Simplify. |

The slope of the tangent line is $m_{\tan} = f'(1) = 1 + \ln 1 = 1 + 0 = 1$

An equation of the tangent line is $y - 0 = 1(x - 1)$ or $y = x - 1$.

63. Parallel lines have the same slope. Since the slope of the line $y = x$ is 1, the slope of the tangent line we seek is also 1. Moreover, the slope of a tangent line is given by the derivative of the function. So we need $m_{\tan} = f'(x) = 1$.
   $f'(x) = e^x = 1$ when $x = 0$, and
   $f(0) = e^0 = 1$
which means an equation of a tangent line to the function is $y - 1 = 1(x - 0)$ or $y = x + 1$.

65. (a) The reaction rate for a dose of 5 units is given by the derivative of R, $R'(x)$ evaluated at $x = 5$ units.

$$R'(x) = \frac{d}{dx}(5.5 \ln x + 10) = \frac{5.5}{x}$$

$$R'(5) = \frac{5.5}{5} = 1.1$$

(b) The reaction rate for a dose of 10 units is given by $R'(10)$.

$$R'(10) = \frac{5.5}{10} = 0.55$$

67. The rate of change in atmospheric pressure is given by the derivative of the atmospheric pressure P.

$$P'(x) = \frac{d}{dx}\left(10^4 e^{-0.00012x}\right) = 10^4 \frac{d}{dx} e^{-0.00012x} = 10^4 \cdot (-0.00012) e^{-0.00012x} = -1.2e^{-0.00012x}$$

The rate of change of atmospheric pressure at $x = 500$ meters is

$$P'(500) = -1.2e^{(-0.00012)(500)} = -1.2e^{-0.060} = -1.130 \text{ kilograms per square meter.}$$

The rate of change of atmospheric pressure at $x = 700$ meters is

$$P'(700) = -1.2e^{(-0.00012)(700)} = -1.2e^{-0.060} = -1.103 \text{ kilograms per square meter.}$$

69. (a) The rate of change of A with respect to time is the derivative $A'(t)$.

$$A'(t) = \frac{d}{dx}\left(102 - 90e^{-0.21t}\right) = \frac{d}{dx}102 - 90 \cdot \frac{d}{dx}e^{-0.21t} = -90 \cdot (-0.21)e^{-0.21t} = 18.9e^{-0.21t}$$

(b) At $t = 5$, $A'(t) = A'(5) = 18.9e^{-0.21(5)} = 18.9e^{-1.05} = 6.614$

(c) At $t = 10$, $A'(t) = A'(10) = 18.9e^{-0.21(10)} = 18.9e^{-2.1} = 2.314$

(d) At $t = 30$, $A'(t) = A'(30) = 18.9e^{-0.21(30)} = 18.9e^{-6.3} = 0.035$

71. (a) The rate of change of S with respect to x is the derivative $S'(x)$.

$$S'(x) = \frac{d}{dx}\left(100{,}000 + 400{,}000 \ln x\right) = \frac{d}{dx}100{,}000 + 400{,}000\frac{d}{dx}\ln x$$

$$= 400{,}000 \cdot \frac{1}{x} = \frac{400{,}000}{x}$$

(b) $S'(10) = \dfrac{400{,}000}{10} = 40{,}000$

(c) $S'(20) = \dfrac{400{,}000}{20} = 20{,}000$

73. (a) For $x = 1000$, the price p is

$$p = 50 - 4\ln\left(\frac{1000}{100} + 1\right) = 50 - 4\ln 11 = \$40.41$$

(b) For $x = 5000$, the price p is

$$p = 50 - 4\ln\left(\frac{5000}{100} + 1\right) = 50 - 4\ln 51 = \$34.27$$

(c) The marginal demand is the derivative of the function p.

$$\frac{d}{dx}p = \frac{d}{dx}\left[50 - 4\ln\left(\frac{x}{100} + 1\right)\right] = \frac{d}{dx}50 - 4\frac{d}{dx}\ln\left(\frac{x}{100} + 1\right)$$

↑

The derivative of a difference.

$$= \frac{d}{dx}50 - 4\frac{d}{dx}\ln\left(\frac{x + 100}{100}\right) = -4\frac{d}{dx}\left[\ln(x + 100) - \ln 100\right]$$

↑   ↑

Write $\dfrac{x}{100} + 1$ as a single fraction.   The logarithm of a quotient is the difference of the logarithms.

$$= -4\left[\frac{d}{dx}\ln(x + 100) - \frac{d}{dx}\ln 100\right] = -4 \cdot \frac{1}{x + 100} = -\frac{4}{x + 100}$$

↑   ↑   ↑

The derivative of a difference.   Differentiate.   Simplify.

The marginal demand for 1000 t-shirts is

$$p'(1000) = -\frac{4}{1000 + 100} = -\frac{4}{1100} = -0.0036$$

(d) The marginal demand for 5000 t-shirts is

$$p'(5000) = -\frac{4}{5000 + 100} = -\frac{4}{5100} = -0.00078$$

(e) The revenue function $R(x) = p \cdot x = \left[50 - 4\ln\left(\frac{x}{100} + 1\right)\right] \cdot x = 50x - 4x\ln\left(\frac{x}{100} + 1\right)$.

(f) The marginal revenue is the derivative of the function R.

$$R'(x) = \frac{d}{dx}\left[50x - 4x\ln\left(\frac{x}{100} + 1\right)\right] = \frac{d}{dx}(50x) - \frac{d}{dx}\left[4x\ln\left(\frac{x}{100} + 1\right)\right]$$

$$= 50\frac{d}{dx}x - 4\left[x\frac{d}{dx}\ln\left(\frac{x}{100} + 1\right) + \ln\left(\frac{x}{100} + 1\right)\frac{d}{dx}x\right]$$

$$= 50 - 4\left[x \cdot \frac{1}{x + 100} + \ln\left(\frac{x}{100} + 1\right)\right] = 50 - 4\left[\frac{x}{x + 100} + \ln\left(\frac{x}{100} + 1\right)\right]$$

When $x = 1000$ t-shirts are sold the marginal revenue is

$$R'(1000) = 50 - 4\left[\frac{1000}{1000 + 100} + \ln\left(\frac{1000}{100} + 1\right)\right] = 50 - 4\left[\frac{10}{11} + \ln 11\right] = \$36.77$$

(g) When $x = 5000$ t-shirts are sold the marginal revenue is

$$R'(5000) = 50 - 4\left[\frac{5000}{5000 + 100} + \ln\left(\frac{5000}{100} + 1\right)\right] = 50 - 4\left[\frac{50}{51} + \ln 51\right] = \$30.35$$

(h) Profit is the difference between revenue and cost.

$$P(x) = R(x) - C(x)$$

$$= \left[50x - 4x\ln\left(\frac{x}{100}+1\right)\right] - [4x] = 46x - 4x\ln\left(\frac{x}{100}+1\right)$$

(i) If 1000 t-shirts are sold the profit is

$$P(1000) = 46(1000) - 4(1000)\ln\left(\frac{1000}{100}+1\right) = 46,000 - 4000\ln 11 = \$36,408.42$$

(j) If 5000 t-shirts are sold the profit is

$$P(5000) = 46(5000) - 4(5000)\ln\left(\frac{5000}{100}+1\right) = 230,000 - 20,000\ln 51 = \$151,363.49$$

(k) To use TABLE to find the quantity x that maximizes profit, we enter the profit function into Y_1, then in TBLSET we select a large value for x, choose ΔTbl, and select the Auto option.

X	Y₁
3.1E6	1.44E7
3.2E6	1.44E7
3.3E6	1.45E7
3.4E6	1.45E7
3.5E6	1.45E7
3.6E6	1.45E7
3.7E6	1.45E7

$Y_1 = 14525251.2745$

We chose $x = 3,000,000$ and ΔTbl $= 1000$. Then using TABLE, we increased x until the profit function stopped increasing and began decreasing in magnitude. The quantity (to the nearest thousand) that maximizes profit is 3,632,000 t-shirts.
The maximum profit is \$14,525,251.

(l) $p(3,632,000) = p = 50 - 4\ln\left(\frac{3,632,000}{100}+1\right) = 50 - 4\ln(36,321) = 8.00$

To maximize profit, the t-shirts should be sold for \$8.00 each.

75. (a) The rate of change of p with respect to t is the derivative of p.

$$\frac{d}{dt}p = \frac{d}{dt}(0.470 + 0.026\ln t) = \frac{d}{dt}0.470 + 0.026\frac{d}{dt}\ln t = 0.026 \cdot \frac{1}{t} = \frac{0.026}{t}$$

(b) In 2002 the rate of change of p was

$$p'(5) = \frac{0.026}{5} = 0.052$$

(c) In 2007 the rate of change of p was

$$p'(10) = \frac{0.026}{10} = 0.026$$

77. Prove: $\dfrac{d}{dx}\ln g(x) = \dfrac{\dfrac{d}{dx}g(x)}{g(x)} = \dfrac{g'(x)}{g(x)}$

Proof: Let $y = f(u) = \ln u$ and $u = g(x)$.

Then $y' = f'(u) = \dfrac{dy}{du} = \dfrac{d}{du} \ln u = \dfrac{1}{u}$ and $u' = \dfrac{du}{dx} = g'(x)$

According to the Chain Rule if both f and g are differentiable functions, then

$$\frac{dy}{dx} = \frac{dy}{du} \cdot \frac{du}{dx} = \frac{1}{u} g'(x) = \frac{1}{g(x)} \cdot g'(x) = \frac{g'(x)}{g(x)}$$

13.6 Higher-Order Derivatives $766 \mid 1-53 \ (ODD)$

1. $f(x) = 2x + 5$
$f'(x) = 2$
$f''(x) = 0$

3. $f(x) = 3x^2 + x - 2$
$f'(x) = 6x + 1$
$f''(x) = 6$

5. $f(x) = -3x^4 + 2x^2$
$f'(x) = -12x^3 + 4x$
$f''(x) = -36x^2 + 4$

7. $f(x) = \dfrac{1}{x} = x^{-1}$

$f'(x) = -x^{-2} = -\dfrac{1}{x^2}$

$f''(x) = 2x^{-3} = \dfrac{2}{x^3}$

9. $f(x) = x + \dfrac{1}{x} = x + x^{-1}$

$f'(x) = 1 - x^{-2} = 1 - \dfrac{1}{x^2}$

$f''(x) = 2x^{-3} = \dfrac{2}{x^3}$

11. $f(x) = \dfrac{x}{x+1}$

$$f'(x) = \frac{d}{dx}\left(\frac{x}{x+1}\right) = \frac{(x+1)\dfrac{d}{dx}x - x\dfrac{d}{dx}(x+1)}{(x+1)^2} = \frac{(x+1)\cdot 1 - x\cdot 1}{(x+1)^2} = \frac{1}{(x+1)^2} = (x+1)^{-2}$$

    ↑   ↑   ↑   ↑

  The derivative of a quotient. Differentiate. Simplify. Write with a
                           negative exponent.

$$f''(x) = -2(x+1)^{-3}\frac{d}{dx}(x+1) = -2(x+1)^{-3}\cdot 1 = -\frac{2}{(x+1)^3}$$

  ↑   ↑

 Use the Power Rule.   Simplify.

13. $f(x) = e^x$
$f'(x) = e^x$
$f''(x) = e^x$

15. $f(x) = \left(x^2 + 4\right)^3$

$f'(x) = 3\left(x^2 + 4\right)^2 \dfrac{d}{dx}\left(x^2 + 4\right) = 3\left(x^2 + 4\right)^2 \cdot 2x = 6x\left(x^2 + 4\right)^2$

    ↑   ↑

Use the Power Rule.   Simplify.

$$f''(x) = \dfrac{d}{dx}\left[6x\left(x^2 + 4\right)^2\right] = 6x\dfrac{d}{dx}\left(x^2 + 4\right)^2 + \left(x^2 + 4\right)^2 \dfrac{d}{dx}(6x) \quad \text{Derivative of a product.}$$

$$= 6x \cdot 2\left(x^2 + 4\right)\dfrac{d}{dx}\left(x^2 + 4\right) + \left(x^2 + 4\right)^2 \cdot 6 \quad \text{Use the Power Rule.}$$

$$= 6x \cdot 2\left(x^2 + 4\right) \cdot 2x + \left(x^2 + 4\right)^2 \cdot 6 \quad \text{Differentiate.}$$

$$= 24x^2\left(x^2 + 4\right) + 6\left(x^2 + 4\right)^2 \quad \text{Simplify.}$$

$$= 6\left(x^2 + 4\right)\left(4x^2 + x^2 + 4\right) \quad \text{Factor.}$$

$$= 6\left(x^2 + 4\right)\left(5x^2 + 4\right) \quad \text{Simplify.}$$

17. $f(x) = \ln x$

$f'(x) = \dfrac{1}{x} = x^{-1}$

$f''(x) = -x^{-2} = -\dfrac{1}{x^2}$

19. $f(x) = xe^x$

$f'(x) = x\dfrac{d}{dx}e^x + e^x\dfrac{d}{dx}x = xe^x + e^x = e^x(x+1)$

    ↑   ↑   ↑

Derivative of a product.   Differentiate.   Factor.

$f''(x) = \dfrac{d}{dx}\left(xe^x + e^x\right) = \dfrac{d}{dx}xe^x + \dfrac{d}{dx}e^x = xe^x + e^x + e^x = e^x(x+2)$

                                ↑   ↑   ↑

Derivative of a sum.   Differentiate; use $f'(x)$.   Factor.

21. $f(x) = \left(e^x\right)^2$

$f'(x) = 2\left(e^x\right)\dfrac{d}{dx}e^x = 2\left(e^x\right)^2 = 2e^{2x}$ Use the Chain Rule.

$f''(x) = 4\left(e^x\right)\dfrac{d}{dx}e^x = 4\left(e^x\right)^2 = 4e^{2x}$ Use the Chain Rule.

23. $f(x) = \dfrac{1}{\ln x} = (\ln x)^{-1}$

$$f'(x) = -\left(\ln x\right)^{-2}\frac{d}{dx}\ln x = -\left(\ln x\right)^{-2}\cdot\frac{1}{x} = -\frac{1}{x\left(\ln x\right)^2}$$

↑   ↑
Use the Chain Rule.   Simplify.

$$f''(x) = -\frac{x\left(\ln x\right)^2\dfrac{d}{dx}1 - \dfrac{d}{dx}\left[x\left(\ln x\right)^2\right]}{\left[x\left(\ln x\right)^2\right]^2} = \frac{x\dfrac{d}{dx}\left(\ln x\right)^2 + \left(\ln x\right)^2\dfrac{d}{dx}x}{x^2\left(\ln x\right)^4}$$

↑   ↑

The derivative of a quotient. $\dfrac{d}{dx}1 = 0$; the derivative of a product.

$$= \frac{2x\ln x\cdot\dfrac{1}{x} + \left(\ln x\right)^2}{x^2\left(\ln x\right)^4} = \frac{2 + \ln x}{x^2\left(\ln x\right)^3}$$

↑   ↑
Use the Power Rule.   Simplify.

25.   (a) The function f is a polynomial, so the domain of f is all real numbers.

     (b) $f'(x) = 2x$

     (c) $f'(x)$ is a monomial, so the domain of f' is all real numbers.

     (d) $f'(x) = 0$ when $2x = 0$, or when $x = 0$.

     (e) There are no numbers in the domain of f for which $f'(x)$ does not exist.

     (f) $f''(x) = 2$

     (g) The domain of f'' is all real numbers.

27.   (a) The function f is a polynomial, so the domain of f is all real numbers.

     (b) $f'(x) = 3x^2 - 18x + 27$

     (c) $f'(x)$ is a polynomial, so the domain of f' is all real numbers.

     (d) $f'(x) = 0$ when $3x^2 - 18x + 27 = 0$, or when $x = 3$.

$$
\begin{array}{ll}
3x^2 - 18x + 27 = 0 & f'(x) = 0 \\
x^2 - 6x + 9 = 0 & \text{Divide both sides by 3.} \\
(x - 3)^2 = 0 & \text{Factor.} \\
x = 3 & \text{Use the square root method.}
\end{array}
$$

     (e) There are no numbers in the domain of f for which $f'(x)$ does not exist.

(f) $f''(x) = 6x - 18$

(g) The domain of f'' is all real numbers.

29. (a) The function f is a polynomial, so the domain of f is all real numbers.

(b) $f'(x) = 12x^3 - 36x^2$

(c) $f'(x)$ is a polynomial, so the domain of f' is all real numbers.

(d) $f'(x) = 0$ when $12x^3 - 36x^2 = 0$, or when $x = 0$ or $x = 3$.

$$
\begin{aligned}
12x^3 - 36x^2 &= 0 && f'(x) = 0 \\
x^3 - 3x^2 &= 0 && \text{Divide both sides by 12.} \\
x^2(x-3)^2 &= 0 && \text{Factor.} \\
x = 0 \quad \text{or} \quad x &= 3 && \text{Use the square root method.}
\end{aligned}
$$

(e) There are no numbers in the domain of f for which $f'(x)$ does not exist.

(f) $f''(x) = 36x^2 - 72x$

(g) f'' is a polynomial, so the domain of f'' is all real numbers.

31. (a) The domain of the function f is all real numbers except $x = 2$ and $x = -2$.

(b) $f'(x) = \dfrac{\left(x^2-4\right)\dfrac{d}{dx}x - x\dfrac{d}{dx}\left(x^2-4\right)}{\left(x^2-4\right)^2} = \dfrac{\left(x^2-4\right)-x\cdot 2x}{\left(x^2-4\right)^2} = -\dfrac{x^2+4}{\left(x^2-4\right)^2}$

(c) The domain of the function $f'(x)$ is all real numbers except $x = 2$ and $x = -2$.

(d) $f'(x)$ is never equal to zero.

(e) There are no numbers in the domain of f for which $f'(x)$ does not exist.

(f) $f''(x) = -\left[\dfrac{\left(x^2-4\right)^2\dfrac{d}{dx}\left(x^2+4\right) - \left(x^2+4\right)\dfrac{d}{dx}\left(x^2-4\right)^2}{\left(x^2-4\right)^4}\right]$

$= -\left[\dfrac{\left(x^2-4\right)^2\cdot(2x) - \left(x^2+4\right)\cdot 2\left(x^2-4\right)(2x)}{\left(x^2-4\right)^4}\right]$

$= -\left[\dfrac{2x\left(x^2-4\right)^2 - 4x\left(x^2+4\right)\left(x^2-4\right)}{\left(x^2-4\right)^3}\right]$

$$= -\left[\frac{2x^3 - 8x - 4x^3 - 16x}{\left(x^2 - 4\right)^3}\right]$$

$$= -\left[\frac{-2x^3 - 24x}{\left(x^2 - 4\right)^3}\right] = \frac{2x^3 + 24x}{\left(x^2 - 4\right)^3}$$

(g) The domain of the function $f''(x)$ is all real numbers except $x = 2$ and $x = -2$.

33. The function f is a polynomial of degree 3, so the fourth derivative is zero.

35. The function f is a polynomial of degree 19, so the twentieth derivative is zero.

37. The function f is a polynomial of degree 8, so the eighth derivative is equal to the constant

$$8! \cdot \frac{1}{8} = 7! = 5040$$

39. $v = s'(t) = 32t + 20$
$a = v'(t) = s''(t) = 32$

41. $v = s'(t) = 9.8\, t + 4$
$a = v'(t) = s''(t) = 9.8$

43. To find a formula for the nth derivative, we take successive derivatives until we see a pattern.

$$f(x) = e^x$$
$$f'(x) = e^x$$
$$f''(x) = e^x$$

We see that each order derivative is e^x, so we conclude that a formula for $f^{(n)}$ is
$$f^{(n)}(x) = e^x$$

45. To find a formula for the nth derivative, we take successive derivatives until we see a pattern.

$$f(x) = \ln x$$
$$f'(x) = \frac{1}{x} = x^{-1}$$
$$f''(x) = -x^{-2} = -\frac{1}{x^2}$$
$$f'''(x) = 2x^{-3} = \frac{2}{x^3}$$
$$f^{(4)}(x) = -3 \cdot 2x^{-4} = -\frac{3!}{x^4}$$
$$f^{(5)}(x) = (-4) \cdot \left(-3! x^{-5}\right) = \frac{4!}{x^5}$$

$$f^{(6)}(x)=(-5)\cdot\left(4!x^{-5}\right)=-\frac{5!}{x^6}$$

We see a pattern, noticing that the sign of the derivative alternates from positive to negative and conclude the formula for $f^{(n)}$ is

$$f^{(n)}(x)=(-1)^{(n-1)}\cdot\frac{(n-1)!}{x^n}$$

47. To find a formula for the nth derivative, we take successive derivatives until we see a pattern.

$$f(x)=x\ln x$$

$$f'(x)=x\cdot\frac{d}{dx}\ln x+\ln x\cdot\frac{d}{dx}x=x\cdot\frac{1}{x}+\ln x\cdot 1$$

$$=1+\ln x$$

$$f''(x)=0+\frac{1}{x}=\frac{1}{x}=x^{-1}$$

$$f'''(x)=-x^{-2}=-\frac{1}{x^2}$$

$$f^{(4)}(x)=2x^{-3}=\frac{2}{x^3}$$

$$f^{(5)}(x)=-3\cdot 2x^{-4}=-\frac{3!}{x^4}$$

$$f^{(6)}(x)=(-4)\cdot\left(-3!x^{-5}\right)=\frac{4!}{x^5}$$

$$f^{(7)}(x)=(-5)\cdot\left(4!x^{-5}\right)=-\frac{5!}{x^6}$$

We see a pattern, noticing that the sign of the derivative alternates from positive to negative and conclude the formula for $f^{(n)}$ is

$$f^{(n)}(x)=(-1)^{(n)}\cdot\frac{(n-2)!}{x^{n-1}} \text{ provided } n>1.$$

49. To find a formula for the nth derivative, we take successive derivatives until we see a pattern.

$$f(x)=(2x+3)^n$$

$$f'(x)=n(2x+3)^{n-1}\frac{d}{dx}(2x+3)=2n(2x+3)^{n-1} \qquad \text{Use the Power Rule.}$$

$$f''(x)=2n\cdot(n-1)(2x+3)^{n-2}\frac{d}{dx}(2x+3)=4n(n-1)(2x+3)^{n-2} \qquad \text{Use the Power Rule.}$$

$$=2^2 n(n-1)(2x+3)^{n-2}$$

$$f'''(x)=2^2 n(n-1)\cdot(n-2)(2x+3)^{n-3}\frac{d}{dx}(2x+3)$$

$$= 2^3 n(n-1)(n-2)(2x+3)^{n-3}$$
$$f^{(4)}(x) = 2^4 n(n-1)(n-2)(n-3)(2x+3)^{n-4}$$

We see a pattern and conclude the formula for $f^{(n)}$ is
$$f^{(n)}(x) = 2^n \cdot n!$$

51. To find a formula for the nth derivative, we take successive derivatives until we see a pattern.

$$f(x) = e^{ax}$$

$$f'(x) = e^{ax} \frac{d}{dx}(ax) = ae^{ax}$$

$$f''(x) = ae^{ax} \frac{d}{dx}(ax) = a^2 e^{ax}$$

$$f'''(x) = a^2 e^{ax} \frac{d}{dx}(ax) = a^3 e^{ax}$$

We see a pattern and conclude the formula for $f^{(n)}$ is
$$f^{(n)}(x) = a^n \cdot e^{ax}$$

53. To find a formula for the nth derivative, we take successive derivatives until we see a pattern.

$$f(x) = \ln(ax)$$

$$f'(x) = \frac{a}{ax} = \frac{1}{x} = \frac{0!}{x^1}$$

$$f''(x) = -\frac{1}{x^2} = -\frac{1!}{x^2}$$

$$f'''(x) = \frac{2}{x^3} = \frac{2!}{x^3}$$

$$f^{(4)}(x) = -\frac{6}{x^4} = -\frac{3!}{x^4}$$

$$f^{(5)}(x) = \frac{24}{x^5} = \frac{4!}{x^5}$$

We see a pattern and conclude the formula for $f^{(n)}$ is
$$f^{(n)}(x) = (-1)^{n-1} \cdot \frac{(n-1)!}{x^n}$$

55. $y = e^{2x}$

$$y' = \frac{d}{dx} e^{2x} = e^{2x} \frac{d}{dx}(2x) = e^{2x} \cdot 2 = 2e^{2x}$$

$$y'' = \frac{d}{dx} y' = \frac{d}{dx}(2e^{2x}) = 2e^{2x} \frac{d}{dx}(2x) = 2e^{2x} \cdot 2 = 4e^{2x}$$

So, $y'' - 4y = 4e^{2x} - 4e^{2x} = 0$

57. $f(x) = x^2 g(x)$

$$f'(x) = x^2 \frac{d}{dx} g(x) + g(x) \frac{d}{dx} x^2 = x^2 g'(x) + 2x g(x)$$

$\uparrow$ Derivative of a product $\uparrow$ Differentiate.

$$f''(x) = \frac{d}{dx} x^2 g'(x) + \frac{d}{dx} 2x g(x)$$

$$= \left[x^2 \frac{d}{dx} g'(x) + g'(x) \frac{d}{dx} x^2 \right] + \left[2x \frac{d}{dx} g(x) + g(x) \frac{d}{dx}(2x) \right]$$

$$= \left[x^2 g''(x) + 2x g'(x) \right] + \left[2x\, g'(x) + 2g(x) \right]$$

$$= x^2 g''(x) + 4x g'(x) + 2g(x)$$

59. (a) The velocity is

$$v = s'(t) = \frac{d}{dx}\left(6 + 80t - 16t^2\right) = 80 - 32t$$

At $t = 2$ seconds the velocity is $v(2) = 80 - 32(2) = 16$ feet per second.

(b) The ball reaches its maximum height when $v = 0$.

$$v = 80 - 32t = 0 \ \text{ when } t = \frac{80}{32} = 2.5 \text{ seconds}$$

The ball reaches its maximum height 2.5 seconds after it is thrown.

(c) At $t = 2.5$, $s(2.5) = 6 + 80(2.5) - 16(2.5^2) = 106$ feet. The ball reaches a maximum height of 106 feet.

(d) The acceleration is $a = v'(t) = -32$ feet per second per second.

(e) The ball strikes the ground when $s(t) = 0$. That is when

$$6 + 80t - 16t^2 = 0$$

Using the quadratic formula to solve for t, we find

$$t = \frac{-80 \pm \sqrt{(80)^2 - 4(-16)(6)}}{2(-16)}$$

$$t = \frac{80 \pm \sqrt{6784}}{32}$$

We need only the positive answer, since t represents time, and we get $t = 5.0739$. So the ball is in the air for 5.0739 seconds.

(f) The ball hits the ground at $t = 5.0739$ seconds, the velocity is

$$v(5.0739) = 80 - 32(5.0739) = -82.365 \text{ feet per second.}$$

The ball is moving at a speed of 82.365 feet per second in a downward direction.

(g) The total distance traveled by the ball is the distance up plus the distance down or $(106 - 6) + 106 = 206$ feet.

61. The velocity of the bullet is

$$v = s'(t) = \frac{d}{dx}\left[8 - (2-t)^3\right] = \frac{d}{dx}8 - \frac{d}{dx}(2-t)^3 = -3(2-t)^2\frac{d}{dx}(2-t) = 3(2-t)^2$$

meters per second. After 1 second, the bullet is traveling at a velocity

$$v(1) = 3(2-1)^2 = 3 \cdot 1^2 = 3 \text{ meters per second.}$$

The acceleration is $a = v'(t) = -6(2-t) = 6t - 12$ meters per second per second.

63. (a) The rock hits the ground when its height is zero. Since the rock started from a height of 88.2 meters, when it hits the ground the rock has traveled 88.2 meters.

$$4.9t^2 = 88.2$$
$$t^2 = \frac{88.2}{4.9} = 18$$
$$t = 3\sqrt{2} \approx 4.24$$

It takes the rock approximately 4.24 seconds to hit the ground.

(b) The average velocity is

$$\frac{ds}{dt} = \frac{s\left(3\sqrt{2}\right) - s(0)}{3\sqrt{2} - 0} = \frac{\left(88.2 - 4.9 \cdot 3\sqrt{2}^2\right) - \left(88.2 - 4.9 \cdot 0^2\right)}{3\sqrt{2}} = \frac{-20.8}{3\sqrt{2}} = 4.9$$

meters per second. That is, the rock is moving at an average speed of 4.9 meters per second in the downward direction.

(c) The average velocity in the first 3 seconds is 4.9 meters per second in a downward direction. (See part (b).)

(d) The velocity of the rock is $v = s'(t) = -9.8t$ meters per second. The rock hits the ground at $t = 4.24$ seconds. The velocity is $v(4.24) = -9.8(4.24) = -41.6$ meters per second.

13.7 Implicit Differentiation   773/1-42 (odd)

1. $$\frac{d}{dx}\left(x^2 + y^2\right) = \frac{d}{dx}4$$

$$2x + 2y\frac{dy}{dx} = 0$$

This is a linear equation in $\frac{dy}{dx}$. Solving for $\frac{dy}{dx}$, we have

$$2y\frac{dy}{dx} = -2x$$
$$\frac{dy}{dx} = \frac{-2x}{2y} = -\frac{x}{y} \quad \text{provided } y \neq 0.$$

3.
$$\frac{d}{dx}\left(x^2 y\right) = \frac{d}{dx} 8$$

$$x^2 \frac{d}{dx} y + y \frac{d}{dx} x^2 = 0$$

$$x^2 \frac{dy}{dx} + y \cdot 2x = 0$$

This is a linear equation in $\frac{dy}{dx}$. Solving for $\frac{dy}{dx}$, we have

$$x^2 \frac{dy}{dx} = -2xy$$

$$\frac{dy}{dx} = \frac{-2xy}{x^2} = -\frac{2y}{x} \text{ provided } x \neq 0.$$

5.
$$\frac{d}{dx}\left(x^2 + y^2 - xy\right) = \frac{d}{dx} 2$$

$$2x + 2y\frac{dy}{dx} - \left(x\frac{d}{dx}y + y\frac{d}{dx}x\right) = 0$$

$$2x + 2y\frac{dy}{dx} - x\frac{dy}{dx} - y = 0$$

This is a linear equation in $\frac{dy}{dx}$. Solving for $\frac{dy}{dx}$, we have

$$2x + (2y - x)\frac{dy}{dx} - y = 0$$

$$(2y - x)\frac{dy}{dx} = y - 2x$$

$$\frac{dy}{dx} = \frac{y - 2x}{2y - x} \text{ provided } 2y - x \neq 0.$$

7.
$$\frac{d}{dx}\left(x^2 + 4xy + y^2\right) = \frac{d}{dx} y$$

$$\frac{d}{dx}x^2 + 4x\frac{d}{dx}y + y\frac{d}{dx}(4x) + \frac{d}{dx}y^2 = \frac{d}{dx} y$$

$$2x + 4x\frac{dy}{dx} + y \cdot 4 + 2y\frac{dy}{dx} = \frac{dy}{dx}$$

This is a linear equation in $\frac{dy}{dx}$. Solving for $\frac{dy}{dx}$, we have

$$2x + 4y + (4x + 2y)\frac{dy}{dx} = \frac{dy}{dx}$$

$$(4x + 2y)\frac{dy}{dx} - \frac{dy}{dx} = -2x - 4y$$

$$(4x + 2y - 1)\frac{dy}{dx} = -2x - 4y$$

$$\frac{dy}{dx} = \frac{-2x - 4y}{4x + 2y - 1} \quad \text{provided } 4x + 2y - 1 \neq 0.$$

9. $$\frac{d}{dx}(3x^2 + y^3) = \frac{d}{dx}1$$

$$6x + 3y^2\frac{dy}{dx} = 0$$

This is a linear equation in $\frac{dy}{dx}$. Solving for $\frac{dy}{dx}$, we have

$$3y^2\frac{dy}{dx} = -6x$$

$$\frac{dy}{dx} = \frac{-6x}{3y^2} = -\frac{2x}{y^2} \quad \text{provided } y \neq 0.$$

11. $$\frac{d}{dx}(4x^3 + 2y^3) = \frac{d}{dx}x^2$$

$$12x^2 + 6y^2\frac{dy}{dx} = 2x$$

This is a linear equation in $\frac{dy}{dx}$. Solving for $\frac{dy}{dx}$, we have

$$6y^2\frac{dy}{dx} = 2x - 12x^2$$

$$\frac{dy}{dx} = \frac{2x - 12x^2}{6y^2} = \frac{x - 6x^2}{3y^2} \quad \text{provided } y \neq 0.$$

13. $$\frac{d}{dx}\left(\frac{1}{x^2} - \frac{1}{y^2}\right) = \frac{d}{dx}\left(x^{-2} - y^{-2}\right) = \frac{d}{dx}4$$

$$-2x^{-3} + 2y^{-3}\frac{dy}{dx} = 0$$

This is a linear equation in $\frac{dy}{dx}$. Solving for $\frac{dy}{dx}$, we have

$$2y^{-3}\frac{dy}{dx} = 2x^{-3}$$

$$\frac{dy}{dx} = \frac{2x^{-3}}{2y^{-3}} = \frac{y^3}{x^3} \quad \text{provided } x \neq 0.$$

15. $\dfrac{d}{dx}\left(\dfrac{1}{x}+\dfrac{1}{y}\right)=\dfrac{d}{dx}\left(x^{-1}+y^{-1}\right)=\dfrac{d}{dx}2$

$$-x^{-2}-y^{-2}\dfrac{dy}{dx}=0$$

This is a linear equation in $\dfrac{dy}{dx}$. Solving for $\dfrac{dy}{dx}$, we have

$$-y^{-2}\dfrac{dy}{dx}=x^{-2}$$

$$\dfrac{dy}{dx}=\dfrac{x^{-2}}{-y^{-2}}=-\dfrac{y^2}{x^2}\quad\text{provided }x\neq 0.$$

17. $\dfrac{d}{dx}\left(x^2+y^2\right)=\dfrac{d}{dx}\left(ye^x\right)$

$$\dfrac{d}{dx}x^2+\dfrac{d}{dx}y^2=y\dfrac{d}{dx}e^x+e^x\dfrac{d}{dx}y$$

$$2x+2y\dfrac{dy}{dx}=ye^x+e^x\dfrac{dy}{dx}$$

This is a linear equation in $\dfrac{dy}{dx}$. Solving for $\dfrac{dy}{dx}$, we have

$$2y\dfrac{dy}{dx}-e^x\dfrac{dy}{dx}=ye^x-2x$$

$$\left(2y-e^x\right)\dfrac{dy}{dx}=ye^x-2x$$

$$\dfrac{dy}{dx}=\dfrac{ye^x-2x}{2y-e^x}\quad\text{provided }2y-e^x\neq 0.$$

19. $\dfrac{d}{dx}\left(\dfrac{x}{y}+\dfrac{y}{x}\right)=\dfrac{d}{dx}\left(6e^x\right)$

$$\left[\dfrac{y\dfrac{d}{dx}x-x\dfrac{d}{dx}y}{y^2}\right]+\left[\dfrac{x\dfrac{d}{dx}y-y\dfrac{d}{dx}x}{x^2}\right]=6e^x$$

$$\left[\dfrac{y-x\dfrac{dy}{dx}}{y^2}\right]+\left[\dfrac{x\dfrac{dy}{dx}-y}{x^2}\right]=6e^x$$

$$\dfrac{x^2y-x^3\dfrac{dy}{dx}+xy^2\dfrac{dy}{dx}-y^3}{x^2y^2}=6e^x$$

$$x^2y-x^3\dfrac{dy}{dx}+xy^2\dfrac{dy}{dx}-y^3=6x^2y^2e^x$$

This is a linear equation in $\dfrac{dy}{dx}$. Solving for $\dfrac{dy}{dx}$, we have

$$-\left(x^3 - xy^2\right)\frac{dy}{dx} = 6x^2 y^2 e^x - x^2 y + y^3$$

$$\frac{dy}{dx} = \frac{6x^2 y^2 e^x - x^2 y + y^3}{-x^3 + xy^2} \quad \text{provided } -x^3 + xy^2 \neq 0.$$

21. $\dfrac{d}{dx} x^2 = \dfrac{d}{dx}\left(y^2 \ln x\right)$

$$2x = y^2 \frac{d}{dx} \ln x + \ln x \frac{d}{dx} y^2$$

$$2x = y^2 \cdot \frac{1}{x} + \ln x \cdot 2y \frac{dy}{dx}$$

This is a linear equation in $\dfrac{dy}{dx}$. Solving for $\dfrac{dy}{dx}$, we have

$$2x = \frac{y^2}{x} + 2y \ln x \frac{dy}{dx}$$

$$2x - \frac{y^2}{x} = 2y \ln x \frac{dy}{dx}$$

$$\frac{dy}{dx} = \frac{2x - \dfrac{y^2}{x}}{2y \ln x} = \frac{2x^2 - y^2}{2xy \ln x} \quad \text{provided } 2xy \ln x \neq 0.$$

23.
$$\frac{d}{dx}(2x+3y)^2 = \frac{d}{dx}\left(x^2 + y^2\right)$$

$$2(2x+3y)\frac{d}{dx}(2x+3y) = 2x + 2y\frac{dy}{dx}$$

$$2(2x+3y)\left(2 + 3\frac{dy}{dx}\right) = 2x + 2y\frac{dy}{dx}$$

$$\cancel{2}(2x+3y)\left(2 + 3\frac{dy}{dx}\right) = \cancel{2}x + \cancel{2}y\frac{dy}{dx}$$

$$4x + 6x\frac{dy}{dx} + 6y + 9y\frac{dy}{dx} = x + y\frac{dy}{dx}$$

This is a linear equation in $\dfrac{dy}{dx}$. Solving for $\dfrac{dy}{dx}$, we have

$$6x\frac{dy}{dx} + 9y\frac{dy}{dx} - y\frac{dy}{dx} = x - 4x - 6y$$

$$(6x + 9y - y)\frac{dy}{dx} = x - 4x - 6y$$

$$\frac{dy}{dx} = \frac{-3x - 6y}{6x + 8y} \quad \text{provided } 6x + 8y \neq 0.$$

25.

$$\frac{d}{dx}\left(x^2+y^2\right)^2 = \frac{d}{dx}\left(x-y\right)^3$$

$$2\left(x^2+y^2\right)\frac{d}{dx}\left(x^2+y^2\right) = 3\left(x-y\right)^2\frac{d}{dx}\left(x-y\right)$$

$$2\left(x^2+y^2\right)\left(2x+2y\frac{dy}{dx}\right) = 3\left(x-y\right)^2\left(1-\frac{dy}{dx}\right)$$

$$4x^3+4x^2y\frac{dy}{dx}+4xy^2+4y^3\frac{dy}{dx} = 3x^2-6xy+3y^2-\left(3x^2-6xy+3y^2\right)\frac{dy}{dx}$$

This is a linear equation in $\dfrac{dy}{dx}$. Solving for $\dfrac{dy}{dx}$, we have

$$4x^2y\frac{dy}{dx}+4y^3\frac{dy}{dx}+\left(3x^2-6xy+3y^2\right)\frac{dy}{dx} = 3x^2-6xy+3y^2-4x^3-4xy^2$$

$$\left(4x^2y+4y^3+3x^2-6xy+3y^2\right)\frac{dy}{dx} = 3x^2-6xy+3y^2-4x^3-4xy^2$$

$$\frac{dy}{dx} = \frac{3x^2-6xy+3y^2-4x^3-4xy^2}{4x^2y+4y^3+3x^2-6xy+3y^2}$$

provided $4x^2y+4y^3+3x^2-6xy+3y^2 \neq 0$.

27.

$$\frac{d}{dx}\left(x^3+y^3\right)^2 = \frac{d}{dx}\left(x^2y^2\right)$$

$$2\left(x^3+y^3\right)\frac{d}{dx}\left(x^3+y^3\right) = x^2\frac{d}{dx}y^2+y^2\frac{d}{dx}x^2$$

$$2\left(x^3+y^3\right)\left(3x^2+3y^2\frac{dy}{dx}\right) = x^2\cdot 2y\frac{dy}{dx}+y^2\cdot 2x$$

$$\not{2}\left(x^3+y^3\right)\left(3x^2+3y^2\frac{dy}{dx}\right) = \not{2}x^2y\frac{dy}{dx}+\not{2}xy^2$$

$$3x^5+3x^3y^2\frac{dy}{dx}+3x^2y^3+3y^5\frac{dy}{dx} = x^2y\frac{dy}{dx}+xy^2$$

This is a linear equation in $\dfrac{dy}{dx}$. Solving for $\dfrac{dy}{dx}$, we have

$$3x^3y^2\frac{dy}{dx}+3y^5\frac{dy}{dx}-x^2y\frac{dy}{dx} = xy^2-3x^5-3x^2y^3$$

$$\left(3x^3y^2+3y^5-x^2y\right)\frac{dy}{dx} = xy^2-3x^5-3x^2y^3$$

$$\frac{dy}{dx} = \frac{xy^2-3x^5-3x^2y^3}{3x^3y^2+3y^5-x^2y} \quad \text{provided } 3x^3y^2+3y^5-x^2y \neq 0.$$

29.

$$\frac{d}{dx}y = \frac{d}{dx}e^{x^2+y^2}$$

$$\frac{dy}{dx} = e^{x^2+y^2}\frac{d}{dx}\left(x^2+y^2\right)$$

$$\frac{dy}{dx} = e^{x^2+y^2}\left(2x + 2y\frac{dy}{dx}\right)$$

$$\frac{dy}{dx} = 2xe^{x^2+y^2} + 2ye^{x^2+y^2}\frac{dy}{dx}$$

This is a linear equation in $\frac{dy}{dx}$. Solving for $\frac{dy}{dx}$, we have

$$\frac{dy}{dx} - 2ye^{x^2+y^2}\frac{dy}{dx} = 2xe^{x^2+y^2}$$

$$\left(1 - 2ye^{x^2+y^2}\right)\frac{dy}{dx} = 2xe^{x^2+y^2}$$

$$\frac{dy}{dx} = \frac{2xe^{x^2+y^2}}{1 - 2ye^{x^2+y^2}} \quad \text{provided } 1 - 2ye^{x^2+y^2} \neq 0.$$

31. The first derivative is $y' = -\dfrac{x}{y}$, provided $y \neq 0$ (from Problem 1).

The second derivative is

$$\frac{d}{dx}\left(2x + 2y\frac{dy}{dx}\right) = \frac{d}{dx}0$$

$$2\frac{d}{dx}x + 2\frac{dy}{dx}(yy') = 0$$

$$2 + 2\left[y\frac{d}{dx}y' + y'\frac{d}{dx}y\right] = 0$$

$$\cancel{2} + \cancel{2}\left[yy'' + y'y'\right] = 0$$

$$1 + yy'' + (y')^2 = 0$$

$$y'' = \frac{-(y')^2 - 1}{y} = \frac{-\left(-\dfrac{x}{y}\right)^2 - 1}{y} = -\frac{\dfrac{x^2}{y^2} + 1}{y} = -\frac{x^2 + y^2}{y^3}$$

provided $y \neq 0$.

33. The first derivative is

$$\frac{d}{dx}\left(xy + yx^2\right) = \frac{d}{dx}2$$

$$\left[x\frac{d}{dx}y + y\frac{d}{dx}x\right] + \left[y\frac{d}{dx}x^2 + x^2\frac{d}{dx}y\right] = 0$$

$$\left[xy' + y\right] + \left[y\cdot 2x + x^2y'\right] = 0$$

$$xy' + y + 2xy + x^2y' = 0$$

$$\left(x + x^2\right)y' + \left(1 + 2x\right)y = 0$$

$$\left(x + x^2\right)y' = -\left(1 + 2x\right)y$$

$$y' = -\frac{(1+2x)y}{x^2+x} \quad \text{provided } x \neq 0, \ x \neq -1.$$

Using the fifth line from above, we find the second derivative is

$$\frac{d}{dx}\left[(x+x^2)y' + (1+2x)y\right] = \frac{d}{dx}0$$

$$\left[(x+x^2)\frac{d}{dx}y' + y'\frac{d}{dx}(x+x^2)\right] + \left[(1+2x)\frac{d}{dx}y + y\frac{d}{dx}(1+2x)\right] = 0$$

$$\left[(x+x^2)y'' + y'(1+2x)\right] + \left[(1+2x)y' + y\cdot 2\right] = 0$$

$$(x+x^2)y'' + y'(1+2x) + (1+2x)y' + 2y = 0$$

$$(x+x^2)y'' + 2(1+2x)y' + 2y = 0$$

$$(x+x^2)y'' = -2(1+2x)y' - 2y$$

$$y'' = -\frac{2(1+2x)y' + 2y}{x^2+x}$$

$$y'' = -\frac{2(1+2x)\left(\dfrac{y+2xy}{x^2+x}\right) + 2y}{x^2+x}$$

$$y'' = -\frac{2(1+2x)(y+2xy) + 2y(x^2+x)}{(x^2+x)^2}$$

$$y'' = -\frac{2y+8xy+8x^2y+2x^2y+2xy}{(x^2+x)^2}$$

$$y'' = -\frac{2y+10xy+10x^2y}{(x^2+x)^2}$$

provided $x \neq 0, \ x \neq -1$.

35. The slope of the tangent line is $\dfrac{dy}{dx}$, which is

$$\frac{d}{dx}(x^2+y^2) = \frac{d}{dx}5$$

$$2x + 2y\frac{dy}{dx} = 0$$

Solving for $\dfrac{dy}{dx}$, we have

$$2y\frac{dy}{dx} = -2x$$

$$\frac{dy}{dx} = \frac{-2x}{2y} = -\frac{x}{y} \quad \text{provided } y \neq 0.$$

The slope of the tangent line at the point $(1, 2)$ is $m_{\tan} = -\dfrac{1}{2}$. The equation of the tangent line is

$$y - y_1 = m(x - x_1)$$

$$y - 2 = -\frac{1}{2}(x - 1)$$

$$y - 2 = -\frac{1}{2}x + \frac{1}{2}$$

$$y = -\frac{1}{2}x + \frac{5}{2}$$

37. (Note: The problem as printed in the first printing of the text is incorrect. That equation has no tangent line at $(0, 0)$.) The slope of the tangent line is $\dfrac{dy}{dx}$, which is

$$\frac{d}{dx}e^{xy} = \frac{d}{dx}x$$

$$e^{xy}\frac{d}{dx}(xy) = 1$$

$$e^{xy}\left(x\frac{d}{dx}y + y\frac{d}{dx}\right) = 1$$

$$e^{xy}\left(x\frac{dy}{dx} + y\right) = 1$$

$$e^{xy}x\frac{dy}{dx} + ye^{xy} = 1$$

$$\frac{dy}{dx} = \frac{1 - ye^{xy}}{xe^{xy}}$$

The slope of the tangent line at the point $(1, 0)$ is $m_{\tan} = \dfrac{1 - 0 \cdot e^0}{1e^0} = 1$. The equation of the tangent line is

$$y - y_1 = m(x - x_1)$$

$$y - 0 = 1(x - 1)$$

$$y = x - 1$$

39. The tangent line is horizontal when the slope is zero, that is when $\dfrac{dy}{dx} = 0$.

$$\frac{d}{dx}(x^2 + y^2) = \frac{d}{dx}4$$

$$2x + 2y\frac{dy}{dx} = 0$$

$$\frac{dy}{dx} = \frac{-2x}{2y} = -\frac{x}{y} \quad \text{provided } y \neq 0.$$

$\dfrac{dy}{dx} = 0$ when $x = 0$.

When $x = 0$, $y^2 = 4$, or $y = \pm 2$. So there are horizontal tangent lines at the points $(0, 2)$ and $(0, -2)$.

41. The tangent line is horizontal when the slope is zero, that is when $\dfrac{dy}{dx} = 0$.

$$\frac{d}{dx}\left(y^2 + 4x^2\right) = \frac{d}{dx}16$$

$$2y\frac{dy}{dx} + 8x = 0$$

$$\frac{dy}{dx} = \frac{-8x}{2y} = -\frac{4x}{y} \quad \text{provided } y \neq 0.$$

$\dfrac{dy}{dx} = 0$ when $x = 0$.

When $x = 0$, $y^2 = 16$, or $y = \pm 4$. So there are horizontal tangent lines at the points $(0, 4)$ and $(0, -4)$.

43. (a) The slope of the tangent line is $\dfrac{dy}{dx}$, which is

$$\frac{d}{dx}\left(x + xy + 2y^2\right) = \frac{d}{dx}6$$

$$\frac{d}{dx}x + x\frac{d}{dx}y + y\frac{d}{dx}x + \frac{d}{dx}\left(2y^2\right) = 0$$

$$1 + x\frac{dy}{dx} + y + 4y\frac{dy}{dx} = 0$$

$$\left(x + 4y\right)\frac{dy}{dx} = -y - 1$$

$$\frac{dy}{dx} = \frac{-y-1}{x+4y} = -\frac{y+1}{x+4y} \quad \text{provided } x + 4y \neq 0.$$

(b) At the point $(2, 1)$ the slope of the tangent line is $m_{\text{tan}} = -\dfrac{1+1}{2+4\cdot 1} = -\dfrac{2}{6} = -\dfrac{1}{3}$, and an equation of the tangent line is

$$y - y_1 = m\left(x - x_1\right)$$

$$y - 1 = -\frac{1}{3}(x - 2)$$

$$y = -\frac{1}{3}x + \frac{5}{3}$$

(c) To find the coordinates of the points (x, y) at which the slope of the tangent line equals the slope of the tangent line at $(2, 1)$, we need to solve the system of equations

$$x + xy + 2y^2 = 6 \qquad (1)$$

$$-\frac{y+1}{x+4y} = -\frac{1}{3} \qquad (2)$$

Beginning with equation (2), we get $3y + 3 = x + 4y$ or $3 = x + y$ or $x = 3 - y$.
Substituting $x = 3 - y$ into equation (1), we get

$$(3 - y) + (3 - y)y + 2y^2 = 6$$

$$3 - y + 3y - y^2 + 2y^2 = 6$$

$$3 + 2y + y^2 = 6$$

$$y^2 + 2y - 3 = 0$$

$$(y - 1)(y + 3) = 0$$

$$y = 1 \qquad y = -3$$

Back substituting $y = 1$ into equation (2) we get $x = 2$.
Back substituting $y = -3$ into equation (2) we get $x = 6$.

So the coordinates of the point which has the same slope as the tangent line at (2, 1) is
(6, −3).

45.

$$\frac{d}{dP}\left(P + \frac{a}{V^2}\right) = \frac{d}{dP}\frac{C}{V - b}$$

$$\frac{d}{dP}P + \frac{d}{dP}aV^{-2} = C\frac{d}{dP}(V - b)^{-1}$$

$$1 - 2aV^{-3}\frac{dV}{dP} = -C(V - b)^{-2}\frac{d}{dP}(V - b)$$

$$1 - \frac{2a}{V^3}\frac{dV}{dP} = -\frac{C}{(V - b)^2}\frac{dV}{dP}$$

$$1 = \left(\frac{2a}{V^3} - \frac{C}{(V - b)^2}\right)\frac{dV}{dP}$$

$$\left(\frac{2a(V - b)^2 - CV^3}{V^3(V - b)^2}\right)\frac{dV}{dP} = 1$$

$$\frac{dV}{dP} = \frac{V^3(V - b)^2}{2a(V - b)^2 - CV^3}$$

47.

(a) $$\frac{d}{dt}e^{N(t)} = \frac{d}{dt}\left(430{,}163t + \frac{3t}{t^2 + 2}\right)$$

$$e^{N(t)}\frac{d}{dt}N(t) = \frac{d}{dt}430{,}163t + \frac{(t^2 + 2)\frac{d}{dt}(3t) - 3t\frac{d}{dt}(t^2 + 2)}{(t^2 + 2)^2}$$

$$e^{N(t)}\frac{dN}{dt} = 430{,}163 + \frac{3(t^2 + 2) - 6t^2}{(t^2 + 2)^2}$$

$$e^{N(t)} \frac{dN}{dt} = \frac{430,163(t^2+2)^2 + 3(t^2+2) - 6t^2}{(t^2+2)^2}$$

$$\frac{dN}{dt} = \frac{430,163(t^2+2)^2 + 3(t^2+2) - 6t^2}{e^{N(t)}(t^2+2)^2} = \frac{430,163(t^2+2)^2 + 3(t^2+2) - 6t^2}{\left(430,163t + \dfrac{3t}{t^2+2}\right)(t^2+2)^2}$$

$$= \frac{430,163(t^2+2)^2 + 3(t^2+2) - 6t^2}{430,163t(t^2+2)^2 + 3t(t^2+2)}$$

(b) $N(2) = \dfrac{430,163(2^2+2)^2 + 3(2^2+2) - 6 \cdot 2^2}{(430,163 \cdot 2)(2^2+2)^2 + (3 \cdot 2)(2^2+2)} = 0.500$

$N(4) = \dfrac{430,163(4^2+2)^2 + 3(4^2+2) - 6 \cdot 4^2}{(430,163 \cdot 4)(4^2+2)^2 + (3 \cdot 4)(4^2+2)} = 0.250$

13.8 The Derivative of $x^{p/q}$

1. $f'(x) = \dfrac{d}{dx} x^{4/3} = \dfrac{4}{3} x^{(4/3)-1} = \dfrac{4}{3} x^{1/3}$

3. $f'(x) = \dfrac{d}{dx} x^{2/3} = \dfrac{2}{3} x^{(2/3)-1} = \dfrac{2}{3} x^{-1/3} = \dfrac{2}{3x^{1/3}}$

5. $f'(x) = \dfrac{d}{dx} \dfrac{1}{x^{1/2}} = \dfrac{d}{dx} x^{-1/2} = -\dfrac{1}{2} x^{(-1/2)-1} = -\dfrac{1}{2} x^{-3/2} = -\dfrac{1}{2x^{3/2}}$

7. $f'(x) = \dfrac{d}{dx}(2x+3)^{3/2} = \dfrac{3}{2}(2x+3)^{(3/2)-1} \dfrac{d}{dx}(2x+3) = \dfrac{3}{2}(2x+3)^{1/2} \cdot 2 = 3(2x+3)^{1/2}$

9. $f'(x) = \dfrac{d}{dx}(x^2+4)^{3/2} = \dfrac{3}{2}(x^2+4)^{(3/2)-1} \dfrac{d}{dx}(x^2+4) = \dfrac{3}{2}(x^2+4)^{1/2} \cdot 2x = 3x(x^2+4)^{1/2}$

11. We first change each radical to its fractional exponent equivalent.

$f(x) = (2x+3)^{1/2}$

$f'(x) = \dfrac{d}{dx}(2x+3)^{1/2} = \dfrac{1}{2}(2x+3)^{(1/2)-1} \dfrac{d}{dx}(2x+3) = \dfrac{1}{2}(2x+3)^{-1/2} \cdot 2 = \dfrac{1}{(2x+3)^{1/2}}$

$$= \dfrac{1}{\sqrt{2x+3}}$$

13. We first change each radical to its fractional exponent equivalent.

$$f(x) = \left(9x^2 + 1\right)^{1/2}$$

$$f'(x) = \frac{d}{dx}\left(9x^2 + 1\right)^{1/2} = \frac{1}{2}\left(9x^2 + 1\right)^{(1/2)-1}\frac{d}{dx}\left(9x^2 + 1\right) = \frac{1}{2}\left(9x^2 + 1\right)^{-1/2} \cdot 18x$$

$$= \frac{9x}{\left(9x^2 + 1\right)^{1/2}} = \frac{9x}{\sqrt{9x^2 + 1}}$$

15.
$$f'(x) = \frac{d}{dx}\left(3x^{5/3} - 6x^{1/3}\right) = 3\frac{d}{dx}x^{5/3} - 6\frac{d}{dx}x^{1/3}$$

$$= 3 \cdot \frac{5}{3}x^{(5/3)-1} - 6 \cdot \frac{1}{3}x^{(1/3)-1}$$

$$= 5x^{2/3} - 2x^{-2/3} = 5x^{2/3} - \frac{2}{x^{2/3}}$$

17. To find the derivative of function f we can either use the formula for the derivative of a product or multiply the factors and find the derivative of the sum. We chose to multiply first.

$$f(x) = x^{1/3}\left(x^2 - 4\right) = x^{7/3} - 4x^{1/3}$$

$$f'(x) = \frac{d}{dx}\left(x^{7/3} - 4x^{1/3}\right) = \frac{d}{dx}x^{7/3} - 4\frac{d}{dx}x^{1/3} = \frac{7}{3}x^{(7/3)-1} - 4 \cdot \frac{1}{3}x^{(1/3)-1}$$

$$= \frac{7}{3}x^{4/3} - \frac{4}{3}x^{-2/3} = \frac{7}{3}x^{4/3} - \frac{4}{3x^{2/3}}$$

19. We first change each radical to its fractional exponent equivalent.

$$f(x) = \frac{x}{\sqrt{x^2 - 4}} = \frac{x}{\left(x^2 - 4\right)^{1/2}}$$

$$f'(x) = \frac{d}{dx}\frac{x}{\left(x^2 - 4\right)^{1/2}} = \frac{\left(x^2 - 4\right)^{1/2}\frac{d}{dx}x - x\frac{d}{dx}\left(x^2 - 4\right)^{1/2}}{\left[\left(x^2 - 4\right)^{1/2}\right]^2}$$

Use the formula for the derivative of a quotient.

$$= \frac{\left(x^2 - 4\right)^{1/2} - x \cdot \frac{1}{2}\left(x^2 - 4\right)^{(1/2)-1}\frac{d}{dx}\left(x^2 - 4\right)}{\left(x^2 - 4\right)}$$

$$= \frac{\left(x^2 - 4\right)^{1/2} - x \cdot \frac{1}{2}\left(x^2 - 4\right)^{-1/2} \cdot 2x}{\left(x^2 - 4\right)}$$

$$= \frac{\left(x^2 - 4\right)^{1/2} - x^2\left(x^2 - 4\right)^{-1/2}}{\left(x^2 - 4\right)}$$

$$= \frac{\left(x^2 - 4\right) - x^2}{\left(x^2 - 4\right)^{3/2}} = -\frac{4}{\left(x^2 - 4\right)^{3/2}}$$

Multiply by $\dfrac{\left(x^2 - 4\right)^{1/2}}{\left(x^2 - 4\right)^{1/2}}$; simplify.

21. We first change each radical to its fractional exponent equivalent.

$$f(x) = \sqrt{e^x} = \left(e^x\right)^{1/2}$$

$$f'(x) = \frac{d}{dx}\left(e^x\right)^{1/2} = \frac{1}{2}\left(e^x\right)^{(1/2)-1}\frac{d}{dx}e^x = \frac{1}{2}\left(e^x\right)^{-1/2}e^x = \frac{1}{2}\left(e^x\right)^{1/2} = \frac{\sqrt{e^x}}{2}$$

23. We first change each radical to its fractional exponent equivalent.

$$f(x) = \sqrt{\ln x} = \left(\ln x\right)^{1/2}$$

$$f'(x) = \frac{d}{dx}\left(\ln x\right)^{1/2} = \frac{1}{2}\left(\ln x\right)^{(1/2)-1}\frac{d}{dx}\ln x = \frac{1}{2}\left(\ln x\right)^{-1/2}\cdot\frac{1}{x} = \frac{1}{2x\left(\ln x\right)^{1/2}} = \frac{1}{2x\sqrt{\ln x}}$$

25. We first change each radical to its fractional exponent equivalent.

$$f(x) = e^{\sqrt[3]{x}} = e^{x^{1/3}}$$

$$f'(x) = \frac{d}{dx}e^{x^{1/3}} = e^{x^{1/3}}\frac{d}{dx}x^{1/3} = e^{x^{1/3}}\cdot\frac{1}{3}x^{(1/3)-1} = \frac{1}{3}x^{-2/3}e^{x^{1/3}} = \frac{e^{x^{1/3}}}{3x^{2/3}} = \frac{e^{\sqrt[3]{x}}}{3x^{2/3}} = \frac{e^{\sqrt[3]{x}}}{3\sqrt[3]{x^2}}$$

27. We first change each radical to its fractional exponent equivalent.

$$f(x) = \sqrt[3]{\ln x} = \left(\ln x\right)^{1/3}$$

$$f'(x) = \frac{d}{dx}\left(\ln x\right)^{1/3} = \frac{1}{3}\left(\ln x\right)^{(1/3)-1}\frac{d}{dx}\ln x = \frac{1}{3}\left(\ln x\right)^{-2/3}\cdot\frac{1}{x} = \frac{1}{3x\left(\ln x\right)^{2/3}} = \frac{1}{3x\sqrt[3]{\left(\ln x\right)^2}}$$

29. We first change each radical to its fractional exponent equivalent.

$$f(x) = \sqrt{x}\,e^x = x^{1/2}e^x$$

$$f'(x) = \frac{d}{dx}\left(x^{1/2}e^x\right) = x^{1/2}\frac{d}{dx}e^x + e^x\frac{d}{dx}x^{1/2} \qquad \text{Use the formula for the derivative of a product.}$$

$$= x^{1/2}e^x + e^x\cdot\frac{1}{2}x^{(1/2)-1}$$

$$= x^{1/2}e^x + \frac{1}{2}x^{-1/2}e^x$$

$$= x^{1/2}e^x + \frac{e^x}{2x^{1/2}}$$

$$= \frac{2xe^x + e^x}{2x^{1/2}} = \frac{e^x(2x+1)}{2\sqrt{x}}$$

31. We first change each radical to its fractional exponent equivalent.

$$f(x) = e^{2x}\sqrt{x^2+1} = e^{2x}\left(x^2+1\right)^{1/2}$$

$$f'(x) = \frac{d}{dx}\left(e^{2x}\left(x^2+1\right)^{1/2}\right) = e^{2x}\frac{d}{dx}\left(x^2+1\right)^{1/2} + \left(x^2+1\right)^{1/2}\frac{d}{dx}e^{2x} \qquad \text{Use the formula for the derivative of a quotient.}$$

$$= e^{2x}\frac{1}{2}\left(x^2+1\right)^{(1/2)-1}\frac{d}{dx}\left(x^2+1\right) + \left(x^2+1\right)^{1/2}e^{2x}\frac{d}{dx}(2x) \qquad \text{Use the Chain Rule.}$$

$$= e^{2x} \frac{1}{2}\left(x^2+1\right)^{-1/2} \cdot 2x + \left(x^2+1\right)^{1/2} e^{2x} \cdot 2 \qquad \text{Differentiate.}$$

$$= xe^{2x}\left(x^2+1\right)^{-1/2} + 2e^{2x}\left(x^2+1\right)^{1/2} \qquad \text{Simplify.}$$

$$= \frac{xe^{2x}}{\left(x^2+1\right)^{1/2}} + 2e^{2x}\left(x^2+1\right)^{1/2} = \frac{xe^{2x} + 2e^{2x}\left(x^2+1\right)}{\left(x^2+1\right)^{1/2}} = \frac{e^{2x}\left(2x^2+x+2\right)}{\sqrt{x^2+1}}$$

33.

$$\frac{d}{dx}\left(\sqrt{x}+\sqrt{y}\right) = \frac{d}{dx}4$$

$$\frac{1}{2\sqrt{x}} + \frac{1}{2\sqrt{y}}\frac{dy}{dx} = 0$$

$$\frac{1}{2\sqrt{y}}\frac{dy}{dx} = -\frac{1}{2\sqrt{x}}$$

$$\frac{dy}{dx} = -\frac{2\sqrt{y}}{2\sqrt{x}} = -\frac{\sqrt{y}}{\sqrt{x}}$$

35.

$$\sqrt{x^2+y^2} = \left(x^2+y^2\right)^{1/2}$$

$$\frac{d}{dx}\left(x^2+y^2\right)^{1/2} = \frac{d}{dx}x$$

$$\frac{1}{2}\left(x^2+y^2\right)^{(1/2)-1}\frac{d}{dx}\left(x^2+y^2\right) = 1$$

$$\frac{1}{2}\left(x^2+y^2\right)^{-1/2}\left(2x+2y\frac{dy}{dx}\right) = 1$$

$$2x + 2y\frac{dy}{dx} = 2\left(x^2+y^2\right)^{1/2}$$

$$y\frac{dy}{dx} = \left(x^2+y^2\right)^{1/2} - x$$

$$\frac{dy}{dx} = \frac{\left(x^2+y^2\right)^{1/2} - x}{y} = \frac{\sqrt{x^2+y^2} - x}{y}$$

37.

$$\frac{d}{dx}\left(x^{1/3}+y^{1/3}\right) = \frac{d}{dx}1$$

$$\frac{1}{3}x^{(1/3)-1} + \frac{1}{3}y^{(1/3)-1} = 0$$

$$x^{-2/3} + y^{-2/3}\frac{dy}{dx} = 0$$

$$y^{-2/3}\frac{dy}{dx} = -x^{-2/3}$$

$$\frac{dy}{dx} = -x^{-2/3}y^{2/3} = -\frac{y^{2/3}}{x^{2/3}}$$

39.

$$\frac{d}{dx}\left(e^{\sqrt{x}}+e^{\sqrt{y}}\right)=\frac{d}{dx}4$$

$$e^{\sqrt{x}}\frac{d}{dx}\sqrt{x}+e^{\sqrt{y}}\frac{d}{dx}\sqrt{y}=0 \qquad \text{Use the Chain Rule.}$$

$$e^{\sqrt{x}}\cdot\frac{1}{\cancel{2}\sqrt{x}}+e^{\sqrt{y}}\cdot\frac{1}{\cancel{2}\sqrt{y}}\frac{dy}{dx}=0 \qquad \text{Differentiate.}$$

$$\frac{e^{\sqrt{x}}}{\sqrt{x}}+\frac{e^{\sqrt{y}}}{\sqrt{y}}\frac{dy}{dx}=0 \qquad \text{Simplify.}$$

$$\frac{e^{\sqrt{y}}}{\sqrt{y}}\frac{dy}{dx}=-\frac{e^{\sqrt{x}}}{\sqrt{x}}$$

$$\frac{dy}{dx}=-\frac{e^{\sqrt{x}}}{\sqrt{x}}\cdot\frac{\sqrt{y}}{e^{\sqrt{y}}}=-\frac{\sqrt{y}\,e^{\sqrt{x}}}{\sqrt{x}\,e^{\sqrt{y}}}=-\frac{\sqrt{y}\,e^{\sqrt{x}-\sqrt{y}}}{\sqrt{x}}$$

41. (a) The domain of f is $\{x \mid x \ge 0\}$ or on the interval $[0, \infty)$..

(b) $f'(x)=\dfrac{1}{2\sqrt{x}}$

(c) The domain of $f'(x)$ is $\{x \mid x > 0\}$ or on the interval $(0, \infty)$.

(d) $f'(x)$ is never equal to 0.

(e) $x = 0$ is in the domain of f, but not in the domain of $f'(x)$.

(f) $f''(x)=\dfrac{d}{dx}\left(\dfrac{1}{2\sqrt{x}}\right)=\dfrac{1}{2}\dfrac{d}{dx}x^{-1/2}=\dfrac{1}{2}\cdot\left(-\dfrac{1}{2}\right)x^{(-1/2)-1}=-\dfrac{1}{4}x^{-3/2}=-\dfrac{1}{4x^{3/2}}$

(g) The domain of $f''(x)$ is $\{x \mid x > 0\}$ or on the interval $(0, \infty)$.

43. (a) The domain of f is all real numbers or the interval $(-\infty, \infty)$.

(b) $f'(x)=\dfrac{d}{dx}x^{2/3}=\dfrac{2}{3}x^{(2/3)-1}=\dfrac{2}{3}x^{-1/3}=\dfrac{2}{3x^{1/3}}=\dfrac{2}{3\sqrt[3]{x}}$

(c) The domain of $f'(x)$ is all real numbers except $x = 0$, that is the set $\{x \mid x \ne 0\}$.

(d) $f'(x)$ is never equal to 0.

(e) $x = 0$ is in the domain of f, but not in the domain of $f'(x)$.

(f) $f''(x)=\dfrac{d}{dx}\left(\dfrac{2}{3}x^{-1/3}\right)=\dfrac{2}{3}\cdot\left(-\dfrac{1}{3}\right)x^{(-1/3)-1}=-\dfrac{2}{9}x^{-4/3}=-\dfrac{2}{9x^{4/3}}$

(g) The domain of $f''(x)$ is all real numbers except $x = 0$, that is the set $\{x \mid x \ne 0\}$.

45. (a) The domain of f is all real numbers or the interval $(-\infty, \infty)$.

(b) $f'(x) = \dfrac{d}{dx}\left(x^{2/3} + 2x^{1/3}\right) = \dfrac{2}{3}x^{(2/3)-1} + 2 \cdot \dfrac{1}{3}x^{(1/3)-1}$

$\qquad\qquad = \dfrac{2}{3}x^{-1/3} + \dfrac{2}{3}x^{-2/3} = \dfrac{2}{3x^{1/3}} + \dfrac{2}{3x^{2/3}}$

(c) The domain of $f'(x)$ is all real numbers except $x = 0$, that is the set $\{x \mid x \neq 0\}$.

(d) $f'(x)$ is never equal to zero.

(e) $x = 0$ is in the domain of f, but not in the domain of $f'(x)$.

(f) $f''(x) = \dfrac{d}{dx}\left(\dfrac{2}{3}x^{-1/3} + \dfrac{2}{3}x^{-2/3}\right) = \dfrac{2}{3} \cdot \left(-\dfrac{1}{3}\right)x^{(-1/3)-1} + \dfrac{2}{3} \cdot \left(-\dfrac{2}{3}\right)x^{(-2/3)-1}$

$\qquad\qquad = -\dfrac{2}{9}x^{-4/3} - \dfrac{4}{9}x^{-5/3} = -\dfrac{2}{9x^{4/3}} - \dfrac{4}{9x^{5/3}}$

(g) The domain of $f''(x)$ is all real numbers except $x = 0$, that is the set $\{x \mid x \neq 0\}$.

47. (a) The domain of f is all real numbers or the interval $(-\infty, \infty)$.

(b) $f'(x) = \dfrac{d}{dx}\left(x^2 - 1\right)^{2/3} = \dfrac{2}{3}\left(x^2 - 1\right)^{(2/3)-1}\dfrac{d}{dx}\left(x^2 - 1\right) = \dfrac{2}{3}\left(x^2 - 1\right)^{-1/3} \cdot 2x$

$\qquad\qquad = \dfrac{4x}{3\left(x^2 - 1\right)^{1/3}}$

(c) The domain of $f'(x)$ is all real numbers except $x = 1$ and $x = -1$, that is the set $\{x \mid x \neq 1 \text{ and } x \neq -1\}$.

(d) $f'(x) = 0$ when $x = 0$.

(e) $x = 1$ and $x = -1$ are in the domain of f, but not in the domain of $f'(x)$.

(f) $f''(x) = \dfrac{d}{dx}\dfrac{4x}{3\left(x^2 - 1\right)^{1/3}} = \dfrac{4}{3}\left[\dfrac{\left(x^2 - 1\right)^{1/3}\dfrac{d}{dx}x - x\dfrac{d}{dx}\left(x^2 - 1\right)^{1/3}}{\left(x^2 - 1\right)^{2/3}}\right]$

$\qquad\qquad = \dfrac{4}{3}\left[\dfrac{\left(x^2 - 1\right)^{1/3} - x \cdot \dfrac{1}{3}\left(x^2 - 1\right)^{(1/3)-1}\dfrac{d}{dx}\left(x^2 - 1\right)}{\left(x^2 - 1\right)^{2/3}}\right]$

$$= \frac{4}{3}\left[\frac{\left(x^2-1\right)^{1/3}-\dfrac{x}{3}\left(x^2-1\right)^{-2/3}\cdot 2x}{\left(x^2-1\right)^{2/3}}\right]$$

$$= \frac{4}{3}\left[\frac{\left(x^2-1\right)^{1/3}-\dfrac{2x^2}{3\left(x^2-1\right)^{2/3}}}{\left(x^2-1\right)^{2/3}}\right]$$

$$= \frac{4}{3}\left[\frac{3\left(x^2-1\right)-2x^2}{3\left(x^2-1\right)^{4/3}}\right] = \frac{4\left(x^2-3\right)}{9\left(x^2-1\right)^{4/3}}$$

(g) The domain of $f''(x)$ is all real numbers except $x=1$ and $x=-1$, that is the set $\{x\,|\,x\neq 1 \text{ and } x\neq -1\}$.

49. (a) Since $\sqrt{1-x^2}\geq 0$, $1-x^2\geq 0$ or $x^2\leq 1$. Solving for x, we get $-1\leq x\leq 1$, so the domain of f is the interval $[-1, 1]$.

(b) $f'(x)=\dfrac{d}{dx}\left(x\sqrt{1-x^2}\right)=\dfrac{d}{dx}\left[x\left(1-x^2\right)^{1/2}\right]=x\dfrac{d}{dx}\left(1-x^2\right)^{1/2}+\left(1-x^2\right)^{1/2}\dfrac{d}{dx}x$

$$= x\cdot\frac{1}{2}\left(1-x^2\right)^{(1/2)-1}\frac{d}{dx}\left(1-x^2\right)+\left(1-x^2\right)^{1/2}\cdot 1$$

$$= \frac{x}{2}\left(1-x^2\right)^{-1/2}\left(-2x\right)+\left(1-x^2\right)^{1/2}$$

$$= \frac{-x^2}{\left(1-x^2\right)^{1/2}}+\left(1-x^2\right)^{1/2}=\frac{-x^2+\left(1-x^2\right)}{\left(1-x^2\right)^{1/2}}=\frac{1-2x^2}{\left(1-x^2\right)^{1/2}}$$

(c) Since $\left(1-x^2\right)^{1/2}>0$, $1-x^2>0$ or $x^2<1$. Solving for x, we get $-1<x<1$, so the domain of $f'(x)$ is all real numbers or the interval $(-1, 1)$.

(d) $f'(x)=0$ when $1-2x^2=0$, or when $x^2=\dfrac{1}{2}$, or $x=-\sqrt{\dfrac{1}{2}}$ or $x=\sqrt{\dfrac{1}{2}}$.

(e) The points $x=-1$ and $x=1$ are in the domain of f, but are not part of the domain of f'.

(f) $f''(x)=\dfrac{d}{dx}\dfrac{1-2x^2}{\left(1-x^2\right)^{1/2}}=\dfrac{\left(1-x^2\right)^{1/2}\dfrac{d}{dx}\left(1-2x^2\right)-\left(1-2x^2\right)\dfrac{d}{dx}\left(1-x^2\right)^{1/2}}{\left(\left(1-x^2\right)^{1/2}\right)^2}$

$$= \frac{\left(1-x^2\right)^{1/2}(-4x)-\left(1-2x^2\right)\frac{1}{2}\left(1-x^2\right)^{(1/2)-1}\frac{d}{dx}\left(1-x^2\right)}{\left(1-x^2\right)}$$

$$= \frac{-4x\left(1-x^2\right)^{1/2}-\left(1-2x^2\right)\frac{1}{2}\left(1-x^2\right)^{-1/2}(-2x)}{\left(1-x^2\right)}$$

$$= \frac{-4x\left(1-x^2\right)^{1/2}+\left(x-2x^3\right)\left(1-x^2\right)^{-1/2}}{\left(1-x^2\right)}$$

$$= \frac{-4x\left(1-x^2\right)+\left(x-2x^3\right)}{\left(1-x^2\right)^{3/2}}$$

$$= \frac{-4x+4x^3+x-2x^3}{\left(1-x^2\right)^{3/2}}=\frac{2x^3-3x}{\left(1-x^2\right)^{3/2}}$$

(g) Since $\left(1-x^2\right)^{3/2}>0$, $1-x^2>0$ or $x^2<1$. Solving for x, we get $-1<x<1$, so the domain of $f''(x)$ is all real numbers or the interval $(-1, 1)$.

51.

(a) $N'(t)=\dfrac{d}{dx}\left(-\dfrac{10,000}{\sqrt{1+0.1t}}+11,000\right)=\dfrac{d}{dx}\left(-10,000(1+0.1t)^{-1/2}+11,000\right)$

$$= -10,000\frac{d}{dx}(1+0.1t)^{-1/2}+\frac{d}{dx}11,000$$

$$= -10,000\cdot\left(-\frac{1}{2}\right)(1+0.1t)^{(-1/2)-1}\frac{d}{dx}(1+0.1t)$$

$$= 5,000(1+0.1t)^{-3/2}(0.1)=\frac{500}{\left(1+0.1t\right)^{3/2}}$$

(b) In 10 years, $t = 10$.

$$N'(10)=\frac{500}{\left(1+0.1(10)\right)^{3/2}}=176.777=177\text{ students}$$

53.   Since z is a constant, we write $z = K$. Then the production function becomes
$$K=x^{0.5}y^{0.4}$$

We will find $\dfrac{dy}{dx}$ using implicit differentiation.

$$\frac{d}{dx}K=\frac{d}{dx}\left(x^{0.5}y^{0.4}\right)$$

$$0=x^{0.5}\frac{d}{dx}y^{0.4}+y^{0.4}\frac{d}{dx}x^{0.5}$$

$$0=x^{0.5}\cdot 0.4y^{0.4-1}\frac{dy}{dx}+y^{0.4}\cdot 0.5x^{0.5-1}$$

$$0 = 0.4x^{0.5}y^{-0.6}\frac{dy}{dx} + 0.5y^{0.4}x^{-0.5}$$

$$0 = \frac{0.4x^{0.5}}{y^{0.6}}\frac{dy}{dx} + \frac{0.5y^{0.4}}{x^{0.5}}$$

$$\frac{0.4x^{0.5}}{y^{0.6}}\frac{dy}{dx} = -\frac{0.5y^{0.4}}{x^{0.5}}$$

$$\frac{dy}{dx} = -\frac{0.5y^{0.4}}{x^{0.5}}\cdot\frac{y^{0.6}}{0.4x^{0.5}} = -\frac{0.5y^{0.4+.06}}{0.4x^{0.5+0.5}}\cdot\frac{10}{10} = -\frac{5y}{4x}$$

55. (a) The instantaneous rate of pollution is the derivative of the function A.

$$A'(t) = \frac{d}{dt}\left(t^{1/4}+3\right)^3 = 3\left(t^{1/4}+3\right)^2\frac{d}{dt}\left(t^{1/4}+3\right) = 3\left(t^{1/4}+3\right)^2\cdot\frac{1}{4}t^{(1/4)-1} = \frac{3\left(t^{1/4}+3\right)^2}{4t^{3/4}}$$

(b) After 16 years the rate of

$$A'(16) = \frac{3\left(16^{1/4}+3\right)^2}{4\cdot16^{3/4}} = \frac{3\cdot5^2}{32} = 2.344 \text{ units per year.}$$

57. (a) Velocity is the derivative of the distance function s.

$$v = s'(t) = \frac{d}{dt}t^{3/2} = \frac{3}{2}t^{(3/2)-1} = \frac{3t^{1/2}}{2} \text{ feet per second.}$$

After 1 second the child has a velocity of $v(1) = \dfrac{3\cdot1^{1/2}}{2} = \dfrac{3}{2} = 1.5$ feet per second.

(b) If the slide is 8 feet long, it takes

$$t^{3/2} = 8$$

$$\left(t^{3/2}\right)^{2/3} = (8)^{2/3} = 4 \text{ seconds to get down the slide and strike the ground.}$$

The velocity the child when striking the ground is

$$v(4) = \frac{3\cdot4^{1/2}}{2} = 3 \text{ feet per second.}$$

Chapter 13 Review

TRUE-FALSE ITEMS

1. True   **3.** True   **5.** False   **7.** True

FILL-IN-THE-BLANKS

1. tangent   **3.** Power Rule; Chain Rule   **5.** zero

REVIEW EXERCISES

1. $f'(x) = 2;\; f'(2) = 2$

3. $f'(x) = 2x;\; f'(2) = 2 \cdot 2 = 4$

5. $f'(x) = 2x - 2; f'(1) = 2 \cdot 1 - 2 = 0$

7. $f'(x) = e^{3x} \dfrac{d}{dx}(3x) = 3e^{3x}; f'(0) = 3e^0 = 3$

9. $f'(x) = \lim\limits_{h \to 0} \dfrac{\left[4(x+h)+3\right] - \left[4x+3\right]}{h} = \lim\limits_{h \to 0} \dfrac{4x + 4h + 3 - 4x - 3}{h} = \lim\limits_{h \to 0} \dfrac{4\cancel{h}}{\cancel{h}} = \lim\limits_{h \to 0} 4 = 4$

11. $f'(x) = \lim\limits_{h \to 0} \dfrac{\left[2(x+h)^2 + 1\right] - \left[2x^2 + 1\right]}{h} = \lim\limits_{h \to 0} \dfrac{2x^2 + 4xh + 2h^2 - 2x^2 - 1}{h}$

$\qquad = \lim\limits_{h \to 0} \dfrac{\cancel{h}(4x + 2h)}{\cancel{h}} = \lim\limits_{h \to 0}(4x + 2h) = \lim\limits_{h \to 0} 4x + \lim\limits_{h \to 0} 2h = 4x + 0 = 4x$

13. $f'(x) = 5x^4$

15. $f'(x) = \dfrac{1}{4} \dfrac{d}{dx} x^4 = \dfrac{1}{4} \cdot 4x^3 = x^3$

17. $f'(x) = 2 \cdot 2x - 3 = 4x - 3$

19. $f'(x) = 7 \dfrac{d}{dx}(x^2 - 4) = 7 \cdot 2x = 14x$

21. $f'(x) = 5\left[(x^2 - 3x)\dfrac{d}{dx}(x-6) + (x-6)\dfrac{d}{dx}(x^2 - 3x)\right] = 5\left[x^2 - 3x + (x-6)(2x-3)\right]$

$\qquad = 5(x^2 - 3x + 2x^2 - 15x + 18) = 5(3x^2 - 18x + 18) = 15(x^2 - 6x + 6)$

23. $f'(x) = \dfrac{d}{dx}\left[12x(8x^3 + 2x^2 - 5x + 2)\right] = 12\dfrac{d}{dx}(8x^4 + 2x^3 - 5x^2 + 2x)$

$\qquad = 12(32x^3 + 6x^2 - 10x + 2) = 24(16x^3 + 3x^2 - 5x + 1)$

25. $f'(x) = \dfrac{d}{dx}\dfrac{2x+2}{5x-3} = \dfrac{(5x-3)\dfrac{d}{dx}(2x+2) - (2x+2)\dfrac{d}{dx}(5x-3)}{(5x-3)^2}$

$\qquad = \dfrac{(5x-3) \cdot 2 - (2x+2) \cdot 5}{(5x-3)^2} = \dfrac{10x - 6 - 10x - 10}{(5x-3)^2} = \dfrac{-16}{(5x-3)^2}$

27. $f'(x) = 2 \cdot (-12)\, x^{-13} = -24\, x^{-13} = -\dfrac{24}{x^{13}}$

29. $f'(x) = \dfrac{d}{dx}\left(2 + \dfrac{3}{x} + \dfrac{4}{x^2}\right) = \dfrac{d}{dx}(2 + 3x^{-1} + 4x^{-2}) = -3x^{-2} - 8x^{-3} = -\dfrac{3}{x^2} - \dfrac{8}{x^3}$

31. $f'(x) = \dfrac{d}{dx}\left(\dfrac{3x-2}{x+5}\right) = \dfrac{(x+5)\dfrac{d}{dx}(3x-2)-(3x-2)\dfrac{d}{dx}(x+5)}{(x+5)^2}$ Derivative of a quotient.

$= \dfrac{(x+5)(3)-(3x-2)(1)}{(x+5)^2} = \dfrac{3x+15-3x+2}{(x+5)^2} = \dfrac{17}{(x+5)^2}$

33. $f'(x) = \dfrac{d}{dx}\left(3x^2-2x\right)^5 = 5\left(3x^2-2x\right)^4 \dfrac{d}{dx}\left(3x^2-2x\right)$ Use the Power Rule.

$= 5\left(3x^2-2x\right)^4(6x-2) = 10\left(3x^2-2x\right)^4(3x-1)$

35. $f'(x) = \dfrac{d}{dx}\left[7x\left(x^2+2x+1\right)^2\right]$

$= 7x\dfrac{d}{dx}\left(x^2+2x+1\right)^2 + \left(x^2+2x+1\right)^2\dfrac{d}{dx}(7x)$ The derivative of a product.

$= 7x\cdot 2\left(x^2+2x+1\right)\dfrac{d}{dx}\left(x^2+2x+1\right) + \left(x^2+2x+1\right)^2\cdot 7$ Use the Power Rule.

$= 14x\left(x^2+2x+1\right)(2x+2) + 7\left(x^2+2x+1\right)^2$ Differentiate.

$= 7\left(x^2+2x+1\right)\left[\,2x(2x+2)+\left(x^2+2x+1\right)\right]$ Factor.

$= 7\left(x^2+2x+1\right)\left[\,4x^2+4x+x^2+2x+1\right]$ Simplify in the brackets.

$= 7\left(x^2+2x+1\right)\left[\,5x^2+6x+1\right]$ Simplify in the brackets.

$= 7(x+1)^2\left[(5x+1)(x+1)\right] = 7(x+1)^3(5x+1)$ Factor.

37. $f'(x) = \dfrac{d}{dx}\left(\dfrac{x+1}{3x+2}\right)^2 = 2\left(\dfrac{x+1}{3x+2}\right)\dfrac{d}{dx}\left(\dfrac{x+1}{3x+2}\right)$ Use the Power Rule.

$= 2\left(\dfrac{x+1}{3x+2}\right)\left[\dfrac{(3x+2)\dfrac{d}{dx}(x+1)-(x+1)\dfrac{d}{dx}(3x+2)}{(3x+2)^2}\right]$ Derivative of a quotient.

$= 2\left(\dfrac{x+1}{3x+2}\right)\left[\dfrac{(3x+2)\cdot 1-(x+1)\cdot 3}{(3x+2)^2}\right]$ Differentiate.

$= \dfrac{2(x+1)(3x+2-3x-3)}{(3x+2)^3}$ Simplify.

$= -\dfrac{2(x+1)}{(3x+2)^3}$

39.

$$f'(x) = \frac{d}{dx} \frac{7}{\left(x^3 + 4\right)^2} = 7\frac{d}{dx}\left(x^3 + 4\right)^{-2}$$ Write the function with a negative exponent.

$$= 7 \cdot (-2)\left(x^3 + 4\right)^{-3} \frac{d}{dx}\left(x^3 + 4\right)$$ Use the Power Rule.

$$= -14\left(x^3 + 4\right)^{-3}\left(3x^2\right) = -\frac{42x^2}{\left(x^3 + 4\right)^3}$$ Differentiate and simplify.

41.

$$f'(x) = \frac{d}{dx}\left(3x + \frac{4}{x}\right)^3 = \frac{d}{dx}\left(\frac{3x^2 + 4}{x}\right)^3$$ Write the function with a common denominator.

$$= 3\left(\frac{3x^2 + 4}{x}\right)^2 \frac{d}{dx}\left(\frac{3x^2 + 4}{x}\right)$$ Use the Power Rule.

$$= 3\left(\frac{3x^2 + 4}{x}\right)^2\left[\frac{x\frac{d}{dx}\left(3x^2 + 4\right) - \left(3x^2 + 4\right)\frac{d}{dx}x}{x^2}\right]$$ The derivative of a quotient.

$$= 3\left(\frac{3x^2 + 4}{x}\right)^2\left[\frac{x(6x) - \left(3x^2 + 4\right)}{x^2}\right] = 3\left(\frac{3x^2 + 4}{x}\right)^2\left[\frac{3x^2 - 4}{x^2}\right]$$ Differentiate; simplify.

$$= \frac{3\left(3x^2 + 4\right)^2\left(3x^2 - 4\right)}{x^4}$$ Simplify.

43.

$$f'(x) = \frac{d}{dx}\left(3e^x + x^2\right) = 3\frac{d}{dx}e^x + \frac{d}{dx}x^2 = 3e^x + 2x$$

45.

$$f'(x) = \frac{d}{dx}e^{3x+1} = e^{3x+1}\frac{d}{dx}(3x + 1) = 3e^{3x+1}$$ Use the Chain Rule.

47.

$$f'(x) = \frac{d}{dx}\left[e^x\left(2x^2 + 7x\right)\right] = e^x\frac{d}{dx}\left(2x^2 + 7x\right) + \left(2x^2 + 7x\right)\frac{d}{dx}e^x$$ The derivative of a product.

$$= e^x(4x + 7) + \left(2x^2 + 7x\right)e^x = e^x\left(2x^2 + 11x + 7\right)$$

49.

$$f'(x) = \frac{d}{dx}\frac{1+x}{e^x} = \frac{e^x\frac{d}{dx}(1+x) - (1+x)\frac{d}{dx}e^x}{\left(e^x\right)^2} = \frac{e^x - (1+x)e^x}{e^{2x}} = \frac{-xe^x}{e^{2x}} = -\frac{x}{e^x}$$

           ↑   ↑   ↑

Derivative of a quotient.   Differentiate.   Simplify.

51. $f'(x) = \dfrac{d}{dx}\left(\dfrac{e^x}{3x}\right)^2 = 2\left(\dfrac{e^x}{3x}\right)\dfrac{d}{dx}\left(\dfrac{e^x}{3x}\right) = 2\left(\dfrac{e^x}{3x}\right)\left[\dfrac{1}{3}\cdot\dfrac{x\dfrac{d}{dx}e^x - e^x\dfrac{d}{dx}x}{x^2}\right]$

$\qquad\qquad\qquad\;\uparrow\qquad\qquad\qquad\qquad\uparrow$

$\qquad$ Use the Power Rule.$\qquad$Use the formula for the derivative of a quotient.

$\qquad\quad = 2\left(\dfrac{e^x}{3x}\right)\left[\dfrac{1}{3}\cdot\dfrac{xe^x - e^x}{x^2}\right] = \dfrac{2e^x\left(xe^x - e^x\right)}{9x^3} = \dfrac{2e^{2x}(x-1)}{9x^3}$

$\quad\;\uparrow\qquad\qquad\qquad\qquad\quad\uparrow\qquad\qquad\qquad\uparrow$

$\quad$ Differentiate.$\qquad\qquad\qquad$Simplify.$\qquad\qquad$Factor.

53. $f'(x) = \dfrac{d}{dx}\ln(4x) = \dfrac{1}{4x}\dfrac{d}{dx}(4x) = \dfrac{4}{4x} = \dfrac{1}{x}$

55. $f'(x) = \dfrac{d}{dx}\left(x^2\ln x\right) = x^2\dfrac{d}{dx}\ln x + \ln x\dfrac{d}{dx}x^2 = x^2\cdot\dfrac{1}{x} + \ln x\cdot 2x = x + 2x\ln x$

$\qquad\qquad\qquad\quad\uparrow\qquad\qquad\qquad\qquad\qquad\uparrow\qquad\qquad\quad\uparrow$

$\qquad\qquad$ Use the formula for the$\qquad\quad$Differentiate.$\qquad$Simplify.
$\qquad\qquad$ derivative of a product.

57. $f'(x) = \dfrac{d}{dx}\ln\left(2x^3 + 1\right) = \dfrac{1}{2x^3 + 1}\dfrac{d}{dx}\left(2x^3 + 1\right) = \dfrac{1}{2x^3 + 1}\cdot\left(6x^2\right) = \dfrac{6x^2}{2x^3 + 1}$

$\qquad\qquad\qquad\qquad\qquad\;\uparrow$

$\qquad\qquad\qquad$ Use the Chain Rule.

59. $f'(x) = \dfrac{d}{dx}\left[2^x + x^2\right] = \dfrac{d}{dx}2^x + \dfrac{d}{dx}x^2 = 2^x\ln 2 + 2x$

61. $f'(x) = \dfrac{d}{dx}\left(x + \log x\right) = \dfrac{d}{dx}x + \dfrac{d}{dx}\log x = 1 + \dfrac{1}{x\ln 10}$

63. $f'(x) = \dfrac{d}{dx}\sqrt{x} = \dfrac{d}{dx}(x)^{1/2} = \dfrac{1}{2}x^{(1/2)-1} = \dfrac{1}{2}x^{-1/2} = \dfrac{1}{2x^{1/2}} = \dfrac{1}{2\sqrt{x}}$

65. $f'(x) = \dfrac{d}{dx}\left(3x^{5/3} + 5\right) = 3\dfrac{d}{dx}x^{5/3} + \dfrac{d}{dx}5 = 3\cdot\dfrac{5}{3}x^{(5/3)-1} = 5x^{2/3}$

67. $f'(x) = \dfrac{d}{dx}\sqrt{x^2 - 3x} = \dfrac{d}{dx}\left(x^2 - 3x\right)^{1/2} = \dfrac{1}{2}\left(x^2 - 3x\right)^{(1/2)-1}\dfrac{d}{dx}\left(x^2 - 3x\right)$

$\qquad\qquad\qquad\qquad\qquad\uparrow\qquad\qquad\qquad\qquad\uparrow$

$\qquad$ Change the radical to an exponent.$\qquad$Use the Power Rule.

$$= \frac{1}{2}\left(x^2 - 3x\right)^{-1/2}(2x - 3) = \frac{2x - 3}{2\left(x^2 - 3x\right)^{1/2}} = \frac{2x - 3}{2\sqrt{x^2 - 3x}}$$

| ↑ | ↑ | ↑ |
| Differentiate. | Simplify. | Alternate form of the solution. |

69. $f'(x) = \dfrac{d}{dx}\dfrac{x+1}{\sqrt{x+5}} = \dfrac{d}{dx}\dfrac{x+1}{(x+5)^{1/2}} = \dfrac{(x+5)^{1/2}\dfrac{d}{dx}(x+1) - (x+1)\dfrac{d}{dx}(x+5)^{1/2}}{\left[(x+5)^{1/2}\right]^2}$

$$= \frac{(x+5)^{1/2} - (x+1)\cdot\dfrac{1}{2}(x+5)^{(1/2)-1}\dfrac{d}{dx}(x+5)}{(x+5)}$$

$$= \frac{(x+5)^{1/2} - (x+1)\cdot\dfrac{1}{2}(x+5)^{-1/2}}{(x+5)}$$

$$= \frac{2(x+5) - (x+1)}{2(x+5)^{3/2}} = \frac{2x + 10 - x - 1}{2(x+5)^{3/2}} = \frac{x+9}{2(x+5)^{3/2}}$$

71. $f'(x) = \dfrac{d}{dx}\left[(1+x)\sqrt{e^x}\right] = \dfrac{d}{dx}\left[(1+x)e^{x/2}\right] = (1+x)\dfrac{d}{dx}e^{x/2} + e^{x/2}\dfrac{d}{dx}(1+x)$

$$= (1+x)e^{x/2}\frac{d}{dx}\left(\frac{x}{2}\right) + e^{x/2}\cdot 1$$

$$= (1+x)e^{x/2}\cdot\frac{1}{2} + e^{x/2}$$

$$= \frac{(1+x)e^{x/2} + 2e^{x/2}}{2}$$

$$= \frac{e^{x/2}(1+x+2)}{2} = \frac{e^{x/2}(x+3)}{2}$$

73. $f'(x) = \dfrac{d}{dx}\left(\sqrt{x}\ln x\right) = \dfrac{d}{dx}\left(x^{1/2}\ln x\right) = x^{1/2}\dfrac{dy}{dx}\ln x + \ln x\dfrac{d}{dx}x^{1/2}$ Derivative of a product.

$$= x^{1/2}\cdot\frac{1}{x} + \ln x\cdot\frac{1}{2}x^{-1/2} \qquad\text{Differentiate.}$$

$$= \frac{1}{x^{1/2}} + \ln x\cdot\frac{1}{2x^{1/2}} = \frac{2 + \ln x}{2x^{1/2}} = \frac{2 + \ln x}{2\sqrt{x}} \qquad\text{Simplify.}$$

75. $f'(x) = \dfrac{d}{dx}\left(x^3 - 8\right) = 3x^2$

$$f''(x) = \frac{d}{dx}\left(3x^2\right) = 6x$$

77. $f'(x) = \dfrac{d}{dx}e^{-3x} = e^{-3x}\dfrac{d}{dx}(-3x) = -3e^{-3x}$

$f''(x) = \dfrac{d}{dx}\left(-3e^{-3x}\right) = -3e^{-3x}\dfrac{d}{dx}(-3x) = -3e^{-3x}\cdot(-3) = 9e^{-3x}$

79. $f'(x) = \dfrac{d}{dx}\dfrac{x}{2x+1} = \dfrac{(2x+1)\dfrac{d}{dx}x - x\dfrac{d}{dx}(2x+1)}{(2x+1)^2}$ Use the formula for the derivative of a quotient.

$= \dfrac{(2x+1)\cdot 1 - x(2)}{(2x+1)^2} = \dfrac{(2x+1)-2x}{(2x+1)^2} = \dfrac{1}{(2x+1)^2}$

$f''(x) = \dfrac{d}{dx}\dfrac{1}{(2x+1)^2} = \dfrac{d}{dx}(2x+1)^{-2} = -2(2x+1)^{-3}\dfrac{d}{dx}(2x+1)$ Use the Power Rule.

$= -2(2x+1)^{-3}\cdot 2 = -\dfrac{4}{(2x+1)^3}$ Simplify.

81.
$$\dfrac{d}{dx}\left(xy+3y^2\right) = \dfrac{d}{dx}(10x)$$

$$\left(x\dfrac{d}{dx}y + y\dfrac{d}{dx}x\right) + 3\dfrac{d}{dx}y^2 = 10$$

$$x\dfrac{dy}{dx} + y + 6y\dfrac{dy}{dx} = 10$$

$$(x+6y)\dfrac{dy}{dx} = 10 - y$$

$$\dfrac{dy}{dx} = \dfrac{10-y}{x+6y}\ \text{ provided }x+6y\neq 0$$

83.
$$\dfrac{d}{dx}\left(xe^{y}\right) = \dfrac{d}{dx}\left(4x^2\right)$$

$$x\dfrac{d}{dx}e^{y} + e^{y}\dfrac{d}{dx}x = 4\cdot 2x$$

$$xe^{y}\dfrac{dy}{dx} + e^{y} = 8x$$

$$xe^{y}\dfrac{dy}{dx} = 8x - e^{y}$$

$$\dfrac{dy}{dx} = \dfrac{8x - e^{y}}{xe^{y}}\ \text{ provided }x\neq 0.$$

85.   The slope of the tangent line to the graph of $f(x) = 2x^2 + 3x - 7$ at the point $(-1,-8)$ is

$$m_{\text{tan}} = \lim_{x\to -1}\dfrac{f(x)-f(-1)}{x-(-1)} = \lim_{x\to -1}\dfrac{(2x^2+3x-7)-(-8)}{x+1} = \lim_{x\to -1}\dfrac{2x^2+3x+1}{x+1}$$

$$= \lim_{x \to -1} \frac{(2x+1)\cancel{(x+1)}}{\cancel{x+1}} = \lim_{x \to -1} (2x+1) = -1$$

An equation of the tangent line is

$$\begin{array}{ll} y-(-8)=(-1)[x-(-1)] & y-f(c)=m_{\tan}(x-c) \\ y+8=-x-1 & \text{Simplify.} \\ y=-x-9 & \end{array}$$

87. The slope of the tangent line to the graph of f is the derivative evaluated at $(0, 1)$.

$$f'(x)=\frac{d}{dx}\left(x^2+e^x\right)=2x+e^x$$

$$m_{\tan}=f'(0)=2\cdot 0+e^0=1$$

An equation of the tangent line is

$$y-1=1(x-0)$$
$$y=x+1$$

89. (a) The average rate of change of f as x changes from 0 to 2 is

$$\frac{\Delta y}{\Delta x}=\frac{f(2)-f(0)}{2-0}=\frac{\left[2^3+3\cdot 2\right]-\left[0^3+3\cdot 0\right]}{2}=\frac{14-0}{2}=7$$

(b) The instantaneous rate of change is the derivative of f evaluated at 2.

$$f'(x)=\frac{d}{dx}\left(x^3+3x\right)=3x^2+3$$

$$f'(2)=3\cdot 2^2+3=15$$

91. (a) When the stone hits the water its height is zero. So we solve the equation $h(t)=0$.

$$-16t^2+100=0$$
$$16t^2=100$$
$$t^2=\frac{100}{16}$$
$$t=\sqrt{\frac{100}{16}}=\frac{10}{4}=2.5$$

It takes the stone 2.5 seconds to hit the water.

(b) The average velocity during its fall is

$$\frac{\Delta y}{\Delta t}=\frac{h(2.5)-h(0)}{2.5-0}=\frac{\left[-16\left(2.5^2\right)+100\right]-\left[-16\left(0^2\right)+100\right]}{2.5}=\frac{-100}{2.5}=-40$$

feet per second in the downward direction.

(c) The instantaneous velocity is the derivative of h.

$$v=h'(t)=\frac{d}{dt}\left(-16t^2+100\right)=-32t$$

When it hits the water the stone's velocity is $v(2.5)=(-32)(2.5)=-80$ feet per second.

93. (a) The ball reaches its highest point when its velocity is zero. The velocity is the derivative of the distance function s.

$$v = s'(t) = \frac{d}{dt}\left(-16t^2+128t+6\right) = -32t+128$$

We set $v = 0$ and solve.

$$-32t + 128 = 0$$
$$32t = 128$$
$$t = 4$$

The ball reach its highest point 4 seconds after it is thrown.

(b) The maximum height of the ball is

$$s(4) = -16(4^2)+128(4)+6 = 262$$

feet above the ground.

(c) The ball travels 262 feet downward and $262 - 6$ feet upward for a total of 518 feet.

(d) The velocity of the ball at time t is $v(t) = -32t+128$ feet per second as we calculated in part (a).

(e) The velocity of the ball is zero at $t = 4$ seconds. At that time it momentarily stops, changing from the upward to the downward direction.

(f) The ball is in the air for 4 seconds on it way up. To find how long it is in the air on the way down, we solve the equation $s(t) = 0$.

$$-16t^2+128t+6 = 0$$

Using the quadratic formula with $a = 16$, $b = -128$, and $c = -6$, we find

$$t = \frac{128 \pm \sqrt{(-128)^2-4(16)(-6)}}{2\cdot 16} = \frac{128 \pm \sqrt{16,768}}{32}$$

We only need the positive answer, and so

$$t = \frac{128 + \sqrt{16,768}}{32} = 8.047 \approx 8.0 \text{ seconds.}$$

(g) The velocity of the ball when it hits the ground is
$v(8.047) = -32(8.047)+128 = -129.504$ feet per second in the downward direction.

(h) The acceleration a is the derivative of the velocity at time t.

$$a = v'(t) = s''(t) = \frac{d}{dx}\left(-32t+128\right) = -32 \text{ feet per second per second.}$$

(i) The velocity of the ball is $v(2) = -32(2)+128 = 64$ feet per second after it is in the air for 2 seconds.

The velocity of the ball is $v(6) = -32(6) + 128 = -64$ feet per second after it is in the air for 6 seconds.

95. (a) Revenue is the product of price and quantity. So the revenue function R is
$$R(x) = px = (-0.50x + 75)x = -0.50x^2 + 75x$$

(b) The marginal revenue function is the derivative of the revenue function R.
$$R'(x) = \frac{d}{dx}(-0.50x^2 + 75x) = -1.00x + 75$$

(c) The marginal cost function is the derivative of the cost function C.
$$C'(x) = \frac{d}{dx}(15x + 550) = 15$$

(d) The break even point is the x-value for which $R(x) = C(x)$.
$$-0.50x^2 + 75x = 15x + 550$$
$$-0.50x^2 + 60x - 550 = 0$$
$$x^2 - 120x + 1100 = 0$$
$$(x - 110)(x - 10) = 0$$
$$(x - 110) = 0 \qquad (x - 10) = 0$$
$$x = 110 \qquad\qquad x = 10$$

(e) The marginal revenue equals the marginal cost when
$$-x + 75 = 15$$
$$x = 60 \text{ units are produced and sold.}$$

CHAPTER 13 PROJECT

1. Since x persons are in the pool, the probability that the pooled test is negative is the same as the probability that each person in the sample tests negative. So, $p_- = q \cdot q \cdot q \cdot \ldots \cdot q = q^x$. On the other hand, the pooled test will be positive if at least 1 person tests positive, so $p_+ = 1 - $ probability that all tests were negative $= 1 - q^x$.

3. Answers will vary. (Note: $\dfrac{N}{x} \cdot x = N$)

5. If $C(x) = KN\left(1 - (0.9944)^x + \dfrac{1}{x}\right)$ then the marginal cost function is the derivative of C.

$$C'(x) = \frac{d}{dx}\left[KN\left(1 - (0.9944)^x + \frac{1}{x}\right)\right] = KN\left[\frac{d}{dx}1 - \frac{d}{dx}(0.9944)^x + \frac{d}{dx}x^{-1}\right]$$

$$= KN\left[-(0.9944)^x \ln 0.9944 - x^{-2}\right] = KN\left[(0.0056)(0.9944)^x - \frac{1}{x^2}\right]$$

7. N will increase (or decrease) the unit marginal cost by a factor of N, the number of persons tested.

9. If each test costs \$5.00 and individual tests are performed, the total cost will be \$5000.

   If the pooling procedure is used the total cost will be $\$5000\left(1-(0.9944)^x+\dfrac{1}{x}\right)$ where x is the number of persons in the pooled group. The savings will be

   $$\$5000-\$5000\left(1-(0.9944)^x+\frac{1}{x}\right)=\$5000\left[(0.9944)^x-\frac{1}{x}\right]$$

   If $x = 13$ persons are in each group, the savings would be approximately

   $$\$5000\left[(0.9944)^{13}-\frac{1}{13}\right]=\$4263.37$$

   If $x = 14$ persons are in each group, the savings would be approximately

   $$\$5000\left[(0.9944)^{14}-\frac{1}{14}\right]=\$4264.81$$

MATHEMATICAL QUESTIONS FROM PROFESSIONAL EXAMS

1. (e) $\quad\dfrac{d}{dx}\left(x^2 e^{x^2}\right)=x^2\dfrac{d}{dx}e^{x^2}+e^{x^2}\dfrac{d}{dx}x^2=x^2 e^{x^2}\dfrac{d}{dx}x^2+e^{x^2}(2x)=2x^3 e^{x^2}+2xe^{x^2}$

3. (d) $\quad\dfrac{d}{dx}\left(be^{c^2+x^2}\right)=be^{c^2+x^2}\dfrac{d}{dx}\left(c^2+x^2\right)=2bxe^{c^2+x^2}$

5. (e) $\quad f''(x)=\dfrac{d}{dx}f'(x)=\dfrac{dy}{dx}\left(x\,f(x)\right)=x\dfrac{d}{dx}\left[f(x)\right]+f(x)\dfrac{d}{dx}x$

   $\qquad\qquad =xf'(x)+f(x)=x\left[x\,f(x)\right]+f(x)=\left(x^2+1\right)f(x)$

   $\quad f''(-2)=\left[(-2)^2+1\right]f(-2)=5\cdot 3=15$

Chapter 14
Applications: Graphing Functions; Optimization

14.1 Horizontal and Vertical Tangent Lines; Continuity and Differentiability   793/1-25 (ODD)

1. $f'(x) = 2x - 4$

    Horizontal tangent lines occur where $f'(x) = 0$.
    $$2x - 4 = 0$$
    $$2x = 4$$
    $$x = 2$$
    Evaluate the function f at $x = 2$: $f(2) = (2)^2 - 4(2) = -4$

    The tangent line to the graph of f is horizontal at the point $(2, f(2)) = (2, -4)$. There is no vertical tangent line since $f'(x)$ is never unbounded.

3. $f'(x) = -2x + 8$

    Horizontal tangent lines occur where $f'(x) = 0$.
    $$-2x + 8 = 0$$
    $$-2x = -8$$
    $$x = 4$$
    Evaluate the function f at $x = 4$: $f(4) = -(4)^2 + 8(4) = 16$

    The tangent line to the graph of f is horizontal at the point $(4, f(4)) = (4, 16)$. There is no vertical tangent line since $f'(x)$ is never unbounded.

5. $f'(x) = -4x + 8$

    Horizontal tangent lines occur where $f'(x) = 0$.
    $$-4x + 8 = 0$$
    $$-4x = -8$$
    $$x = 2$$
    Evaluate the function f at $x = 2$: $f(2) = -2(2)^2 + 8(2) + 1 = -8 + 16 + 1 = 9$

    The tangent line to the graph of f is horizontal at the point $(2, f(2)) = (2, 9)$. There is no vertical tangent line since $f'(x)$ is never unbounded.

7. $f'(x) = \dfrac{2}{3} \cdot 3x^{-1/3} = \dfrac{2}{x^{1/3}}$

    $f'(x)$ is never zero, so there is no horizontal tangent line.

$f'(x)$ is unbounded at $x = 0$. We evaluate f at 0: $f(0) = 3 \cdot 0^{2/3} + 1 = 1$.

The tangent line to the graph of f is vertical at the point $(0, 1)$.

9. $f'(x) = -3x^2 + 3$
    Horizontal tangent lines occur where $f'(x) = 0$.
    $$-3x^2 + 3 = 0$$
    $$x^2 - 1 = 0$$
    $$(x - 1)(x + 1) = 0$$
    $$x - 1 = 0 \quad \text{or} \quad x + 1 = 0$$
    $$x = 1 \quad \text{or} \quad x = -1$$

    Evaluate f at 1 and -1.
    $$f(1) = -1^3 + 3(1) + 1 = 3$$
    $$f(-1) = -(-1)^3 + 3(-1) + 1 = -1$$

    There are two horizontal lines tangent to the graph of f, one at $(1, 3)$ and the other at $(-1, -1)$. There is no vertical tangent line since $f'(x)$ is never unbounded.

11. $$f'(x) = \frac{3}{4} \cdot 4x^{-1/4} = \frac{3}{x^{1/4}}$$
    $f'(x) \neq 0$ so there is no horizontal tangent line.
    $f'(x)$ is unbounded at $x = 0$. We evaluate $f(0) = 4(0)^{3/4} - 2 = -2$.

    The tangent line to the graph of f at $(0, -2)$ is vertical.

13. $f'(x) = 5x^4 - 40x^3$
    Horizontal tangents occur where $f'(x) = 0$.
    $$5x^4 - 40x^3 = 0$$
    $$5x^3(x - 8) = 0$$
    $$5x^3 = 0 \quad \text{or} \quad x - 8 = 0$$
    $$x = 0 \quad \text{or} \quad x = 8$$

    Evaluate f at 0 and 8.
    $$f(0) = (0)^5 - 10(0)^4 = 0 \qquad\qquad f(8) = (8)^5 - 10(8)^4 = -8192$$

    We conclude that the graph of f has 2 horizontal tangent lines, one at $(0, 0)$, the other at $(8, -8192)$. There is no vertical tangent line.

15. $f'(x) = 15x^4 + 60x^2$
    Horizontal tangent lines occur where $f'(x) = 0$.
    $$15x^4 + 60x^2 = 0$$

$$15x^2\left(x^2+4\right)=0$$
$$15x^2=0 \quad \text{or} \quad x^2+4=0$$
$$x=0$$

Evaluate f at 0: $\quad f(0)=3(0)^5+20(0)^3-1=-1$

The graph of f has a horizontal tangent line at $(0,-1)$; f' is never unbounded, so f has no vertical tangent.

17. $\quad f'(x)=\dfrac{2}{3}x^{-1/3}+\dfrac{1}{3}\cdot 2x^{-2/3}=\dfrac{2}{3x^{1/3}}+\dfrac{2}{3x^{2/3}}$

$$=\dfrac{2}{3x^{1/3}}\cdot\dfrac{x^{1/3}}{x^{1/3}}+\dfrac{2}{3x^{2/3}}=\dfrac{2x^{1/3}}{3x^{2/3}}+\dfrac{2}{3x^{2/3}}=\dfrac{2x^{1/3}+2}{3x^{2/3}}$$

Horizontal tangent lines occur where $f'(x)=0$.
$$\dfrac{2x^{1/3}+2}{3x^{2/3}}=0$$
$$2x^{1/3}+2=0$$
$$x^{1/3}=-1$$
$$x=-1$$

Vertical tangent lines occur where $f'(x)$ is unbounded. $f'(x)$ is unbounded at 0.

Evaluate f at 0 and -1:
$$f(0)=0^{2/3}+2(0)^{1/3}=0$$
$$f(-1)=(-1)^{2/3}+2(-1)^{1/3}=-1$$

We conclude that the graph of f has a horizontal tangent line at $(-1,-1)$ and a vertical tangent line at $(0,0)$.

19. $\quad f(x)=x^{2/3}(x-10)=x^{5/3}-10x^{2/3}$

$$f'(x)=\dfrac{5}{3}x^{2/3}-\dfrac{2}{3}\cdot 10x^{-1/3}=\dfrac{5x^{2/3}}{3}-\dfrac{20}{3x^{1/3}}$$

$$=\dfrac{5x^{2/3}}{3}\cdot\dfrac{x^{1/3}}{x^{1/3}}-\dfrac{20}{3x^{1/3}}=\dfrac{5x}{3x^{1/3}}-\dfrac{20}{3x^{1/3}}=\dfrac{5x-20}{3x^{1/3}}$$

Horizontal tangent lines occur where $f'(x)=0$.
$$\dfrac{5x-20}{3x^{1/3}}=0$$
$$5x-20=0$$

$$5x = 20$$
$$x = 4$$

Vertical tangent lines occur where $f'(x)$ is unbounded. $f'(x)$ is unbounded at 0.

Evaluate f at 0 and 4:
$$f(0) = 0^{2/3}(0-10) = 0$$
$$f(4) = 4^{2/3}(4-10) = -6 \cdot 4^{2/3} = -6 \cdot (8 \cdot 2)^{1/3} = -6 \cdot 2 \cdot 2^{1/3} = -12 \cdot 2^{1/3} = -12\sqrt[3]{2}$$

We conclude that the graph of f has a horizontal tangent line at $(4, -12\sqrt[3]{2})$ and a vertical tangent line at $(0, 0)$.

21. $f(x) = x^{2/3}(x^2 - 16) = x^{8/3} - 16x^{2/3}$

$$f'(x) = \frac{8}{3}x^{5/3} - \frac{2}{3} \cdot 16x^{-1/3} = \frac{8x^{5/3}}{3} - \frac{32}{3x^{1/3}}$$

$$= \frac{8x^{5/3}}{3} \cdot \frac{x^{1/3}}{x^{1/3}} - \frac{32}{3x^{1/3}} = \frac{8x^2}{3x^{1/3}} - \frac{32}{3x^{1/3}} = \frac{8x^2 - 32}{3x^{1/3}}$$

Horizontal tangent lines occur where $f'(x) = 0$.
$$\frac{8x^2 - 32}{3x^{1/3}} = 0$$
$$8x^2 - 32 = 0$$
$$8(x^2 - 4) = 0$$
$$(x - 2)(x + 2) = 0$$
$$x - 2 = 0 \quad \text{or} \quad x + 2 = 0$$
$$x = 2 \quad \text{or} \quad x = -2$$

Vertical tangent lines occur where $f'(x)$ is unbounded. $f'(x)$ is unbounded at 0.

Evaluate f at 2, -2, and 0:
$$f(2) = 2^{2/3}(2^2 - 16) = 2^{2/3}(-12) = -12 \cdot 2^{2/3} = -12\sqrt[3]{4}$$
$$f(-2) = (-2)^{2/3}((-2)^2 - 16) = 2^{2/3}(-12) = -12 \cdot 2^{2/3} = -12\sqrt[3]{4}$$
$$f(0) = 0^{2/3}(0^2 - 16) = 0$$

We conclude that the graph of f has horizontal tangent lines at $\left(2, -12\sqrt[3]{4}\right)$ and $\left(-2, -12\sqrt[3]{4}\right)$ and a vertical tangent line at $(0, 0)$.

23.
$$f(x) = \frac{x^{2/3}}{x-2} \qquad x \neq 2$$

$$f'(x) = \frac{\frac{2}{3}x^{-1/3} \cdot (x-2) - x^{2/3} \cdot 1}{(x-2)^2} = \frac{1}{(x-2)^2}\left[\frac{2(x-2)}{3x^{1/3}} - x^{2/3}\right]$$

$$= \frac{1}{(x-2)^2}\left[\frac{2(x-2)}{3x^{1/3}} - x^{2/3} \cdot \frac{3x^{1/3}}{3x^{1/3}}\right]$$

$$= \frac{1}{(x-2)^2}\left[\frac{2(x-2)}{3x^{1/3}} - \frac{3x}{3x^{1/3}}\right]$$

$$= \frac{1}{(x-2)^2}\left[\frac{2x-4-3x}{3x^{1/3}}\right] = \frac{-x-4}{3x^{1/3}(x-2)^2}$$

Horizontal tangent lines occur where $f'(x) = 0$.

$$\frac{-x-4}{3x^{1/3}(x-2)^2} = 0$$

$$-x-4 = 0$$

$$x = -4$$

Vertical tangent lines occur where $f'(x)$ is unbounded. $f'(x)$ is unbounded if $x = 0$ or $x = 2$, but $x = 2$ is not part of the domain of f, so we disregard it.

Evaluate f at -4 and 0:

$$f(-4) = \frac{(-4)^{2/3}}{(-4)-2} = \frac{16^{1/3}}{-6} = -\frac{2 \cdot 2^{1/3}}{6} = -\frac{2^{1/3}}{3} = -\frac{\sqrt[3]{2}}{3}$$

$$f(0) = \frac{0^{2/3}}{0-2} = 0$$

We conclude that the graph of f has a horizontal tangent line at $\left(-4, -\dfrac{\sqrt[3]{2}}{3}\right)$ and a vertical tangent line at $(0, 0)$.

25.

$$f(x) = \frac{x^{1/3}}{x-1} \qquad x \ne 1$$

$$f'(x) = \frac{\frac{1}{3}x^{-2/3}(x-1) - x^{1/3} \cdot 1}{(x-1)^2} = \frac{1}{(x-1)^2}\left[\frac{x-1}{3x^{2/3}} - x^{1/3}\right]$$

$$= \frac{1}{(x-1)^2}\left[\frac{x-1}{3x^{2/3}} - x^{1/3} \cdot \frac{3x^{2/3}}{3x^{2/3}}\right]$$

$$= \frac{1}{(x-1)^2}\left[\frac{x-1}{3x^{2/3}} - \frac{3x}{3x^{2/3}}\right]$$

$$= \frac{1}{(x-1)^2} \left[\frac{x-1-3x}{3x^{2/3}} \right] = \frac{-2x-1}{3x^{2/3}(x-1)^2}$$

Horizontal tangent lines occur where $f'(x) = 0$.

$$\frac{-2x-1}{3x^{2/3}(x-1)^2} = 0$$

$$-2x - 1 = 0$$

$$x = -\frac{1}{2}$$

Vertical tangent lines occur where $f'(x)$ is unbounded. $f'(x)$ is unbounded if $x = 0$ or $x = 1$, but $x = 1$ is not part of the domain of f, so we disregard it.

Evaluate f at $-\frac{1}{2}$ and 0:

$$f\left(-\frac{1}{2}\right) = \frac{\left(-\frac{1}{2}\right)^{1/3}}{-\frac{1}{2}-1} = \frac{\left(-\frac{1}{2}\right)^{1/3}}{-\frac{3}{2}} = \left(-\frac{1}{2}\right)^{1/3} \cdot \left(-\frac{2}{3}\right) = \frac{1}{2^{1/3}} \cdot \frac{2}{3} = \frac{2^{2/3}}{3} = \frac{\sqrt[3]{4}}{3}$$

$$f(0) = \frac{0^{1/3}}{0-1} = 0$$

We conclude that the graph of f has a horizontal tangent line at $\left(-\frac{1}{2}, \frac{\sqrt[3]{4}}{3}\right)$ and a vertical tangent line at $(0, 0)$.

27. (a) $f(0) = 0^{2/3} = 0$. The one-sided limits are

$$\lim_{x \to 0^-} f(x) = \lim_{x \to 0^-} x^{2/3} = 0 \qquad \lim_{x \to 0^+} f(x) = \lim_{x \to 0^+} x^{2/3} = 0$$

Since $\lim_{x \to 0} f(x) = f(0)$ the function is continuous at 0.

(b) The derivative of f at 0 is $f'(0) = \lim_{x \to 0} \frac{f(x) - f(0)}{x - 0} = \lim_{x \to 0} \frac{x^{2/3} - 0}{x - 0}$.

We look at the one-sided limits:

$$\lim_{x \to 0^-} \frac{x^{2/3} - 0}{x - 0} = \lim_{x \to 0^-} \frac{x^{2/3}}{x} = \lim_{x \to 0^-} \frac{1}{x^{1/3}} = -\infty$$

$$\lim_{x \to 0^+} \frac{x^{2/3} - 0}{x - 0} = \lim_{x \to 0^+} \frac{x^{2/3}}{x} = \lim_{x \to 0^+} \frac{1}{x^{1/3}} = \infty$$

Since the one-sided limits are not equal, $f'(0)$ does not exist.

(c) The derivative is unbounded at $x = 0$, so there is a vertical tangent line at 0.

29. (a) $f(1)$ is not defined, so f is not continuous at $x = 1$.

(b) Since f is not continuous at $x = 1$, $f'(1)$ does not exist.

31. (a) $f(0) = 0^2 = 0$. The one sided limits are

$$\lim_{x \to 0^-} f(x) = \lim_{x \to 0^-} 3x = 0 \qquad \lim_{x \to 0^+} f(x) = \lim_{x \to 0^+} x^2 = 0$$

Since $\lim_{x \to 0} f(x) = f(0)$, the function f is continuous at 0.

(b) The derivative of f at 0 is $f'(0) = \lim_{x \to 0} \dfrac{f(x) - f(0)}{x - 0} = \lim_{x \to 0} \dfrac{f(x) - 0}{x - 0}$.

We look at the one-sided limits:

$$\lim_{x \to 0^-} \frac{f(x) - 0}{x - 0} = \lim_{x \to 0^-} \frac{3x - 0}{x - 0} = \lim_{x \to 0^-} \frac{3x}{x} = \lim_{x \to 0^-} 3 = 3$$

$$\lim_{x \to 0^+} \frac{f(x) - 0}{x - 0} = \lim_{x \to 0^+} \frac{x^2 - 0}{x - 0} = \lim_{x \to 0^+} \frac{x^2}{x} = \lim_{x \to 0^+} x = 0$$

Since the one-sided limits are not equal, $f'(0)$ does not exist.

(c) There is no tangent line at $x = 0$.

(d)

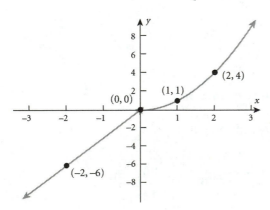

$$f(x) = \begin{cases} 3x & \text{if } x < 0 \\ x^2 & \text{if } x \geq 0 \end{cases}$$

33. (a) $f(2) = 4 \cdot 2 = 8$. The one-sided limits are

$$\lim_{x \to 2^-} f(x) = \lim_{x \to 2^-} 4x = 4 \cdot 2 = 8 \qquad \lim_{x \to 2^+} f(x) = \lim_{x \to 2^+} x^2 = 2^2 = 4$$

Since the one-sided limits are unequal, the $\lim_{x \to 2} f(x)$ doesn't exist and the function is not continuous at $x = 2$.

(b) The function is not continuous at $x = 2$, so $f'(2)$ does not exist.

(c) Does not apply to this problem.

(d)

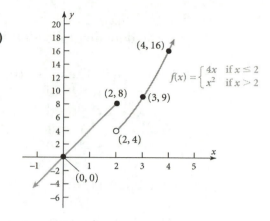

35. (a) $f(0) = 0^3 = 0$. The one-sided limits are:

$$\lim_{x \to 0^-} f(x) = \lim_{x \to 0^-} x^2 = 0^2 = 0 \qquad \lim_{x \to 0^+} f(x) = \lim_{x \to 0^+} x^3 = 0^3 = 0$$

The one-sided limits are equal so $\lim_{x \to 0} f(x)$ exists, and since $\lim_{x \to 0} f(x) = f(0)$, the function is continuous at $x = 0$.

(b) The derivative of f at 0 is $f'(0) = \lim_{x \to 0} \dfrac{f(x) - f(0)}{x - 0} = \lim_{x \to 0} \dfrac{f(x) - 0}{x - 0} = \lim_{x \to 0} \dfrac{f(x)}{x}$.

We look at one-sided limits.

$$\lim_{x \to 0^-} \frac{f(x)}{x} = \lim_{x \to 0^-} \frac{x^2}{x} = \lim_{x \to 0^-} x = 0 \qquad \lim_{x \to 0^+} \frac{f(x)}{x} = \lim_{x \to 0^+} \frac{x^3}{x} = \lim_{x \to 0^+} x^2 = 0$$

We conclude that $f'(0) = 0$, and that there is a horizontal tangent line at 0.

(c) Does not apply to this problem.

(d)

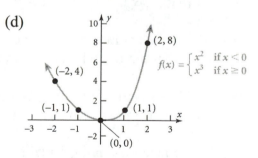

803/2-33 (ODD)

14.2 Increasing and Decreasing Functions; the First Derivative Test

1. The domain of f is $[x_1, x_9]$.

3. The graph of f is increasing on (x_1, x_4), (x_5, x_7), and (x_8, x_9).

5. $f'(x) = 0$ for x_4, x_7, and x_8.

7. f has a local maximum at (x_4, y_4) and at (x_7, y_7).

9. $f(x) = -2x^2 + 4x - 2$
STEP 1 The domain of f is all real numbers.

STEP 2 Let $x = 0$. Then $y = f(0) = -2$. The y-intercept is $(0, -2)$. Now let $y = 0$. Then
$$-2x^2 + 4x - 2 = 0$$
$$x^2 - 2x + 1 = 0$$
$$(x - 1)^2 = 0$$
$$x - 1 = 0$$
$$x = 1$$
The x-intercept is $(1, 0)$.

STEP 3 To find where the graph is increasing or decreasing, we find $f'(x)$:
$$f'(x) = -4x + 4$$

The solution to $f'(x) = 0$ is
$$-4x + 4 = 0$$
$$x = 1$$
Use 1 to separate the number line into 2 parts, and use $x = 0$ and $x = 2$ as test numbers.

$$-\infty < x < 1 \quad \text{and} \quad 1 < x < \infty$$

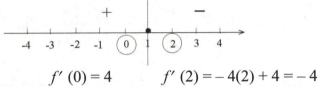

$$f'(0) = 4 \qquad f'(2) = -4(2) + 4 = -4$$

We conclude that the graph of f is increasing on the interval $(-\infty, 1)$; f is decreasing on the interval $(1, \infty)$.

STEP 4 Since f is increasing to the left of 1 and decreasing to the right of 1, we conclude that there is a local maximum at $(1, f(1)) = (1, 0)$.

STEP 5 We found $f'(1) = 0$, indicating that there is a horizontal tangent at $(1, 0)$.
The first derivative is never unbounded, so there is no vertical tangent line.

STEP 6 Since f is a polynomial function, its end behavior is that of $y = -2x^2$.
Polynomial functions have no asymptotes.

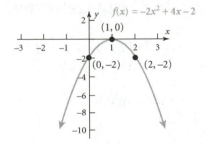

11. $f(x)=x^3-9x^2+27x-27$

STEP 1 The domain of f is all real numbers.

STEP 2 Let $x=0$. Then $y=f(0)=-27$. The y-intercept is $(0,-27)$.
Now let $y=0$. Then $x^3-9x^2+27x-27=0$
$$(x-3)^3=0$$
$$x-3=0$$
$$x=3$$
The x-intercept is $(3,0)$.

STEP 3 To find where the graph is increasing or decreasing, we find $f'(x)$:
$$f'(x)=3x^2-18x+27$$
The solution to $f'(x)=0$ is
$$3x^2-18x+27=0$$
$$3(x^2-6x+9)=0$$
$$3(x-3)^2=0$$
$$x-3=0$$
$$x=3$$
Use 3 to separate the number line into 2 parts, and use 0 and 4 as test numbers.

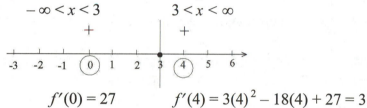

$$-\infty<x<3 \qquad 3<x<\infty$$
$$+ \qquad\qquad +$$

$$f'(0)=27 \qquad f'(4)=3(4)^2-18(4)+27=3$$
We conclude that the function is always increasing that is on the interval $(-\infty,\infty)$.

STEP 4 There are no local extreme points since the first derivative never changes signs.

STEP 5 We found $f'(3)=0$, indicating that there is a horizontal tangent at $(3,f(3))=(3,0)$.
The first derivative is never unbounded, so there is no vertical tangent.

STEP 6 Since f is a polynomial function, its end behavior is that of $y=x^3$. Polynomial functions have no asymptotes.

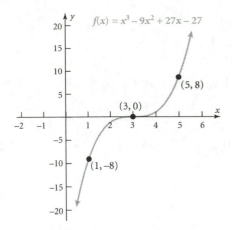

13. $f(x) = 2x^3 - 15x^2 + 36x$

STEP 1 The domain of f is all real numbers.

STEP 2 Let $x = 0$. Then $y = f(0) = 0$. The y-intercept is $(0, 0)$.
Now let $y = 0$. Then $2x^3 - 15x^2 + 36x = 0$
$$x(2x^2 - 15x + 36) = 0$$
$$x = 0$$
The x-intercept is $(0, 0)$.
($2x^2 - 15x + 36 = 0$ has no real solution; its discriminant, $b^2 - 4ac = -63$, is negative.)

STEP 3 To find where the graph is increasing or decreasing, we find $f'(x)$:
$$f'(x) = 6x^2 - 30x + 36$$
The solutions to $f'(x) = 0$ are
$$6x^2 - 30x + 36 = 0$$
$$6(x^2 - 5x + 6) = 0$$
$$6(x - 2)(x - 3) = 0$$
$$x - 2 = 0 \quad \text{or} \quad x - 3 = 0$$
$$x = 2 \quad \text{or} \qquad x = 3$$
Use the numbers to separate the number line into 3 parts, and use 0, 2.5, and 4 as test numbers.

$$-\infty < x < 2 \qquad 2 < x < 3 \qquad 3 < x < \infty$$

$$+ \qquad\qquad - \qquad\qquad +$$

-1 ⓪ 1 2 ③ 4
⟨2.5⟩

$f'(0) = 36 \quad f'(2.5) = 6(2.5)^2 - 30(2.5) + 36 = -1.5 \quad f'(4) = 6(4)^2 - 30(4) + 36 = 12$

We conclude that the function is increasing on the intervals $(-\infty, 2)$ and $(3, \infty)$ and is decreasing on the interval $(2, 3)$.

STEP 4 Since the function is increasing to the left of 2 and decreasing to the right of 2, the point $(2, f(2)) = (2, 28)$ is a local maximum.

The function is decreasing to the left of 3 and increasing to the right of 3, so the point $(3, f(3)) = (3, 27)$ is a local minimum.

STEP 5 We found $f'(2) = 0$ and $f'(3) = 0$, indicating that there are horizontal tangent lines at $(2, f(2)) = (2, 28)$ and $(3, f(3)) = (3, 27)$.

The first derivative is never unbounded, so there is no vertical tangent.

STEP 6 Since f is a polynomial function, its end behavior is that of $y = 2x^3$. Polynomial functions have no asymptotes.

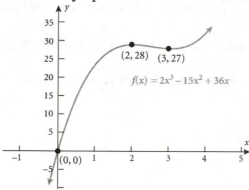

15. $f(x) = -x^3 + 3x - 1$

STEP 1 The domain of f is all real numbers.

STEP 2 Let $x = 0$. Then $y = f(0) = -1$. The y-intercept is $(0, -1)$. The x-intercept is hard to find, so we skip it.

STEP 3 To find where the graph is increasing or decreasing, we find $f'(x)$:
$$f'(x) = -3x^2 + 3$$
The solutions to $f'(x) = 0$ are
$$-3x^2 + 3 = 0$$
$$-3(x^2 - 1) = 0$$
$$-3(x - 1)(x + 1) = 0$$
$$x - 1 = 0 \quad \text{or} \quad x + 1 = 0$$
$$x = 1 \quad \text{or} \quad x = -1$$

We use the numbers 1 and -1 to separate the number line into three parts:
$$-\infty < x < -1 \qquad -1 < x < 1 \qquad 1 < x < \infty$$
and choose a test point from each interval.

For $x = -2$: $\quad f'(-2) = -3(-2)^2 + 3 = -9$

For $x = 0$: $\quad f'(0) = -3(0)^2 + 3 = 3$

For $x = 2$: $f'(2) = -3(2)^2 + 3 = -9$

We conclude that the graph of f is increasing on the interval $(-1, 1)$; f is decreasing on the intervals $(-\infty, -1)$ and $(1, \infty)$.

STEP 4 Since the graph is decreasing to the left of -1 and increasing to the right of -1, the point $(-1, f(-1)) = (-1, -3)$ is a local minimum.

Since the graph is increasing to the left of 1 and decreasing to the right of 1, the point $(1, f(1)) = (1, 1)$ is a local maximum.

STEP 5 $f'(x) = 0$ for $x = 1$ and $x = -1$. The graph of f has horizontal tangent lines at the points $(1, f(1)) = (1, 1)$ and $(-1, f(-1)) = (-1, -3)$.

The first derivative is never unbounded, so there is no vertical tangent.

STEP 6 Since f is a polynomial function, its end behavior is that of $y = -x^3$. Polynomial functions have no asymptotes.

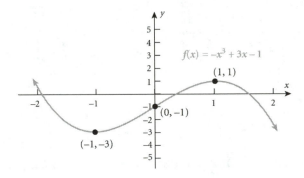

17. $f(x) = 3x^4 - 12x^3 + 2$

STEP 1 The domain of f is all real numbers.

STEP 2 Let $x = 0$. Then $y = f(0) = 2$. The y-intercept is $(0, 2)$.
The x-intercept is hard to find, so we skip it.

STEP 3 To find where the graph is increasing or decreasing, we find $f'(x)$:
$$f'(x) = 12x^3 - 36x^2$$
The solutions to $f'(x) = 0$ are
$$12x^3 - 36x^2 = 0$$
$$12x^2(x - 3) = 0$$
$$12x^2 = 0 \quad \text{or} \quad x - 3 = 0$$
$$x = 0 \quad \text{or} \quad x = 3$$

We use the numbers 0 and 3 to separate the number line into three parts:
$$-\infty < x < 0 \qquad 0 < x < 3 \qquad 3 < x < \infty$$
and choose a test number from each interval.

For $x = -1$: $f'(-1) = 12(-1)^3 - 36(-1)^2 = -48$

For $x = 1$: $f'(1) = 12(1)^3 - 36(1)^2 = -24$

For $x = 4$: $f'(4) = 12(4)^3 - 36(4)^2 = 192$

We conclude that the graph of f is increasing on the interval $(3, \infty)$; f is decreasing on the interval $(-\infty, 3)$.

STEP 4 Since the graph is decreasing to the left of 3 and increasing to the right of 3, the point $(3, f(3)) = (3, -79)$ is a local minimum.
There is no local maximum.

STEP 5 $f'(x) = 0$ for $x = 0$ and $x = 3$. The graph of f has horizontal tangent lines at the points $(0, f(0)) = (0, 2)$ and $(3, f(3)) = (3, -79)$.
The first derivative is never unbounded, so there is no vertical tangent.

STEP 6 Since f is a polynomial function, its end behavior is that of $y = 3x^4$. Polynomial functions have no asymptotes.

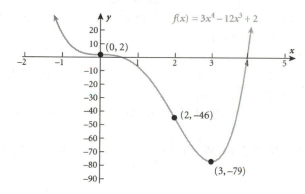

19. $f(x) = x^5 - 5x + 1$

STEP 1 The domain of f is all real numbers.

STEP 2 Let $x = 0$. Then $y = f(0) = 1$. The y-intercept is $(0, 1)$.
The x-intercept is hard to find, so we skip it.

STEP 3 To find where the graph is increasing or decreasing, we find $f'(x)$:
$$f'(x) = 5x^4 - 5$$
The solutions to $f'(x) = 0$ are
$$5x^4 - 5 = 0$$
$$5(x^4 - 1) = 0$$
$$5(x^2 - 1)(x^2 + 1) = 0$$
$$5(x - 1)(x + 1)(x^2 + 1) = 0$$

$$x - 1 = 0 \quad \text{or} \quad x + 1 = 0 \quad \text{or} \quad x^2 + 1 = 0$$
$$x = 1 \quad \text{or} \qquad x = -1$$

The discriminant of $x^2 + 1$ is negative, so $x^2 + 1 = 0$ has no solution.

We use the numbers -1 and 1 to separate the number line into three parts:
$$-\infty < x < -1 \qquad -1 < x < 1 \qquad 1 < x < \infty$$
and choose a test point from each part.

For $x = -2$: $\quad f'(-2) = 5(-2)^4 - 5 = 75$

For $x = 0$: $\quad f'(0) = 5(0)^4 - 5 = -5$

For $x = 2$: $\quad f'(2) = 5(2)^4 - 5 = 75$

We conclude that the graph of f is increasing on the intervals $(-\infty, -1)$ and $(1, \infty)$; f is decreasing on the interval $(-1. 1)$.

STEP 4 Since the graph is increasing to the left of -1 and decreasing to the right of -1, the point $(-1, f(-1)) = (-1, 5)$ is a local maximum.

The graph is decreasing to the left of 1 and increasing to the right of 1, so the point $(1, f(1)) = (1, -3)$ is a local minimum.

STEP 5 $f'(x) = 0$ for $x = -1$ and for $x = 1$. The graph of f has horizontal tangent lines at the points $(-1, f(-1)) = (-1, 5)$ and $(1, f(1)) = (1, -3)$.

The first derivative is never unbounded, so there is no vertical tangent

STEP 6 Since f is a polynomial function, its end behavior is that of $y = x^5$. Polynomial functions have no asymptotes.

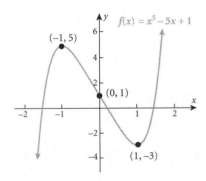

21. $f(x) = 3x^5 - 20x^3 + 1$

STEP 1 The domain of f is all real numbers.

STEP 2 Let $x = 0$. Then $y = f(0) = 1$. The y-intercept is $(0, 1)$.
The x-intercept is hard to find, so we skip it.

STEP 3 To find where the graph is increasing or decreasing, we find $f'(x)$:

$$f'(x) = 15x^4 - 60x^2$$

The solutions to $f'(x) = 0$ are

$$15x^4 - 60x^2 = 0$$
$$15x^2 (x^2 - 4) = 0$$
$$15x^4 (x-2)(x+2) = 0$$
$$15x^4 = 0 \quad \text{or} \quad x - 2 = 0 \quad \text{or} \quad x + 2 = 0$$
$$x = 0 \quad \text{or} \quad x = 2 \quad \text{or} \quad x = -2$$

We use the numbers -2, 0, and 2 to separate the number line into four parts:

$$-\infty < x < -2 \qquad -2 < x < 0 \qquad 0 < x < 2 \qquad 2 < x < \infty$$

and choose a test point from each part.

For $x = -3$: $\quad f'(x)(-3) = 15(-3)^4 - 60(-3)^2 = 675$

For $x = -1$: $\quad f'(x)(-1) = 15(-1)^4 - 60(-1)^2 = -45$

For $x = 1$: $\quad f'(x)(1) = 15(1)^4 - 60(1)^2 = -45$

For $x = 3$: $\quad f'(x)(3) = 15(3)^4 - 60(3)^2 = 675$

We conclude that the graph of f is increasing on the intervals $(-\infty, -2)$ and $(2, \infty)$; f is decreasing on the interval $(-2, 2)$.

STEP 4 Since the graph is increasing to the left of -2 and decreasing to the right of -2, the point $(-2, f(-2)) = (-2, 65)$ is a local maximum.

The graph is decreasing to the left of 2 and increasing to the right of 2, so the point $(2, f(2)) = (2, -63)$ is a local minimum.

STEP 5 $f'(x) = 0$ for $x = -2$, $x = 0$, and $x = 2$. The graph of f has horizontal tangent lines at the points $(-2, f(-2)) = (-2, 65)$, $(0, f(0)) = (0, 1)$, and $(2, f(2)) = (2, -63)$.

The first derivative is never unbounded, so there is no vertical tangent

STEP 6 Since f is a polynomial function, its end behavior is that of $y = 3x^5$. Polynomial functions have no asymptotes.

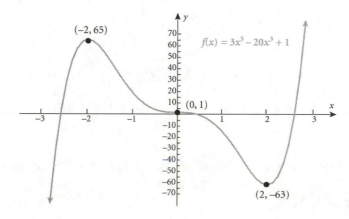

23. $f(x) = x^{2/3} + 2x^{1/3}$

STEP 1 The domain of f is the set of all real numbers.

STEP 2 Let $x = 0$. Then $y = f(0) = 0$. The y-intercept is $(0, 0)$.
Now let $y = 0$. Then $x^{2/3} + 2x^{1/3} = 0$
$$x^{1/3}\left(x^{1/3} + 2\right) = 0$$
$$x^{1/3} = 0 \quad \text{or} \quad x^{1/3} + 2 = 0$$
$$x = 0 \quad \text{or} \quad x^{1/3} = -2$$
$$x = -8$$

The x-intercepts are $(0, 0)$ and $(8, 0)$.

STEP 3 To find where the graph is increasing or decreasing, we find $f'(x)$:
$$f'(x) = \frac{2}{3}x^{-1/3} + \frac{2}{3}x^{-2/3}$$

The solutions to $f'(x) = 0$ are
$$\frac{2}{3}x^{-1/3} + \frac{2}{3}x^{-2/3} = 0$$
$$x^{-1/3} + x^{-2/3} = 0$$
$$\frac{1}{x^{1/3}} + \frac{1}{x^{2/3}} = 0$$
$$\frac{x^{1/3} + 1}{x^{2/3}} = 0$$
$$x^{1/3} = -1$$

$f'(x)$ is not defined when $x = 0$, $f'(x) = 0$ when $x = -1$.
We use these two numbers to separate the number line into three parts:
$$-\infty < x < -1 \qquad -1 < x < 0 \qquad 0 < x < \infty$$
and we choose a test point in each part.

For $x = -8$: $\quad f'(x)(-8) = \frac{2}{3}(-8)^{-1/3} + \frac{2}{3}(-8)^{-2/3} = -\frac{1}{6} \approx -0.167$

For $x = -\frac{1}{8}$: $\quad f'(x)\left(-\frac{1}{8}\right) = \frac{2}{3}\left(-\frac{1}{8}\right)^{-1/3} + \frac{2}{3}\left(-\frac{1}{8}\right)^{-2/3} = \frac{4}{3} \approx 1.333$

For $x = 1$: $\quad f'(x)(1) = \frac{2}{3}(1)^{-1/3} + \frac{2}{3}(1)^{-2/3} = \frac{4}{3} \approx 1.333$

We conclude that the graph of f is increasing on the interval $(-1, \infty)$; f is decreasing on the interval $(-\infty, -1)$.

STEP 4 The graph is decreasing to the left of -1 and increasing to the right of -1, so the point $(-1, f(-1)) = (-1, -1)$ is a local minimum.
There is no local maximum.

STEP 5 $f'(x) = 0$ for $x = -1$. The graph of f has a horizontal tangent line at the point $(-1, f(-1)) = (-1, -1)$.

The first derivative is unbounded at $x = 0$, so there is a vertical tangent line at the point $(0, f(0)) = (0, 0)$.

STEP 6 For the end behavior of f, we look at the two limits at infinity:

$$\lim_{x \to -\infty} f(x) = \lim_{x \to -\infty} \left(x^{2/3} + 2x^{1/3} \right) = \lim_{x \to -\infty} x^{2/3} = \left(\lim_{x \to -\infty} x^{1/3} \right)^2 = \infty$$

$$\lim_{x \to \infty} f(x) = \lim_{x \to \infty} \left(x^{2/3} + 2x^{1/3} \right) = \lim_{x \to \infty} x^{2/3} = \left(\lim_{x \to \infty} x^{1/3} \right)^2 = \infty$$

The graph of f becomes unbounded in the positive direction as $x \to \pm\infty$.

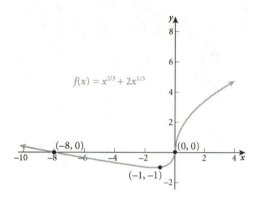

25. $f(x) = \left(x^2 - 1 \right)^{2/3}$

STEP 1 The domain of f is all real numbers.

STEP 2 Let $x = 0$. Then $y = f(0) = (0-1)^{2/3} = (-1)^{2/3} = 1$. The y-intercept is $(0, 1)$.

Now let $y = 0$. Then $\left(x^2 - 1 \right)^{2/3} = 0$

$$x^2 - 1 = 0$$
$$x^2 = 1$$
$$x = \pm 1$$

The x-intercepts are $(-1, 0)$ and $(1, 0)$.

STEP 3 To find where the graph is increasing or decreasing, we find $f'(x)$:

$$f'(x) = \frac{2}{3}\left(x^2 - 1 \right)^{-1/3} \cdot 2x = \frac{4x}{3\left(x^2 - 1 \right)^{1/3}} \qquad x \neq \pm 1$$

The solution to $f'(x) = 0$ is

$$\frac{4x}{3\left(x^2 - 1 \right)^{1/3}} = 0$$
$$4x = 0$$
$$x = 0$$

We use the numbers -1, 0, and 1 to separate the number line into four parts:
$$-\infty < x < -1 \qquad -1 < x < 0 \qquad 0 < x < 1 \qquad 1 < x < \infty$$
and choose a test number from each part.

For $x = -8$: $\quad f'(-8) = \dfrac{4(-8)}{3\left[(-8)^2 - 1\right]^{1/3}} = -2.681$

For $x = -\dfrac{1}{8}$: $\quad f'\left(-\dfrac{1}{8}\right) = \dfrac{4\left(-\dfrac{1}{8}\right)}{3\left[\left(-\dfrac{1}{8}\right)^2 - 1\right]^{1/3}} = 0.168$

For $x = \dfrac{1}{8}$: $\quad f'\left(\dfrac{1}{8}\right) = \dfrac{4\left(\dfrac{1}{8}\right)}{3\left[\left(\dfrac{1}{8}\right)^2 - 1\right]^{1/3}} = -0.168$

For $x = 8$: $\quad f'(8) = \dfrac{4(8)}{3\left[(8)^2 - 1\right]^{1/3}} = 2.681$

We conclude that the graph of f is increasing on the intervals $(-1, 0)$ and $(1, \infty)$; f is decreasing on the intervals $(-\infty, -1)$ and $(0, 1)$.

STEP 4 The graph is decreasing to the left of -1 and increasing to the right of -1, so the point $(-1, f(-1)) = (-1, 0)$ is a local minimum. The graph is also decreasing to the left of 1 and increasing to the right of 1, so the point $(1, f(1)) = (1, 0)$ is another local minimum.

The graph is increasing to the left of 0 and decreasing to the right of 0, so the point $(0, f(0)) = (0, 1)$ is a local maximum.

STEP 5 $f'(x) = 0$ for $x = 0$. The graph of f has a horizontal tangent line at the point $(0, f(0)) = (0, 1)$. The first derivative is unbounded at $x = -1$ and $x = 1$, so there are vertical tangent lines at the points $(-1, 0)$ and $(1, 0)$.

STEP 6 For the end behavior we look at the limits at infinity. Since f is an even function, we need only to consider the limit as $x \to \infty$.
$$\lim_{x \to \infty} \left(x^2 - 1\right)^{2/3} = \lim_{x \to \infty} \left(x^2\right)^{2/3} = \lim_{x \to \infty} x^{4/3} = \infty$$
The graph becomes unbounded at $x \to \pm \infty$.

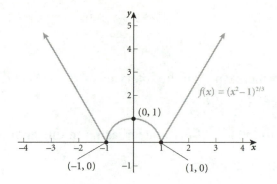

$f(x) = (x^2 - 1)^{2/3}$

(0, 1)

(−1, 0)

(1, 0)

27.
$$f(x) = \frac{8}{x^2 - 16} = \frac{8}{(x-4)(x+4)} \qquad x \neq \pm 4$$

STEP 1 The domain of f is $\{x \mid x \neq -4, x \neq 4\}$.

STEP 2 Let $x = 0$. Then $y = f(0) = \dfrac{8}{0^2 - 16} = \dfrac{8}{-16} = -\dfrac{1}{2}$. The y-intercept is $\left(0, -\dfrac{1}{2}\right)$.

Now let $y = 0$. But $f(x) \neq 0$, so there is no x-intercept.

STEP 3 To find where the graph is increasing or decreasing, we find $f'(x)$:
$$f(x) = \frac{8}{x^2 - 16} = 8(x^2 - 16)^{-1}$$
$$f'(x) = -8(x^2 - 16)^{-2} \cdot 2x = \frac{-16x}{(x^2 - 16)^2} \qquad x \neq \pm 4$$

$f'(x)$ is not defined when $x = -4$ or $x = 4$.
The solution to $f'(x) = 0$ is
$$-16x = 0$$
$$x = 0$$

We use the numbers -4, 0, and 4 to separate the number line into four parts:
$$-\infty < x < -4 \qquad -4 < x < 0 \qquad 0 < x < 4 \qquad 4 < x < \infty$$
and choose a test number from each part.

For $x = -5$: $\quad f'(-5) = -\dfrac{16(-5)}{\left[(-5)^2 - 16\right]^2} = 0.988$

For $x = -1$: $\quad f'(-1) = -\dfrac{16(-1)}{\left[(-1)^2 - 16\right]^2} = 0.071$

For $x = 1$: $\quad f'(1) = -\dfrac{16(1)}{\left[(1)^2 - 16\right]^2} = -0.071$

For $x = 5$: $f'(5) = -\dfrac{16(5)}{\left[(5)^2 - 16\right]^2} = -0.988$

We conclude that the graph of f is increasing on the intervals $(-\infty, -4)$ and $(-4, 0)$; f is decreasing on the intervals $(0, 4)$ and $(4, \infty)$.

STEP 4 The graph is increasing to the left of 0 and decreasing to the right of 0, so the point $(0, f(0)) = \left(0, -\dfrac{1}{2}\right)$ is a local maximum.

There is no local minimum.

STEP 5 $f'(x) = 0$ for $x = 0$. The graph of f has a horizontal tangent line at the point $\left(0, \dfrac{1}{2}\right)$. The first derivative is unbounded at $x = -4$ and $x = 4$, but they are not in the domain of the function so there are no vertical tangent lines.

STEP 6 For the end behavior we look at the limits at infinity. Since f is an even function, we need only to consider the limit as $x \to \infty$.

$$\lim_{x \to \infty} f(x) = \lim_{x \to \infty}\left(\frac{8}{x^2 - 16}\right) = \frac{\lim_{x \to \infty} 8}{\lim_{x \to \infty}(x^2 - 16)} = \frac{8}{\lim_{x \to \infty} x^2} = 0$$

The x-axis $(y = 0)$ is a horizontal asymptote of f as x becomes unbounded in the positive and negative directions.

Since f is a rational function and f is unbounded at $x = -4$ and $x = 4$, the graph of f will have vertical asymptotes at $x = -4$ and $x = 4$.

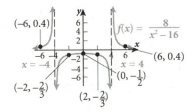

29.

$$f(x) = \frac{x}{x^2 - 9} = \frac{x}{(x-3)(x+3)} \qquad x \neq \pm 3$$

STEP 1 The domain of f is $\{x \mid x \neq -3, x \neq 3\}$.

STEP 2 Let $x = 0$. Then $y = f(0) = \dfrac{0}{0^2 - 9} = 0$. The y-intercept is $(0, 0)$.

Now let $y = 0$. Then $\dfrac{x}{x^2 - 9} = 0$ or $x = 0$. The x-intercept is also $(0, 0)$.

STEP 3 To find where the graph is increasing or decreasing, we find $f'(x)$:

$$f'(x) = \frac{(x^2-9)\cdot 1 - x \cdot 2x}{(x^2-9)^2} = \frac{x^2-9-2x^2}{(x^2-9)^2} = -\frac{x^2+9}{(x^2-9)^2} \qquad x \neq \pm 3$$

$f'(x)$ is not defined when $x = -3$ or $x = 3$.

$f'(x) = 0$ when $-\dfrac{x^2+9}{(x^2-9)^2} = 0$ or when $x^2 + 9 = 0$, which has no solution. So $f'(x) \neq 0$.

We use the numbers -3 and 3 to separate the number line into three parts:
$$-\infty < x < -3 \qquad -3 < x < 3 \qquad 3 < x < \infty$$
and choose a test number from each part.

For $x = -4$: $f'(-4) = -\dfrac{(-4)^2+9}{\left[(-4)^2-9\right]^2} = -0.510$

For $x = 0$: $f'(0) = -\dfrac{0^2+9}{(0^2-9)^2} = -0.111$

For $x = 4$: $f'(4) = -\dfrac{4^2+9}{(4^2-9)^2} = -0.510$

We conclude that the graph of f is decreasing on the intervals $(-\infty, -3)$, $(-3, 3)$, and $(3, \infty)$.

STEP 4 The graph is always decreasing so there is neither a local minimum nor a local maximum.

STEP 5 $f'(x) \neq 0$ so the graph of f has no horizontal tangent line.

The first derivative is unbounded at $x = -3$ and $x = 3$, but these points are not in the domain of the function so there are no vertical tangent lines.

STEP 6 For the end behavior we look at the limits at infinity. Since f is an odd function, we need only to consider the limit as $x \to \infty$.

$$\lim_{x \to \infty} f(x) = \lim_{x \to \infty} \left(\frac{x}{x^2-9}\right) = \lim_{x \to \infty} \left(\frac{x}{x^2}\right) = \lim_{x \to \infty} \left(\frac{1}{x}\right) = 0$$

The x-axis is a horizontal asymptote to the graph as f becomes unbounded.

Since f is a rational function and f is unbounded at $x = -3$ and $x = 3$, the graph of f will have vertical asymptotes at $x = -3$ and $x = 3$.

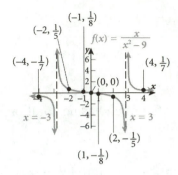

(-1, $\frac{1}{8}$)

(-2, $\frac{1}{5}$)

$f(x) = \dfrac{x}{x^2 - 9}$

(-4, $-\frac{1}{7}$)   (4, $\frac{1}{7}$)

(0, 0)

$x = -3$ $x = 3$

(2, $-\frac{1}{5}$)

(1, $-\frac{1}{8}$)

31.

$$f(x) = \frac{x^2}{x^2 - 4} \qquad x \neq \pm 2$$

STEP 1 The domain of f is $\{x \mid x \neq -2, x \neq 2\}$.

STEP 2 Let $x = 0$. Then $y = f(0) = \dfrac{0^2}{0^2 - 4} = 0$. The y-intercept is $(0, 0)$.

Now let $y = 0$. Then $\dfrac{x^2}{x^2 - 4} = 0$ or $x = 0$. The x-intercept is also $(0, 0)$.

STEP 3 To find where the graph is increasing or decreasing, we find $f'(x)$:

$$f'(x) = \frac{(x^2 - 4) \cdot 2x - x^2 \cdot 2x}{(x^2 - 4)^2} = \frac{2x^3 - 8x - 2x^3}{(x^2 - 4)^2} = -\frac{8x}{(x^2 - 4)^2} \qquad x \neq \pm 2$$

$f'(x)$ is not defined when $x = -2$ or $x = 2$.
The solutions to $f'(x) = 0$ are

$$-\frac{8x}{(x^2 - 4)^2} = 0 \text{ or } x = 0$$

We use the numbers -2, 0, and 2 to separate the number line into four parts:

$$-\infty < x < -2 \qquad -2 < x < 0 \qquad 0 < x < 2 \qquad 2 < x < \infty$$

and choose a test number from each part.

For $x = -3$: $f'(-3) = -\dfrac{8(-3)}{\left[(-3)^2 - 4\right]^2} = 0.96$

For $x = -1$: $f'(-1) = -\dfrac{8(-1)}{\left[(-1)^2 - 4\right]^2} = 0.889$

For $x = 1$: $f'(1) = -\dfrac{8(1)}{\left(1^2 - 4\right)^2} = -0.889$

For $x = 3$: $f'(3) = -\dfrac{8(3)}{\left(3^2 - 4\right)^2} = -0.96$

We conclude that the graph of f is increasing on the intervals $(-\infty, -2)$ and $(-2, 0)$; f is decreasing on the intervals $(0, 2)$ and $(2, \infty)$.

STEP 4 The graph is increasing to the left of 0 and decreasing to the right of 0, so the point $(0, f(0)) = (0, 0)$ is a local maximum. There is no local minimum.

STEP 5 $f'(x) = 0$ for $x = 0$. The graph of f has a horizontal tangent line at the point $(0, f(0)) = (0, 0)$.
    The first derivative is unbounded at $x = -2$ and $x = 2$, but these points are not in the domain of the function so there are no vertical tangent lines.

STEP 6 For the end behavior we look at the limits at infinity. Since f is an even function, we need only to consider the limit as $x \to \infty$.

$$\lim_{x \to \infty} f(x) = \lim_{x \to \infty} \frac{x^2}{x^2 - 4} = \lim_{x \to \infty} \frac{x^2}{x^2} = \lim_{x \to \infty} 1 = 1$$

The line $y = 1$ is a horizontal asymptote to the graph as f becomes unbounded. Since f is a rational function and f is unbounded at $x = -2$ and $x = 2$, the graph of f will have vertical asymptotes at $x = -2$ and $x = 2$.

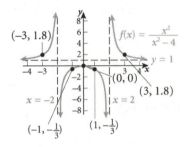

33. $f(x) = x \ln x$

STEP 1 The domain of f is $\{x \mid x > 0\}$.

STEP 2 $x = 0$ is not in the domain of f so there is no y-intercept.
Let $y = 0$. Then $x \ln x = 0$

$$x = 0 \quad \text{or} \quad \ln x = 0$$
$$x = 1$$

We disregard $x = 0$ since it is not in the domain of f. The x-intercept is $(1, 0)$.

STEP 3 To find where the graph is increasing or decreasing, we find $f'(x)$:

$$f'(x) = x \cdot \frac{1}{x} + 1 \cdot \ln x = 1 + \ln x$$

The solution to $f'(x) = 0$ is

$$1 + \ln x = 0$$
$$\ln x = -1$$
$$x = \frac{1}{e} \approx 0.368$$

We use the number 0.368 to separate the positive number line into two parts:
$$0 < x < 0.368 \qquad 0.368 < x < \infty$$
and choose a test number from each part.

For $x = 0.1$: $f'(0.1) = 1 + \ln 0.1 = -1.303$

For $x = 1$: $f'(1) = 1 + \ln 1 = 1$

We conclude that the graph of f is increasing on the interval $\left(\dfrac{1}{e}, \infty\right)$; f is decreasing on

the interval $\left(0, \dfrac{1}{e}\right)$.

STEP 4 The graph is decreasing to the left of $\dfrac{1}{e}$ and increasing to the right of $\dfrac{1}{e}$, so the

point $\left(\dfrac{1}{e}, f\left(\dfrac{1}{e}\right)\right) = \left(\dfrac{1}{e}, -\dfrac{1}{e}\right)$ is a local minimum. There is no local maximum.

STEP 5 $f'(x) = 0$ for $x = \dfrac{1}{e}$. The graph of f has a horizontal tangent line at the point

$\left(\dfrac{1}{e}, -\dfrac{1}{e}\right)$. There are no vertical tangent lines.

STEP 6 For the end behavior we look at the limit at infinity.
$$\lim_{x \to \infty} f(x) = \lim_{x \to \infty} (x \ln x) = \infty$$
The graph of the function becomes unbounded as $x \to \infty$.

Since f becomes unbounded as $x \to 0$, there is a vertical tangent at $x = 0$.

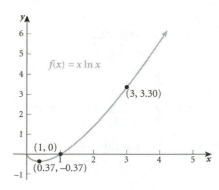

35. S is increasing on the interval where $S'(x) > 0$.
$$S'(x) = 8x + 50$$
$$8x + 50 > 0$$

$$8x > -50$$
$$x > -\frac{50}{8} = -\frac{25}{4} = -6.25$$

Since $S'(x) > 0$ on the domain of S, S is an increasing function.

37. (a) R is increasing on the interval where $R'(x) > 0$ and decreasing where $R'(x) < 0$.
$$R'(x) = -0.010x + 20$$
$$-0.010x + 20 > 0$$

$$-0.010x > -20$$
$$x < 2000$$

The graph of R is increasing on the interval $(0, 2000)$ and it is decreasing on the interval $(2000, \infty)$.

(b) Since the function R is increasing to the left of 2000 and decreasing to the right of 2000, then selling $x = 2000$ trucks will maximize revenue.

(c) The maximum revenue is $R(2000) = -0.005(2000)^2 + 20(2000) = \$20,000$.

(d)

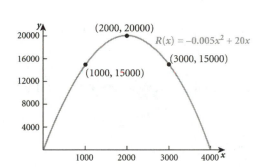

39. (a) A is increasing on the interval where $A'(t) > 0$ and decreasing where $A'(t) < 0$.
$$A'(t) = -238.4t + 113.4$$
$$-238.4t + 113.4 > 0$$
$$-238.4t > -113.4$$
$$t < 0.476$$

The function A is increasing on the interval $(0, 0.476)$.

(b) According to this model the acreage of wheat planted will be decreasing from 2004 to 2008.

41. (a) The yield is increasing when $f'(x) > 0$.
$$f'(x) = -0.417 + \frac{1}{2} \cdot 0.852x^{-1/2} = -0.417 + \frac{0.426}{x^{1/2}} = \frac{-0.417x^{1/2} + 0.426}{x^{1/2}} \qquad x > 0$$

$f'(x) = 0$ when

$$-0.417x^{1/2} + 0.426 = 0$$
$$-0.417x^{1/2} = -0.426$$
$$x^{1/2} = 1.022$$
$$x = 1.044$$

We use 1.044 to separate the positive number line into two parts.

$$0 < x < 1.044 \qquad 1.044 < x < \infty$$

Testing a point in each interval we find

when $x = 1$: $f'(1) = \dfrac{-0.417(1)^{1/2} + 0.426}{(1)^{1/2}} = 0.009$

when $x = 4$: $f'(4) = \dfrac{-0.417(4)^{1/2} + 0.426}{4^{1/2}} = \dfrac{-0.417(2) + 0.426}{2} = -0.204$

The yield will be increasing when the amount of nitrogen is in the interval (0, 1.044).

(b) The yield will be decreasing when the amount of nitrogen is greater than 1.044.

43. $f(x) = 2x^2 - 2x$

1. f is a polynomial function. It is continuous everywhere on its domain. So it is continuous on [0, 1].

2. f is a polynomial function. It is differentiable everywhere on its domain. So f is differentiable on (0, 1).

3. $f(0) = 0,$ $f(1) = 2(1)^2 - 2(1) = 2 - 2 = 0$

$f'(x) = 4x - 2$, $f'(x) = 0$ when
$$4x - 2 = 0$$
$$4x = 2$$
$$x = \frac{1}{2}$$

Since $\dfrac{1}{2}$ is in the interval [0, 1], Rolle's Theorem is verified.

45. $f(x) = x^4 - 1$

1. f is a polynomial function. It is continuous everywhere on its domain. So it is continuous on [– 1, 1].

2. A polynomial function is differentiable everywhere on its domain. So f is differentiable on (– 1, 1).

3. $f(-1) = (-1)^4 - 1 = 1 - 1 = 0$ $\qquad$ $f(1) = 1^4 - 1 = 1 - 1 = 0$

$f'(x) = 4x^3$, $f'(x) = 0$ when $4x^3 = 0$ or when $x = 0$. Since 0 is in the interval $[-1, 1]$, Rolle's Theorem is verified.

47. $f(x) = x^2$

1. f is a polynomial function. It is continuous everywhere on its domain. So it is continuous on $[0, 3]$.

2. A polynomial function is differentiable everywhere on its domain. So f is differentiable on $(0, 3)$.

3. $f(0) = 0^2 = 0$ $\qquad$ $f(3) = 3^2 = 9$

$$\frac{f(b) - f(a)}{b - a} = \frac{f(3) - f(0)}{3 - 0} = \frac{9 - 0}{3 - 0} = \frac{9}{3} = 3$$

$f'(x) = 2x$. $f'(x) = 3$ when $2x = 3$ or when $x = \dfrac{3}{2} = 1.5$. Since $\dfrac{3}{2}$ is in the interval $[0, 3]$,

the Mean Value Theorem is verified.

49. $f(x) = \dfrac{1}{x^2}$ $\quad x \ne 0$

1. f is a rational function. It is continuous everywhere on its domain. So it is continuous on $[1, 2]$.

2. A rational function is differentiable everywhere on its domain. So f is differentiable on $(1, 2)$.

3. $f(1) = \dfrac{1}{1^2} = 1$ $\qquad$ $f(2) = \dfrac{1}{2^2} = \dfrac{1}{4}$

$$\frac{f(b) - f(a)}{b - a} = \frac{f(2) - f(1)}{2 - 1} = \frac{\frac{1}{4} - 1}{1} = -\frac{3}{4}$$

$f'(x) = -2x^{-3} = -\dfrac{2}{x^3}$. $f'(x) = -\dfrac{3}{4}$ when

$$-\frac{2}{x^3} = -\frac{3}{4}$$
$$-3x^3 = -8$$
$$x^3 = \frac{8}{3}$$
$$x = \frac{2}{\sqrt[3]{3}} \approx 1.387$$

Since 1.387 is in the interval $[1, 2]$, the Mean Value Theorem is verified.

14.3 Concavity; the Second Derivative Test 818 | 1-12, 13-27, 29-53(odd)

1. The domain of f is $\{x \mid x_1 \le x < x_4 \text{ or } x_4 < x < x_7\}$ or all the x in the interval $[x_1, x_4)$ or (x_4, x_7).

3. The graph of the function is increasing on the intervals and (x_1, x_3), $(0, x_4)$, and (x_4, x_6).

5. $f'(x) = 0$ when $x = 0$ and when $x = x_6$.

7. f has a local maximum at x_3 and x_6.

9. The graph is concave up on the intervals (x_1, x_3) and (x_3, x_4).

11. There is a vertical asymptote at $x = x_4$.

13. $f(x) = x^3 - 6x^2 + 1$

   $f'(x) = 3x^2 - 12x$

   $f''(x) = 6x - 12$

   $f''(x) > 0$ when $6x - 12 > 0$ or when $6x > 12$ or $x > 2$.
   The graph of f is concave up on the interval $(2, \infty)$. It is concave down on the interval $(-\infty, 2)$. The inflection point is $(2, f(2)) = (2, -15)$.

15. $f(x) = x^4 - 2x^3 + 6x - 1$

   $f'(x) = 4x^3 - 6x^2 + 6$

   $f''(x) = 12x^2 - 12x$

   We solve $f''(x) = 0$ and use test points to determine the intervals for which f is concave up and concave down.

   $$12x^2 - 12x = 0$$
   $$12x(x - 1) = 0$$
   $$12x = 0 \quad \text{or} \quad x - 1 = 0$$
   $$x = 0 \quad \text{or} \quad x = 1$$

   We separate the number line into three parts:
   $$-\infty < x < 0 \quad 0 < x < 1 \quad 1 < x < \infty$$

   For $x = -1$: $\quad f''(-1) = 12(-1)^2 - 12(-1) = 24$

   For $x = 0.5$: $\quad f''(0.5) = 12(0.5)^2 - 12(0.5) = -3$

   For $x = 2$: $\quad f''(2) = 12(2)^2 - 12(2) = 24$

   We conclude that the graph of f is concave up on the intervals $(-\infty, 0)$ and $(1, \infty)$; f is concave down on the interval $(0, 1)$. The inflection points are $(0, f(0)) = (0, -1)$ and $(1, f(1)) = (1, 4)$.

17. $f(x) = 3x^5 - 5x^4 + 60x + 10$

$f'(x) = 15x^4 - 20x^3 + 60$

$f''(x) = 60x^3 - 60x^2$

We solve $f''(x) = 0$ and use test points to determine the intervals for which f is concave up and concave down.

$$60x^3 - 60x^2 = 0$$
$$60x^2(x-1) = 0$$

$$\begin{array}{ccc} 60x^2 = 0 & \text{or} & x - 1 = 0 \\ x = 0 & \text{or} & x = 1 \end{array}$$

We separate the number line into three parts:
$$-\infty < x < 0 \qquad 0 < x < 1 \qquad 1 < x < \infty$$

For $x = -1$: $f''(-1) = 60(-1)^3 - 60(-1)^3 = -120$

For $x = 0.5$: $f''(0.5) = 60(0.5)^3 - 60(0.5)^3 = -7.5$

For $x = 2$: $f''(2) = 60(2)^3 - 60(2)^3 = 240$

We conclude that the graph of f is concave up on the interval $(1, \infty)$; f is concave down on the interval $(-\infty, 1)$. The inflection point is $(1, f(1)) = (1, 68)$.

19. $f(x) = 3x^5 - 10x^3 + 10x + 10$

$f'(x) = 15x^4 - 30x^2 + 10$

$f''(x) = 60x^3 - 60x$

We solve $f''(x) = 0$ and use test points to determine the intervals for which f is concave up and concave down.

$$60x^3 - 60x = 0$$
$$60x(x^2 - 1) = 0$$
$$60x(x-1)(x+1) = 0$$

$$\begin{array}{ccccc} 60x = 0 & \text{or} & x - 1 = 0 & \text{or} & x + 1 = 0 \\ x = 0 & \text{or} & x = 1 & \text{or} & x = -1 \end{array}$$

We separate the number line into four parts:
$$-\infty < x < -1 \qquad -1 < x < 0 \qquad 0 < x < 1 \qquad 1 < x < \infty$$

For $x = -2$: $f''(-2) = 60(-2)^3 - 60(-2) = -360$

For $x = -0.5$: $f''(-0.5) = 60(-0.5)^3 - 60(-0.5) = 22.5$

For $x = 0.5$: $f''(0.5) = 60(0.5)^3 - 60(0.5) = -22.5$

For $x = 2$: $f''(2) = 60(2)^3 - 60(2) = 360$

We conclude that the graph of f is concave up on the intervals $(-1, 0)$ and $(1, \infty)$; f is concave down on the intervals $(-\infty, -1)$ and $(0, 1)$. The inflection points are $(-1, f(-1)) = (-1, 7)$, $(0, f(0)) = (0, 10)$, and $(1, f(1)) = (1, 13)$.

21. $f(x) = x^5 - 10x^2 + 4$

$f'(x) = 5x^4 - 20x$

$f''(x) = 20x^3 - 20 = 20(x^3 - 1)$

We solve $f''(x) = 0$ and use test points to determine the intervals for which f is concave up and concave down.

$$20(x^3 - 1) = 0$$

$$20(x - 1)(x^2 + x + 1) = 0$$

$$x - 1 = 0 \quad \text{or} \quad x^2 + x + 1 = 0$$
$$x = 1$$

The discriminant of $x^2 + x + 1$ is negative, and so $x^2 + x + 1 = 0$ has no real solution. We use 1 to separate the number line into two parts:

$$-\infty < x < 1 \qquad 1 < x < \infty$$

For $x = 0$: $f''(0) = 20(0^3 - 1) = -20$

For $x = 2$: $f''(2) = 20(2^3 - 1) = 140$

We conclude that the graph of f is concave up on the interval $(1, \infty)$; f is concave down on the interval $(-\infty, 1)$. The inflection point is $(1, f(1)) = (1, -5)$.

23. $f(x) = 3x^{1/3} + 9x + 2$

$f'(x) = \dfrac{1}{3} \cdot 3x^{-2/3} + 9 = x^{-2/3} + 9$

$f''(x) = -\dfrac{2}{3}x^{-5/3} = -\dfrac{2}{3x^{5/3}}$

$f''(x)$ is unbounded at $x = 0$. We use 0 to separate the number line into two parts:

$$-\infty < x < 0 \qquad 0 < x < \infty$$

For $x = -1$: $f''(-1) = -\dfrac{2}{3(-1)^{5/3}} = \dfrac{2}{3}$

For $x = 1$: $f''(1) = -\dfrac{2}{3(1)^{5/3}} = -\dfrac{2}{3}$

We conclude that the graph of f is concave up on the interval $(-\infty, 0)$; f is concave down on the interval $(0, \infty)$. The inflection point is $(0, f(0)) = (0, 2)$.

25. $f(x) = x^{2/3}(x-10) = x^{5/3} - 10x^{2/3}$

$$f'(x) = \frac{5}{3}x^{2/3} - \frac{20}{3}x^{-1/3}$$

$$f''(x) = \frac{10}{9}x^{-1/3} + \frac{20}{9}x^{-4/3} = \frac{10}{9x^{1/3}} + \frac{20}{9x^{4/3}} = \frac{10}{9x^{1/3}} \cdot \frac{x}{x^{3/3}} + \frac{20}{9x^{4/3}} = \frac{10x+20}{9x^{4/3}}$$

$f''(x)$ is unbounded at $x = 0$. We solve $f''(x) = 0$ and use test points to determine the intervals for which f is concave up and concave down.

$$\frac{10x+20}{9x^{4/3}} = 0$$
$$10x + 20 = 0$$
$$10x = -20$$
$$x = -2$$

We use 0 and -2 to separate the number line into three parts:
$$-\infty < x < -2 \qquad -2 < x < 0 \qquad 0 < x < \infty$$

For $x = -3$: $f''(-3) = f''(-3) = \dfrac{10(-3)+20}{9(-3)^{4/3}} = -0.257$

For $x = -1$: $f''(-1) = \dfrac{10(-1)+20}{9(-1)^{4/3}} = 1.111$

For $x = 1$: $f''(1) = \dfrac{10(1)+20}{9(1)^{4/3}} = 3.333$

We conclude that the graph of f is concave up on the intervals $(-2, 0)$ and $(0, \infty)$; f is concave down on the interval $(-\infty, -2)$. The inflection point is $(-2, f(-2)) = (-2, -19.049)$.

27. $f(x) = x^{2/3}(x^2 - 16) = x^{8/3} - 16x^{2/3}$

$$f'(x) = \frac{8}{3}x^{5/3} - \frac{32}{3}x^{-1/3}$$

$$f''(x) = \frac{40}{9}x^{2/3} + \frac{32}{9}x^{-4/3} = \frac{40x^{2/3}}{9} + \frac{32}{9x^{4/3}} = \frac{40x^2+32}{9x^{4/3}}$$

$f''(x)$ is unbounded at $x = 0$. We solve $f''(x) = 0$ and use test points to determine the intervals for which f is concave up and concave down.

$$\frac{40x^2+32}{9x^{4/3}} = 0 \quad \text{or} \quad 40x^2 + 32 = 0$$

The discriminant of $40x^2 + 32$ is negative, so $40x^2 + 32 = 0$ has no real solution.

We use 0 to separate the number line into two parts:
$$-\infty < x < 0 \qquad 0 < x < \infty$$

For $x = -1$: $\qquad f''(-1) = \dfrac{40(-1)^2 + 32}{9(-1)^{4/3}} = 8$

For $x = 1$: $\qquad f''(1) = \dfrac{40(1)^2 + 32}{9(1)^{4/3}} = 8$

Since $f''(x)$ is positive for all x, we conclude that the graph of f is concave up on the interval $(-\infty, 0)$ and $(0, \infty)$. There is no inflection point.

29. $f(x) = x^3 - 6x^2 + 1$

STEP 1 Since f is a polynomial, the domain of f is all real numbers.

STEP 2 Let $x = 0$. Then $y = f(0) = 1$. The y-intercept is $(0, 1)$. The x-intercept is hard to find, so we skip it.

STEP 3 To find where the graph is increasing or decreasing, we find $f'(x)$:
$$f'(x) = 3x^2 - 12x$$
The solutions to $f'(x) = 0$ are
$$3x^2 - 12x = 0$$
$$3x(x - 4) = 0$$
$$3x = 0 \quad \text{or} \quad x - 4 = 0$$
$$x = 0 \quad \text{or} \quad x = 4$$
We use the numbers to separate the number line into three parts:
$$-\infty < x < 0 \qquad 0 < x < 4 \qquad 4 < x < \infty$$
and choose a test number from each part.

For $x = -2$: $\qquad f'(-2) = 3(-2)^2 - 12(-2) = 36$

For $x = 2$: $\qquad f'(2) = 3(2)^2 - 12(2) = -12$

For $x = 5$: $\qquad f'(5) = 3(5)^2 - 12(5) = 15$

We conclude that the graph of f is increasing on the intervals $(-\infty, 0)$ and $(4, \infty)$; f is decreasing on the interval $(0, 4)$.

STEP 4 The graph is increasing to the left of 0 and decreasing to the right of 0, so the point $(0, f(0)) = (0, 1)$ is a local maximum.

The graph is decreasing to the left of 4 and increasing to the right of 4, so the point $(4, f(4)) = (4, -31)$ is a local minimum.

STEP 5 $f'(x) = 0$ for $x = 0$ and $x = 4$. The graph of f has a horizontal tangent line at the points $(0, 1)$ and $(4, -31)$.

The first derivative is never unbounded, so there is no vertical tangent.

STEP 6 Since f is a polynomial function, its end behavior is that of $y = x^3$. Polynomial functions have no asymptotes.

STEP 7 To identify the inflection point, if any, we find $f''(x)$

$$f''(x) = 6x - 12$$

$f''(x) = 0$ when $6x = 12$, that is when $x = 2$.

We use the number 2 to separate the number line into two parts:

$$-\infty < x < 2 \qquad\qquad 2 < x < \infty$$

and choose a test number from each part.

For $x = 0$: $\quad f''(0) = 6(0) - 12 = -12$

For $x = 3$: $\quad f''(3) = 6(3) - 12 = 6$

We conclude that the graph of f is concave up on the interval $(2, \infty)$ and is concave down on the interval $(-\infty, 2)$. Since the concavity changes at the point $(2, -15)$, it is an inflection point.

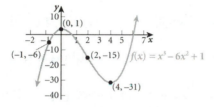

31. $f(x) = x^4 - 2x^2 + 1$

STEP 1 Since f is a polynomial, the domain of f is all real numbers.

STEP 2 Let $x = 0$. Then $y = f(0) = 1$. The y-intercept is $(0, 1)$.
Now let $y = 0$. Then $x^4 - 2x^2 + 1 = 0$

$$\left(x^2 - 1\right)^2 = 0$$

$$x^2 = 1; \; x = \pm 1$$

The x-intercepts are $(1, 0)$ and $(-1, 0)$.

STEP 3 To find where the graph is increasing or decreasing, we find $f'(x)$:

$$f'(x) = 4x^3 - 4x$$

The solutions to $f'(x) = 0$ are

$$4x^3 - 4x = 0$$

$$4x\left(x^2 - 1\right) = 0$$

$$4x(x - 1)(x + 1) = 0$$

$$4x = 0 \quad \text{or} \quad x - 1 = 0 \quad \text{or} \quad x + 1 = 0$$

$$x = 0 \quad \text{or} \quad x = 1 \quad \text{or} \quad x = -1$$

We use the numbers to separate the number line into four parts:
$$-\infty < x < -1 \qquad -1 < x < 0 \qquad 0 < x < 1 \qquad 1 < x < \infty$$
and choose a test number from each part.

For $x = -2$ $\qquad f'(-2) = 4(-2)^3 - 4(-2) = -24$

For $x = -0.5$ $\qquad f'(-0.5) = 4(-0.5)^3 - 4(-0.5) = 1.5$

For $x = 0.5$ $\qquad f'(0.5) = 4(0.5)^3 - 4(0.5) = -1.5$

For $x = 2$ $\qquad f'(2) = 4(2)^3 - 4(2) = 24$

We conclude that the graph of f is increasing on the intervals $(-1, 0)$ and $(1, \infty)$; f is decreasing on the intervals $(-\infty, -1)$ and $(0, 1)$.

STEP 4 The graph is decreasing to the left of -1 and increasing to the right of -1, so the point $(-1, f(-1)) = (-1, 0)$ is a local minimum. The graph is also decreasing to the left of 1 and increasing to the right of 1, so the point $(1, f(1)) = (1, 0)$ is another local minimum.
    The graph is increasing to the left of 0 and decreasing to the right of 0, so the point $(0, f(0)) = (0, 1)$ is a local maximum.

STEP 5 The graph of f has a horizontal tangent line at the points $(-1, f(-1)) = (-1, 0)$, $(1, f(1)) = (1, 0)$, $(0, f(0)) = (0, 1)$.
    The first derivative is never unbounded, so there is no vertical tangent.

STEP 6 Since f is a polynomial function, its end behavior is that of $y = x^4$. Polynomial functions have no asymptotes.

STEP 7 To identify the inflection point, if any, we find $f''(x)$:
$$f''(x) = 12x^2 - 4$$
$f''(x) = 0$ when $12x^2 - 4 = 0$
$$12x^2 = 4$$
$$x^2 = \frac{1}{3} \quad \text{or} \quad x = \pm \frac{1}{\sqrt{3}} \approx \pm 0.577$$

We use the numbers ± 0.577 to separate the number line into three parts
$$-\infty < x < -0.577 \qquad -0.577 < x < 0.577 \qquad 0.577 < x < \infty$$
and choose a test number from each part.

For $x = -1$: $\qquad f''(-1) = 8$

For $x = 0$: $\qquad f''(0) = -4$

For $x = 1$: $\qquad f''(1) = 8$

We conclude that the graph of f is concave up on the intervals $(-\infty, -0.577)$ and $(0.577, \infty)$ and is concave down on the interval $(-0.577, 0.577)$. Since the concavity changes at the points $(-0.577, 0.445)$ and $(0.577, 0.445)$ they are inflection points.

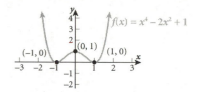

33. $f(x) = x^5 - 10x^4$

STEP 1 Since f is a polynomial, the domain of f is all real numbers.

STEP 2 Let $x = 0$. Then $y = f(0) = 0$. The y-intercept is $(0, 0)$.
Now let $y = 0$. Then $x^5 - 10x^4 = 0$
$$x^4(x - 10) = 0$$
$$x = 0 \qquad \text{or} \qquad x - 10 = 0$$
$$x = 10$$
The x-intercepts are $(0, 0)$ and $(10, 0)$.

STEP 3 To find where the graph is increasing or decreasing, we find $f'(x)$:
$$f'(x) = 5x^4 - 40x^3$$
$f'(x) = 0$ when $5x^4 - 40x^3 = 0$
$$5x^3(x - 8) = 0$$
$$5x^3 = 0 \qquad \text{or} \qquad x - 8 = 0$$
$$x = 0 \qquad \text{or} \qquad x = 8$$
We use the numbers to separate the number line into three parts:
$$-\infty < x < 0 \qquad 0 < x < 8 \qquad 8 < x < \infty$$
and choose a test number from each part.

For $x = -1$: $\qquad f'(-1) = 5(-1)^4 - 40(-1)^3 = 45$

For $x = 1$: $\qquad f'(1) = 5(1)^4 - 40(1)^3 = -35$

For $x = 10$: $\qquad f'(10) = 5(10)^4 - 40(10)^3 = 10,000$

We conclude that the graph of f is increasing on the intervals $(-\infty, 0)$ and $(8, \infty)$; f is decreasing on the interval $(0, 8)$.

STEP 4 The graph is increasing to the left of 0 and decreasing to the right of 0, so the point $(0, f(0)) = (0, 0)$ is a local maximum.
    The graph is decreasing to the left of 8 and increasing to the right of 8, so the point $(8, f(8)) = (8, -8192)$ is a local minimum.

STEP 5 The graph of f has a horizontal tangent line at the points $(0, 0)$ and $(8, -8192)$.
    The first derivative is never unbounded, so there is no vertical tangent.

STEP 6 Since f is a polynomial function, its end behavior is that of $y = x^5$. Polynomial functions have no asymptotes.

STEP 7 To identify the inflection point, if any, we find $f''(x)$:

$$f''(x) = 20x^3 - 120x^2$$
$$f''(x) = 0 \text{ when } 20x^3 - 120x^2 = 0$$
$$20x^2(x-6) = 0$$

$$20x^2 = 0 \quad \text{or} \quad x - 6 = 0$$
$$x = 0 \quad \text{or} \quad x = 6$$

We use the numbers to separate the number line into three parts

$$-\infty < x < 0 \qquad 0 < x < 6 \qquad 6 < x < \infty$$

and choose a test number from each part.

For $x = -1$: $\quad f''(-1) = 20(-1)^3 - 120(-1)^2 = -140$

For $x = 1$: $\quad f''(1) = 20(1)^3 - 120(1)^2 = -100$

For $x = 7$: $\quad f''(7) = 20(7)^3 - 120(7)^2 = 980$

We conclude that the graph of f is concave down on the intervals $(-\infty, 0)$ and $(0, 6)$ and is concave up on the interval $(6, \infty)$. Since the concavity changes at the point $(6, -5184)$, it is an inflection point.

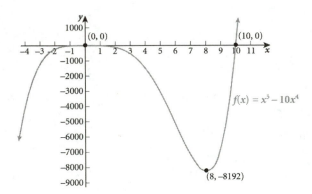

35. $f(x) = x^6 - 3x^5$

STEP 1 Since f is a polynomial, the domain of f is all real numbers.

STEP 2 Let $x = 0$. Then $y = f(0) = 0$. The y-intercept is $(0, 0)$.
Now let $y = 0$. Then $x^6 - 3x^5 = 0$
$$x^5(x-3) = 0$$
$$x^5 = 0 \quad \text{or} \quad x - 3 = 0$$
$$x = 0 \quad \text{or} \quad x = 3$$
The x-intercepts are $(0, 0)$ and $(3, 0)$.

STEP 3 To find where the graph is increasing or decreasing, we find $f'(x)$:
$$f'(x) = 6x^5 - 15x^4$$
$$f'(x) = 0 \text{ when } 6x^5 - 15x^4 = 0$$
$$3x^4(2x-5) = 0$$

$$3x^4 = 0 \quad \text{or} \quad 2x - 5 = 0$$
$$x = 0 \quad \text{or} \quad x = \frac{5}{2}$$

We use the numbers to separate the number line into three parts:

$$-\infty < x < 0 \qquad 0 < x < 2.5 \qquad 2.5 < x < \infty$$

and choose a test number from each part.

For $x = -1$: $\quad f'(-1) = 6(-1)^5 - 15(-1)^4 = -21$

For $x = 1$: $\quad f'(1) = 6(1)^5 - 15(1)^4 = -9$

For $x = 3$: $\quad f'(3) = 6(3)^5 - 15(3)^4 = 243$

We conclude that the graph of f is increasing on the interval $(2.5, \infty)$; f is decreasing on the intervals $(-\infty, 0)$ and $(0, 2.5)$.

STEP 4 The graph is decreasing to the left of 2.5 and increasing to the right of 2.5, so the point $(2.5, f(2.5)) = (2.5, -48.83)$ is a local minimum.
There is no local maximum.

STEP 5 The graph of f has a horizontal tangent line at the point $(2.5, -48.83)$ and $(0, 0)$. The first derivative is never unbounded, so there is no vertical tangent.

STEP 6 Since f is a polynomial function, its end behavior is that of $y = x^6$. Polynomial functions have no asymptotes.

STEP 7 To identify the inflection point, if any, we find $f''(x)$:

$$f''(x) = 30x^4 - 60x^3$$

$f''(x) = 0$ when $30x^4 - 60x^3 = 0$

$$30x^3(x - 2) = 0$$

$$30x^3 = 0 \quad \text{or} \quad x - 2 = 0$$
$$x = 0 \quad \text{or} \quad x = 2$$

We use the numbers to separate the number line into three parts

$$-\infty < x < 0 \qquad 0 < x < 2 \qquad 2 < x < \infty$$

and choose a test number from each part.

For $x = -1$: $\quad f''(-1) = 30(-1)^4 - 60(-1)^3 = 90$

For $x = 1$: $\quad f''(1) = 30(1)^4 - 60(1)^3 = -30$

For $x = 3$: $\quad f''(3) = 30(3)^4 - 60(3)^3 = 810$

We conclude that the graph of f is concave up on the intervals $(-\infty, 0)$ and $(2, \infty)$ and is concave down on the interval $(0, 2)$. Since the concavity changes at the points $(0, 0)$ and $(2, -32)$, they are inflection points.

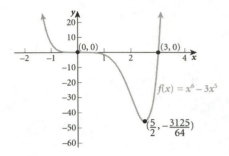

37. $f(x) = 3x^4 - 12x^3$

STEP 1 Since f is a polynomial, the domain of f is all real numbers.

STEP 2 Let $x = 0$. Then $y = f(0) = 0$. The y-intercept is $(0, 0)$.
Now let $y = 0$. Then $3x^4 - 12x^3 = 0$
$$3x^3(x-4) = 0$$
$$3x^3 = 0 \qquad \text{or} \qquad x - 4 = 0$$
$$x = 0 \qquad \text{or} \qquad x = 4$$
The x-intercepts are $(0, 0)$ and $(4, 0)$.

STEP 3 To find where the graph is increasing or decreasing, we find $f'(x)$:
$$f'(x) = 12x^3 - 36x^2$$
$f'(x) = 0$ when $12x^3 - 36x^2 = 0$
$$12x^2(x-3) = 0$$
$$12x^2 = 0 \quad \text{or} \quad x - 3 = 0$$
$$x = 0 \quad \text{or} \quad x = 3$$
We use the numbers to separate the number line into three parts:
$$-\infty < x < 0 \qquad 0 < x < 3 \qquad 3 < x < \infty$$
and choose a test number from each part.

For $x = -1$: $\quad f'(-1) = 12(-1)^3 - 36(-1)^2 = -48$

For $x = 1$: $\quad f'(1) = 12(1)^3 - 36(1)^2 = -24$

For $x = 4$: $\quad f'(4) = 12(4)^3 - 36(4)^2 = 192$

We conclude that the graph of f is increasing on the interval $(3, \infty)$; f is decreasing on the intervals $(-\infty, 0)$ and $(0, 3)$.

STEP 4 The graph is decreasing to the left of 3 and increasing to the right of 3, so the point $(3, f(3)) = (3, -81)$ is a local minimum.
There is no local maximum.

STEP 5 The graph of f has a horizontal tangent line at the points $(0, 0)$ and $(3, -81)$. The first derivative is never unbounded, so there is no vertical tangent.

STEP 6 Since f is a polynomial function, its end behavior is that of $y = 3x^4$. Polynomial functions have no asymptotes.

STEP 7 To identify the inflection point, if any, we find $f''(x)$:
$$f''(x) = 36x^2 - 72x = 36x(x-2)$$

$f''(x) = 0$ when $36x(x-2) = 0$

$$36x = 0 \quad \text{or} \quad x - 2 = 0$$
$$x = 0 \quad \text{or} \quad x = 2$$

We use the numbers 0 and 2 to separate the number line into three parts
$$-\infty < x < 0 \qquad 0 < x < 2 \qquad 2 < x < \infty$$
and choose a test number from each part.

For $x = -1$: $\qquad f''(-1) = 36(-1)(-1-2) = 108$

For $x = 1$: $\qquad f''(1) = 36(1)(1-2) = -36$

For $x = 3$: $\qquad f''(3) = 36(3)(3-2) = 108$

We conclude that the graph of f is concave up on the intervals $(-\infty, 0)$ and $(2, \infty)$ and is concave down on the interval $(0, 2)$. Since the concavity changes at the points $(0, 0)$ and $(2, -48)$, they are inflection points.

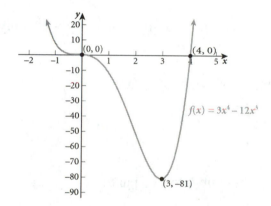

39. $f(x) = x^5 - 10x^2 + 4$

STEP 1 Since f is a polynomial, the domain of f is all real numbers.

STEP 2 Let $x = 0$. Then $y = f(0) = 4$. The y-intercept is $(0, 4)$.
The x-intercept is hard to find, so we skip it.

STEP 3 To find where the graph is increasing or decreasing, we find $f'(x)$:
$$f'(x) = 5x^4 - 20x$$

$f'(x) = 0$ when $5x^4 - 20x = 0$

$$5x(x^3 - 4) = 0$$
$$5x = 0 \quad \text{or} \quad x^3 - 4 = 0$$
$$x = 0 \quad \text{or} \quad x^3 = 4$$
$$x = \sqrt[3]{4} \approx 1.587$$

We use the numbers to separate the number line into three parts:
$$-\infty < x < 0 \qquad 0 < x < 1.587 \qquad 1.587 < x < \infty$$

and choose a test number from each part.

For $x = -1$: $f'(-1) = 5(-1)^4 - 20(-1) = 25$

For $x = 1$: $f'(1) = 5(1)^4 - 20(1) = -15$

For $x = 2$: $f'(2) = 5(2)^4 - 20(2) = 40$

We conclude that the graph of f is increasing on the intervals $(-\infty, 0)$ and $(1.587, \infty)$; f is decreasing on the interval $(0, 1.587)$.

STEP 4 The graph is increasing to the left of 0 and decreasing to the right of 0, so the point $(0, f(0)) = (0, 4)$ is a local maximum.

The graph is decreasing to the left of 1.587 and increasing to the right of 1.587, so the point $(1.587, f(1.587)) = (1.587, -11.119)$ is a local minimum.

STEP 5 The graph of f has a horizontal tangent line at the points $(0, 4)$ and $(1.587, -11.119)$. The first derivative is never unbounded, so there is no vertical tangent.

STEP 6 Since f is a polynomial function, its end behavior is that of $y = x^5$. Polynomial functions have no asymptotes.

STEP 7 To identify the inflection point, if any, we find $f''(x)$:

$$f''(x) = 20x^3 - 20 = 20(x^3 - 1)$$

$$f''(x) = 0 \text{ when } 20(x^3 - 1) = 0$$

$$x^3 - 1 = 0 \quad \text{or} \quad x = 1$$

We use the number 1 to separate the number line into two parts:

$$-\infty < x < 1 \qquad\qquad 1 < x < \infty$$

and choose a test number from each part.

For $x = 0$: $f''(0) = 20[0 - 1] = -20$

For $x = 2$: $f''(2) = 20[2^3 - 1] = 140$

We conclude that the graph of f is concave up on the interval $(1, \infty)$ and is concave down on the interval $(-\infty, 1)$. Since the concavity changes at the point $(1, -5)$, it is an inflection point.

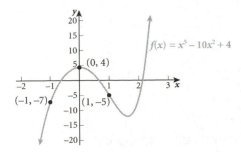

41. $f(x) = x^{2/3}(x-10)$

STEP 1 The domain of f is all real numbers.

STEP 2 Let $x = 0$. Then $y = f(0) = 0$. The y-intercept is $(0, 0)$.
Now let $y = 0$. Then $x^{2/3}(x-10) = 0$

$$x^{2/3} = 0 \quad \text{or} \quad x - 10 = 0$$
$$x = 0 \quad \text{or} \quad x = 10$$

The x-intercepts are $(0, 0)$ and $(10, 0)$.

STEP 3 To find where the graph is increasing or decreasing, we find $f'(x)$:
$$f'(x) = x^{2/3} + \frac{2}{3}x^{-1/3}(x-10) = \frac{3x^{2/3} \cdot x^{1/3}}{3x^{1/3}} + \frac{2x-20}{3x^{1/3}} = \frac{5x-20}{3x^{1/3}}$$

$f'(x)$ is unbounded when $x = 0$. $f'(x) = 0$ when $5x - 20 = 0$ or when $x = 4$.
We use the critical numbers to separate the number line into three parts:
$$-\infty < x < 0 \qquad 0 < x < 4 \qquad 4 < x < \infty$$
and choose a test number from each part.

For $x = -1$: $\qquad f'(-1) = \dfrac{5(-1)-20}{3(-1)^{1/3}} = 8.333$

For $x = 1$: $\qquad f'(1) = \dfrac{5(1)-20}{3(1)^{1/3}} = -5$

For $x = 5$: $\qquad f'(5) = \dfrac{5(5)-20}{3(5)^{1/3}} = 0.975$

We conclude that the graph of f is increasing on the intervals $(-\infty, 0)$ and $(4, \infty)$; f is decreasing on the interval $(0, 4)$.

STEP 4 The graph is increasing to the left of 0 and decreasing to the right of 0, so the point $(0, f(0)) = (0, 0)$ is a local maximum.

The graph is decreasing to the left of 4 and increasing to the right of 4, so the point $(4, f(4)) = (4, -15.12)$ is a local minimum.

STEP 5 The graph of f has a horizontal tangent line at the point $(4, -15.12)$.

The first derivative is unbounded at $x = 0$, so there is a vertical tangent line at the point $(0, f(0)) = (0, 0)$.

STEP 6 For the end behavior of f, we look at the two limits at infinitiy:

As x approaches $-\infty$, f becomes unbounded in the negative direction; as x approaches ∞, f becomes unbounded in the positive direction.

STEP 7 To identify the inflection point, if any, we find $f''(x)$:
$$f''(x) = \frac{10}{9}x^{-1/3} + \frac{20}{9}x^{-4/3} = \frac{10x+20}{9x^{4/3}}$$

$f''(x)$ is unbounded at $x = 0$.

$f''(x) = 0$ when $10x + 20 = 0$, that is when $x = -2$.

We use the numbers 0 and -2 to separate the number line into three parts:

$$-\infty < x < -2 \qquad -2 < x < 0 \qquad 0 < x < \infty$$

and choose a test number from each part.

For $x = -3$: $\quad f''(-3) = \dfrac{10(-3)+20}{9(-3)^{4/3}} = -0.257$

For $x = -1$: $\quad f''(-1) = \dfrac{10(-1)+20}{9(-1)^{4/3}} = 1.111$

For $x = 1$: $\quad f''(1) = \dfrac{10(1)+20}{9(1)^{4/3}} = 3.333$

We conclude that the graph of f is concave up on the intervals $(-2, 0)$ and $(0, \infty)$ and is concave down on the interval $(-\infty, -2)$. Since the concavity changes at the point $(-2, -19.049)$, it is an inflection point.

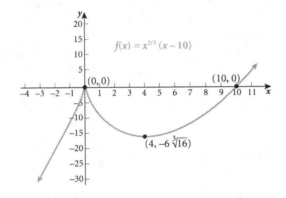

43. $f(x) = x^{2/3}(x^2 - 16)$

STEP 1 The domain of f is all real numbers.

STEP 2 Let $x = 0$. Then $y = f(0) = 0$. The y-intercept is $(0, 0)$.

Now let $y = 0$. Then $x^{2/3}(x^2 - 16) = 0$

$$x^{2/3}(x-4)(x+4) = 0$$

$x^{2/3} = 0 \quad$ or $\quad x - 4 = 0 \quad$ or $\quad x + 4 = 0$

$\qquad x = 0 \quad$ or $\qquad x = 4 \quad$ or $\qquad x = -4$

The x-intercepts $(0, 0)$, $(-4, 0)$, and $(4, 0)$.

STEP 3 To find where the graph is increasing or decreasing, we find $f'(x)$:

$$f'(x) = x^{2/3} \cdot 2x + \frac{2}{3}x^{-1/3}(x^2 - 16) = \frac{2x^{5/2} \cdot 3x^{1/3}}{3x^{1/3}} + \frac{2(x^2 - 16)}{3x^{1/3}}$$

$$= \frac{6x^2 + 2x^2 - 32}{3x^{1/3}} = \frac{8x^2 - 32}{3x^{1/3}}$$

$f'(x)$ is not defined when $x = 0$. $f'(x) = 0$ when $8x^2 - 32 = 0$.

$$8x^2 = 32$$
$$x^2 = 4$$
$$x = \pm 2$$

We use the numbers to separate the number line into four parts:

$$-\infty < x < -2 \qquad -2 < x < 0 \qquad 0 < x < 2 \qquad 2 < x < \infty$$

and choose a test number from each part.

For $x = -3$: $\qquad f'(-3) = \dfrac{8(-3)^2 - 32}{3(-3)^{1/3}} = -9.245$

For $x = -1$: $\qquad f'(-1) = \dfrac{8(-1)^2 - 32}{3(-1)^{1/3}} = 8$

For $x = 1$: $\qquad f'(1) = \dfrac{8(1)^2 - 32}{3(1)^{1/3}} = -8$

For $x = 3$: $\qquad f'(3) = \dfrac{8(3)^2 - 32}{3(3)^{1/3}} = 9.245$

We conclude that the graph of f is increasing on the intervals $(-2, 0)$ and $(2, \infty)$; f is decreasing on the intervals $(-\infty, -2)$ and $(0, 2)$.

STEP 4 The graph is decreasing to the left of -2 and increasing to the right of -2, so the point $(-2, f(-2)) = (-2, -19.05)$ is a local minimum. The graph is also decreasing to the left of 2 and increasing to the right of 2, so the point $(2, f(2)) = (2, -19.05)$ is a local minimum.

   The graph is increasing to the left of 0 and decreasing to the right of 0, so the point $(0, f(0)) = (0, 0)$ is a local maximum.

STEP 5 The graph of f has a horizontal tangent line at the points $(-2, -19.05)$ and $(2, -19.05)$.

   The first derivative is unbounded at $x = 0$, so there is a vertical tangent line at the point $(0, 0)$.

STEP 6 For the end behavior we look at the limits at infinity. Since f is an even function, we need only to consider the limit as $x \to \infty$. As x approaches ∞, f becomes unbounded in the positive direction.

STEP 7 To identify the inflection point, if any, we find $f''(x)$

$$f''(x) = \frac{40}{9}x^{2/3} + \frac{32}{9}x^{-4/3} = \frac{40x^2 + 32}{9x^{4/3}}$$

$f''(x)$ is unbounded at $x = 0$.

$f''(x)$ is never equal to 0.
We use the number 0 to separate the number line into two parts
$$-\infty < x < 0 \qquad 0 < x < \infty$$
and choose a test number from each part.

For $x = -1$: $\qquad f''(-1) = \dfrac{40(-1)^2 + 32}{9(-1)^{4/3}} = 8$

For $x = 1$: $\qquad f''(1) = \dfrac{40(1)^2 + 32}{9(1)^{4/3}} = 8$

We conclude that the graph of f is always concave up, and has no inflection point.

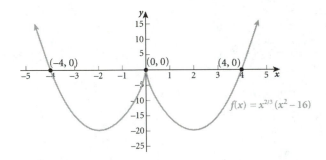

45. $f(x) = xe^x$

STEP 1 The domain of f is all real numbers.

STEP 2 Let $x = 0$. Then $y = f(0) = 0$. The y-intercept is (0, 0).
Now let $y = 0$. Then $xe^x = 0$ or $x = 0$. The x-intercept is (0, 0).

STEP 3 To find where the graph is increasing or decreasing, we find $f'(x)$:
$$f'(x) = xe^x + e^x = e^x(x+1)$$
$f'(x) = 0$ when $e^x(x+1) = 0$
$$x + 1 = 0 \quad \text{or} \quad x = -1$$

We use the numbers to separate the number line into two parts:
$$-\infty < x < -1 \qquad -1 < x < \infty$$
and choose a test number from each part.
For $x = -2$: $\qquad f'(-2) = e^{-2}(-2+1) = -0.135$
For $x = 0$: $\qquad f'(0) = e^0(0+1) = 1$

We conclude that the graph of f is increasing on the interval $(-1, \infty)$; f is decreasing on the interval $(-\infty, -1)$.

STEP 4 The graph is decreasing to the left of -1 and increasing to the right of -1, so the point $(-1, f(-1)) = (-1, -0.368)$ is a local minimum.
There is no local maximum.

STEP 5 The graph of f has a horizontal tangent line at the point $(-1, -0.368)$. There is no vertical tangent.

STEP 6 For the end behavior of f, we look at the two limits at infinitiy:
As x approaches $-\infty$, the graph of f approaches the x-axis. The line $y = 0$ is a horizontal asymptote as x becomes unbounded in the negative direction.
As x approaches ∞, y becomes unbounded in the positive direction. As x becomes unbounded the function f behaves like $y = e^x$.

STEP 7 To identify the inflection point, if any, we find $f''(x)$
$$f''(x) = xe^x + e^x + e^x = e^x(x+2)$$

$f''(x) = 0$ when $e^x(x+2) = 0$, that is when $x = -2$
We use the number -2 to separate the number line into two parts
$$-\infty < x < -2 \qquad\qquad -2 < x < \infty$$
and choose a test number from each part.
For $x = -3$: $\qquad f''(-3) = e^{-3}(-3+2) = -e^{-3} \approx -0.050$

For $x = 0$: $\qquad f''(0) = e^0(2) = 2$

We conclude that the graph of f is concave up on the interval $(-2, \infty)$ and is concave down on the interval $(-\infty, -2)$. Since the concavity changes at the point $(-2, -0.271)$, it is an inflection point.

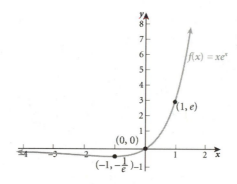

47. First locate the numbers for which $f'(x) = 0$.
$$f'(x) = 3x^2 - 3$$
$$3x^2 - 3 = 0$$
$$3(x^2 - 1) = 0$$
$$3(x-1)(x+1) = 0$$
$$x - 1 = 0 \quad \text{or} \quad x + 1 = 0$$
$$x = 1 \quad \text{or} \quad x = -1$$

Now evaluate $f''(x)$ at these numbers.
$$f''(x) = 6x$$

$f''(-1) = 6(-1) = -6 < 0$. By the Second Derivative Test, f has a local maximum at $(-1, f(-1)) = (-1, 4)$.

$f''(1) = 6(1) = 6 > 0$. By the Second Derivative Test, f has a local minimum at $(1, f(1)) = (1, 0)$.

49. First locate the numbers for which $f'(x) = 0$.
$$f'(x) = 12x^3 + 12x^2$$
$$12x^3 + 12x^2 = 0$$
$$12x^2(x+1) = 0$$
$$x = 0 \quad \text{or} \quad x = -1$$
Now evaluate $f''(x)$ at these numbers.
$$f''(x) = 36x^2 + 24x$$

$f''(0) = 36(0)^2 + 24(0) = 0$. The Second Derivative Test is inconclusive. To determine if a local maximum or minimum exists, we must use the First Derivative Test.
$$f'\left(-\frac{1}{2}\right) = 12\left(-\frac{1}{2}\right)^3 + 12\left(-\frac{1}{2}\right)^2 = -\frac{12}{8} + \frac{12}{4} = \frac{-6+12}{4} = \frac{3}{2} > 0$$
$$f'(1) = 12(1)^3 + 12(1)^2 = 24 > 0$$
Since the first derivative does not change signs, there is no local extreme point at $(0, f(0)) = (0, -3)$.

$f''(-1) = 36(-1)^2 + 24(-1) = 12 > 0$. By the Second Derivative Test, f has a local minimum at $(-1, f(-1)) = (-1, -4)$.

51. First locate the numbers for which $f'(x) = 0$.
$$f'(x) = 5x^4 - 20x^3$$
$$5x^4 - 20x^3 = 0$$
$$5x^3(x-4) = 0$$
$$x = 0 \quad \text{or} \quad x = 4$$

Now evaluate $f''(x)$ at these numbers.
$$f''(x) = 20x^3 - 60x^2$$

$f''(4) = 20(4)^3 - 60(4)^2 = 320$. By the Second Derivative Test, f has a local minimum at $(4, f(4)) = (4, -254)$.

$f''(0) = 20(0)^3 - 60(0)^2 = 0$. The Second Derivative Test is inconclusive. To determine if a local maximum or minimum exists at $x = 0$, we must use the First Derivative Test.
$$f'(-1) = 5(-1)^4 - 20(-1)^3 = 25 > 0$$
$$f'(1) = 5(1)^4 - 20(1)^3 = -15 < 0$$
Since the first derivative is positive to the left of 0 and negative to the right of 0, we conclude that there is a local maximum at $(0, f(0)) = (0, 2)$.

53.
$$f(x) = x + \frac{1}{x} = x + x^{-1} \qquad x \neq 0$$

First locate the numbers for which $f'(x) = 0$.
$$f'(x) = 1 - x^{-2} = 1 - \frac{1}{x^2} = \frac{x^2 - 1}{x^2} \qquad x \neq 0$$
$$\frac{x^2 - 1}{x^2} = 0$$
$$x^2 - 1 = 0$$
$$x = \pm 1$$

Now evaluate $f''(x)$ at these numbers.
$$f''(x) = 2x^{-3} = \frac{2}{x^3} \qquad x \neq 0$$

$f''(-1) = \dfrac{2}{(-1)^3} = -2 < 0$. By the Second Derivative Test, f has a local maximum
at $(-1, -2)$.

$f''(1) = \dfrac{2}{(1)^3} = 2 > 0$. By the Second Derivative Test, f has a local minimum
at $(1, f(1)) = (1, 2)$.

55. Answers will vary.

57. Answers will vary.

59. If the point $(1, 6)$ is an inflection point of the function f, then $f(1) = 6$ and $f''(1) = 0$.
We first find $f''(x)$.
$$f(x) = ax^3 + bx^2 \qquad\qquad f(1) = a + b = 6 \qquad (1)$$
$$f'(x) = 3ax^2 + 2bx$$
$$f''(x) = 6ax + 2b \qquad\qquad f''(1) = 6a + 2b = 0 \qquad (2)$$
We solve the system of equations (1) and (2).
$$a = 6 - b \qquad\qquad (1)$$
$$6(6 - b) + 2b = 0 \qquad (2)$$
$$36 - 6b + 2b = 0$$
$$36 - 4b = 0$$
$$b = 9 \qquad (2)$$
Back-substituting $b = 9$ into equation (1) gives
$$a + 9 = 6 \qquad (1)$$
$$a = -3$$
The point $(1, 6)$ is an inflection point of the function $f(x) = -3x^3 + 9x^2$.

61.

(a) The average cost function is given by $\overline{C}(x) = \dfrac{C(x)}{x}$.

$$\overline{C}(x) = \frac{2x^2 + 50}{x} = 2x + \frac{50}{x}$$

(b) We locate points where $\overline{C}'(x) = 0$.

$$\overline{C}'(x) = 2 - 50x^{-2} = 2 - \frac{50}{x^2} = \frac{2x^2 - 50}{x^2}$$

$$\frac{2x^2 - 50}{x^2} = 0$$

$$2x^2 - 50 = 0$$

$$2(x^2 - 25) = 0$$

$$(x - 5)(x + 5) = 0$$

$$x = 5 \quad \text{or} \quad x = -5$$

We only consider $x = 5$, since $x = -5$ is not part of the domain of the function. We evaluate $\overline{C}''(5)$.

$$\overline{C}''(x) = 100x^{-3} = \frac{100}{x^3}$$

$$\overline{C}''(5) = \frac{100}{5^3} = \frac{100}{125} = \frac{4}{5} > 0$$

By the Second Derivative Test $x = 5$ is a local minimum and the minimum average cost will be $\overline{C}(5) = 2(5) + \dfrac{50}{5} = 10 + 10 = \$20$.

(c) The marginal cost function is given by the derivative $C'(x) = 4x$.

(d)

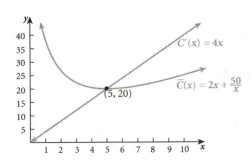

(e) Answers will vary.

63.

(a) The average cost function is given by $\overline{C}(x) = \dfrac{C(x)}{x}$.

$$\overline{C}(x) = \frac{500 + 10x + \frac{x^2}{500}}{x} = \frac{500}{x} + 10 + \frac{x}{500}$$

(b) We locate the points where $\overline{C}'(x) = 0$.

$$\overline{C}'(x) = -500x^{-2} + \frac{1}{500} = -\frac{500}{x^2} + \frac{1}{500} = \frac{x^2 - 500^2}{500x^2}$$

$$\frac{x^2 - 500^2}{500x^2} = 0$$

$$x^2 - 500^2 = 0$$

$$x^2 = 500^2$$

$$x = 500$$

We now evaluate $\overline{C}''(500)$.

$$\overline{C}''(x) = 1000x^{-3} = \frac{1000}{x^3}$$

$$\overline{C}''(500) = \frac{1000}{500^3} = 0.000008 > 0$$

By the Second Derivative Test $x = 500$ is a local minimum and the minimum average cost will be $\overline{C}(500) = 500 + 10(500) + \frac{500^2}{500} = \$6000$.

(c) The marginal cost function is given by the derivative $C'(x) = 10 + \frac{x}{250}$.

(d)

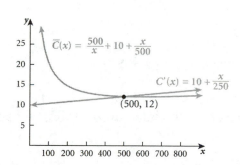

(e) Answers will vary.

65. (a) The domain of the function N is $\{t \mid t \geq 0\}$.

(b) To find the y-intercept we let $t = 0$.

$$N(0) = \frac{50,000}{1 + 49,999e^0} = \frac{50,000}{1 + 49,999} = 1$$

The y-intercept is $(0, 1)$.
There is no t-intercept because $N(t)$ never equals zero.

(c) $N(t) = \dfrac{50,000}{1+49,999e^{-t}} = 50,000\left(1+49,999e^{-t}\right)^{-1}$

$N'(t) = (-1) \cdot 50,000\left(1+49,999e^{-t}\right)^{-2} \cdot (-1) \cdot 49,999e^{-t}$

$N'(t) = \dfrac{50,000 \cdot 49,999e^{-t}}{\left(1+49,999e^{-t}\right)^2} > 0$

since $e^{-t} > 0$. The function is always increasing.

(d)

$N''(t) = \dfrac{-1 \cdot 50,000 \cdot 49,999e^{-t}\left(1+49,999e^{-t}\right)^2 - \left(50,000 \cdot 49,999e^{-t}\right) \cdot 2\left(1+49,999e^{-t}\right)(-1)\,49,999e^{-t}}{\left(1+49,999e^{-t}\right)^4}$

$= \dfrac{\left[-50,000 \cdot 49,999e^{-t}\left(1+49,999e^{-t}\right)\right]\left[\left(1+49,999e^{-t}\right)-2 \cdot 49,999e^{-t}\right]}{\left(1+49,999e^{-t}\right)^4}$

$= \dfrac{-50,000 \cdot 49,999e^{-t}\left(1+49,999e^{-t}\right)\left(1-49,999e^{-t}\right)}{\left(1+49,999e^{-t}\right)^4}$

$= \dfrac{-50,000 \cdot 49,999e^{-t}\left(1-49,999e^{-t}\right)}{\left(1+49,999e^{-t}\right)^3} = \dfrac{50,000 \cdot 49,999e^{-t}\left(49,999e^{-t}-1\right)}{\left(1+49,999e^{-t}\right)^3}$

The sign of N'' is controlled by $49,999e^{-t}-1$ since the rest of the expression is always positive.

$49,999e^{-t}-1 = 0$

$49,999e^{-t} = 1$

$e^{-t} = \dfrac{1}{49,999}$ or when $e^{t} = 49,999$

This occurs when $t = \ln 49,999 = 10.820$. If $t < 10.82$, $49,999e^{-t}-1 > 0$ and $N''(t) > 0$.
We conclude that the function N is concave up on the interval $(0, 10.82)$ and concave down on the interval $(10.82, \infty)$.

(e) The inflection point is $(\ln 49,999, N(\ln 49,999)) \approx (10.82, 25,000)$.

(f)

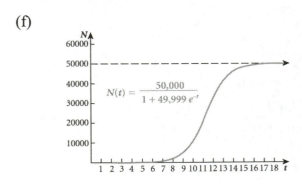

(g) Since $N''(t)$ is positive to the left of 10.82 and negative to the right of 10.82, by the First Derivative Test we conclude that the rumor is spreading at the greatest rate when $t = 10.82$ days.

67. (a) The growth rate of the population is given by $P'(t)$.

$$P(t) = \frac{800e^t}{1+0.1(e^t-1)} = \frac{800e^t}{0.9+0.1e^t} = \frac{8000e^t}{9+e^t}$$

$$P'(t) = \frac{8000e^t(9+e^t) - 8000e^t(e^t)}{(9+e^t)^2}$$

$$= \frac{72,000e^t + 8000e^{2t} - 8000e^{2t}}{(9+e^t)^2} = \frac{72,000e^t}{(9+e^t)^2}$$

(b) To determine where the growth rate is maximum we look at $P''(t)$ and evaluate where it equal zero.

$$P''(t) = \frac{72,000e^t(9+e^t)^2 - 72,000e^t\left[2(9+e^t)(e^t)\right]}{(9+e^t)^4}$$

$$= \frac{72,000e^t(9+e^t)\left[(9+e^t)-2e^t\right]}{(9+e^t)^4}$$

$$= \frac{72,000e^t(9-e^t)}{(9+e^t)^3}$$

$P''(t) = 0$ when $9 - e^t = 0$ since the denominator and $72,000e^t$ are always positive.

$9 - e^t = 0$ when $e^t = 9$ or when $t = \ln 9 \approx 2.197$. Using the First Derivative Test we find that $(\ln 9, P(\ln 9)) \approx (2.197, 4000)$ is a local maximum. So the population is growing the fastest after about 2.197 days.

(c) $\displaystyle\lim_{x \to \infty}\left(\frac{800e^t}{1+0.10(e^t-1)}\right) = \lim_{x \to \infty}\left(\frac{8000e^t}{9+e^t}\right) = 8000 \cdot \lim_{x \to \infty}\left(\frac{e^t}{9+e^t}\right) = 8000 \cdot \lim_{x \to \infty}\left(\frac{e^t}{e^t}\right)$

$$= 8000 \cdot \lim_{x \to \infty}(1) = 8000$$

(d)

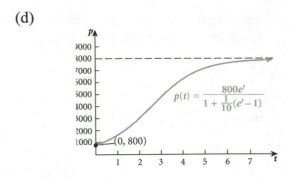

$$p(t) = \frac{800e^t}{1 + \frac{1}{10}(e^t - 1)}$$

(0, 800)

69. (a) The sales rate of the car model is given by $f'(x)$.

$$f(x) = \frac{20,000}{1+50e^{-x}} = 20,000(1+50\,e^{-x})^{-1}$$

$$f'(x) = -20,000(1+50\,e^{-x})^{-2} \cdot (-50\,e^{-x}) = \frac{100,000\,e^{-x}}{(1+50\,e^{-x})^{2}}$$

The sales rate is maximum where $f''(x) = 0$.

$$f''(x) = \frac{-100,000\,e^{-x}(1+50\,e^{-x})^{2} - (100,000\,e^{-x})\left[2(1+50\,e^{-x})(-50\,e^{-x})\right]}{(1+50\,e^{-x})^{4}}$$

$$= \frac{-100,000\,e^{-x}(1+50\,e^{-x})^{2} + 100 \cdot 100,000\,e^{-2x}(1+50\,e^{-x})}{(1+50\,e^{-x})^{4}}$$

$$= \frac{-100,000\,e^{-x}(1+50\,e^{-x}) + 100 \cdot 100,000\,e^{-x}}{(1+50\,e^{-x})^{3}}$$

$$= \frac{-100,000\,e^{-x} - 50 \cdot 100,000\,e^{-2x} + 100 \cdot 100,000\,e^{-2x}}{(1+50\,e^{-x})^{3}}$$

$$= \frac{-100,000\,e^{-x} + 50 \cdot 100,000\,e^{-2x}}{(1+50\,e^{-x})^{3}} = \frac{100,000\,e^{-x}(50\,e^{-x}-1)}{(1+50\,e^{-x})^{3}}$$

$f''(x) = 0$ when $50\,e^{-x} - 1 = 0$ since the denominator and $100,000\,e^{-x}$ are always positive. $50\,e^{-x} - 1 = 0$ when $50\,e^{-x} = 1$ or $e^{x} = 50$ or when $\ln 50 = x$. From the First Derivative Test we see that $(\ln 50, f(\ln 50)) \approx (3.91, 10,000)$ is a relative maximum. The sales rate is a maximum after about 3.91 months of sales.

(b)

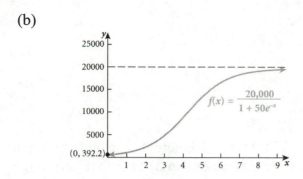

14.4 Optimization $831|1-41$ (ODD)

1. $f(x)=x^2+2x$ $f'(x)=2x+2$

$f'(x)=0$ when $x=-1$

The critical number -1 is in the interval $[-3, 3]$, so we evaluate f at each of the three points.

$$f(-3)=(-3)^2+2(-3)=3$$
$$f(-1)=(-1)^2+2(-1)=-1$$
$$f(3)=(3)^2+2(3)=15$$

The absolute maximum of f on $[-3, 3]$ is 15 and the absolute minimum is -1.

3. $f(x)=1-6x-x^2$ $f'(x)=-6-2x$

$f'(x)=0$ when $x=-3$.

The critical number -3 is not in the interval $[0, 4]$, so we evaluate f only at the endpoints.

$$f(0)=1-6(0)-0^2=1$$
$$f(4)=1-6(4)-4^2=-39$$

The absolute maximum of f on $[0, 4]$ is 1 and the absolute minimum is -39.

5. $f(x)=x^3-3x^2$ $f'(x)=3x^2-6x$

$f'(x)=0$ when $3x^2-6x=0$
$$3x(x-2)=0$$
$$x=0 \quad \text{or} \quad x=2$$

The critical number 0 is not in the interval $[1, 4]$, so we ignore it and evaluate f at each of the other three numbers.

$$f(1)=1^3-3\cdot1^2=-2$$
$$f(2)=2^3-3\cdot2^2=-4$$
$$f(4)=4^3-3\cdot4^2=16$$

The absolute maximum of f on $[1, 4]$ is 16 and the absolute minimum is -4.

7. $f(x) = x^4 - 2x^2 + 1$ $f'(x) = 4x^3 - 4x$

$f'(x) = 0$ when $4x^3 - 4x = 0$

$$4x(x^2 - 1) = 0$$

$$4x(x - 1)(x + 1) = 0$$

$$x = 0 \quad \text{or} \quad x = 1 \quad \text{or} \quad x = -1$$

The critical number -1 is not in the interval [0, 1], so we ignore it. The other two critical numbers are the endpoints of the closed interval, so we evaluate f at each of those two numbers.

$$f(0) = 0^4 - 2(0)^2 + 1 = 1$$

$$f(1) = 1^4 - 2(1)^2 + 1 = 0$$

The absolute maximum of f on [0, 1] is 1 and the absolute minimum is 0.

9. $f(x) = x^{2/3}$ $f'(x) = \dfrac{2}{3}x^{-1/3} = \dfrac{2}{3x^{1/3}}$

$f'(x)$ is not defined at $x = 0$. It is never equal to zero. So the only critical number is 0.

The critical number 0 is in the interval [-1, 1], so we evaluate f at each of the three numbers.

$$f(-1) = (-1)^{2/3} = 1$$

$$f(0) = 0^{2/3} = 0$$

$$f(1) = 1^{2/3} = 1$$

The absolute maximum of f on [-1, 1] is 1 and the absolute minimum is 0.

11. $f(x) = 2\sqrt{x} = 2x^{1/2}$ $f'(x) = \dfrac{1}{2} \cdot 2x^{-1/2} = x^{-1/2} = \dfrac{1}{\sqrt{x}}$

$f'(x)$ is not defined at $x = 0$, and it never equals zero. The critical number 0 is not in the interval [1, 4], so we evaluate f only at the endpoints.

$$f(1) = 2\sqrt{1} = 2$$

$$f(4) = 2\sqrt{4} = 4$$

The absolute maximum of f on [1, 4] is 4 and the absolute minimum is 2.

13. $f(x) = x\sqrt{1 - x^2} = x(1 - x^2)^{1/2}$

$$f'(x) = x \cdot \dfrac{1}{2}(1 - x^2)^{-1/2} \cdot (-2x) + 1 \cdot (1 - x^2)^{1/2}$$

$$= \dfrac{-x^2}{(1 - x^2)^{1/2}} + (1 - x^2)^{1/2} \cdot \dfrac{(1 - x^2)^{1/2}}{(1 - x^2)^{1/2}}$$

$$= \frac{-x^2 + 1 - x^2}{\left(1 - x^2\right)^{1/2}} = \frac{1 - 2x^2}{\left(1 - x^2\right)^{1/2}}$$

$f'(x)$ is not defined at $x = -1$ and $x = 1$; $f'(x) = 0$ when

$$1 - 2x^2 = 0$$
$$x^2 = \frac{1}{2}$$
$$x = \pm \sqrt{\frac{1}{2}}$$

The critical numbers $\sqrt{\frac{1}{2}}$ and $-\sqrt{\frac{1}{2}}$ are in the interval $[-1, 1]$ and the critical numbers -1 and 1 are the endpoints of the interval $[-1, 1]$, so we evaluate f at each of these numbers.

$$f(-1) = -1\sqrt{1 - (-1)^2} = 0$$

$$f\left(-\sqrt{\frac{1}{2}}\right) = -\sqrt{\frac{1}{2}} \cdot \sqrt{1 - \left(-\sqrt{\frac{1}{2}}\right)^2} = -\sqrt{\frac{1}{2}} \cdot \sqrt{1 - \frac{1}{2}} = -\sqrt{\frac{1}{2}} \cdot \sqrt{\frac{1}{2}} = -\frac{1}{2}$$

$$f\left(\sqrt{\frac{1}{2}}\right) = \sqrt{\frac{1}{2}} \cdot \sqrt{1 - \left(\sqrt{\frac{1}{2}}\right)^2} = \sqrt{\frac{1}{2}} \cdot \sqrt{1 - \frac{1}{2}} = \sqrt{\frac{1}{2}} \cdot \sqrt{\frac{1}{2}} = \frac{1}{2}$$

$$f(1) = 1\sqrt{1 - 1^2} = 0$$

The absolute maximum of f on $[-1, 1]$ is $\frac{1}{2}$ and the absolute minimum is $-\frac{1}{2}$.

15.
$$f(x) = \frac{x^2}{x - 1}$$

$$f'(x) = \frac{2x(x-1) - x^2(1)}{(x-1)^2} = \frac{2x^2 - 2x - x^2}{(x-1)^2} = \frac{x^2 - 2x}{(x-1)^2}$$

$f'(x)$ is not defined at $x = 1$; $f'(x) = 0$ when $x^2 - 2x = 0$.

$$x^2 - 2x = 0$$
$$x(x - 2) = 0$$
$$x = 0 \quad \text{or} \quad x = 2$$

The critical number 0 is in the interval $\left[-1, \frac{1}{2}\right]$, the critical numbers 1 and 2 are not in the domain of f. So we evaluate f at 0 and the endpoints.

when $x = -1$:　　$f(-1) = \dfrac{(-1)^2}{(-1)-1} = -\dfrac{1}{2}$

when $x = 0$:　　$f(0) = \dfrac{0^2}{0-1} = 0$

when $x = \dfrac{1}{2}$:　　$f\left(\dfrac{1}{2}\right) = \dfrac{\left(\dfrac{1}{2}\right)^2}{\dfrac{1}{2}-1} = -\dfrac{1}{2}$

The absolute maximum of f on $\left[-1, \dfrac{1}{2}\right]$ is 0, and the absolute minimum is $-\dfrac{1}{2}$.

17.　$f(x) = (x+2)^2 (x-1)^{2/3}$

$f'(x) = (x+2)^2 \cdot \dfrac{2}{3}(x-1)^{-1/3} + 2(x+2)(x-1)^{2/3}$

$= \dfrac{2(x+2)^2}{3(x-1)^{1/3}} + (2x+4)(x-1)^{2/3}$

$= \dfrac{2(x+2)^2}{3(x-1)^{1/3}} + (2x+4)(x-1)^{2/3} \cdot \dfrac{3(x-1)^{1/3}}{3(x-1)^{1/3}}$

$= \dfrac{2(x^2+4x+4)}{3(x-1)^{1/3}} + \dfrac{3(2x+4)(x-1)}{3(x-1)^{1/3}}$

$= \dfrac{2x^2+8x+8+3(2x^2+2x-4)}{3(x-1)^{1/3}} = \dfrac{2x^2+8x+8+6x^2+6x-12}{3(x-1)^{1/3}}$

$= \dfrac{8x^2+14x-4}{3(x-1)^{1/3}}$

$f'(x)$ is not defined at $x = 1$; $f'(x) = 0$ when $8x^2 + 14x - 4 = 0$.

$$8x^2 + 14x - 4 = 2(4x^2 + 7x - 2) = 0$$
$$4x^2 + 7x - 2 = 0$$
$$(4x-1)(x+2) = 0$$
$$x = \dfrac{1}{4} \quad \text{or} \quad x = -2$$

The critical numbers -2, $\dfrac{1}{4}$, and 1 are all in the interval $[-4, 5]$, so we evaluate f at each of the three numbers and at the endpoints.

when $x = -4$: $f(-4) = (-4+2)^2(-4-1)^{2/3} \approx 11.696$

when $x = -2$: $f(-2) = (-2+2)^2(-2-1)^{2/3} = 0$

when $x = \dfrac{1}{4}$: $f\left(\dfrac{1}{4}\right) = \left(\dfrac{1}{4}+2\right)^2\left(\dfrac{1}{4}-1\right)^{2/3} \approx 4.179$

when $x = 1$: $f(1) = (1+2)^2(1-1)^{2/3} = 0$

when $x = 5$: $f(5) = (5+2)^2(5-1)^{2/3} \approx 123.472$

The absolute maximum of f on $[-4, 5]$ is 123.472 and the absolute minimum is 0.

19.

$$f(x) = \frac{(x-4)^{1/3}}{x-1}$$

$$f'(x) = \frac{\dfrac{1}{3}(x-4)^{-2/3}(x-1) - (x-4)^{1/3}(1)}{(x-1)^2}$$

$$= \frac{\dfrac{1}{3(x-4)^{2/3}} \cdot (x-1) - (x-4)^{1/3}}{(x-1)^2} = \frac{\dfrac{1}{3(x-4)^{2/3}} \cdot (x-1) - (x-4)^{1/3} \cdot \dfrac{3(x-4)^{2/3}}{3(x-4)^{2/3}}}{(x-1)^2}$$

$$= \frac{(x-1) - 3(x-4)}{3(x-4)^{2/3}(x-1)^2} = \frac{x-1-3x+12}{3(x-4)^{2/3}(x-1)^2}$$

$$= \frac{-2x+11}{3(x-4)^{2/3}(x-1)^2}$$

$f'(x)$ is not defined at $x = 1$ and $x = 4$; $f'(x) = 0$ when $-2x + 11 = 0$, or $x = \dfrac{11}{2} = 5.5$.

The critical numbers 4 and 5.5 are in the interval $[2, 12]$, so we evaluate f at each of these numbers and at the endpoints.

when $x = 2$: $f(2) = \dfrac{(2-4)^{1/3}}{2-1} \approx -1.260$

when $x = 4$: $f(4) = \dfrac{(4-4)^{1/3}}{4-1} = 0$

when $x = 5.5$: $f(5.5) = \dfrac{(5.5-4)^{1/3}}{5.5-1} \approx 0.254$

when $x = 12$: $f(12) = \dfrac{(12-4)^{1/3}}{12-1} \approx 0.182$

The absolute maximum of f on $[2, 12]$ is 0.254 and the absolute minimum is -1.260.

21. $f(x) = xe^x$

$f'(x) = xe^x + e^x = e^x(x+1)$

$f'(x) = 0$ when $e^x(x+1) = 0$ or when $x = -1$.

The critical number -1 is in the interval $[-10, 10]$, so we evaluate f at it and at the endpoints.

when $x = -10$: $\quad f(-10) = -10e^{-10} \approx -0.0005$

when $x = -1$: $\quad f(-1) = -e^{-1} \approx -0.368$

when $x = 10$: $\quad f(10) = 10e^{10} \approx 220,265$

The absolute maximum of f on $[-10, 10]$ is 220,265 and the absolute minimum is -0.368.

23. $f(x) = \dfrac{\ln x}{x}$

$f'(x) = \dfrac{\dfrac{1}{x} \cdot x - (\ln x) \cdot (1)}{x^2} = \dfrac{1 - \ln x}{x^2}$

$f'(x) = 0$ when $1 - \ln x = 0$ or when $x = e$.

The critical number $e \approx 2.718$ is in the interval $[1, 3]$, so we evaluate f at e and the two endpoints.

when $x = 1$: $\quad f(1) = \dfrac{\ln 1}{1} = 0$

when $x = e$: $\quad f(e) = \dfrac{\ln e}{e} = \dfrac{1}{e} \approx 0.368$

when $x = 3$: $\quad f(3) = \dfrac{\ln 3}{3} \approx 0.366$

The absolute maximum of f on $[1, 3]$ is 0.368 and the absolute minimum is 0.

25. **STEP 1** The quantity to be maximized is volume. Denote it by V.

STEP 2 Denote the dimensions of the side of the small square to be cut out by x. Let the y be the dimension one side of the bottom of box after removing square x.

STEP 3 Then $y = 12 - 2x$

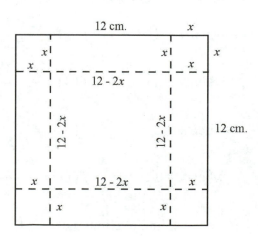

STEP 4 The height of the box is x, while the area of the base of the box is y^2. So,
$$V = xy^2$$
$$V(x) = x(12 - 2x)^2$$
Although the domain of V is the set of real numbers, only values of x between 0 and 6 make sense physically. We need the absolute maximum of V on $[0, 6]$.

STEP 5 Differentiate V and identify the critical numbers.
$$V'(x) = x \cdot 2(12 - 2x) \cdot (-2) + 1 \cdot (12 - 2x)^2$$
$$= -4x(12 - 2x) + (12 - 2x)^2$$
$$= (12 - 2x)[(12 - 2x) - 4x]$$
$$= (12 - 2x)(12 - 6x)$$
If $V'(x) = 0$, then
$$(12 - 2x)(12 - 6x) = 0$$
$$12 - 2x = 0 \quad \text{or} \quad 12 - 6x = 0$$
$$x = 6 \quad \text{or} \quad x = 2$$
We calculate the values of V at the critical number $x = 2$ and at the endpoints 0 and 6.
$$V(0) = 0 \qquad V(2) = 2(12 - 2(2))^2 = 2(8)^2 = 128 \qquad V(6) = 0$$

The maximum volume is 128 cubic centimeters and the dimensions of the box are $x = 2$ centimeters deep by $y = 12 - 2(2) = 8$ centimeters on each side.

27. **STEP 1** We want to minimize the amount of material A which is used to make the box.

    STEP 2 Let x denote dimension of a side of the square bottom of the box, and let y denote the height of the box.

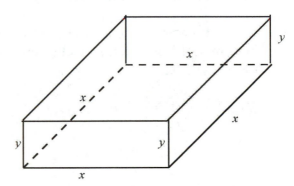

   STEP 3 $\quad V = x^2 y$
   $$8000 = x^2 y$$
   $$y = \frac{8000}{x^2}$$

STEP 4 The amount of material used is the area of the base of the box plus the area of the four sides.
$$A = x^2 + 4xy$$
$$A(x) = x^2 + 4x\left(\frac{8000}{x^2}\right)$$
$$= x^2 + \frac{32,000}{x} = x^2 + 32,000x^{-1}$$
The domain of A is all real numbers other than $x = 0$, but only positive values of x make physical sense.

STEP 5 Differentiate A and find the critical numbers.

$$A'(x) = 2x - 32,000x^{-2} = 2x - \frac{32,000}{x^2} = \frac{2x^3 - 32,000}{x^2} \qquad x \neq 0$$

$x = 0$ is a critical number. We solve $A'(x) = 0$ to find other critical numbers, if they exist.

$$\frac{2x^3 - 32,000}{x^2} = 0$$

$$2x^3 - 32,000 = 0$$

$$x^3 = 16,000$$

$$x = 20\sqrt[3]{2}$$

We use the second derivative test to see if $x = 20\sqrt[3]{2}$ locates a maximum or minimum.

$$A''(x) = 2 + 64,000x^{-3}$$

$$A''\left(20\sqrt[3]{2}\right) = 2 + \frac{64,000}{\left(20\sqrt[3]{2}\right)^3} = 2 + \frac{64,000}{16,000} = 2 + 4 = 6 > 0$$

So we conclude that the least amount of material is used when the base measures $20\sqrt[3]{2} \approx 25.20$ centimeters and the height measures $\dfrac{8000}{\left(20\sqrt[3]{2}\right)^2} \approx 12.60$ centimeters.

29. **STEP 1** We want to minimize the cost of the material A which is used to make the cylinder.

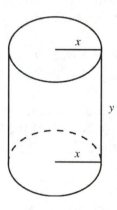

STEP 2 Let x denote the radius of the top and bottom of the cylinder, and let y denote the height of the cylinder.

STEP 3
$$V = \pi x^2 y$$
$$4000 = \pi x^2 y$$
$$y = \frac{4000}{\pi x^2}$$

STEP 4 The cost C of producing the can is \$0.50 times the area of the top and bottom of the cylinder plus \$0.40 times the area of the side of the cylinder.

$$C = 2(0.50)\left(\pi x^2\right) + (0.40)(2\pi xy)$$

$$C(x) = 1.00\left(\pi x^2\right) + 0.80\pi x\left(\frac{4000}{\pi x^2}\right)$$

$$= \pi x^2 + \frac{3200}{x}$$

The domain of C is $\{x \mid x > 0\}$.

STEP 5 Differentiate C and find the critical numbers.

$$C'(x) = 2\pi x - \frac{3200}{x^2} = \frac{2\pi x^3 - 3200}{x^2}$$

We solve $C'(x) = 0$ to find the critical numbers.

$$2\pi x^3 - 3200 = 0$$

$$x^3 = \frac{3200}{2\pi}$$

$$x = \sqrt[3]{\frac{1600}{\pi}} \approx 7.986$$

We use the second derivative test to see if $x = 7.986$ locates a maximum or minimum.

$$C''(x) = 2\pi - (-2)3200x^{-3} = 2\pi + \frac{6400}{x^3}$$

$$C''(7.986) = 2\pi + \frac{6400}{\left(\dfrac{1600}{\pi}\right)} = 2\pi + 4\pi = 6\pi > 0$$

We conclude that the cost is minimized when the can has a radius of 7.986 cm and a height of $\dfrac{4000}{\pi(7.986)^2} \approx 19.965$ centimeters.

31. **STEP 1** Minimize the cost of installing the telephone line.

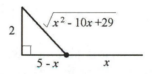

STEP 2 Using the hint, we let x denote the distance between the box and the connection.

STEP 3 Using the Pythagorean theorem, we calculate the distance the line runs off the road.

$$c^2 = a^2 + b^2$$
$$c^2 = (5-x)^2 + 2^2$$
$$= 25 - 10x + x^2 + 4 = x^2 - 10x + 29$$
$$c = \sqrt{x^2 - 10x + 29}$$

STEP 4 The cost C of laying the line is expressed as

$$C(x) = 50x + 60(x^2 - 10x + 29)^{1/2}$$

The domain of C is $\{x \mid 0 \le x \le 5\}$.

STEP 5 The first derivative gives the critical numbers, if they exist.

$$C'(x) = 50 + 60\left[\frac{1}{2}(x^2 - 10x + 29)^{-1/2}(2x-10)\right]$$

$$= 50 + 60\left[\frac{x-5}{(x^2 - 10x + 29)^{1/2}}\right] = \frac{50(x^2 - 10x + 29)^{1/2} + 60x - 300}{(x^2 - 10x + 29)^{1/2}}$$

We solve $C'(x) = 0$.

$$50\left(x^2 - 10x + 29\right)^{1/2} + 60x - 300 = 0$$

$$\left(x^2 - 10x + 29\right)^{1/2} = \frac{300 - 60x}{50} = 6 - 1.2x$$

$$x^2 - 10x + 29 = (6 - 1.2x)^2 = 36 - 14.4x + 1.44x^2$$

$$0.44x^2 - 4.4x + 7 = 0$$

$$x = \frac{4.4 \pm \sqrt{4.4^2 - 4(0.44)(7)}}{2(0.44)} = \frac{4.4 \pm \sqrt{7.04}}{0.88}$$

$$x = \frac{4.4 + \sqrt{7.04}}{0.88} \approx 8.015 \qquad \text{or} \qquad x = \frac{4.4 - \sqrt{7.04}}{0.88} \approx 1..985$$

$x = 8.015$ is not in the domain of the function, so we test only the endpoints and the critical number 1.985.

when $x = 0$: $C(0) = 50(0) + 60\left(0^2 - 10(0) + 29\right)^{1/2} = 60\sqrt{29} \approx 323.11$

when $x = 1.985$: $C(1.985) = 50(1.985) + 60\left(1.985^2 - 10(1.985) + 29\right)^{1/2} \approx 316.33$

when $x = 5$: $C(5) = 50(5) + 60\left(5^2 - 10(5) + 29\right)^{1/2} = 370$

The minimum cost is obtained when the telephone line leaves the road approximately 1.985 kilometers from the box.

33. We want to minimize the cost of operating the truck over the interval [10, 75]. We find the derivative of C and locate any critical numbers in the open interval (10, 75).

$$C'(x) = 1.60\left(-\frac{1600}{x^2} + 1\right) = 1.60\left(\frac{x^2 - 1600}{x^2}\right)$$

The critical numbers are those for which $C'(x) = 0$.

$$1.60\left(\frac{x^2 - 1600}{x^2}\right) = 0$$

$$x^2 - 1600 = 0$$

$$x^2 = 1600$$

$$x = \pm 40$$

Only the critical number $x = 40$ is in the open interval, so we evaluate C at 40 and at the endpoints.

when $x = 10$: $C(10) = 1.60\left(\frac{1600}{10} + 10\right) = 272$

when $x = 40$: $C(40) = 1.60\left(\frac{1600}{40} + 40\right) = 128$

when $x = 75$: $C(75) = 1.60\left(\frac{1600}{75} + 75\right) = 154.13$

The cost of operating the truck is minimized when it is driven at 40 miles per hour.

35.   **STEP 1** Minimize the size S of the page.

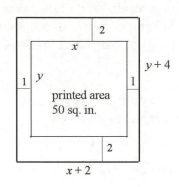

STEP 2 Let x denote the width of the print and y denote the length of the print.

STEP 3 The area of the printed matter is 50 square inches.

$$A = lw$$
$$50 = xy$$
$$y = \frac{50}{x} \text{ inches}$$

STEP 4 The size of the page is expressed by the function

$$S = (y+4)(x+2)$$
$$S(x) = \left(\frac{50}{x}+4\right)(x+2) = 50 + \frac{100}{x} + 4x + 8 = 58 + \frac{100}{x} + 4x$$

The domain of S is $\{x \mid x > 0\}$.

STEP 5 The derivative $S'(x)$ will give critical numbers if they exist.

$$S'(x) = -\frac{100}{x^2} + 4 = \frac{4x^2 - 100}{x^2}$$

We solve the equation $S'(x) = 0$.

$$\frac{4x^2 - 100}{x^2} = 0$$
$$4x^2 - 100 = 0$$
$$x^2 = \frac{100}{4} = 25 \text{ or } x = \pm 5$$

Since $x = -5$ is not in the domain of S we disregard it. We use the second derivative test to see if $x = 5$ locates a maximum or a minimum.

$$S''(x) = 200x^{-3} = \frac{200}{x^3} \qquad\qquad S''(5) = \frac{200}{5^3} = \frac{200}{125} > 0$$

We note that $x = 5$ minimizes S, and we conclude that the most economical page size is one with length $\frac{50}{5} + 4 = 14$ inches and width $5 + 2 = 7$ inches.

37. The relation is justified because an increased tax rate results in a higher price and lower demand. When tax rate is 0, the quantity demanded is 2.45. On the other hand, there is no demand when the tax rate is $t = 18$.

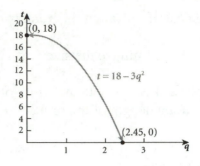

The optimal tax rate maximizes tax revenue. Tax revenue is given by

$$R = qt = q(18 - 3q^2) = 18q - 3q^3$$

The domain of R is $\{q \mid 0 \le q \le \sqrt{6}\}$.

We differentiate R and then set it equal to zero to find the critical numbers.

$$R'(q) = 18 - 9q^2$$
$$18 - 9q^2 = 0$$
$$q^2 = 2$$
$$q = \pm\sqrt{2}$$

The only critical number in the interval $(0, \sqrt{6})$ is $\sqrt{2}$, so we evaluate R at 0, $\sqrt{6}$, and $\sqrt{2}$.

when $q = 0$: $\qquad\qquad R(0) = 18(0) - 3(0)^3 = 0$

when $q = \sqrt{2} \approx 1.414$: $\qquad R(\sqrt{2}) = 18(\sqrt{2}) - 3(\sqrt{2})^3 = 12\sqrt{2} \approx 16.971$

when $q = \sqrt{6}$: $\qquad\qquad R(\sqrt{6}) = 18(\sqrt{6}) - 3(\sqrt{6})^3 = 0$

The revenue is maximized when $q = \sqrt{2}$. The tax rate corresponding to maximum revenue is

$$t = 18 - 3(\sqrt{2})^2 = 12$$

This means that a tax rate of 12% generates a maximum revenue of 16.971 monetary units.

39. The volume of the cylinder is fixed, so

$$V = \pi r^2 h$$

$$h = \frac{V}{\pi r^2}$$

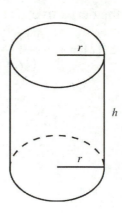

The surface area S of the cylinder is

$$S = \text{(area of side)} + 2\text{(area of the top)}$$

$$S = 2\pi rh + 2\pi r^2$$

$$S(r) = 2\pi r\left(\frac{V}{\pi r^2}\right) + 2\pi r$$

$$= \frac{2V}{r} + 2\pi r^2$$

Differentiating S and finding the critical numbers, if there are any gives:

$$S'(r) = -\frac{2V}{r^2} + 4\pi r = \frac{4\pi r^3 - 2V}{r^2} \qquad r \neq 0$$

Solving $S'(r) = 0$,

$$\frac{4\pi r^3 - 2V}{r^2} = 0$$

$$4\pi r^3 - 2V = 0$$

$$r^3 = \frac{2V}{4\pi} = \frac{V}{2\pi}$$

$$r = \sqrt[3]{\frac{V}{2\pi}}$$

Using the second derivative test, we find that

$$S''(r) = \frac{4V}{r^3} + 4\pi \quad \text{and} \quad S''\left(\sqrt[3]{\frac{V}{2\pi}}\right) = \frac{4V}{\left(\sqrt[3]{\frac{V}{2\pi}}\right)^3} + 4\pi = 12\pi > 0$$

We conclude that $r = \sqrt[3]{\frac{V}{2\pi}}$ will minimize the surface area of a cylinder with volume V.

The height of this cylinder should be

$$h = \frac{V}{\pi r^2} = \frac{V}{\pi\left[\sqrt[3]{\frac{V}{2\pi}}\right]^2} = \frac{V}{\pi\sqrt[3]{\frac{V^2}{4\pi^2}}} = \frac{V}{\pi} \cdot \sqrt[3]{\frac{4\pi^2}{V^2}}$$

$$= \sqrt[3]{\frac{V^3}{\pi^3}} \cdot \sqrt[3]{\frac{4\pi^2}{V^2}} = \sqrt[3]{\frac{4\pi^2 V^3}{V^2 \pi^3}} = \sqrt[3]{\frac{4V}{\pi}} = \sqrt[3]{\frac{8V}{2\pi}} = 2\sqrt[3]{\frac{V}{2\pi}} = 2r$$

41. We use the derivative $C'(t)$ to find when the concentration of the drug is the greatest.

$$C'(t) = \frac{2(16+t^3) - 2t \cdot 3t^2}{(16+t^3)^2} = \frac{32 + 2t^3 - 6t^3}{(16+t^3)^2} = \frac{32 - 4t^3}{(16+t^3)^2}$$

$$\frac{32 - 4t^3}{(16+t^3)^2} = 0$$

$$32 - 4t^3 = 0$$

$$t^3 = \frac{32}{4} = 8$$

$$t = 2$$

Using the first derivative test we see that for $t < 2$, $C'(t) > 0$, and for $t > 2$, $C'(t) < 0$. So we conclude that the concentration of the drug is greatest 2 hours after it is admininstered.

14.5 Elasticity of Demand

1. (a) $\quad x = f(p) = 4000 - 100p$

(b) $\quad f'(p) = -100$

$$E(p) = \frac{pf'(p)}{f(p)} = \frac{p(-100)}{4000 - 100p} = \frac{-100p}{4000 - 100p} = \frac{p}{p - 40}$$

(c) When $p = \$5$, $E(5) = \dfrac{5}{5 - 40} = -0.143$

Increasing the price by 10% to \$5.50, will result in a decrease of approximately 1.43% in quantity demanded.

(d) When $p = \$15$, $E(15) = \dfrac{15}{15 - 40} = -0.6$

Increasing the price by 10% to \$16.50, will result in a decrease of approximately 6 % in quantity demanded.

(e) When $p = \$20$, $E(20) = \dfrac{20}{20 - 40} = -1$

Increasing the price by 10% to \$22.00, will result in a decrease of approximately 10 % in quantity demanded.

3. (a) $\quad x = 200(50 - p) = 10,000 - 200p$

(b) $f'(p) = -200$

$$E(p) = \frac{pf'(p)}{f(p)} = \frac{p(-200)}{10,000-200p} = \frac{-200p}{10,000-200p} = \frac{p}{p-50}$$

(c) When $p = \$10$, $E(10) = \dfrac{10}{10-50} = -0.25$

Increasing the price by 5% will result in a decrease of approximately $(0.25)(5\%) = 0.0125 = 1.25\%$ in quantity demanded.

(d) When $p = \$25$, $E(25) = \dfrac{25}{25-50} = -1$

Increasing the price by 5% will result in a decrease of approximately 5% in quantity demanded.

(e) When $p = \$35$, $E(35) = \dfrac{35}{35-50} = -2.333$

Increasing the price by 5% will result in a decrease of approximately $(2.333)(5\%) = 0.11665 = 11.665\%$ in quantity demanded.

5. $x = f(p) = 600-3p$ $f'(p) = -3$

$$E(p) = \frac{pf'(p)}{f(p)} = \frac{p(-3)}{600-3p} = \frac{-3p}{600-3p} = \frac{p}{p-200}$$

$$E(50) = \frac{50}{50-200} = \frac{50}{-150} = -0.333$$

At $p = \$50$ the demand is inelastic.

7. $x = f(p) = \dfrac{600}{p+4}$ $f'(p) = \dfrac{-600}{(p+4)^2}$

$$E(p) = \frac{pf'(p)}{f(p)} = \frac{p\left[\dfrac{-600}{(p+4)^2}\right]}{\dfrac{600}{p+4}} = \frac{-p}{p+4}$$

$$E(10) = \frac{-10}{10+4} = -0.714$$

At $p = \$10$ the demand is inelastic.

9. $x = f(p) = 10,000-10p^2$ $f'(p) = -20p$

$$E(p) = \frac{pf'(p)}{f(p)} = \frac{p(-20p)}{10,000-10p^2} = \frac{-20p^2}{10,000-10p^2} = \frac{-2p^2}{1000-p^2}$$

$$E(10) = \frac{-2(10^2)}{1000 - 10^2} = -0.222$$

At $p = \$10$ the demand is inelastic.

11. $x = f(p) = \sqrt{100 - p}$ $f'(p) = \frac{1}{2}(100 - p)^{-1/2} \cdot (-1) = \frac{-1}{2\sqrt{100 - p}}$

$$E(p) = \frac{pf'(p)}{f(p)} = \frac{p\left[\dfrac{-1}{2\sqrt{100 - p}}\right]}{\sqrt{100 - p}} = \frac{-p}{2(100 - p)}$$

$$E(10) = \frac{-10}{2(100 - 10)} = -0.056$$

At $p = \$10$ the demand is inelastic.

13. $x = f(p) = 40(4 - p)^3$ $f'(p) = 40 \cdot 3(4 - p)^2 \cdot (-1) = -120(4 - p)^2$

$$E(p) = \frac{pf'(p)}{f(p)} = \frac{p\left[-120(4 - p)^2\right]}{40(4 - p)^3} = -\frac{3p}{4 - p}$$

$$E(2) = -\frac{(3)(2)}{4 - 2} = -3$$

At $p = \$2$ the demand is elastic.

15. $x = f(p) = 20 - 3\sqrt{p}$ $f'(p) = \frac{1}{2} \cdot (-3)p^{-1/2} = -\frac{3}{2\sqrt{p}}$

$$E(p) = \frac{pf'(p)}{f(p)} = \frac{p\left[-\dfrac{3}{2\sqrt{p}}\right]}{20 - 3\sqrt{p}} = -\frac{3p}{2\sqrt{p}\left(20 - 3\sqrt{p}\right)} = -\frac{3p}{40\sqrt{p} - 6p}$$

$$E(4) = -\frac{3(4)}{40\sqrt{4} - 6(4)} = -\frac{12}{80 - 24} = -0.214$$

At $p = \$4$ the demand is inelastic.

17. First we differentiate.
$$\frac{1}{2}x^{-1/2}\frac{dx}{dp} + 2p\frac{dx}{dp} + 2x + 2p = 0$$

$$\frac{dx}{dp} = \frac{-(2x+2p)}{\frac{1}{2}x^{-1/2}+2p} = -\frac{4(x+p)\sqrt{x}}{1+4p\sqrt{x}}$$

$$E(p) = \frac{pf'(p)}{f(p)} = \frac{p}{x} \cdot \left[-\frac{4(x+p)\sqrt{x}}{1+4p\sqrt{x}} \right]$$

When $x = 16$ and $p = 4$,

$$E(4) = \frac{4}{16} \cdot \left[-\frac{4(16+4)\sqrt{16}}{1+4(4)\sqrt{16}} \right] = -\frac{1}{4} \cdot \left[\frac{320}{65} \right] = -\frac{80}{65} = -1.231$$

19. First we differentiate.

$$4x\frac{dx}{dp} + 3p\frac{dx}{dp} + 3x + 20p = 0$$

$$\frac{dx}{dp} = \frac{-(3x+20p)}{4x+3p}$$

$$E(p) = \frac{p}{x} \cdot \left[-\frac{3x+20p}{4x+3p} \right]$$

When $x = 10$ and $p = 5$,

$$E(5) = \frac{5}{10} \cdot \left[-\frac{3 \cdot 10 + 20 \cdot 5}{4 \cdot 10 + 3 \cdot 5} \right] = -\frac{1}{2} \cdot \frac{130}{55} = -1.182$$

21.

$$p = F(x) = 10 - \frac{1}{20}x \qquad\qquad p' = F'(x) = -\frac{1}{20}$$

$$E(x) = \frac{F(x)}{xF'(x)} = \frac{10 - \frac{1}{20}x}{x\left[-\frac{1}{20} \right]} = -\frac{200 - x}{x}$$

$$E(5) = -\frac{200-5}{5} = -39$$

23.

$$p = F(x) = 10 - 2x^2 \qquad\qquad p' = F'(x) = -4x$$

$$E(x) = \frac{F(x)}{xF'(x)} = \frac{10-2x^2}{x(-4x)} = -\frac{10-2x^2}{4x^2} = -\frac{5-x^2}{2x^2}$$

$$E(2) = -\frac{5-2^2}{2(2^2)} = -\frac{1}{8} = -0.125$$

25.

$$p = F(x) = 50 - 2\sqrt{x} = 50 - 2x^{1/2} \qquad\qquad p' = F'(x) = -x^{-1/2} = -\frac{1}{\sqrt{x}}$$

$$E(x) = \frac{F(x)}{xF'(x)} = \frac{50 - 2\sqrt{x}}{x\left(-\dfrac{1}{\sqrt{x}}\right)} = -\frac{50 - 2\sqrt{x}}{\sqrt{x}}$$

$$E(100) = -\frac{50 - 2\sqrt{100}}{\sqrt{100}} = -\frac{50 - 20}{10} = -3$$

27. $\quad x = f(p) = \dfrac{6000}{p} - 500 \qquad\qquad f'(p) = -\dfrac{6000}{p^2}$

$$E(p) = \frac{p \cdot f'(p)}{x} = \frac{p\left(-\dfrac{6000}{p^2}\right)}{\dfrac{6000 - 500p}{p}} = -\frac{6000}{6000 - 500p}$$

When $p = 4$,

$$E(4) = -\frac{6000}{6000 - 500(4)} = -1.5$$

(a) The demand is elastic since $|E(4)| > 1$.
(b) If the price is increased, the revenue will decrease.

29. $\quad x = f(p) = \sqrt{300 - 6p} = (300 - 6p)^{1/2} \quad f'(p) = \dfrac{1}{2}(300 - 6p)^{-1/2} \cdot (-6) = -\dfrac{3}{\sqrt{300 - 6p}}$

$$E(p) = \frac{p \cdot f'(p)}{x} = \frac{p\left[-\dfrac{3}{\sqrt{300 - 6p}}\right]}{\sqrt{300 - 6p}} = -\frac{3p}{300 - 6p} = -\frac{p}{100 - 2p}$$

When $p = 10$,

$$E(10) = -\frac{10}{100 - 2(10)} = -0.125$$

(a) When the price is \$10, the demand is inelastic.
(b) The revenue will decrease if the price is lowered slightly.

31. When $p = 15$, $x = 2000$ and when $p = 18$ then $x = 1800$.

(a) $m = \dfrac{x_2 - x_1}{p_2 - p_1} = \dfrac{1800 - 2000}{18 - 15} = -\dfrac{200}{3} = -66.67$

$$x - 2000 = -\frac{200}{3}(p - 15)$$

$$x = -\frac{200}{3}p + 1000 + 2000 = -\frac{200}{3}p + 3000$$

(b) $f'(p) = -\dfrac{200}{3}$

$$E(p) = \frac{p \cdot f'(p)}{x} = \frac{p\left(-\dfrac{200}{3}\right)}{-\dfrac{200}{3}p + 3000} = \frac{-200p}{-200p + 9000} = -\frac{p}{45 - p}$$

$$E(18) = -\frac{18}{45 - 18} = -0.667$$

(c) If the price is increased by 5%, the demand will decrease by approximately $(0.667)(0.05) = 0.0333 = 3.33\%$.

(d) Since $|E(18)| < 1$ the demand is inelastic, and the increase in price will cause an increase in revenue.

14.6 Related Rates $844 / 1-21 \ (ODD)$

1. $x^2 + y^2 = 13$

$2x\dfrac{dx}{dt} + 2y\dfrac{dy}{dt} = 0$

$\dfrac{dx}{dt} = \dfrac{-2y\dfrac{dy}{dt}}{2x}$

When $x = 2$, $y = 3$, and $\dfrac{dy}{dt} = 2$,

$\dfrac{dx}{dt} = \dfrac{-2 \cdot 3 \cdot 2}{2 \cdot 2} = -3$

3. $x^3 y^2 = 72$

$x^3 \cdot 2y\dfrac{dy}{dt} + 3x^2\dfrac{dx}{dt} \cdot y^2 = 0$

$3x^2 y^2 \dfrac{dx}{dt} = -2x^3 y\dfrac{dy}{dt}$

$\dfrac{dx}{dt} = \dfrac{-2x^3 y\dfrac{dy}{dt}}{3x^2 y^2}$

When $x = 2$, $y = 3$, and $\dfrac{dy}{dt} = 2$,

$\dfrac{dx}{dt} = \dfrac{-2 \cdot 2^3 \cdot 3 \cdot 2}{3 \cdot 2^2 \cdot 3^2} = -\dfrac{8}{9}$

5. $V = 80h^2 \qquad \dfrac{dV}{dt} = 160h\dfrac{dh}{dt}$

If $h = 3$ and $\dfrac{dh}{dt} = \dfrac{1}{12}$, then $\dfrac{dV}{dt} = 160 \cdot 3 \cdot \dfrac{1}{12} = 40$

7. $V = \dfrac{1}{12}\pi h^3 \qquad \dfrac{dV}{dt} = \dfrac{3}{12}\pi h^2\dfrac{dh}{dt} = \dfrac{1}{4}\pi h^2\dfrac{dh}{dt}$

If $h = 8$ and $\dfrac{dh}{dt} = \dfrac{5}{16}\pi$, then $\dfrac{dV}{dt} = \dfrac{1}{4}\pi \cdot 8^2 \cdot \dfrac{5}{16}\pi = 5\pi^2$

9. **STEP 2** The variables are:

    s = length (in centimeters) of a side of the cube

    V = volume (in cubic centimeters) of the cube

    t = time (in seconds).

STEP 1

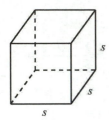

STEP 3 The rates of change are $\dfrac{ds}{dt}$ and $\dfrac{dV}{dt}$.

STEP 4 We know $\dfrac{ds}{dt} = 3$ cm/sec, and we want $\dfrac{dV}{dt}$ when $s = 10$ cm. The volume of a cube is $V = s^3$.

STEP 5 $\dfrac{dV}{dt} = 3s^2 \dfrac{ds}{dt}$

STEP 6 When $s = 10$ and $\dfrac{ds}{dt} = 3$, $\dfrac{dV}{dt} = 3 \cdot 10^2 \cdot 3 = 900$ cubic centimeters per second.

11. **STEP 2** The variables are:

    x = length (in centimeters) of one leg of the triangle

    y = length (in centimeters) of a the other leg

    t = time (in minutes).

STEP 1

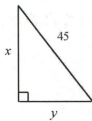

STEP 3 The rates of change are $\dfrac{dx}{dt}$ and $\dfrac{dy}{dt}$.

STEP 4 We know $\dfrac{dx}{dt} = 2$ cm/min, and we want $\dfrac{dy}{dt}$ when $x = 4$ cm. We use the Pythagorean Theorem $x^2 + y^2 = 45^2$, and note that when $x = 4$, $y = \sqrt{45^2 - 4^2} = \sqrt{2009} \approx 44.822$ cm.

STEP 5 $2x\dfrac{dx}{dt} + 2y\dfrac{dy}{dt} = 0$

$$\frac{dy}{dt} = -\frac{2x\dfrac{dx}{dt}}{2y} = -\frac{x}{y}\frac{dx}{dt}$$

STEP 6 When $x = 4$, $y = 44.822$ and $\dfrac{dx}{dt} = 2$, $\dfrac{dy}{dt} = -\dfrac{4}{\sqrt{2009}} \cdot 2 \approx -0.178$ centimeters per minute.

13. **STEP 2** The variables are:

    r = radius (in meters) of the balloon

    V = volume (in cubic meters) of the balloon

    S = surface area (in square meters) of the balloon

    t = time (in minutes).

STEP 1

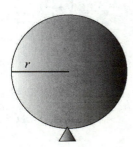

STEP 3 The rates of change are $\dfrac{dr}{dt} \cdot \dfrac{dV}{dt}$, and $\dfrac{dS}{dt}$.

STEP 4 We know $\dfrac{dV}{dt} = -1.5$ m^3 per minute

(negative because the air is leaking out). We want $\dfrac{dS}{dt}$

when $r = 4$ m. The volume of a sphere is $V = \dfrac{4}{3}\pi r^3$ and the surface area is $S = 4\pi r^2$.

STEP 5 $\quad \dfrac{dV}{dt} = 4\pi r^2 \dfrac{dr}{dt} \quad$ and $\quad \dfrac{dS}{dt} = 8\pi r \dfrac{dr}{dt}$

We use $\dfrac{dV}{dt}$ to solve for $\dfrac{dr}{dt}$ and then substitute it into $\dfrac{dS}{dt}$.

$$\frac{dr}{dt} = \frac{1}{4\pi r^2}\frac{dV}{dt}$$

$$\frac{dS}{dt} = 8\pi r \cdot \frac{1}{4\pi r^2}\frac{dV}{dt} = \frac{2}{r}\frac{dV}{dt}$$

STEP 6 When $r = 4$ and $\dfrac{dV}{dt} = -1.5$, $\dfrac{dS}{dt} = \dfrac{2}{4}\cdot(-1.5) = -0.75$ square meters per minute.

15. **STEP 2** The variables are:
    h = water level (in meters) in the deep end.
    V = volume (in cubic meters) of the pool water
    t = time (in minutess).

    STEP 3 The rates of change are $\dfrac{dh}{dt}$ and $\dfrac{dV}{dt}$.

    STEP 4 We know $\dfrac{dV}{dt} = 15$ m^3/min, and we want

    STEP 1

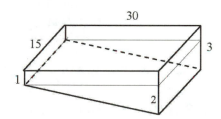

$\dfrac{dh}{dt}$ when $h = 2$ m. Once the water hits the 2 meter mark the volume of the new water is that of a rectangular prism. So $V = lwh$.

STEP 5 $\quad \dfrac{dV}{dt} = lw\dfrac{dh}{dt} \quad$ or $\quad \dfrac{dh}{dt} = \dfrac{1}{lw}\cdot\dfrac{dV}{dt}$

STEP 6 When $l = 30$, $w = 15$ and $\dfrac{dV}{dt} = 15$, $\dfrac{dh}{dt} = \dfrac{1}{30\cdot 15}\cdot 15 = \dfrac{1}{30} \approx 0.033$ meters per minute.

17. **STEP 2** The variables are:

r = radius (in feet) of the oil spill
A = area (in square feet) of the oil spill
t = time (in minutes).

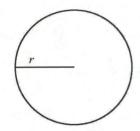

STEP 3 The rates of change are $\dfrac{dr}{dt}$ and $\dfrac{dA}{dt}$.

STEP 4 We know $\dfrac{dr}{dt} = 0.42$ ft/min, and we want $\dfrac{dA}{dt}$ when $r = 120$ feet. The area of a circle is $A = \pi r^2$.

STEP 5 $\dfrac{dA}{dt} = 2\pi r \dfrac{dr}{dt}$

STEP 6 When $r = 120$ and $\dfrac{dr}{dt} = 0.42$, $\dfrac{dA}{dt} = 2\pi \cdot 120 \cdot 0.42 = 100.8\pi \approx 316.673$ square feet per minute.

19. **STEP 2** x = units produced per day
C = cost of producing x units
R = revenue from selling x units
P = profit from selling x units
t = time (in days)

STEP 3 The rates of change are $\dfrac{dx}{dt}$, $\dfrac{dC}{dt}$, $\dfrac{dR}{dt}$, and $\dfrac{dP}{dt}$.

STEP 4 We know $\dfrac{dx}{dt} = 100$ units per day and $x = 1000$ units

(a) **STEP 5** $\dfrac{dC}{dt} = 5\dfrac{dx}{dt}$

STEP 6 Substituting $\dfrac{dx}{dt} = 100$, we find $\dfrac{dC}{dt} = 500$. That is cost is increasing at a rate of $500 per day.

(b) **STEP 5** $\dfrac{dR}{dt} = 15\dfrac{dx}{dt} - \dfrac{x}{5000}\dfrac{dx}{dt}$

STEP 6 When $x = 1000$ and $\dfrac{dx}{dt} = 100$, we find $\dfrac{dR}{dt} = 15 \cdot 100 - \dfrac{1000}{5000} \cdot 100 = 1480$. That is, revenue is increasing at a rate of $1480 per day.

(c) The revenue is increasing when production is 1000 units per day.

(d) Profit is the difference between revenue and cost.
$$P(x) = R(x) - C(x)$$

$$P(x) = \left(15x - \frac{x^2}{10,000}\right) - (5x + 5000) = 10x - \frac{x^2}{10,000} - 5000$$

(e) When $x = 1000$, $\dfrac{dP}{dt} = \dfrac{dR}{dt} - \dfrac{dC}{dt} = 1480 - 500 = 980$. That is the profit is increasing at a rate of $980 per day.

21. STEP 2 q = demand (in thousands) of plasma televisions
p = price of a plasma television
R = revenue derived from selling laptops
t = time (in years)

STEP 3 The rates of change are $\dfrac{dp}{dt}$ and $\dfrac{dR}{dt}$.

STEP 4 We know $\dfrac{dp}{dt} = -100$ dollars per year, and we want $\dfrac{dR}{dt}$ when $p = \$7000$. The quantity demanded is $q = 10,000 - 0.90p$ thousand television, and the revenue (in thousands of dollars) is $R = pq = p(10,000 - 0.90p) = 10,000p - 0.90p^2$.

STEP 5 $\dfrac{dR}{dt} = 10,000\dfrac{dp}{dt} - 1.8p\dfrac{dp}{dt}$

STEP 6 When $p = 7000$ and $\dfrac{dp}{dt} = -100$,

$$\frac{dR}{dt} = 10,000 \cdot (-100) - 1.8 \cdot 7000 \cdot (-100) = 260,000.$$

That is, the revenue is increasing at a rate of $260,000,000 per year.

14.7 The Differential; Linear Approximations

1.
$$y = x^3 - 2x + 1$$
$$f'(x) = \frac{dy}{dx} = 3x^2 - 2$$
$$dy = f'(x)dx = \left(3x^2 - 2\right)dx$$

3.
$$y = \frac{x-1}{x^2 + 2x - 8}$$
$$f'(x) = \frac{dy}{dx} = \frac{1 \cdot \left(x^2 + 2x - 8\right) - (x-1)(2x+2)}{\left(x^2 + 2x - 8\right)^2} = \frac{x^2 + 2x - 8 - 2x^2 - 2x + 2x + 2}{\left(x^2 + 2x - 8\right)^2}$$
$$= \frac{-x^2 + 2x - 6}{\left(x^2 + 2x - 8\right)^2}$$

$$dy = f'(x)dx = \frac{-x^2+2x-6}{(x^2+2x-8)^2}dx$$

5. Take the differential of each side.

$$d(xy) = d(6)$$

$$x\,dy + y\,dx = 0$$

$$dy = -\frac{y}{x}dx \qquad\qquad dx = -\frac{x}{y}dy$$

$$\frac{dy}{dx} = -\frac{y}{x} \qquad\qquad \frac{dx}{dy} = -\frac{x}{y}$$

7. Take the differential of each side.

$$d(x^2+y^2) = d(16)$$

$$2x\,dx + 2y\,dy = 0$$

$$2x\,dx = -2y\,dy$$

$$dy = \frac{2x}{-2y}dx \qquad\qquad dx = \frac{-2y}{2x}dy$$

$$\frac{dy}{dx} = \frac{2x}{-2y} = -\frac{x}{y} \qquad\qquad \frac{dx}{dy} = \frac{-2y}{2x} = -\frac{y}{x}$$

9. Take the differential of each side.

$$d(x^3+y^3) = d(3x^2y)$$

$$3x^2\,dx + 3y^2\,dy = 3x^2 \cdot dy + 6x\,dx \cdot y$$

$$(3y^2-3x^2)dy = (6xy-3x^2)\,dx$$

$$dy = \frac{6xy-3x^2}{3y^2-3x^2}dx \qquad\qquad dx = \frac{3y^2-3x^2}{6xy-3x^2}dy$$

$$\frac{dy}{dx} = \frac{6xy-3x^2}{3y^2-3x^2} = \frac{2xy-x^2}{y^2-x^2} \qquad\qquad \frac{dx}{dy} = \frac{y^2-x^2}{2xy-x^2}$$

11. $$d\left(\sqrt{x-2}\right) = d\left[(x-2)^{1/2}\right] = \frac{1}{2}(x-2)^{-1/2} \cdot dx = \frac{dx}{2(x-2)^{1/2}} = \frac{dx}{2\sqrt{x-2}}$$

13. $$d(x^3-x-4) = (3x^2-1)\,dx$$

15. $f(x) = x^2 - 2x + 1$ and $f(2) = 2^2 - 2(2) + 1 = 1$

$f'(x) = 2x - 2$ and $f'(2) = 2(2) - 2 = 2$

The linear approximation to f near $x = 2$ is
$$f(x) \approx f(2) + f'(2)(x - 2)$$
$$f(x) \approx 1 + 2(x - 2) = 2x - 3$$

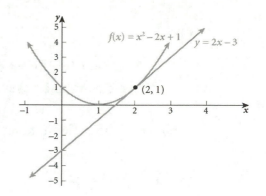

17. $f(x) = \sqrt{x} = x^{1/2}$ and $f(4) = \sqrt{4} = 2$

$f'(x) = \dfrac{1}{2}x^{-1/2} = \dfrac{1}{2\sqrt{x}}$ and $f'(4) = \dfrac{1}{2\sqrt{4}} = \dfrac{1}{4}$

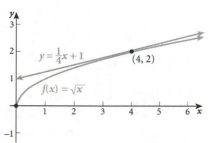

The linear approximation to f near $x = 4$ is
$$f(x) \approx f(4) + f'(4)(x - 4)$$
$$f(x) = 2 + \frac{1}{4}(x - 4) = \frac{1}{4}x + 1$$

19. $f(x) = e^x$ and $f(0) = e^0 = 1$

$f'(x) = e^x$ and $f'(0) = e^0 = 1$

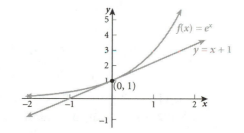

The linear approximation to f near $x = 0$ is
$$f(x) \approx f(0) + f'(0)(x - 0)$$
$$f(x) = 1 + 1(x - 0) = x + 1$$

21. (a) $\Delta y \approx f'(x_0)(x - x_0)$

$\approx 2(3)(3.001 - 3)$

$= 0.006$

(b) $\Delta y \approx f'(x_0)(x - x_0)$

$f'(x) = -\dfrac{1}{(x+2)^2}$

$\Delta y \approx -\dfrac{1}{(2+2)^2}(1.98 - 2) = 0.00125$

23. $A(r) = \pi r^2$ $\qquad\qquad$ $A'(r) = 2\pi r$

$\Delta A \approx dA = A'(r)\, dr = 2\pi r\, dr$

When $r = 10$ centimeters and increases to 10.1 centimeters,

$\Delta A \approx 2\pi \cdot 10 \cdot (0.1) = 2\pi \approx 6.28$

The area increases by approximately 6.28 square centimeters.

25.

$$V(r) = \frac{4}{3}\pi r^3 \qquad\qquad V'(r) = 4\pi r^2$$

$$\Delta V \approx dV = V'(r)\,dr = 4\pi r^2\,dr$$

When $r = 3$ meters and increases to 3.1 meters,

$$\Delta V \approx 4\pi \cdot 3^2 \cdot (0.1) = 3.6\pi \approx 11.310$$

The volume of the balloon increases by approximately 11.310 cubic meters.

27.

$V = s^3$, $V'(s) = 3s^2$ and $\dfrac{\Delta s}{s} = 0.02$. The relative error in the volume,

$$\frac{dV}{V} \approx \frac{dV}{V} = \frac{V'(s)\,ds}{V} = \frac{3s^2\,ds}{s^3} = 3\frac{ds}{s} = 3\frac{\Delta s}{s} = 3 \cdot (0.02) = 0.06$$

The percentage area in volume is 6%.

29. If the diameter d is 4 cm., then the radius r is 2 cm.
Similarly if $d = 3.9$ cm, then $r = 1.95$ cm. We know
that the radius is one-fourth the height, so $h = 4r$.
The volume of a right circular cone is

$$V = \frac{1}{3}\pi r^2 h = \frac{4}{3}\pi r^3$$

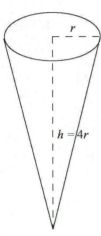

Taking the differential of V,

$$dV = 4\pi r^2 \cdot dr$$

dr is approximately the change in radius or 0.05 cm.
So, $dV = 4\pi\left(2^2\right) \cdot (-0.05) = -2.513$

The cup holds approximately 2.513 cubic
centimeters less than intended.

31. If we let h denote the height of the building and x denote the length of the shadow cast by
the pole, we can use similar triangles to estimate the height of the building.

$$\frac{\text{height}}{\text{base}} = \frac{h}{9+x} = \frac{3}{x}$$

$$h = \frac{27 + 3x}{x} \text{ meters}$$

If the measured shadow of the pole is 1 meter with an error of 1%, then the percentage

error in the height of the building will be $\dfrac{\Delta h}{h} \approx \dfrac{dh}{h}$.

$$dh = f'(x)\,dx = \frac{3 \cdot x - (3x + 27) \cdot 1}{x^2} \cdot dx = -\frac{27}{x^2}\,dx$$

$$\frac{dh}{h} = \frac{f'(x)\,dx}{h} = \frac{-\dfrac{27}{x^2}\,dx}{\dfrac{3x+27}{x}} = -\frac{27}{3x+27} \cdot \frac{dx}{x}$$

When $x = 1$ and $\dfrac{dx}{x} = 0.01$, $\dfrac{\Delta h}{h} \approx -\dfrac{27}{30} \cdot (0.01) = -0.009$

The percentage error in the measured height of the building is approximately 0.9%.

33. If the pendulum is originally 1 meter = 100 centimeters long and increases to 110 centimeters in length, then $\Delta l = 10$ centimeters. From Problem 32, we see that

$\dfrac{\Delta T}{T} \approx \dfrac{1}{2} \cdot \dfrac{dl}{l}$. So, $\dfrac{\Delta T}{T} = \dfrac{1}{2} \cdot \dfrac{10}{100} = \dfrac{1}{20} = 0.05$. The percentage area in the period is approximately 5%.

5% of a day is 0.05(60 min/hour)(24 hours/day) = 72 minutes.

Chapter 14 Review

TRUE-FALSE ITEMS

1.   False   3.   False   5.   True

FILL IN THE BLANKS

1.   decreasing   3.   concave up   5.   concavity

7.   linear approximation

REVIEW EXERCISES

1. $f'(x) = 3x^2 - 2x + 1$

Horizontal tangent lines occur where $f'(x) = 0$. Since the discriminant $b^2 - 4ac = (-2)^2 - 4(3)(1) = -8 < 0$, f has no horizontal tangent line.

There is no vertical tangent line since $f'(x)$ is never unbounded.

3.   The domain of f is $\{x \mid x \neq -4\}$.

$$f'(x) = \frac{\dfrac{1}{3}x^{-2/3}(x+4) - x^{1/3}(1)}{(x+4)^2} = \frac{\dfrac{x+4}{3x^{2/3}} - x^{1/3}}{(x+4)^2} = \frac{x+4-3x}{3x^{2/3}(x+4)^2} = \frac{4-2x}{3x^{2/3}(x+4)^2}$$

Horizontal tangent lines occur where $f'(x) = 0$, that is when $4 - 2x = 0$ or when $x = 2$.

When $x = 2$ $f(2) = \dfrac{2^{1/3}}{2+4} = \dfrac{2^{1/3}}{6}$. So f has a horizontal tangent line at $\left(2, \dfrac{2^{1/3}}{6}\right)$.

Vertical tangent lines occur where $f'(x)$ is unbounded. $f'(x)$ is unbounded at 0 and at -4. Since $x = -4$ is not in the domain of f we disregard it.

When $x = 0$, $f(0) = \dfrac{0^{1/3}}{0+4} = 0$. So f has a vertical tangent line at $(0, 0)$.

5. (a) $f(x) = 3x^{1/5}$　　　　$f(0) = 3 \cdot 0^{1/5} = 0$

The one-sided limits of f near 0 are

$$\lim_{x \to 0^-} f(x) = \lim_{x \to 0^-} 3x^{1/5} = 0 \qquad \lim_{x \to 0^+} f(x) = \lim_{x \to 0^+} 3x^{1/5} = 0$$

Since $\lim_{x \to 0} f(x) = f(0)$ the function is continuous at 0.

(b) The derivative of f at 0 is $f'(0) = \lim_{x \to 0} \dfrac{f(x) - f(0)}{x - 0} = \lim_{x \to 0} \dfrac{3x^{1/5} - 0}{x - 0} = \lim_{x \to 0} \dfrac{3x^{1/5}}{x}$.

$f'(x)$ is unbounded at $x = 0$ and does not exist.

(c) Since the function is continuous at $x = 0$, and the derivative is unbounded at $x = 0$, so there is a vertical tangent line at $(0, 0)$.

7.
$$f(x) = \begin{cases} 3x + 1 & x < 3 \\ x^2 + 1 & x \geq 3 \end{cases}$$

(a) When $x = 3, f(3) = 3^2 + 1 = 10$

The one-sided limits are

$$\lim_{x \to 3^-} f(x) = \lim_{x \to 3^-} (3x + 1) = 10 \qquad \lim_{x \to 3^+} f(x) = \lim_{x \to 3^+} (x^2 + 1) = 10$$

Since $\lim_{x \to 3} f(x) = f(x) = 10$, the function f is continuous at 3.

(b) The derivative of f at 3 is $f'(3) = \lim_{x \to 3} \dfrac{f(x) - f(3)}{x - 3} = \lim_{x \to 3} \dfrac{f(x) - 10}{x - 3}$.

Looking at one-sided limits,

$$\lim_{x \to 3^-} \frac{f(x) - 10}{x - 3} = \lim_{x \to 3^-} \frac{(3x+1) - 10}{x - 3} = \lim_{x \to 3^-} \frac{3x - 9}{x - 3} = \lim_{x \to 3^-} \frac{3(x-3)}{x-3} = \lim_{x \to 3^-} 3 = 3$$

$$\lim_{x \to 3^+} \frac{f(x) - 10}{x - 3} = \lim_{x \to 3^+} \frac{(x^2+1) - 10}{x - 3} = \lim_{x \to 3^+} \frac{x^2 - 9}{x - 3} = \lim_{x \to 3^+} \frac{(x+3)(x-3)}{x-3} = \lim_{x \to 3^+} (x+3) = 6$$

Since the one-sided limits are not equal we concluded that $f'(3)$ does not exist.

(c) The one-sided limits in (b) are unequal, so there is no tangent line at the point $(3, 10)$.

9.
$$f(x) = \frac{1}{5}x^5 - x^3 - 4x$$

(a) **STEP 1**　$f'(x) = x^4 - 3x^2 - 4$

　　STEP 2 Solve $f'(x) = 0$.

$$x^4 - 3x^2 - 4 = 0$$

$$\left(x^2-4\right)\left(x^2+1\right)=0$$
$$\left(x-2\right)\left(x+2\right)\left(x^2+1\right)=0$$
$$x-2=0 \quad \text{or} \quad x+2=0 \quad \text{or} \quad x^2+1=0$$
$$x=2 \quad \text{or} \qquad x=-2$$

These numbers separate the number line into three parts: $-\infty<x<-2$, $-2<x<2$, and $2<x<\infty$, and we use -3, 0, and 3 as test numbers.

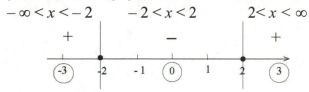

$$f'(-3)=(-3)^4-3(-3)^2-4=50 \quad f'(0)=-4 \quad f'(3)=3^4-3\left(3^2\right)-4=50$$

We conclude that the function is increasing on the intervals $(-\infty,-2)$ and $(2,\infty)$ and is decreasing on the interval $(-2,2)$.

(b) f is increasing for $-\infty<x<-2$ and decreasing for $-2<x<2$. When $x=-2$, $y=f(-2)=9.6$. So by the First Derivative Test, f has a local maximum at the point $(-2, 9.6)$.

f is decreasing for $-2<x<2$ and increasing for $2<x<\infty$. When $x=2$, $y=f(2)=-9.6$, and by the First Derivative Test, f has a local minimum at $(2,-9.6)$.

11.
$$f(x)=\frac{x^2}{x^2-8} \qquad x\neq\pm\sqrt{8}$$

(a) **STEP 1** $\displaystyle f'(x)=\frac{2x\cdot\left(x^2-8\right)-x^2\cdot(2x)}{\left(x^2-8\right)^2}=\frac{2x^3-16x-2x^3}{\left(x^2-8\right)^2}=\frac{-16x}{\left(x^2-8\right)^2}$

STEP 2 $f'(x)$ is not defined at $x=\pm 2\sqrt{2}$

Solve $f'(x)=0$

$$-16x=0 \text{ or } x=0$$

These 3 numbers separate the number line into four parts, we use the test numbers -3, -1, 1, and 3 to see if $f'(x)$ is positive or negative in each part.

$$-\infty<x<-2\sqrt{2} \qquad -2\sqrt{2}<x<0 \qquad 0<x<2\sqrt{2} \qquad 2\sqrt{2}<x<\infty$$

$$f'(-3)=48 \qquad f'(-1)=0.327 \qquad f'(1)=-0.327 \qquad f'(3)=-48$$

We conclude that f is increasing on the intervals $\left(-\infty, -2\sqrt{2}\right)$ and $\left(-2\sqrt{2}, 0\right)$, and that f is decreasing on the intervals $\left(0, 2\sqrt{2}\right)$ and $\left(2\sqrt{2}, \infty\right)$.

(b) f is increasing for $x < 0$ and decreasing for $x > 0$. When $x = 0$, $y = f(0) = 0$. So by the First Derivative Test, there is a local maximum at $(0, 0)$.

13. $f(x) = 1 + 3e^{-x}$

(a) **STEP 1** $f'(x) = -3e^{-x}$

   STEP 2   The domain of f' is all real numbers. $f'(x) \neq 0$.

We select $x = 0$ as a test point $f'(0) = -3e^{0} = -3$. We conclude that f is always decreasing.

(b)   From the First Derivative Test we conclude that f has no local maximum nor local minimum points.

15. $f(x) = x^3 - 3x^2 + 3x - 1$

STEP 1 f is a polynomial, so the domain is all real numbers.

STEP 2   The y-intercept occurs when $x = 0$; $y = f(0) = -1$. The y-intercept is $(0, -1)$. The x-intercept(s) occur when $y = 0$. So we solve $x^3 - 3x^2 + 3x - 1 = 0$. This is hard to do so we skip it.

STEP 3   To find where the graph is increasing or decreasing, we find $f'(x)$:
$$f'(x) = 3x^2 - 6x + 3$$
The solutions to $f'(x) = 0$ are $3x^2 - 6x - 3 = 0$

$$x = \frac{-b \pm \sqrt{b^2 - 4ac}}{2a} = \frac{6 \pm \sqrt{36 - 4(3)(-3)}}{2(3)} = \frac{6 \pm \sqrt{72}}{6} = \frac{6 \pm 6\sqrt{2}}{6} = 1 \pm \sqrt{2}$$

$$x \approx -0.414 \quad \text{or} \quad x \approx 2.414$$

We use these two numbers to separate the number line into three parts:
$$-\infty < x < -0.414 \qquad -0.414 < x < 2.414 \qquad 2.414 < x < \infty$$
and choose a test number from each part.

For $x = -1$: $f'(-1) = 12$

For $x = 0$: $f'(0) = 3$

For $x = 3$ $f'(3) = 12$

We conclude that the graph of f is always increasing.

STEP 4   Since the graph of f is always increasing, there is no relative minimum point or relative maximum point.

STEP 5 $f'(x) = 0$ for $x = -0.414$ and for $x = 2.414$. The graph of f has a horizontal tangent line at the points $(-0.414, f(-0.414)) = (-0.414, -2.828)$ and $(2.414, f(2.414)) = (2.414, 2.828)$

The first derivative is never unbounded, so there is no vertical tangent.

STEP 6 Since f is a polynomial function, its end behavior is that of $y = x^3$. Polynomial functions have no asymptotes.

STEP 7 To identify the inflection point, if any, we find $f''(x)$:

$$f''(x) = 6x - 6$$

$$f''(x) = 0 \text{ when } 6x - 6 = 0 \text{ or when } x = 1.$$

We use the number 1 to separate the number line into two parts:

$$-\infty < x < 1 \qquad \text{and} \qquad 1 < x < \infty$$

and choose a test number from each part.

For $x = 0$: $f''(0) = -6$

For $x = 2$: $f''(2) = 6$

We conclude that the graph of f is concave up on the interval $(1, \infty)$ and is concave down on the interval $(-\infty, 1)$. Since the concavity changes at the point $x = 1$, the point $(1, f(1)) = (1, 0)$ is an inflection point.

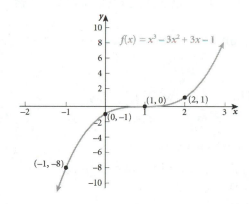

17. $f(x) = x^5 - 5x$

STEP 1 f is a polynomial, so the domain is all real numbers.

STEP 2 The y-intercept occurs when $x = 0$; $y = f(0) = 0$. The y-intercept is $(0, 0)$.
The x-intercept(s) occur when $y = 0$. So we solve $x^5 - 5x = 0$.

$$x(x^4 - 5) = 0$$

$$x = 0 \quad \text{or} \quad x^4 - 5 = 0$$

$$x = \pm \sqrt[4]{5} \approx \pm 1.495$$

The x-intercepts are $(0, 0)$, $(-1.495, 0)$, and $(1.495, 0)$.

STEP 3 To find where the graph is increasing or decreasing, we find $f'(x)$:

$$f'(x) = 5x^4 - 5$$

The solutions to $f'(x) = 0$ are
$$5x^4 - 5 = 0$$
$$x^4 = 1$$
$$x = \pm 1$$

We use the 2 numbers to separate the number line into three parts:
$$-\infty < x < -1 \qquad -1 < x < 1 \qquad 1 < x < \infty$$
and choose a test number from each part.

For $x = -2$: $\qquad f'(x) = 5 \cdot (-2)^4 - 5 = 75$

For $x = 0$: $\qquad f'(0) = 5 \cdot 0 - 5 = -5$

For $x = 2$: $\qquad f'(2) = 5 \cdot 2^4 - 5 = 75$

We conclude that the graph of f is increasing on the intervals $(-\infty, -1)$ and $(1, \infty)$, and f is decreasing on the interval $(-1, 1)$.

STEP 4 The graph is increasing to the left of -1 and decreasing to the right of -1, so the point $(-1, f(-1)) = (-1, 4)$ is a local maximum.

   The graph is decreasing to the left of 1 and increasing to the right of 1, so the point $(1, f(1)) = (1, -4)$ is a local minimum.

STEP 5 $f'(x) = 0$ for $x = -1$ and $x = 1$. The graph of f has a horizontal tangent lines at the points $(-1, f(-1)) = (-1, 4)$ and $(1, f(1)) = (1, -4)$.

   The first derivative is never unbounded, so there is no vertical tangent.

STEP 6 Since f is a polynomial function, its end behavior is that of $y = x^5$. Polynomial functions have no asymptotes.

STEP 7 To identify the inflection point, if any, we find $f''(x)$:
$$f''(x) = 20x^3$$
$$f''(x) = 0 \text{ when } 20x^3 = 0 \text{ or when } x = 0.$$

We use the numbers 0 to separate the number line into two parts:
$$-\infty < x < 0 \qquad \text{and} \qquad 0 < x < \infty$$
and choose a test number from each part.

For $x = -1$: $\qquad f''(-1) = -20$

For $x = 1$: $\qquad f''(1) = 20$

We conclude that the graph of f is concave up on the interval $(0, \infty)$ and is concave down on the interval $(-\infty, 0)$. Since the concavity changes at $x = 0$, the point $(0, 0)$ is an inflection point.

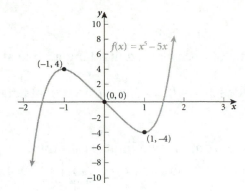

19. $f(x) = x^{4/3} + 4x^{1/3}$

STEP 1 The domain of f is all real numbers.

STEP 2 The y-intercept occurs when $x = 0$; $y = f(0) = 0$. The y-intercept is $(0, 0)$.
The x-intercept(s) occur when $y = 0$. So we solve $x^{4/3} + 4x^{1/3} = 0$.

$$x^{1/3}(x + 4) = 0$$

$$x^{1/3} = 0 \quad \text{or} \quad x + 4 = 0$$
$$x = 0 \quad \text{or} \quad x = -4$$

The x-intercepts are $(0, 0)$ and $(-4, 0)$.

STEP 3 To find where the graph is increasing or decreasing, we find $f'(x)$:

$$f'(x) = \frac{4}{3}x^{1/3} + \frac{4}{3}x^{-2/3} = \frac{4(x+1)}{3x^{2/3}}$$

$f'(x)$ is unbounded when $x = 0$.
The solution to $f'(x) = 0$ is $4(x + 1) = 0$ or $x = -1$.

We use these two numbers to separate the number line into three parts:
$$-\infty < x < -1 \qquad -1 < x < 0 \qquad 0 < x < \infty$$
and choose a test number from each part.

For $x = -8$: $\qquad f'(-8) = \dfrac{4(-8+1)}{3(-8)^{2/3}} = \dfrac{-28}{12} \approx -2.333$

For $x = -\dfrac{1}{8}$: $\qquad f'\left(-\dfrac{1}{8}\right) = \dfrac{4\left(-\dfrac{1}{8}+1\right)}{3\left(-\dfrac{1}{8}\right)^{2/3}} = \dfrac{\dfrac{7}{2}}{\dfrac{3}{4}} = \dfrac{14}{3} \approx 4.667$

For $x = 1$: $\qquad f'(1) = \dfrac{4(1+1)}{3(1)^{2/3}} = \dfrac{8}{3} \approx 2.667$

We conclude that the graph of f is increasing on the intervals $(-1, 0)$ and $(0, \infty)$, and f is decreasing on the interval $(-\infty, -1)$.

STEP 4 The graph is decreasing to the left of -1 and increasing to the right of -1, so the point $(-1, f(-1)) = (-1, -3)$ is a local minimum.

There is no local maximum.

STEP 5 $f'(x) = 0$ for $x = -1$. The graph of f has a horizontal tangent line at the point $(-1, f(-1)) = (-1, -3)$.

The first derivative is unbounded at $x = 0$, so there is a vertical tangent line at the point $(0, f(0)) = (0, 0)$.

STEP 6 For the end behavior of f, we look at the two limits at infinity:

$$\lim_{x \to \infty} f(x) = \lim_{x \to \infty} \left[x^{4/3} + 4x^{1/3} \right] = \infty \qquad \lim_{x \to -\infty} f(x) = \lim_{x \to -\infty} \left[x^{4/3} + 4x^{1/3} \right] = \infty$$

STEP 7 To identify the inflection points, if any, we find $f''(x)$:

$$f''(x) = \frac{4}{9}x^{-2/3} - \frac{8}{9}x^{-5/3} = \frac{4}{9x^{2/3}} - \frac{8}{9x^{5/3}} = \frac{4x - 8}{9x^{5/3}}$$

$f''(x)$ is unbounded at $x = 0$, and $f''(x) = 0$ when $4x - 8 = 0$ or when $x = 2$.
We use the numbers 0 and 2 to separate the number line into three parts:

$$-\infty < x < 0 \qquad 0 < x < 2 \qquad 2 < x < \infty$$

and choose a test number from each part.

For $x = -1$: $\quad f''(-1) = \dfrac{-12}{-9} = \dfrac{4}{3}$

For $x = 1$: $\quad f''(1) = \dfrac{-4}{9}$

For $x = 8$: $\quad f''(8) = \dfrac{24}{288} = \dfrac{1}{12}$

We conclude that the graph of f is concave up on intervals $(-\infty, 0)$ and $(2, \infty)$ and is concave down on interval $(0, 2)$. Since the concavity changes at $x = 0$ and $x = 2$, the points $(0, 0)$ and $(2, 7.560)$ are inflection points.

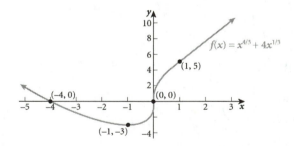

21.
$$f(x) = \frac{2x}{x^2 + 1}$$

STEP 1 f is a rational function. Since the denominator never equals 0, the domain is all real numbers.

STEP 2 The y-intercept occurs when $x = 0$; $y = f(0) = 0$. The y-intercept is $(0, 0)$. The x-intercept(s) occur when $y = 0$. So we solve $2x = 0$ and find $x = 0$. The x-intercept is $(0, 0)$.

STEP 3   To find where the graph is increasing or decreasing, we find $f'(x)$:

$$f'(x) = \frac{2 \cdot (x^2+1) - 2x \cdot 2x}{(x^2+1)^2} = \frac{2x^2+2-4x^2}{(x^2+1)^2} = \frac{2-2x^2}{(x^2+1)^2}$$

The solutions to $f'(x) = 0$ are $2 - 2x^2 = 0$

$$2(1-x)(1+x) = 0$$

$$1 - x = 0 \quad \text{or} \quad 1 + x = 0$$
$$x = 1 \quad \text{or} \quad x = -1$$

We use the numbers to separate the number line into three parts:

$$-\infty < x < -1 \qquad -1 < x < 1 \qquad 1 < x < \infty$$

and choose a test number from each part.

For $x = -2$: $f'(-2) = \dfrac{2-2(-2)^2}{\left[(-2)^2+1\right]^2} = -0.24$

For $x = 0$: $f'(0) = \dfrac{2-2(0)^2}{\left[(0)^2+1\right]^2} = 2$

For $x = 2$: $f'(2) = \dfrac{2-2(2)^2}{\left[2^2+1\right]^2} = -0.24$

We conclude that the graph of f is increasing on the interval $(-1, 1)$, and that the graph of f is decreasing on the intervals $(-\infty, -1)$ and $(1, \infty)$.

STEP 4   The graph is decreasing to the left of -1 and increasing to the right of -1, so the point $(-1, f(-1)) = (-1, -1)$ is a local minimum.

  The graph is increasing to the left of 1 and decreasing to the right of 1, so the point $(1, f(1)) = (1, 1)$ is a local maximum.

STEP 5 $f'(x) = 0$ for $x = 1$ and $x = -1$. The graph of f has a horizontal tangent line at the points $(1, f(1)) = (1, 1)$ and at $(-1, f(-1)) = (-1, -1)$.

  The first derivative is never unbounded, so there is no vertical tangent.

STEP 6   For the end behavior we look at the limits at infinity. Since f is an odd function, we need only to consider the limit as $x \to \infty$, the limit as $x \to -\infty$ will be the opposite.

$$\lim_{x \to \infty} f(x) = \lim_{x \to \infty} \frac{2x}{x^2+1} = \lim_{x \to \infty} \frac{2}{x} = 0$$

As x becomes unbounded in both the positive and negative directions, $y = 0$ is a horizontal asymptote.

STEP 7   To identify the inflection point, if any, we find $f''(x)$:

$$f''(x) = \frac{-4x(x^2+1)^2 - (2-2x^2) \cdot 2(x^2+1)(2x)}{(x^2+1)^4} = \frac{-4x(x^2+1) - 4x(2-2x^2)}{(x^2+1)^3}$$

$$= \frac{-4x(-x^2+3)}{(x^2+1)^3} = \frac{4x(x^2-3)}{(x^2+1)^3}$$

$f''(x) = 0$ when $4x(x^2-3) = 0$

$$4x = 0 \quad \text{or} \quad x^2 - 3 = 0$$
$$x = 0 \quad \text{or} \quad x = \pm\sqrt{3}$$

We use the numbers 0, $\sqrt{3}$, and $-\sqrt{3}$ to separate the number line into four parts:

$$-\infty < x < -\sqrt{3} \qquad -\sqrt{3} < x < 0 \qquad 0 < x < \sqrt{3} \qquad \sqrt{3} < x < \infty$$

and choose a test number from each part.

For $x = -2$: $\quad f''(-2) = -0.064$

For $x = -1$: $\quad f''(-1) = 1$

For $x = 1$: $\quad f''(1) = -1$

For $x = 2$: $\quad f''(2) = 0.064$

We conclude that the graph of f is concave up on the intervals $(-\sqrt{3}, 0)$ and $(\sqrt{3}, \infty)$, and is concave down on the intervals $(-\infty, -\sqrt{3})$ and $(0, \sqrt{3})$. Since the concavity changes at the points $x = 0$, $x = \sqrt{3}$, and $x = -\sqrt{3}$, the points $(0, 0)$, $\left(\sqrt{3}, 0.866\right)$, and $\left(-\sqrt{3}, -0.866\right)$ are inflection points.

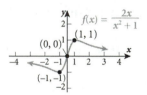

23. First locate the numbers for which $f'(x) = 0$.

$$f'(x) = 12x^2 - 3$$
$$12x^2 - 3 = 0$$
$$3(4x^2 - 1) = 0$$
$$3(2x-1)(2x+1) = 0$$
$$2x - 1 = 0 \quad \text{or} \quad 2x + 1 = 0$$
$$x = \frac{1}{2} \quad \text{or} \quad x = -\frac{1}{2}$$

Now evaluate $f''(x)$ at these numbers.

$$f''(x) = 24x$$

$f''\left(\dfrac{1}{2}\right) = 24\left(\dfrac{1}{2}\right) = 12 > 0$. By the Second Derivative Test, f has a local minimum

at $\left(\dfrac{1}{2}, f\left(\dfrac{1}{2}\right)\right) = \left(\dfrac{1}{2}, -1\right)$.

$f''\left(-\dfrac{1}{2}\right) = 24\left(-\dfrac{1}{2}\right) = -12$. By the Second Derivative Test, f has a local maximum

at $\left(-\dfrac{1}{2}, f\left(-\dfrac{1}{2}\right)\right) = \left(-\dfrac{1}{2}, 1\right)$.

25. First locate the numbers for which $f'(x) = 0$.

$$f'(x) = 4x^3 - 4x$$
$$4x^3 - 4x = 0$$
$$4x(x^2 - 1) = 0$$
$$4x = 0 \quad \text{or} \quad x^2 - 1 = 0$$
$$x = 0 \quad \text{or} \quad x = \pm\sqrt{1} = \pm 1$$

Now evaluate $f''(x)$ at these numbers.

$$f''(x) = 12x^2 - 4$$

$f''(0) = 12(0^2) - 4 = -4 < 0$. By the Second Derivative Test, f has a local maximum
at $(0, f(0)) = (0, 0)$.

$f''(-1) = 12(-1)^2 - 4 = 12 - 4 = 8 > 0$. By the Second Derivative Test, f has a local
minimum at $(-1, f(-1)) = (-1, -1)$.

$f''(1) = 12(1)^2 - 4 = 12 - 4 = 8 > 0$. By the Second Derivative Test, f has a local minimum
at $(1, f(1)) = (1, -1)$.

27. First locate the numbers for which $f'(x) = 0$.

$$f'(x) = xe^x + e^x$$
$$xe^x + e^x = 0$$
$$e^x(x + 1) = 0$$
$$e^x = 0 \quad \text{or} \quad x + 1 = 0$$
$$x = -1$$

Now evaluate $f''(x)$ at this number.

$$f''(x) = xe^x + e^x + e^x = xe^x + 2e^x$$

$f''(-1) = (-1)e^x + 2e^x = e^x > 0$. By the Second Derivative Test, f has a local minimum

at $(-1, f(-1)) = \left(-1, -\dfrac{1}{e}\right) \approx (-1, -0.368)$.

29. $f(x) = x^3 - 3x^2 + 3x - 1$ $f'(x) = 3x^2 - 6x + 3$

$f'(x) = 0$ when $3x^2 - 6x + 3 = 0$.

$$x^2 - 2x + 1 = 0$$
$$(x-1)^2 = 0$$
$$x = 1$$

The critical number 1 is in the interval (0, 3), so we evaluate f at the critical number and the two endpoints.

$$f(0) = 0^3 - 3(0)^2 + 3(0) - 1 = -1$$
$$f(1) = 1^3 - 3(1)^2 + 3(1) - 1 = 0$$
$$f(3) = 3^3 - 3(3)^2 + 3(3) - 1 = 8$$

The absolute maximum of f on [0, 3] is 8 and the absolute minimum is -1.

31. $f(x) = x^4 - 4x^3 + 4x^2$ $f'(x) = 4x^3 - 12x^2 + 8x$

$f'(x) = 0$ when $4x^3 - 12x^2 + 8x = 0$.

$$4x(x^2 - 3x + 2) = 0$$
$$4x(x-2)(x-1) = 0$$

$4x = 0$ or $x - 2 = 0$ or $x - 1 = 0$
  $x = 0$ or $x = 2$ or $x = 1$

The critical number 2 is in the interval (1, 3), so we evaluate f at the critical number and the two endpoints.

$$f(1) = 1^4 - 4(1)^3 + 4(1)^2 = 1$$
$$f(2) = 2^4 - 4(2)^3 + 4(2)^2 = 0$$
$$f(3) = 3^4 - 4(3)^3 + 4(3)^2 = 9$$

The absolute maximum of f on [1, 3] is 9 and the absolute minimum is 0.

33.

$f(x) = x^{4/3} - 4x^{1/3}$ $f'(x) = \dfrac{4}{3}x^{1/3} - \dfrac{4}{3}x^{-2/3} = \dfrac{4x^{1/3}}{3} - \dfrac{4}{3x^{2/3}} = \dfrac{4x-4}{3x^{2/3}}$

$f'(x)$ is unbounded when $x = 0$.
$f'(x) = 0$ when $4x - 4 = 0$ or $x = 1$.

The critical numbers 0 and 1 are in the interval $(-1, 8)$, so we evaluate f at the critical numbers and the two endpoints.

$$f(-1) = (-1)^{4/3} - 4(-1)^{1/3} = 5$$
$$f(0) = (0)^{4/3} - 4(0)^{1/3} = 0$$
$$f(1) = (1)^{4/3} - 4(1)^{1/3} = -3$$

$$f(8) = (8)^{4/3} - 4(8)^{1/3} = 8$$

The absolute maximum of f on $[-1, 8]$ is 8 and the absolute minimum is -3.

35. $x = f(p) = 1000 - 2p^2$ $f'(p) = -4p$

$$E(p) = \frac{pf'(p)}{f(p)} = \frac{p(-4p)}{1000 - 2p^2} = -\frac{4p^2}{1000 - 2p^2} = -\frac{2p^2}{500 - p^2}$$

$$E(20) = -\frac{2(20)^2}{500 - 20^2} = -8$$

At $p = \$20$ the demand is elastic.

37. $x = f(p) = \sqrt{500 - p^2} = (500 - p^2)^{1/2}$

$$f'(p) = \frac{1}{2}(500 - p^2)^{-1/2} \cdot (-2p) = \frac{-2p}{2(500 - p^2)^{1/2}} = -\frac{p}{\sqrt{500 - p^2}}$$

$$E(p) = \frac{pf'(p)}{f(p)} = \frac{p\left[-\dfrac{p}{\sqrt{500 - p^2}}\right]}{\sqrt{500 - p^2}} = -\frac{p^2}{500 - p^2}$$

$$E(10) = -\frac{10^2}{500 - 10^2} = -\frac{100}{400} = -\frac{1}{4} = -0.25$$

At $p = \$10$ the demand is inelastic.

39. $x = f(p) = 40 - 2\sqrt{p} = 40 - 2p^{1/2}$ $f'(p) = -\frac{1}{2} \cdot 2p^{-1/2} = -\frac{1}{\sqrt{p}}$

$$E(p) = \frac{pf'(p)}{f(p)} = \frac{p\left[-\dfrac{1}{\sqrt{p}}\right]}{40 - 2\sqrt{p}} = -\frac{\sqrt{p}}{40 - 2\sqrt{p}}$$

(a) $E(300) = -\dfrac{\sqrt{300}}{40 - 2\sqrt{300}} = -3.232$

At $p = \$300$ the demand is elastic.

(b) Since the demand is elastic, increasing the price to \$310 will cause a decrease in revenue.

41. $y = 3x^4 - 2x^3 + x$ $f'(x) = \dfrac{dy}{dx} = 12x^3 - 6x^2 + 1$

$$dy = f'(x)dx = \left(12x^3 - 6x^2 + 1\right)dx$$

43.
$$y = \frac{3-2x}{1+x}$$

$$f'(x) = \frac{dy}{dx} = \frac{(-2)(1+x) - (3-2x)(1)}{(1+x)^2} = \frac{-2-2x-3+2x}{(1+x)^2} = \frac{-5}{(1+x)^2}$$

$$dy = f'(x)dx = -\frac{5}{(1+x)^2}\,dx$$

45. $f(x) = x^2 - 9$ $f'(x) = 2x$

$f(3) = 3^2 - 9 = 0$ $f'(3) = 2 \cdot 3 = 6$

$f(x) = f(3) + f'(3)(x-3) = 0 + 6(x-3)$

$f(x) = 6x - 18$

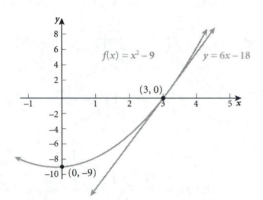

47. $x^2 + y^2 = 8$

$$2x\frac{dx}{dt} + 2y\frac{dy}{dt} = 0$$

$$2x\frac{dx}{dt} = -2y\frac{dy}{dt}$$

$$\frac{dx}{dt} = -\frac{2y}{2x}\frac{dy}{dt} = -\frac{y}{x}\frac{dy}{dt}$$

When $x = 2$, $y = 2$, and $\dfrac{dy}{dt} = 3$,

$$\frac{dx}{dt} = -\frac{2}{2} \cdot 3 = -3$$

49. $xy + 6x + y^3 = -2$

$$x\frac{dy}{dt} + y\frac{dx}{dt} + 6\frac{dx}{dt} + 3y\frac{dy}{dt} = 0$$

$$(x+3y)\frac{dy}{dt} + (y+6)\frac{dx}{dt} = 0$$

$$(x+3y)\frac{dy}{dt} = -(y+6)\frac{dx}{dt}$$

$$\frac{dy}{dt} = -\frac{y+6}{x+3y}\frac{dx}{dt}$$

When $x = 2$, $y = -3$, and $\dfrac{dx}{dt} = 3$, $\dfrac{dy}{dt} = -\dfrac{(-3)+6}{2+3(-3)} \cdot 3 = \dfrac{9}{7}$.

51.   **STEP 2** The variables are: **STEP 1**

   r = radius (in meters) of the balloon
   V = volume (in cubic meters) of the balloon
   S = surface area (in square meters) of the balloon
   t = time (in minutes).

STEP 3 The rates of change are $\dfrac{dr}{dt} \cdot \dfrac{dV}{dt}$, and $\dfrac{dS}{dt}$.

STEP 4 We know $\dfrac{dV}{dt} = 10 \text{ m}^3$ per minute. We want

$\dfrac{dS}{dt}$ when $r = 3$ m. The volume of a sphere is $V = \dfrac{4}{3}\pi r^3$ and the surface area is

$S = 4\pi r^2$.

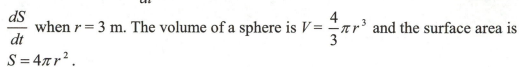

STEP 5 $\dfrac{dV}{dt} = 4\pi r^2 \dfrac{dr}{dt}$ and $\dfrac{dS}{dt} = 8\pi r \dfrac{dr}{dt}$

We use $\dfrac{dV}{dt}$ to solve for $\dfrac{dr}{dt}$ and then substitute it into $\dfrac{dS}{dt}$.

$$\frac{dr}{dt} = \frac{1}{4\pi r^2}\frac{dV}{dt}$$

$$\frac{dS}{dt} = 8\pi r \cdot \frac{1}{4\pi r^2}\frac{dV}{dt} = \frac{2}{r}\frac{dV}{dt}$$

STEP 6 When $r = 3$ and $\dfrac{dV}{dt} = 10$, $\dfrac{dS}{dt} = \dfrac{2}{3} \cdot 10 = \dfrac{20}{3} \approx 6.667$ square meters per minute.

53. $C(x) = 5x^2 + 1125$

(a) $\overline{C}(x) = \dfrac{C(x)}{x} = \dfrac{5x^2 + 1125}{x} = 5x + \dfrac{1125}{x}$

(b) To find the minimum average cost, we check the critical numbers.

$$\overline{C}'(x) = 5 - \frac{1125}{x^2} = \frac{5x^2 - 1125}{x^2}$$

$\overline{C}'(x)$ is unbounded at $x = 0$.

$\overline{C}'(x) = 0$ when $5x^2 - 1125 = 0$

$$x^2 - 225 = 0$$

$$x = \pm\sqrt{225} = \pm 15$$

Since x denotes the number of items produced, we disregard $x = -15$.

$\overline{C}''(x) = 2250x^{-3}$ and $\overline{C}''(15) = \dfrac{2250}{3375} = \dfrac{2}{3}$. By the Second Derivative Test, $x = 15$ locates a

local minimum. The minimum average cost is $\overline{C}(15) = 5(15) + \dfrac{1125}{15} = 150$.

(c) The marginal cost function is the derivative of $C(x)$.
  $C'(x) = 10x$

(d)

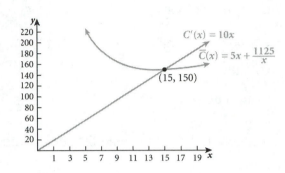

55. Profit = Revenue – Cost, $P(x) = R(x) - C(x)$, where $R(x) = xp$.
$$P(x) = x\left(62{,}402.50 - 0.5x^2\right) - \left(48{,}002.50x + 1500\right)$$
$$= 62{,}402.50x - 0.5x^3 - 48{,}002.50x - 1500$$
$$= -0.5x^3 + 14{,}400x - 1500$$

$$P'(x) = -1.5x^2 + 14{,}400$$
$$P'(x) = 0 \text{ when } -1.5x^2 + 14{,}400 = 0$$
$$x^2 = 9600$$
$$x = \pm\sqrt{9600} = \pm 97.98$$

Since x denotes the units to be sold, we disregard $x = -98$, and use $x = 98$ in the second derivative.
$$P''(x) = -3x \qquad\qquad P''(98) = -3(98) = -294 < 0$$

From the Second Derivative Test $x = 98$ locates a local maximum. 98 units must be sold to maximize profit.

57. **STEP 1** We want to minimize the cost of the material A which is used to make the can.

STEP 2 Let r denote the radius of the top and bottom of the can, and
    let h denote the height of the can.

STEP 3 $V = \pi r^2 h$

$$500 = \pi r^2 h$$

$$h = \frac{500}{\pi r^2}$$

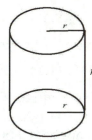

STEP 4 Since the cost of top, bottom and sides are the same, we need to minimize the surface area of the can.

$$S = 2\left(\pi r^2\right) + 2\pi r h$$

$$S(r) = 2\pi r^2 + 2\pi r \left(\frac{500}{\pi r^2}\right)$$

$$= 2\pi r^2 + \frac{1000}{r}$$

The domain of S is $\{r \mid r > 0\}$.

STEP 5 Differentiate S and find the critical numbers.

$$S'(r) = 4\pi r - \frac{1000}{r^2} = \frac{4\pi r^3 - 1000}{r^2}$$

We solve $S'(x) = 0$ to find the critical numbers.

$$4\pi r^3 - 1000 = 0$$

$$r^3 = \frac{1000}{4\pi}$$

$$r = \sqrt[3]{\frac{250}{\pi}} \approx 4.301$$

We use the Second Derivative Test to see if $r = 4.301$ locates a maximum or a minimum.

$$S''(r) = 4\pi - (-2)1000 r^{-3} = 4\pi + \frac{2000}{r^3}$$

$$S''\left(\sqrt[3]{\frac{250}{\pi}}\right) = 4\pi + \frac{2000}{\left(\sqrt[3]{\dfrac{250}{\pi}}\right)^3} = 4\pi + 8\pi = 12\pi > 0$$

We conclude that the cost is minimized when the can has a radius of 4.301 centimeters and a height of $\dfrac{500}{\pi r^2} \approx \dfrac{500}{\pi(4.301)^2} = 8.604$ centimeters.

59. $A = \pi r^2 \qquad dA = 2\pi r\, dr$

If $r = 10$ and changes to 8, then $dr = \Delta r = -2$, and

$$dA = 2\pi(10)(-2) = -40\pi \approx -125.66.$$

The area of the burn decreases by approximately 125.66 square centimeters.

61. $D(x) = -4x^3 - 3x^2 + 2000$ $dD = -(12x^2 + 6x)dx$

(a) If $x = 1.50$ and changes to 2.00, then $dx = \Delta x = 0.50$, and
$$dD = -\left(12(1.5)^2 + 6(1.5)\right)(0.50) = -18$$
The demand for peanuts will decrease by approximately 1800 pounds.

(b) If $x = 2.50$ and changes to 3.50, then $dx = \Delta x = 1.00$, and
$$dD = -\left(12(2.5)^2 + 6(2.5)\right)(1.00) = -90$$
The demand for peanuts will decrease by approximately 9000 pounds.

63. $c(x) = \dfrac{3x}{4 + 2x^2}$

$c'(x) = \dfrac{3(4 + 2x^2) - 3x(4x)}{(4 + 2x^2)^2} = \dfrac{12 + 6x^2 - 12x^2}{(4 + 2x^2)^2} = \dfrac{12 - 6x^2}{(4 + 2x^2)^2}$

$dc = \dfrac{12 - 6x^2}{(4 + 2x^2)^2}\, dx$

(a) If the time is $x = 1.2$ and changes to 1.3, then $dx = \Delta x = 0.1$, and
$$dc = \dfrac{12 - 6(1.2)^2}{\left(4 - 2(1.2)^2\right)^2}(0.1) = 0.0071$$
The concentration of the drug increases by approximately 0.0071.

(b) If the time is $x = 2$ and changes to 2.25, then $dx = \Delta x = 0.25$, and
$$dc = \dfrac{12 - 6(2)^2}{\left(4 - 2(2)^2\right)^2}(0.25) = -0.0208$$
The concentration of the drug decreases by approximately 0.0208.

65.   (b)

CHAPTER 14 PROJECT

(a) Since the demand remains constant throughout the year, on average there are $\dfrac{x}{2}$ vacuum cleaners in the store at any time t. So the average holding costs will be $H \cdot \dfrac{x}{2} = \dfrac{Hx}{2}$.

(b) If the total demand is denoted by D and x vacuum cleaners are shipped per order, x times the number of orders must equal the yearly demand D, or the number of orders must equal $\dfrac{D}{x}$. So the yearly reorder costs will be R times the number of orders placed or $R \cdot \dfrac{D}{x} = \dfrac{RD}{x}$.

(c) Total cost is the sum of ordering cost and holding cost.
$$C(x) = \frac{RD}{x} + \frac{Hx}{2}$$
The domain of C is $\{x \mid 0 < x \le 500\}$.

(d) First we find the derivative $C'(x)$ and the critical numbers.
$$C'(x) = -\frac{RD}{x^2} + \frac{H}{2} = \frac{-2RD + Hx^2}{2x^2}$$
$C'(x)$ is unbounded at $x = 0$, but 0 is not in the domain of C.
$C'(x) = 0$ when $-2RD + Hx^2 = 0$.
$$Hx^2 = 2RD$$
$$x^2 = \frac{2RD}{H}$$
$$x = \sqrt{\frac{2RD}{H}}$$

$$C''(x) = \frac{2RD}{x^3}$$
$$C''\left(\sqrt{\frac{2RD}{H}}\right) = \frac{2RD}{\left(\sqrt{\dfrac{2RD}{H}}\right)^3} = 2RD \cdot \frac{H^{3/2}}{(2RD)^{3/2}} = \frac{H^{3/2}}{\sqrt{2RD}} > 0$$

By the Second Derivative Test, we see that cost C is minimized when $x = \sqrt{\dfrac{2RD}{H}}$.

(e) Assuming $D = 500$ vacuum cleaners, $H = \$10$/vacuum cleaner, and $R = \$40$/order, the lot size that will minimize cost is
$$x = \sqrt{\frac{2RD}{H}} = \sqrt{\frac{2 \cdot 40 \cdot 500}{10}} = \sqrt{4000} \approx 63.24$$
Ordering 63 vacuum cleaners at a time will minimize cost.

If 63 vacuum cleaners are ordered at a time, there needs to be 8 orders placed per year.

(f) If $H = \$3$/vacuum cleaner, the lot size that will minimize cost is
$$x = \sqrt{\frac{2 \cdot 40 \cdot 500}{3}} = \sqrt{\frac{40,000}{3}} \approx 115.47$$
115 vacuum cleaners should be ordered at a time to minimize cost.

If 115 vacuum cleaners are ordered, there needs to be 5 orders placed per year.

(g) If the cost C is revised to reflect shipping costs, then

$$C(x) = \frac{Hx}{2} + \frac{D(R+Sx)}{x} = \frac{Hx}{2} + \frac{DR}{x} + \frac{DSx}{x} = \frac{Hx}{2} + \frac{DR}{x} + DS$$

$$C'(x) = \frac{H}{2} - \frac{DR}{x^2} = \frac{Hx^2 - 2DR}{2x^2}$$

$C'(x) = 0$ when $Hx^2 - 2DR = 0$ or when $x = \sqrt{\dfrac{2RD}{H}}$.

The lot size that will minimize the total cost is still $\sqrt{\dfrac{2RD}{H}}$.

MATHEMATICAL QUESTIONS FROM THE PROFESSIONAL EXAMS

1. **(b)** The marginal cost function is the derivative of C.
$$C'(X) = 6X^2 + 8X + 3$$

3. **(a)**

5. **(a)**

7. **(d)** $D = E^2(100 - I)$

$$D'(t) = -E^2 \frac{dI}{dt} + 2E(100 - I)\frac{dE}{dt} = -(95)^2(3) + 2(95)(100 - 6)(2) = 8645$$

9. **(c)** $\quad f(x) = \dfrac{1}{6}x^3 - 2x \qquad f'(x) = \dfrac{1}{2}x^2 - 2 \qquad f''(x) = x$

f is decreasing when $f'(x) < 0$ or when $-2 < x < 2$.

f is concave up when $f''(x) > 0$ or when $x > 0$.

So f is both decreasing and concave up on the interval $(0, 2)$.

11. **(c)** $\quad f'(x) = \dfrac{(3xe^{3x} + e^{3x})(1+x) - (xe^{3x})(1)}{(1+x)^2}$

$$f'(1) = \frac{(3e^3 + e^3)(1+1) - (e^3)}{(1+1)^2} = \frac{7e^3}{4}$$

13. (a) Using similar triangles, we find

$$\frac{h}{r} = \frac{10}{4}$$

$$r = \frac{2}{5}h$$

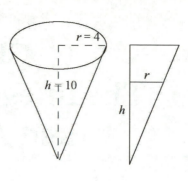

$$V = \frac{1}{3}\pi r^2 \cdot h = \frac{1}{3}\pi\left(\frac{2}{5}h\right)^2 \cdot h = \frac{4}{75}\pi h^3$$

$$\frac{dV}{dt} = \frac{4}{25}\pi h^2 \frac{dh}{dt}$$

$$\frac{dh}{dt} = \frac{25}{4\pi h^2}\frac{dV}{dt}$$

When $h = 5$, $\dfrac{dh}{dt} = \dfrac{25}{4\pi(5)^2} \cdot 2 = \dfrac{1}{2\pi}$

15.

(d) $\quad r = \dfrac{\Delta s}{\Delta t} = \dfrac{ds}{dt}$

We are told that $s_2 = (s_1)^2$, $r_2 = \dfrac{ds_1}{dt} = \dfrac{d}{dx}(s_1^2) = 2s_1 \cdot \dfrac{ds_1}{dt} = 2 \cdot 9 \cdot 3 = 54$

Chapter 15
The Integral of a Function
and Applications

15.1 Antiderivatives; the Indefinite Integral; Marginal Analysis

1.
$$F(x) = \frac{x^4}{4} + K$$

3.
$$F(x) = \frac{2x^2}{2} + 3x + K = x^2 + 3x + K$$

5. $F(x) = 4\ln|x| + K$

7.
$$f(x) = \sqrt[3]{x} = x^{1/3}$$
$$F(x) = \frac{x^{4/3}}{\frac{4}{3}} + K = \frac{3x^{4/3}}{4} + K$$

9. $\int 3\,dx = 3x + K$

11.
$$\int x\,dx = \frac{x^2}{2} + K$$

13.
$$\int x^{1/3}\,dx = \frac{x^{4/3}}{\frac{4}{3}} + K = \frac{3x^{4/3}}{4} + K$$

15.
$$\int x^{-2}\,dx = \frac{x^{-1}}{-1} + K = -\frac{1}{x} + K$$

17.
$$\int x^{-1/2}\,dx = \frac{x^{1/2}}{\frac{1}{2}} + K = 2x^{1/2} + K$$

19.
$$\int (2x^3 + 5x)\,dx = \frac{2x^4}{4} + \frac{5x^2}{2} + K = \frac{x^4}{2} + \frac{5x^2}{2} + K$$

21.
$$\int (x^2 + 2e^x)\,dx = \frac{x^3}{3} + 2e^x + K$$

23.
$$\int (x^3 - 2x^2 + x - 1)\,dx = \frac{x^4}{4} - \frac{2x^3}{3} + \frac{x^2}{2} - x + K$$

25.
$$\int \left(\frac{x-1}{x}\right)dx = \int \left(1 - \frac{1}{x}\right)dx = x - \ln|x| + K$$

27.
$$\int \left(2e^x - \frac{3}{x}\right) dx = 2e^x - 3\ln|x| + K$$

29.
$$\int \left(\frac{3\sqrt{x}+1}{\sqrt{x}}\right) dx = \int (3 + x^{-1/2}) dx = 3x + \frac{x^{1/2}}{\frac{1}{2}} + K = 3x + 2\sqrt{x} + K$$

31.
$$\int \frac{x^2-4}{x+2} dx = \int \frac{(x-2)(x+2)}{x+2} dx = \int (x-2) dx = \frac{x^2}{2} - 2x + K$$

33.
$$\int x(x-1) dx = \int (x^2 - x) dx = \frac{x^3}{3} - \frac{x^2}{2} + K$$

35.
$$\int \left(\frac{3x^5+2}{x}\right) dx = \int \left(3x^4 + \frac{2}{x}\right) dx = \frac{3x^5}{5} + 2\ln|x| + K$$

37.
$$\int \frac{4e^x + e^{2x}}{e^x} dx = \int (4 + e^x) dx = 4x + e^x + K$$

39. $R'(x) = 600 \qquad R(x) = 600x + K$
If $R = 0$ when $x = 0$, then $K = 0$, and $R(x) = 600x$.

41. $R'(x) = 20x + 5 \qquad R(x) = 10x^2 + 5x + K$
If $R = 0$ when $x = 0$, then $K = 0$, and $R(x) = 10x^2 + 5x$.

43.
$$C(x) = \int C'(x) dx = \int (14x - 2800) dx = \frac{14x^2}{2} - 2800x + K$$
Fixed cost $= K = \$4300$ so $C(x) = 7x^2 - 2800x + 4300$.
Cost is minimum when $C'(x) = 0$.
$$14x - 2800 = 0$$
$$14x = 2800$$
$$x = 200$$
Cost is minimum when 200 units are produced.

45.
$$C(x) = \int (20x - 8000) dx = \frac{20x^2}{2} - 8000x + K$$
Fixed cost $= K = \$500$ so $C(x) = 10x^2 - 8000x + 500$.

Cost is minimum when $C'(x) = 0$.
$$20x - 8000 = 0$$
$$20x = 8000$$
$$x = 400$$
Cost is minimum when 400 units are produced.

47.

(a) $C(x) = \int (1000 - 20x + x^2)\,dx = 1000x - \dfrac{20x^2}{2} + \dfrac{x^3}{3} + K$

Fixed cost $= K = \$9000$, so $C(x) = 1000x - 10x^2 + \dfrac{x^3}{3} + 9000$

(b) Revenue is the product of the price and the number of items sold, $R = px$.

$R = R(x) = 3400x$

(c) Profit is the difference between revenue and cost, $P = R - C$.

$$P = P(x) = 3400x - \left(1000x - 10x^2 + \dfrac{x^3}{3} + 9000\right)$$

$$= 3400x - 1000x + 10x^2 - \dfrac{x^3}{3} - 9000 = -\dfrac{x^3}{3} + 10x^2 + 2400x - 9000$$

(d) The maximum profit occurs at a critical number.

$$P'(x) = -x^2 + 20x + 2400$$

$P'(x) = 0$ when $-x^2 + 20x + 2400 = 0$

$$x^2 - 20x - 2400 = 0$$
$$(x - 60)(x + 40) = 0$$
$$x - 60 = 0 \quad \text{or} \quad x + 40 = 0$$
$$x = 60 \quad \text{or} \quad x = -40$$

Since x denotes the number of units produced we disregard -40.

$$P''(x) = -2x + 20 \qquad\qquad P''(60) = -2(60) + 20 = -100 < 0$$

By the Second Derivative Test, selling $x = 60$ units yields maximum profit.

(e) $P(60) = -\dfrac{60^3}{3} + 10 \cdot 60^2 + 2400 \cdot 60 - 9000 = 99{,}000$

The maximum profit is $\$99{,}000$.

(f)

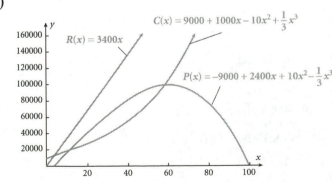

49. Let P denote the prison population. Then

$$P = P(t) = \int (7000t + 20{,}000)\,dt = \frac{7000t^2}{2} + 20{,}000t + K$$

We let t denote time (in years), and letting $t = 0$ represent 1998, we find K.

$$P(0) = 3500(0)^2 + 20{,}000(0) + K = 592{,}462$$
$$K = 592{,}462$$
$$P(t) = 3500t^2 + 20{,}000t + 592{,}462$$

In 2008 $t = 10$, and

$$P(10) = 3500\,(10)^2 + 20{,}000\,(10) + 592{,}462 = 1{,}142{,}462$$

According to this model in 2008 there will be 1,142,462 inmates in the United States.

51. Let P denote the population of the town. Then

$$P'(t) = 2 + t^{4/5} \qquad P(t) = \int P'(t)\,dt = \int (2 + t^{4/5})\,dt = 2t + \frac{t^{9/5}}{\frac{9}{5}} + K$$

$$= 2t + \frac{5t^{9/5}}{9} + K$$

We let t denote time (in months), and let $t = 0$ represent the time when $P = 20{,}000$.
Since $P(0) = 20{,}000$, $K = 20{,}000$. In 10 months $t = 10$, and the population will be

$$P(10) = 2(10) + \frac{5 \cdot 10^{9/5}}{9} + 20{,}000 = 20{,}055 \text{ people.}$$

53. Let P denote the voting population (in thousands). Then

$$P = P(t) = \int (2.2t - 0.8t^2)\,dt = \frac{2.2t^2}{2} - \frac{0.8t^3}{3} + K$$

If t denotes time (in years), and $P(0) = 20$, then we find K

$$P(0) = 1.1 \cdot 0^2 - \frac{0.8}{3} \cdot 0^3 + K = 20$$
$$K = 20$$

In 3 years, $t = 3$, and

$$P(3) = 1.1 \cdot 3^2 - \frac{0.8}{3} \cdot 3^3 + 20 = 22.7$$

In three years the voting population will be 22,700 citizens.

55. Let f denote the amount of end product present in the reaction. Then

$$f = \int \left(\frac{\sqrt{t} - 1}{t} \right) dt = \int \left(\frac{1}{\sqrt{t}} - \frac{1}{t} \right) dt = \int t^{-1/2}\,dt - \int \frac{1}{t}\,dt = \frac{t^{1/2}}{\frac{1}{2}} - \ln|t| + K = 2t^{1/2} - \ln|t| + K$$

Since the reaction started at $t = 1$, there is no end product at that time. So we find K

$$f(1) = 2 \cdot 1^{1/2} - \ln 1 + K = 0$$

$$2 - 0 + K = 0$$
$$K = -2$$

After 4 minutes, $t = 4$, and

$$f(4) = 2 \cdot 4^{1/2} - \ln 4 - 2 = 4 - 2 \ln 2 - 2 = 2 - 2 \ln 2$$

There will be $2 - 2 \ln 2$ milligrams of end product after 4 minutes of reaction.

57.   Let V denote the amount of water in the reservoir at time t. Then

$$V = V(t) = \int \left(15,000 - \frac{5}{2}t\right) dt = \int 15,000 \, dt - \frac{5}{2} \int t \, dt = 15,000t - \frac{5t^2}{4} + K$$

If initially (at $t = 0$) there are 100,000 gallons in the reservoir, we find K

$$V(0) = 15,000 \cdot 0 - \frac{5}{4} \cdot 0^2 + K = 100,000$$

$$K = 100,000$$

$$V(t) = 15,000 \, t - \frac{5}{4} t^2 + 100,000$$

The reservoir will be empty when $V(t) = 0$.

$$15,000 \, t - \frac{5}{4} \, t^2 + 100,000 = 0$$

$$-5t^2 + 60,000 \, t + 400,000 = 0$$

$$t^2 - 12,000 \, t - 80,000 = 0$$

$$t = \frac{12,000 \pm \sqrt{12,000^2 - 4(-80,000)}}{2}$$

$$t \approx 12,006.66$$

(We disregard the negative value of t.)

The reservoir will be empty in approximately 12,007 hours (500.28 days).

15.2 Integration Using Substitution

1.   Let $u = 2x + 1$. Then $du = 2 \, dx$ so $dx = \frac{1}{2} \, du$

$$\int (2x+1)^5 \, dx = \frac{1}{2} \int u^5 \, du = \frac{1}{2} \frac{u^6}{6} + K = \frac{(2x+1)^6}{12} + K$$

3.   Let $u = 2x - 3$. Then $du = 2 \, dx$, so $dx = \frac{1}{2} \, du$.

$$\int e^{2x-3} \, dx = \frac{1}{2} \int e^u \, du = \frac{1}{2} e^u + K = \frac{e^{2x-3}}{2} + K$$

5.
Let $u = -2x + 3$. Then $du = -2\,dx$, so $dx = -\dfrac{1}{2}\,du$.

$$\int (-2x+3)^{-2}\,dx = -\frac{1}{2}\int u^{-2}\,du = -\frac{1}{2}\frac{u^{-1}}{-1} + K$$

$$= \frac{1}{2}(-2x+3)^{-1} + K = \frac{1}{2(-2x+3)} + K = \frac{1}{6-4x} + K$$

7.
Let $u = x^2 + 4$. Then $du = 2x\,dx$, so $x\,dx = \dfrac{1}{2}\,du$.

$$\int x(x^2+4)^2\,dx = \frac{1}{2}\int u^2\,du = \frac{1}{2}\cdot\frac{u^3}{3} + K = \frac{1}{2}\cdot\frac{(x^2+4)^3}{3} + K = \frac{(x^2+4)^3}{6} + K$$

9.
Let $u = x^3 + 1$. Then $du = 3x^2\,dx$, so $x^2\,dx = \dfrac{1}{3}du$.

$$\int e^{x^3+1}x^2\,dx = \frac{1}{3}\int e^u\,du = \frac{1}{3}\cdot e^u + K = \frac{1}{3}\cdot e^{x^3+1} + K = \frac{e^{x^3+1}}{3} + K$$

11.
Let $u = x^2$. Then $du = 2x\,dx$, so $x\,dx = \dfrac{1}{2}\,du$.

$$\int \left(e^{x^2}+e^{-x^2}\right)x\,dx = \int e^{x^2}x\,dx + \int e^{-x^2}x\,dx = \frac{1}{2}\int e^u\,du + \frac{1}{2}\int e^{-u}\,du = \frac{1}{2}\cdot e^u - \frac{1}{2}\cdot e^{-u} + K$$

$$= \frac{e^{x^2}}{2} - \frac{e^{-x^2}}{2} + K = \frac{e^{x^2}-e^{-x^2}}{2} + K$$

13.
Let $u = x^3 + 2$. Then $du = 3x^2\,dx$, so $x^2\,dx = \dfrac{1}{3}du$.

$$\int x^2(x^3+2)^6\,dx = \frac{1}{3}\int u^6\,du = \frac{1}{3}\cdot\frac{u^7}{7} + K = \frac{(x^3+2)^7}{21} + K$$

15.
Let $u = 1 + x^2$. Then $du = 2x\,dx$, so $x\,dx = \dfrac{1}{2}\,du$.

$$\int \frac{x}{\sqrt[3]{1+x^2}}\,dx = \int x(1+x^2)^{-1/3}\,dx = \frac{1}{2}\int u^{-1/3}\,du$$

$$= \frac{1}{2}\cdot\frac{u^{2/3}}{\frac{2}{3}} + K = \frac{3u^{2/3}}{4} + K$$

$$= \frac{3(1+x^2)^{2/3}}{4} + K$$

17. Let $u = x + 3$. Then $du = dx$ and $x = u - 3$.

$$\int x\sqrt{x+3}\, dx = \int (u-3)\sqrt{u}\, du = \int \left(u^{3/2} - 3u^{1/2}\right) du = \frac{u^{5/2}}{\frac{5}{2}} - \frac{3u^{3/2}}{\frac{3}{2}} + K$$

$$= \frac{2(x+3)^{5/2}}{5} - \frac{6(x+3)^{3/2}}{3} + K = \frac{2(x+3)^{5/2}}{5} - 2(x+3)^{3/2} + K$$

19. Let $u = e^x + 1$. Then $du = e^x\, dx$.

$$\int \frac{e^x}{e^x+1}\, dx = \int \frac{1}{u}\, du = \ln|u| + K = \ln\left(e^x + 1\right) + K$$

21. Let $u = \sqrt{x} = x^{1/2}$. Then $du = \frac{1}{2} x^{-1/2}\, dx = \frac{1}{2\sqrt{x}}\, dx$, so $2du = \frac{1}{\sqrt{x}}\, dx$.

$$\int \frac{e^{\sqrt{x}}}{\sqrt{x}}\, dx = 2 \int e^u\, du = 2 \cdot e^u + K = 2 e^{\sqrt{x}} + K$$

23. Let $u = x^{1/3} - 1$. Then $du = \frac{1}{3} x^{-2/3}\, dx = \frac{1}{3x^{2/3}}\, dx$, so $3\, du = = \frac{1}{x^{2/3}}\, dx$.

$$\int \frac{\left(x^{1/3}-1\right)^6}{x^{2/3}}\, dx = 3 \int u^6\, du = 3 \cdot \frac{u^7}{7} + K = \frac{3\left(x^{1/3}-1\right)^7}{7} + K$$

25. Let $u = x^2 + 2x + 3$. Then $du = (2x + 2)\, dx$, so $\frac{1}{2}\, du = (x + 1)\, dx$

$$\int \frac{(x+1)dx}{\left(x^2+2x+3\right)^2} = \frac{1}{2} \int \frac{1}{u^2}\, du = \frac{1}{2} \int u^{-2}\, du = \frac{1}{2} \frac{u^{-1}}{(-1)} + K = -\frac{1}{2u} + K$$

$$= -\frac{1}{2\left(x^2+2x+3\right)} + K = -\frac{1}{2x^2+4x+6} + K$$

27. Let $u = 1 + \sqrt{x}$. Then $du = \frac{1}{2\sqrt{x}}\, dx$, so $2\, du = \frac{1}{\sqrt{x}}\, dx$.

$$\int \frac{dx}{\sqrt{x}\left(1+\sqrt{x}\right)^4} = 2 \int \frac{1}{u^4}\, du = 2 \int u^{-4}\, du = 2 \cdot \frac{u^{-3}}{(-3)} + K = -\frac{2}{3u^3} + K$$

$$= -\frac{2}{3\left(1+\sqrt{x}\right)^3} + K$$

29. Let $u = 2x + 3$. Then $du = 2\, dx$, so $\frac{1}{2}\, du = dx$.

$$\int \frac{dx}{2x+3} = \frac{1}{2} \int \frac{1}{u}\, du = \frac{1}{2} \cdot \ln|u| = \frac{1}{2} \ln|2x+3| + K$$

31.

Let $u = 4x^2 + 1$. Then $du = 8x\,dx$, so $\dfrac{1}{8}\,du = x\,dx$.

$$\int \frac{x\,dx}{4x^2+1} = \frac{1}{8}\int \frac{du}{u} = \frac{1}{8}\cdot \ln|u| + K = \frac{1}{8}\ln(4x^2+1) + K$$

33.

Let $u = x^2 + 2x + 2$. Then $du = (2x+2)dx = 2(x+1)\,dx$, so $\dfrac{1}{2}\,du = (x+1)\,dx$

$$\int \frac{x+1}{x^2+2x+x}\,dx = \frac{1}{2}\int \frac{1}{u}\,du = \frac{1}{2}\ln|u| + K = \frac{1}{2}\ln(x^2+2x+x) + K$$

35. The value V of the car is the antiderivative of the depreciation rate with $V(0) = \$\,27{,}000$.

$$V = \int V'(t)\,dt = \int -6000e^{-0.5t}\,dt = -6000\int e^{-0.5t}\,dt$$

$$= -6000\left(\frac{1}{-0.5}e^{-0.5t}\right) + K = 12{,}000\,e^{-0.5t} + K$$

We use $V(0) = 27{,}000$ to determine K.

$$V(0) = 12{,}000\,e^{-0.5(0)} + K = 27{,}000$$
$$K = 27{,}000 - 12{,}000 = 15{,}000$$
$$V(t) = 12{,}000\,e^{-0.5t} + 15{,}000$$

After 2 years $t = 2$, and the car is worth
$$V(2) = 12{,}000\,e^{-0.5(2)} + 15{,}000 = 12{,}000\,e^{-1} + 15{,}000 = \$19{,}414.55$$
After 4 years $t = 4$, and the car is worth
$$V(4) = 12{,}000\,e^{-0.5(4)} + 15{,}000 = 12{,}000\,e^{-2} + 15{,}000 = \$16{,}624.02$$

37. (a) The budget B is the antiderivative of the growth rate with $B(0) = 68.6$ billion dollars.

$$B = \int B'(t)\,dt = \int 1.715\,e^{0.025t}\,dt = 1.715\int e^{0.025t}\,dt = 1.715\cdot\frac{1}{0.025}e^{0.025t} + K$$

We use $B(0) = 68.6$ to determine K.

$$B(0) = 68.6\,e^{0.025\,(0)} + K = 68.6$$
$$68.6 + K = 68.6$$
$$K = 0$$
$$B(t) = 68.6\,e^{0.025t}$$

(b) We need to find t so that $B(t) > 100$
$$68.6\,e^{0.025t} > 100$$
$$e^{0.025t} > \frac{100}{68.6} \approx 1.4577$$

Changing the exponentional to logarithmic form we find

$$\ln\left(\frac{100}{68.6}\right) = 0.025t$$

$$t = \frac{\ln\left(\dfrac{100}{68.6}\right)}{0.025} \approx 15.075$$

The budget will exceed $ 100 billion in just over 15 years, that is in 2016.

39. (a) The number of employees N is given by the antiderivative of $N'(t)$ with $N(0) = 400$.

$$N = \int N'(t)\, dt = \int 20e^{0.01t}\, dt = 20\int e^{0.01t}\, dt = 20 \cdot \frac{1}{0.01}e^{0.01t} + K = 2000e^{0.01t} + K$$

We use $N(0) = 400$ to determine K.

$$N(0) = 2000\, e^0 + K = 400$$
$$K = 400 - 2000 = -1600$$
$$N(t) = 2000\, e^{0.01t} - 1600$$

(b) We need to find t so that $N(t) = 800$.

$$N(t) = 2000\, e^{0.01t} - 1600 = 800$$
$$2000\, e^{0.01t} = 2400$$
$$e^{0.01t} = \frac{2400}{2000} = 1.2$$
$$\ln 1.2 = 0.01\, t$$
$$t = \frac{\ln 1.2}{0.01} = 18.232$$

It will take about 18.232 years for the number of employees to reach 800.

41. Let $u = ax + b$. Then $du = a\, dx$, so $\dfrac{1}{a}\, du = dx$.

$$\int (ax+b)^n\, dx = \frac{1}{a}\int u^n\, du = \frac{1}{a} \cdot \frac{u^{n+1}}{n+1} + K = \frac{u^{n+1}}{a(n+1)} + K = \frac{(ax+b)^{n+1}}{a(n+1)} + K$$

15.3 Integration by Parts

1. $$\int u\, dv = uv - \int v\, du$$

If $u = x$, then $du = dx$; and if $dv = e^{4x}\, dx$, then $v = \dfrac{1}{4}e^{4x}$.

$$\int x\, e^{4x}\, dx = \frac{1}{4}xe^{4x} - \int \frac{1}{4}e^{4x}\, dx = \frac{1}{4}xe^{4x} - \frac{1}{4} \cdot \frac{1}{4}e^{4x} + K$$

$$= \frac{1}{4}xe^{4x} - \frac{1}{16}e^{4x} + K$$

3. Choose $\quad u = x \quad$ and $\quad dv = e^{2x}\, dx$

Then $\quad du = dx \quad$ and $\quad v = \dfrac{1}{2} e^{2x}$

$$\int u\, dv = uv - \int v\, du$$

$$\int x e^{2x}\, dx = \frac{1}{2} x e^{2x} - \frac{1}{2} \int e^{2x}\, dx = \frac{1}{2} x e^{2x} - \frac{1}{4} e^{2x} + K$$

5. Choose $\quad u = x^2 \quad$ and $\quad dv = e^{-x}\, dx$

Then $\quad du = 2x\, dx \quad$ and $\quad v = -e^{-x}$

$$\int u\, dv = uv - \int v\, du$$

$$\int x^2 e^{-x}\, dx = x^2 \left(-e^{-x} \right) - \int -e^{-x} \cdot 2x\, dx$$

$$= -x^2 e^{-x} + 2 \int x e^{-x}\, dx$$

We use integration by parts once more. This time
choose $\quad u = x \quad$ and $\quad dv = e^{-x}\, dx$

Then $\quad du = dx \quad$ and $\quad v = -e^{-x}$

$$\int x^2 e^{-x}\, dx = -x^2 e^{-x} + 2 \left[x \cdot \left(-e^{-x} \right) - \int -e^{-x} \cdot dx \right]$$

$$= -x^2 e^{-x} - 2x e^{-x} + 2 \int e^{-x}\, dx$$

$$= -x^2 e^{-x} - 2x e^{-x} - 2 e^{-x} + K$$

$$= -e^{-x} \left(x^2 + 2x + 2 \right) + K$$

7. Choose $\quad u = \ln x \quad$ and $\quad dv = \sqrt{x}\, dx$

$\quad du = \dfrac{1}{x}\, dx \quad$ and $\quad v = \dfrac{2}{3} x^{3/2}$

$$\int u\, dv = uv - \int v\, du$$

$$\int \sqrt{x}\, \ln x\, dx = \ln x \cdot \frac{2}{3} x^{3/2} - \int \frac{2}{3} x^{3/2} \cdot \frac{1}{x}\, dx$$

$$= \frac{2}{3} x^{3/2} \ln x - \frac{2}{3} \int x^{1/2}\, dx$$

$$= \frac{2}{3} x^{3/2} \ln x - \frac{2}{3} \cdot \frac{x^{3/2}}{\frac{3}{2}} + K = \frac{2}{3} x^{3/2} \ln x - \frac{4}{9} x^{3/2} + K$$

9. Choose $\quad u = (\ln x)^2 \quad$ and $\quad dv = dx$

$\quad du = 2 \ln x \cdot \dfrac{1}{x}\, dx \quad$ and $\quad v = x$

$$= \frac{2}{x} \ln x \, dx$$

$$\int u \, dv = uv - \int v \, du$$

$$\int (\ln x)^2 \, dx = (\ln x)^2 \cdot x - \int \left(x \cdot \frac{2}{x} \ln x \right) dx$$

$$= x (\ln x)^2 - 2 \int \ln x \, dx$$

We use integration by parts once more.
This time choose $\quad u = \ln x \quad$ and $\quad dv = dx$

$$du = \frac{1}{x} \, dx \qquad \text{and} \qquad v = x$$

$$\int (\ln x)^2 \, dx = x (\ln x)^2 - 2 \left[x \ln x - \int x \cdot \frac{1}{x} \, dx \right]$$

$$= x (\ln x)^2 - 2x \ln x + 2 \int dx$$

$$= x (\ln x)^2 - 2x \ln x + 2x + K$$

11. Choose $\quad u = \ln 3x \qquad$ and $\qquad dv = x^2 \, dx$

$$du = \frac{1}{x} \, dx \qquad \text{and} \qquad v = \frac{x^3}{3}$$

$$\int u \, dv = uv - \int v \, du$$

$$\int x^2 \ln 3x \, dx = \ln 3x \cdot \frac{x^3}{3} - \int \frac{x^3}{3} \cdot \frac{1}{x} \, dx$$

$$= \frac{x^3 \ln 3x}{3} - \frac{1}{3} \int x^2 \, dx$$

$$= \frac{x^3 \ln 3x}{3} - \frac{1}{3} \cdot \frac{x^3}{3} + K$$

$$= \frac{x^3 \ln 3x}{3} - \frac{x^3}{9} + K = \frac{x^3}{9} \left[3 \ln 3x - 1 \right] + K$$

13. Choose $\quad u = (\ln x)^2 \qquad\qquad$ and $\qquad\qquad dv = x^2 \, dx$

$$du = 2 \ln x \cdot \frac{1}{x} \, dx \qquad \text{and} \qquad v = \frac{x^3}{3}$$

$$= \frac{2 \ln x}{x} \, dx$$

$$\int u \, dv = uv - \int v \, du$$

$$\int x^2 (\ln x)^2 \, dx = (\ln x)^2 \cdot \frac{x^3}{3} - \int \frac{x^3}{3} \cdot \frac{2\ln x}{x} \, dx$$

$$= \frac{x^3 (\ln x)^2}{3} - \frac{2}{3} \int x^2 \ln x \, dx$$

We use integration by parts a second time.
This time choose $u = \ln x$ and $dv = x^2 \, dx$

Then $du = \frac{1}{x} \, dx$ and $v = \frac{x^3}{3}$

$$\int x^2 (\ln x)^2 \, dx = \frac{x^3 (\ln x)^2}{3} - \frac{2}{3} \left[\ln x \cdot \frac{x^3}{3} - \int \left(\frac{x^3}{3} \cdot \frac{1}{x} \right) dx \right]$$

$$= \frac{x^3 (\ln x)^2}{3} - \frac{2}{3} \left[\frac{x^3 \ln x}{3} - \frac{1}{3} \int x^2 \, dx \right]$$

$$= \frac{x^3 (\ln x)^2}{3} - \frac{2}{3} \left[\frac{x^3 \ln x}{3} - \frac{1}{3} \cdot \frac{x^3}{3} + K \right]$$

$$= \frac{x^3 (\ln x)^2}{3} - \frac{2x^3 \ln x}{9} + \frac{2x^3}{27} + K$$

15. Choose $u = \ln x$ and $dv = x^{-3} \, dx$

$du = \frac{1}{x} \, dx$ and $v = \frac{x^{-2}}{-2}$

$$\int u \, dv = uv - \int v \, du$$

$$\int \frac{\ln x}{x^3} \, dx = \ln x \cdot \left(-\frac{x^{-2}}{2} \right) - \int -\frac{x^{-2}}{2} \cdot \frac{1}{x} \, dx = -\frac{x^{-2} \ln x}{2} + \frac{1}{2} \int x^{-3} \, dx$$

$$= -\frac{x^{-2} \ln x}{2} + \frac{1}{2} \cdot \frac{x^{-2}}{-2} + K = -\frac{x^{-2} \ln x}{2} - \frac{x^{-2}}{4} + K$$

$$= -\frac{2 \ln x + 1}{4x^2} + K$$

17. The function P is the antiderivative of $P'(t)$ with $P(0) = 5000$.

$$P = \int P'(t) \, dt = \int \left(90\sqrt{t} - 100 t e^{-t} \right) dt = 90 \int \sqrt{t} \, dt - 100 \int \left(t e^{-t} \right) dt$$

We use integration by parts to integrate the second integral.
Choose $u = t$ $dv = e^{-t} \, dt$
$du = dt$ $v = -e^{-t}$

$$P = 90 \frac{t^{3/2}}{\frac{3}{2}} - 100 \left[-t e^{-t} - \int -e^{-t} \, dt \right] = 60 t^{3/2} + 100 t e^{-t} + 100 e^{-t} + K$$

We use $P(0) = 5000$ to determine K.

$$P(0) = 60(0)^{3/2} + 100(0)e^0 + 100e^0 + K = 5000$$
$$100 + K = 5000$$
$$K = 4900$$
$$P(t) = 60t^{3/2} + 100te^{-t} + 100e^{-t} + 4900$$

In 4 days $t = 4$, and $P(4) = 60(4)^{3/2} + 100(4)e^{-4} + 100e^{-4} + 4900 = 5389$ ants.

In one week $t = 7$, and $P(7) = 60(7)^{3/2} + 100(7)e^{-7} + 100e^{-7} + 4900 = 6012$ ants.

15.4 The Definite; Learning Curves; Total Sales Over Time

1.
$$\int_1^2 (3x - 1)\, dx = \left(\frac{3x^2}{2} - x\right)\Bigg|_1^2 = \left(\frac{12}{2} - 2\right) - \left(\frac{3}{2} - 1\right) = 6 - 2 - \frac{3}{2} + 1 = \frac{7}{2}$$

3.
$$\int_0^1 (3x^2 + e^x)\, dx = \frac{3x^3}{3} + e^x\Bigg|_0^1 = (1^3 + e^1) - (0^3 + e^0) = 1 + e - 0 - 1 = e$$

5.

$$\int_0^1 \sqrt{u}\, du = \int_0^1 u^{1/2}\, du = \frac{u^{3/2}}{\frac{3}{2}}\Bigg|_0^1 = \frac{2u^{3/2}}{3}\Bigg|_0^1 = \frac{2(1)^{3/2}}{3} - \frac{2(0)^{2/3}}{3} = \frac{2}{3}$$

7.

$$\int_0^1 \left(t^2 - t^{3/2}\right) dt = \left(\frac{t^3}{3} - \frac{t^{5/2}}{\frac{5}{2}}\right)\Bigg|_0^1 = \left(\frac{t^3}{3} - \frac{2t^{5/2}}{5}\right)\Bigg|_0^1 = \left(\frac{1^3}{3} - \frac{2(1)^{5/2}}{5}\right) - \left(\frac{0^3}{3} - \frac{2(0)^{5/2}}{5}\right)$$

$$= \frac{1}{3} - \frac{2}{5} - 0 = \frac{5-6}{15} = -\frac{1}{15}$$

9.

$$\int_{-2}^3 (x-1)(x+3)\, dx = \int_{-2}^3 \left(x^2 + 2x - 3\right) dx = \left(\frac{x^3}{3} + \frac{2x^2}{2} - 3x\right)\Bigg|_{-2}^3$$

$$= \left[\frac{3^3}{3} + 3^2 - 3(3)\right] - \left[\frac{(-2)^3}{3} + (-2)^2 - 3(-2)\right] = 9 + 9 - 9 + \frac{8}{3} - 4 - 6$$

$$= -1 + \frac{8}{3} = \frac{5}{3}$$

11.

$$\int_1^2 \frac{x^2 - 1}{x^4}\, dx = \int_1^2 \left(x^{-2} - x^{-4}\right) dx = \left(\frac{x^{-1}}{-1} - \frac{x^{-3}}{-3}\right)\Bigg|_1^2 = \left(-\frac{1}{x} + \frac{1}{3x^3}\right)\Bigg|_1^2$$

$$= \left[-\frac{1}{2} + \frac{1}{3(2)^3}\right] - \left[-\frac{1}{1} + \frac{1}{3(1)^3}\right] = -\frac{1}{2} + \frac{1}{24} + 1 - \frac{1}{3} = \frac{5}{24}$$

13.

$$\int_1^8 \left(\sqrt[3]{t^2} + \frac{1}{t}\right) dt = \int_1^8 \left(t^{2/3} + \frac{1}{t}\right) dt = \left(\frac{t^{5/3}}{\frac{5}{3}} + \ln|t|\right)\Bigg|_1^8 = \left(\frac{3t^{5/3}}{5} + \ln|t|\right)\Bigg|_1^8$$

$$= \left[\frac{3(8)^{5/3}}{5} + \ln 8\right] - \left[\frac{3(1)^{5/3}}{5} + \ln 1\right]$$

$$= \frac{3(32)}{5} + \ln 8 - \frac{3}{5} - 0 = \frac{96-3}{5} + \ln 8 = \frac{93}{5} + \ln 8 = 3 \ln 2 + \frac{93}{5}$$

15.

$$\int_1^4 \frac{x+1}{\sqrt{x}}\, dx = \int_1^4 \left(\frac{x}{\sqrt{x}} + \frac{1}{\sqrt{x}}\right) dx = \int_1^4 \left(x^{1/2} + x^{-1/2}\right) dx = \left(\frac{x^{3/2}}{\frac{3}{2}} + \frac{x^{1/2}}{\frac{1}{2}}\right)\Bigg|_1^4 = \left(\frac{2x^{3/2}}{3} + 2x^{1/2}\right)\Bigg|_1^4$$

$$= \left[\frac{2(4)^{3/2}}{3} + 2(4)^{1/2} \right] - \left[\frac{2(1)^{3/2}}{3} + 2(1)^{1/2} \right] = \frac{16}{3} + 4 - \frac{2}{3} - 2 = \frac{14+6}{3} = \frac{20}{3}$$

17. $\int_3^3 \left(5x^4 + 1\right)^{3/2} dx = 0$ Property 4 $\int_a^a f(x)\, dx = 0$

19.
$$\int_{-1}^1 (x+1)^2\, dx = \int_{-1}^1 (x^2 + 2x + 1)\, dx = \left(\frac{x^3}{3} + \frac{2x^2}{2} + x \right)\Bigg|_{-1}^1$$

$$= \left[\frac{1^3}{3} + 1^2 + 1 \right] - \left[\frac{(-1)^3}{3} + (-1)^2 + (-1) \right] = \frac{1}{3} + 2 + \frac{1}{3} - 1 + 1 = \frac{8}{3}$$

21.
$$\int_1^e \left(x - \frac{1}{x} \right) dx = \frac{x^2}{2} - \ln|x| \Bigg|_1^e = \left[\frac{e^2}{2} - \ln e \right] - \left[\frac{1}{2} - \ln 1 \right] = \frac{e^2}{2} - 1 - \frac{1}{2} + 0 = \frac{e^2}{2} - \frac{3}{2} = \frac{e^2 - 3}{2}$$

23.
$$\int_0^1 e^{-x}\, dx = \frac{e^{-x}}{-1} \Bigg|_0^1 = \left[-e^{-1} \right] - \left[-e^0 \right] = -e^{-1} + 1 = 1 - \frac{1}{e}$$

25. We use the method of substitution to evaluate the integral.
Let $u = x + 1$. Then $du = dx$.

We adjust the limits of integration.
When $x = 1$, $u = 1 + 1 = 2$, and when $x = 3$, $u = 3 + 1 = 4$. So
$$\int_1^3 \frac{dx}{x+1} = \int_2^4 \frac{du}{u} = \ln|u| \Big|_2^4 = \ln 4 - \ln 2 = \ln 2$$

27. We use the method of substitution to evaluate the integral.
Let $u = x^{3/2} + 1$. Then $du = \frac{3}{2} x^{1/2}\, dx$. So $\frac{2}{3} du = x^{1/2}\, dx = \sqrt{x}\, dx$

We adjust the limits of integration.
When $x = 0$, $u = 0^{3/2} + 1 = 1$, and when $x = 1$, $u = 1^{3/2} + 1 = 2$. So
$$\int_0^1 \frac{\sqrt{x}}{x^{3/2} + 1}\, dx = \frac{2}{3} \int_1^2 \frac{du}{u} = \frac{2}{3} \ln|u| \Bigg|_1^2 = \frac{2}{3} (\ln 2 - \ln 1) = \frac{2}{3} \ln 2$$

29. We integrate by parts. We choose $u = x$ and $dv = e^{2x}\, dx$. Then $du = dx$ and $v = \frac{1}{2} e^{2x}$.

$$\int_1^3 x \cdot e^{2x}\, dx = \left[x \cdot \frac{1}{2} e^{2x} \right]_1^3 - \int_1^3 \frac{1}{2} e^{2x}\, dx = \frac{xe^{2x}}{2} \Bigg|_1^3 - \frac{1}{4} e^{2x} \Bigg|_1^3$$

$$= \left[\frac{3e^6}{2} - \frac{e^2}{2}\right] - \left[\frac{e^6}{4} - \frac{e^2}{4}\right] = \frac{6e^6 - 2e^2 - e^6 + e^2}{4} = \frac{5e^5 - e^2}{4}$$

31.

We integrate by parts. Choose $u = x$ and $dv = e^{-3x}dx$. Then $du = dx$, and $v = -\frac{1}{3}e^{-3x}$.

$$\int_1^2 xe^{-3x}dx = x\left(-\frac{1}{3}e^{-3x}\right)\bigg|_1^2 - \int_1^2 -\frac{1}{3}e^{-3x}\,dx = -\frac{xe^{-3x}}{3}\bigg|_1^2 + \frac{1}{3}\int_1^2 e^{-3x}\,dx$$

$$= \left[-\frac{2e^{-6}}{3} + \frac{e^{-3}}{3}\right] + \frac{1}{3}\left[-\frac{1}{3}e^{-3x}\bigg|_1^2\right] = -\frac{2e^{-6}}{3} + \frac{e^{-3}}{3} + \frac{1}{3}\left[-\frac{e^{-6}}{3} + \frac{e^{-3}}{3}\right]$$

$$= -\frac{2e^{-6}}{3} + \frac{e^{-3}}{3} - \frac{e^{-6}}{9} + \frac{e^{-3}}{9} = -\frac{6e^{-6}}{9} + \frac{3e^{-3}}{9} - \frac{e^{-6}}{9} + \frac{e^{-3}}{9}$$

$$= -\frac{7e^{-6}}{9} + \frac{4e^{-3}}{9} = -\frac{7}{9e^6} + \frac{4}{9e^3}$$

33.

Integrating by parts, we choose $u = \ln x$ and $dv = dx$. Then $du = \frac{1}{x}dx$ and $v = x$.

$$\int_1^5 \ln x\,dx = \ln x \cdot x\bigg|_1^5 - \int_1^5 x \cdot \frac{1}{x}\,dx = x\ln x\bigg|_1^5 - \int_1^5 dx = x\ln x\bigg|_1^5 - x\bigg|_1^5$$

$$= \left[5\ln 5 - 1\ln 1\right] - \left[5 - 1\right] = 5\ln 5 - 0 - 4 = 5\ln 5 - 4 \approx 4.047$$

35. $\displaystyle\int_2^2 e^{x^2}\,dx = 0$ $\qquad$ Property 4 $\displaystyle\int_a^a f(x)\,dx = 0$

37. $\displaystyle\int_0^1 e^{-x^2}\,dx + \int_1^0 e^{-x^2}\,dx = \int_0^1 e^{-x^2}\,dx - \int_0^1 e^{-x^2}\,dx = 0$ $\quad$ Property 3 $\displaystyle\int_a^b f(x)\,dx = -\int_b^a f(x)\,dx$

39. $\displaystyle\int_1^3 \left[f(x) + g(x)\right]\,dx = \int_1^3 f(x)\,dx + \int_1^3 g(x)\,dx = 4 + (-2) = 2$
$\qquad\qquad\qquad\uparrow$

$\qquad$ (Property 7: $\displaystyle\int_a^b\left[f(x) \pm g(x)\right]\,dx = \int_a^b f(x)\,dx \pm \int_a^b g(x)\,dx$)

41. $\displaystyle\int_3^6 8f(x)\,dx = 8\int_3^6 f(x)\,dx = 8 \cdot 8 = 64$
$\qquad\qquad\uparrow$

$\qquad$ (Property 6: $\displaystyle\int_a^b cf(x)\,dx = c\int_a^b f(x)\,dx$)

43. $\int_3^6 \left[3f(x) + 4g(x) \right] dx = \int_3^6 3f(x)\, dx + \int_3^6 4g(x)\, dx$

(Property 7: $\int_a^b \left[f(x) \pm g(x) \right] dx = \int_a^b f(x)\, dx \pm \int_a^b g(x)\, dx$)

$\qquad = 3\int_3^6 f(x)\, dx + 4\int_3^6 g(x)\, dx = 3 \cdot 8 + 4 \cdot 3 = 36$

(Property 6: $\int_a^b c\, f(x)\, dx = c \int_a^b f(x)\, dx$)

45. $\int_1^6 f(x)\, dx = \int_1^3 f(x)\, dx + \int_3^6 f(x)\, dx = 4 + 8 = 12$

(Property 5: $\int_a^b f(x)\, dx = \int_a^c f(x)\, dx + \int_c^b f(x)\, dx$)

47. $C'(x) = 6x^2 - 100x + 1000$

The increase in cost in raising production from 100 units to 110 units is

$C(110) - C(100) = \int_{100}^{110} C'(x)\, dx$

$\qquad = \int_{100}^{110} \left(6x^2 - 100x + 1000 \right) dx$

$\qquad = \left(\dfrac{6x^3}{3} - \dfrac{100x^2}{2} + 1000x \right) \Big|_{100}^{110}$

$\qquad = \left[2(110)^3 - 50(110)^2 + 1000(110) \right] - \left[2(100)^3 - 50(100)^2 + 1000(100) \right]$

$\qquad = \$\, 567{,}000$

49. (a) $f(x) = 1272 x^{-0.35}$

$\int_{30}^{80} 1272 x^{-0.35}\, dx = \dfrac{1272 x^{0.65}}{0.65} \Big|_{30}^{80} = 1956.92 \left(80^{0.65} - 30^{0.65} \right) = 15{,}921.34$

The total labor hours needed is 15,921.

(b) $f(x) = 1272 x^{-0.15}$

$\int_{30}^{80} 1272 x^{-0.15}\, dx = \dfrac{1272 x^{0.85}}{0.85} \Big|_{30}^{80} = 1496.47 \left(80^{0.85} - 30^{0.85} \right) = 35{,}089.39$

The total labor hours needed is 35,089.

(c) Answers will vary.

51. $D(x) = -8.93x + 70$

$$\int_0^{12}(-8.93x + 70)\, dx = \left(-\frac{8.93x^2}{2} + 70x\right)\Big|_0^{12} = \left[-\frac{8.93(12)^2}{2} + 70(12)\right] - [0] = 197.04$$

The total budget deficit for 2002-2003 is projected to be 197.04 billion dollars.

53. The total sales during the first year is

$$\int_0^{12}\left(1200 - 950e^{-x}\right)dx = \left(1200x + 950e^{-x}\right)\Big|_0^{12}$$
$$= \left[1200(12) + 950e^{-12}\right] - \left[0 + 950e^0\right]$$
$$= \$13,450.01$$

55. $f(x) = 1000x^{-0.5}$

To produce an additional 25 units

$$\int_{35}^{60}1000x^{-0.05}\, dx = \frac{1000\, x^{0.5}}{0.5}\Big|_{35}^{60} = 2000\left[60^{0.5} - 35^{0.5}\right] = 3659.77$$

The job will take 3660 labor hours.

57. (a) $f(x) = x^2$ is an even function because $f(-x) = (-x)^2 = x^2 = f(x)$.

$$\int_{-1}^{1}x^2\, dx = \frac{x^3}{3}\Big|_{-1}^{1} = \frac{1^3}{3} - \frac{(-1)^3}{3} = \frac{1}{3} + \frac{1}{3} = \frac{2}{3}$$

$$2\int_0^1 x^3\, dx = 2\left[\frac{x^3}{3}\Big|_0^1\right] = 2\left[\frac{1^3}{3} - 0\right] = \frac{2}{3}$$

So $\int_{-1}^{1}x^2\, dx = 2\int_0^1 x^3\, dx$.

(b) $f(x) = x^4 + x^2$ is an even function because

$$f(-x) = (-x)^4 + (-x)^2 = x^4 + x^2 = f(x).$$

$$\int_{-1}^{1}\left(x^4 + x^2\right)dx = \left(\frac{x^5}{5} + \frac{x^3}{3}\right)\Big|_{-1}^{1} = \left[\frac{1^5}{5} + \frac{1^3}{3}\right] - \left[\frac{(-1)^5}{5} + \frac{(-1)^3}{3}\right]$$
$$= \frac{1}{5} + \frac{1}{3} + \frac{1}{5} + \frac{1}{3} = \frac{2}{5} + \frac{2}{3} = \frac{6 + 10}{15} = \frac{16}{15}$$

$$2\int_0^1\left(x^4 + x^2\right)dx = 2\left[\left(\frac{x^5}{5} + \frac{x^3}{3}\right)\Big|_0^1\right] = 2\left[\left(\frac{1}{5} + \frac{1}{3}\right) - 0\right] = \frac{16}{15}$$

So $\int_{-1}^{1}\left(x^4 + x^2\right)dx = 2\int_0^1\left(x^4 + x^2\right)dx$.

15.5 Finding Areas; Consumer's Surplus; Producer's Surplus; Maximizing Profit over Time

1.

$$A = \int_2^6 (3x+2)\,dx = \left(\frac{3x^2}{2} + 2x\right)\bigg|_2^6$$

$$= \left[\frac{3 \cdot 6^2}{2} + 2(6)\right] - \left[\frac{3 \cdot 2^2}{2} + 2(2)\right]$$

$$= 54 + 12 - 6 - 4 = 56$$

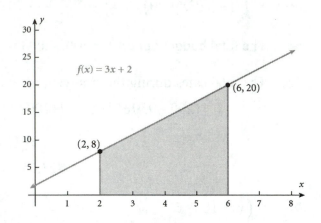

3.

$$A = \int_0^2 x^2\,dx = \frac{x^3}{3}\bigg|_0^2 = \frac{2^3}{3} - 0 = \frac{8}{3}$$

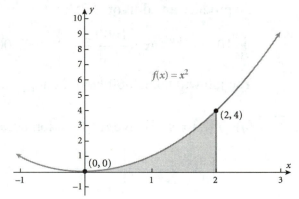

5.

$$A = \int_{-2}^{1}(x^2-1)\,dx - \int_{-1}^{1}(x^2-1)\,dx = \left(\frac{x^3}{3}-x\right)\bigg|_{-2}^{-1} - \left(\frac{x^3}{3}-x\right)\bigg|_{-1}^{1}$$

$$= \left\{\left[\frac{(-1)^3}{3}-(-1)\right]-\left[\frac{(-2)^3}{3}-(-2)\right]\right\} - \left\{\left[\frac{1^3}{3}-1\right]-\left[\frac{(-1)^3}{3}-(-1)\right]\right\}$$

$$= -\frac{1}{3}+1+\frac{8}{3}-2-\frac{1}{3}+1-\frac{1}{3}+1 = \frac{8}{3}$$

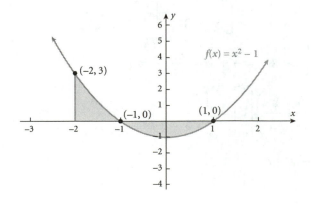

7. $f(x) = \sqrt[3]{x} = x^{1/3}$

$$A = -\int_{-1}^{0} x^{1/3}\, dx + \int_{0}^{8} x^{1/3}\, dx$$

$$= -\frac{x^{4/3}}{\frac{4}{3}}\Bigg|_{-1}^{0} + \frac{x^{4/3}}{\frac{4}{3}}\Bigg|_{0}^{8} = -\frac{3x^{4/3}}{4}\Bigg|_{-1}^{0} + \frac{3x^{4/3}}{4}\Bigg|_{0}^{8}$$

$$= \left\{[-0] - \left[-\frac{3(-1)^{4/3}}{4}\right]\right\} + \left\{\left[\frac{3(8)^{4/3}}{4}\right] - [0]\right\}$$

$$= \frac{3}{4} + 12 = \frac{51}{4}$$

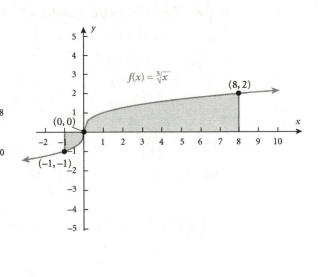

9. $f(x) = e^{x}$

$$A = \int_{0}^{1} e^{x}\, dx = e^{x}\Big|_{0}^{1} = e^{1} - e^{0}$$

$$= e - 1 \approx 1.718$$

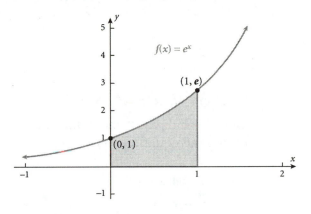

11.
$$A = \int_{0}^{1} g(x)\, dx - \int_{0}^{1} f(x)\, dx$$

$$= \int_{0}^{1} 2x\, dx - \int_{0}^{1} x\, dx$$

$$= \frac{2x^{2}}{2}\Bigg|_{0}^{1} - \frac{x^{2}}{2}\Bigg|_{0}^{1}$$

$$= [1^{2} - 0] - \left[\frac{1^{2}}{2} - 0\right]$$

$$= 1 - \frac{1}{2} = \frac{1}{2}$$

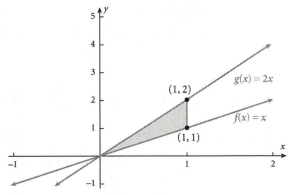

13. First we find where the graphs intersect.

$$f(x) = g(x)$$
$$x^2 = x$$
$$x^2 - x = 0$$
$$x(x-1) = 0$$
$$x = 0 \quad \text{or} \quad x = 1$$
$$f(0) = 0^2 = 0 \qquad f(1) = 1^2 = 1$$
$$g(0) = 0 \qquad g(1) = 1$$

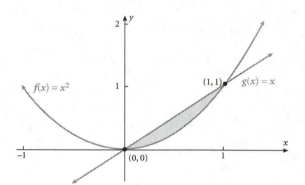

$$A = \int_0^1 g(x)\,dx - \int_0^1 f(x)\,dx = \int_0^1 x\,dx - \int_0^1 x^2\,dx$$

$$= \frac{x^2}{2}\bigg|_0^1 - \frac{x^3}{3}\bigg|_0^1 = \left[\frac{1}{2} - 0\right] - \left[\frac{1}{3} - 0\right]$$

$$= \frac{1}{2} - \frac{1}{3} = \frac{3-2}{6} = \frac{1}{6}$$

15. We graph $f(x) = x^2 + 1$ and $g(x) = x + 1$.
Then we find where the graphs intersect by
solving the equation $f(x) = g(x)$.

$$x^2 + 1 = x + 1$$
$$x^2 - x = 0$$
$$x(x-1) = 0$$
$$x = 0 \quad \text{or} \quad x - 1 = 0$$
$$x = 1$$

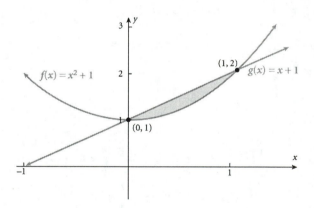

$$A = \int_0^1 \big(g(x) - f(x)\big)\,dx = \int_0^1 \left[(x+1) - (x^2+1)\right] dx$$

$$= \int_0^1 (x - x^2)\,dx = \left(\frac{x^2}{2} - \frac{x^3}{3}\right)\bigg|_0^1$$

$$= \left[\frac{1^2}{2} - \frac{1^3}{3}\right] - [0]$$

$$= \frac{1}{2} - \frac{1}{3} = \frac{1}{6}$$

17. We graph $f(x) = \sqrt{x} = x^{1/2}$ and $g(x) = x^3$. Then we find where the graphs intersect by solving the equation $f(x) = g(x)$.

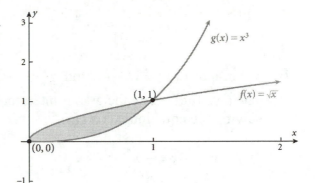

$$\sqrt{x} = x^3$$
$$x = x^6$$
$$x(x^5 - 1) = 0$$
$$x = 0 \quad \text{or} \quad x^5 - 1 = 0$$
$$x = 1$$

$$A = \int_0^1 (f(x) - g(x))\, dx = \int_0^1 (x^{1/2} - x^3)\, dx$$

$$= \left(\frac{x^{3/2}}{\frac{3}{2}} - \frac{x^4}{4} \right) \Bigg|_0^1$$

$$= \left[\frac{2}{3} \cdot 1^{3/2} - \frac{1^4}{4} \right] - [0]$$

$$= \frac{2}{3} - \frac{1}{4} = \frac{8-3}{12} = \frac{5}{12}$$

19. We graph $f(x) = x^2$ and $g(x) = x^4$. Then we find where the graphs intersect by solving the equation $f(x) = g(x)$.

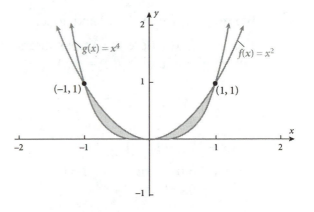

$$x^2 = x^4$$
$$x^4 - x^2 = 0$$
$$x^2(x^2 - 1) = 0$$
$$x^2(x - 1)(x + 1) = 0$$
$$x^2 = 0 \quad \text{or} \quad x - 1 = 0 \quad \text{or} \quad x + 1 = 0$$
$$x = 0 \quad \text{or} \quad x = 1 \quad \text{or} \quad x = -1$$

$$A = \int_{-1}^0 (f(x) - g(x))\, dx + \int_0^1 (f(x) - g(x))\, dx = \int_{-1}^1 (f(x) - g(x))\, dx = \int_{-1}^1 (x^2 - x^4)\, dx$$

$$A = \left(\frac{x^3}{3} - \frac{x^5}{5} \right) \Bigg|_{-1}^1$$

$$= \left[\frac{1}{3} - \frac{1}{5} \right] - \left[\frac{(-1)^3}{3} - \frac{(-1)^5}{5} \right]$$

$$= \frac{1}{3} - \frac{1}{5} + \frac{1}{3} - \frac{1}{5} = \frac{5}{15} - \frac{3}{15} + \frac{5}{15} - \frac{3}{15} = \frac{4}{15}$$

21. We graph $f(x) = x^2 - 4x$ and $g(x) = -x^2$.

Then we find where the graphs intersect by solving the equation $f(x) = g(x)$.

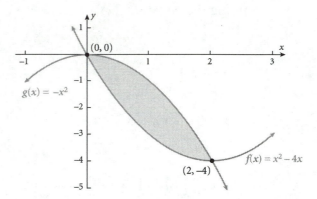

$$x^2 - 4x = -x^2$$
$$2x^2 - 4x = 0$$
$$2x(x-2) = 0$$
$$2x = 0 \quad \text{or} \quad x - 2 = 0$$
$$x = 0 \quad \text{or} \quad x = 2$$

$$A = \int_0^2 (g(x) - f(x))\, dx = \int_0^2 \left[-x^2 - (x^2 - 4x) \right] dx = \int_0^2 (-2x^2 + 4x)\, dx$$

$$= \left(-\frac{2x^3}{3} + \frac{4x^2}{2} \right) \Bigg|_0^2$$

$$= \left[-\frac{2 \cdot 2^3}{3} + 2 \cdot 2^2 \right] - [0]$$

$$= -\frac{16}{3} + 8 = \frac{8}{3}$$

23. We graph $f(x) = 4 - x^2$ and $g(x) = x + 2$.

Then we find where the graphs intersect by solving the equation $f(x) = g(x)$.

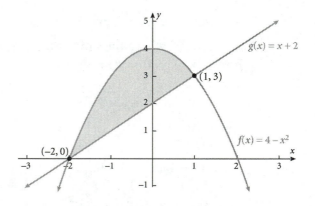

$$4 - x^2 = x + 2$$
$$x^2 + x - 2 = 0$$
$$(x+2)(x-1) = 0$$
$$x + 2 = 0 \quad \text{or} \quad x - 1 = 0$$
$$x = -2 \quad \text{or} \quad x = 1$$

$$A = \int_{-2}^1 \left[f(x) - g(x) \right] dx = \int_{-2}^1 \left[(4 - x^2) - (x + 2) \right] dx = \int_{-2}^1 \left[(2 - x^2 - x) \right] dx$$

$$= \left(2x - \frac{x^3}{3} - \frac{x^2}{2} \right) \Bigg|_{-2}^1$$

$$= \left[2 \cdot 1 - \frac{1^3}{3} - \frac{1^2}{2} \right] - \left[2(-2) - \frac{(-2)^3}{3} - \frac{(-2)^2}{2} \right]$$

$$= 2 - \frac{1}{3} - \frac{1}{2} + 4 - \frac{8}{3} + 2 = \frac{9}{2}$$

25. We graph $f(x) = x^3$ and $g(x) = 4x$.

Then we find where the graphs intersect by solving the equation $f(x) = g(x)$.

$$x^3 = 4x$$
$$x^3 - 4x = 0$$
$$x(x^2 - 4) = 0$$
$$x(x - 2)(x + 2) = 0$$
$$x = 0 \quad \text{or} \quad x - 2 = 0 \quad \text{or} \quad x + 2 = 0$$
$$x = 2 \quad \text{or} \quad x = -2$$

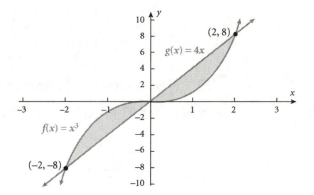

$$A = \int_{-2}^{0} (f(x) - g(x))\, dx + \int_{0}^{2} (g(x) - f(x))\, dx$$

$$= \int_{-2}^{0} (x^3 - 4x)\, dx + \int_{0}^{2} (4x - x^3)\, dx$$

$$= \left(\frac{x^4}{4} - \frac{4x^2}{2} \right) \Bigg|_{-2}^{0} + \left(\frac{4x^2}{2} - \frac{x^4}{4} \right) \Bigg|_{0}^{2}$$

$$= \left\{ [0] - \left[\frac{(-2)^4}{4} - 2(-2)^2 \right] \right\} + \left\{ \left[2(2)^2 - \frac{2^4}{4} \right] - [0] \right\}$$

$$0 - [4 - 8] + [8 - 4] + 0 = 4 + 4 = 8$$

27. We graph $y = x^2$, $y = x$, and $y = -x$.
Then we find where the graphs intersect by solving the equations $x^2 = x$ and $x^2 = -x$.

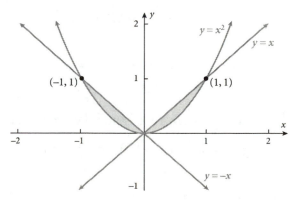

$$\begin{array}{ll}
x^2 = x & x^2 = -x \\
x^2 - x = 0 & x^2 + x = 0 \\
x(x - 1) = 0 & x(x + 1) = 0 \\
x = 0 \quad \text{or} \quad x - 1 = 0 & x = 0 \quad \text{or} \quad x + 1 = 0 \\
\qquad x = 1 & \qquad x = -1
\end{array}$$

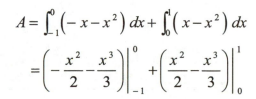

$$A = \int_{-1}^{0} (-x - x^2)\, dx + \int_{0}^{1} (x - x^2)\, dx$$

$$= \left(-\frac{x^2}{2} - \frac{x^3}{3} \right) \Bigg|_{-1}^{0} + \left(\frac{x^2}{2} - \frac{x^3}{3} \right) \Bigg|_{0}^{1}$$

$$= \left\{ [0] - \left[-\frac{(-1)^2}{2} - \frac{(-1)^3}{3} \right] \right\} + \left\{ \left[\frac{1}{2} - \frac{1}{3} \right] - [0] \right\}$$

$$= -\left[\frac{-3+2}{6} \right] + \left[\frac{3-2}{6} \right] = \frac{1}{6} + \frac{1}{6} = \frac{2}{6} = \frac{1}{3}$$

29.

(a) $\int_0^4 (3x+1)\, dx$ represents the area under the graph of $f(x) = 3x+1$ from $x = 0$ to $x = 4$.

(b)

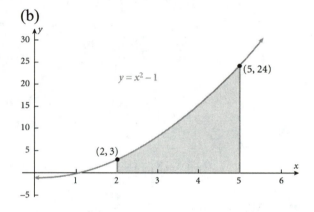

(c) $\int_0^4 (3x+1)\, dx = \left(\frac{3x^2}{2} + x \right) \Big|_0^4$

$$= \left[\frac{3 \cdot 4^2}{2} + 4 \right] - [0]$$

$$= 28$$

31.

(a) $\int_2^5 (x^2 - 1)\, dx$ represents the area under the graph of $f(x) = x^2 - 1$ from $x = 2$ to $x = 5$.

(b)

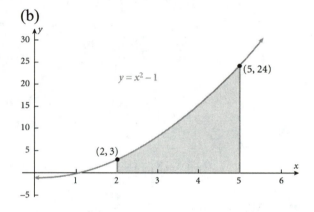

(c) $\int_2^5 (x^2 - 1)\, dx = \left(\frac{x^3}{3} - x \right) \Big|_2^5$

$$= \left[\frac{5^3}{3} - 5 \right] - \left[\frac{2^3}{3} - 2 \right]$$

$$= \frac{125}{3} - 5 - \frac{8}{3} + 2$$

$$= \frac{117}{3} - 3 = 39 - 3 = 36$$

33.

(a) $\int_0^2 e^x\, dx$ represents the area under the graph of $f(x) = e^x$ from $x = 0$ to $x = 2$.

(b)

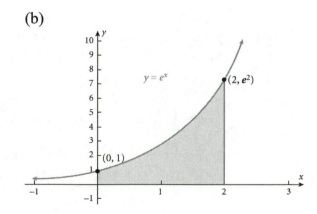

(c) $\int_0^2 e^x\, dx = e^x \Big|_0^2$

$$= e^2 - e^0$$

$$= e^2 - 1$$

35. We first find the equilibrium point (x^*, p^*), by solving the equation $D(x^*) = S(x^*)$.

$$-5x^* + 20 = 4x^* + 8$$
$$-9x^* = -12$$
$$x^* = \frac{12}{9} = \frac{4}{3}$$

and $p^* = D(x^*) = S(x^*)$

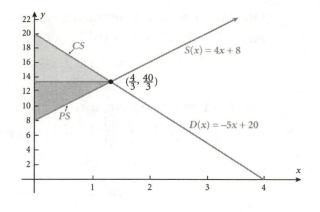

$$= -5\left(\frac{4}{3}\right) + 20 = -\frac{20}{3} + \frac{60}{3} = \frac{40}{3} \approx 13.33$$

The consumer's surplus, $CS = \int_0^{x^*} D(x)\,dx - p^* x^*$

$$CS = \int_0^{4/3}(-5x + 20)\,dx - \left(\frac{4}{3}\right)\left(\frac{40}{3}\right)$$

$$= \left(\frac{-5x^2}{2} + 20x\right)\Bigg|_0^{4/3} - \frac{160}{9}$$

$$= \left[-\frac{5}{2}\cdot\left(\frac{4}{3}\right)^2 + 20\left(\frac{4}{3}\right)\right] - [0] - \frac{160}{9}$$

$$= -\frac{80}{18} + \frac{80}{3} - \frac{160}{9} = \frac{-40 + 240 - 160}{9} = \frac{40}{9} = \$4.44$$

The producer's surplus, $PS = p^* x^* - \int_0^{x^*} S(x)\,dx$

$$PS = \frac{160}{9} - \int_0^{4/3}(4x + 8)\,dx$$

$$= \frac{160}{9} - \left(\frac{4x^2}{2} + 8x\right)\Bigg|_0^{4/3}$$

$$= \frac{160}{9} - \left\{\left[2\left(\frac{4}{3}\right)^2 + 8\left(\frac{4}{3}\right)\right] - [0]\right\}$$

$$\frac{160}{9} - \left[\frac{32}{9} + \frac{32}{3}\right] = \frac{160}{9} - \frac{32}{9} - \frac{96}{9} = \frac{32}{9} = \$3.56$$

37. The best time to terminate operations is when $R'(t) = C'(t)$.

$$19 - t^{1/2} = 3 + 3t^{1/2}$$
$$-4t^{1/2} = -16$$
$$t^{1/2} = 4 \text{ or } t = 16$$

Operations should be terminated after 16 years.

At $t = 16$, $R'(16) = 19 - 16^{1/2} = 19 - 4 = 15$ million dollars.

$$C'(16) = 3 + 3 \cdot 16^{1/2} = 3 + 3 \cdot 4 = 15 \text{ million dollars.}$$

The profit at $t = 16$ years is

$$P(16) = \int_0^{16} \left[\left(19 - t^{1/2} \right) - \left(3 + 3t^{1/2} \right) \right] dt = \int_0^{16} \left[16 - 4t^{1/2} \right] dt$$

$$= \left(16t - \frac{4t^{3/2}}{\frac{3}{2}} \right) \Bigg|_0^{16} = \left[16 \cdot 16 - \frac{8}{3} \cdot 16^{3/2} \right] - [0]$$

$$= 256 - \frac{512}{3} = \frac{768}{3} - \frac{512}{3} = \frac{256}{3} = 85.33 \text{ million dollars.}$$

39. (a) $f(x) = x^2$; $\quad a = 0 \quad b = 1$

$$\int_0^1 x^2 \, dx = \frac{x^3}{3} \Bigg|_0^1 = \frac{1}{3}$$

So, $f(c)(b - a) = \frac{1}{3}$

$$c^2(1 - 0) = \frac{1}{3}$$

$$c^2 = \frac{1}{3} \quad \text{or} \quad c = \frac{1}{\sqrt{3}} = \frac{\sqrt{3}}{3}$$

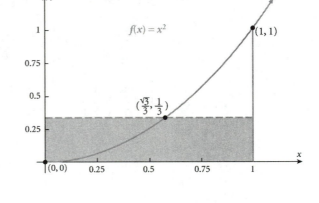

(b) $f(x) = \frac{1}{x^2} = x^{-2}$; $\quad a = 1 \quad b = 4$

$$\int_1^4 x^{-2} \, dx = \frac{x^{-1}}{-1} \Bigg|_1^4 = -\frac{1}{4} + 1 = \frac{3}{4}$$

So, $f(c)(b - a) = \frac{3}{4}$

$$\frac{1}{c^2}(4 - 1) = \frac{3}{4}$$

$$\frac{3}{c^2} = \frac{3}{4}$$

$$c^2 = 4 \quad \text{or} \quad c = 2$$

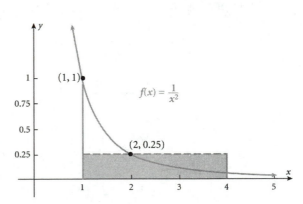

41. $\dfrac{d}{dx} \displaystyle\int_a^x f(t)\, dt = f(x)$

(a) $\dfrac{d}{dx} \displaystyle\int_1^x t^2\, dt = f(x) = x^2$ 　　　　(b) $\dfrac{d}{dx} \displaystyle\int_2^x \sqrt{t^2 - 2}\, dt = f(x) = \sqrt{x^2 - 2}$

(c) $\dfrac{d}{dx} \displaystyle\int_5^x \sqrt{t^t + 2t}\, dt = f(x) = \sqrt{x^x + 2x}$

15.6 Approximating Definite Integrals

1. STEP 1　[1, 3] has been divided into two subintervals of equal length
　　　　　　　[1, 2]　　　　and　　　　[2, 3]
　STEP 2　$f(1) = 1$　　　　　　　$f(2) = 2$

　STEP 3　$\displaystyle\int_1^3 f(x)\, dx \approx f(1)\cdot 1 + f(2)\cdot 1 = 1 + 2 = 3$

3. STEP 1　[0, 8] has been divided into 4 intervals, each of width 2.
　STEP 2　$f(0) = 10$　　　$f(2) = 6$　　　$f(4) = 7$　　　$f(6) = 5$

　STEP 3　$\displaystyle\int_0^8 f(x)\, dx \approx f(0)\cdot 2 + f(2)\cdot 2 + f(4)\cdot 2 + f(6)\cdot 2$
　　　　　　　　$= 10\cdot 2 + 6\cdot 2 + 7\cdot 2 + 5\cdot 2 = 56$

5. (a)

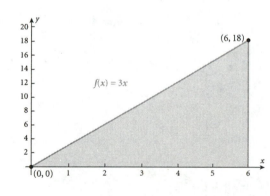

(b) STEP 1　When [0, 6] is separated into three subintervals each will have width 2.
　　　　　　　　[0, 2]　　　[2, 4]　　　[4, 6]

　STEP 2　　$f(0) = 3\cdot 0 = 0$
　　　　　　　　$f(2) = 3\cdot 2 = 6$
　　　　　　　　$f(4) = 3\cdot 4 = 12$

　STEP 3　$A \approx f(0)\cdot 2 + f(2)\cdot 2 + f(4)\cdot 2$
　　　　　　　　$= 2(0 + 6 + 12) = 36$

(c)　We use STEP 1 from part (b).
　STEP 2　$f(2) = 3\cdot 2 = 6$　　　$f(4) = 3\cdot 4 = 12$　　　$f(6) = 3\cdot 6 = 18$

　STEP 3　$A \approx f(2)\cdot 2 + f(4)\cdot 2 + f(6)\cdot 2 = 2(6 + 12 + 18) = 72$

(d) STEP 1　When [0, 6] is separated into six subintervals each will have width 1.
　　　　　　　　[0, 1]　　[1, 2]　　[2, 3]　　[3. 4]　　[4, 5]　　[5, 6]

STEP 2 $f(0) = 0$ $f(1) = 3 \cdot 1 = 3$ $f(2) = 6$ $f(3) = 3 \cdot 3 = 9$

          $f(4) = 12$ $f(5) = 3 \cdot 5 = 15$

STEP 3 $A \approx f(0) \cdot 1 + f(1) \cdot 1 + f(2) \cdot 1 + f(3) \cdot 1 + f(4) \cdot 1 + f(5) \cdot 1$

          $= 1(0 + 3 + 6 + 9 + 12 + 15) = 45$

(e) We use STEP 1 from part (d).

STEP 2 $f(1) = 3 \cdot 1 = 3$ $f(2) = 6$ $f(3) = 3 \cdot 3 = 9$

        $f(4) = 12$ $f(5) = 3 \cdot 5 = 15$ $f(6) = 3 \cdot 6 = 18$

STEP 3 $A \approx f(1) \cdot 1 + f(2) \cdot 1 + f(3) \cdot 1 + f(4) \cdot 1 + f(5) \cdot 1 + f(6) \cdot 1$

          $= 1(3 + 6 + 9 + 12 + 15 + 18) = 63$

(f) Since A is a triangle, with base 6 and altitude 18, the actual area is

$$A = \frac{1}{2}bh = \frac{1}{2} \cdot 6 \cdot 18 = 54$$

7. (a)

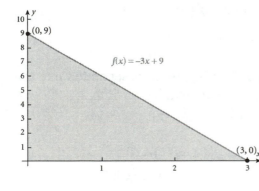

(b) **STEP 1** When [0, 3] is separated into three subintervals each will have width 1.

       [0, 1]   [1, 2]   [2, 3]

STEP 2 $f(0) = -3(0) + 9 = 9$

             $f(1) = -3(1) + 9 = 6$

             $f(2) = -3(2) + 9 = 3$

STEP 3 $A \approx f(0) \cdot 1 + f(1) \cdot 1 + f(2) \cdot 1$

          $= 1(9 + 6 + 3) = 18$

(c) We use STEP 1 from part (b).

STEP 2 $f(3) = -3(3) + 9 = 0$

STEP 3 $A \approx f(1) \cdot 1 + f(2) \cdot 1 + f(3) \cdot 1 = 1(6 + 3 + 0) = 9$

(d) **STEP 1** When [0, 3] is separated into six subintervals each will have width 0.5.

     [0, 0.5]   [0.5, 1]   [1, 1.5]   [1.5, 2]   [2, 2.5]   [2.5, 3]

STEP 2 $f(0) = 9$ $f(0.5) = -3(0.5) + 9 = 7.5$ $f(1) = 6$

     $f(1.5) = -3(1.5) + 9 = 4.5$ $f(2) = 3$ $f(2.5) = -3(2.5) + 9 = 1.5$

STEP 3

$$A \approx f(0) \cdot 0.5 + f(0.5) \cdot 0.5 + f(1) \cdot 0.5 + f(1.5) \cdot 0.5 + f(2) \cdot 0.5 + f(2.5) \cdot 0.5$$
$$= 0.5(9 + 7.5 + 6 + 4.5 + 3 + 1.5) = 0.5(31.5) = 15.75$$

(e) We use STEP 1 from part (d).

STEP 3

$$A \approx f(0.5) \cdot 0.5 + f(1) \cdot 0.5 + f(1.5) \cdot 0.5 + f(2) \cdot 0.5 + f(2.5) \cdot 0.5 + f(3) \cdot 0.5$$
$$= 0.5(7.5 + 6 + 4.5 + 3 + 1.5 + 0) = 0.5(22.5) = 11.25$$

(f) Since A is a triangle, with base 3 and altitude 9, the actual area is
$$A = \frac{1}{2}bh = \frac{1}{2} \cdot 3 \cdot 9 = \frac{27}{2} = 13.5$$

9. (a)

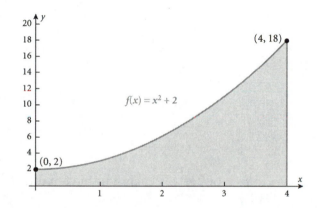

(b) **STEP 1** When [0, 4] is separated into four subintervals each will have width 1.
    [0, 1]   [1, 2]   [2, 3]   [3, 4]

STEP 2 $f(0) = 0^2 + 2 = 2$ $f(1) = 1^2 + 2 = 3$
                  $f(2) = 2^2 + 2 = 6$ $f(3) = 3^2 + 2 = 11$

STEP 3 $A \approx \left[f(0) \cdot 1\right] + \left[f(1) \cdot 1\right] + \left[f(2) \cdot 1\right] + \left[f(3) \cdot 1\right] = 1(2 + 3 + 6 + 11) = 22$

(c) **STEP 1** When [0, 4] is separated into eight subintervals each will have width 0.5.
    [0, 0.5]   [0.5, 1]   [1, 1.5]   [1.5, 2]   [2, 2.5]   [2.5, 3]   [3, 3.5]   [3.5, 4]

STEP 2 $f(0.5) = 0.5^2 + 2 = 2.25$ $f(1.5) = 1.5^2 + 2 = 4.25$
                 $f(2.5) = 2.5^2 + 2 = 8.25$ $f(3.5) = 3.5^2 + 2 = 14.25$

STEP 3 $A \approx \left[f(0) \cdot 0.5 \right] + \left[f(0.5) \cdot 0.5 \right] + \left[f(1) \cdot 0.5 \right] + \left[f(1.5) \cdot 0.5 \right]$
$$+ \left[f(2) \cdot 0.5 \right] + \left[f(2.5) \cdot 0.5 \right] + \left[f(3) \cdot 0.5 \right] + \left[f(3.5) \cdot 0.5 \right]$$
$$= 0.5 \left(2 + 2.25 + 3 + 4.25 + 6 + 8.25 + 11 + 14.25 \right) = 0.5(51)$$
$$A \approx 25.5$$

(d) $A = \int_0^4 \left(x^2 + 2 \right) dx$

(e) $\int_0^4 \left(x^2 + 2 \right) dx = \left(\dfrac{x^3}{3} + 2x \right) \Big|_0^4 = \left[\dfrac{4^3}{3} + 2(4) \right] - [0] = \dfrac{88}{3} \approx 29.333$

11. (a)

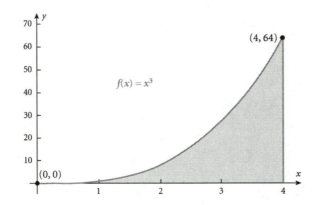

(b) **STEP 1** When [0, 4] is separated into four subintervals each will have width 1.
   [0, 1]   [1, 2]   [2, 3]   [3, 4]

STEP 2 $f(0) = 0^3 = 0$ $\qquad\qquad$ $f(1) = 1^3 = 1$
   $f(2) = 2^3 = 8$ $\qquad\qquad$ $f(3) = 3^3 = 27$

STEP 3 $A \approx \left[f(0) \cdot 1 \right] + \left[f(1) \cdot 1 \right] + \left[f(2) \cdot 1 \right] + \left[f(3) \cdot 1 \right] = 1(0 + 1 + 8 + 27) = 36$

(c) **STEP 1** When [0, 4] is separated into eight subintervals each will have width 0.5.
   [0, 0.5]   [0.5, 1]   [1, 1.5]   [1.5, 2]   [2, 2.5]   [2.5, 3]   [3, 3.5]   [3.5, 4]

STEP 2 $f(0.5) = 0.5^3 = 0.125$ $\qquad\qquad$ $f(1.5) = 1.5^3 = 3.375$
   $f(2.5) = 2.5^3 = 15.625$ $\qquad\qquad$ $f(3.5) = 3.5^3 = 42.875$

STEP 3 $A \approx \left[f(0) \cdot 0.5 \right] + \left[f(0.5) \cdot 0.5 \right] + \left[f(1) \cdot 0.5 \right] + \left[f(1.5) \cdot 0.5 \right]$
$$+ \left[f(2) \cdot 0.5 \right] + \left[f(2.5) \cdot 0.5 \right] + \left[f(3) \cdot 0.5 \right] + \left[f(3.5) \cdot 0.5 \right]$$

$$= 0.5(0 + 0.125 + 1 + 3.375 + 8 + 15.625 + 27 + 42.875) = 0.5(98)$$
$$A \approx 49$$

(d) $A = \displaystyle\int_0^4 x^3 \, dx$

(e) $\displaystyle\int_0^4 x^3 \, dx = \left(\dfrac{x^4}{4} \right)\Bigg|_0^4 = \dfrac{4^4}{4} - 0 = 64$

13. (a)

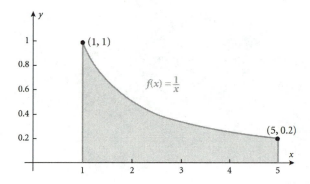

(b) **STEP 1** When [1, 5] is separated into four subintervals each will have width 1.

[1, 2]   [2, 3]   [3, 4]   [4, 5]

STEP 2 $f(1) = \dfrac{1}{1} = 1$ $\qquad\qquad$ $f(2) = \dfrac{1}{2}$

$\qquad\qquad\quad$ $f(3) = \dfrac{1}{3}$ $\qquad\qquad$ $f(4) = \dfrac{1}{4}$

STEP 3 $A \approx \big[f(1) \cdot 1 \big] + \big[f(2) \cdot 1 \big] + \big[f(3) \cdot 1 \big] + \big[f(4) \cdot 1 \big]$

$\qquad\qquad = 1\left(1 + \dfrac{1}{2} + \dfrac{1}{3} + \dfrac{1}{4} \right) = \dfrac{12 + 6 + 4 + 3}{12} = \dfrac{25}{12} \approx 2.083$

(c) **STEP 1** When [1, 5] is separated into eight subintervals each will have width 0.5.

[1, 1.5]   [1.5, 2]   [2, 2.5]   [2.5, 3]   [3, 3.5]   [3.5, 4]   [4, 4.5]   [4.5, 5]

STEP 2 $f\left(\dfrac{3}{2} \right) = \dfrac{1}{\dfrac{3}{2}} = \dfrac{2}{3}$ $\qquad\qquad$ $f\left(\dfrac{5}{2} \right) = \dfrac{1}{\dfrac{5}{2}} = \dfrac{2}{5}$

$\qquad\qquad\quad$ $f\left(\dfrac{7}{2} \right) = \dfrac{1}{\dfrac{7}{2}} = \dfrac{2}{7}$ $\qquad\qquad$ $f\left(\dfrac{9}{2} \right) = \dfrac{1}{\dfrac{9}{2}} = \dfrac{2}{9}$

STEP 3 $A \approx \left[f(1) \cdot \frac{1}{2} \right] + \left[f\left(\frac{3}{2}\right) \cdot \frac{1}{2} \right] + \left[f(2) \cdot \frac{1}{2} \right] + \left[f\left(\frac{5}{2}\right) \cdot \frac{1}{2} \right]$

$$+ \left[f(3) \cdot \frac{1}{2} \right] + \left[f\left(\frac{7}{2}\right) \cdot \frac{1}{2} \right] + \left[f(4) \cdot \frac{1}{2} \right] + \left[f\left(\frac{9}{2}\right) \cdot \frac{1}{2} \right]$$

$$= \frac{1}{2}\left(1 + \frac{2}{3} + \frac{1}{2} + \frac{2}{5} + \frac{1}{3} + \frac{2}{7} + \frac{1}{4} + \frac{2}{9} \right) = \frac{4609}{2520} \approx 1.829$$

(d) $\quad A = \int_{1}^{5} \frac{1}{x}\, dx$
  (e) $\quad \int_{1}^{5} \frac{1}{x}\, dx = \ln x \big|_{1}^{5} = \ln 5 - \ln 1 = \ln 5 \approx 1.609$

15. (a)

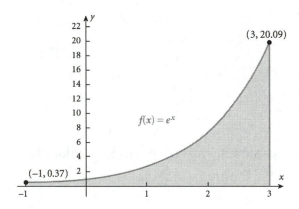

(b) **STEP 1** When $[-1, 3]$ is separated into four subintervals each will have width 1.
$[-1, 0] \quad [0, 1] \quad [1, 2] \quad [2, 3]$

STEP 2 $\quad f(-1) = e^{-1} \qquad f(0) = e^0 = 1$
$\qquad\qquad f(1) = e^1 \qquad\quad f(2) = e^2$

STEP 3
$A \approx \left[f(-1) \cdot 1 \right] + \left[f(0) \cdot 1 \right] + \left[f(1) \cdot 1 \right] + \left[f(2) \cdot 1 \right] = 1\left(e^{-1} + 1 + e + e^2 \right) \approx 11.475$

(c) **STEP 1** When $[0, 4]$ is separated into eight subintervals each will have width 0.5.
$[-1, -0.5] \quad [-0.5, 0] \quad [0, 0.5] \quad [0.5, 1] \quad [1, 1.5] \quad [1.5, 2] \quad [2, 2.5] \quad [2.5, 3]$

STEP 2 $\quad f(-0.5) = e^{-0.5} \qquad f(0.5) = e^{0.5}$
$\qquad\qquad f(1.5) = e^{1.5} \qquad\quad f(2.5) = e^{2.5}$

STEP 3 $\quad A \approx \left[f(-1) \cdot 0.5 \right] + \left[f(-0.5) \cdot 0.5 \right] + \left[f(0) \cdot 0.5 \right] + \left[f(0.5) \cdot 0.5 \right]$

$$+ \left[f(1) \cdot 0.5 \right] + \left[f(1.5) \cdot 0.5 \right] + \left[f(2) \cdot 0.5 \right] + \left[f(2.5) \cdot 0.5 \right]$$

$$= 0.5\left(e^{-1} + e^{-0.5} + 1 + e^{0.5} + e + e^{1.5} + e^2 + e^{2.5}\right) \approx 0.5(30.39465) \approx 15.197$$

(d) $A = \int_1^3 e^x \, dx$

(e) $\int_1^3 e^x \, dx = e^x \Big|_{-1}^3 = e^3 - e^{-1} = \dfrac{e^4 - 1}{e} \approx 19.718$

17.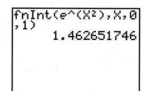

$$\int_0^1 e^{x^2} \, dx \approx 1.46$$

19.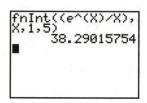

$$\int_1^5 \dfrac{e^x}{x} \, dx \approx 38.29$$

21. (a)

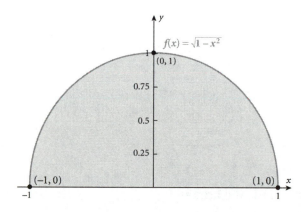

(b) **STEP 1** When [– 1, 1] is separated into five subintervals each will have width 0.4.

[– 1, – 0.6]   [– 0.6, – 0.2]   [– 0.2, 0.2]   [0.2, 0.6]   [0.6, 1]

STEP 2 $f(-1) = \sqrt{1 - (-1)^2} = 0$ $f(-0.6) = \sqrt{1 - (-0.6)^2} = 0.8$

$f(-0.2) = \sqrt{1 - (-0.2)^2} = \sqrt{0.96}$ $f(0.2) = \sqrt{1 - (0.2)^2} = \sqrt{0.96}$

$f(0.6) = \sqrt{1 - (0.6)^2} = 0.8$

STEP 3 $A \approx \left[f(-1) \cdot 0.4\right] + \left[f(-0.6) \cdot 0.4\right] + \left[f(-0.2) \cdot 0.4\right]$

$$+ \left[f(0.2) \cdot 0.4\right] + \left[f(0.6) \cdot 0.4\right]$$

$$= 0.4\left(0 + 0.8 + \sqrt{0.96} + \sqrt{0.96} + 0.8\right) \approx 0.4(3.55959) \approx 1.424$$

(c) **STEP 1** When $[-1, 1]$ is separated into ten subintervals each will have width 0.2.

$[-1, -0.8]$ $[-0.8, -0.6]$ $[-0.6, -0.4]$ $[-0.4, -0.2]$ $[-0.2, 0]$

$[0, 0.2]$ $[0.2, 0.4]$ $[0.4, .06]$ $[0.6, 0.8]$ $[0.8, 1]$

STEP 2 $f(-0.8) = \sqrt{1 - (-0.8)^2} = 0.6$ $f(-0.4) = \sqrt{1 - (-0.4)^2} = \sqrt{0.84}$

$f(0) = \sqrt{1 - 0^2} = 1$ $f(0.4) = \sqrt{1 - (0.4)^2} = \sqrt{0.84}$

$f(0.8) = \sqrt{1 - (0.8)^2} = 0.6$

STEP 3 $A \approx \left[f(-1) \cdot 0.2 \right] + \left[f(-0.8) \cdot 0.2 \right] + \left[f(-0.6) \cdot 0.4 \right] + \left[f(-0.4) \cdot 0.2 \right]$

$+ \left[f(-0.2) \cdot 0.2 \right] + \left[f(0) \cdot 0.2 \right] + \left[f(0.2) \cdot 0.2 \right] + \left[f(0.4) \cdot 0.2 \right]$

$+ \left[f(0.6) \cdot 0.2 \right] + \left[f(0.8) \cdot 0.2 \right]$

$= 0.2 \left(0 + 0.6 + 0.8 + \sqrt{0.84} + \sqrt{0.96} + 1 + \sqrt{0.96} + \sqrt{0.84} + 0.8 + 0.6 \right)$

$= 0.2(7.59262) = 1.519$

(d) $A = \int_{-1}^{1} \sqrt{1 - x^2} \, dx = \int_{1}^{1} \left(1 - x^2 \right)^{1/2} dx$

(e)

```
fnInt(√(1-X²),X,
-1,1)
        1.570796729
```

(f) Since the graph is a semi-circle with radius 1, the area is

$A = \frac{1}{2} \left(\pi r^2 \right) = \frac{1}{2} \pi$

15.7 Differential Equations

1.

The general solution to the differential equation $\dfrac{dy}{dx} = x^2 - 1$ is $y = \dfrac{x^3}{3} - x + K$.

We use $x = 0$ and $y = 0$ to find K.

$0 = \dfrac{0}{3} - 0 + K$

$K = 0$

The particular solution to the differential equation is

$y = \dfrac{x^3}{3} - x$

3.

The general solution to the differential equation $\dfrac{dy}{dx} = x^2 - x$ is $y = \dfrac{x^3}{3} - \dfrac{x^2}{2} + K$

We use $x = 3$ and $y = 3$ to find K.

$$3 = \frac{3^3}{3} - \frac{3^2}{2} + K = 9 - \frac{9}{2} + K$$

$$K = 3 - \frac{9}{2} = -\frac{3}{2}$$

The particular solution to the differential equation is

$$y = \frac{x^3}{3} - \frac{x^2}{2} - \frac{3}{2}$$

5.

The general solution to the equation $\dfrac{dy}{dx} = x^3 - x + 2$ is $y = \dfrac{x^4}{4} - \dfrac{x^2}{2} + 2x + K$.

We use $x = -2$ and $y = 1$ to find K.

$$1 = \frac{(-2)^4}{4} - \frac{(-2)^2}{2} + 2(-2) + K$$

$$K = 1 - 4 + 2 + 4 = 3$$

The particular solution to the differential equation is

$$y = \frac{x^4}{4} - \frac{x^2}{2} + 2x + 3$$

7.

The general solution to the differential equation $\dfrac{dy}{dx} = e^x$ is $y = e^x + K$.

We use $x = 0$ and $y = 4$ to find K.

$$4 = e^0 + K$$

$$K = 4 - 1 = 3$$

The particular solution to the differential equation is

$$y = e^x + 3$$

9.

The general solution to the differential equation $\dfrac{dy}{dx} = \dfrac{x^2 + x + 1}{x} = x + 1 + \dfrac{1}{x}$

is $y = \dfrac{x^2}{2} + x + \ln|x| + K$.

We use $x = 1$ and $y = 0$ to find K.

$$0 = \frac{1^2}{2} + 1 + \ln|1| + K$$

$$K = -\frac{3}{2}$$

The particular solution to the differential equation is

$$y = \frac{x^2}{2} + x + \ln|x| - \frac{3}{2}$$

11.

The differential equation describing the population growth is $\dfrac{dN}{dt} = kN$, where N is the population size and t is the time in minutes.

The general solution of the equation is $N(t) = N_0\, e^{kt}$ where N_0 is the population at $t = 0$ and k is the constant of proportionality. Using $N(5) = 150$ and $N_0 = 100$, we solve for k.

$$150 = 100\, e^{5k}$$
$$e^{5k} = 1.5$$
$$5k = \ln 1.5$$
$$k = \frac{\ln 1.5}{5}$$

After 1 hour $t = 60$ and there will be 12,975 bacteria.

$$N(60) = 100\, e^{(60)\left[\frac{\ln 1.5}{5}\right]} = 100\, e^{(12)\ln 1.5} = 12{,}974.63$$

After 90 minutes, there will be 147,789 bacteria.

$$N(90) = 100\, e^{(90)\left[\frac{\ln 1.5}{5}\right]} = 100\, e^{(18)\ln 1.5} = 147{,}789.188$$

There will be 1,000,000 bacteria after $t = 113.58$ minutes.

$$1{,}000{,}000 = 100\, e^{\left[\frac{\ln 1.5}{5}\right]t}$$
$$10{,}000 = e^{\left[\frac{\ln 1.5}{5}\right]t}$$
$$\ln 10{,}000 = \left[\frac{\ln 1.5}{5}\right]t$$
$$t = \frac{5 \cdot \ln 10{,}000}{\ln 1.5} \approx 113.58$$

13.

The differential equation describing the radioactive decay is $\dfrac{dA}{dt} = kA$, where A is the amount of radium present and t is the time in years.

The general solution of the equation is $A(t) = A_0\, e^{kt}$ where A_0 is the amount at $t = 0$ and k is the constant of proportionality. Using $A_0 = 8$ grams and $A(1690) = 4$ grams, we solve for k.

$$4 = 8\, e^{1690k}$$
$$\frac{1}{2} = e^{1690k}$$
$$\ln \frac{1}{2} = 1690k$$
$$k = \frac{\ln 0.5}{1690} \approx -0.0041015$$

In 100 years there will be 7.679 grams of radium present.

$$N(100) = 8e^{\left[\frac{\ln 0.5}{1690}\right] \cdot 100} = 7.679$$

15. If we begin with 100 grams of carbon and $t = 5600$ years is the half life, then we use the half-life to find k.

$$N(5600) = 50 = 100 e^{5600k}$$

$$\frac{1}{2} = e^{5600k}$$

$$\ln 0.5 = 5600 k$$

$$k = \frac{\ln 0.5}{5600} \approx 0.000124$$

We now find t so that $N(t) = 30$.

$$30 = 100 e^{\left[\frac{\ln 0.50}{5600}\right]t}$$

$$0.3 = e^{\left[\frac{\ln 0.50}{5600}\right]t}$$

$$\ln 0.30 = \frac{\ln 0.50}{5600}t$$

$$t = \frac{5600 \ln(0.30)}{\ln 0.5} \approx 9727$$

The tree is 9727 years old.

17. Since the population obeys the law of uninhibited growth and $N(0) = 1500$, we use $N(24) = 2500$ to find the value of k.

$$N(t) = 1500 e^{kt}$$

$$2500 = 1500 e^{24k}$$

$$\ln\left(\frac{25}{15}\right) = 24k$$

$$k = \frac{1}{24} \cdot \ln\left(\frac{5}{3}\right)$$

After 3 days $t = 72$ hours, and the population is 6944 mosquitos.

$$N(72) = 1500 e^{72\left[\frac{1}{24}\ln\frac{5}{3}\right]} = 6944.44$$

19. The differential equation describing the population growth is

$$\frac{dN}{dt} = 3000 e^{2t/5}$$

We solve the equation to find the function that describes the population.

$$\int dN = \int 3000\, e^{2t/5}\, dt$$

$$N = \frac{3000\, e^{2t/5}}{\dfrac{2}{5}} + K$$

$$N(t) = 7500\, e^{2t/5} + K$$

To find K we use $N(0) = 7500$.

$$7500 = 7500 e^0 + K$$

$$K = 0$$

So $N(t) = 7500 e^{2t/5}$, and when $t = 5$

$$N(5) = 7500\, e^{2\,\cdot\,5/5} = 7500 e^2 = 55{,}417.9$$

There are 55,418 bacteria present.

21. (a) We use $N(0) = 10{,}000$ and $N(t_1) = 20{,}000$ to find the constant of proportionality k.

$$20{,}000 = 10{,}000\, e^{kt_1}$$

$$2 = e^{kt_1}$$

$$\ln 2 = k\, t_1$$

$$k = \frac{\ln 2}{t_1}$$

To find $N(t)$, we use $N(t_1 + 10) = 100{,}000$.

$$100{,}000 = 10{,}000\, e^{\frac{\ln 2}{t_1}(t_1 + 10)}$$

$$10 = e^{\frac{\ln 2}{t_1}(t_1 + 10)}$$

$$\ln 10 = \frac{\ln 2}{t_1}(t_1 + 10) = \ln 2 + \frac{10\ln 2}{t_1}$$

$$\ln 10 - \ln 2 = \frac{10\ln 2}{t_1}$$

$$\ln 5 = \frac{10\ln 2}{t_1}$$

$$t_1 = \frac{10\ln 2}{\ln 5} \approx 4.30677$$

and $k = \dfrac{\ln 2}{\dfrac{10 \ln 2}{\ln 5}} = \dfrac{\ln 5}{10}.$

So $N(t) = 10{,}000\, e^{\left(\frac{\ln 5}{10}\right)t} = 10{,}000(5^{t/10})$

(b) $N(20) = 10{,}000\, e^{\left(\frac{\ln 5}{10}\right)(20)} = 250{,}000$ bacteria

(c) The value of t_1 is 4.30667. So after 4.3 minutes there were 20,000 bacteria present.

23. $\dfrac{dA}{dt} = -\alpha A$ then $A(t) = A_0 e^{-\alpha t}$.

$$\frac{1}{2} = e^{-\alpha t}$$

$$\ln \frac{1}{2} = -\alpha t$$

$$t = -\frac{\ln 0.5}{1.5 \times 10^{-7}} = 4,620,981.2 \text{ years.}$$

25.

(a) The differential equation is $\dfrac{dp}{dx} = k\,p$.

We solve the equation.

$$\frac{dp}{p} = k\,dx$$

$$\ln p = kx + K$$

To find K we use $p(0) = 300$, and find $\ln 300 = 0 + K$ or $K = \ln 300$, and we then find

$$\ln p = kx + \ln 300$$

$$\ln p - \ln 300 = kx$$

$$\ln \frac{p}{300} = kx$$

$$\frac{p}{300} = e^{kx}$$

$$p = 300 e^{kx}$$

To find k we use the second boundary condition $p(200) = 150$.

$$150 = 300 e^{200\,k}$$

$$\frac{1}{2} = e^{200k}$$

$$\ln\left(\frac{1}{2}\right) = 200\,k$$

$$k = \frac{\ln 0.5}{200} \approx -0.0034657$$

So the price-demand equation is $p = 300 e^{\left(\frac{\ln 0.5}{200}\right)x}$.

(b) To sell 300 units, the price p should be

$$p = 300 e^{300\left(\frac{\ln 0.5}{200}\right)} = \$106.07$$

(c) To sell 350 units, the price p should be

$$p = 300e^{350\left(\frac{\ln 0.5}{200}\right)} = \$89.19$$

Chapter 15 Review

TRUE-FALSE ITEMS

1. True   3. False   5. False

7. True   9. False

FILL-IN-THE-BLANKS

1. $F'(x) = f(x)$ 3. integration by parts   5. 0

7. $\displaystyle\int_0^2 \sqrt{x^2 + 1}\, dx$

REVIEW EXERCISES

1. $F(x) = \dfrac{6x^6}{6} + K = x^6 + K$

3. $F(x) = \dfrac{x^4}{4} + \dfrac{x^2}{2} + K$

5. $f(x) = \dfrac{1}{\sqrt{x}} = x^{-1/2}$

   $F(x) = \dfrac{x^{1/2}}{\dfrac{1}{2}} + K = 2x^{1/2} + K = 2\sqrt{x} + K$

7. $\displaystyle\int 7\, dx = 7\int dx = 7x + K$

9. $\displaystyle\int (5x^3 + 2)\, dx = \dfrac{5x^4}{4} + 2x + K$

11. $\displaystyle\int (x^4 - 3x^2 + 6)\, dx = \dfrac{x^5}{5} - \dfrac{3x^3}{3} + 6x + K = \dfrac{x^5}{5} - x^3 + 6x + K$

13. $\displaystyle\int \dfrac{3}{x}\, dx = 3\int \dfrac{1}{x}\, dx = 3\ln|x| + K$

15. We use the method of substitution to evaluate the integral. Let $u = x^2 - 1$, then $du = 2x\,dx$.

$$\int \frac{2x}{x^2 - 1}\,dx = \int \frac{1}{u}\,du = \ln|u| + k = \ln|x^2 - 1| + K$$

17.
$$\int e^{3x}\,dx = \frac{1}{3}e^{3x} + K$$

19. We use the method of substitution to evaluate the integral.

Let $u = x^3 + 3x$. Then $du = (3x^2 + 3)\,dx$

$$du = 3(x^2 + 1)\,dx$$

$$\frac{1}{3}\,du = (x^2 + 1)\,dx$$

$$\int (x^3 + 3x)^5 (x^2 + 1)\,dx = \frac{1}{3}\int u^5\,du = \frac{1}{3} \cdot \frac{u^6}{6} + K = \frac{(x^3 + 3x)^6}{18} + K$$

21.
$$\int 2x(x - 3)\,dx = \int (2x^2 - 6x)\,dx = \frac{2x^3}{3} - \frac{6x^2}{2} + K = \frac{2x^3}{3} - 3x^2 + K$$

23. We use the method of substitution to evaluate the integral.

Let $u = 3x^2 + x$. Then $du = (6x + 1)\,dx$.

$$\int e^{3x^2 + x}(6x + 1)\,dx = \int e^u\,du = e^u + K = e^{3x^2 + x} + K$$

25. We use the method of substitution to evaluate the integral.

Let $u = x - 5$. Then $x = u + 5$ and $du = dx$.

$$\int x\sqrt{x - 5}\,dx = \int (u + 5)\sqrt{u}\,du = \int (u^{3/2} + 5u^{1/2})\,du = \frac{u^{5/2}}{\frac{5}{2}} + \frac{5u^{3/2}}{\frac{3}{2}} + K$$

$$= \frac{2(x - 5)^{5/2}}{5} + \frac{10(x - 5)^{3/2}}{3} + K$$

27. We use the method of integration by parts to evaluate the integral.

Choose $\quad u = x \qquad\qquad dv = e^{4x}\,dx$.

Then $\qquad du = dx \qquad\qquad v = \frac{1}{4}e^{4x}$

$$\int xe^{4x}\,dx = \int u \cdot dv = uv - \int v\,du$$

$$= x \cdot \frac{1}{4}e^{4x} - \int \frac{1}{4}e^{4x}\,dx$$

$$= \frac{xe^{4x}}{4} - \frac{1}{4} \cdot \frac{1}{4} e^{4x} + K = \frac{xe^{4x}}{4} - \frac{e^{4x}}{16} + K$$

29. We use the method of integration by parts to evaluate the integral.

Choose $u = \ln 2x$ $dv = x^{-2} dx$

Then $du = \dfrac{1}{x} dx$ $v = -x^{-1} = -\dfrac{1}{x}$

$$\int x^{-2} \ln 2x \, dx = \int u \cdot dv = uv - \int v \, du$$

$$= \ln 2x \cdot \left(-\frac{1}{x}\right) - \int \left(-\frac{1}{x}\right) \cdot \frac{1}{x} \, dx$$

$$= -\frac{\ln 2x}{x} + \int x^{-2} \, dx$$

$$= -\frac{\ln 2x}{x} - \frac{1}{x} + K$$

$$= -\frac{1}{x}(1 + \ln 2x) + K$$

31.
$$R(x) = \int R'(x) \, dx = \int (5x + 2) \, dx = \frac{5x^2}{2} + 2x + K$$

Since $R(0) = 0$, we have $K = 0$, and the revenue function R is

$$R(x) = \frac{5x^2}{2} + 2x$$

33.
$$C(x) = \int C'(x) \, dx = \int (5x + 120{,}000) \, dx = \frac{5x^2}{2} + 120{,}000x + K$$

Since fixed cost is \$7500, $C(0) = 7500$ and $K = 7500$. The cost function C is

$$C(x) = \frac{5x^2}{2} + 120{,}000x + 7500$$

The minimum cost occurs either at the vertex (since C is a quadratic function) or

at $x = 0$. At the vertex $x = -\dfrac{b}{2a} = -\dfrac{120{,}000}{2\left(\dfrac{5}{2}\right)} = -24{,}000$ which is negative and so not in

the domain. At $x = 0$, $C(0) = 7500$ and is minimum.

35.
(a) $R(x) = \int R'(x) dx = \int (500 - 0.01x) dx = 500x - \dfrac{0.01x^2}{2} + K$

Using the fact that $R(0) = 0$ we solve for K.
$$R(0) = 500(0) - 0.005(0) + K = 0$$
So $K = 0$, and $R(x) = 500x - 0.005x^2$

(b) Since $R(x)$ is a quadratic function, and since $a = 0.005$ is negative, the maximum value occurs at the x-value of the vertex, if it is in the domain.

$$x = -\frac{b}{2a} = -\frac{500}{2(-0.005)} = 50,000$$

Revenue is maximized when 50,000 televisions are sold.

(c) The maximum revenue that can be obtained is
$$R(50,000) = 500(50,000) - 0.005(50,000^2) = \$\,12,500,000$$

(d) If sales increase from 35,000 to 40,000 televisions, the revenue will increase by
$$R(40,000) - R(35,000) = 12,000,000 - 11,375,000 = \$\,625,000$$

37.
$$\int_{-2}^{1}(x^2 + 3x - 1)\,dx = \left(\frac{x^3}{3} + \frac{3x^2}{2} - x\right)\Bigg|_{-2}^{1} = \left(\frac{1}{3} + \frac{3}{2} - 1\right) - \left(\frac{(-2)^3}{3} + \frac{3(-2)^2}{2} - (-2)\right)$$
$$= \frac{1}{3} + \frac{3}{2} - 1 + \frac{8}{3} - 6 - 2 = -\frac{9}{2}$$

39.
$$\int_{4}^{9} 8\sqrt{x}\,dx = 8\int_{4}^{9} x^{1/2}\,dx = 8\left[\left(\frac{x^{3/2}}{\frac{3}{2}}\right)\Bigg|_{4}^{9}\right] = 8\left[\frac{2(9^{3/2})}{3} - 2\left(\frac{4^{3/2}}{3}\right)\right] = \frac{16}{3}[27 - 8] = \frac{304}{3}$$

41.
$$\int_{0}^{1}(e^x - e^{-x})\,dx = (e^x + e^{-x})\Big|_{0}^{1} = (e^1 + e^{-1}) - (e^0 + e^0) = e + e^{-1} - 2 = e + \frac{1}{e} - 2$$

43. We use the method of substitution to evaluate this integral.

Let $u = 3x + 2$. Then $du = 3dx$ and $\frac{1}{3}du = dx$.

When $x = 0$, $u = 3(0) + 2 = 2$, and when $x = 4$, $u = 3(4) + 2 = 14$.

$$\int_{0}^{4} \frac{dx}{(3x + 2)^2} = \frac{1}{3}\int_{2}^{14} \frac{1}{u^2}\,du = \frac{1}{3}\int_{2}^{14} u^{-2}\,du = \frac{1}{3}\left[\frac{u^{-1}}{-1}\Bigg|_{2}^{14}\right]$$
$$= -\frac{1}{3}\left[\frac{1}{14} - \frac{1}{2}\right] = -\frac{1}{3}\left[\frac{1 - 7}{14}\right] = \frac{2}{14} = \frac{1}{7}$$

45.
$$\int_{-2}^{2} e^{3x}\,dx = \frac{1}{3}e^{3x}\Big|_{-2}^{2} = \frac{1}{3}\left[e^6 - e^{-6}\right] \approx 134.475$$

47. We use integration by parts to evaluate this integral by choosing
$$u = x + 2 \qquad\qquad dv = e^{-x}\,dx$$
$$du = dx \qquad\qquad v = -e^{-x}$$

$$\int_0^1 u \, dv = \int_0^1 (x+2)e^{-x}\,dx = (x+2)\left(-e^{-x}\right)\Big|_0^1 - \int_0^1 \left(-e^{-x}\right)dx$$

$$= -(x+2)\,e^{-x}\Big|_0^1 + \int_0^1 e^{-x}\,dx$$

$$= -(x+2)\,e^{-x}\Big|_0^1 - e^{-x}\Big|_0^1$$

$$= \left[-3e^{-1} + 2e^0\right] - \left[e^{-1} - e^0\right]$$

$$= -3e^{-1} + 2 - e^{-1} + 1 = 3 - 4e^{-1} = 3 - \frac{4}{e}$$

49.
$$\int_0^9 f(x)\,dx = \int_0^5 f(x)\,dx + \int_5^9 f(x)\,dx$$

$$= 3 + (-2) = 1$$

Property 5 $\int_a^b f(x)\,dx = \int_a^c f(x)\,dx + \int_c^b f(x)\,dx$

51.
$$\int_9^5 g(x)\,dx = -\int_5^9 g(x)\,dx = -10$$

Property 3 $\int_a^b f(x)\,dx = -\int_b^a f(x)\,dx$

53.
$$A = \int_{-1}^2 (x^2 + 4)\,dx = \left(\frac{x^3}{3} + 4x\right)\Bigg|_{-1}^2$$

$$= \left[\frac{2^3}{3} + 4 \cdot 2\right] - \left[\frac{(-1)^3}{3} + 4 \cdot (-1)\right]$$

$$= \frac{8}{3} + 8 + \frac{1}{3} + 4$$

$$= 15$$

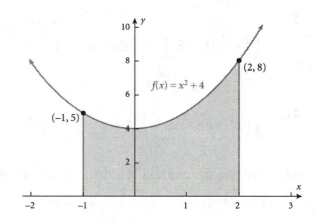

55.
$$A = \int_0^1 (e^x + x)\,dx = \left(e^x + \frac{x^2}{2}\right)\Bigg|_0^1$$

$$= \left[e^1 + \frac{1^2}{2}\right] - \left[e^0 + 0\right]$$

$$= e + \frac{1}{2} - 1$$

$$= e - \frac{1}{2}$$

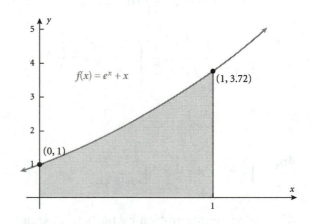

57.

$$A = -\int_0^2 (x^2 - x - 2)\,dx + \int_2^3 (x^2 - x - 2)\,dx = -\left(\frac{x^3}{3} - \frac{x^2}{2} - 2x\right)\Big|_0^2 + \left(\frac{x^3}{3} - \frac{x^2}{2} - 2x\right)\Big|_2^3$$

$$= -\left\{\left[\frac{2^3}{3} - \frac{2^2}{2} - 2(2)\right] - [0]\right\} + \left\{\left[\frac{3^3}{3} - \frac{3^2}{2} - 2(3)\right] - \left[\frac{2^3}{3} - \frac{2^2}{2} - 2(2)\right]\right\}$$

$$= \left[9 - \frac{9}{2} - 6\right] - 2\left[\frac{8}{3} - \frac{4}{2} - 4\right] = \left[3 - \frac{9}{2}\right] - 2\left[\frac{8}{3} - 6\right]$$

$$= 3 - \frac{9}{2} - \frac{16}{3} + 12 = \frac{31}{6}$$

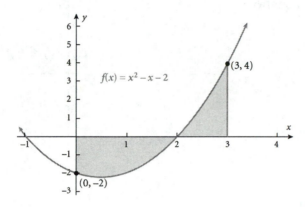

59. The points of intersection of the two graphs are found by solving the system of equations

$$\begin{cases} y = x^2 - 4 \\ x + y = 2 \end{cases}$$

We use substitution.

$$x + (x^2 - 4) = 2$$
$$x^2 + x - 6 = 0$$
$$(x + 3)(x - 2) = 0$$
$$x + 3 = 0 \quad \text{or} \quad x - 2 = 0$$
$$x = -3 \quad \text{or} \quad x = 2$$

When $x = -3$, $y = 5$, and when $x = 2$, $y = 0$. So the points of intersection are $(-3, 5)$ and $(2, 0)$.

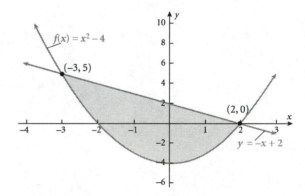

$$A = \int_{-3}^{2}\left[(2-x)-(x^2-4)\right]dx = \int_{-3}^{2}\left(-x^2-x+6\right)dx = \left(-\frac{x^3}{3}-\frac{x^2}{2}+6x\right)\Bigg|_{-3}^{2}$$

$$= \left[-\frac{2^3}{3}-\frac{2^2}{2}+6(2)\right]-\left[-\frac{(-3)^3}{3}-\frac{(-3)^2}{2}+6(-3)\right]$$

$$= -\frac{8}{3}-2+12-9+\frac{9}{2}+18 = \frac{125}{6}$$

61. The points of intersection of the two graphs
are found by solving the system of equations

$$\begin{cases} f(x)=x^3 \\ g(x)=4x \end{cases}$$

$$x^3 = 4x$$
$$x^3 - 4x = 0$$
$$x(x^2-4)=0$$
$$x=0 \quad \text{or} \quad x^2=4$$
$$x=\pm 2$$

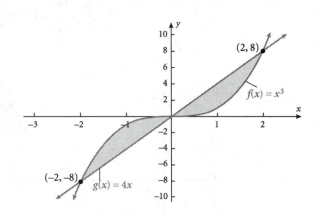

The points of intersection are (0, 0),
(−2, −8), and (2, 8).

$$A = \int_{-2}^{0}\left[f(x)-g(x)\right]dx + \int_{0}^{2}\left[g(x)-f(x)\right]dx = \int_{-2}^{0}\left[x^3-4x\right]dx + \int_{0}^{2}\left[4x-x^3\right]dx$$

$$= \left(\frac{x^4}{4}-\frac{4x^2}{2}\right)\Bigg|_{-2}^{0} + \left(\frac{4x^2}{2}-\frac{x^4}{4}\right)\Bigg|_{0}^{2}$$

$$= \left[0-\left(\frac{(-2)^4}{4}-2(-2)^2\right)\right]+\left[\left(2(-2)^2-\frac{(-2)^4}{4}\right)-0\right]$$

$$= -4+8+8-4 = 8$$

63.

$$\int_{2000}^{2500}P'(x)\,dx = \int_{2000}^{2500}(9-0.004x)\,dx = \left(9x-\frac{0.004x^2}{2}\right)\Bigg|_{2000}^{2500}$$

$$= \left[9(2500)-0.002(2500^2)\right]-\left[9(2000)-0.002(2000)^2\right]=0$$

There is no change in monthly profit obtained by increasing production from 2000 to
2500 pairs of jeans.

65. (a) The time t_{max} of optimal termination is found when

$$R'(x)=C'(x)$$
$$-10t = 2t-12$$

$$-12t = -12$$
$$t_{\max} = 1$$

To maximize profit the owner should keep the machine for 1 time unit.

(b) The total profit that the machine will generate in the unit of time is

$$P(t_{\max}) = \int_0^1 \left[R'(t) - C'(t) \right] dt = \int_0^1 \left[-10t - (2t - 12) \right] dt$$

$$= \left(\frac{-12t^2}{2} + 12t \right) \Bigg|_0^1 = [-6 + 12] - [0] = 6 \text{ monetary units.}$$

67. (a)

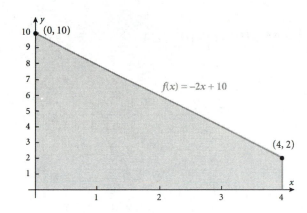

Divide the interval into 4 subintervals of length $\Delta x = 1$.

$$[0, 1] \quad [1, 2] \quad [2, 3] \quad [3, 4]$$

Evaluate f at the endpoints.

$$f(0) = 10$$
$$f(1) = -2 + 10 = 8$$
$$f(2) = -4 + 10 = 6$$
$$f(3) = -6 + 10 = 4$$
$$f(4) = -8 + 10 = 2$$

(b) Using the left endpoints, we have
$$A \approx f(0) \cdot 1 + f(1) \cdot 1 + f(2) \cdot 1 + f(3) \cdot 1$$
$$\approx (10 + 8 + 6 + 4) \cdot 1 = 28$$

(c) Using right endpoints, we have
$$A \approx f(1) \cdot 1 + f(2) \cdot 1 + f(3) \cdot 1 + f(4) \cdot 1$$
$$\approx (8 + 6 + 4 + 2) \cdot 1 = 20$$

(d) Divide the interval into 8 subintervals of length $\Delta x = \dfrac{1}{2}$.

$$\left[0, \frac{1}{2}\right] \quad \left[\frac{1}{2}, 1\right] \quad \left[1, \frac{3}{2}\right] \quad \left[\frac{3}{2}, 2\right] \quad \left[2, \frac{5}{2}\right] \quad \left[\frac{5}{2}, 3\right] \quad \left[3, \frac{7}{2}\right] \quad \left[\frac{7}{2}, 4\right]$$

Evaluate f at the new endpoints.

$$f\left(\frac{1}{2}\right) = -2\left(\frac{1}{2}\right) + 10 = 9 \qquad f\left(\frac{3}{2}\right) = -2\left(\frac{3}{2}\right) + 10 = 7$$

$$f\left(\frac{5}{2}\right) = -2\left(\frac{5}{2}\right) + 10 = 5 \qquad f\left(\frac{7}{2}\right) = -2\left(\frac{7}{2}\right) + 10 = 3$$

Using the left endpoints, we have
$$A \approx f(0) \cdot \frac{1}{2} + f\left(\frac{1}{2}\right) \cdot \frac{1}{2} + f(1) \cdot \frac{1}{2} + f\left(\frac{3}{2}\right) \cdot \frac{1}{2} + f(2) \cdot \frac{1}{2} + f\left(\frac{5}{2}\right) \cdot \frac{1}{2} + f(3) \cdot \frac{1}{2} + f\left(\frac{7}{2}\right) \cdot \frac{1}{2}$$

$$\approx (10+9+8+7+6+5+4+3) \cdot \frac{1}{2} = 26$$

(e) Using the right endpoints, we have

$$A \approx f\left(\frac{1}{2}\right) \cdot \frac{1}{2} + f(1) \cdot \frac{1}{2} + f\left(\frac{3}{2}\right) \cdot \frac{1}{2} + f(2) \cdot \frac{1}{2} + f\left(\frac{5}{2}\right) \cdot \frac{1}{2} + f(3) \cdot \frac{1}{2} + f\left(\frac{7}{2}\right) \cdot \frac{1}{2} + f(4) \cdot \frac{1}{2}$$

$$\approx (9+8+7+6+5+4+3+2) \cdot \frac{1}{2} = 22$$

(f) $A = \int_0^4 f(x)\, dx = \int_0^4 (-2x+10)\, dx$

(g) $\int_0^4 (-2x+10)\, dx = \left. \left(\frac{-2x^2}{2} + 10x \right) \right|_0^4 = \left[-4^2 + 10(4) \right] - [0] = 24$

69. $\displaystyle\int_1^{10} x \ln xx\, dx \approx 90.38$

```
fnInt(Xln(X),X,1
,10)
      90.37925465
```

71. The general solution to the differential equation $\dfrac{dy}{dx} = x^2 + 5x - 10$ is

$$y = \int \left(x^2 + 5x - 10 \right) dx$$

$$y = \frac{x^3}{3} + \frac{5x^2}{2} - 10x + K$$

We use $x = 0$ and $y = 1$ to find K.

$$1 = 0 + 0 - 0 + K$$
$$1 = K$$

The particular solution to the differential equation is

$$y = \frac{x^3}{3} + \frac{5x^2}{2} - 10x + 1$$

73. The general solution to the differential equation $\dfrac{dy}{dx} = e^{2x} - x$ is

$$y = \int \left(e^{2x} - x \right) dx$$

$$y = \frac{1}{2} e^{2x} - \frac{x^2}{2} + K$$

We use $x = 0$ and $y = 3$ to find K.

$$3 = \frac{1}{2}e^0 - 0 + K$$

$$\frac{5}{2} = K$$

The particular solution to the differential equation is

$$y = \frac{1}{2}e^{2x} - \frac{x^2}{2} + \frac{5}{2}$$

75. The general solution to the differential equation $\dfrac{dy}{dx} = 10y$ is

$$\frac{dy}{y} = 10\,dx$$

$$\int \frac{dy}{y} = \int 10\,dx$$

$$\ln|y| = 10x + K$$

$$y = e^{10x} + K$$

We use $x = 0$ and $y = 1$ to find K.

$$\ln 1 = 0 + K$$

$$0 = K$$

The particular solution to the differential equation is

$$y = e^{10x}$$

77. $N(0) = 2000 \qquad N(2) = 3 \cdot 2000 = 6000$

$$N(t) = N(0)\,e^{kt} = 2000\,e^{kt}$$

We use $N(2) = 6000$ to find k.

$$6000 = 2000\,e^{kt}$$

$$3 = e^{k \cdot 2}$$

$$\ln 3 = 2k$$

$$k = \frac{\ln 3}{2}$$

So in $4\frac{1}{2}$ hours $t = \frac{9}{2}$, and there will be

$$N\left(\frac{9}{2}\right) = 2000e^{\left(\frac{\ln 3}{2}\right)\left(\frac{9}{2}\right)} = 23{,}689 \text{ bacteria.}$$

79.

If $A(0) = 1$, then $A(5600) = \dfrac{1}{2}(1) = 0.5$, and $A(t) = 0.4(1) = 0.4$.

$$A(t) = A(0)e^{kt} = e^{kt}$$

We use $A(5600)$ to find k.

$$0.5 = e^{5600k}$$

$$\ln 0.5 = 5600\,k$$

$$\frac{\ln 0.5}{5600} = k$$

We now solve for t, the age of the bones.

$$0.4 = e^{\left(\frac{\ln 0.5}{5600}\right)t}$$

$$\ln 0.4 = \frac{\ln 0.5}{5600}\,t$$

$$t = \frac{5600\ln 0.4}{\ln 0.5} \approx 7402.8$$

The bones are 7403 years old.

81. $E'(t) = 0.02t^2 + t \qquad E(0) = 5$

$$E(t) = \int E'(t)\,dt = \int \left(0.02t^2 + t\right) dt$$

$$= \frac{0.02t^3}{3} + \frac{t^2}{2} + K = = \frac{t^3}{150} + \frac{t^2}{2} + 5$$

Since $E(0) = 5$, then $K = 5$, and

$$E(5) == \frac{0.02(5)^3}{3} + \frac{(5)^2}{2} + 5 = 18.33 \text{ million dollars.}$$

83.

$$\int_8^{18} f(x)\,dx = \int_8^{18} 700x^{-0.1}\,dx = \left(700 \cdot \frac{x^{0.9}}{0.9} \right)\Bigg|_8^{18}$$

$$= \frac{700}{0.9}\left[18^{0.9} - 8^{0.9}\right] = 5431.77$$

Margo should allow an additional 5432 labor hours to produce the 500 additional tennis ball servers.

85. (a) At market equilibrium $D(x) = S(x)$.

$$12 - \frac{x^*}{50} = \frac{x^*}{20} + 5$$

$$1200 - 2x^* = 5x^* + 500$$

$$700 = 7x^*$$

$$x^* = 100$$

The price p^* at $x^* = 100$ is $D(100) = 12 - \dfrac{100}{50} = 12 - 2 = 10$.

So at market equilibrium 100 units will be sold at \$10 each.

(b)   The consumer's surplus $CS = \int_0^{x^*} D(x)\, dx - p^* x^*$

$$CS = \int_0^{100} \left(12 - \frac{x}{50}\right) dx - 10 \cdot 100 = \left(12x - \frac{x^2}{100}\right)\bigg|_0^{100} - 1000$$

$$= 12(100) - \frac{100^2}{100} - 1000 = 1200 - 100 - 1000 = 100$$

The supplier's surplus $PS = p^* s^* - \int_0^{x^*} S(x)\, dx$.

$$PS = 10 \cdot 100 - \int_0^{x^*} \left(\frac{x}{20} + 5\right) dx = 1000 - \left[\left(\frac{x^2}{40} + 5x\right)\bigg|_0^{100}\right]$$

$$= 1000 - \left[\frac{100^2}{40} + 5(100)\right]$$

$$= 1000 - [250 + 500] = \$250$$

(c)

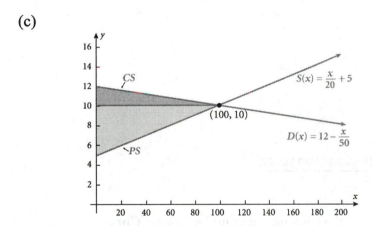

87.

(a)
$$\frac{dp}{dx} = k\, p$$

$$\frac{dp}{p} = k\, dx$$

$$\ln p = k\, x + k$$

We find k by using the facts that $x = 0$ when $p = 800$.

$$\ln 800 = k \cdot 0 + k$$

So, $\ln p = kx + \ln 800$

$$\ln p - \ln 800 = kx$$

$$\ln \frac{p}{800} = kx$$

$$\frac{p}{800} = e^{kx}$$

$$p = 800 e^{kx}$$

We use the fact that $x = 80$ when $p = 600$ to find k.

$$600 = 800 \, e^{80k}$$
$$0.75 = e^{80\,k}$$
$$\ln 0.75 = 80 \, k$$
$$k = \frac{\ln 0.75}{80} \approx -0.003596$$

The demand equation is $p(x) = 800 \, e^{\left(\frac{\ln 0.75}{80}\right)x}$

(b) If the manufacturer produces $x = 200$ units, the price will be

$$p(200) = 800 \, e^{200\left(\frac{\ln 0.75}{80}\right)} = \$389.71$$

(c) If the manufacturer produces $x = 250$ units, the price will be

$$p(250) = 800 \, e^{250\left(\frac{\ln 0.75}{80}\right)} = \$325.58$$

CHAPTER 15 PROJECT

1.
$$A = \int_0^1 x \, dx = \left(\frac{x^2}{2}\right)\bigg|_0^1 = \frac{1}{2}$$

3.
$$G = \frac{\text{area of the top piece}}{\text{area of the triangle}} = \frac{\text{area of the top piece}}{\dfrac{1}{2}}$$

$$= \frac{\text{area of the triangle - area under Lorenz Curve}}{\dfrac{1}{2}}$$

$$= \frac{\dfrac{1}{2} - \int_0^1 L(x)\, dx}{\dfrac{1}{2}}$$

$$= 1 - 2\int_0^1 L(x)\, dx$$

5. Answers will vary.

7. The Gini coefficient for the US economy in 1993 is

$$G = 1 - 2\int_0^1 \left(0.442x^2 + 5.8x^3 - 23.71x^4 + 31.036x^5 + 5.71x^6 - 38.842x^7 + 20.564x^8\right) dx$$

$$= 1 - 2\left(\frac{0.442x^3}{3} + \frac{5.8x^4}{4} - \frac{23.71x^5}{5} + \frac{31.036x^6}{6} + \frac{5.71x^7}{7} - \frac{38.842x^8}{8} + \frac{20.564x^9}{9}\right)\Bigg|_0^1$$

$$= 1 - 2\left[\left(\frac{0.442}{3} + \frac{5.8}{4} - \frac{23.71}{5} + \frac{31.036}{6} + \frac{5.71}{7} - \frac{38.842}{8} + \frac{20.564}{9}\right) - 0\right]$$

$$= 0.453$$

MATHEMATICAL QUESTIONS FROM PROFESSIONAL EXAMS

1. (b)

Evaluate the integral using integration by parts. Choose

$$u = \ln x \qquad\qquad dv = \frac{1}{x}\,dx$$

$$dv = \frac{1}{x}\,dx \qquad\qquad v = \ln x$$

$$\int_1^e \frac{1}{x}\ln x\,dx = (\ln x)^2\Big|_1^e - \int_1^e \frac{1}{x}\ln x\,dx$$

$$2\int_1^e \frac{1}{x}\ln x\,dx = (\ln x)^2\Big|_1^e$$

$$\int_1^e \frac{1}{x}\ln x\,dx = \frac{1}{2}\left[(\ln x)^2\Big|_1^e\right] = \frac{1}{2}\left[1^2 - 0\right] = \frac{1}{2}$$

3. (c)

If $\int_1^b f(x)\,dx = b^2 e^b - e$, then $\int f(x)\,dx = x^2 e^x + K$, and the derivative of $x^2 e^x$ is $f(x)$.

$$\frac{d}{dx}\left(x^2 e^x\right) = x^2 e^x + 2x e^x = f(x).$$

5. (c)

Let $t = 0$ denote 4 years ago. Then $P(0) = 25{,}000$ and $P(4) = 36{,}000$.

$$P(t) = P(0)e^{kt}$$

$$P(t) = 25{,}000\,e^{kt}$$

We use $P(4)$ to find k.

$$36{,}000 = 25{,}000\,e^{4k}$$

$$1.44 = e^{4k}$$

$$\ln 1.44 = 4k$$

$$k = \frac{\ln 1.44}{4}$$

Sis years from now $t = 10$, and

$$P(10) = 25{,}000\,e^{10\left(\frac{\ln 1.44}{4}\right)} = 62{,}208$$

Chapter 16
Other Applications and
Extensions of the Integral

16.1 Improper Integrals

1.
$\displaystyle\int_0^\infty x^2\,dx$ is an improper integral because the upper limit of integration is not finite.

3.
$\displaystyle\int_0^1 \frac{1}{x}\,dx$ is an improper integral because the function is not continuous at $x = 0$.

5.
$\displaystyle\int_1^2 \frac{dx}{x-1}$ is an improper integral because the function is not continuous at $x = 1$.

7.
$$\int_1^\infty e^{-4x}\,dx = \lim_{b\to\infty}\int_1^b e^{-4x}\,dx = \lim_{b\to\infty}\left[-\frac{1}{4}e^{-4x}\,\Big|_1^b\right]$$

$$= -\frac{1}{4}\lim_{b\to\infty}\left(e^{-4b}\right) + \frac{1}{4}\lim_{b\to\infty}\left(e^{-4}\right) = -\frac{1}{4}\lim_{b\to\infty}\left(\frac{1}{e^{4b}}\right) + \frac{1}{4}e^{-4}$$

$$= \frac{1}{4}e^{-4}$$

9.
$$\int_0^\infty \sqrt{x}\,dx = \lim_{b\to\infty}\int_0^\infty x^{1/2}\,dx = \lim_{b\to\infty}\left[\frac{x^{3/2}}{\dfrac{3}{2}}\,\Bigg|_0^b\right]$$

$$= \lim_{b\to\infty}\left[\frac{2b^{3/2}}{3} - \frac{2\left(0^{3/2}\right)}{3}\right] = \frac{2}{3}\lim_{b\to\infty} b^{2/3} = \infty$$

Since the limit is infinite, the integral has no value.

11.
$$\int_{-1}^0 \frac{1}{\sqrt[5]{x}}\,dx = \lim_{b\to 0^-}\int_{-1}^b x^{-1/5}\,dx = \lim_{b\to 0^-}\left[\frac{x^{4/5}}{\dfrac{4}{5}}\,\Bigg|_{-1}^b\right]$$

$$= \lim_{b\to 0^-}\left[\frac{5b^{4/5}}{4} - \frac{5(-1)^{4/5}}{4}\right]$$

$$= \lim_{b\to 0^-}\frac{5b^{4/5}}{4} - \lim_{b\to 0^-}\frac{5}{4}$$

$$= 0 - \frac{5}{4} = -\frac{5}{4}$$

13.
$$\int_0^1 \frac{1}{x}\, dx = \lim_{a \to 0^+} \int_a^1 \frac{1}{x}\, dx = \lim_{a \to 0^+} \left[\ln x \Big|_a^1 \right]$$
$$= \lim_{a \to 0^+} \left[\ln 1 - \ln a \right] = - \lim_{a \to 0^+} \left[\ln a \right] = \infty$$

Since the limit is infinite, the integral has no value.

15.
$$A = \int_0^1 f(x)\, dx = \int_0^1 \frac{1}{\sqrt{x}}\, dx = \lim_{a \to 0^+} \int_a^1 x^{-1/2}\, dx = \lim_{a \to 0^+} \left[\frac{x^{1/2}}{\frac{1}{2}} \Big|_a^1 \right]$$
$$= 2 \left[\lim_{a \to 0^+} \left(1 - a^{1/2} \right) \right] = 2(1) = 2$$

17.
$$V(t) = \int_0^\infty 5124 e^{-0.05t}\, dt = \lim_{t \to b} \int_0^b 5124 e^{-0.05t}\, dt = 5124 \lim_{t \to b} \frac{e^{-0.05t}}{-0.05} \Big|_0^b$$
$$= -\frac{5124}{0.05} \lim_{t \to b} \left[e^{-0.05b} - e^0 \right]$$
$$= -\frac{5124}{0.05} \lim_{t \to b} \left[\frac{1}{e^{0.05b}} - 1 \right]$$
$$= \frac{5124}{0.05} = \$102{,}480$$

19. (a) Answers will vary.

(b) Total reaction is
$$\int_0^\infty t e^{-t^2}\, dt = \lim_{b \to \infty} \int_0^b t e^{-t^2}\, dt$$

We use the method of substitution to evaluate the integral.

Let $u = -t^2$. Then $du = -2t\, dt$, and $-\frac{1}{2} du = t\, dt$.

When $t = 0$, $u = 0$, and when $t = b$, $u = -b^2$.
$$\int_0^\infty t e^{-t^2}\, dt = \lim_{b \to \infty} \left[-\frac{1}{2} \int_0^{-b^2} e^u\, du \right] = -\frac{1}{2} \lim_{b \to \infty} \left[e^u \Big|_0^{-b^2} \right]$$
$$= -\frac{1}{2} \lim_{b \to \infty} \left[e^{-b^2} - e^0 \right] = -\frac{1}{2}[0 - 1] = \frac{1}{2}$$

21. (a)
$$\int_{-1}^1 \frac{1}{x^2}\, dx = \int_{-1}^0 x^{-2}\, dx + \int_0^1 x^{-2}\, dx = \lim_{c \to 0^-} \int_{-1}^c x^{-2}\, dx + \lim_{c \to 0^+} \int_c^1 x^{-2}\, dx$$

$$= \lim_{c \to 0^-}\left[-x^{-1}\Big|_{-1}^{c}\right] + \lim_{c \to 0^+}\left[-x^{-1}\Big|_{c}^{1}\right]$$

$$= \lim_{c \to 0^-}\left[-\frac{1}{c}+1\right] + \lim_{c \to 0^+}\left[-1+\frac{1}{c}\right] = \infty$$

Since the limit is infinite, the integral has no value.

(b) $$\int_0^4 \frac{x\,dx}{\sqrt[3]{x^2-4}} = \int_0^2 \frac{x\,dx}{\sqrt[3]{x^2-4}} + \int_2^4 \frac{x\,dx}{\sqrt[3]{x^2-4}} = \lim_{c \to 2^-}\int_0^c \frac{x\,dx}{\sqrt[3]{x^2-4}} + \lim_{c \to 2^+}\int_c^4 \frac{x\,dx}{\sqrt[3]{x^2-4}}$$

We use the method of substitution to evaluate the integrals.

Let $u = x^2 - 4$. Then $du = 2x\,dx$ and $\frac{1}{2}du = x\,dx$.

When $x = 0$, $u = -4$; when $x = 2$, $u = 0$; and when $x = 4$, $u = 12$.

$$\int_0^4 \frac{x\,dx}{\sqrt[3]{x^2-4}} = \lim_{c \to 2^-}\left[\frac{1}{2}\int_{-4}^{c^2-4} u^{-1/3}\,du\right] + \lim_{c \to 2^+}\left[\frac{1}{2}\int_{c^2-4}^{12} u^{-1/3}\,du\right]$$

$$= \frac{1}{2}\lim_{c \to 2-}\left[\frac{3u^{2/3}}{2}\Big|_{-4}^{c^2-4}\right] + \frac{1}{2}\lim_{c \to 2+}\left[\frac{3u^{2/3}}{2}\Big|_{c^2-4}^{12}\right]$$

$$= \frac{3}{4}\left\{\lim_{c \to 2^-}\left[(c^2-4)^{2/3}-(-4)^{2/3}\right] + \lim_{c \to 2^+}\left[12^{2/3}-(c^2-4)^{2/3}\right]\right\}$$

$$= \frac{3}{4}\left\{\left[0^{2/3}-4^{2/3}\right] + \left[12^{2/3}-0\right]\right\}$$

$$= \frac{3}{4}\left[-4^{2/3}+12^{2/3}\right] \approx 2.041$$

16.2 Average Value of a Function

1.

$$AV = \frac{1}{1-0}\int_0^1 x^2\,dx = \frac{x^3}{3}\Big|_0^1 = \frac{1}{3}-0 = \frac{1}{3}$$

3.

$$AV = \frac{1}{1-(-1)}\int_{-1}^1 (1-x^2)\,dx = \frac{1}{2}\left[\left(x-\frac{x^3}{3}\right)\Big|_{-1}^1\right]$$

$$= \frac{1}{2}\left[\left(1-\frac{1}{3}\right)-\left(-1-\frac{(-1)^3}{3}\right)\right]$$

$$= \frac{1}{2}\left[1-\frac{1}{3}+1-\frac{1}{3}\right] = \frac{1}{2}\left[\frac{4}{3}\right] = \frac{2}{3}$$

5.
$$AV = \frac{1}{5-1} \int_1^5 3x\, dx = \frac{1}{4}\left[\frac{3x^2}{2}\Bigg|_1^5\right] = \frac{1}{4}\left[\frac{3(5^2)}{2} - \frac{3(1^2)}{2}\right] = \frac{1}{4}\left[\frac{75}{2} - \frac{3}{2}\right] = \frac{72}{8} = 9$$

7.
$$AV = \frac{1}{2-(-2)} \int_{-2}^2 \left(-5x^4 + 4x - 10\right) dx = \frac{1}{4}\left[\left(\frac{-5x^5}{5} + \frac{4x^2}{2} - 10x\right)\Bigg|_{-2}^2\right]$$
$$= \frac{1}{4}\left[\left(-2^5 + 2^3 - 10 \cdot 2\right) - \left(-(-2)^5 + 2(-2)^2 - 10(-2)\right)\right]$$
$$= \frac{1}{4}\left[-32 + 8 - 20 - 32 - 8 - 20\right] = \frac{1}{4}\left[-104\right] = -26$$

9.
$$AV = \frac{1}{1-0} \int_0^1 e^x\, dx = e^x\Big|_0^1 = e^1 - e^0 = e - 1$$

11. The average value of the population during the next 20 years is
$$AV = \frac{1}{20-0} \int_0^{20} \left(6 \cdot 10^9\right) e^{0.03t}\, dt = \frac{6 \cdot 10^9}{20} \int_0^{20} e^{0.03t}\, dt = 3 \cdot 10^8 \frac{e^{0.03t}}{0.03}\Bigg|_0^{20}$$
$$= 3 \cdot 10^8 \cdot \frac{\left(e^{0.6} - 1\right)}{0.03} \approx 8.22 \cdot 10^9$$

13. The average temperature AT of the rod is
$$AT = \frac{1}{3-0} \int_0^3 25x\, dx = \frac{1}{3}\left[\frac{25x^2}{2}\Bigg|_0^3\right] = \frac{1}{3}\left[\frac{25 \cdot 3^2}{2} - 0\right] = \frac{75}{2} = 37.5^\circ\, C.$$

15. If the car is accelerating at a rate of 3 meters per second per second, its velocity is
$$v = \int 3\, dt = 3t + K$$

Since at time $t = 0$, the car is at rest, we have $v(0) = 0$. We use this condition to solve for K.
$$v(0) = 3(0) + K = 0$$
$$K = 0$$

So the average speed during the first 8 seconds of acceleration is
$$AV = \frac{1}{8-0} \int_0^8 3t\, dt = \frac{1}{8}\left[\frac{3t^2}{2}\Bigg|_0^8\right] = \frac{1}{8}\left[\frac{3 \cdot 8^2}{2} - 0\right] = 12 \text{ meters per}$$
second.

17.
$$AR = \frac{1}{6-1} \int_1^6 \left(-4.43x^3 + 46.17x^2 - 132.5x + 290\right) dx$$
$$= \frac{1}{5}\left[\left(\frac{-4.43x^4}{4} + \frac{46.17x^3}{3} - \frac{132.5x^2}{2} + 290x\right)\Bigg|_1^6\right]$$

$$= \frac{1}{5}\left[\left(\frac{-4.43\left(6^4\right)}{4} + \frac{46.17\left(6^3\right)}{3} - \frac{132.5\left(6^2\right)}{2} + 290(6)\right) - \left(\frac{-4.43}{4} + \frac{46.17}{3} - \frac{132.5}{2} + 290\right)\right]$$

$$= \frac{1}{5}\left[1243.92 - 238.0325\right] = 201.1775$$

The average annual revenue of Exxon-Mobil Corporation between 1997 and 2002 is 201.1775 billion dollars.

19.
$$AR = \frac{1}{90-0} \int_0^{90} \left(-0.000414x + 0.206748\right)\, dx$$

$$= \frac{1}{90}\left[\left(-0.000414\,\frac{x^2}{2} + 0.206748x\right)\Big|_0^{90}\right]$$

$$= \frac{1}{90}\left[\left(-0.000414\left(\frac{90^2}{2}\right) + 0.206748(90)\right) - 0\right]$$

$$= -0.000414\,(45) + 0.206748(1) = 0.188118$$

There was an average of 0.188 inches of rain per day during the first 90 days of the year.

16.3 Continuous Probability Functions

1. We show the two conditions are satisfied.

Condition 1: $f(x) = \frac{1}{2} > 0$ everywhere.

Condition 2: $\int_0^2 f(x)\, dx = 1$

$$\int_0^2 \frac{1}{2}\, dx = \frac{1}{2}x\Big|_0^2 = \frac{1}{2}(2-0) = 1$$

So f is a probability density function.

3. We show the two conditions are satisfied.

Condition 1: $f(x) = 2x \geq 0$ when $x \geq 0$. So f is nonnegative on the interval [0, 1].

Condition 2: $\int_0^1 f(x)\, dx = 1$

$$\int_0^1 2x\, dx = \frac{2x^2}{2}\Big|_0^1 = 1^2 - 0 = 1$$

So f is a probability density function.

5.   We show the two conditions are satisfied.

Condition 1: $f(x) = \dfrac{3}{250}(10x - x^2)$. We find where $f = 0$, use the points to separate the number line into 3 parts and test a point from each part for its sign.

$$\frac{3}{250}(10x - x^2) = 0$$

$$10x - x^2 = 0$$

$$x(10 - x) = 0$$

$$x = 0 \quad \text{or} \quad 10 - x = 0$$

$$x = 10$$

When $x = -1, f(-1) = -0.132$
When $x = 1, f(1) = 0.108$
When $x = 11, f(11) = -0.132$

We conclude that f is positive on the interval $(0, 10)$. So f is nonnegative on the interval $[0, 5]$.

Condition 2: $\displaystyle\int_0^5 f(x)\, dx = 1$

$$\int_0^5 \frac{3}{250}(10x - x^2)\, dx = \frac{3}{250}\left[\left(\frac{10x^2}{2} - \frac{x^3}{3}\right)\Bigg|_0^5\right]$$

$$= \frac{3}{250}\left[\left(\frac{250}{2} - \frac{125}{3}\right) - (0)\right] = \frac{3}{250}\left[\frac{750}{6} - \frac{250}{6}\right]$$

$$= \frac{3}{250} \cdot \frac{500}{6} = 1$$

So f is a probability density function.

7.   We show the two conditions are satisfied.

Condition 1: $f(x) = \dfrac{1}{x} > 0$ when $x > 0$. So f is nonnegative on the interval $[1, e]$.

Condition 2: $\displaystyle\int_1^e f(x)\, dx = 1$

$$\int_1^e \frac{1}{x}\, dx = \ln x\Big|_1^e = \ln e - \ln 1 = 1 - 0 = 1$$

So f is a probability density function.

9. $\displaystyle\int_0^3 k\, dx = kx\Big|_0^3 = 3k - 0 = 1.$ So $k = \dfrac{1}{3}$.

11. $\displaystyle\int_0^2 kx\, dx = k \cdot \frac{x^2}{2}\Bigg|_0^2 = k(2 - 0) = 1.$ So $k = \dfrac{1}{2}$.

13.

$$\int_0^5 k\left(10x - x^2\right) dx = k \int_0^5 \left(10x - x^2\right) dx = k\left[\left(\frac{10x^2}{2} - \frac{x^3}{3}\right)\Big|_0^5\right] = k\left[125 - \frac{125}{3}\right] = \frac{250}{3}k = 1$$

So $k = \dfrac{3}{250}$.

15.

$$\int_1^2 \frac{k}{x}\, dx = k \int_1^2 \frac{1}{x}\, dx = k\left[\ln x \Big|_1^2\right] = k\left[\ln 2 - \ln 1\right] = k \ln 2 = 1. \text{ So } k = \frac{1}{\ln 2}.$$

17.

$$E(x) = \int_0^2 x f(x)\, dx = \int_0^2 \frac{1}{2} x\, dx = \frac{x^2}{4}\Big|_0^2 = \frac{1}{4}(4 - 0) = 1$$

19.

$$E(x) = \int_0^1 x f(x)\, dx = \int_0^1 2x^2\, dx = \frac{2x^3}{3}\Big|_0^1 = \frac{2}{3}(1 - 0) = \frac{2}{3}$$

21.

$$E(x) = \int_0^5 x f(x)\, dx = \int_0^5 \frac{3}{250}\left(10x^2 - x^3\right) dx = \frac{3}{250}\left[\left(\frac{10x^3}{3} - \frac{x^4}{4}\right)\Big|_0^5\right]$$

$$= \frac{3}{250}\left(\frac{10\left(5^3\right)}{3} - \frac{5^4}{4}\right) = \frac{3}{250}\left(\frac{1250}{3} - \frac{625}{4}\right)$$

$$= \frac{3}{250}\left(\frac{5000 - 1875}{12}\right) = \frac{1}{250} \cdot \frac{3125}{4} = \frac{25}{8}$$

23.

$$E(x) = \int_1^e x f(x)\, dx = \int_1^e dx = x\Big|_1^e = e - 1$$

25.

$$P(1 \le X \le 3) = \int_1^3 \frac{1}{5}\, dx = \frac{1}{5} x\Big|_1^3 = \frac{1}{5}(3 - 1) = \frac{2}{5}$$

There is a 40% probability that the number selected is from the interval [1, 3].

27. Let the random variable X denote the interval between incoming calls.

$$P(X \ge 6) = 1 - P(X < 6) = 1 - \int_0^6 0.5\, e^{-0.5x}\, dx$$

$$= 1 - \left[-e^{-0.5x}\Big|_0^6\right]$$

$$= 1 + e^{-3} - e^0 = e^{-3} \approx 0.0498$$

The probability of waiting at least 6 minutes for the next call is 4.98%.

29.

$$P(X < 5) = \int_0^5 0.4e^{-0.4x}\,dx = -e^{-0.4x}\Big|_0^5 = -e^{-2} + e^0 = 1 - e^{-2} \approx 0.86466$$

The probability a subject makes a choice in fewer than 5 seconds is 0.865.

31.

Since the average life of the light bulb is 2000 hours, $\lambda = \dfrac{1}{2000}$, and

$$f(x) = \begin{cases} \dfrac{1}{2000}e^{-x/2000} & x \ge 0 \\ 0 & x < 0 \end{cases}$$

(a) If the random variable X denotes the length of the light bulb's life, then

$$P(1800 \le X \le 2200) = \int_{1800}^{2200} \frac{1}{2000}e^{-x/2000}\,dx$$

$$= -e^{-x/2000}\Big|_{1800}^{2200}$$

$$= -e^{-22/20} + e^{-18/20} \approx 0.07370$$

The probability a light bulb lasts between 1800 and 2200 hours is 0.074.

(b) $P(X \ge 2500) = 1 - P(X \le 2500) = 1 - \displaystyle\int_0^{2500} \frac{1}{2000}e^{-x/2000}\,dx$

$$= 1 - \left[-e^{-x/2000}\Big|_0^{2500} \right]$$

$$= 1 - \left[-e^{-25/20} + e^0 \right]$$

$$= 1 + e^{-25/20} - 1 \approx 0.2865$$

There is a 28.7% probability that a light bulb will last at least 2500 hours.

33.

Since the average life of the light bulb is 2000 hours, $\lambda = \dfrac{1}{2000}$, and

$$f(x) = \begin{cases} \dfrac{1}{2000}e^{-x/2000} & x \ge 0 \\ 0 & x < 0 \end{cases}$$

Let the random variable X denote the length of the light bulb's life.

(a) $P(X < 1500) = \displaystyle\int_0^{1500} \frac{1}{2000}e^{-x/2000}\,dx = -e^{-x/2000}\Big|_0^{1500}$

$$= -e^{-15/20} + e^0 = 1 - e^{-3/4} \approx 0.52763$$

The probability the light burns out in under 1500 hours is 0.528.

(b) $P(1750 \le X \le 2000) = \int_{1750}^{2000} \frac{1}{2000} e^{-x/2000} \, dx = -\left. e^{-x/2000} \right|_{1750}^{2000}$

$$= -e^{-1} + e^{-175/200} \approx 0.04898$$

The probability the light bulb lasts between 1750 and 2000 hours is 0.049.

(c) $P(X > 1900) = 1 - P(X \le 1900) = 1 - \int_{0}^{1900} \frac{1}{2000} e^{-x/2000} \, dx$

$$= 1 - \left[-\left. e^{-x/2000} \right|_{0}^{1900} \right]$$

$$= 1 - \left[-e^{-19/20} + e^{0} \right]$$

$$= 1 + e^{-19/20} - 1 = e^{-19/20} \approx 0.38674$$

The probability that a light bulb lasts longer than 1900 hours is 0.387.

(d)   Answers may vary.

35. $P(X > 10) = 1 - P(X \le 10) = 1 - \int_{0}^{10} \frac{1}{8} e^{-x/8} \, dx = 1 - \left[-\left. e^{-x/8} \right|_{0}^{10} \right]$

$$= 1 - \left[-e^{-10/8} + e^{0} \right]$$

$$= 1 + e^{-5/4} - 1 = e^{-5/4} = 0.28650$$

A customer waits longer than 10 minutes for lunch 28.7% of the time.

37.   (a) Since the average wait time is 10 minutes, $\lambda = \dfrac{1}{10}$.

$$f(x) = \begin{cases} \dfrac{1}{10} e^{-x/10} & x \ge 0 \\ 0 & x < 0 \end{cases}$$

(b) $P(7 \le x \le 12) = \int_{7}^{12} \frac{1}{10} e^{-x/10} \, dx = -\left. e^{-x/10} \right|_{7}^{12} = -e^{-12/10} + e^{-7/10} \approx 0.19539$

There is a 19.5% probability of waiting in line between 7 and 12 minutes.

(c) $P(X > 15) = 1 - P(X \le 15) = 1 - \int_{0}^{15} \frac{1}{10} e^{-x/10} \, dx$

$$= 1 - \left[-\left. e^{-x/10} \right|_{0}^{15} \right]$$

$$= 1 + e^{-15/10} - 1 \approx 0.22313$$

The probability a customer waits longer than 15 minutes is 0.223.

(d) The mother will be on time to the bus stop if she waits less than 15 minutes on line.

$$P(X<15)=e^{-15/10}-1=0.77687$$

The probability of her getting out in less than 15 minutes is 0.777.

39. The expected waiting time between tee-offs is the average time between tee-offs. So it will be 9 minutes.

41.

$$E(x)=\int_0^4 x f(x)\,dx=\int_0^4 \frac{3}{56}\left(5x^2-x^3\right)dx$$

$$=\frac{3}{56}\left[\left(\frac{5x^3}{3}-\frac{x^4}{4}\right)\Big|_0^4\right]$$

$$=\frac{3}{56}\left[\frac{320}{3}-\frac{256}{4}\right]=2.286$$

We can expect the contractor's cost estimate to be off by 2.286%.

43. (a) We want the probability the pregnancy lasts longer than 287 days. We use the fact that f is a probability density function, and the properties of the definite integral. That is, $\int_a^b f(x)\,dx=1$ and $\int_a^b f(x)\,dx=\int_a^c f(x)\,dx+\int_c^b f(x)\,dx$ to determine the probability.

$$P(X>287)=1-P(X\le287)$$

$$P(X>287)=1-\int_0^{287}\frac{1}{10\sqrt{2\pi}}e^{-(x-280)^2/200}\,dx$$

$$=1-\frac{1}{10\sqrt{2\pi}}\int_0^{287}e^{-(x-280)^2/200}\,dx$$

We evaluate the integral using a graphing utility. (We used a TI-83 Plus.)
In the math subroutine we go to 9: fnInt (and enter the integral in the following way.

fnInt(function to be integrated, the variable, the lower limit, the upper limit)
fnInt($e^{\wedge}(-(x-280)^2/200),x,0,287$)

Then we calculated $1-\dfrac{1}{10\sqrt{2\pi}}$ * ans. (See the screen shot below.)

```
fnInt(e^(-(X-280
)²/200),X,0,287)
        19.00115343
1-(1/(10√(2π)))*
Ans
       .2419636522
■
```

The probability a pregnancy lasts more than one week beyond the mean is 0.242.

(b) We want the probability the pregnancy lasts between 273 and 287 days. That is,

$$P(273 < X < 287) = \int_{273}^{287} \frac{1}{10\sqrt{2\pi}} e^{-(x-280)^2/200} \, dx = \frac{1}{10\sqrt{2\pi}} \int_{273}^{287} e^{-(x-280)^2/200} \, dx$$

Again we use a graphing utility. This time the lower limit of integration is 273.

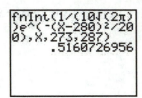

The probability a baby is born within one week of the mean gestation period is 0.516.

45. (a) Since the numbers are so large, we will use the cost of the automobiles in thousands of dollars.

$$c = 17 \qquad f(c) = f(17) = \frac{2}{20-10} = \frac{1}{5}$$

We first find $m_1 x + b_1$.

$$m_1 = \frac{y_2 - y_1}{x_2 - x_1} = \frac{\frac{1}{5} - 0}{17 - 10} = \frac{1}{35}$$

$$y = m_1 x + b_1$$

$$0 = \frac{1}{35}(10) + b_1$$

$$b_1 = -\frac{2}{7}$$

$$y = \frac{1}{35} x - \frac{2}{7}$$

Then we find $m_2 x + b_2$

$$m_2 = \frac{y_2 - y_1}{x_2 - x_1} = \frac{\frac{1}{5} - 0}{17 - 20} = -\frac{1}{15}$$

$$y = m_2 x + b_2$$

$$0 = -\frac{1}{15}(20) + b_2$$

$$b_2 = \frac{4}{3}$$

$$y = -\frac{1}{15} x + \frac{4}{3}$$

So the probability density function is

$$f(x) = \begin{cases} \dfrac{1}{35} x - \dfrac{2}{7} & \text{if} \quad 10 \le x \le 17 \\[2mm] -\dfrac{1}{15} x + \dfrac{4}{3} & \text{if} \quad 17 < x \le 20 \end{cases}$$

(b) The probability the car will cost less than $15,000 is 0.357.

$$P(X < 15) = \int_{10}^{15}\left(\frac{1}{35}x - \frac{2}{7}\right)dx = \left(\frac{x^2}{70} - \frac{2x}{7}\right)\Bigg|_{10}^{15}$$

$$= \left(\frac{225}{70} - \frac{300}{70}\right) - \left(\frac{100}{70} - \frac{200}{70}\right) = \frac{25}{70} = \frac{5}{14} = 0.35714$$

(c)

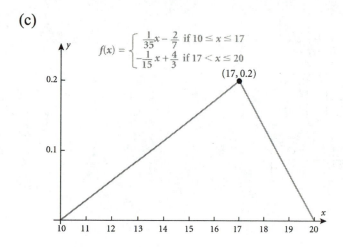

$$f(x) = \begin{cases} \frac{1}{35}x - \frac{2}{7} & \text{if } 10 \le x \le 17 \\ -\frac{1}{15}x + \frac{4}{3} & \text{if } 17 < x \le 20 \end{cases}$$

(17, 0.2)

(d) Answers will vary.

(e) $$E(x) = \int_{10}^{20} x\, f(x)\, dx = \int_{10}^{17}\left(\frac{x^2}{35} - \frac{2x}{7}\right)dx + \int_{17}^{20}\left(-\frac{x^2}{15} + \frac{4x}{3}\right)dx$$

$$= \left(\frac{x^3}{105} - \frac{x^2}{7}\right)\Bigg|_{10}^{17} + \left(-\frac{x^3}{45} + \frac{2x^2}{3}\right)\Bigg|_{17}^{20}$$

$$=$$

$$\left[\left(\frac{17^3}{105} - \frac{17^2}{7}\right) - \left(\frac{10^3}{105} - \frac{10^2}{7}\right)\right] + \left[\left(-\frac{20^3}{45} + \frac{2(20^2)}{3}\right) - \left(-\frac{17^3}{45} + \frac{2(17^2)}{3}\right)\right]$$

$$= 10.26667 + 5.4 = 15.66667$$

The expected price of the car is $15,666.67

(f) Answers will vary.

47. (a) Results will vary.

(b) $$P(0.6 \le X < 0.9) = \int_{0.6}^{0.9} dx = x\Big|_{0.6}^{0.9} = 0.9 - 0.6 = 0.3$$

49. The uniform probability density function is

$$f(x) = \begin{cases} \dfrac{1}{b-a} & \text{if} \quad a \le x \le b \\ 0 & \text{if} \quad a < x \text{ or } x > b \end{cases}$$

$$\sigma^2 = \int_a^b x^2 f(x)\, dx - \left[E(x)\right]^2$$

$$= \int_a^b x^2 \cdot \frac{1}{b-a}\, dx - \left[\int_a^b x \cdot \frac{1}{b-a}\, dx\right]^2$$

$$= \frac{1}{b-a}\left(\frac{x^3}{3}\bigg|_a^b\right) - \left[\frac{1}{b-a}\left(\frac{x^2}{2}\bigg|_a^b\right)\right]^2$$

$$= \frac{1}{b-a}\left[\frac{b^3-a^3}{3}\right] - \frac{1}{(b-a)^2}\left[\frac{b^2-a^2}{2}\right]^2$$

$$= \frac{b^3-a^3}{3(b-a)} - \frac{\left(b^2-a^2\right)^2}{4(b-a)^2}$$

$$= \frac{4(b-a)\left(b^3-a^3\right) - 3\left(b^2-a^2\right)^2}{12(b-a)^2}$$

$$= \frac{4\cancel{(b-a)}\cancel{(b-a)}\left(b^2+ab+a^2\right) - 3\cancel{(b-a)^2}(b+a)^2}{12\cancel{(b-a)^2}}$$

$$= \frac{4\left(b^2+ab+a^2\right) - 3(b+a)^2}{12}$$

$$= \frac{4b^2+4ab+4a^2 - 3b^2 - 6ab - 3a^2}{12}$$

$$= \frac{b^2+a^2-2ab}{12} = \frac{(b-a)^2}{12}$$

Chapter 16 Review

TRUE-FALSE ITEMS

1. False **3.** False

FILL-IN-THE-BLANKS

1. average value **3.** probability density **5.** $\displaystyle\lim_{t \to 2^-} \int_0^t f(x)\, dx$

REVIEW EXERCISES

1.
$$\int_0^\infty 20e^{-20x}\,dx = \lim_{b\to\infty}\int_0^b 20e^{-20x}\,dx = \lim_{b\to\infty}\left(-e^{-20x}\right)\bigg|_0^b$$
$$= \lim_{b\to\infty}\left(-e^{-20b}+1\right)$$
$$= \lim_{b\to\infty}1 - \lim_{b\to\infty}\frac{1}{e^{20b}} = 1-0 = 1$$

3.
$$\int_0^8\frac{1}{\sqrt[3]{x}}\,dx = \lim_{t\to0^+}\int_t^8\frac{1}{\sqrt[3]{x}}\,dx = \lim_{t\to0^+}\int_t^8 x^{-1/3}\,dx = \lim_{t\to0^+}\left[\frac{x^{2/3}}{\frac{2}{3}}\bigg|_t^8\right]$$
$$= \lim_{t\to0^+}\left[\frac{3\left(8^{2/3}\right)}{2}-\frac{3t^{2/3}}{2}\right]$$
$$= \lim_{t\to0^+}\left(\frac{3\left(8^{2/3}\right)}{2}\right) - \lim_{t\to0^+}\left(\frac{3t^{2/3}}{2}\right)$$
$$= 6-0 = 6$$

5.
$$\int_0^1\frac{x+1}{x}\,dx = \lim_{t\to0^+}\left[\int_t^1\frac{x+1}{x}\,dx\right] = \lim_{t\to0^+}\left[\int_t^1\left(1+\frac{1}{x}\right)dx\right] = \lim_{t\to0^+}\left[\left(x+\ln|x|\right)\bigg|_t^1\right]$$
$$= \lim_{t\to0^+}\left[(1+\ln 1)-(t+\ln|t|)\right]$$
$$= \lim_{t\to0^+}1 - \lim_{t\to0^+}\ln|t|$$
$$= -\lim_{t\to0^+}\ln|t| = \infty$$

Since the limit is infinite, the integral has no value.

7. Since the graph of f lies below the x-axis, the area is given by
$$A = -\int_0^\infty f(x)\,dx = \lim_{b\to\infty}\left[-\int_0^b f(x)\,dx\right] = \lim_{b\to\infty}\left[-\int_0^b -e^{-x}\,dx\right] = \lim_{b\to\infty}\left[\int_0^b e^{-x}\,dx\right]$$
$$= \lim_{b\to\infty}\left[\left(-e^{-x}\right)\bigg|_0^b\right]$$
$$= \lim_{b\to\infty}\left[\left(-e^{-b}\right)-\left(-e^0\right)\right]$$
$$= \lim_{b\to\infty}\left[\frac{1}{e^b}+1\right] = 1$$

The area is 1 square unit.

9.

$$AV = \frac{1}{3-(-1)} \int_{-1}^{3} x^3 \, dx = \frac{1}{4} \left[\frac{x^4}{4} \right]_{-1}^{3} = \frac{1}{16} \left[3^4 - (-1)^4 \right] = \frac{80}{16} = 5$$

11.

$$AV = \frac{1}{6-2} \int_{2}^{6} \left(x^2 + x \right) dx = \frac{1}{4} \left[\left(\frac{x^3}{3} + \frac{x^2}{2} \right) \bigg|_{2}^{6} \right]$$

$$= \frac{1}{4} \left[\left(\frac{6^3}{3} + \frac{6^2}{2} \right) - \left(\frac{2^3}{3} + \frac{2^2}{2} \right) \right]$$

$$= \frac{1}{4} \left[72 + 18 - \frac{8}{3} - 2 \right] = 22 - \frac{2}{3} = \frac{64}{3}$$

13.

$$AV = \frac{1}{2-(-2)} \int_{-2}^{2} 3x^2 \, dx = \frac{1}{4} \left[\frac{3x^3}{3} \bigg|_{-2}^{2} \right] = \frac{1}{4} \left[2^3 - (-2)^3 \right] = \frac{1}{4} [8+8] = 4$$

15. (a) We show the two conditions are satisfied.

Condition 1: $f(x) \geq 0$

$$f(x) = \frac{8}{9} x$$

$\frac{8}{9} x \geq 0$ when $x \geq 0$. So $f(x) \geq 0$ for all x in the interval $\left[0, \frac{3}{2} \right]$.

Condition 2: $\int_{0}^{3/2} f(x) \, dx = 1$

$$\int_{0}^{3/2} \frac{8}{9} x \, dx = \frac{8}{9} \left[\frac{x^2}{2} \right]_{0}^{3/2} = \frac{8}{9} \left[\frac{\left(\frac{3}{2} \right)^2}{2} - 0 \right] = \frac{8}{9} \cdot \frac{1}{2} \cdot \frac{9}{4} = 1$$

So f is a probability density function.

(b) $E(x) = \int_{a}^{b} x f(x) \, dx = \int_{0}^{3/2} \frac{8}{9} x^2 \, dx = \frac{8}{9} \left[\frac{x^3}{3} \right]_{0}^{3/2} = \frac{8}{27} \left[\left(\frac{3}{2} \right)^3 - 0 \right] = \frac{8}{27} \cdot \frac{27}{8} = 1$

The expected value is 1.

17. (a) We show the two conditions are satisfied.

Condition 1: $f(x) \geq 0$

$$f(x) = 12x^3 \left(1 - x^2 \right)$$

To determine where $f(x) \geq 0$ we solve $12x^3 \left(1 - x^2 \right) = 0$ and choose test points.

$$12x^3(1-x)(1+x)=0$$

$$12x^3 = 0 \quad \text{or} \quad 1-x = 0 \quad \text{or} \quad 1+x=0$$
$$x = 0 \quad \text{or} \quad x = 1 \quad \text{or} \quad x = -1$$

These numbers separate the number line into 4 parts. We are only interested in the interval [0, 1], and we choose $x = \dfrac{1}{2}$ and test it. $f\left(\dfrac{1}{2}\right) = \dfrac{9}{8} = 1.125$. So we conclude $f(x) \geq 0$ on the interval [0, 1].

Condition 2: $\displaystyle\int_0^1 f(x)\,dx = 1$

$$\int_0^1 \left[12x^3(1-x^2)\right] dx = \int_0^1 (12x^3 - 12x^5)\,dx = 12\left[\left(\frac{x^4}{4} - \frac{x^6}{6}\right)\Big|_0^1\right]$$

$$= 12\left[\left(\frac{1^4}{4} - \frac{1^6}{6}\right) - 0\right] = 12\left(\frac{3-2}{12}\right) = 1$$

So f is a probability density function.

(b) $\displaystyle E(x) = \int_a^b x\,f(x)\,dx = \int_0^1 (12x^4 - 12x^6)\,dx = 12\left[\left(\frac{x^5}{5} - \frac{x^7}{7}\right)\Big|_0^1\right]$

$$= 12\left[\left(\frac{1^5}{5} - \frac{1^7}{7}\right) - 0\right] = 12\left(\frac{7-5}{35}\right) = \frac{24}{35} = 0.6857$$

The expected value is $\dfrac{24}{35}$.

19.
$$AS = \frac{1}{10-0}\int_0^{10}(1340 - 850\,e^{-t})\,dt = \frac{1}{10}\left[(1340t + 850e^{-t})\Big|_0^{10}\right]$$

$$= \frac{1}{10}\left[(13{,}400 + 850e^{-10}) - (0 + 850e^0)\right]$$

$$= \frac{1}{10}(13{,}400 + 850e^{-10} - 850) = 1255.0$$

On the average, 1255 units are sold each year.

21.
$$AP = \frac{1}{150-100}\int_{100}^{150} 50e^{-0.01x}\,dx = \frac{1}{50}\cdot 50 \int_{100}^{150} e^{-0.01x}\,dx = -\frac{1}{0.01}\,e^{-0.01x}\,\Big|_{100}^{150}$$

$$= -100\left(e^{-1.5} - e^{-1}\right) = 14.47$$

The average price of the sandals is \$14.47.

23.

(a) $P(X \leq 1) = \int_{-2}^{1} \frac{1}{12} dx = \frac{1}{12} x \Big|_{-2}^{1} = \frac{1}{12}[1 - (-2)] = \frac{1}{4}$

(b) $P(X \geq 5) = \int_{5}^{10} \frac{1}{12} dx = \frac{1}{12} x \Big|_{5}^{10} = \frac{1}{12}[10 - 5] = \frac{5}{12}$

(c) The expected value of X is

$$E(x) = \int_{-2}^{10} \frac{1}{12} x \, dx = \frac{1}{12} \cdot \frac{x^2}{2} \Big|_{-2}^{10} = \frac{1}{24}[10^2 - (-2)^2] = \frac{96}{24} = 4$$

25. (a) We show the two conditions are satisfied.

Condition 1: $f(x) \geq 0$

$$\frac{3}{635,840}(x^2 - 28x + 196) \geq 0 \quad \text{when} \quad x^2 - 28x + 196 \geq 0$$

$x^2 - 28x + 196 = (x - 14)^2$ is always nonnegative. So condition 1 is satisfied.

Condition 2: $\int_{20}^{100} f(x) \, dx = 1$

$$\int_{20}^{100} \frac{3}{635,840}(x^2 - 28x + 196) \, dx = \frac{3}{635,840} \int_{20}^{100}(x^2 - 28x + 196) \, dx$$

$$= \frac{3}{635,840}\left[\frac{x^3}{3} - 14x^2 + 196x \Big|_{20}^{100}\right]$$

$$= \frac{3}{635,840}\left[\left(\frac{100^3}{3} - 14(100^2) + 196(100)\right) - \left(\frac{20^3}{3} - 14(20^2) + 196(20)\right)\right]$$

$$= 1$$

Both conditions are satisfied, making f a probability density function.

(b) $P(X \leq 40) = \int_{20}^{40} \frac{3}{635,840}(x^2 - 28x + 196) \, dx = \frac{3}{635,840} \int_{20}^{40}(x^2 - 28x + 196) \, dx$

$$= \frac{3}{635,840}\left[\frac{x^3}{3} - 14x^2 + 196x \Big|_{20}^{40}\right]$$

$$= \frac{3}{635,840}\left[\left(\frac{40^3}{3} - 14(40^2) + 196(40)\right) - \left(\frac{20^3}{3} - 14(20^2) + 196(20)\right)\right]$$

$$= = \frac{3}{635,840}[373.33333 - 986.66667] = \frac{217}{7948} = 0.0273$$

The probability a man dies at or before age 40 is 0.032.

(c) $P(X \leq 60) = \displaystyle\int_{20}^{60} \frac{3}{635,840}\left(x^2 - 28x + 196\right) dx = \frac{3}{635,840} \int_{20}^{60}\left(x^2 - 28x + 196\right) dx$

$= \dfrac{3}{635,840}\left[\dfrac{x^3}{3} - 14x^2 + 196x \,\Big|_{20}^{60}\right]$

$= \dfrac{3}{635,840}\left[\left(\dfrac{60^3}{3} - 14\left(60^2\right) + 196(60)\right) - \left(\dfrac{20^3}{3} - 14\left(20^2\right) + 196(20)\right)\right]$

$= \dfrac{3}{635,840}\left[33,360 - 986.66667\right] = \dfrac{607}{3974} = 0.15274$

The probability a man dies at or before age 60 is 0.153.

(d) The expected age of death is

$E(x) = \displaystyle\int_{20}^{100} x f(x) \, dx = \int_{20}^{100} \frac{3}{635,840}\left(x^3 - 28x^2 + 196x\right) dx$

$= \dfrac{3}{635,840} \displaystyle\int_{20}^{100}\left(x^3 - 28x^2 + 196x\right) dx$

$= \dfrac{3}{635,840}\left[\dfrac{x^4}{4} - \dfrac{28x^3}{3} + 98x^2 \,\Big|_{20}^{100}\right]$

$= \dfrac{3}{635,840}\left[\left(\dfrac{100^4}{4} - \dfrac{28\left(100^3\right)}{3} + 98\left(100^2\right)\right) - \left(\dfrac{20^4}{4} - \dfrac{28\left(20^3\right)}{3} + 98\left(20^2\right)\right)\right]$

$= 78.52$ years.

27. (a) We show the two conditions are satisfied.
Condition 1: $f(x) \geq 0$

$\qquad f(x) = \dfrac{1}{2}x \geq 0$ whenever $x \geq 0$

So f is nonnegative on the interval $[0, 2]$.

Condition 2: $\displaystyle\int_0^2 f(x)\, dx = 1$

$\displaystyle\int_0^2 \frac{1}{2} x \, dx = \frac{1}{2}\left[\frac{x^2}{2}\,\Big|_0^2\right] = \frac{1}{2}[2 - 0] = 1$

Both conditions are satisfied, making f a probability density function.

(b) The probability X is less than one is

$P(X < 1) = \displaystyle\int_0^1 \frac{1}{2} x \, dx = \frac{1}{2}\left[\frac{x^2}{2}\,\Big|_0^1\right] = \frac{1}{2}\left[\frac{1}{2} - 0\right] = \frac{1}{4} = 0.25$

(c) The probability X is between 1 and 1.5 is

$$P(1 < X < 1.5) = \int_1^{1.5} \frac{1}{2} x \, dx = \frac{1}{2} \left[\frac{x^2}{2} \right]_1^{1.5} = \frac{1}{4}[2.25 - 1] = \frac{1.25}{4} = \frac{5}{16} = 0.3125$$

(d) The probability X is greater than 1.5 is

$$P(X > 1.5) = \int_{1.5}^2 \frac{1}{2} x \, dx = \frac{1}{2} \left[\frac{x^2}{2} \right]_{1.5}^2 = \frac{1}{4}[4 - 2.25] = \frac{1.75}{4} = \frac{7}{16} = 0.4375$$

(e) The expected value of X is

$$E(x) = \int_0^2 x f(x) \, dx = \int_0^2 \frac{1}{2} x^2 \, dx = \frac{1}{2} \left[\frac{x^3}{3} \right]_0^2 = \frac{1}{6}[8 - 0] = \frac{8}{6} = \frac{4}{3}$$

29. The probability density function f is

$$f(x) = \begin{cases} \dfrac{1}{15} & \text{if} \quad 0 \le x \le 15 \\ 0 & \text{if} \quad x < 0 \text{ or } x > 15 \end{cases}$$

(a) The probability the tourist waits fewer than 3 minutes to hear the chimes is

$$P(X < 3) = \int_0^3 \frac{1}{15} \, dx = \frac{1}{15} x \Big|_0^3 = \frac{1}{15}[3 - 0] = \frac{1}{5} = 0.20$$

(b) The probability the tourist must wait more than 10 minutes to hear the clock is

$$P(X > 10) = \int_{10}^{15} \frac{1}{15} \, dx = \frac{1}{15} x \Big|_{10}^{15} = \frac{1}{15}[15 - 10] = \frac{5}{15} = \frac{1}{3}$$

(c) The expected time a person will wait to hear the chimes is

$$E(X) = \int_0^{15} x f(x) \, dx = \int_0^{15} \frac{1}{15} x \, dx = \frac{1}{15} \cdot \frac{x^2}{2} \Big|_0^{15} = \frac{1}{15} \left[\frac{15^2}{2} - 0 \right] = \frac{15}{2} = 7.5$$

minutes.

31. Let the random variable X denote the time one waits for a call. The probability density function is

$$f(x) = \begin{cases} 2.5 e^{-2.5x} & \text{if} \quad x \ge 0 \\ 0 & \text{if} \quad x < 0 \end{cases}$$

The probability the switchboard is idle for more than one minute is

$$P(X \ge 1) = 1 - P(X < 1) = 1 - \int_0^1 2.5 e^{-2.5x} \, dx = 1 + \left[e^{-2.5x} \Big|_0^1 \right]$$

$$= 1 + \left[e^{-2.5} - e^0 \right] = e^{-2.5} \approx 0.0821$$

33. Let the random variable X denote the life of the light bulb. If the bulb has an average life of 1750 hours, then $\lambda = \dfrac{1}{1750}$, and the probability density function is

$$f(x) = \begin{cases} \dfrac{1}{1750}e^{-x/1750} & \text{if } x \geq 0 \\ 0 & \text{if } x < 0 \end{cases}$$

(a) The probability the light bulb lasts between 1500 and 2000 hours is

$$P(1500 \leq X \leq 2000) = \int_{1500}^{2000} \frac{1}{1750}e^{-x/1750}\,dx = -\left. e^{-x/1750}\right|_{1500}^{2000}$$
$$= -e^{-2000/1750} + e^{-1500/1750} = 0.1055$$

(b) The probability the light bulb burns for more than 2000 hours is

$$P(X > 2000) = 1 - P(X \leq 2000) = 1 - \int_{0}^{2000} \frac{1}{1750}e^{-x/1750}\,dx$$
$$= 1 - \left[-\left. e^{-x/1750}\right|_{0}^{2000} \right] = 1 + e^{-2000/1750} - e^{0}$$
$$= e^{-2000/1750} \approx 0.3189$$

35. The probability the toll collector waits more than 1 minute for the first car is

$$P(X > 1) = 1 - P(X \leq 1) = 1 - \left[\int_{0}^{1} \frac{2}{3}e^{-2t/3}\,dt \right] = 1 - \left[-\left. e^{-2t/3}\right|_{0}^{1} \right]$$
$$= 1 + e^{-2/3} - 1 = 0.5134$$

CHAPTER 16 PROJECT

1.

Interval	Width	Tally	Frequency	Relative Frequency	Probability
1	0 – 5	⊮ ⊮ ⊮ ⊮	21	0.23596	0.26288
2	5 – 10	⊮ ⊮ ⊮ |||	18	0.20225	0.19377
3	10 – 15	⊮ ⊮ ||	12	0.13483	0.14283
4	15 – 20	⊮ ⊮	10	0.11236	0.10529
5	20 – 25	⊮ ⊮ ||	12	0.13483	0.07761
6	25 – 30	||	2	0.02247	0.05721
7	30 – 35	||	2	0.02247	0.04217
8	35 – 40	|||	3	0.03371	0.03108
9	40 – 45	||	2	0.02247	0.02291
10	45 – 50	||||	4	0.04494	0.01689
11	50 – 55			1	0.01124	0.01245
12	55 – 60		0	0	0.00918
13	60 – 65			1	0.01124	0.00676
14	65 – 70		0	0	0.00499
15	70 – 75		0	0	0.00368
16	75 – 80		0	0	0.00271
17	80 – 85		0	0	0.00200
18	85 – 90			1	0.01124	0.00147

3.   Answers will vary.

5.

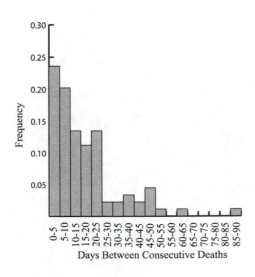

7. See column 6 of the table in Problem 1.

9.

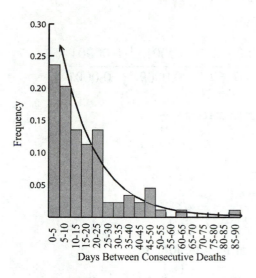

Days Between Consecutive Deaths

Comments will vary.

MATHEMATICAL QUESTIONS FROM PROFESSIONAL EXAMS

1. **(d)**

$$\int_0^1 x \ln x \, dx = \lim_{t \to 0^+} \left[\int_t^1 x \ln x \, dx \right]$$

We evaluate the integral by using integration by parts. We choose

$$u = \ln x \qquad\qquad dv = x \, dx$$
$$du = \frac{1}{x} \, dx \qquad\qquad v = \frac{x^2}{2}$$

$$\lim_{t \to 0^+} \left[\int_t^1 x \ln x \, dx \right] = \lim_{t \to 0^+} \left[uv \Big|_t^1 - \int_t^1 v \, du \right]$$

$$= \lim_{t \to 0^+} \left[\frac{x^2 \ln x}{2} \Big|_t^1 - \int_t^1 \frac{x}{2} \, dx \right]$$

$$= \lim_{t \to 0^+} \left[\left(\frac{x^2 \ln x}{2} \Big|_t^1 \right) - \left(\frac{x^2}{4} \Big|_t^1 \right) \right]$$

$$= \lim_{t \to 0^+} \left[\left(\frac{1 \cdot \ln 1}{2} - \frac{t \cdot \ln t}{2} \right) - \left(\frac{1}{4} - \frac{t^2}{4} \right) \right]$$

$$= \lim_{t \to 0^+} \left[-\frac{t \cdot \ln t}{2} - \frac{1}{4} + \frac{t^2}{4} \right]$$

$$= -\lim_{t \to 0^+} \left(\frac{t \cdot \ln t}{2} \right) - \lim_{t \to 0^+} \left(\frac{1}{4} \right) + \lim_{t \to 0^+} \left(\frac{t^2}{4} \right)$$

$$= -\lim_{t \to 0^+} \left(\frac{t \cdot \ln t}{2} \right) - \frac{1}{4}$$

To determine $\lim_{t \to 0^+} (t \ln t)$ we use a table.

t	1	0.1	0.01	0.001	0.0001	0.00001
$t \ln t$	0	-0.2303	-0.0461	-0.0069	-0.0009	-0.0001

We conclude $\lim_{t \to 0^+} (t \ln t) = 0$, and $\int_0^1 x \ln x \, dx = -\frac{1}{4}$

3. (b)

$$\int_0^\infty \frac{x+1}{\left(x^2 + 2x + 2\right)^2} \, dx = \lim_{b \to \infty} \left[\int_0^b \frac{x+1}{\left(x^2 + 2x + 2\right)^2} \, dx \right]$$

We use the method of substitution to evaluate the integral.

Let $u = x^2 + 2x + 2$. Then $du = 2(x+1) \, dx$, and $\frac{1}{2} du = (x+1) \, dx$.

When $x = 0$, $u = 2$, and when $x = b$, $u = b^2 + 2b + 2$.

$$\lim_{b \to \infty} \left[\int_0^b \frac{x+1}{\left(x^2 + 2x + 2\right)^2} \, dx \right] = \lim_{b \to \infty} \left[\frac{1}{2} \int_2^{b^2 + 2b + 2} u^{-2} \, du \right]$$

$$= \frac{1}{2} \lim_{b \to \infty} \left[-u^{-1} \Big|_2^{b^2 + 2b + 2} \right]$$

$$= \frac{1}{2} \lim_{b \to \infty} \left[-\frac{1}{b^2 + 2b + 2} + \frac{1}{2} \right] = \frac{1}{4}$$

5. (d)

We are only concerned with the time after 8:30.

The probability X arrives after 8:30 is $\frac{1}{2}$.

The probability Y arrives after 8:30 is 1.

As long as X arrives after 8:30, Y arrives first $\frac{1}{2}$ the time. So the probability Y arrives

first if $\frac{1}{2} \cdot \frac{1}{2} = \frac{1}{4}$.

Chapter 17
Calculus of Functions of
Two or More Variables

17.1 Rectangular Coordinates in Space

1.

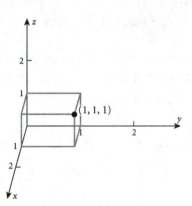

3.

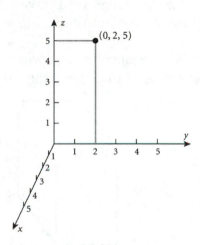

5.

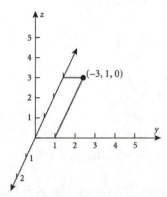

7. Since one vertex is the origin and the edges of the box are parallel to the axes, three of the remaining vertices are also on an axis.

$(2, 0, 0), (0, 1, 0), (0, 0, 3)$

Because the figure is a rectangular box, the remaining three corners will have one coordinate that is 0.

$(2, 1, 0), (2, 0, 3), (0, 1, 3)$

9. Draw a rectangular box (prism) and label the two given vertices. The other 6 vertices are determined by changing the value of one coordinate to that of the other given vertex.

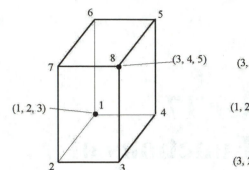
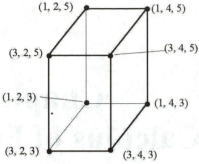

Vertex 2: change $x = 1$ in vertex 1 to $x = 3$ giving (3, 2, 3).
Vertex 3: change $y = 2$ in vertex 2 to $y = 4$ giving (3, 4, 3).
Vertex 4: change $x = 3$ in vertex 3 to $x = 1$ giving (1, 4, 3).
Vertex 5: change $z = 3$ in vertex 4 to $z = 5$ giving (1, 4, 5).
Vertex 6: change $y = 4$ in vertex 5 to $y = 2$ giving (1, 2, 5).
Vertex 7: change $x = 1$ in vertex 6 to $x = 3$ giving (3, 2, 5).
Changing $y = 2$ in vertex 7 to $y = 4$ will give you (3, 4, 5) which is vertex 8.

11. Draw a rectangular box (prism) and label the two given vertices. The other 6 vertices are determined by changing the value of one coordinate to that of the other given vertex.

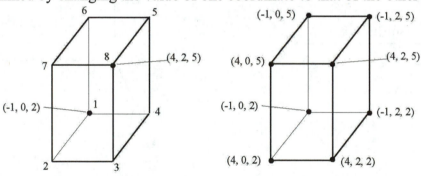

Vertex 2: change $x = -1$ in vertex 1 to $x = 4$ giving (4, 0, 2).
Vertex 3: change $y = 0$ in vertex 2 to $y = 2$ giving (4, 2, 2).
Vertex 4: change $x = 4$ in vertex 3 to $x = -1$ giving $(-1, 2, 2)$.
Vertex 5: change $z = 2$ in vertex 4 to $z = 5$ giving $(-1, 2, 5)$.
Vertex 6: change $y = 2$ in vertex 5 to $y = 0$ giving $(-1, 0, 5)$.
Vertex 7: change $x = -1$ in vertex 6 to $x = 4$ giving (4, 0, 5).
Changing $y = 0$ in vertex 7 to $y = 2$ will give you (4, 2, 5) which vertex 8.

13. $y = 3$ describes a plane parallel to the xz-plane and passing through the point (0, 3, 0).

15. $x = 0$ describes the yz-plane.

17. $z = 5$ describes a plane parallel to the xy-plane and passing through the point (0, 0, 5).

19. We use the distance formula with $\left(x_1, y_1, z_1 \right) = (1, 3, 0)$ and $\left(x_2, y_2, z_2 \right) = (4, 1, 2)$.

$$d = \sqrt{(x_2 - x_1)^2 + (y_2 - y_1)^2 + (z_2 - z_1)^2}$$
$$d = \sqrt{(4-1)^2 + (1-3)^2 + (2-0)^2}$$
$$d = \sqrt{9+4+4} = \sqrt{17}$$

21. We use the distance formula with $(x_1, y_1, z_1) = (-1, 2, -3)$ and $(x_2, y_2, z_2) = (4, -2, 1)$.

$$d = \sqrt{(x_2 - x_1)^2 + (y_2 - y_1)^2 + (z_2 - z_1)^2}$$
$$d = \sqrt{(4-(-1))^2 + (-2-2)^2 + (1-(-3))^2}$$
$$d = \sqrt{25+16+16} = \sqrt{57}$$

23. We use the distance formula with $(x_1, y_1, z_1) = (4, -2, -2)$ and $(x_2, y_2, z_2) = (3, 2, 1)$.

$$d = \sqrt{(x_2 - x_1)^2 + (y_2 - y_1)^2 + (z_2 - z_1)^2}$$
$$d = \sqrt{(3-4)^2 + (2-(-2))^2 + (1-(-2))^2}$$
$$d = \sqrt{1+16+9} = \sqrt{26}$$

25. The equation of the sphere whose center is $(3, 1, 1)$ and whose radius is 1 is
$$(x-3)^2 + (y-1)^2 + (z-1)^2 = 1$$

27. The equation of the sphere whose center is $(-1, 1, 2)$ and whose radius is 3 is
$$(x+1)^2 + (y-1)^2 + (z-2)^2 = 9$$

29. Complete the squares.
$$x^2 + y^2 + z^2 + 2x - 2y = 2$$
$$(x^2 + 2x) + (y^2 - 2y) + (z^2) = 2$$
$$(x^2 + 2x + 1) + (y^2 - 2y + 1) + (z^2) = 2 + 1 + 1$$
$$(x+1)^2 + (y-1)^2 + z^2 = 4$$
The center of the sphere is $(-1, 1, 0)$ and the radius is 2.

31. Complete the squares.
$$x^2 + y^2 + z^2 + 4x + 4y + 2z = 0$$
$$(x^2 + 4x) + (y^2 + 4y) + (z^2 + 2z) = 0$$
$$(x^2 + 4x + 4) + (y^2 + 4y + 4) + (z^2 + 2z + 1) = 0 + 4 + 4 + 1$$
$$(x+2)^2 + (y+2)^2 + (z+1)^2 = 9$$
The center of the sphere is $(-2, -2, -1)$, and the radius is 3.

33. Complete the squares.
$$2x^2 + 2y^2 + 2z^2 - 8x + 4z = -2$$
$$x^2 + y^2 + z^2 - 4x + 2z = -1$$
$$\left(x^2 - 4x\right) + \left(y^2\right) + \left(z^2 + 2z\right) = -1$$
$$\left(x^2 - 4x + 4\right) + \left(y^2\right) + \left(z^2 + 2z + 1\right) = -1 + 4 + 1$$
$$(x-2)^2 + y^2 + (z+1)^2 = 4$$
The center of the sphere is $(2, 0, -1)$, and the radius is 2.

35. To determine the equation of the sphere, we first need to find the center and the radius of the sphere.
The center is the midpoint of the diameter.

$$x_m = \frac{x_1 + x_2}{2} = \frac{-2+2}{2} = 0 \qquad\qquad y_m = \frac{y_1 + y_2}{2} = \frac{0+6}{2} = 3$$

$$z_m = \frac{z_1 + z_2}{2} = \frac{4+8}{2} = 6$$

The center of the sphere is $(0, 3, 6)$.

The radius is the distance from the midpoint to one of endpoints. We use $(2, 6, 8)$.
$$r = \sqrt{(2-0)^2 + (6-3)^2 + (8-6)^2} = \sqrt{4+9+4} = \sqrt{17}$$

The equation of the sphere is
$$x^2 + (y-3)^2 + (z-6)^2 = 17$$

17.2 Functions and Their Graphs

1. $f(2,1) = 2^2 + 1 = 5$

3. $f(2,1) = \sqrt{2 \cdot 1} = \sqrt{2}$

5. $f(2,1) = \dfrac{1}{2 \cdot 2 + 1} = \dfrac{1}{5}$

7. $f(2,1) = \dfrac{2^2 - 1}{2-1} = 3$

9. $f(2,1) = \sqrt{4 - 2^2 \cdot 1^2} = \sqrt{4-4} = 0$

11. $f(x,y) = 3x + 2y + xy$

(a) $f(1,0) = 3 \cdot 1 + 2 \cdot 0 + 1 \cdot 0 = 3$

(b) $f(0,1) = 3 \cdot 0 + 2 \cdot 1 + 0 \cdot 1 = 2$

(c) $f(2,1) = 3 \cdot 2 + 2 \cdot 1 + 2 \cdot 1 = 10$

(d) $f(x+\Delta x, y)=3(x+\Delta x)+2y+(x+\Delta x)y$

$\qquad = 3x+3\Delta x+2y+xy+\Delta xy$

(e) $f(x,y+\Delta y)=3x+2(y+\Delta y)+x(y+\Delta y)$

$\qquad = 3x+2y+2\Delta y+xy+x\Delta y$

13. $f(x, y)=\sqrt{xy}+x$

(a) $f(0, 0)=\sqrt{0\cdot 0}+0=0$ 　　　　　(b) $f(0,1)=\sqrt{0\cdot 1}+0=0$

(c) $f(a^2, t^2)=\sqrt{a^2\cdot t^2}+a^2=\sqrt{a^2}\cdot\sqrt{t^2}+a^2=at+a^2$

(d) $f(x+\Delta x, y)=\sqrt{(x+\Delta x)y}+x+\Delta x$

(e) $f(x, y+\Delta y)=\sqrt{x(y+\Delta y)}+x$

15. $f(x, y, z)=x^2y+y^2z$

(a) $f(1, 2, 3)=1^2\cdot 2+2^2\cdot 3=2+12=14$

(b) $f(0, 1, 2)=0^2\cdot 1+1^2\cdot 2=0+2=2$

(c) $f(-1,-2,-3)=(-1)^2\cdot(-2)+(-2)^2\cdot(-3)=-2-12=-14$

17. $z=f(x, y)=\sqrt{x}\sqrt{y}$

Since the radicand must be nonnegative, we
have $x\geq 0$ and $y\geq 0$. The domain is the set
$\{(x, y)\mid x\geq 0 \text{ and } y\geq 0\}$. That is, the domain is
the first quadrant and the positive x- and y- axes.

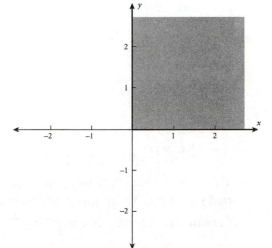

19. $z = f(x, y) = \sqrt{9 - x^2 - y^2}$

Since the radicand must be nonnegative,

$$9 - x^2 - y^2 \geq 0$$
$$x^2 + y^2 \leq 9$$

This inequality describes the domain of f. That is, the domain of f is the points (x, y) that are either on the circle $x^2 + y^2 = 9$ or inside the circle.

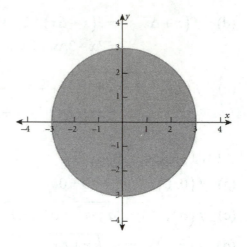

21. $z = f(x, y) = \dfrac{\ln x}{\ln y}$

The logarithmic function is defined only for positive numbers, so x and y must be positive. However, $\ln 1 = 0$, so we must eliminate $y = 1$ from the domain. The domain of f is the set $\{(x, y) \mid x > 0 \text{ and } y > 0, \text{ but } y \neq 1\}$.

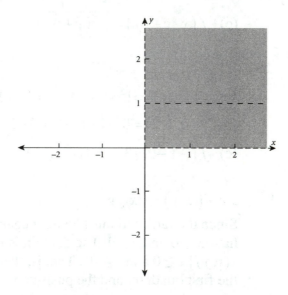

23. $z = f(x, y) = \dfrac{3}{x^2 + y^2 - 4}$

This is a rational function, so its domain is all real numbers except those which make the denominator zero. That is, $x^2 + y^2 - 4 \neq 0$.

$$x^2 + y^2 \neq 4$$

The domain of f is the set of all real numbers excluding the boundary of the circle $x^2 + y^2 = 4$.

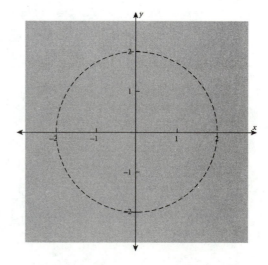

25. $z = f(x, y) = \ln\left(x^2 + y^2\right)$

Logarithm functions are defined only for positive numbers. $x^2 + y^2 > 0$ provided both x and y are not both equal to zero. The domain of f is the set $\{(x, y) \mid (x, y) \neq (0, 0)\}$.

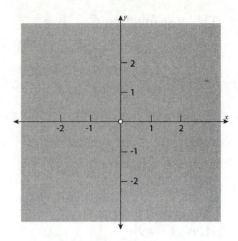

27. $w = f(x, y, z) = \sqrt{x^2 + y^2 + z^2 - 16}$

Since the radicand must be nonnegative, $x^2 + y^2 + z^2 - 16 \geq 0$. That is, $x^2 + y^2 + z^2 \geq 16$. This describes the surface of a sphere of radius 4 and all the space outside it. The domain of f is the set $\left\{(x, y, z) \mid x^2 + y^2 + z^2 \geq 16\right\}$.

29. $w = f(x, y, z) = \dfrac{4}{x^2 + y^2 + z^2}$

f is a rational function, so the domain is all real numbers except those that make the denominator zero. That is, $x^2 + y^2 + z^2 \neq 0$. The domain of f is the set $\{(x, y, z) \mid (x, y, z) \neq (0, 0, 0)\}$.

31. $z = f(x, y) = 3x + 4y$

(a) $f(x + \Delta x, y) = 3(x + \Delta x) + 4y$

(b) $f(x + \Delta x, y) - f(x, y) = \left[3(x + \Delta x) + 4y\right] - \left[3x + 4y\right]$
$$= 3x + 3\Delta x + 4y - 3x - 4y = 3\Delta x$$

(c) $\dfrac{f(x + \Delta x, y) - f(x, y)}{\Delta x} = \dfrac{3\Delta x}{\Delta x}$

(d) $\displaystyle\lim_{\Delta x \to 0} \dfrac{f(x + \Delta x, y) - f(x, y)}{\Delta x} = \lim_{\Delta x \to 0} \dfrac{3\Delta x}{\Delta x} = \lim_{\Delta x \to 0} 3 = 3$

33. Let r denote the radius of the tank, and h denote the height of the tank.

The area A_1 of the top (or bottom) of the tank is given by
$$A_1 = \pi r^2$$

The area A_2 of the side of the tank is given by
$$A_2 = 2\pi r \cdot h = 2\pi rh$$

The cost function C is then
$$C = 2\left[300A_1\right] + 500A_2$$
$$C(r,h) = 600\pi r^2 + 1000\pi rh \text{ dollars.}$$

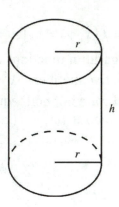

35.
$$A(N, I) = 9\left(\frac{N}{I}\right)$$

(a) $A(3, 4) = 9\left(\dfrac{3}{4}\right) = \dfrac{27}{4} = 6.75$

(b) $A(6, 3) = 9\left(\dfrac{6}{3}\right) = 18.0$

(c) $A(2, 9) = 9\left(\dfrac{2}{9}\right) = 2.0$

(d) $A(3, 18) = 9\left(\dfrac{3}{18}\right) = \dfrac{3}{2} = 1.5$

37. Here $x = 650 - 500 = 150$, and $y = 1600 - 1500 = 100$. So,
$$B(150, 100) = 79.99 + 0.4(150) + 0.02(100) = 79.99 + 60 + 2 = \$141.99$$

39. $H = -42.379 + 2.04901523t + 10.14333127r - 0.22475541tr - 0.00683783t^2$
$\qquad - 0.05481717r^2 + 0.00122874t^2r + 0.00085282tr^2 - 0.00000199t^2r^2$

(a) When $t = 95°$ and $r = 50$, then $H(t, r) = 105.216°$ F.

(b) When $t = 97°$ F, the function $H = H(r)$.

$H = \left[-42.379 + 2.04901523t - 0.00683783t^2\right] + \left[10.14333127 - 0.22475541t + 0.00122874t^2\right]r$
$\qquad + \left[-0.05481717 + 0.00085282t - 0.00000199t^2\right]r^2$
$\quad = 0.00918246r^2 - 0.09672884r + 92.03833484$

We want to find r that will make $H = 105$. That is, we want to solve
$$0.00918246r^2 - 0.09672884r + 92.03833484 = 105$$
$$0.00918246r^2 - 0.09672884r + 92.03833484 - 105 = 0$$
$$0.00918246r^2 - 0.09672884r - 12.96166516 = 0$$

This is a quadratic function in one variable r and can be solved using the quadratic formula or a graphing utility. We used a TI-83 Plus by graphing
$$Y1 = 0.00918246x^2 - 0.09672884x - 12.96166516$$
and using 2nd CALC 2: zero.

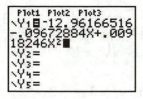 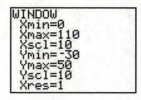 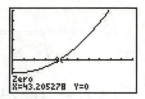

43% relative humidity results in a heat index of 105° F

(c) If $t = 102°$, then
$$H = 0.01146651r^2 + 0.00209041r + 95.47977014$$
We want to find r that will make $H = 130$. That is, we want to solve
$$0.01146651r^2 + 0.00209041r + 95.47977014 = 130$$
$$0.01146651r^2 + 0.00209041r + 95.47977014 - 130 = 0$$
$$0.01146651r^2 + 0.00209041r - 34.52022986 = 0$$
Again using a TI-83. Plus, we graph
$$Y1 = 0.01146651x^2 + 0.00209041x - 34.52022986$$

 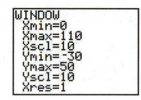 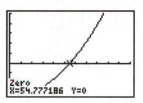

55% relative humidity will result in a heat index of 130° F.

17.3 Partial Derivatives

1. $f_x(x,y) = 3$ 　　　　　　　　　　$f_y(x,y) = -2 + 9y^2$

　　　$f_x(2,-1) = 3$ 　　　　　　　　　$f_y(-2,3) = -2 + 9 \cdot 3^2 = 79$

3. $f_x(x,y) = 2(x-y)$ 　　　　　　　$f_y(x,y) = 2(x-y)(-1) = -2(x-y)$

　　　$f_x(2,-1) = 2(2-(-1)) = 2 \cdot 3 = 6$ 　　$f_y(-2,3) = -2[(-2)-3] = 10$

5. $f_x(x,y) = \dfrac{1}{2}(x^2+y^2)^{-1/2} \cdot 2x$ 　　$f_y(x,y) = \dfrac{1}{2}(x^2+y^2)^{-1/2} \cdot 2y$

　　　　　　$= \dfrac{x}{\sqrt{x^2+y^2}}$ 　　　　　　　　$= \dfrac{y}{\sqrt{x^2+y^2}}$

$$f_x(2,-1)=\frac{2}{\sqrt{2^2+(-1)^2}}=\frac{2}{\sqrt{5}}=\frac{2\sqrt{5}}{5} \qquad f_y(-2,3)=\frac{3}{\sqrt{(-2)^2+3^2}}=\frac{3}{\sqrt{13}}=\frac{3\sqrt{13}}{13}$$

7. $\quad f_x(x,y)=-2y-24x \qquad\qquad\qquad f_y(x,y)=3y^2-2x+2y$

$\qquad f_{xx}(x,y)=-24 \qquad\qquad\qquad\qquad f_{yy}(x,y)=6y+2$

$\qquad f_{xy}(x,y)=-2 \qquad\qquad\qquad\qquad\quad f_{yx}(x,y)=-2$

9. $\quad f_x(x,y)=e^y+ye^x+1 \qquad\qquad\quad f_y(x,y)=xe^y+e^x$

$\qquad f_{xx}(x,y)=ye^x \qquad\qquad\qquad\qquad f_{yy}(x,y)=xe^y$

$\qquad f_{xy}(x,y)=e^y+e^x \qquad\qquad\qquad f_{yx}(x,y)=e^y+e^x$

11. $\quad f(x,y)=\dfrac{x}{y}=x\cdot y^{-1}$

$\qquad f_x(x,y)=\dfrac{1}{y} \qquad\qquad\qquad\qquad\quad f_y(x,y)=-x\cdot y^{-2}=-\dfrac{x}{y^2}$

$\qquad f_{xx}(x,y)=0 \qquad\qquad\qquad\qquad\quad f_{yy}(x,y)=-x\cdot(-2)y^{-3}=\dfrac{2x}{y^3}$

$\qquad f_{xy}(x,y)=-y^{-2}=-\dfrac{1}{y^2} \qquad\qquad f_{yx}(x,y)=-1\cdot y^{-2}=-\dfrac{1}{y^2}$

13. $\quad f(x,y)=\ln\left(x^2+y^2\right)$

$\qquad f_x(x,y)=\dfrac{1}{x^2+y^2}\cdot 2x=\dfrac{2x}{x^2+y^2} \qquad f_y(x,y)=\dfrac{1}{x^2+y^2}\cdot 2y=\dfrac{2y}{x^2+y^2}$

$\qquad f_{xx}(x,y)=\dfrac{2\left(x^2+y^2\right)-2x(2x)}{\left(x^2+y^2\right)^2}=\dfrac{2x^2+2y^2-4x^2}{\left(x^2+y^2\right)^2}=\dfrac{2y^2-2x^2}{\left(x^2+y^2\right)^2}$

$\qquad f_{yy}(x,y)=\dfrac{2\left(x^2+y^2\right)-2y(2y)}{\left(x^2+y^2\right)^2}=\dfrac{2x^2+2y^2-4y^2}{\left(x^2+y^2\right)^2}=\dfrac{2x^2-2y^2}{\left(x^2+y^2\right)^2}$

$\qquad f_{xy}(x,y)=\dfrac{-2x(2y)}{\left(x^2+y^2\right)^2}=\dfrac{-4xy}{\left(x^2+y^2\right)^2}$

$\qquad f_{yx}(x,y)=\dfrac{-2y(2x)}{\left(x^2+y^2\right)^2}=\dfrac{-4xy}{\left(x^2+y^2\right)^2}$

15.

$$f(x, y) = \frac{10 - x + 2y}{xy}$$

$$f_x(x, y) = \frac{(-1) \cdot xy - (10 - x + 2y) \cdot y}{(xy)^2} = \frac{-xy - 10y + xy - 2y^2}{x^2 y^2} = \frac{-10\cancel{y} - 2y^{\cancel{2}}}{x^2 y^{\cancel{2}}}$$

$$= \frac{-10 - 2y}{x^2 y}$$

$$f_y(x, y) = \frac{2 \cdot xy - (10 - x + 2y) \cdot x}{(xy)^2} = \frac{2xy - 10x + x^2 - 2xy}{x^2 y^2} = \frac{-10\cancel{x} + x^{\cancel{2}}}{x^{\cancel{2}} y^2} = \frac{-10 + x}{xy^2}$$

$$f_{xx}(x, y) = \frac{0 \cdot x^2 y - (-10 - 2y)(2xy)}{(x^2 y)^2} = \frac{20\cancel{xy} + 4\cancel{xy}^{\cancel{2}}}{x^{\cancel{3}} y^{\cancel{2}}} = \frac{20 + 4y}{x^3 y}$$

$$f_{yy}(x, y) = \frac{-(-10 + x)(2xy)}{(xy^2)^2} = \frac{(20 - 2x)\cancel{xy}}{x^{\cancel{2}} y^{\cancel{3}}} = \frac{20 - 2x}{xy^3}$$

$$f_{xy}(x, y) = \frac{-2(x^2 y) - (-10 - 2y) \cdot x^2}{(x^2 y)^2} == \frac{-2x^2 y + 10x^2 + 2x^2 y}{x^4 y^2} = \frac{10x^2}{x^4 y^2} = \frac{10}{x^2 y^2}$$

$$f_{yx}(x, y) = \frac{1 \cdot (xy^2) - (-10 + x) \cdot y^2}{(x^2 y)^2} == \frac{\cancel{xy^2} + 10y^2 \cancel{-xy^2}}{x^4 y^2} = \frac{10x^{\cancel{2}}}{x^{\cancel{4}2} y^2} = \frac{10}{x^2 y^2}$$

17.

$$f(x, y) = x^3 + y^2$$

$$f_x(x, y) = 3x^2 \qquad\qquad f_y(x, y) = 2y$$

$$f_{xy}(x, y) = 0 \qquad\qquad f_{yx}(x, y) = 0$$

19.

$$f(x, y) = 3x^4 y^2 + 7x^2 y$$

$$f_x(x, y) = 12x^3 y^2 + 14xy \qquad\qquad f_y(x, y) = 6x^4 y + 7x^2$$

$$f_{xy}(x, y) = 24x^3 y + 14x \qquad\qquad f_{yx}(x, y) = 24x^3 y + 14x$$

21.

$$f(x, y) = \frac{y}{x^2} = y \cdot x^{-2}$$

$$f_x(x, y) = -2yx^{-3} = \frac{-2y}{x^3} \qquad\qquad f_y(x, y) = x^{-2} = \frac{1}{x^2}$$

$$f_{xy}(x, y) = -2x^{-3} = \frac{-2}{x^3} \qquad\qquad f_{yx}(x, y) = -2x^{-3} = \frac{-2}{x^3}$$

23. $f_x(x, y, z) = 2xy - 3yz$

$f_y(x, y, z) = x^2 - 3xz$

$f_z(x, y, z) = -3xy + 3z^2$

25. $f_x(x, y, z) = e^y$

$f_y(x, y, z) = xe^y + e^z$

$f_z(x, y, z) = ye^z$

27. $f_x(x, y, z) = \ln(yz) + y \cdot \dfrac{1}{x} = \ln(yz) + \dfrac{y}{x}$

$f_y(x, y, z) = x \cdot \dfrac{1}{y} + \ln(xz) = \dfrac{x}{y} + \ln(xz)$

$f_z(x, y, z) = x \cdot \dfrac{1}{z} + y \cdot \dfrac{1}{z} = \dfrac{x}{z} + \dfrac{y}{z} = \dfrac{x+y}{z}$

29. $f_x(x, y, z) = \dfrac{2x}{x^2 + y^2 + z^2}$

$f_y(x, y, z) = \dfrac{2y}{x^2 + y^2 + z^2}$

$f_z(x, y, z) = \dfrac{2z}{x^2 + y^2 + z^2}$

31. $z = f(x, y) = 5x^2 + 3y^2$

$f_x(x, y) = 10x$

$f_x(2, 3) = 10 \cdot 2 = 20$

The slope of the tangent line to the curve of intersection of the surface z and the plane $y = 3$ is 20.

33. $z = f(x, y) = \sqrt{16 - x^2 - y^2} = \left(16 - x^2 - y^2\right)^{1/2}$

$f_y(x, y) = \dfrac{1}{2}\left(16 - x^2 - y^2\right)^{-1/2} \cdot (-2y) = \dfrac{-y}{\left(16 - x^2 - y^2\right)^{-1/2}} = \dfrac{-y}{\sqrt{16 - x^2 - y^2}}$

$f_y(1, 2) = \dfrac{-2}{\sqrt{16 - 1^2 - 2^2}} = \dfrac{-2}{\sqrt{11}} = -\dfrac{2\sqrt{11}}{11}$

The slope of the tangent line to the curve of intersection of the surface z and the plane $x = 1$ is $-\dfrac{2\sqrt{11}}{11}$.

35. $z = f(x, y) = e^x \ln y$

$$f_y(x, y) = e^x \cdot \frac{1}{y} = \frac{e^x}{y}$$

$$f_y(0, 1) = \frac{e^0}{1} = 1$$

The slope of the tangent line to the curve of intersection of the surface z and the plane $x = 0$ is 1.

37. $z = f(x, y) = 2\ln \sqrt{x^2 + y^2} = 2\ln \left(x^2 + y^2\right)^{1/2} = \ln\left(x^2 + y^2\right)$

$$f_y(x, y) = \frac{2y}{x^2 + y^2}$$

$$f_y(1, 1) = \frac{2 \cdot 1}{1^2 + 1^2} = \frac{2}{2} = 1$$

The slope of the tangent line to the curve of intersection of the surface z and the plane $x = 1$ is 1.

39. $z = x^2 + 4y^2$

$$\frac{\partial z}{\partial x} = 2x \qquad\qquad\qquad\qquad \frac{\partial z}{\partial y} = 8y$$

$$x\frac{\partial z}{\partial x} + y\frac{\partial z}{\partial y} = x \cdot 2x + y \cdot 8y = 2x^2 + 8y^2 = 2\left(x^2 + 4y^2\right) = 2z$$

41.
$$z = \ln \sqrt{x^2 + y^2} = \ln\left(x^2 + y^2\right)^{1/2} = \frac{1}{2}\ln\left(x^2 + y^2\right)$$

$$\frac{\partial z}{\partial x} = \frac{1}{2} \cdot \frac{2x}{x^2 + y^2} = \frac{x}{x^2 + y^2} \qquad\qquad \frac{\partial z}{\partial y} = \frac{1}{2} \cdot \frac{2y}{x^2 + y^2} = \frac{y}{x^2 + y^2}$$

$$\frac{\partial^2 z}{\partial x^2} = \frac{1 \cdot \left(x^2 + y^2\right) - x \cdot 2x}{\left(x^2 + y^2\right)^2} \qquad\qquad \frac{\partial^2 z}{\partial y^2} = \frac{1 \cdot \left(x^2 + y^2\right) - y \cdot 2y}{\left(x^2 + y^2\right)^2}$$

$$= \frac{y^2 - x^2}{\left(x^2 + y^2\right)^2} \qquad\qquad\qquad\qquad = \frac{x^2 - y^2}{\left(x^2 + y^2\right)^2}$$

$$\frac{\partial^2 z}{\partial x^2} + \frac{\partial^2 z}{\partial y^2} = \frac{y^2 - x^2}{\left(x^2 + y^2\right)^2} + \frac{x^2 - y^2}{\left(x^2 + y^2\right)^2} = \frac{y^2 - x^2 + x^2 - y^2}{\left(x^2 + y^2\right)^2} = 0$$

43. $z = 1000 - 20x - 50y$

(a) $\dfrac{\partial z}{\partial x} = -20 \qquad\qquad\qquad\qquad \dfrac{\partial z}{\partial y} = -50$

(b) Answers will vary.

45.
$$A = 9\left(\frac{N}{I}\right)$$

(a) $\dfrac{\partial A}{\partial N} = \dfrac{9}{I}$ $\dfrac{\partial A}{\partial I} = -9\left(\dfrac{N}{I^2}\right)$

(b) If $N = 78$ and $I = 217$, then

$$\dfrac{\partial A}{\partial N} = \dfrac{9}{217} = 0.0415$$ $$\dfrac{\partial A}{\partial I} = -9\left(\dfrac{78}{217^2}\right) = -0.0149$$

(c) Answers will vary.

47.
(a) $\dfrac{\partial H}{\partial t} = 2.049015323 - 0.22475541r - 0.01367566t + 0.00245748\ tr + 0.00085282r^2$
$- 0.00000398tr^2$

(b) Answers will vary.

(c) $\dfrac{\partial H}{\partial r} = 10.14333127 - 0.22475541t - 0.10963434r + 0.00122874t^2 + 0.00170564tr$
$- 0.00000398t^2r$

(d) Answers will vary.

49.   No, you should not believe it. Explanations will vary.

17.4 Local Maxima and Local Minima

1.   We find the partial derivatives f_x and f_y, set each equal to zero, and solve the system of equations.

$f_x(x, y) = 4x^3 - 4x = 0$ $f_y(x, y) = 2y = 0$
$\quad\quad 4x(x^2 - 1) = 0$ $\quad\quad\quad y = 0$
$\quad 4x(x - 1)(x + 1) = 0$
$\quad 4x = 0$ or $x - 1 = 0$ or $x + 1 = 0$
$\quad\ x = 0$ or $\quad x = 1$ or $\quad x = -1$
The critical points are $(0, 0)$, $(1, 0)$, and $(-1, 0)$.

3.   We find the partial derivatives f_x and f_y, set each equal to zero, and solve the system of equations.

$f_x(x, y) = 4y - 4x^3 = 0$ $f_y(x, y) = 4x - 4y^3 = 0$
$\quad\quad\quad y = x^3$ $\quad\quad\quad x - (x^3)^3 = 0$
$\quad\quad\quad\quad\quad\quad\quad x(1 - x^8) = 0$
$\quad\quad\quad\quad\quad\quad x = 0$ or $1 - x^8 = 0$
$\quad\quad\quad\quad\quad\quad\quad\quad\quad\quad\quad x = \pm 1$
The critical points are $(x, x^3) = (0, 0)$, $(1, 1)$, and $(-1, -1)$.

5. We find the partial derivatives f_x and f_y, set each equal to zero, and solve the system of equations.

$$f_x(x, y) = 4x^3 = 0 \qquad\qquad f_y(x, y) = 4y^3 = 0$$
$$x = 0 \qquad\qquad\qquad y = 0$$

The critical point is $(0, 0)$.

7. We find the partial derivatives f_x and f_y, set each equal to zero, and solve the system of equations.

$$f_x(x, y) = 6x - 2y = 0 \qquad\qquad f_y(x, y) = -2x + 2y = 0$$

Adding $\quad 6x - 2y = 0$
and $\quad\underline{-2x + 2y = 0}$
we get $\quad 4x \qquad = 0 \quad$ or $\quad x = 0.$

Substituting $x = 0$ into the first equation, we find $-2y = 0$ or $y = 0$.
The critical point is $(0, 0)$.

To determine the character of the critical point we find the two second order partial derivatives and the mixed partial derivative, evaluate them at the critical point, and calculate D.

$$f_{xx}(x, y) = 6 \qquad\qquad f_{yy}(x, y) = 2 \qquad\qquad f_{xy}(x, y) = -2$$
$$f_{xx}(0, 0) = 6 \qquad\qquad f_{yy}(0, 0) = 2 \qquad\qquad f_{xy}(0, 0) = -2$$

$$D = f_{xx}(x_0, y_0) \cdot f_{yy}(x_0, y_0) - \left[f_{xy}(x_0, y_0)\right]^2 = 6 \cdot 2 - (-2)^2 = 8 > 0$$

Since $f_{xx}(0, 0) > 0$ and $D > 0$, f has a local minimum at $(x_0, y_0, z_0) = (0, 0, 0)$.

9. We find the partial derivatives f_x and f_y, set each equal to zero, and solve the system of equations.

$$f_x(x, y) = 2x - 3 = 0 \qquad\qquad f_y(x, y) = 2y = 0$$

$$x = \frac{3}{2} = 1.5 \qquad\qquad\qquad y = 0$$

The critical point is $\left(\dfrac{3}{2}, 0\right)$.

To determine the character of the critical point we find the two second order partial derivatives and the mixed partial derivative, evaluate them at the critical point, and calculate D.

$$f_{xx}(x, y) = 2 \qquad\qquad f_{yy}(x, y) = 2 \qquad\qquad f_{xy}(x, y) = 0$$
$$f_{xx}\left(\frac{3}{2}, 0\right) = 2 \qquad\qquad f_{yy}\left(\frac{3}{2}, 0\right) = 2 \qquad\qquad f_{xy}\left(\frac{3}{2}, 0\right) = 0$$

$$D = f_{xx}(x_0, y_0) \cdot f_{yy}(x_0, y_0) - \left[f_{xy}(x_0, y_0)\right]^2 = 2 \cdot 2 - 0^2 = 4 > 0$$

Since $f_{xx}\left(\dfrac{3}{2}, 0\right) > 0$ and $D > 0$, f has a local minimum at $\left(x_0, y_0, z_0\right) = \left(\dfrac{3}{2}, 0, \dfrac{39}{4}\right)$.

11. We find the partial derivatives f_x and f_y, set each equal to zero, and solve the system of equations.

$$f_x(x, y) = 2x + 4 = 0 \qquad\qquad f_y(x, y) = -2y + 8 = 0$$
$$x = -2 \qquad\qquad\qquad\qquad y = 4$$

The critical point is $(-2, 4)$.

To determine the character of the critical point we find the two second order partial derivatives and the mixed partial derivative, evaluate them at the critical point, and calculate D.

$$f_{xx}(x, y) = 2 \qquad\qquad f_{yy}(x, y) = -2 \qquad\qquad f_{xy}(x, y) = 0$$
$$f_{xx}(-2, 4) = 2 \qquad\qquad f_{yy}(-2, 4) = -2 \qquad\qquad f_{xy}(-2, 4) = 0$$
$$D = f_{xx}(x_0, y_0) \cdot f_{yy}(x_0, y_0) - \left[f_{xy}(x_0, y_0)\right]^2 = 2 \cdot (-2) - 0^2 = -4$$

Since $D < 0$, $\left(x_0, y_0, z_0\right) = (-2, 4, 12)$ is a saddle point.

13. We find the partial derivatives f_x and f_y, set each equal to zero, and solve the system of equations.

$$f_x(x, y) = 2x - 4 = 0 \qquad\qquad f_y(x, y) = 8y + 8 = 0$$
$$x = 2 \qquad\qquad\qquad\qquad y = -1$$

The critical point is $(2, -1)$.

To determine the character of the critical point we find the two second order partial derivatives and the mixed partial derivative, evaluate them at the critical point, and calculate D.

$$f_{xx}(x, y) = 2 \qquad\qquad f_{yy}(x, y) = 8 \qquad\qquad f_{xy}(x, y) = 0$$
$$D = f_{xx}(x_0, y_0) \cdot f_{yy}(x_0, y_0) - \left[f_{xy}(x_0, y_0)\right]^2 = 2 \cdot 8 - 0^2 = 16$$

Since $f_{xx}(2, -1) > 0$ and $D > 0$, $\left(x_0, y_0, z_0\right) = (2, -1, -9)$ is a local minimum.

15. We find the partial derivatives f_x and f_y, set each equal to zero, and solve the system of equations.

$$f_x(x, y) = 2x + y - 6 = 0 \qquad\qquad f_y(x, y) = 2y + x = 0$$
$$x = -2y$$

$$2(-2y) + y - 6 = 0$$
$$-3y = 6 \text{ or } y = -2 \qquad\qquad x = -2(-2) = 4$$

The critical point is $(4, -2)$.

To determine the character of the critical point we find the two second order partial derivatives and the mixed partial derivative, evaluate them at the critical point, and calculate D.

$$f_{xx}(x, y) = 2 \qquad\qquad f_{yy}(x, y) = 2 \qquad\qquad f_{xy}(x, y) = 1$$

$$D = f_{xx}(x_0, y_0) \cdot f_{yy}(x_0, y_0) - \left[f_{xy}(x_0, y_0) \right]^2 = 2 \cdot 2 - 1^2 = 3$$

Since $f_{xx}(4, -2) > 0$ and $D > 0$, $(x_0, y_0, z_0) = (4, -2, -6)$ is a local minimum.

17. We find the partial derivatives f_x and f_y, set each equal to zero, and solve the system of equations.

$$f_x(x, y) = 2x + y = 0 \qquad\qquad f_y(x, y) = -2y + x = 0$$
$$y = -2x \qquad\qquad\qquad -2(-2x) + x = 0$$
$$5x = 0$$
$$x = 0 \text{ and } y = 0$$

The critical point is $(0, 0)$.

To determine the character of the critical point we find the two second order partial derivatives and the mixed partial derivative, evaluate them at the critical point, and calculate D.

$$f_{xx}(x, y) = 2 \qquad\qquad f_{yy}(x, y) = -2 \qquad\qquad f_{xy}(x, y) = 1$$

$$D = f_{xx}(x_0, y_0) \cdot f_{yy}(x_0, y_0) - \left[f_{xy}(x_0, y_0) \right]^2 = 2(-2) - 1^2 = -5$$

Since $D < 0$, $(x_0, y_0, z_0) = (0, 0, 2)$ is a saddle point.

19. We find the partial derivatives f_x and f_y, set each equal to zero, and solve the system of equations.

$$f_x(x, y) = 3x^2 - 6y = 0 \qquad\qquad f_y(x, y) = -6x + 3y^2 = 0$$

$$y = \frac{1}{2} x^2 \qquad\qquad\qquad -6x + 3\left(\frac{x^2}{2} \right)^2 = -2x + \frac{x^4}{4} = 0$$

$$x\left(\frac{x^3}{4} - 2 \right) = 0$$

$$x = 0 \quad \text{or} \quad \frac{x^3}{4} - 2 = 0$$
$$x^3 = 8$$
$$x = 2$$

When $x = 0$, $y = 0$, and when $x = 2$, $y = 2$. So the critical points are $(0, 0)$ and $(2, 2)$.

To determine the character of the critical points we find the two second order partial derivatives and the mixed partial derivative, evaluate them at the critical points, and calculate D.

$$f_{xx}(x, y) = 6x \qquad\qquad f_{yy}(x, y) = 6y \qquad\qquad f_{xy}(x, y) = -6$$

$$f_{xx}(0, 0) = 0 \qquad f_{yy}(0, 0) = 0 \qquad f_{xy}(0, 0) = -6$$

$$D = f_{xx}(0, 0) \cdot f_{yy}(0, 0) - \left[f_{xy}(0, 0) \right]^2 = 0 \cdot 0 - (-6)^2 = -36$$

Since $D < 0$ $(x_0, y_0, z_0) = (0, 0, 0)$ is a saddle point.

$$f_{xx}(2, 2) = 6 \cdot 2 = 12 \qquad f_{yy}(2, 2) = 6 \cdot 2 = 12 \qquad f_{xy}(2, 2) = -6$$

$$D = f_{xx}(2, 2) \cdot f_{yy}(2, 2) - \left[f_{xy}(2, 2) \right]^2 = 12 \cdot 12 - (-6)^2 = 108$$

Since $f_{xx}(2, 2) > 0$ and $D > 0$, $(x_0, y_0, z_0) = (2, 2, -8)$ is a local minimum.

21. We find the partial derivatives f_x and f_y, set each equal to zero, and solve the system of equations.

$$f_x(x, y) = 3x^2 + 2xy = 0 \qquad\qquad f_y(x, y) = x^2 + 2y = 0$$

$$y = -\frac{1}{2}x^2$$

$$3x^2 + 2x\left(-\frac{1}{2}x^2 \right) = 0$$

$$3x^2 - x^3 = 0$$

$$x^2(3 - x) = 0$$

$$x^2 = 0 \quad \text{or} \quad 3 - x = 0$$

$$x = 0 \quad \text{or} \qquad x = 3$$

When $x = 0$, then $y = 0$, and when $x = 3$, then $y = -\dfrac{9}{2}$. So the critical points are $(0, 0)$

and $\left(3, -\dfrac{9}{2} \right)$.

To determine the character of the critical points we find the two second order partial derivatives and the mixed partial derivative, evaluate them at the critical points, and calculate D.

$$f_{xx}(x, y) = 6x + 2y \qquad\qquad f_{yy}(x, y) = 2 \qquad\qquad f_{xy}(x, y) = 2x$$

$$f_{xx}(0, 0) = 0 \qquad\qquad f_{yy}(0, 0) = 2 \qquad\qquad f_{xy}(0, 0) = 0$$

$$D = f_{xx}(x_0, y_0) \cdot f_{yy}(x_0, y_0) - \left[f_{xy}(x_0, y_0) \right]^2 = 0 \cdot 2 - 0^2 = 0.$$

Since $D = 0$, no information is given by the test.

$$f_{xx}\left(3, -\frac{9}{2} \right) = 6 \cdot 3 + 2\left(-\frac{9}{2} \right) = 9 \qquad f_{yy}\left(3, -\frac{9}{2} \right) = 2 \qquad f_{xy}\left(3, -\frac{9}{2} \right) = 2 \cdot 3 = 6$$

$$D = f_{xx}\left(3, -\frac{9}{2} \right) \cdot f_{yy}\left(3, -\frac{9}{2} \right) - \left[f_{xy}\left(3, -\frac{9}{2} \right) \right]^2 = 9 \cdot 2 - 6^2 = -18$$

Since $D < 0$, $(x_0, y_0, z_0) = \left(3, -\dfrac{9}{2}, \dfrac{27}{4}\right)$ is a saddle point.

23. We find the partial derivatives f_x and f_y, set each equal to zero, and solve the system of equations.

$$f_x(x, y) = \frac{0 - y(1)}{(x+y)^2} = -\frac{y}{(x+y)^2} = 0 \qquad f_y(x, y) = \frac{1 \cdot (x+y) - y \cdot 1}{(x+y)^2} = \frac{x}{(x+y)^2} = 0$$

$$y = 0 \hspace{7cm} x = 0$$

but the point $(0, 0)$ is not in the domain of f. So f has no local maximum or local minimum point.

25. If we let R denote the revenue function, then
$$R(x) = xp = x(12 - x) \text{ and } R(y) = y(8 - y).$$
The joint revenue function is given by $R(x, y) = R(x) + R(y)$.
$$R(x, y) = 12x - x^2 + 8y - y^2$$
The joint profit function is given by $P(x, y) = R(x, y) - C(x, y)$.
$$P(x, y) = \left(12x - x^2 + 8y - y^2\right) - \left(x^2 + 2xy + 3y^2\right)$$
$$= 12x - 2x^2 + 8y - 4y^2 - 2xy$$
The first-order partial derivatives are
$$P_x(x, y) = 12 - 4x - 2y \qquad\qquad P_y(x, y) = 8 - 8y - 2x$$
The solution to the system is the critical point of P.
$$4x + 2y = 12$$
$$2x + 8y = 8$$
Using the first equation, we find that $y = 6 - 2x$. Substituting for y in the second equation, we have
$$2x + 8(6 - 2x) = 8$$
$$x + 24 - 8x = 4$$
$$20 = 7x \ \text{ or } x = \frac{20}{7}$$

Back-substituting into the first equation we get $y = 6 - \dfrac{40}{7} = \dfrac{2}{7}$. So $\left(\dfrac{20}{7}, \dfrac{2}{7}\right)$ is the critical point.

The second order partial derivatives of P are
$$P_{xx}(x, y) = -4 \qquad\qquad P_{yy}(x, y) = -8 \qquad\qquad P_{xy}(x, y) = -2$$
and $D = (-4)(-8) - (-2)^2 = 32 - 4 = 28$.

Since $P_{xx}\left(\dfrac{20}{7}, \dfrac{2}{7}\right) = -4 < 0$ and $D > 0$, there is a local maximum at $\left(\dfrac{20}{7}, \dfrac{2}{7}\right)$.

When $x = \dfrac{20}{7} \approx 2.857$, $p = 12 - 2.857 = 9.14286$, and when $y = \dfrac{2}{7} \approx 0.286$, $q = 7.71429$

So to maximize profit, the company should sell 2857 units of product x at $9142.86 per unit and 286 units of product y at $7714.29 per unit. The maximum profit is $18,285.71.

27.
$$P(x, y) = 2000x - \frac{1}{5}x^2 + 1150y - 2y^2 + xy + 10,000$$

We find the partial derivatives P_x and P_y, set each equal to zero, and solve the system of equations.

$$P_x(x, y) = 2000 - \frac{2}{5}x + y = 0 \qquad\qquad P_y(x, y) = 1150 - 4y + x = 0$$

$$y = \frac{2}{5}x - 2000 \qquad\qquad 1150 - 4\left(\frac{2}{5}x - 2000\right) + x = 0$$

$$1150 - \frac{8}{5}x + 8000 + x = 0$$

$$9150 - \frac{3}{5}x = 0$$

$$x = \frac{5}{3} \cdot 9150 = 15,250$$

$$y = \frac{2}{5} \cdot 15,250 - 2000 = 4100$$

The critical point is (15250, 4100).

The second order partial derivatives are

$$P_{xx}(x, y) = -\frac{2}{5} \qquad\qquad P_{yy}(x, y) = -4 \qquad\qquad P_{xy}(x, y) = 1$$

$$D = \left(-\frac{2}{5}\right) \cdot (-4) - 1^2 = \frac{8}{2} - 1 = \frac{3}{5}$$

Since $P_{xx}(15250, 4100) = -\frac{2}{5} < 0$ and $D > 0$, there is a local maximum at (15250, 4100).

The maximum profit of $17,617,500 is obtained when 15,250 tons of grade A steel and 4100 tons of grade B steel is produced.

29. (a) If x is constant, then
$$R_y(x, y) = 2x^2 y(a-x)(b-y) + x^2 y^2(a-x)(-1) = 0$$
$$\left[(a-x)x^2 y\right]\left[2(b-y) - y\right] = 0$$
$$\left[(a-x)x^2 y\right] = 0 \quad \text{or} \quad \left[2b - 3y\right] = 0$$
$$y = 0 \quad \text{or} \quad y = \frac{2b}{3}$$

We use the second derivative test to see which value of y locates a relative maximum.
$$R_{yy}(x, y) = \left[(a-x)x^2\right]\left[2b - 3y\right] - \left[3(a-x)x^2 y\right]$$
$$R_{yy}(x, 0) = \left[(a-x)x^2\right]\left[2b\right] > 0 \text{ since } x < a \text{ and } b > 0.$$

$$R_{yy}\left(x, \frac{2b}{3}\right) = \left[(a-x)x^2\right]\left[2b-2b\right] - \left[2(a-x)x^2b\right] = 0 - 2(a-x)x^2b < 0$$

When $y = \dfrac{2b}{3}$, the reaction to the drug is maximized.

(b) If the amount y of the second drug is held constant, then

$$R_x(x, y) = 2xy^2(a-x)(b-y) - x^2y^2(b-y) = 0$$

$$\left[(b-y)xy^2\right]\left[2(a-x)-x\right] = 0$$

$$(b-y)xy^2 = 0 \quad \text{or} \quad 2a - 3x = 0$$

$$x = 0 \quad \text{or} \quad x = \frac{2a}{3}$$

We use the second derivative test to see which value of x locates a relative maximum.

$$R_{xx}(x, y) = -3\left[(by^2 - y^3)x\right] + (by^2 - y^3)(2a - 3x)$$

$$R_{xx}\left(\frac{2a}{3}, y\right) = -2(by^2 - y^3)a + (by^2 - y^3)(2a - 2a) = -2y^2(b-y)a < 0$$

$$R_{xx}(0, y) = -3\left[(by^2 - y^3)0\right] + (by^2 - y^3)(2a - 3(0)) = (by^2 - y^3)(2a) > 0$$

When $x = \dfrac{2a}{3}$ the reaction to the drug is maximized.

(c) If both x and y are variable, then

$$R_{xy}(x, y) = (2a - 3x)(2ybx - 3xy^2) = (2a - 3x)(2b - 3y)xy$$

$$R_{xy}\left(\frac{2a}{3}, \frac{2b}{3}\right) = \left[2a + 3\left(\frac{2a}{3}\right)\right]\left[2b - 3\left(\frac{2b}{3}\right)\right]\left[\left(\frac{2a}{3}\right)\left(\frac{2b}{3}\right)\right] = 0$$

$$R_{xx}\left(\frac{2a}{3}, \frac{2b}{3}\right) = \left[b - \frac{2b}{3}\right]\left[\frac{2b}{3}\right]^2\left[2a - 6\left(\frac{2a}{3}\right)\right]$$

$$= \left[\frac{3b - 2b}{3}\right]\left[\frac{4b^2}{9}\right]\left[-2a\right] = \frac{b}{3} \cdot \frac{4b^2}{9} \cdot (-2a) = -\frac{8ab^3}{27}$$

$$R_{yy}\left(\frac{2a}{3}, \frac{2b}{3}\right) = \left[a - \frac{2a}{3}\right]\left[\frac{2a}{3}\right]^2\left[2b - 6\left(\frac{2b}{3}\right)\right]$$

$$= \left[\frac{3a - 2a}{3}\right]\left[\frac{4a^2}{9}\right]\left[-2b\right] = \frac{a}{3} \cdot \frac{4a^2}{9} \cdot (-2b) = -\frac{8a^3b}{27}$$

$$D = \left[-\frac{8ab^3}{27}\right]\left[-\frac{8a^3b}{27}\right] - 0 = \frac{64a^4b^4}{27^2} > 0$$

Since $R_{xx}\left(\dfrac{2a}{3}, \dfrac{2b}{3}\right) < 0$ and $D > 0$, then the reaction to the drug is a maximum.

31. To find the maximum reaction to the drug, we first find the critical points of y.
$$y = x^2 [a-x] t = ax^2 t - x^3 t$$

$$\frac{\partial y}{\partial x} = 2axt - 3x^2 t = 0 \qquad\qquad \frac{\partial y}{\partial t} = ax^2 - x^3 = 0$$

$$x^2(a-x) = 0$$
$$x = 0 \quad \text{or} \quad x = a$$

We find the second order partial derivatives and test to see if the critical point locates a relative maximum point.

$$\frac{\partial^2 y}{\partial x^2} = 2at - 6xt \qquad\qquad \frac{\partial^2 y}{\partial t^2} = 0 \qquad\qquad \frac{\partial^2 y}{\partial x \partial t} = 2ax - 3x^2$$

$$D = \frac{\partial^2 y}{\partial x^2} \cdot \frac{\partial^2 y}{\partial t^2} - \left[\frac{\partial y}{\partial x \partial t}\right]^2 = 0 \cdot 0 - \left(2a^2 - 3a^2\right)^2 = -a^4$$

Since $D < 0$, there is a saddle point at $(a, 0, 0)$, and there is no maximum reaction to the drug.

33. (a) Let w denote the width, h denote the depth, and l denote the length of the box.
There are two restrictions $l \le 108$, and $l + 2w + 2h \le 130$ inches. So $l \le 130 - 2w - 2h$.
$$V = lwh = (130 - 2w - 2h)wh = 130wh - 2w^2 h - 2wh^2$$
To find the dimensions of the box that meet regulations while maximizing volume, we find the critical points of V.

$$V_h = 130w - 2w^2 - 4wh = 0 \qquad\qquad V_w = 130h - 4wh - 2h^2 = 0$$
$$2w(65 - w - 2h) = 0 \qquad\qquad\qquad = h[130 - 4w - 2h]$$
$$w = 0 \quad \text{or} \quad 65 - w - 2h = 0$$
$$w = 65 - 2h$$

When $w = 0$, then $130h - 2h^2 = 0$ or $h = 0$ or $h = 65$, but both of these measurements result in $V = 0$ cubic inches, which is a minimum.

When $w = 65 - 2h$, then
$$h[130 - 4(65 - 2h) - 2h] = 0$$
$$h[-130 + 6h] = 0$$
$$h = 0 \quad \text{or} \quad 6h - 130 = 0$$
$$h = \frac{65}{3} \qquad \text{and so} \quad w = 65 - 2\left(\frac{65}{3}\right) = \frac{65}{3} = 21.667$$

To check to see if these values result in the maximum volume, we find the second order partials and evaluate D.

$$V_{hh} = -4w \qquad\qquad V_{ww} = -4h \qquad\qquad V_{hw} = 130 - 4w - 4h$$

$$V_{hh}\left(\frac{65}{3}, \frac{65}{3}\right) = -4\left(\frac{65}{3}\right) = -\frac{260}{3} \qquad\qquad V_{ww}\left(\frac{65}{3}, \frac{65}{3}\right) = -4\left(\frac{65}{3}\right) = -\frac{260}{3}$$

$$V_{lw}\left(\frac{65}{3},\frac{65}{3}\right)=130-4\left(\frac{65}{3}\right)-4\left(\frac{65}{3}\right)=130-8\left(\frac{65}{3}\right)=-\frac{130}{3}$$

$$D=\left(-\frac{260}{3}\right)\left(-\frac{260}{3}\right)-\left(-\frac{130}{3}\right)^2=\frac{4\cdot130^2}{9}-\frac{130^2}{9}=\frac{3\cdot130^2}{9}>0$$

Since $V_{hh}\left(\frac{65}{3},\frac{65}{3}\right)<0$ and $D>0$, the volume is maximum when the width = 21.667

inches, the depth is 21.667 inches and the length is $130-4(21.667)=43.332$ inches. The maximum volume is 20,342.59 cubic inches.

(b) If we let r denote the radius of the cylinder and h denote the height, then we have the restriction $h+2\pi r\le130$ or $h\le130-2\pi r$. The volume of the cylinder is

$$V=\pi r^2 h=\pi r^2(130-2\pi r)=130\pi r^2-2\pi^2 r^3$$

To find the dimensions of the cylinder that meet regulations while maximizing volume, we find the critical points of V.

$$\frac{dV}{dr}=260\pi r-6\pi^2 r^2=2\pi r(130-3\pi r)=0$$

$$r=0 \quad\text{or}\quad 130-3\pi r=0$$

$$r=\frac{130}{3\pi}$$

We use the second derivative test to see if the volume is maximized.

$$\frac{d^2V}{dr^2}=260\pi-12\pi^2 r$$

When $r=\frac{130}{3\pi}$, then $\frac{d^2V}{dr^2}=260\pi-12\left(\frac{130}{3\pi}\right)\pi^2=-260\pi<0$. So the volume is

maximized if the cylinder has a radius of $\frac{130}{3\pi}\approx13.79$ inches and a height of 43.33

inches.

35. We find the critical points of W.

$$W_x=\frac{1}{100}\left[\frac{1}{10}x-y-4\right]=0 \qquad\qquad W_y=\frac{1}{100}[50y-x]=0$$

$$x-10y-40=0 \qquad\qquad\qquad 50y-x=0 \quad\text{or}\quad x=50y$$

$$50y-10y=40$$
$$40y=40$$
$$y=1 \qquad\text{and}\qquad x=50$$

The critical point is (50, 1). We find the second order partial derivatives to test if the critical point locates a minimum value.

$$W_{xx} = \frac{1}{100} \cdot \frac{1}{10} = \frac{1}{1000} \quad W_{yy} = \frac{1}{100}[50] = \frac{1}{2} \quad W_{xy} = \frac{1}{100}[-1] = -\frac{1}{100}$$

$$D = \left[\frac{1}{1000}\right]\left[\frac{1}{2}\right] - \left[-\frac{1}{100}\right]^2 = \frac{1}{2000} - \frac{1}{10,000} = \frac{5-1}{10,000} = \frac{4}{10,000} > 0$$

Since $W_{xx}(50, 1) > 0$ and $D > 0$, waste is minimized when the manufacturer uses 50 tons of steel at the rate of 1 ton per week.

17.5 Lagrange Multipliers

1. **STEP 1** Find the maximum of $z = f(x, y) = 3x + 4y$
    Subject to the constraint $g(x, y) = x^2 + y^2 - 9 = 0$

    STEP 2 Construct the function $F(x, y, \lambda) = 3x + 4y + \lambda(x^2 + y^2 - 9)$.

    STEP 3 Set up the system of equations

    $$\frac{\partial F}{\partial x} = 3 + 2\lambda x = 0 \qquad (1)$$

    $$\frac{\partial F}{\partial y} = 4 + 2\lambda y = 0 \qquad (2)$$

    $$\frac{\partial F}{\partial \lambda} = x^2 + y^2 - 9 = 0 \qquad (3)$$

    STEP 4 Solve the system of equations for x, y, and λ.

    $$\lambda = -\frac{3}{2x} \qquad (1)$$

    $$4 + 2\left(-\frac{3}{2x}\right)y = 0 \qquad (2)$$

    $$4 - \frac{3y}{x} = 0$$

    $$y = \frac{4x}{3}$$

    $$x^2 + \left(\frac{4x}{3}\right)^2 - 9 = 0 \qquad (3)$$

    $$9x^2 + 16x^2 = 81$$

    $$25x^2 = 81$$

    $$x = \pm\sqrt{\frac{81}{25}} = \pm\frac{9}{5}$$

    $$y = \frac{4}{3} \cdot \left(\pm\frac{9}{5}\right) = \pm\frac{12}{5}, \text{ and } \lambda = -\frac{3}{2} \cdot \left(\pm\frac{5}{9}\right) = \pm\frac{5}{6}$$

STEP 5 Evaluate $z = f(x, y)$.

$$z = f\left(\frac{9}{5}, \frac{12}{5}\right) = 3 \cdot \frac{9}{5} + 4 \cdot \frac{12}{5} = \frac{27 + 48}{5} = \frac{75}{5} = 15$$

$$z = f\left(-\frac{9}{5}, -\frac{12}{5}\right) = 3 \cdot \left(-\frac{9}{5}\right) + 4 \cdot \left(-\frac{12}{5}\right) = \frac{-27 - 48}{5} = \frac{-75}{5} = -15$$

The maximum value of z subject to constraint g, is 15.

3. **STEP 1** Find the minimum of $z = f(x, y) = x^2 + y^2$

Subject to the constraint $\quad g(x, y) = x + y - 1 = 0$

STEP 2 Construct the function $F(x, y, \lambda) = f(x, y) + \lambda\, g(x, y)$.

$$F(x, y, \lambda) = x^2 + y^2 + \lambda(x + y - 1) = 0$$

STEP 3 Set up the system of equations

$$F_x(x, y, \lambda) = 2x \quad + \lambda = 0 \qquad \Rightarrow \qquad \lambda = -2x \qquad (1)$$
$$F_y(x, y, \lambda) = \quad 2y + \lambda = 0 \qquad \Rightarrow \qquad \lambda = -2y \qquad (2)$$
$$F_\lambda(x, y, \lambda) = x + y - 1 = 0 \qquad\qquad\qquad\qquad\qquad (3)$$

STEP 4 Solve the system of equations for x, y, and λ.

From (1) and (2) we find that $-2x = -2y$ or $x = y$.

Substituting this result in (3) gives $y + y = 1$ or $2y = 1$ or $y = \frac{1}{2}$ and $x = \frac{1}{2}$.

STEP 5 $\quad z = f\left(\frac{1}{2}, \frac{1}{2}\right) = \left(\frac{1}{2}\right)^2 + \left(\frac{1}{2}\right)^2 = \frac{1}{2}$

We conclude that z attains its minimum value at $\left(\frac{1}{2}, \frac{1}{2}\right)$. The minimum value is $\frac{1}{2}$.

5. **STEP 1** Find the maximum of $z = f(x, y) = 12xy - 3y^2 - x^2$

Subject to the constraint $\quad g(x, y) = x + y - 16 = 0$

STEP 2 Construct the function $F(x, y, \lambda) = f(x, y) + \lambda\, g(x, y)$

$$F(x, y, \lambda) = 12xy - 3y^2 - x^2 + \lambda(x + y - 16)$$

STEP 3 Set up the system of equations

$$F_x(x, y, \lambda) = 12y - 2x + \lambda = 0 \qquad (1)$$
$$F_y(x, y, \lambda) = 12x - 6y + \lambda = 0 \qquad (2)$$
$$F_\lambda(x, y, \lambda) = \quad x + y - 16 = 0 \qquad (3)$$

STEP 4 Solve the system of equations for x and y.

Subtracting (2) from (1) we get $18y - 14x = 0$

$$x = \frac{18}{14}y = \frac{9}{7}y$$

Substituting for x in (3), gives $\dfrac{9}{7}y + y = 16$

$$\dfrac{16}{7}y = 16 \text{ or } y = 7 \text{ and } x = 9.$$

STEP 5 Evaluate $z = f(x, y)$ at $(9, 7)$.

$$z = 12(9)(7) - 3(7^2) - 9^2 = 528$$

We conclude that z attains its maximum value at $(9, 7)$. The maximum value is 528.

7. **STEP 1** Find the minimum of $z = f(x, y) = 5x^2 + 6y^2 - xy$
   Subject to the constraint $g(x, y) = x + 2y - 24 = 0$
   STEP 2 Construct the function $F(x, y, \lambda) = f(x, y) + \lambda\, g(x, y)$
   $$F(x, y, \lambda) = 5x^2 + 6y^2 - xy + \lambda(x + 2y - 24)$$
   STEP 3 Set up the system of equations
   $$\begin{aligned} F_x(x,y,\lambda) &= 10x - y + \lambda = 0 &\quad (1)\\ F_y(x,y,\lambda) &= 12y - x + 2\lambda = 0 &\quad (2)\\ F_\lambda(x,y,\lambda) &= x + 2y - 24 = 0 &\quad (3) \end{aligned}$$

   STEP 4 Solve the system of equations for x and y.
   Subtracting twice (1) from (2) gives $-21x + 14y = 0$
   $$y = \dfrac{21x}{14} = \dfrac{3}{2}x$$
   Substituting for y in (3) we get $x + 2\left(\dfrac{3}{2}x\right) = 24$

   $$4x = 24 \text{ or } x = 6 \text{ and } y = \dfrac{3}{2}\cdot 6 = 9$$

   STEP 5 Evaluate $z = f(x, y)$ at $(6, 9)$.
   $$z = 5(6^2) + 6(9^2) - (6)(9) = 612$$
   We conclude that z attains its minimum value at $(6, 9)$. The minimum value is 612.

9. **STEP 1** Find the maximum of $w = f(x, y, z) = xyz$
   Subject to the constraint $g(x, y, z) = x + 2y + 2z - 120 = 0$
   STEP 2 Construct the function $F(x, y, z, \lambda) = f(x, y, z) + \lambda\, g(x, y, z)$
   $$= xyz + x(x + 2y + 2z - 120)$$

   STEP 3 Set up the system of equations

$$F_x(x, y, z, \lambda) = yz + \lambda = 0 \qquad \Rightarrow \qquad \lambda = -yz \qquad (1)$$

$$F_y(x, y, z, \lambda) = xz + 2\lambda = 0 \qquad \Rightarrow \qquad \lambda = -\frac{xz}{2} \qquad (2)$$

$$F_z(x, y, z, \lambda) = xy + 2\lambda = 0 \qquad \Rightarrow \qquad \lambda = -\frac{xy}{2} \qquad (3)$$

$$F_\lambda(x, y, z, \lambda) = x + 2y + 2z - 120 = 0 \qquad (4)$$

STEP 4 Solve the system of equations for x, y, and z.

From (1) and (2) we find $-yz = -\frac{xz}{2}$ or $y = \frac{x}{2}$.

From (1) and (3) we find $-yz = -\frac{xy}{2}$ or $z = \frac{x}{2}$.

Substituting for y and z in (4) gives $x + 2\left(\frac{x}{2}\right) + 2\left(\frac{x}{2}\right) = 120$ or $3x = 120$ or $x = 40$.

Back substituting we find $y = 20$ and $z = 20$.
STEP 5 Evaluate $w = f(x, y, z)$ at $(40, 20, 20)$.
$$w = (40)(20)(20) = 16{,}000$$
We conclude that w attains its maximum value at $(40, 20, 20)$. The maximum is 16,000.

11. **STEP 1** Find the minimum of $w = f(x, y, z) = x^2 + y^2 + z^2 - x - 3y - 5z$
     Subject to the constraint $g(x, y, z) = x + y + 2z - 20 = 0$
STEP 2 Construct the function $F(x, y, z, \lambda) = f(x, y, z) + \lambda g(x, y, z)$
$$F(x, y, z, \lambda) = x^2 + y^2 + z^2 - x - 3y - 5z + \lambda(x + y + 2z - 20)$$
STEP 3 Set up the system of equations
$$F_x(x, y, z, \lambda) = 2x - 1 + \lambda = 0 \qquad \Rightarrow \qquad \lambda = 1 - 2x \qquad (1)$$
$$F_y(x, y, z, \lambda) = 2y - 3 + \lambda = 0 \qquad \Rightarrow \qquad \lambda = 3 - 2y \qquad (2)$$

$$F_z(x, y, z, \lambda) = 2z - 5 + 2\lambda = 0 \qquad \Rightarrow \qquad \lambda = \frac{5 - 2z}{2} \qquad (3)$$

$$F_\lambda(x, y, z, \lambda) = x + y + 2z - 20 = 0 \qquad (4)$$

STEP 4 Solve the system of equations for x, y, and z.
From (1) and (2) we find $1 - 2x = 3 - 2y$
$$-2x = 2 - 2y \quad \text{or} \quad x = y - 1$$

From (2) and (3) we find $3 - 2y = \frac{5 - 2z}{2}$
$$6 - 4y = 5 - 2z$$
$$1 - 4y = -2z$$
$$-\frac{1}{2} + 2y = z$$

Substituting for x and z in (4) gives
$$y - 1 + y + 2\left(-\frac{1}{2} + 2y\right) = 20$$

$$y - 1 + y - 1 + 4y = 20$$

$$6y = 22 \quad \text{or} \quad y = \frac{22}{6} = \frac{11}{3}$$

Back substituting we get $x = y - 1 = \dfrac{8}{3}$ and $z = -\dfrac{1}{2} + 2y = -\dfrac{1}{2} + \dfrac{22}{3} = \dfrac{-3 + 44}{6} = \dfrac{41}{6}.$

STEP 5 Evaluate $w = f(x, y, z)$ at $\left(\dfrac{8}{3}, \dfrac{11}{3}, \dfrac{41}{6} \right).$

$$w = \left(\frac{8}{3} \right)^2 + \left(\frac{11}{3} \right)^2 + \left(\frac{41}{6} \right)^2 - \frac{8}{3} - 3 \left(\frac{11}{3} \right) - 5 \left(\frac{41}{6} \right) = \frac{233}{12} \approx 19.417$$

We conclude that w attains its minimum at $\left(\dfrac{8}{3}, \dfrac{11}{3}, \dfrac{41}{6} \right).$ The minimum value is $\dfrac{233}{12}.$

13. **STEP 1**   Maximize $z = f(x, y) = xy$
Subject to the constraint $g(x, y) = x + y - 100 = 0$
STEP 2 Construct the function $F(x, y, \lambda) = f(x, y) + \lambda g(x, y)$
$$F(x, y, \lambda) = xy + \lambda(x + y - 100)$$
STEP 3 Set up the system of equations
$$
\begin{aligned}
F_x(x, y, \lambda) &= y + \lambda = 0 &&\Rightarrow &&\lambda = -y &&(1) \\
F_y(x, y, \lambda) &= x + \lambda = 0 &&\Rightarrow &&\lambda = -x &&(2) \\
F_\lambda(x, y, \lambda) &= x + y - 100 = 0 &&&&&&(3)
\end{aligned}
$$

STEP 4 Solve the system of equations for x and y.
From (1) and (2) we find $-y = -x$ or $y = x$.
Substituting for y in (3) gives $x + x = 100$
$$2x = 100 \quad \text{or} \quad x = 50$$
Since $x = 50$, $y = 50$.

STEP 5 Evaluate $z = f(x, y)$ at $(50, 50)$.
$$z = (50)(50) = 2500$$
The two numbers whose sum is 100 and whose product is a maximum are 50 and 50.

15. **STEP 1** Find the maximum of $w = f(x, y, z) = x + y + z$
Subject to the constraint $g(x, y) = x^2 + y^2 + z^2 - 25 = 0$
STEP 2 Construct the function $F(x, y, z, \lambda) = f(x, y, z) + \lambda g(x, y, z)$
$$F(x, y, z, \lambda) = x + y + z + \lambda(x^2 + y^2 + z^2 - 25)$$
STEP 3 Set up the system of equations

$$F_x(x, y, z, \lambda) = 1 + 2x\lambda = 0 \qquad \Rightarrow \qquad \lambda = -\frac{1}{2x} \qquad (1)$$

$$F_y(x, y, z, \lambda) = 1 + 2y\lambda = 0 \qquad \Rightarrow \qquad \lambda = -\frac{1}{2y} \qquad (2)$$

$$F_z(x, y, z, \lambda) = 1 + 2z\lambda = 0 \qquad \Rightarrow \qquad \lambda = -\frac{1}{2z} \qquad (3)$$

$$F_\lambda(x, y, z, \lambda) = x^2 + y^2 + z^2 - 25 = 0 \qquad (4)$$

STEP 4 Solve the system of equations for x, y, and z.

From (1) and (2) we find $-\dfrac{1}{2x} = -\dfrac{1}{2y}$

$$2y = 2x \quad \text{or} \quad y = x$$

From (1) and (3) we find $-\dfrac{1}{2x} = -\dfrac{1}{2z}$

$$2z = 2x \quad \text{or} \quad z = x$$

Substituting for y and z in (4) gives

$$x^2 + x^2 + x^2 = 25$$
$$3x^2 = 25$$
$$x^2 = \frac{25}{3} \quad \text{or} \quad x = \pm\frac{5}{\sqrt{3}} = \pm\frac{5\sqrt{3}}{3}$$

When $x = \dfrac{5\sqrt{3}}{3}$, $y = x = \dfrac{5\sqrt{3}}{3}$ and $z = x = \dfrac{5\sqrt{3}}{3}$.

When $x = -\dfrac{5\sqrt{3}}{3}$, $y = x = -\dfrac{5\sqrt{3}}{3}$ and $z = x = -\dfrac{5\sqrt{3}}{3}$.

STEP 5 Evaluate $w = f(x, y, z)$ at each solution (x_0, y_0, z_0).

At $\left(\dfrac{5\sqrt{3}}{3}, \dfrac{5\sqrt{3}}{3}, \dfrac{5\sqrt{3}}{3} \right)$, $w = \dfrac{5\sqrt{3}}{3} + \dfrac{5\sqrt{3}}{3} + \dfrac{5\sqrt{3}}{3} = 5\sqrt{3}$.

At $\left(-\dfrac{5\sqrt{3}}{3}, -\dfrac{5\sqrt{3}}{3}, -\dfrac{5\sqrt{3}}{3} \right)$, $w = \left(-\dfrac{5\sqrt{3}}{3} \right) + \left(-\dfrac{5\sqrt{3}}{3} \right) + \left(-\dfrac{5\sqrt{3}}{3} \right) = -5\sqrt{3}$

We conclude the sum of the three numbers that satisfy the constraint is maximum when each of the numbers equals $\dfrac{5\sqrt{3}}{3}$.

17. We solve the problem using the method of Lagrange multipliers.

STEP 1 Find the minimum cost $C = C(x, y) = 18x^2 + 9y^2$

Subject to the constraint $g(x, y) = x + y - 54 = 0$

STEP 2 Construct the function $F(x, y, \lambda) = C(x, y) + \lambda\, g(x, y)$

$$F(x, y, \lambda) = 18x^2 + 9y^2 + \lambda(x + y - 54)$$

STEP 3 Set up the system of equations

$$F_x(x,y,\lambda) = 36x + \lambda = 0 \qquad \Rightarrow \qquad \lambda = -36x \qquad (1)$$
$$F_y(x,y,\lambda) = 18y + \lambda = 0 \qquad \Rightarrow \qquad \lambda = -18y \qquad (2)$$
$$F_\lambda(x,y,\lambda) = x + y - 54 = 0 \qquad\qquad\qquad\qquad\qquad (3)$$

STEP 4 Solve the system of equations for x and y.
From (1) and (2) we find $-36x = -18y$ or $y = 2x$.

Substituting for y in (3) gives $x + 2x = 54$
$$3x = 54 \quad \text{or} \quad x = 18 \quad \text{and} \quad y = 2x = 2(18) = 36$$

STEP 5 Evaluate $C = f(x, y)$ at (18, 36).
$$C = 18 \cdot 18^2 + 9 \cdot 36^2 = 17{,}496$$

Cost is minimized when 18 units of product x and 36 units of product y are produced. The minimum cost is \$17,496.

19.   We solve the problem using the method of Lagrange multipliers.
STEP 1 Find the maximum of $w = V(l, w, h) = lwh$
  Subject to the constraint $g(l, w, h) = l + w + h - 62 = 0$
STEP 2 Construct the function $F(l, w, h, \lambda) = f(l, w, h) + \lambda g(l, w, h)$
$$F(l, w, h, \lambda) = lwh + \lambda(l + w + h - 62)$$

STEP 3 Set up the system of equations

$$F_l(l,w,h,\lambda) = wh + \lambda = 0 \qquad \Rightarrow \qquad \lambda = -wh \qquad (1)$$
$$F_w(l,w,h,\lambda) = lh + \lambda = 0 \qquad \Rightarrow \qquad \lambda = -lh \qquad (2)$$
$$F_h(l,w,h,\lambda) = wl + \lambda = 0 \qquad \Rightarrow \qquad \lambda = -wl \qquad (3)$$
$$F_\lambda(l,w,h,\lambda) = l + w + h - 62 = 0 \qquad\qquad\qquad\qquad (4)$$

STEP 4 Solve the system of equations for x, y, and z.
From (1) and (2) we find $wh = lh$ or $w = l$.
From (1) and (3) we find $wh = wl$ or $h = l$.
Substituting for w and h in (4) gives
$$l + l + l - 62 = 0$$
$$3l = 62$$
$$l = \frac{62}{3} = 20.667$$
Then $w = l = 20.667$ and $h = l = 20.667$.

STEP 5 Evaluate $w = V(l, w, h) = \left(\dfrac{62}{3}\right)^3 = \dfrac{238328}{27} \approx 8826.96$ cubic inches. The box of

greatest volume has dimensions 20.667 inches by 20.667 inches by 20.667 inches.

21.   (a) Maximize production $P(K, L) = 1.01K^{0.25}L^{0.75}$

  Subject to the constraint $g(K, L) = 175K + 125L - 125{,}000 = 0$

  We use the Method of Lagrange Multipliers and construct the function

$$F(K, L, \lambda) = 1.01K^{0.25}L^{0.75} + \lambda(175K + 125L - 125{,}000)$$

The system of equations formed by the partial derivatives is

$$F_K(K,L,\lambda) = (1.01)(.25)K^{-.75}L^{.75} + 175\lambda = 0 \quad \Rightarrow \quad \lambda = -\frac{(1.01)(.25)K^{-.75}L^{.75}}{175} \quad (1)$$

$$F_L(K,L,\lambda) = (1.01)(.75)K^{.25}L^{-.25} + 125\lambda = 0 \quad \Rightarrow \quad \lambda = -\frac{(1.01)(.75)K^{.25}L^{-.25}}{125} \quad (2)$$

$$F_\lambda(K,L,\lambda) = 175K + 125L - 125{,}000 = 0 \quad (3)$$

From (1) and (2) we find

$$\frac{\cancel{(1.01)}(.25)K^{-.75}L^{.75}}{175} = \frac{\cancel{(1.01)}(.75)K^{.25}L^{-.25}}{125}$$

$$\frac{1}{\cancel{4}} \cdot \frac{L^{.75}}{K^{.75}} \cdot \frac{1}{175} = \frac{3}{\cancel{4}} \cdot \frac{K^{.25}}{L^{.25}} \cdot \frac{1}{125}$$

$$\frac{L}{175} = \frac{3K}{125}$$

$$L = 4.2K$$

Substitute for L in (3)

$$175K + 125(4.2K) = 125{,}000$$

$$175K + 525K = 125{,}000$$

$$700K = 125{,}000$$

$$K = \frac{1250}{7} = 178.57$$

Back-substituting for K, we get $L = 4.2K = 750$.
To maximize total production, the company should use 178.87 units of capital and 750 units of labor.

(b) The maximum production is $P\left(\dfrac{1250}{7}, 750\right) = (1.01)\left(\dfrac{1250}{7}\right)^{.25}(750)^{.75} = 529.140$.

23. Minimize the amount of material used $A(l, w, h) = 2lw + 2lh + 2wh$
     Subject to the constraint $g(l, w, h) = lwh - 175 = 0$
We use the Method of Lagrange Multipliers and construct the function
$$F(l, w, z, \lambda) = 2lw + 2lh + 2wh + \lambda(lwh - 175)$$
The system of equations formed by the partial derivatives is

$$F_l(l,w,h,\lambda)=2w+2h+wh\lambda=0 \quad \Rightarrow \quad \lambda=-\frac{2w+2h}{wh} \qquad (1)$$

$$F_w(l,w,h,\lambda)=2l+2h+lh\lambda=0 \quad \Rightarrow \quad \lambda=-\frac{2l+2h}{lh} \qquad (2)$$

$$F_h(l,w,h,\lambda)=2l+2w+lw\lambda=0 \quad \Rightarrow \quad \lambda=-\frac{2l+2w}{lw} \qquad (3)$$

$$F_\lambda(l,w,h,\lambda)=lwh-175=0 \qquad (4)$$

To solve the system of equations we first use (1) and (2)

$$\frac{2w+2h}{wh}=\frac{2l+2h}{lh}$$

$$\frac{w+h}{w}=\frac{l+h}{l}$$

$$lw+lh=lw+wh$$

$$lh=wh$$
$$l=w$$

We use (2) and (3) and find

$$\frac{2l+2h}{lh}=\frac{2l+2w}{lw}$$

$$\frac{l+h}{h}=\frac{l+w}{w}$$

$$lw+wh=lh+wh$$

$$lw=lh$$

$$w=h$$

We substitute for h and l in (4), and get $w^3=175$ or $w=\sqrt[3]{175}\approx 5.593$. Since $w=l=h$, we have the dimensions of the container that uses the least material and holds 175 cubic feet are $\sqrt[3]{175}\approx 5.593$ feet by $\sqrt[3]{175}\approx 5.593$ feet by $\sqrt[3]{175}\approx 5.593$ feet.

25. Minimize the cost C of making a box
$$C(l, \text{w}, h)=2lw+lw+2lh+2wh=3lw+2lh+2wh$$
Subject to the constraint $V(l, \text{w}, h)=lwh-18=0$
We use the Method of Lagrange Multipliers and construct the function
$$F(l, w, z, \lambda)=3lw+2lh+2wh+\lambda(lwh-18)$$

The system of equations formed by the partial derivatives is

$$F_l(l,w,h,\lambda)=3w+2h+wh\lambda=0 \quad \Rightarrow \quad \lambda=-\frac{3w+2h}{wh} \qquad (1)$$

$$F_w(l,w,h,\lambda)=3l+2h+lh\lambda=0 \quad \Rightarrow \quad \lambda=-\frac{3l+2h}{lh} \qquad (2)$$

$$F_h(l,w,h,\lambda)=2l+2w+lw\lambda=0 \quad \Rightarrow \quad \lambda=-\frac{2l+2w}{lw} \qquad (3)$$

$$F_\lambda(l,w,h,\lambda)=lwh-18=0 \qquad (4)$$

We solve the system of equations.

From (1) and (2) we find $\dfrac{3w+2h}{wh} = \dfrac{3l+2h}{lh}$

$$\dfrac{3w+2h}{w} = \dfrac{3l+2h}{l}$$

$$3lw + 2lh = 3lw + 2wh$$

$$2lh = 2wh$$

$$l = w$$

From (2) and (3) we find $\dfrac{3l+2h}{lh} = \dfrac{2l+2w}{lw}$

$$\dfrac{3l+2h}{h} = \dfrac{2l+2w}{w}$$

$$3lw + 2wh = 2lh + 2wh$$

$$3w = 2h \quad \text{or} \quad h = \dfrac{3}{2}w$$

We substitute for l and h in (4), and get $w \cdot w \cdot \left(\dfrac{3w}{2}\right) = 18$

$$w^3 = 12 \quad \text{or} \quad w = \sqrt[3]{12} \approx 2.289$$

So we find $l = w = 2.289$ and $h = \dfrac{3}{2}w = \dfrac{3}{2}\sqrt[3]{12} \approx 3.434$.

The cost is minimized if the box has a bottom measuring 2.289 feet by 2.289 feet and a height measuring 3.343 feet.

17.6 The Double Integral

1.
$$\int_0^2 (xy^3 + x^2)\,dx = \left(\dfrac{x^2 y^3}{2} + \dfrac{x^3}{3}\right)\Bigg|_0^2 = \left[\dfrac{4y^3}{2} + \dfrac{8}{3}\right] - [0] = 2y^3 + \dfrac{8}{3}$$

3.
$$\int_2^4 (3x^2 y + 2x)\,dy = \left(3x^2 \cdot \dfrac{y^2}{2} + 2xy\right)\Bigg|_2^4 = \left[3x^2 \cdot \dfrac{16}{2} + 2x \cdot 4\right] - \left[3x^2 \cdot \dfrac{4}{2} + 2x \cdot 2\right]$$
$$= 24x^2 + 8x - 6x^2 - 4x = 18x^2 + 4x$$

5.
$$\int_2^3 (x + 3y)\,dx = \left(\dfrac{x^2}{2} + 3xy\right)\Bigg|_2^3 = \left[\dfrac{9}{2} + 9y\right] - \left[\dfrac{4}{2} + 6y\right] = \dfrac{5}{2} + 3y$$

7.
$$\int_2^4 (4x - 6y + 7)\,dy = \left(4xy - \dfrac{6y^2}{2} + 7y\right)\Bigg|_2^4 = [16x - 3 \cdot 16 + 28] - [8x - 3 \cdot 4 + 14] = 8x - 22$$

9.
$$\int_0^1 \frac{x^2}{\sqrt{1+y^2}}\, dx = \frac{1}{\sqrt{1+y^2}}\int_0^1 x^2\, dx = \frac{1}{\sqrt{1+y^2}}\cdot\frac{x^3}{3}\bigg|_0^1 = \frac{1}{\sqrt{1+y^2}}\left[\frac{1}{3}-0\right] = \frac{1}{3\sqrt{1+y^2}}$$

11.
$$\int_0^2 e^{x+y}\, dx = e^y \int_0^2 e^x\, dx = e^y\left[e^x\big|_0^2\right] = e^y\left[e^2-e^0\right] = e^y\left(e^2-1\right)$$

13.
$$\int_0^4 e^{x-4y}\, dx = e^{-4y}\int_0^4 e^x\, dx = e^{-4y}\left[e^x\big|_0^4\right] = e^{-4y}\left(e^4-1\right)$$

15.
$$\int_0^2 \frac{x}{\sqrt{y+6}}\, dx = \frac{1}{\sqrt{y+6}}\int_0^2 x\, dx = \frac{1}{\sqrt{y+6}}\left[\frac{x^2}{2}\bigg|_0^2\right] = \frac{1}{\sqrt{y+6}}(2-0) = \frac{2}{\sqrt{y+6}}$$

17.
$$\int_0^2\left[\int_0^4 y\, dx\right] dy$$

Evaluating the inner integral first.

$$\int_0^4 y\, dx = y\int_0^4 dx = y\cdot\left(x\big|_0^4\right) = y[4-0] = 4y$$

Then

$$\int_0^2\left[\int_0^4 y\, dx\right] dy = \int_0^2 4y\, dy = \frac{4y^2}{2}\bigg|_0^2 = 8-0 = 8$$

19.
$$\int_1^2\left[\int_1^3 \left(x^2+y\right) dx\right] dy$$

Evaluating the inner integral first.

$$\int_1^3 \left(x^2+y\right) dx = \left(\frac{x^3}{3}+yx\right)\bigg|_1^3 = \left[(9+3y)-\left(\frac{1}{3}+y\right)\right] = \frac{26}{3}+2y$$

Then

$$\int_1^2\left[\int_1^3 \left(x^2+y\right) dx\right] dy = \int_1^2\left(\frac{26}{3}+2y\right) dy = \left(\frac{26}{3}y+\frac{2y^2}{2}\right)\bigg|_1^2$$

$$= \left(\frac{52}{3}+4\right)-\left(\frac{26}{3}+1\right) = \frac{26}{3}+3 = \frac{35}{3}$$

21.
$$\int_0^1\left[\int_1^2 \left(x^2+y\right) dx\right] dy$$

Evaluating the inner integral first.

$$\int_1^2 \left(x^2+y\right) dx = \left(\frac{x^3}{3}+yx\right)\bigg|_1^2 = \left(\frac{8}{3}+2y\right)-\left(\frac{1}{3}+y\right) = \frac{7}{3}+y$$

Then

$$\int_0^1 \left[\int_1^2 (x^2+y)dx \right] dy = \int_0^1 \left(\frac{7}{3}+y \right) dy = \left(\frac{7}{3}y+\frac{y^2}{2} \right) \Bigg|_0^1 = \left(\frac{7}{3}+\frac{1}{2} \right) - (0) = \frac{14+3}{6} = \frac{17}{6}$$

23.

$$\int_1^2 \left[\int_3^4 (4x+2y+5)dx \right] dy$$

Evaluating the inner integral first.

$$\int_3^4 (4x+2y+5)dx = \left(\frac{4x^2}{2}+2yx+5x \right) \Bigg|_3^4 = (32+8y+20)-(18+6y+15) = 2y+19$$

Then

$$\int_1^2 \left[\int_3^4 (4x+2y+5)dx \right] dy = \int_1^2 (2y+19)\, dy = \left(\frac{2y^2}{2}+19y \right) \Bigg|_1^2 = (4+38)-(1+19) = 22$$

25.

$$\int_2^4 \left[\int_0^1 (6xy^2-2xy+3)dy \right] dx$$

Evaluating the inner integral first.

$$\int_0^1 (6xy^2-2xy+3)dy = \left(\frac{6xy^3}{3}-\frac{2xy^2}{2}+3y \right) \Bigg|_0^1 = (2x-x+3)-(0) = x+3$$

Then

$$\int_2^4 \left[\int_0^1 (6xy^2-2xy+3)dy \right] dx = \int_2^4 (x+3)dx = \left(\frac{x^2}{2}+3x \right) \Bigg|_2^4 = (8+12)-(2+6) = 12$$

27.

$$\iint_R (y+3x^2)\, dx\, dy = \int_1^3 \left[\int_0^2 (y+3x^2)\, dx \right] dy = \int_1^3 \left[\left(yx+\frac{3x^3}{3} \right) \Bigg|_0^2 \right] dy$$

$$= \int_1^3 (2y+8)\, dy = \left(\frac{2y^2}{2}+8y \right) \Bigg|_1^3$$

$$= (9+24)-(1+8) = 24$$

29.

$$\iint_R (x+y)\, dy\, dx = \int_0^2 \left[\int_1^4 (x+y)\, dy \right] dx = \int_0^2 \left[\left(xy+\frac{y^2}{2} \right) \Bigg|_1^4 \right] dx$$

$$= \int_0^2 \left[(4x+8)-\left(x+\frac{1}{2} \right) \right] dx = \int_0^2 \left(3x+\frac{15}{2} \right) dx$$

$$= \left(\frac{3x^2}{2} + \frac{15}{2}x \right) \Bigg|_0^2 = (6+15)-(0) = 21$$

31.

$$V = \iint_R f(x, y)\, dy\, dx = \int_1^2 \left[\int_3^4 (2x+3y+4)\, dy \right] dx = \int_1^2 \left[\left(2xy + \frac{3y^2}{2} + 4y \right) \Bigg|_3^4 \right] dx$$

$$= \int_1^2 \left[(8x+24+16) - \left(6x + \frac{27}{2} + 12 \right) \right] dx$$

$$= \int_1^2 \left(2x + \frac{29}{2} \right) dx = \left(\frac{2x^2}{2} + \frac{29}{2}x \right) \Bigg|_1^2$$

$$= (4+29) - \left(1 + \frac{29}{2} \right)$$

$$= 32 - \frac{29}{2} = \frac{64-29}{2} = \frac{35}{2} \text{ cubic units}$$

Chapter 17 Review

TRUE-FALSE ITEMS

1. True   **3.** False

FILL-IN-THE-BLANKS

1. surface   **3.** $x = x_0$

REVIEW EXERCISES

1. We use the distance formula with $(x_1, y_1, z_1) = (1, 6, -2)$ and $(x_2, y_2, z_2) = (2, 4, 0)$.

$$d = \sqrt{(x_2 - x_1)^2 + (y_2 - y_1)^2 + (z_2 - z_1)^2}$$

$$d = \sqrt{(2-1)^2 + (4-6)^2 + (0-(-2))^2}$$

$$d = \sqrt{1+4+4} = \sqrt{9} = 3$$

3. We use the distance formula with $(x_1, y_1, z_1) = (4, 6, 8)$ and $(x_2, y_2, z_2) = (6, 2, 1)$.

$$d = \sqrt{(x_2 - x_1)^2 + (y_2 - y_1)^2 + (z_2 - z_1)^2}$$

$$d = \sqrt{(6-4)^2 + (2-6)^2 + (1-8)^2}$$
$$d = \sqrt{4+16+49} = \sqrt{69}$$

5. We use the distance formula with $(x_1, y_1, z_1) = (-3, 7, -1)$ and $(x_2, y_2, z_2) = (0, 3, -1)$.

$$d = \sqrt{(x_2 - x_1)^2 + (y_2 - y_1)^2 + (z_2 - z_1)^2}$$
$$d = \sqrt{(0-(-3))^2 + (3-7)^2 + (-1-(-1))^2}$$
$$d = \sqrt{9+16+0} = \sqrt{25} = 5$$

7. The radius is the distance between the center and a point on the edge of the sphere.

$$r = \sqrt{(x_1 - x_0)^2 + (y_1 - y_0)^2 + (z_1 - z_0)^2}$$
$$r = \sqrt{(3-2)^2 + (4-2)^2 + (0-2)^2}$$
$$r = \sqrt{1+4+4} = \sqrt{9} = 3$$

9. The standard equation of the sphere with radius $r = 2$ and center at $(x_0, y_0, z_0) = (-6, 3, 1)$ is

$$(x-x_0)^2 + (y-y_0)^2 + (z-z_0)^2 = r^2$$
$$(x-(-6))^2 + (y-3)^2 + (z-1)^2 = 2^2$$
$$(x+6)^2 + (y-3)^2 + (z-1)^2 = 4$$

11.
$$(x-1)^2 + (y+3)^2 + (z+8)^2 = 25$$

Compare the equation to the standard equation of the sphere

$$(x-x_0)^2 + (y-y_0)^2 + (z-z_0)^2 = r^2$$
$$(x-1)^2 + (y-(-3))^2 + (z-(-8))^2 = 5^2$$

The center of the sphere is $(1, -3, -8)$ and its radius is 5.

13. (a) Complete the squares and put the equation into standard form.
$$x^2 + y^2 + z^2 - 2x + 8y - 6z = 10$$
$$(x^2 - 2x) + (y^2 + 8y) + (z^2 - 6z) = 10$$
$$(x^2 - 2x + 1) + (y^2 + 8y + 16) + (z^2 - 6z + 9) = 10 + 1 + 16 + 9$$
$$(x-1)^2 + (y+4)^2 + (z-3)^2 = 36$$

(b) The center of the sphere is $(1, -4, 3)$ and its radius is 6.

15. $f(x, y) = 2x^2 + 6xy - y^3$

(a) $f(1, -3) = 2(1^2) + 6(1)(-3) - (-3)^3 = 2 - 18 + 27 = 11$

(b) $f(4, -2) = 2(4^2) + 6(4)(-2) - (-2)^3 = 32 - 48 + 8 = -8$

17. $f(x, y) = \dfrac{x + 2y}{x - 3y}$

(a) $f(1, -3) = \dfrac{1 + 2(-3)}{1 - 3(-3)} = \dfrac{-5}{10} = -\dfrac{1}{2}$ (b) $f(4, -2) = \dfrac{4 + 2(-2)}{4 - 3(-2)} = 0$

19. $x^2 + 3y + 5$ is a polynomial, and polynomials are defined for all real numbers. So the domain of $z = f(x, y)$ is the entire xy-plane.

21. Since only logarithms of positive numbers are allowed, $y - x^2 - 4 > 0$ or $y > x^2 + 4$. The domain of $z = f(x, y)$ is the set or ordered pairs $\{(x, y) \mid y > x^2 + 4\}$. This describes the set of points (x, y) inside the parabola $y = x^2 + 4$.

23. Since only square roots of nonnegative numbers are allowed in the real number system,
$$x^2 + y^2 + 4x - 5 \geq 0$$
$$(x^2 + 4x) + y^2 \geq 5$$
$$(x^2 + 4x + 4) + y^2 \geq 5 + 4$$
$$(x + 2)^2 + y^2 \geq 9$$

The domain of $z = f(x, y)$ is the set or ordered pairs $\{(x, y) \mid (x + 2)^2 + y^2 \geq 9\}$. This describes the set of points either on the circle centered at $(-2, 0)$ and having a radius of 3 or outside the circle.

25. $z = f(x, y) = x^2 y + 4x$

$\qquad f_x(x, y) = 2xy + 4 \qquad\qquad f_y(x, y) = x^2$

$\qquad f_{xx}(x, y) = 2y \qquad\qquad\quad f_{yy}(x, y) = 0$

$\qquad f_{xy}(x, y) = 2x \qquad\qquad\quad f_{yx}(x, y) = 2x$

27. $z = f(x, y) = y^2 e^x + x \ln y$

$\qquad f_x(x, y) = y^2 e^x + \ln y \qquad\qquad f_y(x, y) = 2y e^x + \dfrac{x}{y}$

$\qquad f_{xx}(x, y) = y^2 e^x \qquad\qquad\qquad f_{yy}(x, y) = 2 e^x - \dfrac{x}{y^2}$

$$f_{xy}(x,y) = 2y\, e^x + \frac{1}{y} \qquad\qquad f_{yx}(x,y) = 2y\, e^x + \frac{1}{y}$$

29. $z = f(x,y) = \sqrt{x^2+y^2} = \left(x^2+y^2\right)^{1/2}$

$$f_x(x,y) = \frac{1}{2}\left(x^2+y^2\right)^{-1/2} \cdot 2x = \frac{x}{\left(x^2+y^2\right)^{1/2}} = \frac{x}{\sqrt{x^2+y^2}}$$

$$f_y(x,y) = \frac{1}{2}\left(x^2+y^2\right)^{-1/2} \cdot 2y = \frac{y}{\left(x^2+y^2\right)^{1/2}} = \frac{y}{\sqrt{x^2+y^2}}$$

$$f_{xx}(x,y) = \frac{1\cdot\left(x^2+y^2\right)^{1/2} - x\cdot\dfrac{1}{2}\left(x^2+y^2\right)^{-1/2}\cdot 2x}{x^2+y^2} = \frac{\left(x^2+y^2\right)-x^2}{\left(x^2+y^2\right)^{3/2}} = \frac{y^2}{\left(x^2+y^2\right)^{3/2}}$$

$$f_{yy}(x,y) = \frac{\left[1\cdot\left(x^2+y^2\right)^{1/2}\right] - \left[y\cdot\dfrac{1}{2}\left(x^2+y^2\right)^{-1/2}\cdot 2y\right]}{x^2+y^2}$$

$$= \frac{\left[x^2+y^2\right] - \left[y^2\right]}{\left(x^2+y^2\right)^{3/2}} = \frac{x^2}{\left(x^2+y^2\right)^{3/2}}$$

$$f_{xy}(x,y) = -\frac{1}{2}x\left(x^2+y^3\right)^{-3/2} \cdot 2y = -\frac{xy}{\left(x^2+y^3\right)^{3/2}}$$

$$f_{yx}(x,y) = -\frac{1}{2}y\left(x^2+y^3\right)^{-3/2} \cdot 2x = -\frac{xy}{\left(x^2+y^3\right)^{3/2}}$$

31. $z = f(x,y) = e^x \ln(5x+2y)$

$$f_x(x,y) = e^x \cdot \frac{5}{5x+2y} + e^x \ln(5x+2y) = \frac{5\,e^x}{5x+2y} + e^x \ln(5x+2y)$$

$$f_y(x,y) = e^x \cdot \frac{2}{5x+2y} = \frac{2e^x}{5x+2y}$$

$$f_{xx}(x,y) = \frac{5e^x(5x+2y)-25e^x}{(5x+2y)^2} + \frac{5e^x}{5x+2y} + e^x \ln(5x+2y)$$

$$= \frac{10e^x(5x+2y)-25e^x}{(5x+2y)^2} + e^x \ln(5x+2y)$$

$$= \frac{5e^x(10x+4y-5)}{(5x+2y)^2} + e^x \ln(5x+2y)$$

$$f_{yy}(x, y) = \frac{-2e^x \cdot (2)}{(5x+2y)^2} = -\frac{4e^x}{(5x+2y)^2}$$

$$f_{xy}(x,y) = \frac{-5e^x(2)}{(5x+2y)^2} + e^x \cdot \frac{2}{5x+2y} = \frac{-10e^x + 2e^x(5x+2y)}{(5x+2y)^2}$$

$$= \frac{2e^x(10x+2y-5)}{(5x+2y)^2}$$

$$f_{yx}(x, y) = \frac{2e^x(5x+2y) - 2e^x(5)}{(5x+2y)^2} = \frac{2e^x(5x+2y-5)}{(5x+2y)^2}$$

33. $f(x, y, z) = 3xe^y + xye^z - 12x^2y$

$$f_x(x, y, z) = 3e^y + ye^z - 24xy$$

$$f_y(x, y, z) = 3xe^y + xe^z - 12x^2$$

$$f_z(x, y, z) = xye^z$$

35. The slope of the line tangent to the intersection of z and the plane $y = 2$ is

$$f_x(x, y) = 3y^2$$

At the point (1, 2, 12) the slope is

$$f_x(1, 2) = 3 \cdot 2^2 = 12$$

37. The slope of the line tangent to the intersection of z and the plane $x = 1$ is

$$f_y(x, y) = x^2 e^{xy}$$

At the point (1, 0, 1) the slope is

$$f_y(1, 0) = 1^2 \cdot e^{1 \cdot 0} = 1$$

39. $z = f(x, y) = xy - 6x - x^2 - y^2$

(a) Find the partial derivatives of z, set each equal to zero, and solve the system of equations.

$$f_x(x, y) = y - 6 - 2x = 0 \qquad f_y(x, y) = x - 2y = 0$$

$$\begin{cases} y - 6 - 2x = 0 & (1) \\ x - 2y = 0 \quad \Rightarrow \quad x = 2y & (2) \end{cases}$$

Substituting (2) into (1) gives $y - 6 - 2(2y) = 0$

$$y - 4y = 6$$

$$-3y = 6$$

$$y = -2$$

Solving for x in (2) we get $x = 2(-2) = -4$.

The critical point is $(-4, -2)$.

(b) Find the second-order partial derivatives, and find the value of D.

$$f_{xx}(x, y) = -2 \qquad f_{yy}(x, y) = -2 \qquad f_{xy}(x, y) = 1$$

$$D = f_{xx}(-4, -2) \cdot f_{yy}(-4, -2) - \left[f_{xy}(-4, -2)\right]^2 = (-2)(-2) - 1^2 = 3 > 0$$

Since $f_{xx}(-4, -2) < 0$ and $D > 0$, the function z has a local maximum at $(-4, -2)$. The value of the local maximum is $z = 12$.

41. $z = f(x, y) = 2x - x^2 + 4y - y^2 + 10$

(a) Find the partial derivatives of z, set each equal to zero, and solve the system of equations.

$$f_x(x, y) = 2 - 2x = 0 \qquad\qquad f_y(x, y) = 4 - 2y = 0$$
$$x = 1 \qquad\qquad\qquad\qquad y = 2$$

The critical point is $(1, 2)$.

(b) Find the second-order partial derivatives, and find the value of D.

$$f_{xx}(x, y) = -2 \qquad\qquad f_{yy}(x, y) = -2 \qquad\qquad f_{xy}(x, y) = 0$$
$$D = f_{xx}(1, 2) \cdot f_{yy}(1, 2) - \left[f_{xy}(1, 2)\right]^2 = (-2)(-2) - 0^2 = 4 > 0$$

Since $f_{xx}(1, 2) < 0$ and $D > 0$, the function z has a local maximum at $(1, 2)$. The value of the local maximum is $z = 15$.

43. $z = f(x, y) = x^2 - 9y + y^2$

(a) Find the partial derivatives of z, set each equal to zero, and solve the system of equations.

$$f_x(x, y) = 2x = 0 \qquad\qquad f_y(x, y) = -9 + 2y = 0$$
$$x = 0 \qquad\qquad\qquad\qquad y = \frac{9}{2}$$

The critical point is $\left(0, \dfrac{9}{2}\right)$.

(b) Find the second-order partial derivatives, and find the value of D.

$$f_{xx}(x, y) = 2 \qquad\qquad f_{yy}(x, y) = 2 \qquad\qquad f_{xy}(x, y) = 0$$
$$D = f_{xx}\left(0, \frac{9}{2}\right) \cdot f_{yy}\left(0, \frac{9}{2}\right) - \left[f_{xy}\left(0, \frac{9}{2}\right)\right]^2 = 2 \cdot 2 - 0^2 = 4 > 0$$

Since $f_{xx}\left(0, \dfrac{9}{2}\right) > 0$ and $D > 0$, the function z has a local minimum at $\left(0, \dfrac{9}{2}\right)$. The value of the local minimum is $z = -\dfrac{81}{4}$.

45. **STEP 1** Find the maximum of $z = f(x, y) = 5x^2 + 3y^2 + xy$
Subject to the constraint $g(x, y) = 2x - y - 20 = 0$

STEP 2 Construct the function $F(x, y, \lambda) = f(x, y) + \lambda\, g(x, y)$
$$F(x, y, \lambda) = 5x^2 + 3y^2 + xy + \lambda(2x - y - 20)$$

STEP 3 Set up the system of equations
$$F_x(x, y, \lambda) = 10x + y + 2\lambda = 0 \qquad (1)$$
$$F_y(x, y, \lambda) = 6y + x - \lambda = 0 \qquad (2)$$
$$F_\lambda(x, y, \lambda) = 2x - y - 20 = 0 \qquad (3)$$

STEP 4 Solve the system of equations for x and y.
Add twice (2) to (1) to eliminate λ.
$$(10x + y + 2\lambda) + 2(6y + x - \lambda) = 0$$
$$10x + y + 2\lambda + 12y + 2x - 2\lambda = 0$$
$$12x + 13y = 0 \qquad (1)$$
Subtract six times (3) from (1) to eliminate x.
$$(12x + 13y) - 6(2x - y - 20) = 0$$
$$12x + 13y - 12x + 6y + 120 = 0$$
$$13y + 6y = -120$$
$$19y = -120$$
$$y = \frac{-120}{19}$$

Substituting for y in (3) we get
$$2x - \left(\frac{-120}{19}\right) - 20 = 0 \qquad (3)$$

$$2x = \frac{260}{19} \quad \text{or} \quad x = \frac{130}{19}$$

STEP 5 Evaluate $z = f(x, y)$ at $\left(\dfrac{130}{19}, \dfrac{-120}{19}\right)$.

$$z = = 5\left(\frac{130}{19}\right)^2 + 3\left(\frac{-120}{19}\right)^2 + \left(\frac{130}{19}\right)\left(\frac{-120}{19}\right) = \frac{112{,}100}{19^2} \approx 310.526$$

The maximum value of z subject to the condition is 310.526.

47. **STEP 1** Find the minimum of $z = f(x, y) = x^2 + y^2$
Subject to the constraint $g(x, y) = 2x + y - 4 = 0$

STEP 2 Construct the function $F(x, y, \lambda) = f(x, y) + \lambda\, g(x, y)$
$$F(x, y, \lambda) = x^2 + y^2 + \lambda(2x + y - 4)$$

STEP 3 Set up the system of equations
$$F_x(x, y, \lambda) = 2x + 2\lambda = 0 \quad \Rightarrow \quad \lambda = -x \qquad (1)$$
$$F_y(x, y, \lambda) = 2y + \lambda = 0 \quad \Rightarrow \quad \lambda = -2y \qquad (2)$$
$$F_\lambda(x, y, \lambda) = 2x + y - 4 = 0 \qquad (3)$$

STEP 4 Solve the system of equations for x and y.
From (1) and (2) we find $x = 2y$.

Substituting for x in (3) gives $2(2y) + y - 4 = 0$

$$5y = 4$$

$$y = \frac{4}{5} \quad \text{and} \quad x = 2y = \frac{8}{5}$$

STEP 5 Evaluate $z = f(x, y)$ at $\left(\frac{8}{5}, \frac{4}{5}\right)$.

$$z = \left(\frac{8}{5}\right)^2 + \left(\frac{4}{5}\right)^2 = \frac{64 + 16}{25} = \frac{80}{25} = \frac{16}{5} = 3.2$$

The minimum value of z subject to the constraint is 3.2.

49.

$$\int_0^2 \left(4x^2 y - 12y\right) dx = \left(\frac{4x^3 y}{3} - 12xy\right)\Bigg|_0^2 = \frac{4 \cdot 2^3 y}{3} - 12 \cdot 2y = \frac{32}{3} y - 24y = -\frac{40}{3} y$$

51.

$$\int_{-1}^3 \left(6x^2 y + 2y\right) dy = \left(\frac{6x^2 y^2}{2} + \frac{2y^2}{2}\right)\Bigg|_{-1}^3 = \left(3x^2 \cdot 3^2 + 3^2\right) - \left(3x^2 (-1)^2 + (-1)^2\right)$$

$$= 27x^2 + 9 - 3x^2 - 1 = 24x^2 + 8$$

53.

$$\int_1^2 \left[\int_0^3 \left(6x^2 + 2x\right) dy\right] dx$$

Evaluating the inner integral first,

$$\int_0^3 \left(6x^2 + 2x\right) dy = \left(6x^2 y + 2xy\right)\Bigg|_0^3 = 18x^2 + 6x$$

Then

$$\int_1^2 \left[\int_0^3 \left(6x^2 + 2x\right) dy\right] dx = \int_1^2 \left(18x^2 + 6x\right) dx = \left(\frac{18x^3}{3} + \frac{6x^2}{2}\right)\Bigg|_1^2$$

$$= \left(6 \cdot 2^3 + 3 \cdot 2^2\right) - \left(6 + 3\right) = 48 + 12 - 9 = 51$$

55.

$$\int_0^2 \left[\int_1^8 \left(x^2 + 2xy - y^2\right) dx\right] dy$$

Evaluating the inner integral first,

$$\int_1^8 \left(x^2 + 2xy - y^2\right) dx = \left(\frac{x^3}{3} + \frac{2x^2 y}{2} - xy^2\right)\Bigg|_1^8$$

$$= \left(\frac{8^3}{3} + 64y - 8y^2\right) - \left(\frac{1}{3} + y - y^2\right) = \frac{511}{3} + 63y - 7y^2$$

Then

$$\int_0^2 \left[\int_1^8 \left(x^2 + 2xy - y^2 \right) dx \right] dy = \int_0^2 \left(\frac{511}{3} + 63y - 7y^2 \right) dy = \left(\frac{511}{3} y + \frac{63y^2}{2} - \frac{7y^3}{3} \right) \Bigg|_0^2$$

$$= \frac{1022}{3} + 126 - \frac{56}{3} = \frac{966}{3} + 126 = 448$$

57.

$$\iint\limits_R f(x, y)\, dy\, dx = \int_{-1}^1 \int_1^3 (2x + 4y)\, dy\, dx = \int_{-1}^1 \left[\left(2xy + \frac{4y^2}{2} \right) \Bigg|_1^3 \right] dx$$

$$= \int_{-1}^1 \left[(6x + 18) - (2x + 2) \right] dx$$

$$= \int_{-1}^1 (4x + 16)\, dx = \left(\frac{4x^2}{2} + 16x \right) \Bigg|_{-1}^1$$

$$= (2 + 16) - (2 - 16) = 32$$

59.

$$\iint\limits_R f(x, y)\, dy\, dx = \int_0^3 \int_1^2 (2xy)\, dy\, dx = \int_0^3 \left[\left(\frac{2xy^2}{2} \right) \Bigg|_1^2 \right] dx = \int_0^3 (4x - x)\, dx$$

$$= \int_0^3 3x\, dx = \frac{3x^2}{2} \Bigg|_0^3 = \frac{27}{2}$$

61.

$$V = \iint\limits_R f(x, y)\, dy\, dx = \int_1^8 \int_0^6 (2x + 2y + 1)\, dy\, dx = \int_1^8 \left[\left(2xy + \frac{2y^2}{2} + y \right) \Bigg|_0^6 \right] dx$$

$$= \int_1^8 (12x + 36 + 6)\, dx = \int_1^8 (12x + 42)\, dx$$

$$= \left(\frac{12x^2}{2} + 42x \right) \Bigg|_1^8$$

$$= 6 \cdot 64 + 42 \cdot 8 - 6 - 42 = 672 \text{ cubic units}$$

63. (a) $\dfrac{\partial z}{\partial K} = 80 \cdot \dfrac{1}{4} K^{-3/4} L^{3/4} = 20 \left(\dfrac{L}{K} \right)^{3/4}$ $\qquad$ $\dfrac{\partial z}{\partial L} = 80 \cdot \dfrac{3}{4} K^{1/4} L^{-1/4} = 60 \left(\dfrac{K}{L} \right)^{1/4}$

(b) When $K = \$800{,}000$ and $L = 20{,}000$ labor hours

$$\frac{\partial z}{\partial K} = 20 \left(\frac{20{,}000}{800{,}000} \right)^{3/4} = 20 \left(\frac{1}{40} \right)^{3/4} \approx 1.257$$

$$\frac{\partial z}{\partial L} = 60\left(\frac{800,000}{20,000}\right)^{1/4} = 60(40)^{1/4} \approx 150.892$$

(c) The factory should increase the use of labor. Explanations will vary.

65. $C(x, y) = 1050 + 40x + 45y$

$C_x(x, y) = 40$ $\qquad\qquad\qquad\qquad$ $C_y(x, y) = 45$

Explanations will vary.

67. (a) $R(x, y) = px + qy$

$$= (350 - 6x + y)x + (400 + 2x - 8y)y$$

$$= 350x - 6x^2 + xy + 400y + 2xy - 8y^2$$

$$= 350x - 6x^2 + 3xy + 400y - 8y^2$$

(b) $R_x(x, y) = 350 - 12x + 3y$

$R_y(x, y) = 3x + 400 - 16y$

Explanations will vary.

69. (a) $P(x, y) = R(x, y) - C(x, y)$

$$= \left[350x - 6x^2 + 3xy + 400y - 8y^2\right] - \left[1050 + 40x + 45y\right]$$

$$= 310x - 6x^2 + 3xy + 355y - 8y^2 - 1050$$

(b) $P_x(x, y) = 310 - 12x + 3y$

$P_x(50, 30) = 310 - 12(50) + 3(30) = -200$

$P_y(x, y) = 3x + 355 - 16y$

$P_y(50, 30) = 3(50) + 355 - 16(30) = 25$

Explanations will vary.

71. (a) $P(x, y) = R(x, y) - C(x, y)$

First we find R. $R(x, y) = px + qy = (9 - x)x + (21 - 2y)y = 9x - x^2 + 21y - 2y^2$

$$P(x, y) = \left[9x - x^2 + 21y - 2y^2\right] - (x + y + 225)$$

$$= 8x - x^2 + 20y - 2y^2 - 225$$

We find the critical points of function P.

$P_x(x, y) = 8 - 2x = 0$ $\qquad\qquad\qquad$ $P_y(x, y) = 20 - 4y = 0$

$x = 4$ $\qquad\qquad\qquad\qquad\qquad\qquad$ $y = 5$

The critical point is (4, 5).

We find the second-order partial derivatives and D to determine the character of the critical point.

$$P_{xx}(x, y) = -2 \qquad P_{yy}(x, y) = -4 \qquad P_{xy}(x, y) = 0$$

$$D = P_{xx}(4, 5) \cdot P_{yy}(4, 5) - \left[P_{xy}(4, 5) \right]^2 = (-2)(-4) - 0 = 8$$

Since $P_{xx}(4, 5) < 0$ and $D > 0$, there is a local maximum at $(4, 5)$. The supermarket should sell 4000 units of juice x and 5000 units of juice y to maximize profit.

(b) The maximum profit attainable from the orange juice sales is given by $P(4, 5)$.

$$P(4, 5) = 8(4) - 4^2 + 20(5) - 2(5^2) - 225 = -159$$

Since x and y are in thousands, the maximum profit is a loss of $159,000.

73. We want to find the maximum production $P(K, L) = 10 K^{0.3} L^{0.7}$

$$\text{subject to the condition} \qquad g(K, L) = 50K + 100L - 51,000 = 0$$

We use the method of Lagrange multipliers, and construct the function

$$F(K, L, \lambda) = 10 K^{0.3} L^{0.7} + \lambda(50K + 100L - 51,000)$$

The system of equations formed by the partial derivatives is

$$F_K(K, L, \lambda) = 3K^{-0.7} L^{0.7} + 50\lambda = 0 \qquad \Rightarrow \qquad \lambda = -\frac{3K^{-0.7} L^{0.7}}{50} \quad (1)$$

$$F_L(K, L, \lambda) = 7K^{0.3} L^{-0.3} + 100\lambda = 0 \qquad \Rightarrow \qquad \lambda = -\frac{7K^{0.3} L^{-0.3}}{100} \quad (2)$$

$$F_\lambda(K, L, \lambda) = 50K + 100L - 51,000 = 0 \qquad (3)$$

From (1) and (2) we find

$$\frac{3K^{-0.7} L^{0.7}}{50} = \frac{7K^{0.3} L^{-0.3}}{100}$$

$$\frac{300 L^{0.7}}{K^{0.7}} = \frac{350 K^{0.3}}{L^{0.3}}$$

$$300L = 350K$$

$$L = \frac{350}{300} K = \frac{7}{6} K$$

Substitute for L in (3)

$$50K + 100 \left(\frac{7}{6} K \right) = 51,000$$

$$\frac{1000}{6} K = 51,000$$

$$K = 306$$

Substituting back and solving for L gives $L = \frac{7}{6} K = \frac{7}{6} \cdot 306 = 357$

To maximize productivity $50(306) = \$15,300$ should be allotted for capital and $100(357) = \$35,700$ should be allotted for labor.

(b) The maximum number of units that can be produced is 3409 units.
$$P(306, 357) = 10\left(306^{0.3}\right)\left(357^{0.7}\right) = 3408.66$$

CHAPTER 17 PROJECT

1. Total cost C is the sum of holding cost and reorder cost. If we let x denote the lot size of the vacuum cleaners and y denote the lot size of the microwave ovens, we find
$$C(x, y) = \left(\frac{30x}{2} + \frac{15y}{2}\right) + \left(40\left(\frac{500}{x}\right) + 60\left(\frac{800}{y}\right)\right)$$
$$= 15x + 7.5y + \frac{20,000}{x} + \frac{48,000}{y}$$

3. Assuming that there are $\dfrac{37}{2} = 19$ vacuum cleaners in the store and $\dfrac{80}{2} = 40$ microwave ovens in the store, you would need
$$19\,(20) + 40\,(10) = 780 \text{ cubic feet of storage.}$$

5. The problem is to now minimize cost $C(x, y) = 15x + 7.5y + \dfrac{20,000}{x} + \dfrac{48,000}{y}$

    subject to the constraint $\qquad g(x, y) = 20x + 10y - 1000 = 0$

    To use the method of Lagrange multipliers we need to construct function F.
$$F(x, y, \lambda) = 15x + 7.5y + \frac{20,000}{x} + \frac{48,000}{y} + \lambda(20x + 10y - 1000)$$
    The system of equations that are used to determine the minimum C is formed by the partial derivatives of F.
$$F_x(x, y, \lambda) = 15 - \frac{20,000}{x^2} + 20\lambda = 0 \qquad \Rightarrow \qquad \lambda = \frac{20,000 - 15x^2}{20x^2} \qquad (1)$$
$$F_y(x, y, \lambda) = 7.5 - \frac{48,000}{y^2} + 10\lambda = 0 \qquad \Rightarrow \qquad \lambda = \frac{48,000 - 7.5y^2}{10y^2} \qquad (2)$$
$$F_\lambda(x, y, \lambda) = 20x + 10y - 1000 = 0 \qquad (3)$$

7. If the demand for vacuum cleaners is 500 units, and they are ordered in lots of 24, then there will be 21 orders placed per year. $\left(\dfrac{500}{24} = 20.833\right)$

If the demand for microwave ovens is 800 units, and they are ordered in lots of 52, then there will be 15 orders placed per year (with the last order of size 20).

On the average there will be 12 vacuum cleaners and 26 microwave ovens in the store. The space occupied by these items is 20 (12) + 10(26) = 500 cubic feet.

Appendix A Review

A.1 Real Numbers

1. (a) Natural Numbers: 2, 5

   (b) Integers: $-6, 2, 5$

   (c) Rational numbers: $-6, \frac{1}{2}, -1.333\ldots, 2, 5$

   (d) Irrational numbers: π

   (e) Real numbers: $-6, \frac{1}{2}, -1.333\ldots, \pi, 2, 5$

3. (a) Natural Numbers: 1

   (b) Integers: 0, 1

   (c) Rational numbers: $0, 1, \frac{1}{2}, \frac{1}{3}, \frac{1}{4}$

   (d) Irrational numbers: none

   (e) Real numbers: $0, 1, \frac{1}{2}, \frac{1}{3}, \frac{1}{4}$

5. (a) Natural Numbers: none
   (b) Integers: none
   (c) Rational numbers: none

   (d) Irrational numbers: $\sqrt{2}, \pi, \sqrt{2}+1, \pi+\frac{1}{2}$

   (e) Real numbers: $\sqrt{2}, \pi, \sqrt{2}+1, \pi+\frac{1}{2}$

7. Number: 18.9526
   Rounded: 18.953
   Truncated: 18.952

9. Number: 28.65319
   Rounded: 28.653
   Truncated: 28.653

11. Number: 0.06291
    Rounded: 0.063
    Truncated: 0.062

13. Number: 9.9985
    Rounded: 9.999
    Truncated: 9.998

15. Number: $\frac{3}{7} = 0.428571\ldots$

    Rounded: 0.429
    Truncated: 0.428

17. Number: $\frac{521}{15} = 34.73333\ldots$

    Rounded: 34.733
    Truncated: 34.733

19. $3 + 2 = 5$

21. $x + 2 = 3 \cdot 4$

23. $3y = 1 + 2$

25. $x - 2 = 6$

27. $\dfrac{x}{2} = 6$

29. $9 - 4 + 2 = 5 + 2 = 7$

31. $-6 + 4 \cdot 3 = -6 + 12 = 6$

33. $4 + 5 - 8 = 9 - 8 = 1$

35. $4 + \dfrac{1}{3} = \dfrac{12}{3} + \dfrac{1}{3} = \dfrac{13}{3}$

37.
$$
\begin{aligned}
6 - [3 \cdot 5 + 2 \cdot (3 - 2)] &= 6 - [15 + 2 \cdot 1] \\
&= 6 - [15 + 2] \\
&= 6 - 17 \\
&= -11
\end{aligned}
$$

39.
$$
\begin{aligned}
2 \cdot (3 - 5) + 8 \cdot 2 - 1 &= 2 \cdot (-2) + 16 - 1 \\
&= -4 + 16 - 1 \\
&= 12 - 1 = 11
\end{aligned}
$$

41.
$$
\begin{aligned}
10 - [6 - 2 \cdot 2 + (8 - 3)] \cdot 2 &= 10 - [6 - 4 + 5] \cdot 2 \\
&= 10 - [7] \cdot 2 \\
&= 10 - 14 \\
&= -4
\end{aligned}
$$

43. $(5 - 3)\,\dfrac{1}{2} = (2)\,\dfrac{1}{2} = 1$

45. $\dfrac{4 + 8}{5 - 3} = \dfrac{12}{2} = 6$

47. $\dfrac{3}{5} \cdot \dfrac{10}{21} = \dfrac{30}{105} = \dfrac{2}{7}$

49. $\dfrac{6}{25} \cdot \dfrac{10}{27} = \dfrac{2 \cdot 3 \cdot 2 \cdot 5}{5 \cdot 5 \cdot 3 \cdot 9} = \dfrac{4}{45}$

51. $\dfrac{3}{4} + \dfrac{2}{5} = \dfrac{3 \cdot 5}{4 \cdot 5} + \dfrac{2 \cdot 4}{4 \cdot 5} = \dfrac{15 + 8}{20} = \dfrac{23}{20}$

53. $\dfrac{5}{6} + \dfrac{9}{5} = \dfrac{5 \cdot 5}{6 \cdot 5} + \dfrac{9 \cdot 6}{5 \cdot 6} = \dfrac{25 + 54}{30} = \dfrac{79}{30}$

55. $\dfrac{5}{18} + \dfrac{1}{12} = \dfrac{5 \cdot 2}{18 \cdot 2} + \dfrac{1 \cdot 3}{12 \cdot 3} = \dfrac{10 + 3}{36} = \dfrac{13}{36}$

57.
$$
\begin{aligned}
\frac{1}{30} - \frac{7}{18} &= \frac{1 \cdot 3}{30 \cdot 3} - \frac{7 \cdot 5}{18 \cdot 5} \\
&= \frac{3 - 35}{90} = \frac{-32}{90} = -\frac{16}{45}
\end{aligned}
$$

59. $\dfrac{3}{20} - \dfrac{2}{15} = \dfrac{3 \cdot 3}{20 \cdot 3} - \dfrac{2 \cdot 4}{15 \cdot 4} = \dfrac{9 - 8}{60} = \dfrac{1}{60}$

61. $\dfrac{\dfrac{5}{18}}{\dfrac{11}{27}} = \dfrac{5}{18} \cdot \dfrac{27}{11} = \dfrac{5 \cdot 9 \cdot 3}{9 \cdot 2 \cdot 11} = \dfrac{15}{22}$

63. $6(x + 4) = 6x + 24$

65. $x(x - 4) = x^2 - 4x$

67.
$$
\begin{aligned}
(x + 2)(x + 4) &= (x + 2)x + (x + 2)4 \\
&= x^2 + 2x + 4x + 8 \\
&= x^2 + 6x + 8
\end{aligned}
$$

69.
$$
\begin{aligned}
(x - 2)(x + 1) &= (x - 2)x + (x - 2)1 \\
&= x^2 - 2x + x - 2 \\
&= x^2 - x - 2
\end{aligned}
$$

71. $(x - 8)(x - 2) = (x - 8)x + (x - 8)(-2)$
$\qquad = x^2 - 8x - 2x + 16$
$\qquad = x^2 - 10x + 16$

73. $(x + 2)(x - 2) = (x + 2)x + (x + 2)(-2)$
$\qquad = x^2 + 2x - 2x - 4$
$\qquad = x^2 - 4$

75. Answers will vary.

77. Answers will vary.

79. Subtraction is not commutative. Examples will vary.

81. Division is not commutative. Examples will vary.

83. Explanations will vary.

85. There are no real numbers that are both rational and irrational.
There are no real numbers that are neither rational nor irrational.
Explanations will vary.

87. $0.9999 \ldots = 1$
To show that $0.9999 \ldots = 1$, we let $n = 0.9999 \ldots$, then $10n = 9.9999 \ldots$

$\qquad 10n = 9.9999 \ldots \qquad (1)$
$\qquad \underline{n = 0.9999 \ldots \qquad (2)}$
$\qquad 9n = 9.0000 \ldots \qquad$ Subtract (2) from (1).
$\qquad n = 1 \qquad$ Divide both sides by 9.

A.2 Algebra Review

1.

3. $\dfrac{1}{2} > 0$

5. $-1 > -2$

7. $\pi > 3.14$

9. $\dfrac{1}{2} = 0.5$

11. $\dfrac{2}{3} < 0.67$

13. $x > 0$

15. $x < 2$

17. $x \leq 1$

19. $x \geq -2$

21. $x > -1$

23. $d(C, D) = |C - D| = |0 - 1| = |-1| = 1$

25. $d(D, E) = |D - E| = |1 - 3| = |-2| = 2$

27. $d(A, E) = |A - E| = |-3 - 3| = |-6| = 6$

29. If $x = -2$ and $y = 3$, then $x + 2y = (-2) + 2(3) = -2 + 6 = 4$

31. If $x = -2$ and $y = 3$, then $5xy + 2 = 5(-2)(3) + 2 = -30 + 2 = -28$

33. If $x = -2$ and $y = 3$, then $\dfrac{2x}{x - y} = \dfrac{2(-2)}{(-2) - 3} = \dfrac{-4}{-5} = \dfrac{4}{5}$

35. If $x = -2$ and $y = 3$, then $\dfrac{3x + 2y}{2 + y} = \dfrac{3(-2) + 2(3)}{2 + 3} = \dfrac{-6 + 6}{5} = 0$

37. If $x = 3$ and $y = -2$, then $|x + y| = |3 + (-2)| = |1| = 1$

39. If $x = 3$ and $y = -2$, then $|x| + |y| = |3| + |-2| = 3 + 2 = 5$

41. If $x = 3$ and $y = -2$, then $\dfrac{|x|}{x} = \dfrac{|3|}{3} = \dfrac{3}{3} = 1$

43. If $x = 3$ and $y = -2$, then $|4x - 5y| = |4(3) - 5(-2)| = |12 - (-10)| = |22| = 22$

45. If $x = 3$ and $y = -2$,
then $\big||4x| - |5y|\big| = \big||4(3)| - |5(-2)|\big| = \big||12| - |-10|\big| = |12 - 10| = |2| = 2$

47. We must exclude values of x that would cause the denominator to equal zero.
$x \neq 0$ **(c)**

49. We must exclude values of x that would cause the denominator to equal zero.
$x^2 - 9 \neq 0$
$(x - 3)(x + 3) \neq 0$
$x \neq 3;\ x \neq -3$ **(a)**

51. We must exclude values of x that would cause the denominator to equal zero, but $x^2 + 1$ can never equal zero, so no values are excluded.

53. We must exclude values of x that would cause the denominator to equal zero.
$x^3 - x \neq 0$
$x(x^2 - 1) \neq 0$
$x(x - 1)(x + 1) \neq 0$
$x \neq 0;\ x \neq 1;\ x \neq -1$ **(b), (c),** and **(d)**

55. The domain of the variable x is $\{x \mid x \neq 5\}$.

57. The domain of the variable x is $\{x \mid x \neq -4\}$.

59.
$$C = \frac{5}{9}(F - 32) \quad \text{If } F = 32°, \text{ then } C = \frac{5}{9}(32 - 32) = 0°.$$

61.
$$C = \frac{5}{9}(F - 32) \quad \text{If } F = 77°, \text{ then } C = \frac{5}{9}(77 - 32) = \frac{5}{9}(45) = 25°.$$

63. $(-4)^2 = 16$

65.
$$4^{-2} = \frac{1}{4^2} = \frac{1}{16}$$

67. $3^{-6} \cdot 3^4 = 3^{-6+4} = 3^{-2} = \frac{1}{3^2} = \frac{1}{9}$

69. $(3^{-2})^{-1} = 3^{(-2) \cdot (-1)} = 3^2 = 9$

71. $\sqrt{25} = 5$

73. $\sqrt{(-4)^2} = \sqrt{16} = 4$

75. $\left(8x^3\right)^2 = 8^2 \cdot x^{3 \cdot 2} = 64x^6$

77. $\left(x^2 y^{-1}\right)^2 = x^{2 \cdot 2} \cdot y^{(-1) \cdot 2} = x^4 y^{-2} = \frac{x^4}{y^2}$

79.
$$\frac{x^2 y^3}{xy^4} = \frac{x^2}{x} \cdot \frac{y^3}{y^4} = x^{2-1} \cdot y^{3-4} = x \cdot y^{-1} = \frac{x}{y}$$

81.
$$\frac{(-2)^3 x^4 (yz)^2}{3^2 xy^3 z} = \frac{-8x^4 y^2 z^2}{9xy^3 z} = -\frac{8}{9} \cdot \frac{x^4}{x} \cdot \frac{y^2}{y^3} \cdot \frac{z^2}{z} = -\frac{8}{9} \cdot x^{4-1} \cdot y^{2-3} \cdot z^{2-1}$$
$$= -\frac{8}{9} \cdot x^3 \cdot y^{-1} \cdot z = -\frac{8}{9} \cdot \frac{x^3 z}{y} = -\frac{8x^3 z}{9y}$$

83.
$$\left(\frac{3x^{-1}}{4y^{-1}}\right)^{-2} = \left(\frac{3y}{4x}\right)^{-2} = \left(\frac{4x}{3y}\right)^2 = \frac{4^2 \cdot x^2}{3^2 \cdot y^2} = \frac{16x^2}{9y^2}$$

85. If $x = 2$ and $y = -1$, then $2xy^{-1} = 2 \cdot 2 \cdot (-1)^{-1} = -4$.

87. If $x = 2$ and $y = -1$, then $x^2 + y^2 = 2^2 + (-1)^2 = 4 + 1 = 5$.

89. If $x = 2$ and $y = -1$, then $(xy)^2 = (2 \cdot (-1))^2 = (-2)^2 = 4$.

91. If $x = 2$ and $y = -1$, then $\sqrt{x^2} = \sqrt{2^2} = \sqrt{4} = 2$.

93. If $x = 2$ and $y = -1$, then $\sqrt{x^2 + y^2} = \sqrt{2^2 + (-1)^2} = \sqrt{4 + 1} = \sqrt{5}$.

95. If $x = 2$ and $y = -1$, then $x^y = 2^{-1} = \frac{1}{2}$.

97. If $x = 2$, then $2x^3 - 3x^2 + 5x - 4 = 2(2)^3 - 3(2)^2 + 5(2) - 4 = 10$.

If $x = 1$, then $2x^3 - 3x^2 + 5x - 4 = 2(1)^3 - 3(1)^2 + 5(1) - 4 = 0$.

99. $\dfrac{(666)^4}{(222)^4} = \left(\dfrac{666}{222}\right)^4 = 3^4 = 81$

101. $(8.2)^6 = 304{,}006.6714 = 304{,}006.671$

103. $(6.1)^{-3} = 0.0044057 = 0.004$

105. $(-2.8)^6 = 481.890304 = 481.890$

107. $(-8.11)^{-4} = 0.00023116 = 0.000$

109. $454.2 = 4.542 \times 10^2$

111. $0.013 = 1.3 \times 10^{-2}$

113. $32{,}155 = 3.2155 \times 10^4$

115. $0.000423 = 4.23 \times 10^{-4}$

117. $6.15 \times 10^4 = 61{,}500$

119. $1.214 \times 10^{-3} = 0.001214$

121. $1.1 \times 10^8 = 110{,}000{,}000$

123. $8.1 \times 10^{-2} = 0.081$

125. $A = lw$;   domain: $A > 0, l > 0, w > 0$

127. $C = \pi d$;   domain: $C > 0, d > 0$

129. $A = \dfrac{\sqrt{3}}{4} \cdot x^2$;   domain: $A > 0, x > 0$

131. $V = \dfrac{4}{3}\pi r^3$;   domain: $V > 0, r > 0$

133. $V = x^3$;   domain: $V > 0, x > 0$

135. $C = 4000 + 2x$
(a) If $x = 1000$ watches are produced, it will cost
  $C = 4000 + 2(1000) = \$6000.00$

(b) If $x = 2000$ watches are produced, it will cost
  $C = 4000 + 2(2000) = \$8000.00$

137. (a) If actual voltage is $x = 113$ then
    $|113 - 115| = |-2| = 2$
  Since $2 < 5$, an actual voltage of 113 is acceptable.

(b) If actual voltage is $x = 109$ then
    $|109 - 115| = |-6| = 6$

Since $6 > 5$, an actual voltage of 109 is not acceptable.

139. (a) If the radius is $x = 2.999$, then
$$|x - 3| = |2.999 - 3| = |-0.001| = 0.001$$
Since $0.001 < 0.010$ the ball bearing is not acceptable.
(b) If the radius is $x = 2.89$, then
$$|x - 3| = |2.89 - 3| = |-0.11| = 0.11$$
Since $0.11 > 0.01$, the ball bearing is not acceptable.

141. $\dfrac{1}{3} \neq 0.333; \quad \dfrac{1}{3} > 0.333$

$\dfrac{1}{3} = 0.333\ldots; \quad 0.333\ldots - 0.333 = 0.000333\ldots$

143. The answer is no. Student answers should justify and explain why not.

145. Answers will vary.

A.3 Exponents and Logarithms   A-32/1-49 (ODD)

1. $4^3 = 4 \bullet 4 \bullet 4 = 64$

3. $2^{-3} = \dfrac{1}{2^3} = \dfrac{1}{2 \cdot 2 \cdot 2} = \dfrac{1}{8}$

5. $8^0 = 1$

7. $\sqrt{16} = 4$

9. $\sqrt[3]{27} = 3$

11. $\sqrt[4]{16} = 2$

13. $8^{\frac{2}{3}} = \left(\sqrt[3]{8}\right)^2 = 2^2 = 4$

15. $16^{-\frac{3}{2}} = \dfrac{1}{16^{\frac{3}{2}}} = \dfrac{1}{\left(\sqrt{16}\right)^3} = \dfrac{1}{4^3} = \dfrac{1}{64}$

17. $-8^{-\frac{2}{3}} = \dfrac{1}{-8^{\frac{2}{3}}} = \dfrac{1}{\left(\sqrt[3]{-8}\right)^2} = \dfrac{1}{(-2)^2} = \dfrac{1}{4}$

19. (a) $3^{2.2} = 11.2116$
(b) $3^{2.23} = 11.5873$
(c) $3^{2.236} = 11.6639$
(d) $3^{\sqrt{5}} = 11.6648$

21. (a) $2^{3.14} = 8.8152$
(b) $2^{3.141} = 8.8214$
(c) $2^{3.1415} = 8.8244$
(d) $2^{\pi} = 8.8250$

23. (a) $3.1^{2.7} = 21.2166$
(b) $3.14^{2.71} = 22.2167$
(c) $3.141^{2.718} = 22.4404$
(d) $\pi^e = 22.4592$

25. $3^y = 27$
$y = 3$

27. $2^y = \dfrac{1}{2}$
$y = -1$

29. $2^3 = N$
$N = 8$

31. $3^{-1} = N$
$N = \dfrac{1}{3}$

33. $a^3 = 8$
$a = 2$

35. $a^2 = 9$
$a = 3$

37. $\log_2 5 = x$

39. $\log_{1.1} 10 = t$

41. $\log_{1.1} 200 = \dfrac{\log 200}{\log 1.1} = 55.5903$

43. $\log_{1.005} 1000 = \dfrac{\log 1000}{\log 1.005} = 1385.0021$

45. $\log_{1.002} 20 = \dfrac{\log 20}{\log 1.002} = 1499.3635$

47. $\log_{1.0005} 500 = \dfrac{\log 500}{\log 1.0005} = 12,432.3232$

49. $\log_{1.003} 500 = \dfrac{\log 500}{\log 1.003} = 2074.6418$

A.4 Recursively Defined Sequences; Geometric Sequences

1. $\{n\} = \{1, 2, 3, 4, 5\}$

3. $\left\{\dfrac{n}{n+1}\right\}: \quad \dfrac{1}{1+1} = \dfrac{1}{2}$

$\dfrac{2}{2+1} = \dfrac{2}{3}$

$\dfrac{3}{3+1} = \dfrac{3}{4}$

$\dfrac{4}{4+1} = \dfrac{4}{5}$

$\dfrac{5}{5+1} = \dfrac{5}{6}$

$\left\{\dfrac{1}{2}, \dfrac{2}{3}, \dfrac{3}{4}, \dfrac{4}{5}, \dfrac{5}{6}\right\}$

5. $\{(-1)^{n+1} n^2\}: \quad (-1)^{(1+1)} \cdot 1^2 = 1$
$(-1)^{(2+1)} \cdot 2^2 = -4$
$(-1)^{(3+1)} \cdot 3^2 = 9$
$(-1)^{(4+1)} \cdot 4^2 = -16$
$(-1)^{(5+1)} \cdot 5^2 = 25$

$\{1, -4, 9, -16, 25\}$

7. $\left\{\dfrac{2^n}{3^n+1}\right\}$: $\dfrac{2^1}{3^1+1}=\dfrac{2}{4}=\dfrac{1}{2}$

$$\dfrac{2^2}{3^2+1}=\dfrac{4}{10}=\dfrac{2}{5}$$

$$\dfrac{2^3}{3^3+1}=\dfrac{8}{28}=\dfrac{2}{7}$$

$$\dfrac{2^4}{3^4+1}=\dfrac{16}{82}=\dfrac{8}{41}$$

$$\dfrac{2^5}{3^5+1}=\dfrac{32}{244}=\dfrac{8}{61}$$

$$\left\{\dfrac{1}{2},\ \dfrac{2}{5},\ \dfrac{2}{7},\ \dfrac{8}{41},\ \dfrac{8}{61}\right\}$$

9. $\left\{\dfrac{(-1)^n}{(n+1)(n+2)}\right\}$: $\dfrac{(-1)^1}{(1+1)(1+2)}=-\dfrac{1}{6}$

$$\dfrac{(-1)^2}{(2+1)(2+2)}=\dfrac{1}{12}$$

$$\dfrac{(-1)^3}{(3+1)(3+2)}=-\dfrac{1}{20}$$

$$\dfrac{(-1)^4}{(4+1)(4+2)}=\dfrac{1}{30}$$

$$\dfrac{(-1)^5}{(5+1)(5+2)}=-\dfrac{1}{42}$$

$$\left\{-\dfrac{1}{6},\ \dfrac{1}{12},\ -\dfrac{1}{20},\ \dfrac{1}{30},\ -\dfrac{1}{42}\right\}$$

11. $\left\{\dfrac{n}{e^n}\right\}$: $\left\{\dfrac{1}{e^1},\ \dfrac{2}{e^2},\ \dfrac{3}{e^3},\ \dfrac{4}{e^4},\ \dfrac{5}{e^5}\right\}$

13. $a_1=1$
$a_2=2+a_1=2+1=3$
$a_3=2+a_2=2+3=5$
$a_4=2+a_3=2+5=7$
$a_5=2+a_4=2+7=9$

15. $a_1=-2$
$a_2=1+a_1=1+(-2)=-1$
$a_3=2+a_2=2+(-1)=\;\;1$
$a_4=3+a_3=3+(1)=\;\;4$
$a_5=4+a_4=4+(4)=\;\;8$

17. $a_1 = 5$
$a_2 = 2a_1 = 2 \cdot 5 = 10$
$a_3 = 2a_2 = 2 \cdot 10 = 20$
$a_4 = 2a_3 = 2 \cdot 20 = 40$
$a_5 = 2a_4 = 2 \cdot 40 = 80$

19. $a_1 = 3$
$a_2 = \dfrac{a_1}{1} = 3$
$a_3 = \dfrac{a_2}{2} = \dfrac{3}{2}$
$a_4 = \dfrac{a_3}{3} = \dfrac{\frac{3}{2}}{3} = \dfrac{1}{2}$
$a_5 = \dfrac{a_4}{4} = \dfrac{\frac{1}{2}}{4} = \dfrac{1}{8}$

21. $a_1 = 1$
$a_2 = 2$
$a_3 = a_1 \cdot a_2 = 1 \cdot 2 = 3$
$a_4 = a_2 \cdot a_3 = 2 \cdot 3 = 6$
$a_5 = a_3 \cdot a_4 = 3 \cdot 6 = 18$

23. $a_1 = A$
$a_2 = a_1 + d = A + d$
$a_3 = a_2 + d = (A + d) + d = A + 2d$
$a_4 = a_3 + d = (A + 2d) + d = A + 3d$
$a_5 = a_4 + d = (A + 3d) + d = A + 4d$

25. $a_1 = \sqrt{2}$
$a_2 = \sqrt{2 + a_1} = \sqrt{2 + \sqrt{2}}$
$a_3 = \sqrt{2 + a_2} = \sqrt{2 + \sqrt{2 + \sqrt{2}}}$
$a_4 = \sqrt{2 + a_3} = \sqrt{2 + \sqrt{2 + \sqrt{2 + \sqrt{2}}}}$
$a_5 = \sqrt{2 + a_4} = \sqrt{2 + \sqrt{2 + \sqrt{2 + \sqrt{2 + \sqrt{2}}}}}$

27. $\{2^n\}$
(a) First term: $a = 2$; common ratio: $r = 2$
(b) $2, \; 2^2, \; 2^3, \; 2^4$ or $2, \; 4, \; 8, \; 16$
(c) $S_n = 2\left(\dfrac{1 - 2^n}{1 - 2}\right) = -2\left(1 - 2^n\right) = 2^{n+1} - 2$

29. $\left\{ -3\left(\dfrac{1}{2}\right)^n \right\}$

(a) First term: $a = -3\left(\dfrac{1}{2}\right) = -\dfrac{3}{2}$; common ratio: $r = \dfrac{1}{2}$

(b) $-\dfrac{3}{2}, \; -\dfrac{3}{4}, \; -\dfrac{3}{8}, \; -\dfrac{3}{16}$

(c) $S_n = \left(-\dfrac{3}{2}\right)\left[\dfrac{1 - \left(\frac{1}{2}\right)^n}{1 - \frac{1}{2}}\right] = -3\left[1 - \left(\dfrac{1}{2}\right)^n\right] = -3 + \dfrac{3}{2^n}$

31. $\left\{ \dfrac{2^{n-1}}{4} \right\}$

(a) First term: $a = \dfrac{2^0}{4} = \dfrac{1}{4}$; common ratio: $r = 2$

(b) $\dfrac{1}{4}$, $\dfrac{2}{4} = \dfrac{1}{2}$, $\dfrac{2^2}{4} = \dfrac{4}{4} = 1$, $\dfrac{2^3}{4} = \dfrac{8}{4} = 2$

(c) $S = \dfrac{1}{4}\left[\dfrac{1 - 2^n}{1 - 2} \right] = -\dfrac{1}{4}\left(1 - 2^n\right) = 2^{n-2} - \dfrac{1}{4}$

33. $\left\{ 2^{\frac{n}{3}} \right\}$

(a) First term: $a = 2^{\frac{1}{3}}$; common ratio: $r = 2^{\frac{1}{3}}$

(b) $2^{\frac{1}{3}}$, $2^{\frac{2}{3}}$, $2^{\frac{3}{3}} = 2$, $2^{\frac{4}{3}}$

(c) $S = 2^{\frac{1}{3}}\left(\dfrac{1 - 2^{\frac{n}{3}}}{1 - 2^{\frac{1}{3}}} \right)$

35. $\left\{ \dfrac{3^{n-1}}{2^n} \right\}$

(a) First term: $a = \dfrac{3^0}{2^1} = \dfrac{1}{2}$; common ratio: $r = \dfrac{3}{2}$

(b) $\dfrac{1}{2}$, $\dfrac{3^{2-1}}{2^2} = \dfrac{3}{4}$, $\dfrac{3^{3-1}}{2^3} = \dfrac{9}{8}$, $\dfrac{3^{4-1}}{2^4} = \dfrac{27}{16}$

(c) $S = \dfrac{1}{2}\left[\dfrac{1 - \left(\dfrac{3}{2}\right)^n}{1 - \dfrac{3}{2}} \right] = -1\left[1 - \left(\dfrac{3}{2}\right)^n \right]$

A.5 Polynomials and Rational Expressions

1. $\left(10x^5 - 8x^2\right) + \left(3x^3 - 2x^2 + 6\right) = 10x^5 + 3x^3 - 10x^2 + 6$

3. $(x + a)^2 - x^2 = \left[x^2 + 2ax + a^2\right] - x^2$
$= 2ax + a^2$

5. $(x + 8)(2x + 1) = 2x^2 + x + 16x + 8 = 2x^2 + 17x + 8$

7. $\left(x^2 + x - 1\right)\left(x^2 - x + 1\right) = x^2\left(x^2 - x + 1\right) + x\left(x^2 - x + 1\right) - 1\left(x^2 - x + 1\right)$
$= x^4 - x^3 + x^2 + x^3 - x^2 + x - x^2 + x - 1$
$= x^4 - x^2 + 2x - 1$

9.
$$(x+1)^3 - (x-1)^3 = (x^3 + 3x^2 + 3x + 1) - (x^3 - 3x^2 + 3x - 1)$$
$$= x^3 + 3x^2 + 3x + 1 - x^3 + 3x^2 - 3x + 1$$
$$= 6x^2 + 2$$

11. This is the difference of 2 squares.
$$x^2 - 36 = (x-6)(x+6)$$

13. This is the difference of 2 squares.
$$1 - 4x^2 = (1-2x)(1+2x)$$

15. $x^2 + 7x + 10 = (x+5)(x+2)$ The product of 5 and 2 is 10; the sum of 5 and 2 is 7.

17. This polynomial cannot be factored; it is prime.

19. This polynomial cannot be factored; it is prime.

21. $15 + 2x - x^2 = (5-x)(3+x)$ The product $5 \cdot 3 = 15$; the sum $5 \cdot 1 + 3 \cdot (-1) = 2$.

23. $3x^2 - 12x - 36 = 3(x^2 - 4x - 12)$ Factor out the common factor of 3.
$$= 3(x-6)(x+2)$$ The product $-6 \cdot 2 = -12$; the sum $-6 + 2 = -4$.

25. $y^4 + 11y^3 + 30y^2 = y^2(y^2 + 11y + 30)$ Factor out the common factor of y^2.
$$= y^2(y+5)(y+6)$$ The product $5 \cdot 6 = 30$; the sum $5 + 6 = 11$.

27. This polynomial is a perfect square.
$$4x^2 + 12x + 9 = (2x+3)(2x+3) = (2x+3)^2$$

29. $3x^2 + 4x + 1 = (3x+1)(x+1)$ The products $3 \cdot 1 = 3$ and $1 \cdot 1 = 1$; the sum $3 \cdot 1 + 1 \cdot 1 = 4$.

31. $x^4 - 81 = (x^2 - 9)(x^2 + 9)$ Treat $x^4 - 81$ as the difference of 2 squares, and factor.
$$= (x-3)(x+3)(x^2 + 9)$$ $x^2 - 9$ is the difference of 2 squares.

33. Let $u = x^3$. Then
$$x^6 - 2x^3 + 1 = u^2 - 2u + 1 = (u-1)(u-1) = (u-1)^2$$
Now substitute back for u. That is, replace u with x^3.
$$x^6 - 2x^3 + 1 = (x^3 - 1)^2$$

35. $x^7 - x^5 = x^5(x^2 - 1)$ Factor out the common factor of x^5.
$$= x^5(x-1)(x+1)$$ $x^2 - 1$ is the difference of 2 squares.

37. $5+16x-16x^2=(1+4x)(5-4x)$ The products $4\cdot(-4)=-16$ and $1\cdot 5=5$; the sum $1\cdot(-4)+4\cdot 5=16$.

39. $4y^2-16y+15=(2y-5)(2y-3)$ The products $2\cdot 2=4$ and $-5\cdot(-3)=15$; the sum $2\cdot(-3)+(-5)\cdot 2=-16$.

41. Let $u=x^2$, then
$$1-8x^2-9x^4=1-8u-9u^2=(1-9u)(1+u)$$
Now substitute back for u. That is, replace u with x^2.
$$1-8x^2-9x^4=(1-9x^2)(1+x^2)$$
$1-9x^2$ is the difference of 2 squares.
$$1-8x^2-9x^4=(1-9x^2)(1+x^2)=(1-3x)(1+3x)(1+x^2)$$

43. $x(x+3)-6(x+3)=(x+3)(x-6)$ Factor out the common factor $x+3$.

45. $(x+2)^2-2(x+2)=(x+2)\big[(x+2)-5\big]$ Factor out the common factor $x+2$.
$$=(x+2)(x-3)$$ Simplify.

47. $6x(2-x)^4-9x^2(2-x)^3=3x(2-x)^3\big[2(2-x)-3x\big]$ Factor out the common factors.
$$=3x(2-x)^3(4-2x-3x)$$ Simplify.
$$=3x(2-x)^3(4-5x)$$ Simplify.
$$=3x(x-2)^3(5x-4)$$ Multiply $(x-2)^3$ by $(-1)^3$ and $(5x-4)$ by -1. $[(-1)^3\cdot(-1)=1]$.

49. $x^3+2x^2-x-2=(x^3+2x^2)-(x+2)$ Group the polynomial into the difference of 2 binomials.
$$=x^2(x+2)-1(x+2)$$ Factor the common factors from each binomial.
$$=(x^2-1)(x+2)$$ Factor the common factor $x+2$.
$$=(x-1)(x+1)(x+2)$$ Factor the difference of 2 squares.

51. $x^4-x^3+x-1=(x^4-x^3)+(x-1)$ Group the polynomial into the sum of 2 binomials.
$$=x^3(x-1)+(x-1)$$ Factor the common factors from each binomial.
$$=(x-1)(x^3+1)$$ Factor the common factor $x-1$.
$$=(x-1)(x+1)(x^2-x+1)$$ Factor the sum of 2 cubes.

53. $\dfrac{3x-6}{5x}\cdot\dfrac{x^2-x-6}{x^2-4}=\dfrac{3\cancel{(x-2)}}{5x}\cdot\dfrac{(x-3)\cancel{(x+2)}}{\cancel{(x-2)}\cancel{(x+2)}}$ Factor.

$$= \frac{3(x-3)}{5x} \qquad \text{Simplify.}$$

55.
$$\frac{4x^2-1}{x^2-16} \cdot \frac{x^2-4x}{2x+1} = \frac{(2x-1)\cancel{(2x+1)}}{\cancel{(x-4)}(x+4)} \cdot \frac{x\cancel{(x-4)}}{\cancel{2x+1}} = \frac{(2x-1)x}{x+4}$$

57.
$$\frac{x}{x^2-7x+6} - \frac{x}{x^2-2x-24} = \frac{x}{(x-6)(x-1)} - \frac{x}{(x-6)(x+4)}$$

$$= \frac{x(x+4)}{(x-6)(x-1)(x+4)} - \frac{x(x-1)}{(x-6)(x+4)(x-1)}$$

$$= \frac{x^2+4x}{(x-6)(x-1)(x+4)} - \frac{x^2-x}{(x-6)(x+4)(x-1)}$$

$$= \frac{x^2+4x-x^2+x}{(x-6)(x-1)(x+4)}$$

$$= \frac{5x}{(x-6)(x-1)(x+4)}$$

59.
$$\frac{4}{x^2-4} - \frac{2}{x^2+x-6} = \frac{4}{(x-2)(x+2)} - \frac{2}{(x+3)(x-2)}$$

$$= \frac{4(x+3)}{(x-2)(x+2)(x+3)} - \frac{2(x+2)}{(x+3)(x-2)(x+2)}$$

$$= \frac{4(x+3)-2(x+2)}{(x-2)(x+2)(x+3)}$$

$$= \frac{4x+12-2x-4}{(x-2)(x+2)(x+3)}$$

$$= \frac{2x+8}{(x-2)(x+2)(x+3)} = \frac{2(x+4)}{(x-2)(x+2)(x+3)}$$

61.
$$\frac{1}{x} - \frac{2}{x^2+x} + \frac{3}{x^3-x^2} = \frac{1}{x} - \frac{2}{x(x+1)} + \frac{3}{x^2(x-1)}$$

$$= \frac{x(x+1)(x-1)}{x^2(x+1)(x-1)} - \frac{2x(x-1)}{x^2(x+1)(x-1)} + \frac{3(x+1)}{x^2(x-1)(x+1)}$$

$$= \frac{x(x+1)(x-1)-2x(x-1)+3(x+1)}{x^2(x+1)(x-1)}$$

$$= \frac{x^3-x-2x^2+2x+3x+3}{x^2(x+1)(x-1)}$$

$$= \frac{x^3-2x^2+4x+3}{x^2(x+1)(x-1)}$$

63.
$$\frac{1}{h}\left(\frac{1}{x+h}-\frac{1}{x}\right)=\frac{1}{h}\left(\frac{x}{(x+h)x}-\frac{1(x+h)}{x(x+h)}\right)$$
$$=\frac{1}{h}\left(\frac{x-x-h}{(x+h)x}\right)$$
$$=\frac{1}{\cancel{h}}\left(\frac{-\cancel{h}}{(x+h)x}\right)$$
$$=-\frac{1}{x(x+h)}$$

65.
$$2(3x+4)^2+(2x+3)\cdot2(3x+4)\cdot3=(3x+4)\left[2(3x+4)+6(2x+3)\right]$$
$$=(3x+4)(6x+8+12x+18)$$
$$=(3x+4)(18x+26)$$
$$=2(3x+4)(9x+13)$$

67.
$$2x(2x+5)+x^2\cdot2=2x(2x+5+x)$$
$$=2x(3x+5)$$

69.
$$2(x+3)(x-2)^3+(x+3)^2\cdot3(x-2)^2=(x+3)(x-2)^2\left[2(x-2)+3(x+3)\right]$$
$$=(x+3)(x-2)^2(2x-4+3x+9)$$
$$=(x+3)(x-2)^2(5x+5)$$
$$=5(x+3)(x-2)^2(x+1)$$

71.
$$(4x-3)^2+x\cdot2(4x-3)\cdot4=(4x-3)\left[(4x-3)+8x\right]$$
$$=(4x-3)(12x-3)$$
$$=3(4x-3)(4x-1)$$

73.
$$2(3x-5)\cdot3(2x+1)^3+(3x-5)^2\cdot3(2x+1)^2\cdot2=6(3x-5)(2x+1)^2\left[(2x+1)+(3x-5)\right]$$
$$=6(3x-5)(2x+1)^2(5x-4)$$

75.
$$\frac{(2x+3)\cdot3-(3x-5)\cdot2}{(3x-5)^2}=\frac{6x+9-6x+10}{(3x-5)^2}=\frac{19}{(3x-5)^2}$$

77.
$$\frac{x\cdot2x-(x^2+1)\cdot1}{(x^2+1)^2}=\frac{2x^2-x^2-1}{(x^2+1)^2}=\frac{x^2-1}{(x^2+1)^2}=\frac{(x-1)(x+1)}{(x^2+1)^2}$$

79.
$$\frac{(3x+1)\cdot 2x - x^2\cdot 3}{(3x+1)^2} = \frac{6x^2+2x-3x^2}{(3x+1)^2} = \frac{3x^2+2x}{(3x+1)^2}$$

81.
$$\frac{(x^2+1)\cdot 3-(3x+4)\cdot 2x}{(x^2+1)^2} = \frac{3x^2+3-6x^2-8x}{(x^2+1)^2} = \frac{-3x^2-8x+3}{(x^2+1)^2}$$

A.6 Solving Equations

1. $3x = 21$
$x = 7$

3. $5x+15 = 0$
$5x = -15$
$x = -3$

5. $2x-3 = 5$
$2x = 8$
$x = 4$

7. $\frac{1}{3}x = \frac{5}{12}$
$x = \frac{5}{\cancel{12}_4}\cdot \frac{\cancel{3}}{1}$
$x = \frac{5}{4}$

9. $6-x = 2x+9$
$-3x = 3$
$x = -1$

11. $2(3+2x) = 3(x-4)$
$6+4x = 3x-12$
$x = -18$

13. $8x-(2x+1) = 3x-10$
$6x-1 = 3x-10$
$3x = -9$
$x = -3$

15. $\frac{1}{2}x-4 = \frac{3}{4}x$
$4\cdot\left(\frac{1}{2}x-4\right) = 4\cdot\left(\frac{3}{4}x\right)$
$2x-16 = 3x$
$x = -16$

17. $0.9t = 0.4+0.1t$
$0.8t = 0.4$
$t = \frac{0.4}{0.8} = 0.5$

19. $\frac{2}{y}+\frac{4}{y} = 3$
$\frac{6}{y} = 3$
$3y = 6$
$y = 2$

21.
$$(x+7)(x-1)=(x+1)^2$$
$$x^2-x+7x-7=x^2+2x+1$$
$$x^2+6x-7=x^2+2x+1$$
$$6x-7=2x+1$$
$$4x=8$$
$$x=2$$

23.
$$z(z^2+1)=3+z^3$$
$$z^3+z=3+z^3$$
$$z=3$$

25.
$$x^2=9x$$
$$x^2-9x=0$$
$$x(x-9)=0$$
$$x=0 \quad \text{or} \quad x-9=0$$
$$x=9$$

The solution set is $\{0, 9\}$.

27.
$$t^3-9t^2=0$$
$$t^2(t-9)=0$$
$$t^2=0 \quad \text{or} \quad t-9=0$$
$$t=0 \quad \text{or} \quad t=9$$

The solution set is $\{0, 9\}$.

29.
$$\frac{3}{2x-3}=\frac{2}{x+5}$$
$$3(x+5)=2(2x-3)$$
$$3x+15=4x-6$$
$$x=21$$

31.
$$(x+2)(3x)=(x+2)(6)$$
$$3x^2+6x=6x+12$$
$$3x^2-12=0$$
$$x^2-4=0$$
$$x^2=4$$
$$x=-2 \quad \text{or} \quad x=2$$

The solution set is $\{-2, 2\}$.

33.
$$\frac{2}{x-2}=\frac{3}{x+5}+\frac{10}{(x+5)(x-2)}$$

L.C.D.: $(x+5)(x-2)$

$$\frac{2(x+5)}{(x-2)(x+5)}=\frac{3(x-2)}{(x+5)(x-2)}+\frac{10}{(x+5)(x-2)}$$

Write with the common denominator.

$$2(x+5)=3(x-2)+10$$
$$2x+10=3x-6+10$$
$$x=6$$

Solve the equation formed by the numerators.

Check the answer for extraneous solutions.

The solution set is $\{6\}$.

35.
$$|2x|=6$$
Either
$$2x=6 \quad \text{or} \quad 2x=-6$$
$$x=3 \quad \text{or} \quad x=-3$$
The solution set is $\{-3, 3\}$.

37.
$$|2x+3|=5$$
Either
$$2x+3=5 \quad \text{or} \quad 2x+3=-5$$
$$2x=2 \quad \text{or} \quad 2x=-8$$
$$x=1 \quad \text{or} \quad x=-4$$
The solution set is $\{-4, 1\}$.

39. $|1-4t|=5$

Either
$$1-4t=5 \quad \text{or} \quad 1-4t=-5$$
$$-4t=4 \quad \text{or} \quad -4t=-6$$
$$t=-1 \quad \text{or} \quad t=\frac{3}{2}$$

The solution set is $\left\{-1, \frac{3}{2}\right\}$.

41. $|-2x|=8$

Either
$$-2x=8 \quad \text{or} \quad -2x=-8$$
$$x=-4 \quad \text{or} \quad x=4$$

The solution set is $\{-4, 4\}$.

43. $|-2|x=4$
$$2x=4$$
$$x=2$$

45. $|x-2|=-\frac{1}{2}$

This equation has no solution. Absolute values are always nonnegative.

47. $|x^2-4|=0$
$$x^2-4=0$$
$$x^2=4$$
$$x=\pm 2$$
The solution set is $\{-2, 2\}$.

49. $|x^2-2x|=3$

Either
$$x^2-2x=3 \qquad\qquad \text{or} \qquad\qquad x^2-2x=-3$$
$$x^2-2x-3=0 \qquad\qquad\qquad x^2-2x+3=0 \qquad a=1, b=-2, c=3$$
$$(x-3)(x+1)=0 \qquad\qquad \text{The discriminant,} \quad b^2-4ac=4-12=-8 \text{ is}$$
$$x-3=0 \quad \text{or} \quad x+1=0 \qquad \text{negative; the equation has no real solutions.}$$
$$x=3 \quad \text{or} \qquad x=-1$$

The solution set is $\{-1, 3\}$.

51. $|x^2+x-1|=1$

Either
$$x^2+x-1=1 \qquad\qquad \text{or} \qquad\qquad x^2+x-1=-1$$
$$x^2+x-2=0 \qquad\qquad\qquad\qquad x^2+x=0$$
$$(x-1)(x+2)=0 \qquad\qquad\qquad\qquad x(x+1)=0$$
$$x-1=0 \quad \text{or} \quad x+2=0 \quad \text{or} \qquad x=0 \quad \text{or} \quad x+1=0$$
$$x=1 \quad \text{or} \qquad x=-2 \quad \text{or} \qquad x=0 \quad \text{or} \quad x=-1$$

The solution set is $\{-2, -1, 0, 1\}$.

53.
$$x^2 = 4x$$
$$x^2 - 4x = 0$$
$$x(x - 4) = 0$$
$$x = 0 \quad \text{or} \quad x - 4 = 0$$
$$x = 4$$
The solution set is $\{0, 4\}$.

55.
$$z^2 + 4z - 12 = 0$$
$$(z - 2)(z + 6) = 0$$
$$z - 2 = 0 \quad \text{or} \quad z + 6 = 0$$
$$z = 2 \quad \text{or} \quad z = -6$$
The solution set is $\{-6, 2\}$.

57.
$$2x^2 - 5x - 3 = 0$$
$$(2x + 1)(x - 3) = 0$$
$$2x + 1 = 0 \quad \text{or} \quad x - 3 = 0$$
$$x = -\frac{1}{2} \quad \text{or} \quad x = 3$$
The solution set is $\left\{-\frac{1}{2}, 3\right\}$.

59.
$$x(x - 7) + 12 = 0$$
$$x^2 - 7x + 12 = 0$$
$$(x - 3)(x - 4) = 0$$
$$x - 3 = 0 \quad \text{or} \quad x - 4 = 0$$
$$x = 3 \quad \text{or} \quad x = 4$$
The solution set is $\{3, 4\}$.

61.
$$4x^2 + 9 = 12x$$
$$4x - 12x + 9 = 0$$
$$(2x - 3)(2x - 3) = 0$$
$$2x - 3 = 0$$
$$x = \frac{3}{2}$$
The solution set is $\left\{\frac{3}{2}\right\}$.

63.
$$6x - 5 = \frac{6}{x}$$
$$6x^2 - 5x = 6$$
$$6x^2 - 5x - 6 = 0$$
$$(3x + 2)(2x - 3) = 0$$
$$3x + 2 = 0 \quad \text{or} \quad 2x - 3 = 0$$
$$x = -\frac{2}{3} \quad \text{or} \quad x = \frac{3}{2}$$
The solution set is $\left\{-\frac{2}{3}, \frac{3}{2}\right\}$.

65.
$$\frac{4(x - 2)}{x - 3} + \frac{3}{x} = \frac{-3}{x(x - 3)}$$

The lowest common denominator is $x(x - 3)$.

$$\frac{4x(x - 2)}{x(x - 3)} + \frac{3(x - 3)}{x(x - 3)} = \frac{-3}{x(x - 3)}$$

Write the equation with the common denominator.

$$4x(x - 2) + 3(x - 3) = -3$$

Consider the equation formed by the numerator.

$$4x^2 - 8x + 3x - 9 = -3$$

Simplify.

$$4x^2 - 5x - 6 = 0$$

Put the quadratic equation in standard form.

$$(4x + 3)(x - 2) = 0$$

Factor.

$$4x + 3 = 0 \quad \text{or} \quad x - 2 = 0$$

Use the Zero-Product Property.

$$x = -\frac{3}{4} \quad \text{or} \quad x = 2$$

Solve; be sure to check for extraneous solutions.

The solution set is $\left\{-\frac{3}{4}, 2\right\}$.

67.
$$x^2 = 25$$
$$x = \pm \sqrt{25}$$
$$x = \pm 5$$
The solution set is $\{-5, 5\}$.

69.
$$(x-1)^2 = 4$$
$$x - 1 = \pm \sqrt{4}$$
$$x - 1 = \pm 2$$
$$x = 2 + 1 \quad \text{or} \quad x = -2 + 1$$
$$x = 3 \qquad \text{or} \quad x = -1$$
The solution set is $\{-1, 3\}$.

71.
$$(2x+3)^2 = 9$$
$$2x + 3 = \pm \sqrt{9}$$
$$2x + 3 = \pm 3$$
$$2x = -3 + 3 \quad \text{or} \quad 2x = -3 - 3$$
$$2x = 0 \qquad \text{or} \quad 2x = -6$$
$$x = 0 \qquad \text{or} \quad x = -3$$
The solution set is $\{-3, 0\}$.

73.
$$x^2 + 8x$$
Add $\left(\dfrac{8}{2}\right)^2 = 16$.

Result $x^2 + 8x + 16$

75.
$$x^2 + \frac{1}{2}x$$
Add $\left(\dfrac{1}{4}\right)^2 = \dfrac{1}{16}$

Result $x^2 + \dfrac{1}{2}x + \dfrac{1}{16}$

77.
$$x^2 - \frac{2}{3}x$$
Add $\left(\dfrac{1}{3}\right)^2 = \dfrac{1}{9}$

Result $x^2 - \dfrac{2}{3}x + \dfrac{1}{9}$

79.
$$x^2 + 4x = 21$$
$$x^2 + 4x + 4 = 21 + 4 \qquad \text{Add 4 to both sides.}$$
$$(x+2)^2 = 25 \qquad \text{Factor.}$$
$$x + 2 = \pm 5 \qquad \text{Use the Square Root Method.}$$
$$x = -2 \pm 5$$
The solution set is $\{-7, 3\}$.

81.
$$x^2 - \frac{1}{2}x - \frac{3}{16} = 0$$
$$x^2 - \frac{1}{2}x = \frac{3}{16}$$
$$x^2 - \frac{1}{2}x + \frac{1}{16} = \frac{3}{16} + \frac{1}{16} \qquad \text{Add } \frac{1}{16} \text{ to both sides.}$$
$$\left(x - \frac{1}{4}\right)^2 = \frac{4}{16}$$
$$x - \frac{1}{4} = \pm \sqrt{\frac{4}{16}}$$
$$x = \frac{1}{4} \pm \frac{2}{4}$$

The solution set is $\left\{-\dfrac{1}{4}, \dfrac{3}{4}\right\}$.

83.
$$3x^2 + x - \dfrac{1}{2} = 0$$
$$x^2 + \dfrac{1}{3}x - \dfrac{1}{6} = 0$$
$$x^2 + \dfrac{1}{3}x = \dfrac{1}{6}$$
$$x^2 + \dfrac{1}{3}x + \dfrac{1}{36} = \dfrac{1}{6} + \dfrac{1}{36}$$
$$\left(x + \dfrac{1}{6}\right)^2 = \dfrac{7}{36}$$
$$x + \dfrac{1}{6} = \pm\sqrt{\dfrac{7}{36}}$$
$$x = -\dfrac{1}{6} \pm \dfrac{\sqrt{7}}{6}$$

The solution set is $\left\{\dfrac{-1-\sqrt{7}}{6}, \dfrac{-1+\sqrt{7}}{6}\right\}$.

85. $x^2 - 4x + 2 = 0 \qquad a = 1, b = -4, \text{ and } c = 2$

The discriminant $b^2 - 4ac = (-4)^2 - 4(1)(2) = 16 - 8 = 8$ is positive, so there are 2 real solutions to the equation.
$$x = \dfrac{-b \pm \sqrt{b^2 - 4ac}}{2a} = \dfrac{4 \pm \sqrt{8}}{2} = \dfrac{4 \pm 2\sqrt{2}}{2} = 2 \pm \sqrt{2}$$

The solution set is $\left\{2 - \sqrt{2}, 2 + \sqrt{2}\right\}$.

87. $x^2 - 5x - 1 = 0 \qquad a = 1, b = -5, \text{ and } c = -1$

The discriminant $b^2 - 4ac = (-5)^2 - 4(1)(-1) = 25 + 4 = 29$ is positive, so there are 2 real solutions to the equation.
$$x = \dfrac{-b \pm \sqrt{b^2 - 4ac}}{2a} = \dfrac{5 \pm \sqrt{29}}{2}$$

The solution set is $\left\{\dfrac{5 - \sqrt{29}}{2}, \dfrac{5 + \sqrt{29}}{2}\right\}$.

89. $2x^2 - 5x + 3 = 0 \qquad a = 2, b = -5, \text{ and } c = 3$

The discriminant $b^2 - 4ac = (-5)^2 - 4(2)(3) = 25 - 24 = 1$ is positive, so there are 2 real solutions to the equation.
$$x = \dfrac{-b \pm \sqrt{b^2 - 4ac}}{2a} = \dfrac{5 \pm \sqrt{1}}{4} = \dfrac{5 \pm 1}{4}$$

The solution set is $\left\{1, \dfrac{3}{2}\right\}$.

91. $\quad 4y^2 - y + 2 = 0 \qquad a = 4, b = -1, \text{ and } c = 2$

The discriminant $b^2 - 4ac = (-1)^2 - 4(4)(2) = 1 - 32 = -31$ is negative, so the equation has no real solution.

93. $\quad 4x^2 = 1 - 2x \quad$ First we rewrite the equation in standard form.

$\quad 4x^2 + 2x - 1 = 0 \qquad a = 4, b = 2, \text{ and } c = -1$

The discriminant $b^2 - 4ac = (2)^2 - 4(4)(-1) = 4 + 16 = 20$ is positive, so there are 2 real solutions to the equation.

$$x = \frac{-b \pm \sqrt{b^2 - 4ac}}{2a} = \frac{-2 \pm \sqrt{20}}{8} = \frac{-2 \pm 2\sqrt{5}}{8} = \frac{-1 \pm \sqrt{5}}{4}$$

The solution set is $\left\{\dfrac{-1 - \sqrt{5}}{4}, \dfrac{-1 + \sqrt{5}}{4}\right\}$.

95. $\quad x^2 + \sqrt{3}x - 3 = 0 \qquad a = 1, b = \sqrt{3}, \text{ and } c = -3$

The discriminant $b^2 - 4ac = \left(\sqrt{3}\right)^2 - 4(1)(-3) = 3 + 12 = 15$ is positive, so there are 2 real solutions to the equation.

$$x = \frac{-b \pm \sqrt{b^2 - 4ac}}{2a} = \frac{-\sqrt{3} \pm \sqrt{15}}{2}$$

The solution set is $\left\{\dfrac{-\sqrt{3} - \sqrt{15}}{2}, \dfrac{-\sqrt{3} + \sqrt{15}}{2}\right\}$.

97. $\quad x^2 - 5x + 7 = 0 \qquad a = 1, b = -5, \text{ and } c = 7$

The discriminant $b^2 - 4ac = (-5)^2 - 4(1)(7) = 25 - 28 = -3$ is negative, so the equation has no real solution.

99. $\quad 9x^2 - 30x + 25 = 0 \qquad a = 9, b = -30, \text{ and } c = 25$

The discriminant $b^2 - 4ac = (-30)^2 - 4(9)(25) = 900 - 900 = 0$, so the equation has a repeated solution, a root of multiplicity 2.

101. $\quad 3x^2 + 5x - 8 = 0 \qquad a = 3, b = 5, \text{ and } c = -8$

The discriminant $b^2 - 4ac = (5)^2 - 4(3)(-8) = 25 + 96 = 121$ is positive, so there are 2 real solutions to the equation.

103. $\quad ax - b = c$

$\qquad\quad ax = b + c$

$$x = \frac{b+c}{a}$$

105.

$$\frac{x}{a} + \frac{x}{b} = c$$

$$\frac{bx}{ab} + \frac{ax}{ab} = \frac{abc}{ab}$$

$$bx + ax = abc$$

$$(b + a)x = abc$$

$$x = \frac{abc}{a+b}$$

107.

$$\frac{1}{x-a} + \frac{1}{x+a} = \frac{2}{x-1}$$

$$\frac{(x+a)(x-1)}{(x-a)(x+a)(x-1)} + \frac{(x-a)(x-1)}{(x-a)(x+a)(x-1)} = \frac{2(x-a)(x+a)}{(x-1)(x-a)(x+a)}$$

$$(x+a)(x-1) + (x-a)(x-1) = 2(x-a)(x+a)$$

$$x^2 + ax - x - a + x^2 - x - ax + a = 2x^2 + 2ax - 2ax - 2a^2$$

$$2x^2 - 2x = 2x^2 - 2a^2$$

$$-2x = -2a^2$$

$$x = a^2$$

109.

$$\frac{1}{R} = \frac{1}{R_1} + \frac{1}{R_2}$$

$$\frac{R_1 R_2}{RR_1 R_2} = \frac{RR_2}{RR_1 R_2} + \frac{RR_1}{RR_1 R_2}$$

$$R_1 R_2 = RR_2 + RR_1$$

$$R_1 R_2 = R\left(R_2 + R_1\right)$$

$$R = \frac{R_1 R_2}{R_1 + R_2}$$

111.

$$F = \frac{mv^2}{R}$$

$$FR = mv^2$$

$$R = \frac{mv^2}{F}$$

113.

$$S = \frac{a}{1-r}$$

$$S(1-r) = a$$

$$S - Sr = a$$

$$Sr = S - a$$

$$r = \frac{S-a}{S}$$

115. The roots of the quadratic function $ax^2 + bx + c = 0$ are

$$x_1 = \frac{-b - \sqrt{b^2 - 4ac}}{2a} \quad \text{and} \quad x_2 = \frac{-b + \sqrt{b^2 - 4ac}}{2a}$$

The sum $x_1 + x_2 = \dfrac{-b - \sqrt{b^2 - 4ac}}{2a} + \dfrac{-b + \sqrt{b^2 - 4ac}}{2a}$

$$= \frac{-b - \sqrt{b^2 - 4ac} + (-b) + \sqrt{b^2 - 4ac}}{2a}$$

$$= \frac{-2b}{2a} = -\frac{b}{a}$$

117. If $kx^2 + x + k = 0$ has a repeated real solution, then its discriminant is zero.
$a = k$, $b = 1$, and $c = k$

$$\text{discriminant:} \quad b^2 - 4ac = 1^2 - 4(k)(k) = 0$$
$$1 - 4k^2 = 0$$
$$4k^2 = 1$$
$$2k = \pm 1$$

So the equation has one repeated root if $k = \dfrac{1}{2}$ or $k = -\dfrac{1}{2}$.

119. The real solutions of the equation $ax^2 + bx + c = 0$ are

$$x_1 = \frac{-b - \sqrt{b^2 - 4ac}}{2a} \quad \text{and} \quad x_2 = \frac{-b + \sqrt{b^2 - 4ac}}{2a}$$

$$= -\frac{b + \sqrt{b^2 - 4ac}}{2a} \qquad\qquad = -\frac{b - \sqrt{b^2 - 4ac}}{2a}$$

The real solutions of the equation $ax^2 - bx + c = 0$ are

$$x_3 = \frac{b - \sqrt{b^2 - 4ac}}{2a} \quad \text{and} \quad x_4 = \frac{b + \sqrt{b^2 - 4ac}}{2a}$$

So $x_1 = -x_4$ and $x_2 = -x_3$.

121. (a) $x^2 = 9$ and $x = 3$ are not equivalent. The solution set of $x^2 = 9$ is $\{-3, 3\}$, but the solution set of $x = 3$ is $\{3\}$.

(b) $x = \sqrt{9}$ and $x = 3$ are equivalent since they both have the same solution set, $\{3\}$.

(c) $(x - 1)(x - 2) = (x - 1)^2$ and $x - 2 = x - 1$ are not equivalent. The solution of the first equation is $\{1\}$, but the second equation has no solution.

123. – 127. Answers will vary.

A.7 Intervals; Solving Inequalities

1. The graph represents $[0, 2]$ or $0 \le x \le 2$.

3. The graph represents $(-1, 2)$ or $-1 < x < 2$.

5. The graph represents $[0, 3)$ or $0 \le x < 3$.

7. $[0, 4]$

9. $[4, 6)$

11. $[4, \infty)$

13. $(-\infty, -4)$

15. $2 \le x \le 5$

17. $-3 < x < -2$

19. $4 \le x < \infty$

21. $-\infty < x < -3$

23. $3 < 5$

(a) $3 + 3 < 5 + 3$
$\quad 6 < 8$

(b) $3 - 5 < 5 - 5$
$\quad -2 < 0$

(c) $(3)(3) < (3)(5)$
$\quad 9 < 15$

(d) $(-2)(3) > (-2)(5)$
$\quad -6 > -10$

25. $4 > -3$

(a) $4 + 3 > -3 + 3$
$\quad 7 > 0$

(b) $4 - 5 > -3 - 5$
$\quad -1 > -8$

(c) $(3)(4) > (3)(-3)$
$\quad 12 > -9$

(d) $(-2)(4) < (-2)(-3)$
$\quad -8 < 6$

27. $2x + 1 < 2$

(a) $(2x + 1) + 3 < 2 + 3$
$\quad 2x + 4 < 5$

(b) $(2x + 1) - 5 < 2 - 5$
$\quad 2x - 4 < -3$

(c) $(3)(2x + 1) < (3)(2)$
$\quad 6x + 3 < 6$

(d) $(-2)(2x + 1) > (-2)(2)$
$\quad -4x - 2 > -4$

29. $<$
$$x < 5$$
$$x - 5 < 5 - 5$$
$$x < 0$$

31. $>$
$$x > -4$$
$$x + 4 > -4 + 4$$
$$x + 4 > 0$$

33. $\geq$
$$x \geq -4$$
$$3x \geq (3)(-4)$$
$$3x \geq -12$$

35. $<$
$$x > 6$$
$$-2x < (-2)(6)$$
$$-2x < -12$$

37. $\leq$
$$x \geq 5$$
$$-4x \leq (-4)(5)$$
$$-4x \leq -20$$

39. $>$
$$2x > 6$$
$$\frac{2x}{2} > \frac{6}{2}$$
$$x > 3$$

41. $\geq$
$$-\frac{1}{2}x \leq 3$$
$$(-2) \cdot \left(-\frac{1}{2}x\right) \geq (-2)(3)$$
$$x \geq -6$$

43.
$$x + 1 < 5$$
$$x + 1 - 1 < 5 - 1$$
$$x < 4$$

The solution set is $\{x \mid x < 4\}$ or the interval $(-\infty, 4)$.

45.
$$1 - 2x \leq 3$$
$$1 - 2x - 1 \leq 3 - 1$$
$$-2x \leq 2$$
$$\frac{-2x}{-2} \geq \frac{2}{-2}$$
$$x \geq -1$$

The solution set is $\{x \mid x \geq -1\}$ or the interval $[-1, \infty)$.

47.
$$3x - 7 > 2$$
$$3x - 7 + 7 > 2 + 7$$
$$3x > 9$$
$$\frac{3x}{3} > \frac{9}{3}$$
$$x > 3$$

The solution set is $\{x \mid x > 3\}$ or the interval $(3, \infty)$.

49.
$$3x - 1 \geq 3 + x$$
$$3x - 1 + 1 \geq 3 + x + 1$$
$$3x \geq 4 + x$$
$$3x - x \geq 4 + x - x$$
$$2x \geq 4$$
$$x \geq 2$$

The solution set is $\{x \mid x \geq 2\}$ or the interval $[2, \infty)$.

51.
$$-2(x + 3) < 8$$
$$\frac{-2(x + 3)}{-2} > \frac{8}{-2}$$
$$x + 3 > -4$$
$$x + 3 - 3 > -4 - 3$$
$$x > -7$$

The solution set is $\{x \mid x > -7\}$ or the interval $(-7, \infty)$.

53.
$$4 - 3(1 - x) \le 3$$
$$4 - 3 + 3x \le 3$$
$$1 + 3x \le 3$$
$$1 + 3x - 1 \le 3 - 1$$
$$3x \le 2$$
$$x \le \frac{2}{3}$$

The solution set is $\left\{ x \mid x \le \frac{2}{3} \right\}$ or the

interval $\left(-\infty, \frac{2}{3} \right]$.

55.
$$\frac{1}{2}(x - 4) > x + 8$$
$$2 \cdot \left[\frac{1}{2}(x - 4) \right] > 2 \cdot (x + 8)$$
$$x - 4 > 2x + 16$$
$$x - 4 - x > 2x + 16 - x$$
$$-4 > x + 16$$
$$-4 - 16 > x + 16 - 16$$
$$-20 > x \quad \text{or} \quad x < -20$$

The solution set is $\{x \mid x < -20\}$ or the
interval $(-\infty, -20)$.

57.
$$\frac{x}{2} \ge 1 - \frac{x}{4}$$
$$4 \cdot \left(\frac{x}{2} \right) \ge 4 \cdot \left(1 - \frac{x}{4} \right)$$
$$2x \ge 4 - x$$
$$2x + x \ge 4 - x + x$$
$$3x \ge 4$$
$$x \ge \frac{4}{3}$$

The solution set is $\left\{ x \mid x \ge \frac{4}{3} \right\}$ or the

interval $\left[\frac{4}{3}, \infty \right)$.

59. $0 \le 2x - 6 \le 4$ is equal to the two inequalities

$$0 \le 2x - 6 \qquad \text{and} \qquad 2x - 6 \le 4$$
$$0 + 6 \le 2x - 6 + 6 \qquad\qquad 2x - 6 + 6 \le 4 + 6$$
$$6 \le 2x \qquad\qquad\qquad 2x \le 10$$
$$3 \le x \qquad\qquad\qquad x \le 5$$

The solution set consists of all x for which $x \ge 3$ and $x \le 5$ which is written either as
$\{x \mid 3 \le x \le 5\}$ or as the interval $[3, 5]$.

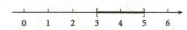

61. $-5 \le 4 - 3x \le 2$ is equal to the two inequalities

$$-5 \le 4 - 3x \qquad\qquad \text{and} \qquad\qquad 4 - 3x \le 2$$
$$-9 \le -3x \qquad\qquad\qquad\qquad\qquad\qquad -3x \le -2$$
$$3 \ge x \qquad\qquad\qquad\qquad\qquad\qquad\qquad x \ge \frac{2}{3}$$

The solution set consists of all x for which $x \ge \frac{2}{3}$ and $x \le 3$ which is written either as

$\left\{ x \mid \dfrac{2}{3} \le x \le 3 \right\}$ or as the interval $\left[\dfrac{2}{3}, 3 \right]$.

63. $-3 < \dfrac{2x - 1}{4} < 0$ is equal to the two inequalities

$$-3 < \frac{2x - 1}{4} \qquad\qquad \text{and} \qquad\qquad \frac{2x - 1}{4} < 0$$
$$-12 < 2x - 1 \qquad\qquad\qquad\qquad\qquad\qquad 2x - 1 < 0$$
$$-11 < 2x \qquad\qquad\qquad\qquad\qquad\qquad\qquad 2x < 1$$
$$-\frac{11}{2} < x \qquad\qquad\qquad\qquad\qquad\qquad\qquad x < \frac{1}{2}$$

The solution set consists of all x for which $x \ge -\dfrac{11}{2}$ and $x \le \dfrac{1}{2}$ which is written either as

$\left\{ x \mid -\dfrac{11}{2} < x < \dfrac{1}{2} \right\}$ or as the interval $\left(-\dfrac{11}{2}, \dfrac{1}{2} \right)$.

65. $1 < 1 - \dfrac{1}{2}x < 4$ is equal to the two inequalities

$$1 < 1 - \frac{1}{2}x \qquad\qquad \text{and} \qquad\qquad 1 - \frac{1}{2}x < 4$$
$$0 < -\frac{1}{2}x \qquad\qquad\qquad\qquad\qquad\qquad -\frac{1}{2}x < 3$$
$$0 > x \qquad\qquad\qquad\qquad\qquad\qquad\qquad x > -6$$

The solution set consists of all x for which $x > -6$ and $x < 0$ which is written either as $\{x \mid -6 < x < 0\}$ or as the interval $(-6, 0)$.

67. $(4x + 2)^{-1} = \dfrac{1}{4x + 2} < 0$ is satisfied if $4x + 2 < 0$. That is, when $4x < -2$ or $x < -\dfrac{1}{2}$.

The solution set is written either as $\left\{ x \mid x < -\dfrac{1}{2} \right\}$ or as the interval $\left(-\infty, -\dfrac{1}{2} \right)$.

69. $0 < \dfrac{2}{x} < \dfrac{3}{5}$ is equal to the two inequalities

$$0 < \dfrac{2}{x} \qquad\qquad \text{and} \qquad\qquad \dfrac{2}{x} < \dfrac{3}{5}$$

$$\qquad\qquad\qquad\qquad\qquad\qquad\qquad 10 < 3x$$

$0 < \dfrac{2}{x}$ when $x > 0$. $\qquad\qquad\qquad\qquad\qquad \dfrac{10}{3} < x$

The solution set consists of all x for which $x > 0$ and $x > \dfrac{10}{3}$ which is written either as

$\left\{ x \mid x > \dfrac{10}{3} \right\}$ or as the interval $\left(\dfrac{10}{3}, \infty \right)$.

71. First we solve the equation $(x - 3)(x + 1) = 0$ and use the solutions to separate the real number line.

$$(x - 3)(x + 1) = 0$$
$$x - 3 = 0 \quad \text{or} \quad x + 1 = 0$$
$$x = 3 \quad \text{or} \quad x = -1$$

We separate the number line into the intervals

$$(-\infty, -1) \qquad (-1, 3) \qquad (3, \infty)$$

In each interval we select a number and evaluate the expression $(x - 3)(x + 1)$ at that value. We choose $-5, 0, 5$.

For $x = -5$: $(-5 - 3)(-5 + 1) = 32$, a positive number.
For $x = 0$: $(0 - 3)(0 + 1) = -3$, a negative number.
For $x = 5$: $(5 - 3)(5 + 1) = 12$, a positive number.

Since $(x - 3)(x + 1) < 0$ for $-1 < x < 3$, we write the solution set either as $\{x \mid -1 < x < 3\}$ or as the interval $(-1, 3)$.

73. First we solve the equation $-x^2 + 9 = 0$ and use the solutions to separate the real number line.

$$-x^2 + 9 = 0$$
$$(-x + 3)(x + 3) = 0$$
$$-x + 3 = 0 \quad \text{or} \quad x + 3 = 0$$
$$x = 3 \quad \text{or} \quad x = -3$$

We separate the number line into the intervals

$$(-\infty, -3) \qquad (-3, 3) \qquad (3, \infty)$$

We select a number in each interval and evaluate the expression $(-x + 3)(x + 3)$ at that value. We choose $-5, 0, 5$.

For $x = -5$: $(-(-5) + 3)(-5 + 3) = -16$, a negative number.

For $x = 0$: $(0 + 3)(0 + 3) = 9$, a positive number.

For $x = 5$: $(-5 + 3)(5 + 3) = -16$, a negative number.

Since the expression $-x^2 + 9 > 0$ for $-3 < x < 3$, we write the solution set either as $\{x \mid -3 < x < 3\}$ or as the interval $(-3, 3)$.

75. First we solve the equation $x^2 + x = 12$ and use the solutions to separate the real number line.

$$x^2 + x = 12$$
$$x^2 + x - 12 = 0$$
$$(x - 3)(x + 4) = 0$$

$$x - 3 = 0 \quad \text{or} \quad x + 4 = 0$$
$$x = 3 \quad \text{or} \quad x = -4$$

We separate the number line into the intervals

$$(-\infty, -4) \qquad (-4, 3) \qquad (3, \infty)$$

We select a number in each interval and evaluate the expression $x^2 + x - 12$ at that value. We choose $-5, 0, 5$.

For $x = -5$: $(-5)^2 + (-5) - 12 = 8$, a positive number.

For $x = 0$: $0^2 + 0 - 12 = -12$, a negative number.

For $x = 5$: $5^2 + 5 - 12 = 18$, a positive number.

The expression $x^2 + x - 12 > 0$ for $x < -4$ or $x > 3$. We write the solution set either as $\{x \mid x < -4 \text{ or } x > 3\}$ or as all x in the interval $(-\infty, -4)$ or $(3, \infty)$.

77. First we solve the equation $x(x - 7) = -12$ and use the solutions to separate the real number line.

$$x(x - 7) = -12$$
$$x^2 - 7x = -12$$
$$x^2 - 7x + 12 = 0$$
$$(x - 3)(x - 4) = 0$$
$$x - 3 = 0 \quad \text{or} \quad x - 4 = 0$$
$$x = 3 \quad \text{or} \quad x = 4$$

We separate the number line into the intervals

$$(-\infty, 3) \qquad (3, 4) \qquad (4, \infty)$$

We select a number in each interval and evaluate the expression $x^2 - 7x + 12$ at that value. We choose 0, 3.5, and 5.

For $x = 0$: $0^2 - 7(0) + 12 = 12$, a positive number.

For $x = 3.5$: $(3.5)^2 - 7(3.5) + 12 = -0.25$, a negative number.

For $x = 5$: $5^2 - 7(5) + 12 = 2$, a positive number.

The expression $x^2 - 7x + 12 > 0$ for $x < 3$ or $x > 4$. We write the solution set either as $\{x \mid x < 3 \text{ or } x > 4\}$ or as all x in the interval $(-\infty, 3)$ or $(4, \infty)$.

79. First we solve the equation $4x^2 + 9 = 6x$ and use the solutions to separate the real number line.

$$4x^2 + 9 = 6x$$
$$4x^2 - 6x + 9 = 0$$

This equation has no real solutions. Its discriminant, $b^2 - 4ac = 36 - 144 = -108$, is negative. The value of $4x^2 - 6x + 9$ either is always positive or always negative. To see which is true, we test $x = 0$. Since $4(0)^2 - 6(0) + 9 = 9$ is positive, we conclude that expression is always positive, and the inequality

$$4x^2 + 9 < 6x \text{ has no solution.}$$

81. First we solve the equation $(x-1)(x^2 + x + 1) = 0$ and use the solutions to separate the real number line.

$$(x-1)(x^2 + x + 1) = 0$$
$$x - 1 = 0 \quad \text{or} \quad x^2 + x + 1 = 0$$
$$x = 1$$

$x = 1$ is the only solution, since the equation $x^2 + x + 1 = 0$ has a negative discriminant. We use $x = 1$ to separate the number line into two parts:

$$-\infty < x < 1 \qquad \text{and} \qquad 1 < x < \infty$$

In each part select a test number and evaluate the expression $(x-1)(x^2 + x + 1)$.

For $x = 0$: $(0-1)(0^2 + 0 + 1) = -1$, a negative number.

For $x = 2$: $(2-1)(2^2 + 2 + 1) = 7$, a positive number.

The expression $(x-1)(x^2 + x + 1) > 0$ for $x > 1$. The solution set is $\{x \mid x > 1\}$, or for all x in the interval $(1, \infty)$.

83. First we solve the equation $(x-1)(x-2)(x-3) = 0$ and use the solutions to separate the real number line.

$$(x-1)(x-2)(x-3) = 0$$
$$x-1=0 \quad \text{or} \quad x-2=0 \quad \text{or} \quad x-3=0$$
$$x=1 \quad \text{or} \quad x=2 \quad \text{or} \quad x=3$$

We separate the number line into the following 4 parts, choose a test number in each part, and evaluate the expression $(x-1)(x-2)(x-3)$ at each test number.

Parts: $\quad -\infty < x < 1 \qquad 1 < x < 2 \qquad 2 < x < 3 \qquad 3 < x < \infty$

For $x = 0$: $(0-1)(0-2)(0-3) = -6$, a negative number.

For $x = 1.5$ $(1.5-1)(1.5-2)(1.5-3) = 0.375$, a positive number.

For $x = 2.5$: $(2.5-1)(2.5-2)(2.5-3) = -0.375$, a negative number.

For $x = 5$: $(5-1)(5-2)(5-3) = 24$, a positive number.

The expression $(x-1)(x-2)(x-3) < 0$ for $x < 1$ or for $2 < x < 3$. The solution set is $\{x \mid x < 1 \text{ or } 2 < x < 3\}$, or for all x in the interval $(-\infty, 1)$ or $(2, 3)$.

85. First we solve the equation $-x^3 + 2x^2 + 8x = 0$ and use the solutions to separate the real number line.

$$-x^3 + 2x^2 + 8x = 0$$
$$-x(x^2 - 2x - 8x) = 0$$
$$-x(x+2)(x-4) = 0$$
$$-x=0 \quad \text{or} \quad x+2=0 \quad \text{or} \quad x-4=0$$
$$x=0 \quad \text{or} \quad x=-2 \quad \text{or} \quad x=4$$

We separate the number line into the following 4 parts, choose a test number in each part, and evaluate the expression $-x^3 + 2x^2 + 8x$ at each test number.

Parts: $\quad -\infty < x < -2 \qquad -2 < x < 0 \qquad 0 < x < 4 \qquad 4 < x < \infty$

For $x = -3$: $\quad -(-3)^3 + 2(-3)^2 + 8(-3) = 21$, a positive number.

For $x = -1$: $\quad -(-1)^3 + 2(-1)^2 + 8(-1) = -5$, a negative number.

For $x = 1$: $\quad -(1)^3 + 2(1)^2 + 8(1) = 9$, a positive number.

For $x = 10$: $\quad -(10)^3 + 2(10)^2 + 8(10) = -720$, a negative number.

The expression $-x^3 + 2x^2 + 8x < 0$ for $-2 < x < 0$ or for $x > 4$. The solution set is $\{x \mid -2 < x < 0 \text{ or } x > 4\}$, or for all x in the interval $(-2, 0)$ or $(4, \infty)$.

87. First we solve the equation $x^3 = x$ and use the solutions to separate the real number line.

$$x^3 = x$$
$$x^3 - x = 0$$
$$x(x^2 - 1) = 0$$
$$x(x-1)(x+1) = 0$$
$$x = 0 \quad \text{or} \quad x - 1 = 0 \quad \text{or} \quad x + 1 = 0$$
$$x = 0 \quad \text{or} \quad x = 1 \quad \text{or} \quad x = -1$$

We separate the number line into the following 4 parts, choose a test number in each part, and evaluate the expression $x^3 - x$ at each test number.

Parts: $-\infty < x < -1$ $-1 < x < 0$ $0 < x < 1$ $1 < x < \infty$

For $x = -2$: $(-2)^3 - (-2) = -6$, a negative number.

For $x = -0.5$: $(-0.5)^3 - (-0.5) = 0.375$, a positive number.

For $x = 0.5$: $(0.5)^3 - (0.5) = -0.375$, a negative number.

For $x = 2$: $(2)^3 - (2) = 6$, a positive number.

The expression $x^3 - x > 0$ for $-1 < x < 0$ or for $x > 1$. The solution set is $\{x \mid -1 < x < 0 \text{ or } x > 1\}$, or for all x in the interval $(-1, 0)$ or $(1, \infty)$.

89. First we solve the equation $x^3 = x^2$ and use the solutions to separate the real number line.

$$x^3 = x^2$$
$$x^3 - x^2 = 0$$
$$x^2(x-1) = 0$$
$$x^2 = 0 \quad \text{or} \quad x - 1 = 0$$
$$x = 0 \quad \text{or} \quad x = 1$$

We separate the number line into the following 3 parts, choose a test number in each part, and evaluate the expression $x^3 - x^2$ at each test number.

Parts: $-\infty < x < 0$ $0 < x < 1$ $1 < x < \infty$

For $x = -1$: $(-1)^3 - (-1)^2 = -2$, a negative number.

For $x = 0.5$: $0.5^3 - 0.5^2 = -0.125$, a negative number.

For $x = 2$: $2^3 - 2^2 = 4$, a positive number.

The expression $x^3 - x^2 > 0$ for $x > 1$. The solution set is $\{x \,|\, x > 1\}$, or for all x in the interval $(1, \infty)$.

91. $\dfrac{x+1}{1-x}$ is not defined when $1 - x = 0$ or when $x = 1$.

$\dfrac{x+1}{1-x} = 0$ when $x + 1 = 0$ or when $x = -1$.

We use these two numbers to separate the number line into three parts. We then choose a test number in each part and evaluate the expression $\dfrac{x+1}{1-x}$ at the test number.

Parts: $-\infty < x < -1$ $-1 < x < 1$ $1 < x < \infty$

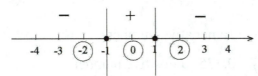

For $x = -2$: $\dfrac{(-2)+1}{1-(-2)} = -\dfrac{1}{3}$, which is a negative number.

For $x = 0$: $\dfrac{0+1}{1-0} = 1$, which is a positive number.

For $x = 2$: $\dfrac{2+1}{1-2} = -3$, which is a negative number.

The expression $\dfrac{x+1}{1-x} < 0$ for $x < -1$ or for $x > 1$. The solution set is $\{x \,|\, x < -1 \text{ or } x > 1\}$, or for all x in the interval $(-\infty, -1)$ or $(1, \infty)$.

93. $\dfrac{(x-1)(x+1)}{x}$ is not defined for $x = 0$. $\dfrac{(x-1)(x+1)}{x} = 0$ for $x = 1$ or $x = -1$.

We use these three numbers to separate the number line into four parts.

Parts: $-\infty < x < -1$ $-1 < x < 0$ $0 < x < 1$ $1 < x < \infty$

For $x = -2$: $\dfrac{[(-2)-1][(-2)+1]}{-2} = -1.5$, which is a negative number.

For $x = -\dfrac{1}{2}$: $\dfrac{\big[(-0.5)-1\big]\big[(-0.5)+1\big]}{-0.5} = 1.5$, which is a positive number.

For $x = \dfrac{1}{2}$: $\dfrac{(0.5-1)(0.5+1)}{0.5} = -1.5$, which is a negative number.

For $x = 2$: $\dfrac{(2-1)(2+1)}{2} = 1.5$, which is a positive number.

The expression $\dfrac{(x-1)(x+1)}{x} < 0$ for $x < -1$ or for $0 < x < 1$. The solution set is

$\{x \mid x < -1 \text{ or } 0 < x < 1\}$, or for all x in the interval $(-\infty, -1)$ or $(0, 1)$.

$$\xleftarrow{\hspace{2cm}} \quad \underset{-2}{|} \quad \overset{)}{\underset{-1}{|}} \quad \underset{0}{|} \quad \overset{(}{\underset{1}{|}} \quad \overset{)}{\underset{2}{|}} \quad \underset{3}{|} \quad \underset{4}{|} \xrightarrow{\hspace{1cm}}$$

95. $\dfrac{x-2}{x^2-1} = \dfrac{x-2}{(x-1)(x+1)}$ is not defined for $x = 1$ or $x = -1$. $\dfrac{x-2}{x^2-1} = 0$ for $x = 2$.

We use these three numbers to separate the real number line into four parts.

Parts: $-\infty < x < -1 \qquad -1 < x < 1 \qquad 1 < x < 2 \qquad 2 < x < \infty$

$$-\quad\big|\quad +\quad\big|\quad -\quad\big|\quad +$$
$$\xrightarrow{\hspace{5cm}}$$
$$\underset{\boxed{-2}}{} \quad \underset{-1}{} \quad \underset{\boxed{0}}{} \quad \underset{1\boxed{\frac{3}{2}}\,2}{} \quad \underset{\boxed{3}}{}$$

For $x = -2$: $\dfrac{(-2)-2}{(-2)^2-1} = -\dfrac{4}{3}$, which is a negative number.

For $x = 0$: $\dfrac{0-2}{0^2-1} = 2$, which is a positive number.

For $x = \dfrac{3}{2} = 1.5$: $\dfrac{1.5-2}{1.5^2-1} = -0.4$, which is a negative number.

For $x = 3$: $\dfrac{3-2}{3^2-1} = 0.125$, which is a positive number.

The expression $\dfrac{x-2}{x^2-1} \geq 0$ for $-1 < x < 1$ or for $2 \leq x < \infty$. The solution set is

$\{x \mid -1 < x < 1 \text{ or } x \geq 2\}$, or for all x in the interval $(-1, 1)$ or $[2, \infty)$.

$$\xleftarrow{\hspace{2cm}} \quad \underset{-2}{|} \quad \overset{(}{\underset{-1}{|}} \quad \underset{0}{|} \quad \overset{)}{\underset{1}{|}} \quad \overset{[}{\underset{2}{|}} \quad \underset{3}{|} \quad \underset{4}{|} \xrightarrow{\hspace{1cm}}$$

97. First we rewrite $\dfrac{x+4}{x-2} \leq 1$ so it has a 0 on the right.

$$\dfrac{x+4}{x-2} - 1 \leq 0$$

or $\qquad \dfrac{x+4}{x-2} - \dfrac{x-2}{x-2} = \dfrac{6}{x-2} \leq 0$

The expression $\dfrac{6}{x-2}$ is not defined for $x = 2$; it is never zero.

We use $x = 2$ to separate the number line into two parts,
$$-\infty < x < 2 \quad \text{and} \quad 2 < x < \infty$$

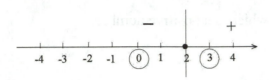

For $x = 0$: $\dfrac{0+4}{0-2} - 1 = -3$, which is a negative number.

For $x = 3$: $\dfrac{3+4}{3-2} - 1 = 6$, which is a positive number.

The expression $\dfrac{x+4}{x-2} - 1 \le 0$ for $x < 2$. The solution set is $\{x \mid x < 2\}$, or for all x in the interval $(-\infty, 2)$.

99. First we rewrite $\dfrac{2x+5}{x+1} > \dfrac{x+1}{x-1}$ so it has a 0 on the right.
$$\dfrac{2x+5}{x+1} - \dfrac{x+1}{x-1} > 0$$

Then we write the expression with a single denominator.
$$\dfrac{(2x+5)(x-1)}{(x+1)(x-1)} - \dfrac{(x+1)(x+1)}{(x-1)(x+1)} = \dfrac{(2x+5)(x-1) - (x+1)(x+1)}{(x+1)(x-1)}$$

$$= \dfrac{2x^2 + 3x - 5 - (x^2 + 2x + 1)}{(x+1)(x-1)} = \dfrac{x^2 + x - 6}{(x+1)(x-1)} = \dfrac{(x+3)(x-2)}{(x+1)(x-1)} > 0$$

The expression $\dfrac{(x+3)(x-2)}{(x+1)(x-1)}$ is not defined for $x = 1$ or $x = -1$. $\dfrac{(x+3)(x-2)}{(x+1)(x-1)} = 0$

for $x = 2$ or $x = -3$. We use these numbers to separate the real number line into 5 parts
$$-\infty < x < -3 \qquad -3 < x < -1 \qquad -1 < x < 1 \qquad 1 < x < 2 \qquad 2 < x < \infty$$

For $x = -4$: $\dfrac{[(-4)+3][(-4)-2]}{[(-4)+1][(-4)-1]} = \dfrac{2}{5}$

For $x = -2$: $\dfrac{[(-2)+3][(-2)-2]}{[(-2)+1][(-2)-1]} = -\dfrac{4}{3}$

For $x = 0$: $\dfrac{(0+3)(0-2)}{(0+1)(0-1)} = 6$

For $x = \dfrac{3}{2} = 1.5$: $\dfrac{(1.5+3)(1.5-2)}{(1.5+1)(1.5-1)} = -1.8$

For $x = 3$: $\dfrac{(3+3)(3-2)}{(3+1)(3-1)} = 0.75$

The expression $\dfrac{2x+5}{x+1} - \dfrac{x+1}{x-1} > 0$ for $x < -3$ or $-1 < x < 1$ or $2 < x < \infty$. The solution set is $\{x \mid x < -3 \text{ or } -1 < x < 1 \text{ or } x > 2\}$ or for all x in the interval $(-\infty, -3)$ or $(-1, 1)$ or $(2, \infty)$.

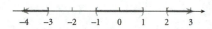

101. Let x represent the score on the last test. To get a B you need $80 < \text{average} < 90$.

The average is $\dfrac{68 + 82 + 87 + 89 + x}{5} = \dfrac{326 + x}{5}$.

We will solve the inequality

$$80 \leq \dfrac{326 + x}{5} < 90$$

which is equivalent to the two inequalities

$$80 \leq \dfrac{326 + x}{5} \quad \text{and} \quad \dfrac{326 + x}{5} < 90$$

Solving each inequality we find

$$80 \leq \dfrac{326 + x}{5} \qquad\qquad \dfrac{326 + x}{5} < 90$$
$$400 \leq 326 + x \qquad\qquad 326 + x < 450$$
$$74 \leq x \qquad\qquad\qquad\quad x < 124$$

The solution set to the combined inequality is $\{x \mid 74 \leq x < 124\}$, but since 100 is usually the highest score possible, you need to score between a 74 and 100 to get a B.

103. If we let x represent the selling price of the property, then we can write an equation relating the commission C to the selling price.

$$C = 45{,}000 + 0.25(x - 900{,}000)$$
$$= 45{,}000 + 0.25x - 225{,}000$$
$$= 0.25x - 180{,}000$$

We are told that $900{,}000 \leq x \leq 1{,}100{,}000$, so the commission varies between

$$0.25(900{,}000) - 180{,}000 \leq C \leq 0.25(1{,}100{,}000) - 180{,}000$$
$$225{,}000 - 180{,}000 \leq C \leq 275{,}000 - 180{,}000$$
$$45{,}000 \leq C \leq 95{,}000$$

The commission on the sale varies between \$45,000 and \$95,000 inclusive.

If the apartment complex sells for $900,000, then the $45,000 commission represents

$$\frac{45,000}{900,000} = 0.05 = 5\% \text{ of the sale.}$$

If the apartment complex sells for $1,100,000, then the $95,000 commission represents

$$\frac{95,000}{1,100,000} = 0.086 = 8.6\% \text{ of the sale.}$$

The sale's commission varies between 5% and 8.6% of the sale.

105. If we let x represent the weekly wages, then an equation relating the withholding W to x is

$$\begin{aligned} W &= 74.35 + 0.25(x - 592) \\ &= 74.35 + 0.25x - 148 \\ &= 0.25x - 73.65 \end{aligned}$$

If wages are between $600 and $800 inclusive, then $600 \leq x \leq 800$, and

$600 \leq x$	and	$x \leq 800$
$0.25(600) - 73.65 \leq 0.25x - 73.65$		$0.25x - 73.65 \leq 0.25(800) - 73.65$
$150 - 73.65 \leq W$		$W \leq 200 - 73.65$
$76.35 \leq W$		$W \leq 126.35$

The tax withholdings are between $76.35 and $126.35 inclusive.

107. If x represents the monthly electric usage, then an equation relating the monthly cost C to x is

$$C = 0.08275x + 7.58$$

If the monthly bills ranged between $63.47 and $214.53, then

$63.47 \leq 0.08275x + 7.58$	and	$0.08275x + 7.58 \leq 214.53$
$55.89 \leq 0.08275x$		$0.08275x \leq 206.95$
$675.41 \leq x$		$x \leq 2500.91$

The monthly electricity usage ranged between 675.41 kilowatt-hours and 2500.91 kilowatt-hours.

109. The price of a car is determined by the dealer's cost x plus the markup. We are told the price is $8800, and that the markup ranges between 12% of the dealer's cost and 18% of the dealer's cost.
If the markup is 12%, we have

$$\begin{aligned} 8800 &= x + 0.12x \\ 8800 &= 1.12x \\ x &= 7857.14 \end{aligned}$$

If the markup is 18%, we have

$$\begin{aligned} 8800 &= x + 0.18x \\ 8800 &= 1.18x \\ x &= 7457.63 \end{aligned}$$

So the dealer's cost varies between $7457.63 and $7857.14 inclusive.

111. – 113. Answers will vary.

A.8 *n*th Roots; Rational Exponents

1. $\sqrt[3]{27} = \sqrt[3]{3^3} = 3$

3. $\sqrt[3]{-8} = \sqrt[3]{(-2)^3} = -2$

5. $\sqrt{8} = \sqrt{4} \cdot \sqrt{2} = 2\sqrt{2}$

7.
$$\sqrt[3]{-8x^4} = \sqrt[3]{(-2x)^3 \, x}$$
$$= \sqrt[3]{(-2x)^3} \cdot \sqrt[3]{x}$$
$$= -2x \cdot \sqrt[3]{x}$$

9.
$$\sqrt[4]{x^{12}y^8} = \sqrt[4]{(x^3)^4 \cdot (y^2)^4}$$
$$= \sqrt[4]{(x^3)^4} \cdot \sqrt[4]{(y^2)^4}$$
$$= x^3 y^2$$

11.
$$\sqrt[4]{\frac{x^9 y^7}{xy^3}} = \sqrt[4]{x^{9-1} \, y^{7-3}}$$
$$= \sqrt[4]{x^8 \, y^4}$$
$$= \sqrt[4]{(x^2)^4} \cdot \sqrt[4]{y^4}$$
$$= x^2 y$$

13. $\sqrt{36x} = \sqrt{6^2 \cdot x} = 6\sqrt{x}$

15.
$$\sqrt{3x^2}\,\sqrt{12x} = \sqrt{36x^3}$$
$$= \sqrt{6^2 \cdot x^2 \cdot x}$$
$$= \sqrt{(6x)^2} \cdot \sqrt{x}$$
$$= 6x \cdot \sqrt{x}$$

17.
$$\left(\sqrt{5}\,\sqrt[3]{9}\right)^2 = \sqrt{5^2} \cdot \sqrt[3]{81}$$
$$= 5 \cdot \sqrt[3]{3^3 \cdot 3}$$
$$= 5 \cdot 3\sqrt[3]{3}$$
$$= 15\sqrt[3]{3}$$

19.
$$\left(3\sqrt{6}\right)\left(2\sqrt{2}\right) = (3 \cdot 2)\left(\sqrt{6} \cdot \sqrt{2}\right)$$
$$= 6\sqrt{12}$$
$$= 6\sqrt{4 \cdot 3}$$
$$= 6 \cdot 2\sqrt{3}$$
$$= 12\sqrt{3}$$

21.
$$\left(\sqrt{3}+3\right)\left(\sqrt{3}-1\right) = \left(\sqrt{3}\right)^2 - \sqrt{3} + 3\sqrt{3} - 3$$
$$= 3 + 2\sqrt{3} - 3$$
$$= 2\sqrt{3}$$

23.
$$\left(\sqrt{x}-1\right)^2 = \left(\sqrt{x}-1\right)\left(\sqrt{x}-1\right) = \left(\sqrt{x}\right)^2 - \sqrt{x} - \sqrt{x} + 1$$
$$= x - 2\sqrt{x} + 1$$

25.
$$3\sqrt{2} - 4\sqrt{8} = 3\sqrt{2} - 4\sqrt{4 \cdot 2}$$
$$= 3\sqrt{2} - 4 \cdot 2\sqrt{2}$$
$$= 3\sqrt{2} - 8\sqrt{2}$$
$$= -5\sqrt{2}$$

27.
$$\sqrt[3]{16x^4} - \sqrt[3]{2x} = \sqrt[3]{(2x)^4} - \sqrt[3]{2x}$$
$$= \sqrt[3]{(2x)^3 \cdot 2x} - \sqrt[3]{2x}$$
$$= \sqrt[3]{(2x)^3} \cdot \sqrt[3]{2x} - \sqrt[3]{2x}$$
$$= 2x\sqrt[3]{2x} - \sqrt[3]{2x}$$
$$= (2x - 1)\sqrt[3]{2x}$$

29.
$$\frac{1}{\sqrt{2}} \cdot \frac{\sqrt{2}}{\sqrt{2}} = \frac{\sqrt{2}}{\sqrt{2^2}} = \frac{\sqrt{2}}{2}$$

31.
$$\frac{-\sqrt{3}}{\sqrt{5}} \cdot \frac{\sqrt{5}}{\sqrt{5}} = \frac{-\sqrt{15}}{\sqrt{5^2}} = -\frac{\sqrt{15}}{5}$$

33.
$$\frac{\sqrt{3}}{5 - \sqrt{2}} \cdot \frac{5 + \sqrt{2}}{5 + \sqrt{2}} = \frac{5\sqrt{3} + \sqrt{6}}{25 - 2} = \frac{5\sqrt{3} + \sqrt{6}}{23}$$

35.
$$\frac{2 - \sqrt{5}}{2 + 3\sqrt{5}} \cdot \frac{2 - 3\sqrt{5}}{2 - 3\sqrt{5}} = \frac{4 - 6\sqrt{5} - 2\sqrt{5} + 3\left(\sqrt{5}\right)^2}{4 - 9\left(\sqrt{5}\right)^2}$$
$$= \frac{4 - 8\sqrt{5} + 15}{4 - 45}$$
$$= -\frac{19 - 8\sqrt{5}}{41} = \frac{8\sqrt{5} - 19}{41}$$

37.
$$\frac{5}{\sqrt[3]{2}} \cdot \frac{\sqrt[3]{2^2}}{\sqrt[3]{2^2}} = \frac{5\sqrt[3]{4}}{\sqrt[3]{2^3}} = \frac{5\sqrt[3]{4}}{2}$$

39.
$$\frac{\sqrt{x+h} - \sqrt{x}}{\sqrt{x+h} + \sqrt{x}} \cdot \frac{\sqrt{x+h} - \sqrt{x}}{\sqrt{x+h} - \sqrt{x}} = \frac{\left(\sqrt{x+h} - \sqrt{x}\right)^2}{\left(\sqrt{x+h}\right)^2 - \left(\sqrt{x}\right)^2} = \frac{\left(\sqrt{x+h} - \sqrt{x}\right)^2}{x + h - x}$$
$$= \frac{\left(\sqrt{x+h} - \sqrt{x}\right)^2}{h} = \frac{(x+h) - 2\sqrt{(x+h)x} + x}{h}$$
$$= \frac{2x + h - 2\sqrt{(x^2 + xh)}}{h}$$

41.
$$\sqrt[3]{2t-1} = 2$$
$$\left(\sqrt[3]{2t-1}\right)^3 = (2)^3$$
$$2t - 1 = 8$$
$$2t = 9$$
$$t = \frac{9}{2}$$

43.
$$\sqrt{15-2x} = x$$
$$\left(\sqrt{15-2x}\right)^2 = x^2$$
$$15 - 2x = x^2$$
$$x^2 + 2x - 15 = 0$$
$$(x+5)(x-3) = 0$$

$$x + 5 = 0 \quad \text{or} \quad x - 3 = 0$$
$$x = -5 \qquad\qquad x = 3$$

The solution set is {3}.

Check: $x = -5$
$$\sqrt{15-2(-5)} = \sqrt{15+10} = \sqrt{25} = 5 \neq x$$
So $x = -5$ is not a solution.

$x = 3$
$$\sqrt{15-2(3)} = \sqrt{15-6} = \sqrt{9} = 3 = x$$
So $x = 3$ is a solution.

45.
$$8^{2/3} = \left(\sqrt[3]{8}\right)^2 = 2^2 = 4$$

47.
$$(-27)^{1/3} = \sqrt[3]{-27} = -3$$

49.
$$16^{3/2} = \left(\sqrt{16}\right)^3 = 4^3 = 64$$

51.
$$9^{-3/2} = \frac{1}{9^{3/2}} = \frac{1}{\left(\sqrt{9}\right)^3} = \frac{1}{3^3} = \frac{1}{27}$$

53.
$$\left(\frac{9}{8}\right)^{3/2} = \left(\sqrt{\frac{9}{8}}\right)^3 = \left(\frac{\sqrt{9}}{\sqrt{8}}\right)^3 = \frac{3^3}{\left(2\sqrt{2}\right)^3} = \frac{27}{8 \cdot \left(\sqrt{2}\right)^2 \cdot \sqrt{2}}$$
$$= \frac{27}{16\sqrt{2}} \cdot \frac{\sqrt{2}}{\sqrt{2}} = \frac{27\sqrt{2}}{16 \cdot 2} = \frac{27\sqrt{2}}{32}$$

55.
$$\left(\frac{8}{9}\right)^{-3/2} = \left(\frac{9}{8}\right)^{3/2} = \frac{27\sqrt{2}}{32}$$
(See problem 53.)

57.
$$x^{3/4} x^{1/3} x^{-1/2} = x^{9/12} x^{4/12} x^{-6/12}$$
$$= x^{9/12 + 4/12 - 6/12}$$
$$= x^{7/12}$$

59.
$$\left(x^3 y^6\right)^{1/3} = \left(x^3\right)^{1/3} \left(y^6\right)^{1/3} = x y^2$$

61.
$$\left(x^2 y\right)^{1/3} \left(x y^2\right)^{2/3} = x^{2/3} y^{1/3} x^{2/3} y^{4/3} = \left(x^{2/3} x^{2/3}\right)\left(y^{1/3} y^{4/3}\right) = x^{4/3} y^{5/3}$$

63.
$$\left(16 x^2 y^{-1/3}\right)^{3/4} = 16^{3/4} x^{6/4} y^{-1/4} = 2^3 x^{3/2} \cdot \frac{1}{y^{1/4}} = \frac{8 x^{3/2}}{y^{1/4}}$$

65.

$$\frac{x}{(1+x)^{1/2}} + 2(1+x)^{1/2} = \frac{x}{(1+x)^{1/2}} + \frac{2(1+x)^{1/2}(1+x)^{1/2}}{(1+x)^{1/2}}$$

$$= \frac{x + 2(1+x)}{(1+x)^{1/2}}$$

$$= \frac{3x + 2}{(1+x)^{1/2}}$$

67.

$$2x(x^2+1)^{1/2} + x^2 \cdot \frac{1}{\cancel{2}}(x^2+1)^{-1/2} \cdot \cancel{2}x = 2x(x^2+1)^{1/2} + x^2 \cdot \frac{1}{(x^2+1)^{1/2}} \cdot x$$

$$= 2x(x^2+1)^{1/2} + \frac{x^3}{(x^2+1)^{1/2}}$$

$$= \frac{2x(x^2+1)^{1/2} \cdot (x^2+1)^{1/2}}{(x^2+1)^{1/2}} + \frac{x^3}{(x^2+1)^{1/2}}$$

$$= \frac{2x(x^2+1)}{(x^2+1)^{1/2}} + \frac{x^3}{(x^2+1)^{1/2}}$$

$$= \frac{2x^3 + 2x + x^3}{(x^2+1)^{1/2}} = \frac{3x^3 + 2x}{(x^2+1)^{1/2}}$$

69.

$$\sqrt{4x+3} \cdot \frac{1}{2\sqrt{x-5}} + \sqrt{x-5} \cdot \frac{1}{5\sqrt{4x+3}} = \frac{\sqrt{4x+3}}{2\sqrt{x-5}} + \frac{\sqrt{x-5}}{5\sqrt{4x+3}}$$

$$= \frac{\left(\sqrt{4x+3}\right)\left(5\sqrt{4x+3}\right)}{\left(2\sqrt{x-5}\right)\left(5\sqrt{4x+3}\right)} + \frac{\left(2\sqrt{x-5}\right)\left(\sqrt{x-5}\right)}{\left(2\sqrt{x-5}\right)\left(5\sqrt{4x+3}\right)}$$

$$= \frac{\left(\sqrt{4x+3}\right)\left(5\sqrt{4x+3}\right) + \left(2\sqrt{x-5}\right)\left(\sqrt{x-5}\right)}{\left(2\sqrt{x-5}\right)\left(5\sqrt{4x+3}\right)}$$

$$= \frac{5(4x+3) + 2(x-5)}{10\sqrt{(x-5)(4x+3)}}$$

$$= \frac{20x+15+2x-10}{10\sqrt{(x-5)(4x+3)}}$$

$$= \frac{22x+5}{10\sqrt{(x-5)(4x+3)}} = \frac{22x+5}{10\sqrt{4x^2-17x-15}}$$

71.

$$\frac{\sqrt{1+x} - x \cdot \dfrac{1}{2\sqrt{1+x}}}{1+x} = \frac{\dfrac{2\left(\sqrt{1+x}\right)\left(\sqrt{1+x}\right) - x}{2\sqrt{1+x}}}{1+x} = \frac{\dfrac{2(1+x) - x}{2\sqrt{1+x}}}{1+x}$$

$$= \frac{2+x}{2(1+x)\sqrt{1+x}} = \frac{2+x}{2(1+x)^{3/2}}$$

73.

$$\frac{(x+4)^{1/2} - 2x(x+4)^{-1/2}}{x+4} = \frac{(x+4)^{1/2} - \dfrac{2x}{(x+4)^{1/2}}}{x+4} = \frac{1}{x+4}\left[(x+4)^{1/2} - \frac{2x}{(x+4)^{1/2}}\right]$$

$$= \frac{1}{x+4}\left[\frac{(x+4)^{1/2} \cdot (x+4)^{1/2}}{(x+4)^{1/2}} - \frac{2x}{(x+4)^{1/2}}\right]$$

$$= \frac{1}{x+4}\left[\frac{x+4}{(x+4)^{1/2}} - \frac{2x}{(x+4)^{1/2}}\right]$$

$$= \frac{1}{x+4}\left[\frac{x+4-2x}{(x+4)^{1/2}}\right] = \frac{4-x}{(x+4)^{3/2}}$$

75.

$$\frac{\dfrac{x^2}{(x^2-1)^{1/2}} - (x^2-1)^{1/2}}{x^2} = \frac{1}{x^2}\left[\frac{x^2}{(x^2-1)^{1/2}} - (x^2-1)^{1/2}\right]$$

$$= \frac{1}{x^2}\left[\frac{x^2}{(x^2-1)^{1/2}} - \frac{(x^2-1)^{1/2} \cdot (x^2-1)^{1/2}}{(x^2-1)^{1/2}}\right]$$

$$= \frac{1}{x^2}\left[\frac{x^2}{(x^2-1)^{1/2}} - \frac{x^2-1}{(x^2-1)^{1/2}}\right]$$

$$= \frac{1}{x^2}\left[\frac{x^2 - x^2 + 1}{(x^2-1)^{1/2}}\right] = \frac{1}{x^2(x^2-1)^{1/2}}$$

77.

$$\frac{\dfrac{1+x^2}{2\sqrt{x}} - 2x\sqrt{x}}{(1+x^2)^2} = \frac{1}{(1+x^2)^2} \cdot \left[\frac{1+x^2}{2\sqrt{x}} - 2x\sqrt{x}\right] = \frac{1}{(1+x^2)^2} \cdot \left[\frac{1+x^2}{2\sqrt{x}} - \frac{2x\sqrt{x} \cdot 2\sqrt{x}}{2\sqrt{x}}\right]$$

$$= \frac{1}{(1+x^2)^2} \cdot \left[\frac{1+x^2}{2\sqrt{x}} - \frac{4x \cdot x}{2\sqrt{x}}\right]$$

$$= \frac{1}{(1+x^2)^2} \cdot \left[\frac{1+x^2-4x^2}{2\sqrt{x}}\right]$$

$$= \frac{1}{(1+x^2)^2} \cdot \left[\frac{1-3x^2}{2\sqrt{x}}\right] = \frac{1-3x^2}{2\sqrt{x}(1+x^2)^2}$$

79.
$$(x+1)^{3/2} + x \cdot \frac{3}{2}(x+1)^{1/2} = (x+1)^{1/2}\left[x+1+\frac{3}{2}x\right]$$
$$= (x+1)^{1/2}\left(\frac{5}{2}x+1\right) = \frac{1}{2}(x+1)^{1/2}(5x+2)$$

81.
$$6x^{1/2}(x^2+x) - 8x^{3/2} - 8x^{1/2} = 2x^{1/2}\left[3(x^2+x) - 4x - 4\right]$$
$$= 2x^{1/2}(3x^2 - x - 4)$$
$$= 2x^{1/2}(3x-4)(x+1)$$

83.
$$3(x^2+4)^{4/3} + x \cdot 4(x^2+4)^{1/3} \cdot 2x = 3(x^2+4)^{4/3} + 8x^2(x^2+4)^{1/3}$$
$$= (x^2+4)^{1/3}\left[3(x^2+4) + 8x^2\right]$$
$$= (x^2+4)^{1/3}\left[3x^2 + 12 + 8x^2\right]$$
$$= (x^2+4)^{1/3}(11x^2 + 12)$$

85.
$$4(3x+5)^{1/3}(2x+3)^{3/2} + 3(3x+5)^{4/3}(2x+3)^{1/2}$$
$$= (3x+5)^{1/3}(2x+3)^{1/2}\left[4(2x+3) + 3(3x+5)\right]$$
$$= (3x+5)^{1/3}(2x+3)^{1/2}\left[8x+12+9x+15\right]$$
$$= (3x+5)^{1/3}(2x+3)^{1/2}(17x+27)$$

87.
$$3x^{-1/2} + \frac{3}{2}x^{1/2} = \frac{3}{x^{1/2}} + \frac{3}{2}x^{1/2}$$
$$= \frac{3}{x^{1/2}} + \frac{3}{2}\frac{x^{1/2} \cdot x^{1/2}}{x^{1/2}}$$
$$= \frac{2 \cdot 3}{2x^{1/2}} + \frac{3x}{2x^{1/2}}$$
$$= \frac{3(2+x)}{2x^{1/2}}$$

89.
$$x\left(\frac{1}{2}\right)(8-x^2)^{-1/2}(-2x) + (8-x^2)^{1/2} = -x^2(8-x^2)^{-1/2} + (8-x^2)^{1/2}$$
$$= \frac{-x^2}{(8-x^2)^{1/2}} + (8-x^2)^{1/2}$$
$$= \frac{-x^2}{(8-x^2)^{1/2}} + \frac{(8-x^2)^{1/2}(8-x^2)^{1/2}}{(8-x^2)^{1/2}}$$

$$= \frac{-x^2}{\left(8-x^2\right)^{1/2}} + \frac{8-x^2}{\left(8-x^2\right)^{1/2}}$$

$$= \frac{-x^2+8-x^2}{\left(8-x^2\right)^{1/2}}$$

$$= \frac{-2x^2+8}{\left(8-x^2\right)^{1/2}} = \frac{2(2-x)(2+x)}{\left(8-x^2\right)^{1/2}}$$

A.9 Geometry Review

1. $a = 5,\ b = 12$
$c^2 = a^2 + b^2$
$c^2 = 5^2 + 12^2 = 25 + 144 = 169$
$c = \sqrt{169} = 13$

3. $a = 10,\ b = 24$
$c^2 = a^2 + b^2$
$c^2 = 10^2 + 24^2 = 100 + 576 = 676$
$c = \sqrt{676} = 26$

5. $a = 7,\ b = 24$
$c^2 = a^2 + b^2$
$c^2 = 7^2 + 24^2 = 49 + 576 = 625$
$c = \sqrt{625} = 25$

7. Square the sides of the triangle. $3^2 = 9$ $4^2 = 16$ $5^2 = 25$
Since $9 + 16 = 25$, the triangle is a right triangle. The hypotenuse is 5 (the longest side).

9. Square the sides of the triangle. $4^2 = 16$ $5^2 = 25$ $6^2 = 36$
The sum $16 + 25 = 41 \neq 36$, so the triangle is not a right triangle.

11. Square the sides of the triangle. $7^2 = 49$ $24^2 = 576$ $25^2 = 625$
Since $49 + 576 = 625$, the triangle is a right triangle. The hypotenuse is 25 (the longest side).

13. Square the sides of the triangle. $6^2 = 36$ $4^2 = 16$ $3^2 = 9$
The sum $16 + 9 = 25 \neq 36$, the triangle is not a right triangle.

15. The area A of a rectangle is $A = lw = 4 \cdot 2 = 8$ square inches.

17. The area A of a triangle is $A = \dfrac{1}{2}\,bh = \dfrac{1}{2} \cdot 2 \cdot 4 = 4$ square inches.

19. The area A of a circle is $A = \pi r^2 = \pi \cdot 5^2 = 25\pi$ square meters.
The circumference of a circle is $C = 2\pi r = 2\pi \cdot 5 = 10\pi$ meters.

21. The volume V of a rectangular box is $V = lwh = 8 \cdot 4 \cdot 7 = 224$ cubic feet.

The surface area SA of a rectangular box is
$$SA = 2lw + 2lh + 2wh = 2 \cdot 8 \cdot 4 + 2 \cdot 8 \cdot 7 + 2 \cdot 4 \cdot 7$$
$$= 64 + 112 + 56 = 232 \text{ square feet.}$$

23.

The volume V of a sphere is $V = \dfrac{4}{3}\pi r^3 = \dfrac{4}{3}\pi \cdot 4^3 = \dfrac{4^4}{3}\pi = \dfrac{256}{3}\pi$ centimeters cubed.

The surface area S of a sphere is $4\pi r^2 = 4\pi 4^2 = 4^3\pi = 64\pi$ centimeters squared.

25. The volume V of a right circular cylinder is $V = \pi r^2 h = \pi \cdot 9^2 \cdot 8 = 648\pi$ cubic inches.

The surface area S of a right circular cylinder is $S = 2\pi r^2 + 2\pi rh$.
$$S = 2\pi \cdot 9^2 + 2\pi \cdot 9 \cdot 8 = 162\pi + 144\pi = 306\pi \text{ square inches}$$

27. The shaded region is a circle of radius $r = 1$.

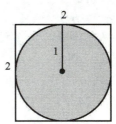

$$A = \pi r^2$$
$$A = \pi \cdot 1^2$$
$$A = \pi \text{ square units}$$

29. The shaded region is the area of the circle. First we use the Pythagorean theorem to find its diameter.

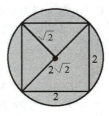

$$c^2 = a^2 + b^2$$
$$c^2 = 2^2 + 2^2 = 4 + 4 = 8$$
$$c = \sqrt{8} = 2\sqrt{2}$$

The radius is half the diameter, so $r = \sqrt{2}$, and
$$A = \pi r^2$$
$$A = \pi \cdot \left(\sqrt{2}\right)^2$$
$$A = 2\pi \text{ square units}$$

31. In 1 revolution the wheel travels its circumference C. So in 4 revolutions it travels a distance $D = 4C$.
$$D = 4C = 4\left(\pi d\right)$$
$$= 4\left(\pi \cdot 16\right) = 64\pi \approx 201 \text{ inches}$$

33. The area of the border is the difference between the area A of the outer square which has a side $S = 10 = 6 + 2 + 2$ feet and the area a of the inner square which has a side $s = 6$ feet.

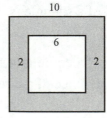

$$A - a = S^2 - s^2$$
$$= 10^2 - 6^2$$
$$= 100 - 36$$
$$= 64 \text{ square feet}$$

35. The area of the window is the sum of the area of half the circle and the area of the rectangle.

$$\text{Area} = \frac{1}{2}\left(\pi r^2\right) + lw$$

$$\text{Area} = \frac{1}{2} \cdot \pi \cdot 2^2 + 6 \cdot 4$$

$$= 2\pi + 24 \approx 30.28 \text{ square feet.}$$

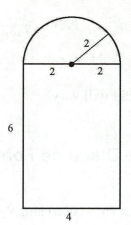

The amount of wood frame is measured by the perimeter of the window. The perimeter is the sum of the half the circumference of the circle and the 3 outer sides of the rectangle.

$$\text{Perimeter} = \frac{1}{2}\left(\pi d\right) + 2l + w$$

$$= \frac{1}{2} \cdot \pi \cdot 4 + 2 \cdot 6 + 4$$

$$= 2\pi + 16 \approx 22.28 \text{ feet}$$

37. Since 1 mile = 5280 feet, 20 feet = $\dfrac{20}{5280}$ mile.

We use the Pythagorean theorem to find the distance d that we can see.

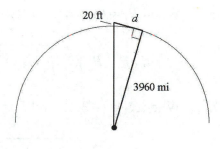

$$d^2 + (3960)^2 = \left(3960 + \frac{20}{5280}\right)^2$$

$$d^2 = \left(3960 + \frac{20}{5280}\right)^2 - (3960)^2 \approx 30.0000$$

$$d \approx 5.48 \text{ miles}$$

39. Since 1 mile = 5280 feet, 150 feet = $\dfrac{150}{5280}$ mile.

We use the Pythagorean theorem to find the distance d that we can see.
From the deck:

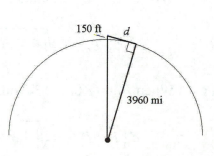

$$d^2 + (3960)^2 = \left(3960 + \frac{100}{5280}\right)^2$$

$$d^2 = \left(3960 + \frac{100}{5280}\right)^2 - (3960)^2 \approx 150.0004$$

$$d \approx 12.25 \text{ miles}$$

From the bridge:

$$d^2 + (3960)^2 = \left(3960 + \frac{150}{5280}\right)^2$$

$$d^2 = \left(3960 + \frac{150}{5280}\right)^2 - (3960)^2 \approx 225.0008$$

$$d \approx 15.00 \text{ miles}$$

41. Answers will vary.

A.10 The Distance Formula

1. When $P_1 = (0, 0)$ and $P_2 = (2, 1)$ the distance is

$$d(P_1, P_2) = \sqrt{(1-0)^2 + (2-0)^2} = \sqrt{1+4} = \sqrt{5} \approx 2.24$$

3. When $P_2 = (-2, 2)$ and $P_1 = (1, 1)$ the distance is

$$d(P_1, P_2) = \sqrt{[1-(-2)]^2 + (1-2)^2} = \sqrt{3^2 + (-1)^2} = \sqrt{9+1} = \sqrt{10} \approx 3.16$$

5. When $P_1 = (3, -4)$ and $P_2 = (5, 4)$, the distance is

$$d(P_1, P_2) = \sqrt{(5-3)^2 + [4-(-4)]^2} = \sqrt{2^2 + 8^2} = \sqrt{4+64} = \sqrt{68} \approx 8.25$$

7. When $P_1 = (-3, 2)$ and $P_2 = (6, 0)$, the distance is

$$d(P_1, P_2) = \sqrt{[6-(-3)]^2 + (0-2)^2} = \sqrt{9^2 + (-2)^2} = \sqrt{81+4} = \sqrt{85} \approx 9.22$$

9. When $P_1 = (4, -3)$ and $P_2 = (6, 4)$, the distance is

$$d(P_1, P_2) = \sqrt{(6-4)^2 + [4-(-3)]^2} = \sqrt{2^2 + 7^2} = \sqrt{4+49} = \sqrt{53} \approx 7.28$$

11. When $P_1 = (-0.2, 0.3)$ and $P_2 = (2.3, 1.1)$, the distance is

$$d(P_1, P_2) = \sqrt{[2.3-(-0.2)]^2 + (1.1-0.3)^2}$$
$$= \sqrt{2.5^2 + 0.8^2} = \sqrt{6.25+0.64} = \sqrt{6.89} \approx 2.62$$

13. When $P_1 = (a, b)$ and $P_2 = (0, 0)$, the distance is

$$d(P_1, P_2) = \sqrt{(0-a)^2 + (0-b)^2} = \sqrt{(-a)^2 + (-b)^2} = \sqrt{a^2 + b^2}$$

15. We first find the length of each side of the triangle.

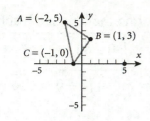

$$d(A, B) = \sqrt{[1-(-2)]^2 + (3-5)^2} = \sqrt{9+4} = \sqrt{13}$$

$$d(B, C) = \sqrt{[(-1)-1]^2 + (0-3)^2} = \sqrt{4+9} = \sqrt{13}$$

$$d(C, A) = \sqrt{[(-2)-(-1)]^2 + (5-0)^2} = \sqrt{1+25}$$
$$= \sqrt{26}$$

To verify that the triangle is right, we show

$$[d(A, B)]^2 + [d(B, C)]^2 = \left(\sqrt{13}\right)^2 + \left(\sqrt{13}\right)^2$$

$$= 13 + 13 = 26 = \left[\sqrt{26}\right]^2 = [d(C, A)]^2$$

From the converse of the Pythagorean Theorem, triangle ABC is a right triangle.

The area of a triangle Area $= \dfrac{1}{2}bh = \dfrac{1}{2} \cdot \sqrt{13} \cdot \sqrt{13} = \dfrac{13}{2} = 7.5$ square units.

17. We first find the length of each side of the triangle.

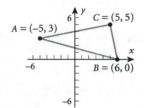

$$d(A, B) = \sqrt{[6-(-5)]^2 + (0-3)^2}$$
$$= \sqrt{11^2 + (-3)^2} = \sqrt{130}$$

$$d(B, C) = \sqrt{(5-6)^2 + (5-0)^2} = \sqrt{(-1)^2 + 5^2}$$
$$= \sqrt{1+25} = \sqrt{26}$$

$$d(C, A) = \sqrt{(-5-5)^2 + (3-5)^2}$$
$$= \sqrt{(-10)^2 + (-2)^2} = \sqrt{100+4} = \sqrt{104}$$

To verify that the triangle is right, we show

$$[d(B, C)]^2 + [d(C, A)]^2 = \left(\sqrt{26}\right)^2 + \left(\sqrt{104}\right)^2 = 26 + 104 = 130 = \left(\sqrt{130}\right)^2 = [d(A, B)]^2$$

From the converse of the Pythagorean Theorem, triangle ABC is a right triangle.

The area of the triangle $= \dfrac{1}{2}bh = \dfrac{1}{2}\left(\sqrt{26}\right)\left(\sqrt{104}\right) = \dfrac{1}{2}\sqrt{2704} = \dfrac{1}{2} \cdot 52 = 26$ square units.

19. We first find the length of each side of the triangle.

$$d(A, B) = \sqrt{(0-4)^2 + \left[-3-(-3)\right]^2}$$
$$= \sqrt{(-4)^2 + 0} = \sqrt{16} = 4$$

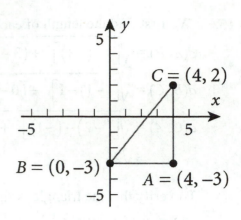

$$d(B, C) = \sqrt{(4-0)^2 + \left[2-(-3)\right]^2} = \sqrt{4^2 + 5^2}$$
$$= \sqrt{16 + 25} = \sqrt{41}$$

$$d(C, A) = \sqrt{(4-4)^2 + (-3-2)^2} = \sqrt{0 + (-5)^2}$$
$$= \sqrt{25} = 5$$

To verify that the triangle is right, we show
$$[d(A, B)]^2 + [d(C, A)]^2 = 4^2 + 5^2 = 16 + 25 = 41 = [d(B, C)]^2$$
From the converse of the Pythagorean Theorem, triangle ABC is a right triangle.

The area of the triangle $= \dfrac{1}{2} bh = \dfrac{1}{2}(4)(5) = 10$ square units.

21. We want the points $(2, y)$ for which the distance between $(2, y)$ and $(-2, -1)$ is 5
$$d^2 = \left[2-(-2)\right]^2 + \left[y-(-1)\right]^2 = 5^2$$
$$4^2 + y^2 + 2y + 1 = 25$$
$$y^2 + 2y - 8 = 0$$
$$(y-2)(y+4) = 0$$
$$y - 2 = 0 \quad \text{or} \quad y + 4 = 0$$
$$y = 2 \quad \text{or} \quad y = -4$$
The points that are 5 units from $(-2, -1)$ are $(2, 2)$ and $(2, -4)$.

23. We want the points $(x, 0)$ that are 5 units away from $(4, -3)$.
$$d^2 = (x-4)^2 + \left[0-(-3)\right]^2 = 5^2$$
$$x^2 - 8x + 16 + 9 = 25$$
$$x^2 - 8x = 0$$
$$x(x-8) = 0$$
$$x = 0 \quad \text{or} \quad x - 8 = 0$$
$$x = 8$$
The points on the x-axis that are a distance of 5 from $(4, -3)$ are $(0, 0)$ and $(8, 0)$.

25. When the points on the y-axis that are 5 units from (4.4) are (0.1) and (0.7), $P_1 = (1, 3)$ and $P_2 = (5, 15)$, then the length of the line segment is

$$d(P_1, P_2) = \sqrt{(5-1)^2 + (15-3)^2} = \sqrt{(4)^2 + (12)^2} = \sqrt{16 + 144} = \sqrt{160} \approx 12.65$$

27. When $P_1 = (-4, 6)$ and $P_2 = (4, -8)$, then the length of the line segment is

$$d\left(P_1, P_2\right) = \sqrt{\left[4-(-4)\right]^2 + (-8-6)^2} = \sqrt{8^2 + (-14)^2} = \sqrt{64+196} = \sqrt{260} \approx 16.12$$

29. Since the baseball "diamond" is a square, the baselines meet at right angles, and the triangle formed by home plate, first base, and second base is a right triangle. The distance from home plate to second base is the hypotenuse of the right triangle.

$$c^2 = a^2 + b^2$$
$$c^2 = 90^2 + 90^2 = 8100 + 8100 = 16,200$$
$$c = \sqrt{16,200} \approx 127.28 \text{ feet}$$

31.

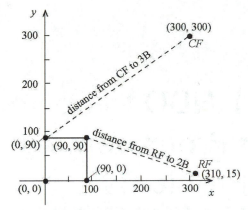

(a) The coordinates of first base are (90 feet, 0 feet), second base are (90 feet, 90 feet), third base are (0 feet, 90 feet).

(b) The distance between the right fielder and second base is the distance between the points (310, 15) and (90, 90).

$$d = \sqrt{(90-310)^2 + (90-15)^2} = \sqrt{220^2 + 75^2} = \sqrt{54,025} \approx 232.42 \text{ feet}$$

(c) The distance between the center fielder and third base is the distance between the points (300, 300) and (0, 90).

$$d = \sqrt{(300-0)^2 + (300-90)^2} = \sqrt{300^2 + 210^2} = \sqrt{134,100} \approx 366.20 \text{ feet}$$

33. After t hours the Intrepid has traveled $30t$ miles to the east, and the truck has traveled $40t$ miles south. Since east and south are 90° apart, the distance between the car and the truck is the hypotenuse of a right triangle. See the diagram.

$$d = \sqrt{(30t)^2 + (40t)^2}$$
$$= \sqrt{900t^2 + 1600t^2}$$
$$= \sqrt{2500t^2}$$
$$= 50t \text{ miles}$$

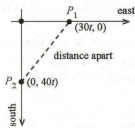

Appendix B

Using LINDO to Solve Linear Programming Problems

1. Enter on the LINDO blank window.

max $3x1 + 2x2 + x3$
subject to
$3x1 + x2 + x3 < 30$
$5x1 + 2x2 + x3 < 24$
$x1 + x2 + 4x3 < 20$
end

Click solve;
click no, when asked for reports;
click close.
Move aside the window on which you entered the problem to see the solution on the right.

LP OPTIMUM FOUND AT STEP   1

OBJECTIVE FUNCTION VALUE

1)   24.00000

VARIABLE	VALUE	REDUCED COST
X1	0.000000	2.000000
X2	12.000000	0.000000
X3	0.000000	0.000000

ROW	SLACK OR SURPLUS	DUAL PRICES
2)	18.000000	0.000000
3)	0.000000	1.000000
4)	8.000000	0.000000

NO. ITERATIONS = 1

The maximum value of $P = 24$, and it is attained when $x_1 = 0$, $x_2 = 12$, and $x_3 = 0$.

3. Enter on the LINDO blank window:

max $3x1 + x2 + x3$
subject to
$x1 + x2 + x3 < 6$
$2x1 + 3x2 + 4x3 < 10$
end

Click solve;
click no, when asked for reports;
click close.
Move aside the window on which you entered the problem to see the solution on the right.

LP OPTIMUM FOUND AT STEP   1

OBJECTIVE FUNCTION VALUE

1)   15.00000

VARIABLE	VALUE	REDUCED COST
X1	5.000000	0.000000
X2	0.000000	3.500000
X3	0.000000	5.000000

ROW	SLACK OR SURPLUS	DUAL PRICES
2)	1.000000	0.000000
3)	0.000000	1.500000

NO. ITERATIONS = 1

5. Enter on the LINDO blank window.

max $2x1 + x2 + 3x3$
subject to
$x1 + x2 - x3 < 10$
$x2 + x3 < 4$
end

Click solve;
click no, when asked for reports;
click close.
Move aside the window on which you entered the problem to see the solution on the right.

LP OPTIMUM FOUND AT STEP   1

OBJECTIVE FUNCTION VALUE

1)   40.00000

VARIABLE	VALUE	REDUCED COST
X1	14.000000	0.000000
X2	0.000000	6.000000
X3	4.000000	0.000000

ROW	SLACK OR SURPLUS	DUAL PRICES
2)	0.000000	2.000000
3)	0.000000	5.000000

NO. ITERATIONS = 1

The maximum value of $P = 40$, and it is attained when $x_1 = 14$, $x_2 = 0$, and $x_3 = 4$.

7. Enter on the LINDO blank window,

LP OPTIMUM FOUND AT STEP   1

max $x1 + x2 + x3$
subject to
$x1 + x2 + x3 < 6$
$4x1 + x2 > 12$
end

OBJECTIVE FUNCTION VALUE

1)   6.000000

VARIABLE	VALUE	REDUCED COST
X1	6.000000	0.000000
X2	0.000000	0.000000
X3	0.000000	0.000000

Click solve;
click no, when asked for reports;
click close.
Move aside the window on
which you entered the problem
to see the solution on the right.

ROW	SLACK OR SURPLUS	DUAL PRICES
2)	0.000000	1.000000
3)	12.000000	0.000000

NO. ITERATIONS = 1

The maximum value of $P = 6$ and it is attained when $x_1 = 6$, $x_2 = 0$, and $x_3 = 0$.

9. Enter on the LINDO blank window Enter on the LINDO blank window.

LP OPTIMUM FOUND AT STEP   3

max $2x1 + x2 + 3x3$
subject to
$5x1 + 2x2 + x3 < 20$
$6x1 + x2 + 4x3 < 24$
$x1 + x2 + 4x3 < 16$
end

OBJECTIVE FUNCTION VALUE

1)   15.20000

VARIABLE	VALUE	REDUCED COST
X1	1.600000	0.000000
X2	4.800000	0.000000
X3	2.400000	0.000000

Click solve;
click no, when asked for reports;
click close.
Move aside the window on
which you entered the problem
to see the solution on the right.

ROW	SLACK OR SURPLUS	DUAL PRICES
2)	0.000000	0.142857
3)	0.000000	0.114286
4)	0.000000	0.600000

NO. ITERATIONS = 3

The maximum value of $P = 15.2$, and it is attained when $x_1 = 1.6$, $x_2 = 4.8$, and $x_3 = 2.4$.

11. max $2x1 + 3x2 + x3$
subject to
$x1 + x2 + x3 < 50$
$3x1 + 2x2 + x3 < 10$
end

Click solve;
click no, when asked for reports;
click close.
Move aside the window on
which you entered the problem
to see the solution on the right.

LP OPTIMUM FOUND AT STEP   1

OBJECTIVE FUNCTION VALUE

1)   15.00000

VARIABLE	VALUE	REDUCED COST
X1	0.000000	2.500000
X2	5.000000	0.000000
X3	0.000000	0.500000

ROW	SLACK OR SURPLUS	DUAL PRICES
2)	45.000000	0.000000
3)	0.000000	1.500000

NO. ITERATIONS = 1

The maximum value of $P = 15$, and it is attained when $x_1 = 0$, $x_2 = 5$, and $x_3 = 0$.

13. Enter on the LINDO blank
window.

max $2x1 + x2 + x3$
subject to
$-2x1 + x2 - 2x3 < 4$
$x1 - 2x2 + x3 < 2$
end

Click solve;
An error message appears:
Unbounded solution at STEP 1.

Close the boxes to look at the
report on the right

UNBOUNDED VARIABLES ARE:
  X3
SLK   3
  X2

OBJECTIVE FUNCTION VALUE

1)   0.9999990E+08

VARIABLE	VALUE	REDUCED COST
X1	0.000000	10.000000
X2	4.000000	0.000000
X3	99999904.000000	7.000000

ROW	SLACK OR SURPLUS	DUAL PRICES
2)	0.000000	−3.000000
3)	10.000000	−2.000000

NO. ITERATIONS = 1

This problem is unbounded and has no maximum solution.

15. Enter on the LINDO blank window,

max $2x1 + x2 + 3x3$
subject to
$x1 + 2x2 + x3 < 25$
$3x1 + 2x2 + 3x3 < 30$
end

Click solve;
click no, when asked for reports;
click close.
Move aside the window on
which you entered the problem
to see the solution on the right.

LP OPTIMUM FOUND AT STEP   1

OBJECTIVE FUNCTION VALUE

1)   30.00000

VARIABLE	VALUE	REDUCED COST
X1	0.000000	1.000000
X2	0.000000	1.000000
X3	10.000000	0.000000

ROW	SLACK OR SURPLUS	DUAL PRICES
2)	15.000000	0.000000
3)	0.000000	1.000000

NO. ITERATIONS = 1

The maximum value of $P = 30$, and it is attained when $x_1 = 0$, $x_2 = 0$, and $x_3 = 10$.

17. Enter on the LINDO blank window.

max $2x1 + 4x2 + x3 + x4$
subject to
$2x1 + x2 + 2x3 + 3x4 < 12$
$2x2 + x3 + 2x4 < 20$
$2x1 + x2 + 4x3 < 16$
end

Click solve;
click no, when asked for reports;
click close.
Move aside the window on
which you entered the problem
to see the solution on the right.

LP OPTIMUM FOUND AT STEP   2

OBJECTIVE FUNCTION VALUE

1)   42.00000

VARIABLE	VALUE	REDUCED COST
X1	1.000000	0.000000
X2	10.000000	0.000000
X3	0.000000	2.500000
X4	0.000000	5.000000

ROW	SLACK OR SURPLUS	DUAL PRICES
2)	0.000000	1.000000
3)	0.000000	1.500000
4)	4.000000	0.000000

NO. ITERATIONS = 2

The maximum value of $P = 42$ and it is attained when $x_1 = 1$, $x_2 = 10$, $x_3 = 0$, and $x_4 = 0$.

19. Enter on the LINDO blank window.

max $2x1 + x2 + x3$
subject to
$x1 + 2x2 + 4x3 < 20$
$2x1 + 4x2 + 4x3 < 60$
$3x1 + 4x2 + x3 < 90$
end

Click solve;
click no, when asked for reports;
click close.
Move aside the window on which you entered the problem to see the solution on the right.

LP OPTIMUM FOUND AT STEP   1

OBJECTIVE FUNCTION VALUE

1)   40.00000

VARIABLE	VALUE	REDUCED COST
X1	20.000000	0.000000
X2	0.000000	3.000000
X3	0.000000	7.000000

ROW	SLACK OR SURPLUS	DUAL PRICES
2)	0.000000	2.000000
3)	20.000000	0.000000
4)	30.000000	0.000000

NO. ITERATIONS = 1

The maximum value of $P = 40$ and it is attained when $x_1 = 20$, $x_2 = 0$, and $x_3 = 0$.

21. Enter on the LINDO blank window.

max $x1 + 2x2 + 4x3 - x4$
subject to
$5x1 + 4x3 + 6x4 < 20$
$4x1 + 2x2 + 2x3 + 8x4 < 40$
end

Click solve;
click no, when asked for reports;
click close.
Move aside the window on which you entered the problem to see the solution on the right.

LP OPTIMUM FOUND AT STEP   2

OBJECTIVE FUNCTION VALUE

1)   50.00000

VARIABLE	VALUE	REDUCED COST
X1	0.000000	5.500000
X2	15.000000	0.000000
X3	5.000000	0.000000
X4	0.000000	12.000000

ROW	SLACK OR SURPLUS	DUAL PRICES
2)	0.000000	0.500000
3)	0.000000	1.000000

NO. ITERATIONS = 2

The maximum value of $P = 50$ and it is attained when $x_1 = 0$, $x_2 = 15$, $x_3 = 5$, and $x_4 = 0$.

23.   Enter on the LINDO blank window.

min $x1 + x2 + x3 + x4 + x5 + x6 + x7$
subject to
$4x1 + 2x2 + x3 + 2x5 + x6 > 75$
$x2 + 2x3 + 3x4 + x6 > 110$
$x5 + x6 + 2x7 > 50$
end

Click solve;
click no, when asked for reports;
click close.
Move aside the window on which you entered the problem to see the solution below.

LP OPTIMUM FOUND AT STEP   3

OBJECTIVE FUNCTION VALUE

1)   76.25000

VARIABLE	VALUE	REDUCED COST
X1	6.250000	0.000000
X2	0.000000	0.166667
X3	0.000000	0.083333
X4	20.000000	0.000000
X5	0.000000	0.083333
X6	50.000000	0.000000
X7	0.000000	0.166667

ROW	SLACK OR SURPLUS	DUAL PRICES
2)	0.000000	−0.250000
3)	0.000000	−0.333333
4)	0.000000	−0.416667

NO. ITERATIONS = 3

The minimum value of $P = 76.25$ and it is attained when $x_1 = 6.25$, $x_4 = 20$, $x_6 = 50$, and $x_2 = x_3 = x_5 = x_7 = 0$.

Appendix C
Graphing Utilities

C.1   The Viewing Rectangle

1.   Using the window shown, X-scale = 1, and Y-scale = 2, we get the point (-1, 4).

3.   Using the window shown, X-scale = 1, and Y-scale = 1, we get the point (3, 1).

5. Xmin = -6 Ymin = -4   7. Xmin = -6 Ymin = -1
     Xmax = 6 Ymax = 4 Xmax = 6 Ymax = 3
     Xscl = 2 Yscl = 2 Xscl = 2 Yscl = 1

9. Xmin = 3 Ymin = 2
     Xmax = 9 Ymax = 10
     Xscl = 1 Yscl = 2

11. Answers will vary, but an appropriate viewing window would be
     Xmin = -12 Ymin = -4
     Xmax = 6 Ymax = 8
     Xscl = 1 Yscl = 1

13. Answers will vary, but an appropriate viewing window would be
     Xmin = -30 Ymin = -100
     Xmax = 50 Ymax = 50
     Xscl = 10 Yscl = 10

15. Answers will vary, but an appropriate viewing window would be
     Xmin = -10 Ymin = -20
     Xmax = 110 Ymax = 180
     Xscl = 10 Yscl = 20

C.2   Using a Graphing Utility to Graph Equations

1.   (a)   (b)   (c)   (d)

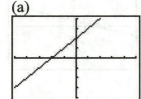 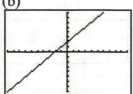 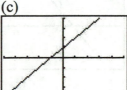 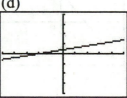

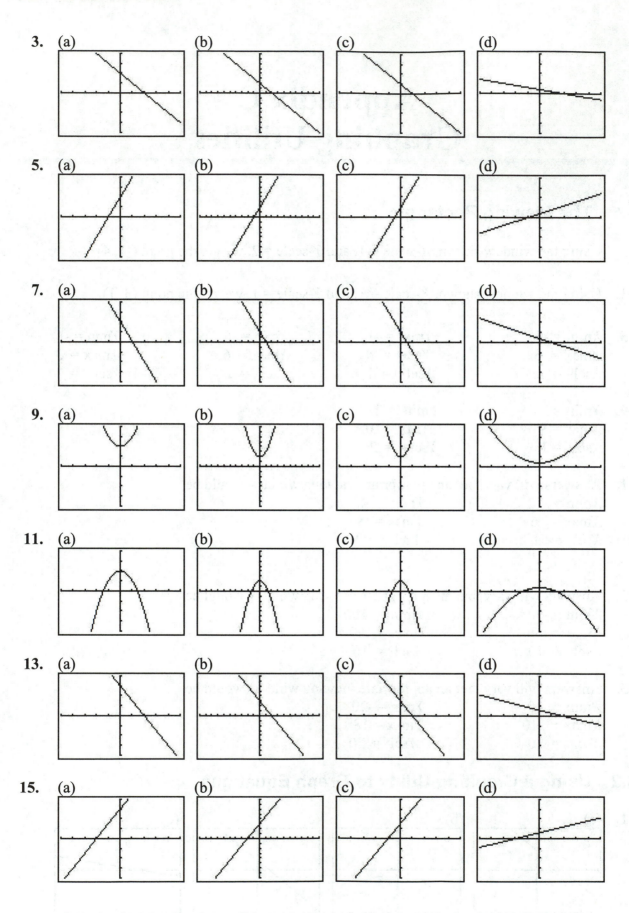

17. $y = x + 2$

X	Y1	
-3	-1	
-2	0	
-1	1	
0	2	
1	3	
2	4	
3	5	

X= -3

19. $y = -x + 2$

X	Y3	
-3	5	
-2	4	
-1	3	
0	2	
1	1	
2	0	
3	-1	

X= -3

21. $y = 2x + 2$

X	Y5	
-3	-4	
-2	-2	
-1	0	
0	2	
1	4	
2	6	
3	8	

X= -3

23. $y = -2x + 2$

X	Y7	
-3	8	
-2	6	
-1	4	
0	2	
1	0	
2	-2	
3	-4	

X= -3

25. $y = x^2 + 2$

X	Y9	
-3	11	
-2	6	
-1	3	
0	2	
1	3	
2	6	
3	11	

X= -3

27. $y = -x^2 + 2$

X	Y1	
-3	-7	
-2	-2	
-1	1	
0	2	
1	1	
2	-2	
3	-7	

X= -3

29. $3x + 2y = 6$

X	Y3	
-3	7.5	
-2	6	
-1	4.5	
0	3	
1	1.5	
2	0	
3	-1.5	

X= -3

31. $-3x + 2y = 6$

X	Y5	
-3	-1.5	
-2	0	
-1	1.5	
0	3	
1	4.5	
2	6	
3	7.5	

X= -3

C.3 Square Screens

1. A square screen results if $2(X\text{max} - X\text{min}) = 3(Y\text{max} - Y\text{min})$
In this window we test
$$2(3 - (-3)) \ ? \ 3(2 - (-2))$$
$$2 \cdot 6 \qquad 3 \cdot 4$$
$$12 \ = \ 12$$
The window is square.

3. A square screen results if $2(X\text{max} - X\text{min}) = 3(Y\text{max} - Y\text{min})$
In this window we test
$$2(9 - 0) \ ? \ 3(4 - (-2))$$
$$2 \cdot 9 \qquad 3 \cdot 6$$
$$18 \ = \ 18$$
The window is square.

5. A square screen results if $2(X\text{max} - X\text{min}) = 3(Y\text{max} - Y\text{min})$
   In this window we test

$$2(6 - (-6)) \; ? \; 3(2 - (-2))$$
$$2 \cdot 12 \qquad 3 \cdot 4$$
$$24 \; \neq \; 12$$

   The window is not square.

7. A square screen results if $2(X\text{max} - X\text{min}) = 3(Y\text{max} - Y\text{min})$
   In this window we test

$$2(9 - 0) \; ? \; 3(3 - (-3))$$
$$2 \cdot 9 \qquad 3 \cdot 6$$
$$18 \; = \; 18$$

   The window is square.

9. Answers may vary.
   If $X\text{min} = -4$ and $X\text{max} = 8$, then $2(X\text{max} - X\text{min}) = 2(8 - (-4)) = 2 \cdot 12 = 24$.

   To make the screen square, the difference between $Y\text{min}$ and $Y\text{max}$ must be $24 \div 3 = 8$.

   To include the point $(4, 8)$, the y-axis must go at least as high as 8. A possible choice is $Y\text{min} = 1$ and $Y\text{max} = 9$.

C.4 Using a Graphing Utility to Solve Inequalities

Note: There are no problems for this section.

C.5 Using a Graphing Utility to Locate Intercepts and Check for Symmetry

1.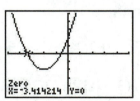

   The smaller x-intercept is $(-3.41, 0)$.

3.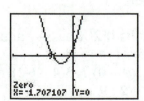

   The smaller x-intercept is $(-1.71, 0)$.

5.

The smaller x-intercept is $(-0.28, 0)$.

7.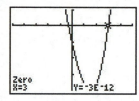

The positive x-intercept is $(3, 0)$.

9.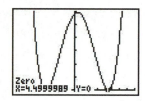

The positive x-intercept is $(4.50, 0)$.

11.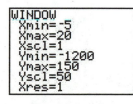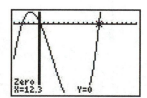

There are two positive x-intercepts; they are $(0.32, 0)$ and $(12.3, 0)$.

13.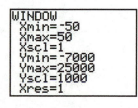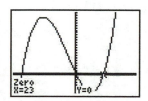

There are two positive x-intercepts; they are $(1, 0)$ and $(23, 0)$.

15. (a) The x-intercepts are $(-1, 0)$ and $(1, 0)$. The y-intercept is $(0, -1)$.
(b) The graph is symmetric with respect to the y-axis.

17. (a) There is no x-intercept and there is no y-intercept.
(b) The graph is symmetric with respect to the origin.

C.6 Using a Graphing Utility to Solve Equations

Note: There are no problems for this section.